T0311961

Assisted Phytoremediation

Assisted Phytoremediation

Edited by

Vimal Chandra Pandey

Department of Environmental Science,
Babasaheb Bhimrao Ambedkar University, Lucknow, India

Elsevier
Radarweg 29, PO Box 211, 1000 AE Amsterdam, Netherlands
The Boulevard, Langford Lane, Kidlington, Oxford OX5 1GB, United Kingdom
50 Hampshire Street, 5th Floor, Cambridge, MA 02139, United States

British Library Cataloguing-in-Publication Data
A catalogue record for this book is available from the British Library

Library of Congress Cataloging-in-Publication Data
A catalog record for this book is available from the Library of Congress

ISBN: 978-0-12-822893-7

For Information on all Elsevier publications visit our website at
https://www.elsevier.com/books-and-journals

Publisher: Candice Janco
Acquisitions Editor: Maris LaFleur
Editorial Project Manager: Michelle Fisher
Production Project Manager: Kumar Anbazhagan
Cover Designer: Matthew Limbert
Cover Credit: Vimal Chandra Pandey

Typeset by Aptara, New Delhi, India

Contents

Contributors

Giorgia Aimola
Water Research Institute-Italian National Research Council, Bari, Italy.

Khalid J. Alzahrani
Department of Clinical Laboratories Sciences, College of Applied Medical Sciences, Taif University, Saudi Arabia.

Adenike Eunice Amoo
Food Security and Safety Niche, Faculty of Natural and Agricultural Sciences, North-West University, Mmabatho, South Africa.

Valeria Ancona
Water Research Institute-Italian National Research Council, Bari, Italy.

Sanem Argin
Department of Agricultural Trade and Management, School of Applied Science, Yeditepe University, Istanbul, Turkey.

Tuba Arjumend
Department of Plant Protection, Faculty of Agriculture, Usak University, Uşak, Turkey.

Anam Ashraf
School of Environment, Tsinghua University, Beijing, China.

Ayansina Segun Ayangbenro
Food Security and Safety Niche, Faculty of Natural and Agricultural Sciences, North-West University, Mmabatho, South Africa.

Gargi Bhattacharjee
Department of Biosciences, School of Science, Indrashil University, Rajpur, Mehsana, Gujarat, India.

Irshad Bibi
Institute of Soil and Environmental Sciences, University of Agriculture Faisalabad, Faisalabad, Pakistan.

Dilara Birinci
Department of Genetics and Bioengineering, Faculty of Engineering, Yeditepe University, Istanbul, Turkey.

Parisa Bolouri
Department of Genetics and Bioengineering, Faculty of Engineering, Yeditepe University, Istanbul, Turkey.

Domenico Borello
Water Research Institute-Italian National Research Council, Bari, Italy; Department of Mechanical and Aerospace Engineering, Sapienza University of Rome, Italy.

Sylvain Bourgerie
University of Orleans, Orléans, France.

Anna Barra Caracciolo
Water Research Institute-Italian National Research Council, Monterotondo, Rome, Italy

Mukkaram Ejaz
School of Environmental and Municipal Engineering, Lanzhou Jiaotong University, Lanzhou, PR China.

Melek Ekinci
Department of Horticulture, Faculty of Agriculture, Atatürk University, Erzurum, Turkey.

Nilda Ersoy
Department of Organic Agriculture, Vocational School of Technical Sciences, Akdeniz University, Antalya, Turkey.

Joël Fontaine
Univ. Littoral Côte d'Opale, UR 4492, UCEIV, Unit of environmental chemistry and interactions with living organisms, SFR Condorcet FR CNRS 3417, Dunkerque, France.

Gordana Gajić
Department of Ecology, Institute of Biological Research "Siniša Stanković", National Institute of Republic of Serbia, University of Belgrade, Belgrade, Serbia.

Nisarg Gohil
Department of Biosciences, School of Science, Indrashil University, Rajpur, Mehsana, Gujarat, India.

Paola Grenni
Water Research Institute-Italian National Research Council, Monterotondo, Rome, Italy.

Ankita Gupta
Plant Stress Biology Laboratory, Institute of Environment and Sustainable Development, Banaras Hindu University, Varanasi, India.

Adem Güneş
Department of Soil Science, Faculty of Agriculture, Erciyes University, Kayseri, Turkey.

Fasih U. Haider
College of Resources and Environmental Sciences, Gansu Agricultural University, Lanzhou, China.

Sunila Hooda
Ram Lal Anand College, University of Delhi, India.

Sajid Husain
Hainan Key Laboratory for Sustainable Utilization of Tropical Bioresource, College of Tropical Crops, Hainan University, Haikou, Hainan, China.

Azhar Hussain
Department of Soil Science, the Islamia University of Bahawalpur, Pakistan.

David Okeh Igwe
Department of Biotechnology, Ebonyi State University, Abakaliki, Ebonyi State, Nigeria; Section of Plant Pathology, Boyce Thompson Institute for Plant Research, Ithaca, New York, United StatesPlant Pathology and Plant-Microbe Biology, School of Integrated Plant Sciences, Cornell University, Ithaca, NY, USA.

N.F. Islam
Department of Botany, Nanda Nath Saikia College, Titabar, Assam, India.

Ksenija Jakovljević
University of Belgrade, Faculty of Biology, Institute of Botany and Botanical Garden, Belgrade, Serbia.

Sonia Labidi
Université de Carthage, Institut National Agronomique de Tunisie, Laboratoire des Sciences Horticoles, LR13AGR01, Tunis, Mahrajène, Tunisia.

Manhattan Lebrun
University of Orleans, Orléans, France.

Cheng Liu
College of Environment, Hohai University Nanjing, China.

Francisco J. López-Bellido
Department of Plant Production and Agricultural Technology, School of Agricultural Engineering, University of Castilla-La Mancha, Ciudad Real (Spain).

Sahrish Majeed
Ram Lal Anand College, University of Delhi, India.

Arnab Majumdar
Department of Earth Sciences, Indian Institute of Science Education and Research (IISER) Kolkata, Mohanpur, West Bengal, India.

Garima Malik
Raghunath Girls' Post Graduate College, C.C.S. University, Meerut, India.

Rupesh Maurya
Department of Biosciences, School of Science, Indrashil University, Rajpur, Mehsana, Gujarat, India.

Hacène Meglouli
Institut de Recherche en Biologie Végétale (IRBV), de l'Université de Montréal, Canada.

Tariq Mehmood
College of Environment, Hohai University Nanjing, China.

Florie Miard
University of Orleans, Orléans, France.

Miroslava Mitrović
Department of Ecology, Institute of Biological Research "Siniša Stanković", National Institute of Republic of Serbia, University of Belgrade, Belgrade, Serbia.

Domenico Morabito
University of Orleans, Orléans, France.

Romain Nandillon
University of Orleans, Orléans, France.

Nabeel Khan Niazi
Institute of Soil and Environmental Sciences, University of Agriculture Faisalabad, Faisalabad, Pakistan.

Omena Bernard Ojuederie
Department of Biological Sciences, Faculty of Science, Kings University, Odeomu, Osun State, Nigeria; Food Security and Safety Niche, Faculty of Natural and Agricultural Sciences, North-West University, Mmabatho, South Africa.

Shesan John Owonubi
Department of Chemistry, University of Zululand, KwaDlangezwa, KwaZulu-Natal, South Africa.

Janhvi Pandey
Academy of Scientific and Innovative Research (AcSIR), India; Division of Agronomy and Soil Science, CSIR-Central Institute of Medicinal and Aromatic Plants, Lucknow, India.

Vimal Chandra Pandey
Department of Environmental Science, Babasaheb Bhimrao Ambedkar University, Lucknow, India.

Henny Patel
Department of Biosciences, School of Science, Indrashil University, Rajpur, Mehsana, Gujarat, India.

Rupshikha Patowary
Centre for the Environment, Indian Institute of Technology Guwahati, Guwahati, Assam, India.

Pavle Pavlović
Department of Ecology, Institute of Biological Research "Siniša Stanković", National Institute of Republic of Serbia, University of Belgrade, Belgrade, Serbia.

Jacob Olagbenro Popoola
Department of Biological Sciences, Covenant University, Ota, Ogun State, Nigeria.

Dragana Ranđelović
Institute for Technology of Nuclear and Other Mineral Raw Materials, Belgrade, Serbia.

Ida Rascio
Water Research Institute-Italian National Research Council, Bari, Italy; Department of Soil, Plant and Food Sciences, University of Bari, Italy.

Umair Riaz
Soil and Water Testing Laboratory for Research Bahawalpur, Pakistan.

Luis Rodríguez
Department of Chemical Engineering, School of Civil Engineering, University of Castilla-La Mancha, Ciudad Real (Spain).

Sumeera Asghar
Department College of Horticulture, China Agricultural University, Beijing, China.

Anissa Lounès-Hadj Sahraoui
Univ. Littoral Côte d'Opale, UR 4492, UCEIV, Unit of environmental chemistry and interactions with living organisms, SFR Condorcet FR CNRS 3417, Dunkerque, France.

Maryline Calonne-Salmon
Earth and Life Institute, Applied Microbiology, Mycology, Université catholique de Louvain, Louvain-la-Neuve, Belgium.

Sougata Sarkar
Division of Agronomy and Soil Science, CSIR-Central Institute of Medicinal and Aromatic Plants, Lucknow, India; Genetic Resources and Agro-Technology Division, CSIR-Indian Institute of Integrative Medicine, Jammu, India.

Hemen Sarma
Department of Botany, Nanda Nath Saikia College, Titabar, Assam, India.

M. Shahid
Department of Environmental Sciences, COMSATS University Islamabad, Vehari, Pakistan.

Shreya Shakhreliya
Department of Biosciences, School of Science, Indrashil University, Rajpur, Mehsana, Gujarat, India.

Mehak Shaz
Department of Environmental Sciences and Engineering, Government College University, Faisalabad, Pakistan.

Vijai Singh
Department of Biosciences, School of Science, Indrashil University, Rajpur, Mehsana, Gujarat, India.

Sudhakar Srivastava
Plant Stress Biology Laboratory, Institute of Environment and Sustainable Development, Banaras Hindu University, Varanasi, India.

Virtudes Sánchez
Department of Chemical Engineering, School of Civil Engineering, University of Castilla-La Mancha, Ciudad Real (Spain).

Metin Turan
Department of Genetics and Bioengineering, Faculty of Engineering, Yeditepe University, Istanbul, Turkey.

Vito F. Uricchio
Water Research Institute-Italian National Research Council, Bari, Italy.

Ertan Yıldırım
Department of Horticulture, Faculty of Agriculture, Atatürk University, Erzurum, Turkey.

Tijana Zeremski
Institute of Field and Vegetable Crops, National Institute of the Republic of Serbia, Novi Sad, Serbia.

About the Editor

Vimal Chandra Pandey

Dr. Pandey featured in the world's top 2% scientists by Stanford University, United States. He is a leading researcher in the field of environmental engineering, especially phytomanagement of polluted sites. His research focuses mainly on the remediation and management of degraded lands, including heavy metal-polluted lands and postindustrial lands such as fly ash, red mud, mine spoil, and others for regaining ecosystem services and supporting a bio-based economy with phytoproducts through the affordable green technology (phytoremediation). His research interest also lies in exploring industrial crop-based phytoremediation to attain bioeconomy security and restoration, adaptive phytoremediation practices, phytoremediation based biofortification, carbon sequestration in waste dumpsites, fostering bioremediation for utilizing polluted lands, and attaining UN-Sustainable Development Goals. Recently, Dr. Pandey worked as a CSIR-Pool Scientist (Senior Research Associate) in the Department of Environmental Science at Babasaheb Bhimrao Ambedkar University, Lucknow, India. Dr. Pandey also worked as Consultant at Council of Science and Technology, Uttar Pradesh, DST-Young Scientist in Plant Ecology and Environmental Science Division at CSIR-National Botanical Research Institute, Lucknow and DS Kothari Postdoctoral fellow in the Department of Environmental Science at Babasaheb Bhimrao Ambedkar University, Lucknow. He is a recipient of a number of awards/honours/fellowships, and a member of National Academy of Sciences India. Dr. Pandey serves as a subject expert and the panel member for the evaluation of research and professional activities in India and abroad for fostering environmental sustainability. He has published over 100 scientific articles/book chapters in peer reviewed journals/books. Dr. Pandey is also the author and editor of seven books published by Elsevier with several more forthcoming. He is Associate Editor of *Land Degradation and Development* (Wiley), Editor of *Restoration Ecology* (Wiley), Associate Editor of *Environment, Development and Sustainability* (Springer), Associate Editor of *Ecological Processes* (Springer Nature), Advisory Board Member of Ambio (Springer), Editorial Board Member of *Environmental Management* (Springer), Editorial Board Member of *Bulletin of Environmental Contamination and Toxicology* (Springer). He also works/worked as Guest Editor for *Energy, Ecology and Environment* (Springer); *Bulletin of Environmental Contamination and Toxicology* (Springer); *Sustainability* (*MDPI*). Email address: vimalcpandey@gmail.com, ORCID: https://orcid.org/0000-0003-2250-6726, Google Scholar: https://scholar.google.co.in/citations?user=B-5sDCoAAAAJ&hl.

Foreword

Environmental pollution is one of the most serious challenges worldwide. It is caused by uncontrolled release of wide-ranging pollutants on the earth. As a result, their impact on human health has been well recognized via contaminated air and water or food chain. Therefore, a holistic approach is an urgent need globally to remove the pollutants in affordable way. In this direction, the plant-based remediation is appropriate to reduce contaminants. The present book, *Assisted Phytoremediation*, is aimed to offer such promising and potential tools to enhance plant performance. It can be assisted by altering the environmental conditions and stimulate the microbial or plants grown for degradation or reduction of pollutants. The phytoremediation efficiency can be enhanced either by amendments or microorganisms; chelate, additional nutrients, genetic engineering, and maintenance of pH are required. The present book focuses on a wide range of strategies such as fungi-assisted, biochar-assisted, chelate-assisted, nanoparticles-assisted, transgenic plant-mediated, CRISPR-assisted, compost-assisted, PGPR-assisted, phosphate-assisted, electrokinetic-assisted, and biosurfactant-assisted approaches that have been applied to enhance plant performance for degradation and uptake of pollutants.

The main idea behind the compilation of this book is to draw together chapters from eminent professors and scientists worldwide and benefit by their established expertise in phytoremediation. Currently, there is no such kind of book that is available in the market that can cover a broad spectrum of assisted phytoremediation. This book provides an up-to-date applied knowledge of assisted phytoremediation that will be useful for utilizing polluted lands for phytoproducts production such as fibers, timber, biofuels, biomass, dye, essential oils, etc. I appreciate the efforts of Editor Dr. Vimal Chandra Pandey, in bringing out this valuable edition through the leading global publisher, Elsevier Publishing, with 15 chapters covering various aspects of the assisted phytoremediation. I hope the book will be a notable asset for researchers, PhD students and plant scientists, stakeholders, policy makers, practitioners, entrepreneurs, and stakeholders alike.

Manju Sharma

Prof. (Mrs.) Manju Sharma
Distinguished Woman Scientist Chair
Past President, NASI
Former Secretary to Govt. of India
September 7, 2021

Preface

Assisted Phytoremediation is a new book that provides potential tools to enhance plant performance in phytoremediation. This book briefed different aspects of assisted phytoremediation by altering the environmental conditions and stimulated the microbial or plants grown for degrading or decreasing the pollutants. To increase the phytoremediation efficiency, a better understanding of the mechanisms underlying pollutant accumulation and degradation in plant is very indispensable for addressing the most critical and complex pollution challenges. Therefore, this book presents all aspects of soil remediation offering a great prospect in development of enhanced phytoremediation using different strategies namely, fungi-assisted, biochar-assisted, chelate-assisted, nanoparticles-assisted, transgenic plant-mediated, CRISPR-assisted, compost-assisted, PGPR-assisted, phosphate-assisted, electrokinetic-assisted and biosurfactant-assisted approaches that is useful to make a pollution-free earth. The book was aimed to provide an up-to-date knowledge on phytoremediation that can be used to increase plant performance and their adaptation against harsh conditions. This valuable book will support students, researchers, environmentalists, ecological engineers, practitioners, regulatory agencies, policy makers, and stakeholders.

The first chapter provides general and brief information about important aspects of assisted phytoremediation to help readers better understand the potential tools to enhance plant performance and their better use in future for utilizing polluted sites for phytoproducts production, that is, fibers, essential oils, timber, dye, fuel-wood, biomass, and biofuels. The second chapter focuses on the plant-assisted bioremediation strategy, relying on the synergistic actions between plant root system and natural microbes (bacteria and fungi), can be effective for stabilizing, storing and degrading soil pollutants. Third chapter covers a description of the arbuscular mycorrhizal fungi diversity as well as their potential role in phytomanagement of persistent organic pollutants and trace elements-polluted soils and the impact of mycorrhizal inoculation on soil refunctionalization. The fourth chapter describes biochar-assisted phytoremediation for metal(loid) polluted soils, besides the biochar also help to reduce soil acidity, increase organic matter and nutrient content, thereby improving plant growth. The fifth chapter covers chelate-assistant phytoremediation and their limitations and drawbacks. The detailed review about nanoparticles-assisted phytoremediation and their advances and applications are presented in the sixth chapter, whereas the transgenic plant-mediated phytoremediation for heavy metals, metalloids and xenobiotics and their applications, challenges, and prospects are described in the seventh chapter. The eighth chapter discusses recent developments in CRISPR-Cas9 based microorganisms and plant genome editing for bioremediation of pollutants in order to clean our environment for healthy human and animal life on the earth. Some potential approaches for assisted phytoremediation of arsenic contaminated sites are focused in the 9th chapter. The 10th chapter summarizes compost-assisted phytoremediation as well as their impact on metal(loids) mobility in soil/plant systems, soil microbial activity and plants. The role of plant growth-promoting rhizobacteria as biological control agents in the biosorption of soil contaminants is focused in the 11th chapter, whereas the 12th chapter describes the joint action of plant/bacteria and fungi in removal of metal(loid)s, radionuclides, and chlorinated compounds. The 13th chapter delivers a complete review of phosphate-assisted phytoremediation of potentially toxic metal(loid)s in soil. The multiple aspects about the applicability of electrokinetic-assisted phytoremediation, such as the electrochemical processes and the physicochemical changes occurring in the soil, along with the effects of electric current on plant biomass and some practical issues of the technique are extensively discussed in the 14th chapter. The 15th chapter focuses on different aspects of biosurfactants-assisted phytoremediation and their role in the management of heavy metals and petroleum-contaminated soils.

Vimal Chandra Pandey
Editor

Acknowledgments

I sincerely wish to thank Maris LaFleur (Acquisitions Editor), Michelle Fisher (Editorial Project Manager), and Bhaskaran Srinivasan (Copyrights Coordinator) and Kumar Anbazhagan (Production Project Manager) from Elsevier for their excellent support, guidance, and coordination of this fascinating project. I would like to thank all the authors for their excellent chapter contributions. I would like to thank all the reviewers for their time and expertise to review the chapters of this book. Special thanks go to Prof. (Mrs.) Manju Sharma, Distinguished Woman Scientist Chair, Past President, NASI, and Former Secretary to Govt. of India for providing the "Foreword" of this book. Finally, I want to thank my beloved wife and sons for their unending support, interest, and encouragement, and apologize for the many missed dinners!

CHAPTER

1

Understanding assisted phytoremediation: Potential tools to enhance plant performance

Garima Malik[a], Sunila Hooda[b], Sahrish Majeed[b], Vimal Chandra Pandey[c]

[a]*Raghunath Girls' Post Graduate College, C.C.S. University, Meerut, India*

[b]*Ram Lal Anand College, University of Delhi, India*

[c]*Department of Environmental Science, Babasaheb Bhimrao Ambedkar University, Lucknow, India*

1.1 Introduction

Land and soil degradation is one of the global problems that humanity is facing today. The severity of soil degradation has affected ecosystem functions and services. Globally, more than three billion people are suffering by land degradation, especially small farmers and poor people. Scientists have warned that land degradation is happening at an alarming pace and if the trend continues ~90% of world land could become degraded by 2050. Worldwide, governments are investing billions of dollars to restore/reclaim polluted and degraded lands. Sustainable land management has become a focal area of policy makers and efforts are being made to adopt eco-friendly methods for consistent restoration of polluted lands at global scale.

Phytoremediation, the utilization of plants to eradicate, degrade, or stabilize organic and inorganic pollutants from the environment, is a promising, profitable, and eco-friendly bioremediation method (Pandey and Singh, 2020). The basic idea that vegetation (trees, shrubs, grasses, and aquatic plants) can be used for soil, air, and water remediation is primitive; however, several novel scientific studies along with an interdisciplinary research approach has led to the expansion of this knowledge into a global method for restoration of ecological environment (Gajić et al., 2019; Pandey and Bauddh, 2018; Pandey and Souza-Alonso, 2019; Gajić et al., 2020a; Grbović et al., 2019; Grbović et al., 2020; Pathak et al., 2020; Pandey, 2020). Apart from contaminant removal, there are added benefits of opting phytoremediation, such as soil quality enhancement, soil carbon sequestration, biomass production, and aesthetically pleasing (Pandey and Souza-Alonso, 2019). Numerous pollutants, including heavy metals, organic compounds, pesticides, and xenobiotic can be effectively remediated by plants.

Plant-assisted bioremediation, a kind of phytoremediation, includes the collaborative action of plant roots and the microbes residing in the rhizosphere to remediate soils containing high concentrations of pollutants. Some "hyperaccumulator" plants have the capacity to accumulate huge amounts of metals in their shoots, many of these metals do not appear to be necessary for plant functioning. A large number of plant taxa belong to this category of metal hyperaccumulators; *Alyssum* and *Thlaspi* species both

Assisted Phytoremediation. DOI: https://doi.org/10.1016/B978-0-12-822893-7.00015-X

from the Brassicaceae family are the most commonly known hyperaccumulators species (Baker et al., 2000). The choice of process, phytodegradation (usage of plants and allied microbes to degrade organic contaminants), phytoextraction (utilization of pollutant-amassing plants to eliminate contaminants from soil by accumulating them in the plant parts which can be easily harvested), phytostabilization (plants decrease the bioavailability of contaminants), rhizofiltration (the utilization of roots of aquatic plants for absorption and adsorption of contaminants), phytovolatilization (plants volatilize pollutants), to be selected for cleaning depends upon the habitats, types of contaminant, and climatic conditions (Gajić et al., 2018; Gajić and Pavlović, 2018; Pandey and Bajpai, 2019).

Generally, the plants applied in phytoremediation have high contaminant tolerance, high biomass and growth rate, extensive root systems, and capability to either degrade or accumulate large amount of contaminant. The plants have the intrinsic capacity to detoxify pollutants, but they by and large do not have the necessary catabolic pathway, like microbes, for the complete degradation of contaminants (Yan et al., 2020). Moreover, phytoremediation is considered to be a slow process, and hindrance in achieving contaminant removal goals in reasonable time frame due to plant seasonality issues is an added pressure. Additionally, there are apprehensions over the probable introduction of pollutants into the food chain, and there are further problems associated with the ways in which plants that amass xenobiotics are discarded. Due to these biosafety issues, the possibility of phytoremediation as a global method to remediate environmental pollutants is still under scanner. In the last decade, information of the physiological and molecular mechanisms involved in phytoremediation began to emerge accompanied by precise and simple genetic engineering approaches intended to augment and expand phytoremediation (Gajić et al., 2016; Gajić et al., 2020b; Pandey and Singh, 2020). In order to enhance the workability of phytoremediation on ecological restoration, research is going on to assess the effects of diverse types of "catalysts" on the competency of phytoremediation. The combined use of these catalysts' approaches, for example, inclusion of genetic engineering, microbial-assisted, biochar-assisted, chelate-assisted, compost-assisted, electrokinetic-assisted, etc., to improve phytoremediation efficiency may address the weakness of plant-based remediation methods.

Recently, phytomanagement has emerged as an excellent approach for sustainable reclamation of polluted land resources. Phytomanagement is defined as the use of commercially and economically important plants to remediate soil pollution so as to make a beneficial and sustainable use of land resource by generating marketable products. The idea of phytomanagement correlates with the fact that many polluted soils can still symbolize a valuable resource that should be utilized sustainably. An additional income along with ecosystem restoration is a winning combination that will ensure benefits to the involved stakeholders and the dependent local community. The broad objective of the present chapter is to explore the different concepts of assisted phytoremediation, types, and applications in addressing aspects that enhance plant's remediation potential and future directions toward sustainable phytomanagement.

1.2 Assisted phytoremediation

To overcome the limitations of selected plant species with phytoremediation potential or to improve plant's remediation efficacy in cleaning up polluted sites, strategy to combine different types of assistance should be pursued to attain the goal of transforming phytoremediation into widely accepted technique. The combination of different approaches (discussed below) is essential for the remediation of certain polluted sites and could become a highly efficient, eco-friendly, and low-input bioremediation technology for the future (Fig. 1.1). However, successful implementation of a

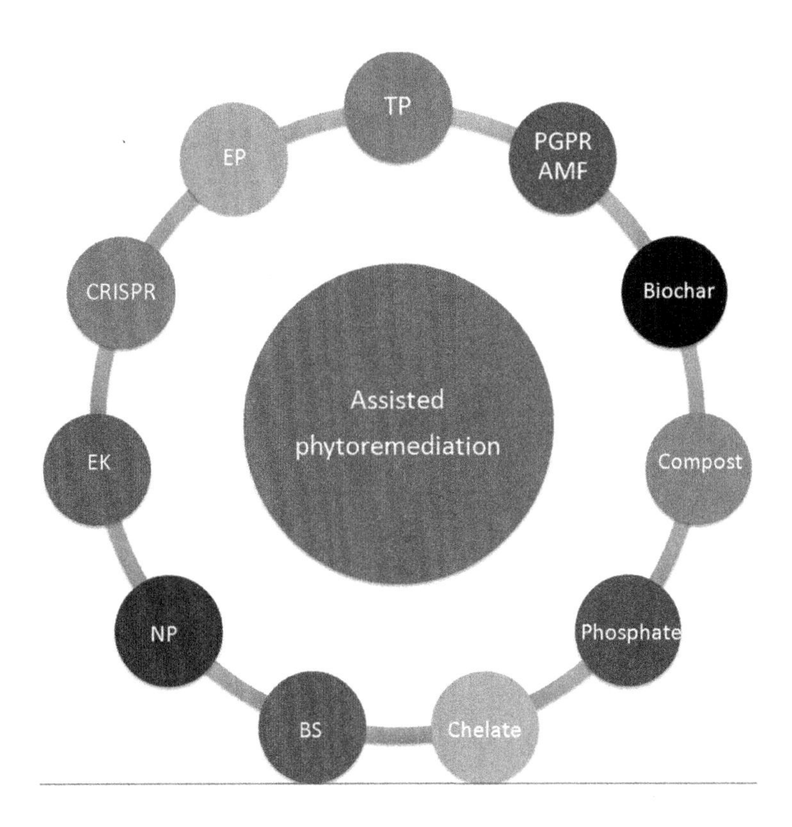

FIG. 1.1

Different approaches of assisted phytoremediation: transgenic plants (TP)/plant growth-promoting rhizobacteria (PGPR) and arbuscular mycorrhizal fungi (AMF)/biochar/compost/phosphate/chelate/biosurfactant (BS)/ nanoparticles (NPs)/electrokinetic (EK)/CRISPR/economic plants (EP).

multi-approach remediation method requires a comprehensive understanding of intricate interactions/ crosstalk at different levels.

1.2.1 Transgenic plant mediated phytoremediation

Genetic engineering or use of various molecular biology tools has helped plant biologist to understand various regulatory mechanisms that control different biochemical and physiological processes in plants. Using this knowledge, scientists have developed transgenic plants that along with substantially improved remediation competencies have broader and safer application in detoxification of polluted land and water bodies. Scientists have attempted to design plants by genetic manipulation of genes involved in uptake, breakdown, or transport of individual pollutants by manipulation of metabolic and enzymatic pathways involved in accumulation, chemical transformation, and detoxification of targeted metals and metalloids and by introducing microbial and mammalian catabolic genes to complement and enhance plant metabolic ability and accomplish nearly complete degradation of organic pollutants. The usage of genetically engineered plants for phytoremediation purpose has been reviewed by several scientists, including Van Aken (2008), Doty (2008), Cherian and Oliveira (2005).

Table 1.1 Assisted phytoremediation of contaminated sites.

Assisted phytoremediation	Contaminants	References
Transgenic plants		
Pityrogramma calomelanos *Isatis cappadocia* *Brassica juncea* *Brassica carinata* *Mimosa pudica* *Heliathus annus* *Pteris vittata* *Arabidopsis* sp.	As	Chen et al. (2013) Mecwan et al. (2018)
Plant/bacteria/fungi		
Shewanella sp. ANA-3 *Geobacter* *Bacillus* selenatarsenatis SF-1 *Sporosarcina* ginsengisoli CR5 *Candida glabrata* *Pteris vittata/Methylobacterium* *Ricinus communis/Acaulospora* sp./*Funneliformis mosseae/Gigaspora gigantea* *Zea mays/Claroideoglomus etunicatum* *Pseudomonas aeruginosa*- HMR1	As As Lead-acid batteries Lanthanum Zn	Roy et al. (2015) Rathinasabapathi et al. (2006) Gonzales – Chavez et al. (2019) Hao et al. (2021) Bhojiya et al. (2021)
Humic substances		
Humic acids	Cd Industrial waters Antibiotics Mine soil	Evangelou et al. (2004) Lipczynska-Kochany and Kochany (2008) Porras et al. (2011) Vargas et al. (2016)
Microbial enzymes		
Laccases, oxidases, peroxidases, lipase, MMO	Heavy metals, PAH, phenolic	Park et al. (2006), Korcan et al. (2013), Singh and Singh (2017)
Biochar		
Plant biochar (carbonates, organic anions) Biochar/*Amaranthus tricolor* Animal biochar	Heavy metals Cd Cr	Paz-Ferreiro et al. (2014), Wang et al. (2019) Lu et al. (2015) Choppala et al. (2012), Zhang et al. (2013)
Compost		
Municipal solid waste compost Green waste Compost/*M. oleifera*	Polluted soil Hg Cr, Cu, Pb, Zn Pb	Garau et al. (2019) Smolinska (2015) Brunetti et al. (2012) Ogundiran et al. (2018)

Table 1.1 Assisted phytoremediation of contaminated sites. ***Continued***

Assisted phytoremediation	Contaminants	References
Phosphate solubilizing bacteria		
Bacillus, Pseudomonas, Enterobacter, Pantoea, Burkholderia, Acinetobacter, Rhizobium, Flavibacterium, Penicillium, Aspergillus	Heavy metals	Ahemad (2015)
Chelate		
EDTA, DTPA, EDDS, NTA, HEDTA, CDTA, EGTA, EDDHA	Heavy metals	Huang et al. (1997), Vassil et al. (1998), Chen and Cutright (2001), Meers et al. (2005), Quartacci et al. (2005), Evangelou et al. (2007)
Biosurfactants		
Rhamnolipids, surfactin Sophorolipids, lipopeptides	Oil Pharmaceutical, cosmetic compounds Cd	Clien et al. (2007), Wang et al. (2008), Sheng et al. (2008)
Nanoparticles		
Fe nanoparticles Ti nanodioxide	TCE, PCE, c-DCE, atrazine, molinate, chlorpyrifos, TNT, DDT, chloroform Cd	Zhang et al. (2003), Ansari et al. (2019) Singh and Lee, (2016)
Electrokinetic		
EK/potato/rapeseed/tobacco EK/maize EK/maize/ryegrass	Pb, Zn, Cd, Cu Petroleum spiked soils Atrazin	Aboughlama et al. (2008) Bi et al. (2011) Rocha et al. (2019) Sanchez et al. (2018, 2019,a,b)
CRISPR		
CRISPR/poplar/maize	Heavy metals, pesticides	Tang et al. (2017), Basharat et al. (2018), Basu et al. (2018)
Economically valuable plants		
Energy plants: *Brassica napus, Brassica carinata, Cynara cardunculus, Arundo donax, Panicum virgatum, Phalaris arundinace, Miscanthus, Jatropha curcas, Ricinus communis, Sorghum, Salix, Populus, Betula, Robinia, Acer, Pyrus* Fiber crops: *Linum usitatissimum, Cannabis sativa, Gossypium hirsutum, Corchorus capsularis, Hibiscus cannabinus, Boehmeria nivea* Aromatic plants: *Chrysopogon zizanioides, Cymbopogon flexuosus, Cymbopogon* sp., *Ocimum*, mentha, lavender, *Salvia*, rosemary, chamomile, geranium	Heavy metals, mining and industrial sites Polluted sites	Pandey et al. (2015), Sathya et al. (2016), Pogrzeba et al. (2018), Hauptvogl et al. (2020), Li et al. (2014), Uddin et al. (2016), Ludvikova and Griga (2019), Guo et al. (2020), Pandey et al. (2022) Verma et al. (2014), Pandey et al. (2019), Pandey and Praveen (2020), Pandey et al. (2020)

Presence of elevated levels of heavy metals in the agricultural soil has posed a severe risk to environmental health in addition to the negative health impacts on human and animals, over the last few decades. Arsenic (As) is one such toxic element that cannot be degraded, although transformation between different oxidation states is possible. As pollution is a worldwide menace, especially in South Asia, including Bangladesh and eastern parts of India. Various plants, including *Pityrogramma calomelanos, Isatis cappadocica, Brassica juncea, Brassica carinata, Mimosa pudica*, and *Helianthus annus* have been shown either to amass tremendously high concentrations of As or show hypertolerance (Mecwan et al., 2018) (Table 1.1). Chinese brake fern, *Pteris vittata* L., was the first plant to be reported as As hyperaccumulator. Several other fern species in the Pteridales have been identified to be As hyperaccumulator. Two genes of *P. vittate* namely, PvACR2 (encoding for arsenate reductase) and PvACR3 (encoding for arsenite efflux/arsenite [As(III)] antiporter) were identified, characterized, and reported to have a critical role in detoxification and storage of As (Chen et al., 2013; Mecwan et al., 2018). Transgenic *Arabidopsis* expressing PvACR3 accumulated ~7.5-fold more As and showed greatly enhanced As tolerance in comparison to wild type plants (Chen et al., 2013) (Table 1.1). Specific studies to identify and characterize genes involved in hyperaccumulation in terms of uptake, transport, and sequestration may help scientist to improve the natural mechanism in plants and thereby in generating transgenic plants. Using genetic engineering tools to develop transgenic plants having high level of metal-binding proteins or transgenics that may release specific ligands for metal selection into the rhizosphere and thus solubilize elements on the contaminated site may help in fixing the problem of heavy metal contaminated soil in near future.

The specific improvement obtained in the plants *via* transgenic approaches has opened up new opportunities in the field of phytoremediation; however, not enough research statistics is available to comprehend the phytoremediation potential of these genetically engineered plants on the basis of field performance. Therefore, focused field trials studies are needed to promote transgenic plant mediated phytoremediation a commercially feasible and suitable technology. Successful generation of transgenic plants having varied remediation properties, wide applicability and desired field-testing results will provide green and sustainable prospects for environmental clean-up, leading to enhanced air, soil, and water qualities.

1.2.2 Phytobial remediation by bacteria and fungi

Microbial-assisted phytoremediation is one of the most widely used methods for the removal of pollutants from heavily contaminated soils and water bodies (Gerhardt et al., 2009; Reddy, 2010; Bhojiya et al., 2021; Majhi et al., 2021). It is an *in-situ* biological remediation method involving bacteria and fungi. It is also known as phytobial remediation as it combines both bioremediation and phytoremediation to decrease the level of inorganic and organic pollutants from soil. Since microorganisms play a significant role in biogeochemical cycles, whether its mineralization or biotransformation, they become the key component that cannot be neglected in any phytoremediation process. The microbial communities found in the soil are further divided into three types depending on their association and individual roles: (1) Free living microbes in soil, (2) microbes associated with rhizosphere, and (3) microbes present within plants or endophytes.

The mechanism used by free living microorganism includes immobilization, mobilization, biotransformation reactions. The microorganism like *Shewanella* sp. strain ANA-3, *Geobacter, Bacillus selenatarsenatis* SF-1 use mobilization mechanism for As removal, while *Sporosarcina ginsengisoli*

CR5, *Candida glabrata, Schizosaccharomyces pombe* use immobilization mechanism as reviewed by Roy et al. (2015) (Table 1.1). Biotransformation of metal ions is done by many bacteria, fungi and algae. The significant contribution of arbuscular fungi in enhancement of metal tolerance by phytoextraction and phytostabilization has been reviewed recently (Janeeshma and Puthur, 2020; Rozpądek et al., 2019). Rhizosphere associated microorganism are mainly known as plant growth promoting rhizobacteria (PGPR), which aid in the growth of plants by solubilization of P and chelation of heavy metals. They secrete carboxylic acids, siderophores, phytohormones, which increase availability and mobilization of heavy metals and support plant growth. The endophytes are known to have a significant role in plant growth; they are also explored for their role in degradation of xenobiotics like volatile aromatic hydrocarbons. Many heavy metal tolerant endophytic fungi like *Phomopsis* and *Bipolaris* isolated from many plants growing in polluted sites may prove to be useful in microbial-assisted phytoremediation as reviewed by Rozpądek et al. (2019). *Methylobacterium,* which is an endophyte associated with *Pteris vittate,* was shown to have As tolerance (Rathinasabapathi et al., 2006).

Phytobial remediation offers various advantages as the *in situ*, cleanest, and economical method that can be used on a wider scale for decontamination of polluted soil, sediments, and groundwater. Further, it helps in soil preservation and nutrient enrichment of the soil. However, this method has its own limitation in terms of long duration and need of regular monitoring for heavy metals entry into food chain.

1.2.3 Arbuscular mycorrhizal fungi-assisted phytoremediation

In natural conditions, arbuscular mycorrhizal fungi (AMF) are known to form symbiotic alliance with the roots of majority of vascular plant and offer protection against abiotic and biotic stress. AMF mycelium creates an extensive below-ground complex which serves as a conduit between soil, plant roots, and soil microbes. The extra matrical hyphae form hyphosphere by extending the rhizosphere, and in turn significantly increase plant's nutrients and pollutants acquisition. Many AMF (*Glomus* sp., *Gigaspora* sp., *Entrophospora* sp., etc.) have been shown to be present in close association with roots of plants growing in heavy metal contaminated sites indicating that these fungi may have evolved heavy metal tolerance mechanism and therefore may be effectively utilized in mitigation of heavy metal toxicity in contaminated sites (Mathur et al., 2007). The augmentation of heavy metals (such as Zn, Cd, As, and Se) phytoaccumualtion was proved by inoculation of plant roots with AMF (Giasson et al., 2005).

A number of research concluded that AMF inoculation have better outcome when native fungi and early seral fungi are used as a consortium instead of single/few species (Asmelash et al., 2016). Recently, *Ricinus communis* plants inoculated with AMF (*Acaulospora* sp., *Funneliformis mosseae*, and *Gigaspora gigantea*) showed 100% survival as compared to non-inoculated plants (57%) on modified soil severely polluted by lead-acid batteries (LAB) in Mexico (González-Chávez et al., 2019) (Table 1.1). Similarly, AMF (*Claroideoglomus etunicatum*) inoculated maize augment maize tolerance to lanthanum (La) and ease La phytotoxicity due to the prominent impact of AMF on the development of plants and microorganisms present in the rhizosphere soils (Hao et al., 2021) (Table 1.1). Various scientific studies indicate that AMF have the potential to enhance the phytoremediation capabilities of plants in conditions where multiple kinds of contaminants are present. AMF-assisted phytoremediation is therefore a valuable approach for the remediation of heavy metals contaminated soils and other inorganic pollutants, restoration of degraded lands and should be considered for forthcoming studies. Future research efforts should focus on formulating economical inocula creation approaches and proper AMF management which is still in its infancy. Progress in these fields may be utilized for proper plant and its fungal symbionts selection and in turn will be helpful in augmenting necessary conditions required for effective AMF-mediated phytoremediation.

1.2.4 Bioremediation with PGPR, humic substances, and enzyme combination

Increased concentration of heavy metals in agriculture soil interferes with plant growth and metabolism, affect activity of soil microbes, and reduce soil fertility and plant yield. Plants used for phytoextraction have limited metal tolerance, as increase in metal concentration beyond a certain threshold can be toxic and result in slow growth rate and decreased biomass production (Jing et al., 2007). Interactions between soil, microorganisms, metals, and plant roots are controlled by environmental conditions, characteristic of soil and microbes, which in turn determine the extent of phytoremedition (Glick, 1995). Rhizosphere, region around plant roots, contains diverse microbial population, including PGPR that are important in promoting plant growth and nutrient recycling by direct methods (e.g., N fixation, releasing chemical substances like siderophores, plant hormones like auxin and cytokinins that helps in cell division, differentiation, and root development) and indirect methods (including synthesis of antibiotic to control pathogens and prevention of diseases) and affect the availability of soluble metals to plants using various strategies (Hayat et al., 2010). Iron-siderophores formed by rhizobacteria provide iron to chlorotic plants growing in heavy metal concentrated soils which are deficient in iron content (Reid et al., 1986; Imsande, 1998). The potential use of PGPR along with phytoextractor plants is a novel approach towards phytoremediation of metal polluted soils. The addition of PGPR, *K. ascorbata* SUD165/26, has increased production and plant size of Indian mustard by decreasing Ni concentration in the soil (Burd et al.,1998).

Potential use of humic substances (HS) has become an interesting alternative to increase the efficiency of phytoextraction to researchers. HS, organic decomposition products of terrestrial and aquatic biomass, are known to stimulate both plants and microbial activities through various mechanisms (Ekin 2019). Humic acids, fulvic acid, and humin are the three types of HS depending upon their solubility. HS decrease the bioavailability of organic contaminants and hence their toxicity (Tranvik, 2014). Significant increase in Cd levels in the shoots of plants in Cd contaminated soil was marked with the application of humic acid, thereby indicating enhanced metal uptake by plants due to addition of humic acid (Evangelou et al., 2004). Multiple researches have been done to study the effect of HS on bioremediation of heavily contaminated industrial waters (Lipczynska-Kochany and Kochany, 2008), phytodegradation of soils contaminated with petroleum hydrocarbons (Park et al., 2011), antibiotics like ciprofloxacin which poses threat to aquatic ecosystem (Porras et al., 2016), and combined use of HS and vetiver grass for phytoremediation of mine soils (Vargas et al., 2016) (Table 1.1).

Microbial enzymes are known to degrade various pollutants, restore and remediate the ecosystem, and regulate various biochemical processes. Extracellular enzymes like laccases, oxidases, peroxidases secreted by white rot fungi can degrade PAHs and lignins (Korcan et al., 2013) (Table 1.1). Oxidoreductases can detoxify phenolic pollutants (Park et al., 2006). Methane monooxygenase from methanotrophs can metabolize aromatic and heavy metals and is thus used as a bioremediation tool (Pandey et al., 2014). Lipase producing microorganisms are often exploited for bioremediation of oil polluted environments (Basha, 2021). Use of enzymes is more feasible and efficient as compared to using a whole cell which requires proper environmental conditions and nutrition to survive. The large-scale production of metal degrading enzymes from various microorganisms and their *in-situ* application in polluted sites can be included in the strategies to be investigated further.

1.2.5 Biochar assisted phytoremediation

Biochar amended phytoremediation has increasingly proven to be a promising approach, which can be utilized to eliminate different contaminants in soils. Biochar is a C-rich product made by pyrolyzing biomass wastes obtained from agriculture and forestry operations (Liu et al., 2011; Wang et al., 2010). Bio-oil and syngas are also obtained as byproducts of biochar production. Depending upon temperature, residence time of biomass and heating rates, pyrolysis is of two types: fast and slow. High-temperature pyrolysis and water vapor activation generally yields biochar with high pH (Hass et al., 2012) and reduced cation exchange capacity. Biochar has larger surface area which controls the chemical adsorption onto biochar. Inorganic carbonates and organic anions add to the alkalinity of biochars that partly lowers the concentrations of accessible heavy metals in biochar assisted phytoremediation in agricultural soils (Paz-Ferreiro et al., 2014; Wang et al., 2020) (Table 1.1). Several characteristics of biochar, such as specific surface area, elemental composition, surface chemical property, pH etc., vary, depending upon pyrolysis conditions, residence time, type, and moisture content of biomass used (Wang et al., 2020; Zhang et al., 2013). Carbon content is low in animal biochar whereas protein and inorganic substances are present in large amount. Moreover, animal biochar shows higher content of ash and phosphorus (P) as compared to plant biochar, which has the potential to boost the soil productivity (O'Kelly, 2014). Therefore, properties of soil and biochar are important to consider while selecting biochar for remediation process (Paz-Ferreiro et al., 2014).

Various researches have revealed biochar application effectively immobilize heavy metals which lowers the phytotoxicity and bioavailability of these metals. The mechanism of biochar assisted remediation in metal polluted soils include ion exchange, electrostatic interaction, physical adsorption, complex formation and precipitation, thereby decreasing the bioavailability of the pollutants and significantly reducing the chances of entry of the toxic substances into biological food chain or leaching to groundwater. Biochar obtained from chicken manure when applied to chromate polluted soils immobilized Cr and prevented its leaching by reducing Cr(VI) to Cr(III), which has comparatively lesser mobility (Choppala et al., 2012; Zhang et al., 2013) (Table 1.1). Biochar-amended phytoremediation has huge potential for immobilization of heavy metal cations in tailings and mining affected soils, specifically in acidic ones. Biochar-amended phytoextraction approach using *Amaranthus tricolor L.* was applied in the remediation of Cd contaminated soil (Lu et al., 2015). Biochar also facilitates the application of phytostabilization. In another study, effectiveness of biochar is also documented using AMF (Lehmann et al., 2011). Additional benefit includes improvement of plant response to diseases in biochar assisted soils (Graber et al., 2010). Combination of biochar and phytoextractors seems to be the effective methodology in remediating multimetal contaminated soils, where each of them independently work to eliminate two elements at the same time.

1.2.6 Compost-assisted phytoremediation

Compost is a type of organic amendment which is added to agricultural soil to improve physicochemical and biological condition. It facilitates phytostabilization in which plants are used to stabilize and immobilize wastes, thereby alleviating erosion of top soil, leaching of toxic pollutants into groundwater (Sinhal et al., 2015). The common types of compost that have been used in various studies are: domestic waste, green waste, sewage sludge, cattle manure, poultry manure, etc. It is composed of nutrients, mineral ions, humic matter, microbes, and enzymes. Compost can further improve the efficiency of phytostabilization either alone or in combination with biochar or other biosorbents (Garau et al., 2019;

Mary et al., 2016). Compost-assisted phytoremediation not only helps in removal of soil pollutants but also enhances the physical, chemical, and biological properties of soil along with microbial activity and plant growth (Bacchetta et al., 2015; Garau et al., 2014; Manzano et al., 2016). It is a cost-effective, green alternative to restore heavy metal-contaminated agricultural soil and has generated considerable interest in the last decade (Alvarenga et al., 2008; Castaldi et al., 2018).

The biostimulatory effect of municipal solid waste compost (MSWC) with cardoon plants in polluted soils was reported by Garau et al. (2019). Recently, Garau et al. (2021) compared the different grass and legume species with MSWC in assisted phytoremediation. Green waste compost was reported to efficiently extract Hg (Smolinska, 2015), while Brunetti et al. (2012) showed that it helps in the increased build-up of heavy metals such as Cr, Cu, Pb, and Zn (Table 1.1). In a similar study, immobilization of Ni, Pb, and Cu and phytoextraction of Ni by mustards was reported by Rodríguez-Vila et al. (2015). The enhanced phytoremediation capability was reported when compost and *M. oleifera* were used in combination in Pb contaminated soil (Ogundiran et al., 2018).

The use of compost mediated phytoremediation is an effective strategy for repair of multiple heavy metal contaminated soils. Moreover, it will also reduce long-lasting impacts associated with compost application. However, detailed studies are required to further understand the underlying mechanism of interaction of microbes present in compost and their significance in plant rhizosphere in agricultural soils as well as metal polluted sites.

1.2.7 Phosphate-assisted phytoremediation

Phosphate solubilizing bacteria (PSB) facilitates the remediation of polluted soils by converting insoluble P into soluble organic phosphates, which can be readily metabolized by plants. In heavy metal contaminated environments, PSB aid in phytostabilization and phytoextraction of metal species (Ahemad, 2015). Therefore, PSB assisted phytoremediation approach is gaining more popularity in the heavy metal contaminated soils. Majority of PSB belong to *Bacillus, Pseudomonas, Enterobacter, Pantoea, Burkholderia, Acinetobacter, Rhizobium,* and *Flavibacterium* genus. *Penicillium* and *Aspergillus* species have also been used for phosphate assisted phytoremediation recently. These microorganisms can be either applied as a pure culture or as consortium (Jia et al., 2016; Gupta and Kumar, 2017) (Table 1.1). The practice of using microbial consortium is more effective and gaining more popularity worldwide as it increases metal uptake by plants by converting these heavy metals into soluble and bioavailable forms, which also enhances the crop growth and thus facilitate efficient phytoremediation. The bioavailability of organic phosphorus depends on the solubilization, mineralization, and immobilization processes. Microbes provide organic acids, indole acetic acid, siderophores, and ACC deaminase, which in turn contribute a lot in increasing the phytoremediation capability of plants.

The key role of PSB consortium in enhancement of phytoremediation potential of various crops plants has been studied. However, it needs to be explored in more detail considering the different plants species, bacterial species, concentration of contaminants, soil physicochemical characteristics, and other environmental factors.

1.2.8 Chelate assisted phytoremediation

In recent years, to increase the metal uptake capacities of hyperaccumulating plants, different phytoremediation strategies were adopted, including addition of chelating substances like ethylene diamine tetraacetic acid (EDTA), ethylene diaminedisuccinate (EDDS), and nitrilotriacetic acid (NTA).

Chelating agents bind with metals to form soluble complexes, hereby enhancing their bioavailability in soil. EDTA was suggested to be used as chelating agent in late 20th century to enhance the effect of phytoextraction (Evangelou et al., 2007) (Table 1.1). A lot of studies have been done on the effect of application of EDTA as chelator in different metal polluted soils using different plant species, its effect on the mobilization of heavy metals, solubilization, and bioavailability in soil, thereby its implication in phytoremediation. The addition of EDTA resulted in the facilitated Cd and Ni translocation across the plant but no translocation was observed in case of Cr (Chen and Cutright, 2001) (Table 1.1). Other studies reported that EDTA forms soluble complex with Pb which facilitates its uptake by the plant (Vassil et al., 1998). EDTA has been demonstrated to significantly improve phytoextraction but due to its poor degradability, it has become a major environmental concern (Oviedo and Rodríguez, 2003).

Other synthetic APCAs include hydroxyl ethylenediaminetetraacetic acid (HEDTA), trans-1,2-cyclohexylenedinitrilotetraacetic acid (CDTA), diethylenetriaminopentaacetic acid (DTPA), ethylene bis[oxyethylenetrinitrilo]tetraacetic acid (EGTA), ethylenediamine-N, N'bis(O-hydroxyphenyl)acetic acid (EDDHA) (Evangelou et al., 2007) (Table 1.1). EDTA was reported to have maximum effect of them all in Pb accumulation in pea and corn (Huang et al., 1997). EDDS and NTA are natural biodegradable APCAs which were studied for their phytoremediation potential in the last few years. Meers et al. (2005) reported that EDDS was more efficient than EDTA in the extraction of Cu, Ni, and Zn, whereas Pb and Cd mobilization occurred more by EDTA than EDDS (Table 1.1). Similar results were reported by Luo et al. (2005). NTA was found to be more effective in As and Zn extraction than other synthetic APCAs in metal contaminated soils using vetiver and maize (Chiu et al., 2005). Research has also been conducted investigating the role of organic acids in metal chelation and as an alternative to EDTA. Low molecular weight organic acids are weak organic acids like citric acid, tartaric acid, malic acid, etc., secreted by microorganisms and plant roots in rhizosphere which influence the uptake, bioavailability, solubility of metal ions, microbial activity, and other physical structure of the soil (Evangelou et al., 2007; Agnello et al., 2014). In a study on Cd extraction by Indian mustard by Quartacci et al. 2005, it was observed that the shoots showed enhanced cadmium accumulation by a factor of 2.6 in NTA (20 mmol kg^{-1}) amended soil and 2.3 in citrate (20 mmol kg^{-1}) amended soil. Chen et al. (2006) reported the enhancement in Cu uptake and accumulation in shoots when treated with citric acid and glucose. It is to be noted that the ability of chelating agents to bind with metals is largely dependent on the type of heavy metal and the plant species used (Evangelou et al., 2007).

1.2.9 Biosurfactant assisted phytoremediation

Biosurfactants consist of both hydrophobic and hydrophilic moieties that is, they are amphipathic surface-active molecules. These compounds act at liquid interfaces (oil/water or water/oil) reducing surface tension and increasing the surface of contact of the hydrophobic molecules, thereby enhancing their solubility and bioavailability (Pacwa-Płociniczak et al., 2011; Santos et al., 2016; Silva et al., 2014). As biosurfactants can enhance the mobility and biodegradation of organic compounds, they are suitable for bioremediation purposes. Low molecular weight compounds effectively reduce surface and interfacial tensions, while high molecular weight compounds can efficiently do emulsion-stabilization (Rosenberg and Ron, 1999). Efficiency of biosurfactants is measured using critical micelle concentration. Biosurfactants are less toxic, biodegradable, highly selective, can detoxify pollutants, and operational in wide range of temperature, pH, and saline conditions and thus favored over synthetic surfactants (Silva et al., 2014). These properties have gained much attention of scientists and environmentalists.

Biosurfactants forms complex with heavy metals at the rhizosphere soil interface, inducing desorption of metals and increasing their bioavailability in soil for better uptake by plant roots. Use of biosurfactants can enhance phytoremediation process and many studies have been conducted to assess this new approach. Agnello et al. (2014) mentions two ways to biosurfactant assisted phytoremediation namely, (1) application of biosurfactants to the contaminated site and (2) inoculation of biosurfactant producing microorganisms to the contaminated site. Bacteria *Pseudomonas aeruginosa* and *Bacillus subtilis* producing rhamnolipids and surfactin, respectively, have been studied extensively as the large producers of biosurfactants. Species of the genus *Candida* are commonly studied yeasts in the biosurfactant production. Rhamnolipid is efficient in the bioremediation of oil contaminate sites (Clien et al., 2007) (Table 1.1). Wang et al. (2008) identified and isolated a surfactin producing strain of *Bacillus subtitlis* and showed that it can be used to remove contaminants in pharmaceutics, environment, cosmetic, and oil recovery. Sophorolipids, lipopeptides, and other biosurfactants have also been evaluated for their bioremediation potential in the hydrocarbon contaminated marine and terrestrial locations. Zhu and Zhang (2008) in their hydroponic experiment reported that the application of rhamnolipid enhanced the uptake of PAHs by ryegrass (*Lolium multiflorum*) (Table 1.1).

Using bioaugmentation strategy in phytoremediation, the positive influence of biosurfactant producing *Bacillus* strain on biomass of tomato and its Cd uptake was evaluated (Sheng et al., 2008). The practicability of biosurfactant producing microbial strains in *in situ* bioremediation strategies of highly polluted environments is a novel and eco-feasible approach. But little is known about this as most of the research is done in laboratory conditions. Further efforts are to be made to evaluate other potential applications of these surface-active molecules in the clean-up of toxic contaminants from the environment.

1.2.10 Nanoparticle assisted phytoremediation

Nanoparticles (NPs) can easily penetrate and adsorb contaminants because of their large surface area to volume ratio (Ansari et al., 2019), which is basis of nanotech phytoremediation techniques to degrade and reduce various contaminations in soils. Fe based NPs, for example, nanoscalezerovalent iron (nZVI) has been widely used for degrading the pollutants like trichloroethylene (TCE), tetrachloroethylene (PCE), cis-1,2-dichloroethylene (c-DCE) (EPA, 2011). Studies have shown that nanosized zerovalent ions can degrade organic pollutants like atrazine, molinate, and chlorpyrifos (Ansari et al., 2019) (Table 1.1). Cd extraction was improved in soyabean plants by the addition of nano titanium dioxide (Singh and Lee, 2016). The various contaminants present in environment like chlorinated organic compounds, including organochlorine pesticides, and PCBs can be detoxified using nano iron particles (Zhang, 2003) (Table 1.1).

NPs have also been used in enzyme-assisted phytoremediation techniques. Pollutants such as TNT, chloroform, DDT, As etc., are potentially remediated by nano irons. NPs can be obtained from plants, fungi, bacteria and used for the clean-up of heavily contaminated environments (Yadav et al., 2017). The feasibility and efficiency of nanophytoremediation is based on certain factors. These include the physical and chemical properties of NPs as well as pollutants, environmental conditions, and plant dynamics (Ansari et al., 2019). Nanotechnology has not only proven to be efficient in the remediation of contaminated land but also a promising technique in facilitating the removal of contaminants in wastewater treatment processes. For example, CNTs functionalized with polymers can potentially facilitate the removal to heavy metal ions from water (Theron et al., 2008). NPs also have the ability to promote the growth of plants and increase biomass. The accumulation and uptake of NPs by plants determine the extent of nanophytoremediation processes (Ebrahimbabaie et al., 2020). Souri et al. (2017)

demonstrated the combined application of salicylic acid nanoparticles enhanced root-shoot elongation and development of plant under As stress conditions. Application of NPs along with soil amendments like biochar has also proven to efficiently increase plant growth and biomass. Zand et al. (2020) showed that the combined application of nZVI and biochar on metal contaminated soil resulted in the enhanced plant germination and biomass. Though application of nanotechnology assisted phytoremediation strategies has shown favorable results, yet there are very few studies in this field and much more is left to be explored further.

1.2.11 CRISPR-assisted strategies for futuristic phytoremediation

A range of genes and regulators have been identified in bacterial and plant species with potential in phytoremediation using multi-omics approach and *in silico* tools. As mentioned above, studies were focused on improving the genetic makeup of organisms using transgenic approaches for phytoremediation. Recently, CRISPR/Cas9 system has been developed as a simple, efficient, relatively inexpensive and easy to carry out approach for genome editing (Jinek et al., 2012). It is being proposed as a more precise approach for plant genome engineering as compared to the transgenics based approach (Wada et al., 2020). Though initially proposed as a flexible nuclease mediated gene editing toolbox to bring out changes at the target site, it can be used to carry out substitutions in genomes and targeted epigenetic modifications (Gallego-Bartolomé et al., 2018). It is used widely across a range of organisms including microorganisms, animals and model plants in addition to crop plants (Montecillo et al., 2020). The versatile CRIPR/cas9 technology has also found various applications in environmental and agricultural research. Yet, CRIPSR mediated genome editing of plants to be used to remediate polluted soils and water bodies remain to be explored fully.

The CRIPSR mediated gene editing is demonstrated for expression of particular genes in *Pseudomonas* (Chen et al., 2018), *Escherichia coli* (Marshall et al., 2018), *Achromobacter* species (Liang et al., 2020), etc. Similarly, CRIPSR/Cas9 technique can be applied to increase the phytoremediation potential of plants to a broad range of organic as well as inorganic contaminants in soil (Basharat et al., 2018). In plants, CRISPRi (interference) and CRISPRa (activation) which modulates gene expression by the use of gRNA-guided dCas9 has been proven useful for phytoremediation (Lowder et al., 2015). In rice, CRISPR mediated deletion of metal transporter gene (OsNramp5) resulted in reduced Cd accumulation (Tang et al., 2017) (Table 1.1). Poplar and maize genomes have been engineered using CRIPSR/Cas9 technique for phytoremediation use (Basharat et al., 2018). Moreover, the significant contribution of PGPRs in phytoremediation can also be explored using Cas9/sgRNA system to get customized desired genomic modifications (Basu et al., 2018).

1.2.12 Electrokinetic assisted phytoremediation

Electrokinetic assisted phytoremediation process includes the insertion of electrodes in the metal polluted soils. The ions present in the soil under treatment are mobilized between the two electrodes: anode and cathode under the low intensity current, facilitating solubilization of metals by the mechanism of electro-osmosis and electrophoresis (Cameselle, 2012). It can be either carried out as sequential electrokinetic assisted phytoremediation or as coupled EK-phytoremediation. Sequential electrokinetic assisted phytoremediation involves the electrokinetic remediation (EKR) in the first step and phytoremediation in later stage. It is suitable to be applied in heavy metal contaminated soils, which in turn may be followed by phytoremediation. It aids in the clean-up of residual contaminants and improvement in soil physicochemical properties (Wan et al., 2012). In coupled electrokinetic-phytoremediation technique, a low intensity current is applied to the growing plants in metal contaminated soil. It increases the bioavailability of the metals to the plants; therefore, more suitable for hyperaccumulator plants (Vamerali et al., 2010). It can be used in two different ways. It can be used for EKR to improve soil physicochemical properties. Secondly, it can also be used for electrokinetic stabilization to improve the availability of contaminants to the growing plants. EK-phytoremediation process has been applied to bioremediation of barren acidic soils (Faizun, 2014). Cang et al., 2010; 2011 compared the biomass production and heavy metal removal in the Indian mustard under low and high voltage conditions in soils contaminated with two or more metals. The efficacy to remove Pb, Zn, Cd, and Cu ions by EK-phytoremediation was studied using potato plants (Aboughlama et al., 2008) (Table 1.1). In another study, influence of electric field on phytoremediation potential was checked in metal polluted soils with rapeseed and tobacco plants (Bi et al., 2011). There are several other published reports on use of electrokinetic phytoremediation to remove both organic and inorganic contaminants (Acosta-Santoyo et al., 2017; Chirakkara et al., 2015; Sanchez et al., 2019a,b) (Table 1.1). Recently, Rocha et al. (2019) reported coupled EKR was shown to enhance maize plant biomass and remediation efficiency in petroleum spiked soils (Table 1.1). Recently, effectiveness of EK assisted phytoremediation was used in soils contaminated with atrazine using maize and ryegrass (Sanchez et al., 2018; 2019a,b) (Table 1.1).

The EKR depends on various factors like electrode type, electrolyte used, and voltage and duration of current applied. The limitation of this method is that it is not suitable for all soil types. Most of the studies have been done in metal polluted soils and there is an increasing need to study the application of EKR to remove organic pollutants. The advantages associated with EKR include increase the soil pH, soil strength, and availability of nutrients into the soil. The addition of chelators like EDTA, ammonium sulfate, critic acid, and organic compost can further improve phytoextraction in contaminated soils. This process is yet to be applied at a large scale, but it seems to be an attractive alternative along with other phytoremediation methods.

1.3 Potential possibilities of application of assisted phytoremediation for utilizing polluted sites using economically valuable plants: economic and environmental sustainability

The above-described assisted remediation approaches in combination with economically valuable plants having remediation potential would significantly augment the current status of eco-friendly phytoremediation technology into a valuable tool for converting polluted lands into profitable and sustainable land resources by producing marketable products from phytoremediated biomass and yields. In this direction, studies on economically valuable plants that have colonizing nature, are perennial and unpalatable by livestock, and yet have socioecological value is urgently required. The above-described tools to enhance plant performance should be harnessed for establishment, growth, and yield of economically valuable plants in phytoremediation programs.

Various economically valuable plant species, such as energy crops, fiber plants, and aromatic plants, are known to have the capability to endure and remediate contaminants from the polluted and degraded lands (Pandey et al., 2015; Pandey et al., 2016; Tripathi et al., 2016; Grzegórska et al., 2020; Luo et al., 2020). Many more economically important plants are being explored to identify, characterize, and categorize their potential for commercial and sustainable phytoremediation practices. Assisted remediation using economically valuable plant has emerged as a holistic and multipurpose approach for improving quality of soil, restoring degraded lands, positive impacts on the local environment along with monetary benefits by producing marketable products. The large-scale production of such plants may play a big role in achieving the idea of phytomanagement in near future. Some economically valuable plants that must be used intensely in assisted phytoremediation programs are discussed below.

Various energy plants (such as *B. napus*, *B. carinata*, *Cynara cardunculus*, *Arundo donax*, *Panicum virgatum*, *Phalaris arundinacea*, etc.) provide high biomass yields, are resistant against abiotic stress conditions, helps to restore the desirable soil properties by accumulating toxic substances (especially heavy metals and metalloids), and thus, are suitable for sustainable phytoremediation purposes as they can be utilized for generation of local energy supplies that cuts reliance on outdoor energy source (Table 1.1). A recent study on energy crops revealed that *P. virgatum* (switch grass) accumulate lot of Zn and Pb, whereas *S. hermaphrodita* (virginia mallow) accumulate high amount of Cd and *Miscanthus* amass large amount of Pb (Pogrzeba et al., 2018) (Table 1.1). Some energy plant such as *Salix*, *Populus*, *Betula*, *Robinia*, *Acer*, and *Pyrus* are known to naturally grow in degraded and contaminated land especially, in areas affected by mining and industrial activities (Hauptvogl et al., 2020) (Table 1.1). Among the energy crop, sorghum, cultivated for production of bioethanol can be easily grown in poor quality soil and has substantial accumulation capability for heavy metals Pb, Ni, and Cu (Sathya et al. 2016) (Table 1.1). Pandey et al. (2015), emphasized on the utilization of naturally colonizing, economically treasured perennial energy crops such as *Ricinus communis*, *Jatropha curcas*, and *Miscanthus giganteus* for sustainable phytoremediation of contaminated soils (Table 1.1). Hauptvogl et al. (2020) reviewed the phytoremediation potential of energy woody crops (*Salix* and *Populus*), and energy grasses (*Miscanthus* and *Arundo*), and pointed out their role in phytoextraction, phytostabilization, and rhizofiltration of lethal contaminants from the soil by amassing them in their aboveground biomass and widespread root system. This multipurpose approach of phytoremediation along with renewable bioenergy generation from crops thriving in contaminated land resources will lead to drop in CO_2 generation, less dependency on fossil fuels, social benefits, ease the pressure on ecosystem, and keep a check on human's carbon foot printing.

Fiber crops are and will be the future source of ecofriendly, biodegradable, and recyclable raw materials and marketable products. They have a large harvestable biomass, which can be used as a commercial and industrial resource, along with generation of biofuels and energy. By 2050, the total worldwide demand for fiber is estimated to be ~130 million tonnes per year. Apart from their textile usage, fiber crops such as flax (*Linum usitatissimum*), industrial hemp (*Cannabis sativa*), cotton (*Gossypium hirsutum*), jute (*Corchorus capsularis*), kenaf (*Hibiscus cannabinus*), mesta (*Hibiscus sabdariffa*), and ramie (*Boehmeria nivea*) are well known for their phytoremediation potential (Ludvíková and Griga, 2019; Pandey et al., 2022) (Table 1.1). These fiber plants are apt for growing in industrially polluted regions, as they extract substantial quantities of heavy metals from the soil. Flax has been attaining a rising interest for prospective use in phytoremediation of Cd polluted soils due to its Cd-accumulating and tolerance capability (Guo et al., 2020). Similarly, jute, kenaf, and mesta are reported to accumulate considerable amounts of Pb (Uddin et al., 2016), whereas ramie seems to be a hopeful contender for phytoremediation of Cd-contaminated farmland (Li et al., 2014). Cultivation of fiber crops on degraded or contaminated lands is a relevant phytomanagement option as it will not only extend the productivity of the marginal land and economic gains but also mitigate the land-use conflict between the call for food and the growing requirement for plant based raw material.

Some plants are cultivated especially for their secondary metabolites and aromatic plants are one of them. They are grown for the production of essential oils that are used in cosmetics and food processing industries. In the recent years, a large number of aromatic plants have been studied to analyze their phytoremediation potential. Aromatic grasses such as vetiver (*Chrysopogon zizanioides*), lemon grass (*Cymbopogon flexuosus*), palmarosa (*Cymbopogon martinii*), and citronella (*Cymbopogon winterianus Jowitt*) from family – Poaceae and plants like *Ocimum*, mentha, lavender, *Salvia*, rosemary (family – Lamiaceae), chamomile (family – Asteraceae), and geranium (family – Geraniaceae) have been found to be most promising candidates for phytoremediation of heavy metal contaminated sites (Verma et al., 2014; Pandey and Singh, 2015; Pandey et al., 2019; Pandey and Praveen, 2020; Pandey et al., 2020) (Table 1.1). As aromatic plants are not directly linked to food chain and their essential oils are free from the threat of heavy metals build up, they hold an upper hand in comparison to food crops for phytoremediation purpose. Aromatic plant resources are very abundant, and are profitable and feasible option for phytomanagement of contaminated land with low cost, minimum risk, and several benefits.

Taken together, combined utilization of assisted phytoremediation approaches and economically valuable plants can offer great prospects for not only enhancing phytoremediation technologies to clean-up contaminated soils but also profitable and sustainable use of land resources.

1.4 Conclusion

The need to stabilize degraded and polluted land resources is even more critical now than in the past. Phytoremediation is a very promising and green approach for remediation of polluted sites with some limitations. Assistance *via* microbes, chelates, enzymes, genetic engineering, NPs, etc., to enhance natural capabilities of plants to perform remediation functions is a holistic novel approach. In-depth investigations and field trials should be conducted focussing these strategies to establish the optimized assisted phytoremediation technique to aid ecological restoration. The strategy we proposed for assisted phytoremediation based on economically valuable plants is indeed a safe and potential approach to attain bioeconomy security and soil restoration. Plant scientists and native farmers both are required to work together to achieve this goal by producing scientific data and using that data to make the difference on ground-level. However, in spite of the emergence of various food crops, fiber plants, and aromatic plants as phytomanagement tools throughout the world, still there is a vast gap between experimental researches in controlled conditions of labs and greenhouses and under actual conditions in the field where plant have to survive under various abiotic and biotic environmental pressures and polluted environment. Furthermore, comprehensive research on commercially important native plant species that are the best adapted to local conditions, require fewer inputs and check invasive species spread, is required to understand their potential in terms of future phytoremediation prospects.

References

Aboughalma, H., Bi, R., Schlaak, M., 2008. Electrokinetic enhancement on phytoremediation in Zn, Pb, Cu and Cd contaminated soil using potato plants. J. Environ. Sci. Health. 43 (8), 926–933.

Acosta-Santoyo, G., Cameselle, C., Bustos, E., 2017. Electrokinetic–enhanced ryegrass cultures in soils polluted with organic and inorganic compounds. Environ. Res. 158, 118–125.

Agnello, A.C., Huguenot, D., Van Hullebusch, E.D., Esposito, G., 2014. Enhanced phytoremediation: a review of low molecular weight organic acids and surfactants used as amendments. Crit. Rev. Environ. Sci. Technol. 44, 2531–2576.

Ahemad, M., 2015. Enhancing phytoremediation of chromium-stressed soils through plant-growth-promoting bacteria. J. Genet. Engg. Biotechno. 13 (1), 51–58.

Alvarenga, P., Gonçalves, A.P., Fernandes, R.M., de Varennes, A., Vallini, G., Duarte, E., Cunha-Queda, A.C., 2008. Evaluation of composts and liming materials in the phytostabilization of a mine soil using perennial ryegrass. Sci. Total Environ. 406, 43–56.

Ansari, A.A., Gill, S.S., Gill, R., Lanza, G.R., Newman, L., 2019. Phytoremediation: management of environmental contaminants. Phytoremediation Manag. Environ. Contam. 6, 1–476.

Asmelash, F., Bekele, T., Birhane, E., 2016. The potential role of arbuscular mycorrhizal fungi in the restoration of degraded lands. Front. Microbiol. 7, 1095.

Bacchetta, G., Cappai, G., Carucci, A., Tamburini, E., 2015. Use of native plants for the remediation of abandoned mine sites in Mediterranean semiarid environments. B. Environ. Contam.Tox. 94 (3), 326–333.

Baker, A.J.M., McGrath, S.P., Reeves, R.D., Smith, J.A.C., 2000. Metal hyperaccumulator plants: a review of the ecology and physiology of a biological resource for phytoremediation of metal-polluted soils. In: Terry, N., Banuelos, G. (Eds.), Phytoremediation of Contaminated Soil and Water. Lewis Publishers, Boca Raton, pp. 85–108.

Basha, P.A., 2021. Oil degrading lipases and their role in environmental pollutionRecent Developments in Applied Microbiology and Biochemistry, 2. Elsevier, pp. 269–277. https://doi.org/10.1016/b978-0-12-821406-0.00025-4.

Basharat, Z., Novo, L.A.B., Yasmin, A., 2018. Genome editing weds CRISPR: what is in it for phytoremediation? Plants 7 (3), 51.

Basu, S., Rabara, R.C., Negi, S., Shukla, P., 2018. Engineering PGPMOs through gene editing and systems biology: a solution for phytoremediation? Trends Biotechnol. 36 (5), 499–510.

Bi, R., Schlaak, M., Siefert, E., Lord, R., Connolly, H., 2011. Influence of electrical fields (AC and DC) on phytoremediation of metal polluted soils with rapeseed (*Brassica napus*) and tobacco (*Nicotiana tabacum*). Chemosphere 83, 318–326.

Brunetti, G., Farrag, K., Soler-Rovira, P., Ferrara, M., Nigro, F., Senesi, N., 2012. The effect of compost and Bacillus licheniformis on the phytoextraction of Cr, Cu, Pb and Zn by three brassicaceae species from contaminated soils in the Apulia region, Southern Italy. Geoderma 170, 322–330.

Bhojiya, A.A., Joshi, H., Upadhyay, S.K., Srivastava, A.K., Pathak, V.V., Pandey, V.C., Jain, D., 2021. Screening and optimization of zinc removal potential in *Pseudomonas aeruginosa*- HMR1 and its plant growth-promoting attributes. Bull. Environm. Contam. Toxicol. doi:10.1007/s00128-021-03232-5.

Burd, G.I., Dixon, D.G., Glick, B.R., 1998. A plant growth-promoting bacterium that decreases nickel toxicity in seedlings. Appl. Environ. Microbiol. 64, 3663–3668.

Cameselle, C., 2012. Development and enhancement of electro-osmotic flow for the removal of contaminants from soils. Electrochim. Acta 86, 10–22.

Cang, L., Wang, Q.Y., Zhou, D.M., Xu, H., 2011. Effects of electrokinetic-assisted phytoremediation of a multiple-metal contaminated soil on soil metal bioavailability and uptake by Indian mustard. Sep. Purif. Technol. 79, 246–253.

Cang, L., Zhou, D.M., Wang, Q.Y., Fan, G.P., 2010. Impact of electrokinetic-assisted phytoremediation of heavy metal contaminated soil on its physicochemical properties, enzymatic and microbial activities. Electrochim. Acta 86, 41–48.

Castaldi, P., Silvetti, M., Manzano, R., Brundu, G., Roggero, P.P., Garau, G., 2018. Mutual effect of Phragmites australis Arundo donax and immobilization agents on arsenic and trace metals phytostabilization in polluted soils. Geoderma 314, 63–72.

Chen, et al., 2013. Engineering arsenic tolerance and hyperaccumulation in plants for phytoremediation by a PvACR3 transgenic approach. Environ. Sci. Technol. 47 (16), 9355–9362.

Chen, H., Cutright, T., 2001. EDTA and HEDTA effects on Cd, Cr, and Ni uptake by *Helianthus annuus*. Chemosphere 45, 21–28.

Chen, W., Zhang, Y., Zhang, Y., Pi, Y., Gu, T., Song, L., et al., 2018. CRISPR/Cas9-based genome editing in *Pseudomonas aeruginosa* and cytidinedeaminase-mediated base editing in *Pseudomonas* species. Science 6, 222–231.

Chen, Y.X., Wang, Y.P., Wu, W.X., Lin, Q., Xue, S.G., 2006. Impacts of chelate-assisted phytoremediation on microbial community composition in the rhizosphere of a copper accumulator and non-accumulator. Sci. Total Environ. 356, 247–255.

Cherian, S., Oliveira, M.M., 2005. Transgenic plants in phytoremediation: recent advances and new possibilities. Environ. Sci. Technol. 39, 9377–9390.

Chirakkara, R.A., Reddy, K.R., Cameselle, C., 2015. Electrokinetic amendment in phytoremediation of mixed contaminated soil. Electrochim. Acta 181, 179–191.

Chiu, K.K., Ye, Z.H., Wong, M.H., 2005. Enhanced uptake of As, Zn, and Cu by *Vetiveria zizanioide*s and *Zea mays* using chelating agents. Chemosphere 60, 1365–1375.

Choppala, G.K., Bolan, N.S., Megharaj, M., Chen, Z., Naidu, R., 2012. The influence of biochar and black carbon on reduction and bioavailability of chromate in soils. J. Environ. Qual. 41, 1175–1184.

Clien, S.-Y., Lu, W.-B., Wei, Y.-H., Chen, W.-M., Chang, J.-S., 2007. Improved production of biosurfactant with newly isolated Pseudomonas aeruginosa S2. Biotechnol. Prog. 23, 661–666.

Doty, S.L., 2008. Enhancing phytoremediation through the use of transgenics and endophytes. New Phytol. 179, 318–333.

Ebrahimbabaie, P., Meeinkuirt, W., Pichtel, J., 2020. Phytoremediation of engineered nanoparticles using aquatic plants: Mechanisms and practical feasibility. J. Environ. Sci. China 93, 151–163.

Ekin, Z., 2019. Integrated use of humic acid and plant growth promoting rhizobacteria to ensure higher potato productivity in sustainable agriculture. Sustainability 11, 3417.

EPA, 2011. Nanotechnology: applications for environmental remediation, CLU-IN technology focus area fact sheet. https://clu-in.org/download/remed/nano-fact-sheet-2011.pdf. (Accessed 2021-05-21).

Evangelou, M.W.H., Daghan, H., Schaeffer, A., 2004. The influence of humic acids on the phytoextraction of cadmium from soil. Chemosphere 57, 207–213.

Evangelou, M.W.H., Ebel, M., Schaeffer, A., 2007. Chelate assisted phytoextraction of heavy metals from soil. Effect, mechanism, toxicity, and fate of chelating agents. Chemosphere 68, 989–1003.

Faizun, H., 2014. Phytoremediation Technique to Remediate the Acidic Soil Sampled at Ayer Hitam, Johor. UniversitiTunHussienOnn Malaysia, pp. 2–10.

Gallego-Bartolomé, J., Gardiner, J., Liu, W., Papikian, A., Ghoshal, B., Kuo, H.Y., Zhao, J.M., Segal, D.J., Jacobsen, S.E., 2018. Targeted DNA demethylation of the Arabidopsis genome using the human TET1 catalytic domain, Proceedings of the National Academy of Sciences of the United States of America, 115, E2125–E2134.

Gajić, G., Djurdjević, L., Kostić, O., Jarić, S., Mitrović, M., Stevanović, B., Pavlović, P., 2016. Assessment of the phytoremediation potential and an adaptive response of *Festuca rubra* L. sown on fly ash deposits: native grass has a pivotal role in ecorestoration management. Ecol. Eng. 93, 250–261.

Gajić, G., Djurdjević, L., Kostić, O., Jarić, S., Mitrović, M., Stevanović, B., Mitrović, M., Pavlović, P., 2020a. Phytoremediation potential, photosinthetic and antioxidant response to arsenic-induced stress of *Dactylis glomerata* L. sown on the fly ash deposits. Plants-Basel 9 (5), 657.

Gajić, G., Djurđević, L., Kostić, O., Jarić, S., Mitrović, M., Pavlović, P., 2018. Ecological potential of plants for phytoremediation and ecorestoration of fly ash deposits and mine wastes. Front. Environ. Sci. 6, 124.

Gajić, G., Mitrović, M., Pavlović, P., 2019. Ecorestoration of fly ash deposits by native plant species at thermal power stations in Serbia. In: Pandey, V.C., Bauddh, K. (Eds.), Phytomanagement of Polluted Sites: Market Opportunities in Sustainable Phytoremediation. Elsevier, Amsterdam, Netherlands, pp. 113–177.

Gajić, G., Mitrović, M., Pavlović, P., 2020. Feasibility of *Festuca rubra* L. native grass in phytoremediation. In: Pandey, V.C., Singh, D.P. (Eds.), Phytoremediation Potential of Perennial Grasses. Elsevier, Amsterdam, Netherlands, pp. 115–164.

Gajić, G., Pavlović, P., 2018. The role of vascular plants in the phytoremediation of fly ash deposits. In: Matichenkov, V. (Ed.), Phytoremediation: Methods, Management and Assessment. Nova Science Publishers Inc, New York, USA, pp. 151–236.

Garau, G., Silvetti, M., Castaldi, P., Mele, E., Deiana, P., Deiana, S., 2014. Stabilising metal(loid)s in soil with iron and aluminium–based products: microbial biochemical and plant growth impact. J Environ. Manag. 139, 146–153.

Garau, M., Castaldi, P., Patteri, G., et al., 2021. Evaluation of Cynara cardunculus L. and municipal solid waste compost for aided phytoremediation of multi potentially toxic element–contaminated soils. Environ. Sci. Pollut. Res. 28, 3253–3265.

Garau, M., Garau, G., Diquattro, S., Roggero, P.P., Castaldi, P., 2019. Mobility bioaccessibility and toxicity of potentially toxic elements in a contaminated soil treated with municipal solid waste compost. Ecotox. Environ. Safe. 186, 109766.

Gerhardt., K.E., Huang, X.D., Glick, B.R., Greenberg, R.M., 2009. Phytoremediation and rhizoremediation of organic soil contaminants: Potential and challenges. Plant Sci 176 (1), 20–30.

Giasson, P., Jaouich, A., Gagne, S.M., 2005. Arbuscular mycorrhizal fungi involvement in zinc and cadmium speciation change and phytoaccumulation. Remediation 15 (2), 75–81.

Glick, B.R., 1995. The enhancement of plant growth by free-living bacteria. Can. J. Microbiol. 41, 109–117.

González-Chávez, M.D.C.A., Carrillo-González, R., Cuellar-Sánchez, A., Delgado-Alvarado, A., Suárez-Espinosa, J., Ríos-Leal, E., Solís-Domínguez, F.A., Maldonado-Mendoza, I.E., 2019. Phytoremediation assisted by mycorrhizal fungi of a Mexican defunct lead-acid battery recycling site. Sci. Total Environ. 650 (2), 3134–3144.

Graber, E.R., Harel, Y.M., Kolton, M., Cytryn, E., Silber, A., David, D.R., Tsechansky, L., Borenshtein, M., Elad, Y., 2010. Biochar impact on development and productivity of pepper and tomato grown in fertigated soilless media. Plant Soil 337, 481–496.

Grbović, F., Gajić, G., Branković, S., Simić, Z., Vuković, N., Pavlović, P., Topuzović, M., 2020. Complex effect of *Robinia pseudoacacia* L.and *Ailanthus altissima* (Mill.) Swingle growing on asbestos deposits: allelopathy and biogeochemistry. J. Serbian Chem. Soc. 85 (1), 141–153.

Grbović, F., Gajić, G., Branković, S., Simić, Z., Ćirić, A., Rakonjac, Lj., Pavlović, P., Topuzović, M., 2019. Allelopathic potential of selected woody species growing on fly-ash deposits. Arch. Biol. Sci. 71 (1), 83–94.

Grzegórska, A., Rybarczyk, P., Rogala, A., Zabrocki, D., 2020. Phytoremediation-From environment cleaning to energy generation-current status and future perspectives. Energies 13 (11), 2905.

Guo, Y., Qiu, C., Long, S., Wang, H., Wang, Y., 2020. Cadmium accumulation, translocation, and assessment of eighteen *Linum usitatissimum* L. cultivars growing in heavy metal contaminated soil. Int. J. Phytoremediation 22 (5), 490–496.

Gupta, P., Kumar, V., 2017. Value added phytoremediation of metal stressed soils using phosphate solubilizing microbial consortium. World J. Microbiol. Biotechnol. 33, 9.

Hao, L., Zhang, Z., Hao, B., Diao, F., Zhang, J., Bao, Z., Guo, W., 2021. Arbuscular mycorrhizal fungi alter microbiome structure of rhizosphere soil to enhance maize tolerance to La. Ecotoxicol. Environ. Saf. 212, 111996.

Hass, A., Gonzalez, J.M., Lima, I.M., Godwin, H.W., Halvorson, J.J., Boyer, D.G., 2012. Chicken manure biochar as liming and nutrient source for acid appalachian soil. J. Environ. Qual. 41, 1096–1106.

Hauptvogl, M., Kotrla, M., Prčík, M., Pauková, Z., Kováčik, M., Lošák, T., 2020. Phytoremediation potential of fast-growing energy plants: challenges and perspectives-a Review. Pol. J. Environ. Stud. 29 (1), 505–516.

Hayat, R., Ali, S., Amara, U., Khalid, R., Ahmed, I., 2010. Soil beneficial bacteria and their role in plant growth promotion: a review. Ann. Microbiol. 60, 579–598.

Huang, J.W., Chen, J., Berti, W.R., Cunningham, S.D., 1997. Phytoremediadon of lead-contaminated soils: role of synthetic chelates in lead phytoextraction. Environ. Sci. Technol. 31, 800–805.

Imsande, J., 1998. Iron, sulfur, and chlorophyll deficiencies: a need for an integrative approach in plant physiology. Physiol. Plant 103, 139–144.

Janeeshma, E., Puthur, J.T., 2020. Direct and indirect influence of arbuscular mycorrhizae on enhancing metal tolerance of plants. Arch. Microbiol. 202, 1–16.

Jia, X., Liu, C., Song, H., Ding, M., Du, J., Ma, Q., Yuan, Y., 2016. Design, analysis and analysis of synthetic microbial consortia. Synth. Syst. Biotechnol. 1, 109–117.

Jinek, M., Chylinski, K., Fonfara, I., Hauer, M., Doudna, J.A., Charpentier, E.A., 2012. Programmable dual-RNA-guided DNA endonuclease in adaptive bacterial immunity. Science 337 (6096), 816–821.

Jing, Y.de, He, Z.li, Yang, X.e., 2007. Role of soil rhizobacteria in phytoremediation of heavy metal contaminated soils. J. Zhejiang Univ. Sci. B. 8 (3), 192–207.

Korcan, S.E., Ciğerci, İ.H., Konuk, M., 2013. White-Rot Fungi in Bioremediation. Springer, Berlin, Heidelberg, pp. 371–390.

Lehmann, J., Rillig, M.C., Thies, J., Masiello, C.A., Hockaday, W.C., Crowley, D., 2011. Biochar effects on soil biota - a review. Soil Biol. Biochem. 43 (9), 1812–1836.

Li, H., Liu, Y., Zeng, G., Zhou, L., Wang, X., Wang, Y., Wang, C., Hu, X., Xu, W., 2014. Enhanced efficiency of cadmium removal by *Boehmeria nivea* (L.) Gaud. in the presence of exogenous citric and oxalic acids. J. Environ. Sci. 26 (12), 2508–2516.

Liang, Y., Jiao, S., Wang, M., Yu, H., Shen, Z., 2020. A CRISPR/Cas9-based genome editing system for Rhodococcus ruber TH. Metab. Eng. 57, 13–22.

Lipczynska-Kochany, E., Kochany, J., 2008. Humic substances in bioremediation of industrial wastewater - mitigation of inhibition of activated sludge caused by phenol and formaldehyde. J. Environ. Sci. Heal. - Part A Toxic/Hazardous Subst. Environ. Eng. 43, 619–626.

Liu, Y., Yang, M., Wu, Y., Wang, H., Chen, Y., Wu, W., 2011. Reducing CH_4 and CO_2 emissions from waterlogged paddy soil with biochar. J. Soils Sediments 11, 930–939.

Lowder, L.G., Zhang, D., Baltes, N.J., Paul, J.W., Tang, X., Zheng, X., Voytas, D.F., Hsieh, T.-F., Zhang, Y., Qi, Y.A., 2015. CRISPR/Cas9 toolbox for multiplexed plant genome editing and transcriptional regulation. Plant Physiol. 169, 971–985.

Lu, H., Li, Z., Fu, S., Méndez, A., Gascó, G., Paz-Ferreiro, J., 2015. Combining phytoextraction and biochar addition improves soil biochemical properties in a soil contaminated with Cd. Chemosphere 119, 209–216.

Ludvíková, M., Griga, M., 2019. Transgenic fiber crops for phytoremediation of metals and metalloids. In: Prasad, M.N.V. (Ed.), Transgenic Plant Technology for Remediation of Toxic Metals and Metalloids. Academic Press, pp. 341–358. eBook ISBN: 9780128143902.

Luo, C., Shen, Z., Li, X., 2005. Enhanced phytoextraction of Cu, Pb, Zn and Cd with EDTA and EDDS. Chemosphere 59, 1–11.

Luo, F., Hu, X.F., Oh, K. et al., 2020. Using profitable chrysanthemums for phytoremediation of Cd- and Zn-contaminated soils in the suburb of Shanghai. J. Soils Sediments 20, 4011–4022.

Manzano, R., Silvetti, M., Garau, G., Deiana, S., Castaldi, P., 2016. Influence of iron–rich water treatment residues and compost on the mobility of metal(loid)s in mine soils. Geoderma 283, 1–9.

Marshall, R., Maxwell, C.S., Collins, S.P., Jacobsen, T., Luo, M.L., Begemann, M.B., et al., 2018. Rapid and scalable characterization of CRISPR technologies using an E. coli cell-free transcription-translation system. Mol. Cell. 69, 146–157.

Mary, G.S., Sugumaran, P., Niveditha, S., Ramalakshmi, B., Ravichandran, P., Seshadri, S., 2016. Production, characterization and evaluation of biochar from pod (*Pisum sativum*), leaf (*Brassica oleracea*) and peel (*Citrus sinensis*) wastes. Int. J. Recycl. Org. Waste Agric. 5, 43–53.

Mathur, N., Bohra, J.S.S., Quaizi, A., Vyas, A., 2007. Arbuscular mycorrhizal fungi: a potential tool for phytoremediation. J. Plant Sci. 2 (2), 127–140.

Mecwan, N.V., Yagnik, B.N., Solanki, H.A., 2018. A review on phytoremediation of arsenic-contaminated soil. Int. Res. J. Environmental Sci. 7 (4), 27–36.

Meers, E., Ruttens, A., Hopgood, M.J., Samson, D., Tack, F.M.G., 2005. Comparison of EDTA and EDDS as potential soil amendments for enhanced phytoextraction of heavy metals. Chemosphere 58, 1011–1022.

Majhi, P.K., Kothari, R., Arora, N.K., Pandey, V.C., Tyagi, V.V., 2021. Impact of pH on pollutional parameters of textile industry wastewater with use of *Chlorella pyrenoidosa* at lab-scale: a green approach. Bull. Enviro. Contam. Toxicol. doi:10.1007/s00128-021-03208-5.

Montecillo, J.A.V., Chu, L.L., Hanhong, B., 2020. CRISPR-Cas9 system for plant genome editing: current approaches and emerging developments. Agronomy 10 (7), 1033.

O'Kelly, B.C., 2014. Drying temperature and water content–strength correlations. Environ. Geotech. 1, 81–95.

Ogundiran, M.B., Mekwunyei, N.S., Adejumo, S.A., 2018. Compost and biochar assisted phytoremediation potentials of Moringa oleifera for remediation of lead contaminated soil. J. Environ. Chem. Engg. 6 (2), 2206–2213.

Oviedo, C., Rodríguez, J., 2003. EDTA: the chelating agent under environmental scrutiny. Quim.Nova 26 (6).

Pacwa-Płociniczak, M., Płaza, G.A., Piotrowska-Seget, Z., Cameotra, S.S., 2011. Environmental applications of biosurfactants: Rrecent advances. Int. J. Mol. Sci. 12, 633–654.

Pandey, V.C., 2020. Phytomanagement of Fly Ash. Elsevier (Authored book). ISBN: 9780128185445. doi:10.1016/C2018-0-01318-3.

Pandey, V.C., Bajpai, O., 2019. Phytoremediation: From Theory towards Practice. In: Pandey, V.C., Bauddh, K. (Eds.), Phytomanagement of Polluted Sites. Elsevier, Amsterdam, pp. 1–49. 10.1016/B978-0-12-813912-7.00001-6.

Pandey, V.C., Bajpai, O., Singh, N., 2016. Energy crops in sustainable phytoremediation. J. Clean. Prod. Renew. Sustain. Energy Rev. 54, 58–73.

Pandey, V.C., Bauddh, K., 2018. Phytomanagement of Polluted Sites. Elsevier (Edited Book). doi:10.1016/C2017-0-00586-4.

Pandey, V.C., Mahajan, P., Saikia, P., Praveen, A., et al., 2022Fibre Crop-Based Phytoremediation: Socio-economic and Environmental Sustainability. Elsevier (Authored Book), Amsterdam. ISBN: 9780128242469. In press.

Pandey, V.C., Pandey, D., Singh, N., 2015. Sustainable phytoremediation based on naturally colonizing and economically valuable plants. J. Clean. Prod. 86, 37.

Pandey, V.C., Praveen, A., 2020. Vetiveria zizanioides (L.) Nash–more than a promising crop in phytoremediation. In: Pandey, V.C., Singh, D.P. (Eds.), Phytoremediation Potential of Perennial Grasses. Elsevier, Amsterdam, pp. 31–62. doi:10.1016/B978-0-12-817732-7.00002-X.

Pandey, V.C., Rai, A., Korstad, J., 2019. Aromatic crops in phytoremediation: from contaminated to waste dumpsites. In: Pandey, V.C., Bauddh, K. (Eds.), Phytomanagement of Polluted Sites. Elsevier, pp. 255–275.

Pandey, V.C., Rai, A., Kumari, A., Singh, D.P., 2020. Cymbopogon flexuosus–an essential oil-bearing aromatic grass for phytoremediation. In: Pandey, V.C., Singh, D.P. (Eds.), Phytoremediation Potential of Perennial Grasses,. Elsevier, Amsterdam, pp. 195–209. doi:10.1016/B978-0-12-817732-7.00009-2.

Pandey, V.C., Singh, J.S., Singh, D.P., Singh, R.P., 2014. Methanotrophs: promising bacteria for environmental remediation. Int. J. Environ. Sci. Technol. 11, 241–250. doi:10.1007/s13762-013-0387-9.

Pandey, V.C., Singh, N., 2015. Aromatic plants versus arsenic hazards in soils. J. Geochem. Explor 157, 77–80. doi:10.1016/j.gexplo.2015.05.017.

Pandey, V.C., Singh, V., 2020. Bioremediation of Pollutants: From Genetic Engineering to Genome Engineering. Elsevier. ISBN: 9780128190258. doi:10.1016/C2018-0-05022-7.

Pandey, V.C., Souza-Alonso, P., 2019. Market opportunities in sustainable phytoremediation. In: Pandey, V.C., Bauddh, K. (Eds.), Phytomanagement of Polluted Sites. Elsevier, Amsterdam, pp. 51–82. doi:10.1016/B978-0-12-813912-7.00002-8.

Pathak, S., Agarwal, A.V., Pandey, V.C., 2020. Phytoremediation—a holistic approach for remediation of heavy metals and metalloids. In: Pandey, V.C., Singh, V. (Eds.), Bioremediation of Pollutants. Elsevier, Amsterdam, pp. 3–14. doi:10.1016/B978-0-12-819025-8.00001-6.

Park, J.W., Park, B.K., Kim, J.E., 2006. Remediation of soil contaminated with 2,4-dichlorophenol by treatment of minced shepherd's purse roots. Arch. Environ. Contam.Toxicol. 50, 191–195.

Park, S., Kim, K.S., Kim, J.T., Kang, D., Sung, K., 2011. Effects of humic acid on phytodegradation of petroleum hydrocarbons in soil simultaneously contaminated with heavy metals. J. Environ. Sci. 23, 2034–2041.

Paz-Ferreiro, J., Lu, H., Fu, S., Méndez, A., Gascó, G., 2014. Use of phytoremediation and biochar to remediate heavy metal polluted soils: a review. Solid Earth 5, 65–75.
Pogrzeba, M., et al., 2018. Possibility of using energy crops for phytoremediation of heavy metals contaminated land—a three-year experience. In: Mudryk, K., Werle, S. (Eds.), Renewable Energy Sources: Engineering, Technology, Innovation. Springer Proceedings in Energy. Springer, Cham.
Porras, J., Bedoya, C., Silva-Agredo, J., Santamaría, A., Fernández, J.J., Torres-Palma, R.A., 2016. Role of humic substances in the degradation pathways and residual antibacterial activity during the photodecomposition of the antibiotic ciprofloxacin in water. Water Res. 94, 1–9.
Quartacci, M.F., Baker, A.J.M., Navari-Izzo, F., 2005. Nitrilotriacetate- and citric acid-assisted phytoextraction of cadmium by Indian mustard (*Brassica juncea* (L.) Czernj, Brassicaceae). Chemosphere 59, 1249–1255.
Rathinasabapathi, B., Raman, S.B., Kertulis, G., Ma, L.Q., 2006. Arsenic resistant proteobacterium in the phyllosphere of the Chinese brake fern (*Pteris vittata* L.) reduces arsenate to arsenite. Can. J. Microbiol. 52, 695–700.
Reddy, K.R., 2010. Technical challenges to in-situ remediation of polluted sites. Geotech. Geol. Eng. 28, 211–221.
Reid, C.P.P., Szaniszlo, P.J., Crowley, D.E., 1986. Siderophore involvement in plant iron nutritionIron, Siderophores, and Plant Diseases. Springer US, pp. 29–42.
Rocha, I.M.V., Silva, K.N.O., Silva, D.R., Martínez-Huitle, C.A., Santos, E.V., 2019. Coupling electrokinetic remediation with phytoremediation for depolluting soil with petroleum and the use of electrochemical technologies for treating the effluent generated. Sep. Purif. Technol. 208, 194–200.
Rodríguez-Vila, A., Asensio, V., Forján, R., Covelo, E.F., 2015. Chemical fractionation of Cu, Ni, Pb and Zn in a mine soil amended with compost and biochar and vegetated with *Brassica juncea* L. J. Geochem. Explor. 158, 74–81.
Rosenberg, E., Ron, E.Z., 1999. High- and low-molecular-mass microbial surfactants. Appl. Microbiol. Biotechnol. 52, 154–162.
Roy, M., Giri, A.K., Dutta, S., Mukherjee, P., 2015. Integrated phytobial remediation for sustainable management of arsenic in soil and water. Environ. Int. 75, 180–198.
Rozpądek, P., Domka, A.M., Turnau, K., 2019. Are fungal endophytes merely mycorrhizal copycats? The role of fungal endophytes in the adaptation of plants to metal toxicity. Front Microbiol. 10, 371.
Sánchez, V., López-Bellido, F.J., Cañizares, P., Rodríguez, L., 2018. Can electrochemistry enhance the removal of organic pollutants by phytoremediation? J. Environ. Manage. 225, 280–287.
Sánchez, V., López-Bellido, F.J., Rodrigo, M.A., Rodríguez, L., 2019. Electrokinetic-assisted phytoremediation of atrazine: differences between electrode and interelectrode soil sections. Sep. Purif. Technol. 211, 19–27.
Sánchez, V., Francisco, López-Bellido, J., Rodrigo, M.A., Rodríguez, L., 2019b. Enhancing the removal of atrazine from soils by electrokinetic-assisted phytoremediation using ryegrass (*Lolium perenne* L.). Chemosphere 232, 204–212.
Santos, D.K.F., Rufino, R.D., Luna, J.M., Santos, V.A., Sarubbo, L.A., 2016. Biosurfactants: multifunctional biomolecules of the 21st century. Int. J. Mol. Sci. 17 (3), 401.
Sathya, A., Kanaganahalli, V., Rao, P.S., Gopalakrishnan, S., 2016. Cultivation of sweet sorghum on heavy metal-contaminated soils by phytoremediation approach for production of bioethanol. In: Prasad, M.N.V. (Ed.), Bioremediation and Bioeconomy. Elsevier, pp. 271–292. https://hdl.handle.net/20.500.11766/6570.
Sheng, X., He, L., Wang, Q., Ye, H., Jiang, C., 2008. Effects of inoculation of biosurfactant-producing Bacillus sp. J119 on plant growth and cadmium uptake in a cadmium-amended soil. J. Hazard. Mater. 155, 17–22.
Silva, R., de, C.F.S., Almeida, D.G., Rufino, R.D., Luna, J.M., Santos, V.A., Sarubbo, L.A., 2014. Applications of biosurfactants in the petroleum industry and the remediation of oil spills. Int. J. Mol. Sci. 15 (7), 12523–12542.
Singh, J., Lee, B.K., 2016. Influence of nano-TiO_2 particles on the bioaccumulation of Cd in soybean plants (*Glycine max*): a possible mechanism for the removal of Cd from the contaminated soil. J. Environ. Manage. 170, 88–96.
Singh, J.S., Singh, D.P., 2017. Methanotrophs: An emerging bioremediation tool with unique broad spectrum methane monooxygenase (MMO) enzyme, in: Agro-Environmental Sustainability. Springer International Publishing, pp. 1–18.
Sinhal, V.K., Srivastava, A., Singh, V.P., 2015. Phytoremediation: a technology to remediate soil contaminated with heavy metals. Int. J. Green Herbal Chem. 4 (3), 439–460.

Smolinska, B., 2015. Green waste compost as an amendment during induced phytoextraction of mercury-contaminated soil. Environmental Sci. Pollution Res. 22 (5), 3528–3537.

Souri, Z., Karimi, N., Sarmadi, M., Rostami, E., 2017. Salicylic acid nanoparticles (SANPs) improve growth and phytoremediation efficiency of Isatis cappadocica Desv., under As stress. IET Nanobiotechnol. 11, 650–655.

Tang, L., Mao, B., Li, Y., Lv, Q., Zhang, L., Chen, C., He, H., Wang, W., Zeng, X., Shao, Y., et al., 2017. Knockout of OsNramp5 using the CRISPR/Cas9 system produces low Cd-accumulating indica rice without compromising yield. Sci. Rep. 7, 14438.

Theron, J., Walker, J.A., Cloete, T.E., 2008. Nanotechnology and water treatment: applications and emerging opportunities. Crit. Rev. Microbiol. 34, 43–69.

Tranvik, L.J., 2014. Dystrophy in freshwater systemsReference Module in Earth Systems and Environmental Sciences. Elsevier. https://doi.org/10.1016/b978-0-12-409548-9.09396-9.

Tripathi, V., Edrisi, S.A., Abhilash, P.C., 2016. Towards the coupling of phytoremediation with bioenergy production. Renew. Sust. Energ. Rev. 57, 1386–1389.

Uddin, M.N., Wahid-Uz-Zaman, M., Rahman, M.M., Islam, M.S., 2016. Phytoremediation potentiality of lead from contaminated soils by fibrous crop varieties. Am. J. Appl. Sci. Res. 2 (5), 22–28.

Vamerali, T., Bandiera, M., Mosca, G., 2010. Field crops for phytoremediation of metal-contaminated land. A review. Environ. Chem. Lett. 8, 1–17.

Van Aken, B., 2008. Transgenic plants for phytoremediation: helping nature to clean up environmental pollution. Trends Biotechnol. 26, 225–227.

Verma, S.K., Singh, K., Gupta, A.K., Pandey, V.C., Trivedi, P., Verma, R .K., Patra, D.D., 2014. Aromatic grasses for phytomanagement of coal fly ash hazards. Ecol. Eng. 73, 425–428. doi:10.1016/j.ecoleng.2014.09.106.

Vargas, C., Pérez-Esteban, J., Escolástico, C., Masaguer, A., Moliner, A., 2016. Phytoremediation of Cu and Zn by vetiver grass in mine soils amended with humic acids. Environ. Sci. Pollut. Res. 23, 13521–13530.

Vassil, A.D., Kapulnik, Y., Raskin, I., Salt, D.E., 1998. The Role of EDTA in lead transport and accumulation by indian mustard. Plant Physiol. 117 (2), 447–453.

Wada, N., Ueta, R., Osakabe, Y., et al., 2020. Precision genome editing in plants: state-of-the-art in CRISPR/Cas9-based genome engineering. BMC Plant Biol. 20, 234.

Wan, Q.F., Deng, D.C., Bai, Y., Xia, C.Q., 2012. Phytoremediation and electrokinetic remediation of uranium contaminated soils: a review. J. Nucl. Radiochem. 34, 148–156.

Wang, D., Liu, Y., Lin, Z., Yang, Z., Hao, C., 2008. Isolation and identification of surfactin producing *Bacillus subtilis* strain and its effect of surfactin on crude oil. Wei Sheng Wu XueBao 48, 304–311.

Wang, H., Lin, K., Hou, Z., Richardson, B., Gan, J., 2010. Sorption of the herbicide terbuthylazine in two New Zealand forest soils amended with biosolids and biochars. J. Soils Sediments 10, 283–289.

Wang, Y., Wang, H.-S., Tang, C.-S., Gu, K., Shi, B., 2019. Remediation of heavy-metal-contaminated soils by biochar: a review. Environ. Geotech. 1–14.

Yadav, K.K., Singh, J.K., Gupta, N., Kumar, V., 2017. A review of nanobioremediation technologies for environmental cleanup: a novel biological approach. JMES 8 (2), 740–757.

Yan, A., Wang, Y., Tan, S.N., Yusof, M.L.M., Ghosh, S., Chen, Z., 2020. Phytoremediation: A Promising Approach for Revegetation of Heavy Metal-Polluted Land. Frontiers in Plant Science. 11, 359. doi: 10.3389/fpls.2020.00359.

Zand, A.D., Tabrizi, A.M., Heir, A.V., 2020. Incorporation of biochar and nanomaterials to assist remediation of heavy metals in soil using plant species. Environ. Technol. Innov. 20, 101134.

Zhang, W., 2003. Nanoscale iron particles for environmental remediation: an overview. J. Nanoparticle Res. 5, 323–332.

Zhang, X., Wang, H., He, L., Lu, K., Sarmah, A., Li, J., Bolan, N.S., Pei, J., Huang, H., 2013. Using biochar for remediation of soils contaminated with heavy metals and organic pollutants. Environ. Sci. Pollut. Res. 20, 8472–8483.

Zhu, L., Zhang, M., 2008. Effect of rhamnolipids on the uptake of PAHs by ryegrass. Environ. Pollut. 156, 46–52.

CHAPTER

2

Plant-assisted bioremediation: Soil recovery and energy from biomass

Valeria Ancona[a], Ida Rascio[a,b], Giorgia Aimola[a], Anna Barra Caracciolo[c], Paola Grenni[c], Vito F. Uricchio[a], Domenico Borello[a,d]

[a]*Water Research Institute-Italian National Research Council, Bari, Italy*
[b]*Department of Soil, Plant and Food Sciences, University of Bari, Italy*
[c]*Water Research Institute-Italian National Research Council, Monterotondo, Rome, Italy*
[d]*Department of Mechanical and Aerospace Engineering, Sapienza University of Rome, Italy*

2.1 Introduction

Plant-assisted bioremediation or phyto-assisted bioremediation (PABR) is a useful, cost-effective and eco-friendly technology aimed at recovering contaminated areas exploiting the capabilities of selected plants and rhizosphere microbial communities to transform and degrade contaminants (Cunningham and Ow, 1996; Ashraf et al., 2019). The synergic interactions between plants and microorganisms make it possible different processes for stabilizing (phytostabilization), storing in specific parts of the plant (phytoextraction and phytoaccumulation) and degrading contaminants not only from soil, but also from water, sludge and sediment (Ramanjaneyulu et al., 2017). This green remediation can be used for recovering impacted soils from both organic (e.g., persistent organic pollutants) and inorganic (e.g., heavy metals [HMs]) contaminants (Vaajasaari and Joutti, 2006). In the case of organic compounds, such as pesticides and polychlorinated biphenyls (PCBs), plants can promote their removal through release of radical exudates, which in turn can also help autochthonous microbial communities to transform these contaminants. Microbial communities are the only organisms able to degrade completely contaminants until their mineralisation. However, recovery from contamination is only possible if toxicity does not hamper microbial activity and plant development. For example, in the case of HMs, it is necessary to select plants able to resist to their toxic effects not only in the surrounding soil, but also inside plant tissues (Jabeen et al., 2016; Sarwar et al., 2017; Farid et al., 2018). Moreover, these plants need to have a rapid growth, obtaining high amounts of biomass to be easily harvested by pruning; if the biomass is recycled in an sustainable way, this process is in line with the circular economy. Fig. 2.1 reports a scheme of the PABR technology for remediating PCB and/or HM polluted soils.

In the following paragraphs, recent developments achieved for improving the efficiency of PABR technology and the potential of PABR biomass for energy purposes, are summarised.

Assisted Phytoremediation. DOI: https://doi.org/10.1016/B978-0-12-822893-7.00012-4

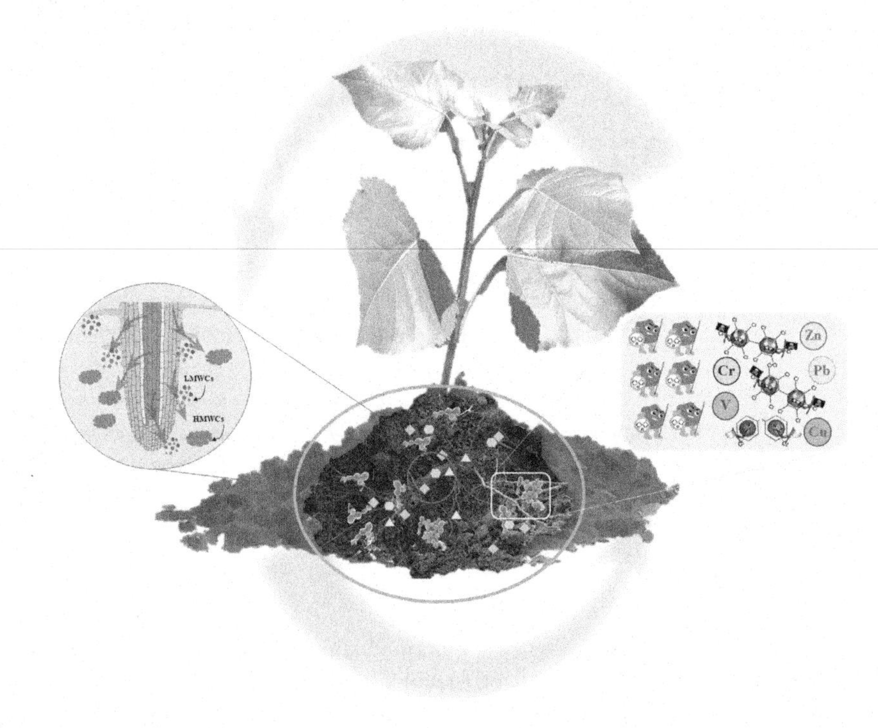

FIG. 2.1

Scheme of the synergistic interactions occurring between plant and microorganisms in PABR strategies for recovering soil from organic (e.g., PCBs) and/or inorganic (HMs) pollutants.

2.2 Soil amendments for enhancing phyto-assisted bioremediation efficiency

Anthropogenically polluted soils are often poor in organic matter content (Kwiatkowska-Malina, 2018). Organic amendments, such as compost and biochar, have been tested for their effectiveness to enhance soil fertility and structure and to help plants to remove contaminants from polluted soil (Sohi et al., 2010; Asensio et al., 2013; Zhou et al., 2017). The performance of an amendment depends on its nature, dose and specific characteristics of polluted soil (Dede et al., 2012; Wei et al., 2014; Parraga-Aguado et al., 2015). Amendments are materials with different origin and physico-chemical properties that can change soil characteristics, such as nutrient binding, cation exchange capacity, soil adsorption. Moreover, they can differently influence microbial community activity (Barra Caracciolo et al., 2015; Di Lenola et al., 2020), including nitrogen fixation, transformation of organic matter and of organic compounds (Asai et al., 2009; Hossain et al., 2010; Laird et al., 2010; Ahmad et al., 2014; Venegas et al., 2016; Li et al., 2019). Different studies conducted in the last decade report that various amendments can improve PABR effectiveness. Really, adding organic amendments can increase plant productivity, crop yield and rhizosphere microbial biomass, with positive effects on the PABR performance (Fowles, 2007; Karami et al., 2011; Wei et al., 2014; Zhang et al., 2014; Forján et al., 2017, 2018; Li et al., 2019; Sigua et al., 2019). Carbonaceous materials are generally underestimated wastes and their use as by-products perfectly fits with the purposes of a circular economy (Lima et al., 2014).

FIG. 2.2

Application of organic amendments (compost and biochar) in a multicontaminated area of Southern Italy recovered with poplar-assisted bioremediation technology.

Fig. 2.2 shows an example of compost or biochar application in a PCB and HM historically contaminated area in Southern Italy.

2.2.1 Biochar

Biochar is a solid carbon-enriched material, a by-product obtained in controlled conditions by pyrolysis treatment of organic matter from a wide range of feedstock (biomass of both vegetable and animal origin, mainly from agricultural activities or woodland).

Thanks to its multiple properties, in the last years biochar is being used for a great number of applications: as an amendment for enhancing soil fertility, in veterinary medicine as an additive for animal feed, in the construction field (insulation material), as an energy source (fuel by producing bio-oil and gases as hydrogen during its production), and finally, in technological contexts (production of super capacitors or pseudo capacitors for accumulation of electrical energy) (Mukherjee and Zimmerman, 2013; Hartley et al., 2009). In addition, as reported by several authors (Schmidt 2012; Sajjad et al., 2020) biochar is being employed in environmental recovery for carbon sequestration, pollutant removal (e.g., HMs, ammonium and nitrate), as an organic compound filter for water and soil purification and, for absorbing foul-smelling volatile organic substances.

Biochar is widely used as a soil amendment because it improves soil quality acting on its structure, pH, cation exchange capacity and density (Nichols et al., 2000). Owing to is highly porous structure, enlarging surface area, it can ameliorate water retention capacity of soil, increasing its porosity and sorption and increasing soil ability to retain and release nutrients (Jefferey et al., 2011; Sajjad et al., 2020, Mukherjee and Zimmerman, 2013; Sajjad et al., 2020).

Beesley et al. (2011) described encouraging results for reducing organic contaminants with activated carbon (Zimmerman et al., 2004), however more studies recommend the use of biochar, as non-activated charcoals, for reducing organic contaminant bioavailability in soil (Beesley et al., 2011; Gomez-Eyles et al., 2011; Qin et al., 2013; Xin et al., 2014). This reduction depends on the type of biochar and specific contaminant to be degraded (Beesley et al., 2011).

The use of biochar for bioremediation of HM-contaminated soils is reported in several works (Singh et al., 2020).

Biochar indeed can form on its surface complexes between functional groups (e.g., oxygen functional groups, Ca^{2+}, Mg^{2+}) and HMs. Sorption mechanisms depend on the cation content in soil and biochar. HMs can also be stabilized by precipitation of some compounds such as sulphates and carbonates present in the ash with some pollutants. Alkaline nature of biochar also helps in lowering bioavailable metal concentrations in biochar-amended land. After biochar addition, HM precipitation in soil can occur at higher pH values (Cantrell et al., 2012; Sajjad et al., 2020).

Differently from HMs removal studies, only few researches are performed on the use of biochar for soil remediation from organic pollutants (Cheng et al., 2017; Hussain et al., 2018; Saum et al., 2018; Zhang et al., 2019). Biochar can adsorb a wide range of organic pollutants, such as pesticides and PAHs, reducing their bioavailability in soil (Evangelou et al., 2015; Zhang et al., 2013; Kookana et al. 2011; Sarmah et al. 2010).

Biochar can increase nutrient availability for plant uptake through an enrichment in carbon source and in soil nutrients (K, Mg, N, P) (Biederman and Stanley Harpole, 2013 ; Nigussie et al., 2012; Rodríguez -Vila et al., 2016; Wu et al., 2019; Bird et al., 2011). Due to its alkaline nature, biochar can be also applied for enhancing pH of acidic soils (Bird et al., 2011; Jefferey et al., 2011; Van Zwieten et al., 2010). Soil microbial biomass can also be positively influenced by biochar; in fact, its porous structure can promote formation of suitable habitats for microbes where they can degrade organic compounds, including pollutants. For the overall reasons, biochar can enhance the efficiency of PABR (Sajjad et al., 2020; Sizmur et al., 2016) and other green technologies (Wu et al., 2019; Kolb et al., 2009; Lehman et al., 2011; Masiello et al., 2013).

The effectiveness of biochar in PABR strategies has been shown by several authors (Beesley et al., 2011; Park et al., 2011; Rees et al., 2015; Novak et al., 2018) and it is recommended due to its low cost and its relative environmental stability (Beesley et al., 2011, Forján et al., 2018). Table 2.1 reports some examples of biochar application in PABR. However, the use of biochar requires specific considerations regarding its characteristics, type of soil contamination and plant selected for PABR. In fact, the variability of feedstock type and some process characteristics (e.g., temperature, residence time, heat transfer rate) give the biochar distinctive physico-chemical properties. The properties of a biochar, such as carbon content, pH, surface area, can be extremely important in influencing soil physico-chemical characteristics and restoration (Lima et al., 2014; Verheijen et al., 2010; Sajjad et al., 2020). Paz-Ferreiro et al. (2014) described in an interesting review several works concerning biochar involved in plant-assisted bioremediation strategy. The paper reported an overall biochar capability to immobilize metals, with different behaviours of the various elements and soil factors that determined HM remediation.

Despite the large number of advantages in using biochar for soil recovering, there are some aspects that need to be further studied and clarified. The possible changing in soil structure can sometimes become a potential drawback associated with biochar soil addition. In a recent review, Kuppusamy et al. (2016) reported a possible binding and deactivation of herbicides and nutrients in the soil, an oversupply of nutrients, an excessive increase in soil EC and pH, an impact on germination and soil biological processes, and in some cases a release of toxic compounds (e.g. HMs and PAHs) from biochar. However, contaminant values were below the legal limit thresholds (Hilber et al., 2012; Kloss et al., 2012; Singh et al., 2010). In any case, before any application the possible content of pollutants in a biochar and the site-specific characteristics of the area to be remediated, needs to be evaluated. A particularly interesting aspect concerns the biochar capability to absorb and immobilize organic and inorganic contaminants in micropores, diminishing their bioavailability and the consequent biodegradation or transformation into less toxic compounds by microorganisms (Rhodes et al., 2008; Beesley et al., 2011; Lou et al., 2011).

Table 2.1 Examples of biochar applications in plant-assisted phytoremediation technologies.

Contaminant	Plant species	Growth condition	Reference
Zn, Cd	*Zea mays L.*	Greenhouse	Sigua et al. (2019)
Cd	*Lactuca sativa L.*	Greenhouse	Zhang et al. (2017)
Cd, Zn, Pb, As	*Oryza sativa L.*	Greenhouse	Zheng et al. (2012)
Cd, Pb, Zn	*Carnavalia ensiformis, Mucuna aterrima*	Greenhouse	Puga et al. (2015)
As, Pb	*Salix purpurea, Populus euramericana*	Greenhouse	Lebrun et al. (2016)
As, Pb	*Solanum lycopersicum L.*	Greenhouse	Beesley et al. (2013)
Fe, Mn	*Brassica juncea L.*	Greenhouse	Forján et al. (2017)
Pb, Zn, Ba, As, Cd	*Salix viminales, Lolium perenne*	Greenhouse	Norini et al. (2019)
Cd	*Crocus sativus L.*		Moradi et al. (2019)
Cd, Zn, Pb	*Brassica Napus L.*	Greenhouse	Houben et al. (2013)
Al, Cd, Cr, Cu, Fe, Mn, Ni, Pb, Tl, Zn	*Anthyllis vulneraria subsp. polyphylla (Dc.) Nyman, Noccaea rotundifolium (L.) Moench subsp. cepaeifolium and Poa alpina L. subsp. Alpine*	Greenhouse	Fellet et al. (2014)
As	*Miscanthus x giganteus*	Greenhouse	Hartley et al. (2009)
Cd	*Beta vulgars*	Greenhouse	Gu et al. (2020)
Petroleum hydrocarbons	*Lolium multiflorum*	Greenhouse	Hussain et al. (2018)
Petroleum hydrocarbons	*Paludicella articulate*	Greenhouse	Saum et al. (2018)
Petroleum hydrocarbons	*Lolium L.*	Greenhouse	Han et al. (2016)
Cd, iprodione	*Medicago sativa L.*	Greenhouse	Zhang et al. (2019)
Difenoconazole	*Tobacco strain K326*	Greenhouse	Cheng et al. (2017)

For example, microbial communities can accumulate HMs by either adsorption or absorption (Jin et al., 2018); however, this possibility can vary if biochar reduces HM bioavailability and mobility.

This phenomenon depends on the sorption strength and the specific characteristics of the contaminant, char and soil which can influence its overall retention over time. Consequently, depending on remediation goals and the characteristics of a pollutant, some aspects of biochar process production have to be taken into consideration (e.g., feedstock and temperature during the pyrolysis). At this regard, it has been demonstrated that, with an increment of temperature in pyrolysis process, there is an increment in carbonization and in the surface area of biochar that enhance the adsorption of compounds (Uchimiya et al., 2010a; Wang et al., 2010; Beesley et al., 2011). Moreover, due to the relative lightness of biochar compared to soil, erosion phenomena and off-site transport could be encountered, with negative ecological consequences in the surrounding environment (Kuppusamy et al., 2016). Furthermore, some studies proved that biochar can undergo an aging process in soil and in presence of other compounds, such as organic matter and minerals. The sorption efficiency of organic pollutants can change through the loss of sorption sites due to the competition with other molecules (Beesley et al., 2011; Gell et al., 2011; Kookana, 2011; Martin et al., 2012; Uchimiya et al., 2010b; Wang et al., 2010).

Really, knowledge on the complete behaviour of biochar especially at field-scale application, need to be improved and further investigations are necessary for better understanding the complete and long-term capabilities of this amendment in enhancing efficiency of soil remediation technologies.

2.2.2 Compost

Compost is the product of artificially controlled bio-oxidation and humification of a mix of organic materials such as solid organic waste from green and woody biodegradable plant residues as pruning waste, manure, sewage waste. Compost is the result of microorganism (mainly of bacteria and fungi) processes (Gandolfi et al., 2010; Huang et al., 2016); compost derives from the organic matter (organic waste) break down by biological processes, producing carbon dioxide, water, heat, and humus, the relatively stable organic end product. Compost is used as an improver for agronomic uses or in horticulture because it can enhance soil structure and nutrient availability (N, P, K, Mg), (Hartl et al., 2003; Gopinath et al., 2008; Luna et al., 2016; Miyasaka et al., 2001), increasing soil fertility and crop growth and yield (Huang et al., 2016). Owing to its action in augmenting soil porous structure, it facilitates roots penetration and improves soil water content and flow of gases (Ferreras et al., 2005; Walker et al., 2004). Furthermore, compost acts as a "glue" for soil, because mixed with it improves physical characteristics contributing to the formation of a good aggregation of soil particles, also for the development of clay mineral flocs (Tejada et al., 2006). Compost has a considerable amount of organic carbon that stimulates activity of soil microorganisms responsible for biogeochemical cycles. It is both partially mineralized and converted in humic and fulvic acids that support an increase in the overall soil organic carbon and structure. Moreover, compost can act as a soil biological activator, introducing microorganisms in soil, increasing biodiversity and activity of microbial communities (Huang et al., 2016; Schmidt 2012, Wu et al., 2016; Barra Caracciolo et al., 2015). In a recent review, Palansooriya et al. (2020) provided detailed information about the capability of different soil amendments, including compost, for immobilization of potentially toxic elements in contaminated soils. These authors affirmed that compost could have different effects on toxic compound immobilization, depending on both its maturity level, composition and soil properties. Moreover, compost obtained from municipal solid waste can sometimes contain organic and inorganic pollutants that might negatively influence soil quality. Therefore, before employing such amendment in soil recovery applications it is advisable to characterize it and certificate its quality.

Several studies have used compost for improving PABR effectiveness. In the case of HMs, some functional groups of compost can form strong complexes with metal cations (Manzano et al., 2016). In the case of organic pollutants, adding compost can accelerate their degradation (Zubillaga et al., 2012). This organic amendment not only improves soil aeration, but also promotes plant growth (Guidi et al., 2012), revegetation (Vaajasaari and Joutti, 2006) and provides a nutrient reserve and structural support for microorganisms involved in soil bioremediation (Robichaud et al., 2019; Wyszkowski and Ziólkowska, 2009). Recent works showed the synergic effects of adding compost and plants (Medicago sativa) in increasing soil quality in terms of microbial activity and structure, together with the degradation of some persistent organic pollutants such as PCBs both in a historically polluted soil and in an Apirolio spiked soil (Di Lenola et al., 2018; Di Lenola et al., 2020). Table 2.2 reports some papers on the use of compost for improving PABR technology.

2.3 Root exudates: Key compounds in driving plant-microbial interactions

2.3.1 Role of root exudates

Plant root exudates include a wide variety of compounds which are released in the surrounding soil, termed as rhizosphere. In this microhabitat, many microbiological, biochemical, chemical and physical processes occur as an effect of root growth and root-exudate release (Rugova et al., 2017). The root

Table 2.2 Examples of compost amendment in plant-assisted bioremediation studies.

Contaminant	Plant species	Origin of compost	Plant growth condition	Reference
Pb	*Moring oleifera*	Wild sunflower (*Tithonia diversifolia*) and poultry manure	Greenhouse	Ogundiran et al. (2019)
Petroleum hydrocarbons	*Lolium multiflorum*	Green garden waste material	Greenhouse	Hussain et al. (2018)
Pb	*Pelargonium hortorum*	Local nursery of Islamabad	Greenhouse	Gul et al. (2020)
Cu	*Oenothera picensis*	Municipal organic waste	Greenhouse	Pérez et al. (2021)
Cd, Zn	*Brassica juncea, Coss*	Pig manure	Rain shelter	Di Guo et al. (2019)
Petroleum hydrocarbons: As, Cd, Cu, Pb, Zn	*Salix planifolia and Saliz alaxensis*	Local compost	Field plot experiments	Robichaud et al. (2019)
PCB; V, Cr, Pb, Sn	*Hybrid poplar genotype Monviso (Populus generosa X Populus nigra)*	Municipal organic waste	Field	Ancona et al. (2017a)
Cd, V, Pb, Sn, Se, Zn, Cr	*Hybrid poplar genotype Monviso (Populus generosa X Populus nigra)*	Municipal organic waste	Field	Ancona et al. (2020)
Apirolio	*Medicago sativa*	Municipal organic waste	Greenhouse	Di Lenola et al. (2018)
PCB	*Medicago sativa*	Municipal organic waste	Greenhouse	Di Lenola et al. (2020)

PCB, polychlorinated biphenyl.

exudates can be classified on their (1) chemical properties (molecular weight and solubility in water fraction), (2) mechanisms of release through the plasma membrane (passive release or active excretion/secretion), and (3) functions (Uren, 2007; Neumann and Römheld, 2007; Valentinuzzi et al., 2015). In accordance with their chemical properties, root exudates can be divided into two major groups: (1) low molecular weight compounds (LMWCs), which include organic acids, amino acids, sugars, phenolic acids, flavonoids, phytosiderophores, and (2) high molecular weight compounds, including mucilage, vitamins and proteins (Chen et al., 2017).

LMWCs, also known as "diffusates," are involved in many processes (Uren, 2007) for their high water-solubility which makes it possible a fast diffusion from roots to soil and microorganisms. The contribution of LMWCs to the total root exudation pattern is quite high if compared to other organic compounds, such as fatty acids, sterols, enzymes, vitamins, and plant growth regulators (i.e., gibberellins, auxins, and cytokinins), which are usually exuded in very small quantities (Oburger et al., 2013). The exudation pattern (quantities and quality of root exudates) strictly depends on plant species, plant age, and environmental conditions (e.g., soil pollution and/or nutrient deficiency), (Bertin et al., 2003; Oburger et al., 2013). For example, an increase in exudation of aromatic and phenolic acids is usually observed in response to symbiotic, pathogenic, or allelopathic interactions (Bertin et al., 2003; Bais et al., 2006; Cesco et al., 2010; Cesco et al., 2012; Oburger et al., 2013). In a similar way, in some detoxification mechanisms (e.g., of Al^{3+} at low soil pH), roots release a higher amount of sugars, phytosiderophores, and phenols. These compounds can promote metal-complexation reactions for their high affinity to some elements. Organic acids play a crucial role in soil nutrient-deficiency conditions and can directly affect nutrient solubility in rhizosphere by ligand-exchange reactions or by ligand-promoted mineral dissolution and metal complexation reactions (Johnson and Loeppert, 2006; Oburger et al., 2011; Oburger et al., 2013). Root exudates can also influence bioavailability and biogeochemical processes of soil organic pollutants. Rodríguez-Garrido et al. (2020) reported the role of 15 compounds (mainly organic acids and phenolic compounds) in favouring the mobilization of hexachlorocyclohexane (HCH) adsorbed to soil, through mechanisms of desorption based on hydrophobic interactions between the exudates and HCH molecules.

All these functions may introduce new and interesting scenarios as these compounds are gaining popularity for their efficiency in promoting PABR processes of polluted soils. A brief description of some exudates and their main functions is reported in Table 2.3.

Table 2.3 Classification of root exudates.

Molecular weight	Class of compounds	Single components	Functions
Low molecular weight compounds (LMWCs)	Carbohydrates (or sugars)	Arabinose, glucose, fructose, galactose, maltose, raffinose, rhamnose, ribose, sucrose and xylose	Provide a favourable environment for microorganism growth
	Amino acids and amides	All 20 proteinogenic amino acids, aminobutyric acid, homoserine, cystathionine, mugineic acid and phytosiderophores	Inhibition of nematodes and root growth of different plant species
	Aliphatic acids (or organic acids)	Formic, acetic, butyric, propionic, maleic, citric, isocitric, oxalic, fumaric, malonic, succinic, maleic, tartaric, oxaloacetic, pyruvic, oxaloglucaric, glycolic, shikimic, acetonic, valeric and gluconic	Plant growth regulation and inhibition
	Aromatic acids	p-hydroxybenzoic, caffeic, p-coumeric, ferulic, gallic, gentisic, protocatechuic, salicylic, sinapic, syringic	Plant growth stimulation depending on concentration
	Miscellaneous phenolics	Flavanol, flavones, flavanones, anthocyanins, isoflavonoids	Plant growth inhibition or stimulation depending on concentration
	Fatty acids	Linoleic, linolenic, oleic, palmitic, stearic	Plant growth regulation
	Sterols	Campestrol, cholesterol, sitosterol, stigmasterol	Plant growth regulation
High molecular weight compounds (HMWCs)	Enzymes and Miscellaneous	-	Usually unknown

From Bertin C, Yang X, Weston, LA: The role of root exudates and allelochemicals in the rhizosphere. Plant and Soil 256:67–83, 2003.

2.3.2 Chemical assessment of root exudates

An accurate evaluation of the dynamic of exudate-driven rhizosphere processes in promoting contaminant removal needs suitable approaches and methodologies (Oburger et al., 2013; Oburger and Jones, 2018).

Since these molecules belong to different chemically classes, their analysis requires different methodological approaches, from the sampling procedure up to detection methods. Mass spectrometry-based approaches are generally the most adequate for determination of exudates patterns because they combine both high sensitivity and wide range of metabolites (Dundek et al., 2011).

Targeted analytical approaches, such as the high-pressure liquid chromatography or gas chromatography coupled with tandem mass spectrometry (HPLC-MS/MS, GC-MS/MS) can be very useful for investigating changes in exudation patterns, under different environmental conditions (Oburger and Jones, 2018). The GC-MS technique is usually used for volatile compounds, whereas HPLC-MS analysis is more suitable for separation of small and non-volatile compounds, both polar and non-polar (Pantigoso et al., 2021). The target analysis could sometimes give limited information as it does not consider the total content of root exudation. For this reason, the non-targeted approach by means of high-resolution mass spectrometry, represents a new and improved methodology for root exudates detection since it allows to cover the entire diversity of molecules released by roots (Fuhrer and Zamboni, 2015). This technique, due to the high sensitivity, is also particularly suitable for low-abundance metabolites. Both targeted and non-targeted approaches require a prior analytes collection from root system (or soil samples) and this can be performed using water or diluted trap solutions and, when necessary, be followed by a subsequent concentration step (Dundek et al., 2011). Despite the numerous sampling techniques available in literature studies, choosing the best strategy for collecting root exudates is always a critical point. Root damage or accidental collection of microbial-produced metabolites and degraded root exudates can affect the sampling of plants in field conditions. On the other hand, the collection from plants grown under laboratory conditions (i.e., hydroponic cultures) cannot be representative of a real environmental exudation profile (Pantigoso et al., 2021; Oburger and Jones, 2018). Unfortunately, root exudate collection can often be costly and time-consuming with sometimes inadequate results.

The spectroscopic techniques such as nuclear magnetic resonance or Fourier-transform infrared spectroscopy are also used in metabolomics studies as they allow to analyse large sample sets in relatively short time and good reproducibility. No sample preparation is needed but, the major limitation is the low sensitivity which inhibits low-abundance molecules detection (Oburger and Jones, 2018).

Other approaches include isotope-based techniques which can give information about the exuded compounds by monitoring the quantity of assimilated-carbon isotopes within the plant-microbe-soil system. As it is very difficult to separate the roots from soil without causing any damage to the root system, the quantity of isotopes present on the root surface is often overestimated (Pausch and Kuzyakov, 2017).

2.3.3 Future perspectives in root exudate investigation

Elucidating the role of root exudates and their contribution to biochemical processes occurring in the rhizosphere is a key point. Despite a widespread attention to these compounds, as well as the analytical and methodological developments reached in the past decades, most of mechanisms involved in root exudates, remain poorly understood (Bais et al., 2006; Barra Caracciolo et al., 2020).

The composition and release rates of root products may be deeply influenced by several conditions, i.e., soil microbial communities and environmental factors, such as pH, soil status, oxygen content and nutrient availability. For example, microbial metabolism can lead to an underestimation of the content of sugar and organic acids from root exudates and this can affect the complete understanding of their role in soil decontamination processes (van Dam and Bouwmeester, 2016). Sterilising agents can help to exclude the effects of microbial activity on root exudate composition, however this practice has also limitations. Consequently, choosing the best collection protocol is crucial for obtaining reliable results on root exudation pattern (Oburger and Jones, 2018; van Dam and Bouwmeester, 2016; Cheng et al., 2017).

An ecological approach is essential to investigate the "rhizosphere effect" in polluted soils. All the "omics" advances (metabolomics, proteomics, meta-transcriptomics) can be useful to elucidate both biochemical and microbiological contributions in rhizosphere (van Dam and Bouwmeester, 2016).

2.4 Investigation of soil microbial community structure and functioning in PABR experiments

In PABR rhizosphere microbial populations are strongly stimulated by exudates from root system. The main positive effects on microbial communities are an increase in their abundance and activity. For example, the dehydrogenase activity (DHA) which is an indicator at broad scale process level is influenced by organic matter content (Kumar et al., 2013), oxygenation, and level of contamination (Grenni et al., 2009; Ancona et al., 2017). DHA is a good indicator of soil quality (Barra Caracciolo et al., 2015) and its increase together with microbial abundance was observed in several PABR studies in both real scenarios (Ancona et al., 2017a) and microcosm experiments (Di Lenola et al., 2018; Di Lenola et al., 2020).

The characterization of the structure and functioning of soil microbial communities is a key point for understanding synergistic interactions occurring between plant species and belowground bacterial populations. This is possible using a combination of various microbial ecology methods. For example, the microbial diversity in soil can be assessed by biochemical methods such as phospholipid-derived fatty-acid and the total ester linked fatty-acid, molecular methods such as polymerase chain reaction-based approaches and epifluorescence microscope-based methods. The recent development of next-generation sequencing technology has further contributed to unveil either microbial or plant functioning in the rhizosphere, to yield a global view of the structure and diversity of the rhizosphere microbiota. Recently, metagenomic investigations have improved an overall knowledge on plant-microbial interactions in PABR studies. A detailed description of all these methods and their advantages is reported in Mercado-Blanco et al. (2018).

The composition of belowground microbiota is the results of the interactions between the initial microbial composition and the plant species used for soil bioremediation. Plant can influence microbial structure through its root morphology, growth and rhizodeposition (Philippot et al., 2013; Sasse et al., 2018). In a recent work carried out by Vergani et al. (2017) using three plant species selected for remediating strongly PCB-polluted soils, the bacterial communities were differentially affected in their composition in the root-associated soil fraction.

Recently, Barra Caracciolo et al. (2020), using DNA and cDNA sequencing and analysis identified the changes of the natural microbial community in the rhizosphere associated to PABR technology applied for recovering soil from PCB and HM pollution. The synergic relationships established between plants and belowground microorganisms were able to increase soil quality, enriching the soil with active bacteria able to degrade and contain contaminants, resist stresses and increase soil nutrient content.

2.5 Energy from phyto-assisted bioremediation biomass

In order to increase the economic feasibility of PABR technology, biomass from PABR applications can be employed for the production of energy or new materials. However, PABR biomass could store toxic compounds (especially potential toxic elements, PTEs) when phytoextraction processes or pollutants air deposition occurs. With a view to use PABR biomass for energy purposes, it is fundamental to assess biomass quality both to investigate its suitability to be treated with effective disposal strategies (pyrolysis, gasification, torrefaction) and also to evaluate the occurrence of possible pollutant residues. Therefore, a detailed chemical characterization of PABR biomass is necessary (Ancona et al., 2017b; Ancona et al., 2019). An important challenge for the research is to develop best practice guidelines for defining PABR biomass characteristics for energy production.

2.5.1 Biomass conversion end energy valorisation

The wide availability of PABR biomass could support the development of interesting opportunities for the energy and transport sector. As a matter of fact, biomass can be processed in combustion/gasification cogeneration plants for producing heat/electricity (Ahrenfeldt et al., 2013), as well as be led to pyrolysis/fermentation plants for generating biofuels (Menon and Rao, 2012).

Distributed small cogeneration plants (up to 1 MW) are widely used where sufficient biomass is available (Kalina, 2011). The produced heat and electricity can feed local communities (electricity can be given to the transmission grid). Furthermore, biomass conversion can generate several by-products that can be advantageously employed. For example, gasification process produces biochar and wood vinegar that can be employed as fertilizers. Biochar, mainly composed by black carbon, is a greenhouse gas (GHG) sink as its use or disposal reduce the CO_2 emissions due to combustion.

On the other hand, biofuel production can have a large impact on decarbonizing of the transport sector. RED II Directive[1] imposes that in 2030 biofuel will feed 14% EU transport sector. Part of such biofuels (3.5% of the total) must be produced by advanced biomass meaning that such feedstock must be secondary biomass or wastes.

This framework supports the quest for the smart use of residual biomass obtained from PABR. In fact, such biomass produced on marginal soils, reduces contamination, and supports soil fertility. PABR biomass can be employed provided that it can be safely converted, disposing eventual contaminants that can be present in the feedstock. The sequestration and disposal of contaminants is always possible, the real challenge is to minimize costs and to make the process as simple as possible.

The set-up of a proper conversion process requires contaminant identification in the feedstock to shape appropriate treatments for sequestrating such substances during the conversion process.

[1]https://eur-lex.europa.eu/legal-content/EN/TXT/?uri=uriserv:OJ.L_.2018.328.01.0082.01.ENG&toc=OJ:L:2018:328:TOC
https://eur-lex.europa.eu/legal-content/EN/TXT/?uri=uriserv:OJ.L_.2018.328.01.0082.01.ENG&toc=OJ:L:2018:328:TOC

PABR biomass obtained from soil contaminated by persistent organic pollutants, such us PCBs, does not accumulate this kind of contaminants; in fact, due to their high hydrophobicity, PCBs cannot translocate in upper part of arboreal plants (leaves, shoots, steams) and tend to be accumulated in the root system (Ancona et al., 2017a; Ancona et al., 2019). Differently, HMs and other inorganic toxic elements can be easily accumulated in PABR biomass, owing to translocation processes. The microorganisms located in the rhizosphere can increase bioavailability of HMs facilitating absorption by plants. This process is enabled by their ability to alter soil pH, release chelating molecules, such as organic acids and siderophores, and develop oxidation and reduction reactions (Gadd, 2000; Khan et al., 2009; Kidd et al., 2009; Ma et al., 2011; Rajkumar et al., 2010; Uroz et al., 2009; Wenzel, 2009).

Depending on the conversion process, several separation devices can be considered. In direct combustion, contaminants can be found in ashes or in flue gas. Separation is possible for the gas at the end of the process (Nzihou and Stanmore, 2013). However, it is preferable to perform the capture process upstream of the energy conversion, as a pre-treatment of biomass. This suggests considering gasification process. In gasification, solid biomass is transformed in a combustible gas (full of CO, H_2) before being burned. Several solutions are possible for biomass gasification (fixed bed – updraft, downdraft; fluidized bed). In all of them there are few intermediate steps operating at different thermo-dynamic conditions where by-products can be captured and eventually disposed. Nzihou and Stanmore (2013) analysed the distribution of HMs among the different available fractions (bed-, bottom-, cyclone-, flying-ashes) examining the different distribution depending on the metal volatility and understanding that the most volatile metals concentrated in the flying ashes, while the less volatiles were the most abundant in the bed- and bottom-ashes. Performing proper bubbling and cooling gas treatment it is possible produce a high-quality syngas to be burned in gas turbines and internal combustion engines. In recent studies (Aghaalikhani et al., 2017; Ancona et al., 2019), an efficient gasification process, considering also the use of Ni-mayenite catalyst for reducing tar content, was adopted for the conversion of PABR biomass produced in a multicontaminated soil in Southern Italy. The experiments were conducted in the experimental gasifier of the Mechanical and Aerospace Engineering Dept. Lab at Rome Sapienza University (Fig. 2.3).

The operating conditions are detailed in Table 2.4.

As for biofuel production, separation of contaminants from produced biofuel could be very complex and this technology is not fully established yet. However, Dastyar et al. (2019), carried out a deep analysis of the different technologies available for biofuel production including pyrolysis, gasification, combustion, and liquefaction/fermentation by considering different PABR biomass. They were able to conclude that pyrolysis seems to represent the best strategy to minimize the transfer of the HMs in the final products, with a reduced complexity of a plant.

2.5.2 Phyto-assisted bioremediation coherence with circular economy

The impact in terms of circularity of the PABR technology is very promising. The PABR biomass valorisation is a sort of "symbiotic process" where the energy/fuel production can be considered as a by-product of the most relevant remediation strategy. Moreover, recovery of contaminated land and use of biomass grown on it avoids the risks of indirect land use change as well as can fight soil degradation processes such as desertification and soil sealing.

The product of the conversion process (syngas, liquid biofuels) substitutes the fossil fuels in a few applications dealing with mobility and energy uses. Furthermore, in gasification process, the produced biochar can be used as soil amendment if no relevant contamination is present, or alternatively easily

FIG. 2.3

Experimental Gasifier at Mechanical and Aerospace Engineering Dept. (DIMA, Sapienza University).

Table 2.4 Gasifier operating conditions.

T (°C)	Air (Nml/min)	H_2O (g/h)	Biomass (g/h)	Steam/biomass	E.R.
800~820	4440~5900	120~150	250~300	0.4~0.5	0.3

From Aghaalikhani A, Savuto E, Di Carlo A, Borello D: Poplar from phytoremediation as a renewable energy source: gasification properties and pollution analysis. Energy Procedia 142: 924–931, 2017.

disposed. In both cases, the net result is that a relevant part (5-10% of biomass weight) of the carbon is sequestrated leading to the implementation of a carbon negative technique, aiming at mitigating the effect of climate changes. Similarly, it is possible to exploit the use of wood vinegar (few % of the biomass weight) for agricultural use (e.g., for increasing plant defence response against biotic and abiotic stress). Presently, no recycle of the extracted contaminants is envisaged.

2.6 Conclusions

The increase in feasibility and success of the PABR technology is achievable through the development and implementation of different site-specific strategies. Adding organic amendments (e.g., compost, biochar) can have the dual advantages of supporting plant growth by ensuring nutrient supply and increasing proliferation and activity of rhizosphere microbial communities. However, the quality and characteristics of an organic amendments needs to be tested before its application.

A major challenge pursued by scientists for improving knowledge on interactions between plants and microorganisms in rhizosphere is the identification of root exudates. These molecules act as key signals underlying plant-microorganism processes which promote decontamination of polluted soils. Moreover, the potential of microorganisms to degrade persistent organic contaminants and detoxify inorganic substances from polluted soil needs to be unveiled by detailed microbial analysis of soil microbial community structure and functioning. Finally, PABR is an environmental-friendly technology and makes also possible to obtain a by-product (biomass) which can be profitably valorised for producing energy or new materials, cutting down costs of this technology, which are lower than those of traditional soil remediation strategies.

Acknowledgments

This chapter was written in the framework of the "Energy for Taranto- Technology And pRocesses for the Abatement of pollutaNts and the remediation of conTaminated sites with raw materials recovery and production of energy tOtally green (TARANTO)" project, funded by the Ministry of Education, University and Research, grant number ARS01_00637.

References

Ancona V., Barra Caracciolo A., Grenni P., Di Lenola M., Campanale C., Calabrese A., Uricchio V.F., Mascolo G., Massacci A., 2017. Plant-assisted bioremediation of a historically PCB and heavy metal-contaminated area in Southern Italy. New Biotechnology, Special Issue: S1 Bioremediation Advances. 38, Part B, 65–73. http://dx.doi.org/10.1016/j.nbt.2016.09.006.

Aghaalikhani, A., Savuto, E., Di, Carlo A., Borello, D., 2017. Poplar from phytoremediation as a renewable energy source: gasification properties and pollution analysis. Energy Procedia 142, 924–931. doi:10.1016/j.egypro.2017.12.148.

Ahmad, R, Waraich, EA, Ashraf, MY, Ahmad, S, Aziz, T, 2014. Does nitrogen fertilization enhance drought tolerance in sunflower? A review. J. Plant Nutr. 37, 942–963. doi:10.1080/01904167.2013.868480.

Ahrenfeldt, J., Thomsen, T.P., Henriksen, U., Clausen, L.R., 2013. Biomass gasification cogeneration - a review of state of the art technology and near future perspectives. Appl. Therm. Eng. 50, 1407–1417. doi:10.1016/j.applthermaleng.2011.12.040.

Ancona, V., Barra Caracciolo, A., Grenni, P., Di Lenola, M., Campanale, C., Calabrese, A., Uricchio, V.F., Mascolo, G., Massacci, A., 2017a. Plant-assisted bioremediation of a historically PCB and heavy metal-contaminated area in Southern Italy. New Biotechnol 38, 65–73. doi:10.1016/j.nbt.2016.09.006.

Ancona, V., Grenni, P., Barra Caracciolo, A., Campanale, C., Di Lenola, M., Rascio, I., Uricchio, V.F., Massacci, A., 2017. Plant-assisted bioremediation: an ecological approach for recovering multi-contaminated areas. In: Lukac, M., Grenni, P., Gamboni, M. (Eds.), Soil Biological Communities and Ecosystem Resilience. Springer International Edition, Berlin, Germany, pp. 291–303.

Ancona, V., Barra Caracciolo, A., Campanale, C., De Caprariis, B., Grenni, P., Uricchio, V.F., Borello, D, 2019. Gasification treatment of poplar biomass produced in a contaminated area restored using plant assisted bioremediation. J. Environ. Manag. 239, 137–141. doi:10.1016/j.jenvman.2019.03.038.

Ancona, V., Barra Caracciolo, A., Campanale, C., Rascio, I., Grenni, P., Di Lenola, M., Bagnuolo, G., Uricchio, V.F., 2020. Heavy metal phytoremediation of a poplar clone in a contaminated soil in southern Italy. J. Chem. Technol. Biotechnol. 95, 940–949. doi:10.1002/jctb.6145.

Asai, H., Samson, B.K., Stephan, H.M., Songyikhangsuthor, K., Homma, K., Kiyono, Y., Inoue, Y., Shiraiwa, T., Horie, T., 2009. Biochar amendment techniques for upland rice production in Northern Laos. 1. Soil physical properties, leaf SPAD and grain yield. Field Crops Res. 111, 81–84. doi:10.1016/j.fcr.2008.10.008.

Asensio, V, Vega, F.A., Andrade, M.L., Covelo, E.F., 2013. Tree vegetation and waste amendments to improve the physical condition of copper mine soils. Chemosphere 90, 603–661. doi:10.1016/j.chemosphere.2012.08.050.

Ashraf, S., Ali, Q., Zahir, Z.A., Ashraf, S., Asghar, H.N, 2019. Phytoremediation: environmentally sustainable way for reclamation of heavy metal polluted soils. Ecotoxicol. Environ. Saf. 174, 714–727. doi:10.1016/j.ecoenv.2019.02.068.

Bais, H.P., Weir, T.L., Perry, L.G., Gilroy, S., Vivanco, J.M, 2006. The role of root exudates in rhizosphere interactions with plants and other organisms. Ann. Rev. Plant Biol. 57, 233–266. doi:10.1146/annurev.arplant.57.032905.105159.

Barra Caracciolo, A., Bustamante, M.A., Nogues, I., Di Lenola, M., Luprano, M.L., Grenni, P, 2015. Changes in microbial community structure and functioning of a semiarid soil due to the use of anaerobic digestate derived composts and rosemary plants. Geoderma 245-246, 89–97. doi:10.1016/j.geoderma.2015.01.021.

Barra Caracciolo, A., Grenni, P., Garbini, G.L., Rolando, L., Campanale, C, Aimola, G., Fernandez-Lopez, M., Fernandez-Gonzalez, A.J., Villadas, P.J., Ancona, V, 2020. Characterization of the Belowground Microbial Community in a Poplar-Phytoremediation Strategy of a Multi-Contaminated Soil. Front. Microbiol. 11, 2073. doi:10.3389/fmicb.2020.02073.

Beesley, L., Marmiroli, M., 2011. The immobilisation and retention of soluble arsenic, cadmium and zinc by biochar. Environ. Pollut. 159, 474–480. doi:10.1016/j.envpol.2010.10.016.

Beesley, L., Marmiroli, M., Pagano, L., Pigoni, V., Fellet, G., Fresno, T., Vamerali, T., Bandiera, M., Marmiroli, N., 2013. Biochar addition to an arsenic contaminated soil increases arsenic concentrations in the pore water but reduces uptake to tomato plants (*Solanum lycopersicum* L.). Sci. Tot. Environ. 454-455, 598–603. doi:10.1016/j.scitotenv.2013.02.047.

Bertin, C., Yang, X., Weston, L.A., 2003. The role of root exudates and allelochemicals in the rhizosphere. Plant Soil 256, 67–83. doi:10.1023/A:1026290508166.

Biederman, L.A., Stanley Harpole, W., 2013. Biochar and its effects on plant productivity and nutrient cycling: a meta-analysis. GCB Bioenergy 5 (2), 202–214. doi:10.1111/gcbb.12037.

Bird, M.I., Wurster, C.M., de, Paula, Silva, P.H, Bass, A.M., de Nys, R., 2011. Algal biochar – production and properties. Bioresour. Technol. 102, 1886–1891. doi:10.1016/j.biortech.2010.07.106.

Cantrell, K.B., Hunt, P.G., Uchimiya, M., Novak, J.M., Ro, K.S., 2012. Impact of pyrolysis temperature and manure source on physicochemical characteristics of biochar. Bioresour. Technol. 107, 419–428. doi:10.1016/j.biortech.2011.11.084.

Cesco, S., Neumann, G., Tomasi, N., Pinton, R., Weisskopf, L., 2010. Release of plant-borne flavonoids into the rhizosphere and their role in plant nutrition. Plant Soil 329, 1–25. doi:10.1007/s11104-009-0266-9.

Cesco, S., Mimmo, T., Tonon, G., et al., 2012. Plant-borne flavonoids released into the rhizosphere: impact on soil bio-activities related to plant nutrition. A review. Biol. Fertil. Soils 48, 123–149. doi:10.1007/s00374-011-0653-2.

Chen, Y.T., Wang, Y., Yeh, K.C., 2017. Role of root exudates in metal acquisition and tolerance. Curr. Opin. Plant Biol. 39, 66–72. doi:10.1016/j.pbi.2017.06.004.

Cheng, J., Lee, X., Gao, W., Chen, Y., Pan, W., Tang, Y., 2017. Effect of biochar on the bioavailability of difenoconazole and microbial community composition in a pesticide-contaminated soil. Appl. Soil Ecol. 121, 185–192. doi:10.1016/j.apsoil.2017.10.009.

Cunningham, S.D., Ow, D.W., 1996. Promises and prospects of phytoremediation. Plant Physiol. 110 (3), 715–771. doi:10.1104/pp.110.3.715.

Dastyar, W., Raheem, A., He, J., Zhao, M., 2019. Biofuel production using thermochemical conversion of heavy metal-contaminated biomass (HMCB) harvested from phytoextraction process. Chem. Eng. J. 358, 759–785. doi:10.1016/j.cej.2018.08.111.

Dede, G., Ozdemir, S., Hulusi Dede, O., 2012. Effect of soil amendments on phytoextraction potential of *Brassica juncea* growing on sewage sludge. Int. J. Environ. Sci. Technol. 9 (3), 559–564. doi:10.1007/s13762-012-0058-2.

Guo, D., Ren, C., Ali, A., Li, R., Du, J., Liu, X., Guan, W., Zhang, Z., 2019. *Streptomyces pactum* combined with manure compost alters soil fertility and enzymatic activities, enhancing phytoextraction of potentially toxic metals (PTMs) in a smelter-contaminated soil. Ecotoxicol. Environ. Saf. 181, 312–320. doi:10.1016/j.ecoenv.2019.06.024.

Di Lenola, M., Barra Caracciolo, A., Grenni, P., Ancona, V., Rauseo, J., Laudicina, V.A., Uricchio, V.F., Massacci, A., 2018. Effects of apirolio addition and alfalfa and compost treatments on the natural microbial community of a historically PCB-contaminated soil. Water Air Soil Pollut 229, 143. doi:10.1007/s11270-018-3803-4.

Di Lenola, M., Barra Caracciolo, A., Ancona, V., Laudicina, V.A., Garbini, G.L., Mascolo, G., Grenni, P., 2020. Combined effects of compost and *Medicago Sativa* in recovery a PCB contaminated soil. Water 12, 860. doi:10.3390/w12030860.

Dundek, P., Holík, L., Rohlík, T., Hromádko, L., Vranová, V., Rejšek, K., Formánek, P., 2011. Methods of plant root exudates analysis: a review. Acta Universitatis Agriculturae et Silviculturae Mendelianae Brunensis 59, 241–246.

Evangelou, M.W.H., Fellet, G., Ji, R., Schulin, R, et al., 2015. Phytoremediation and biochar application as an amendment. In: Ansari, A.A et al (Ed.), Phytoremediation: Management of Environmental Contaminants. Springer International Publishing, Switzerland, 1, pp. 253–263. doi:10.1007/978-3-319-10395-2_17.

Farid, M., Ali, S., Zubair, M., Saeed, R., Rizwan, M., Sallah-Ud-Din, R., Azam, A., Ashraf, R., Ashraf, W., 2018. Glutamic acid assisted phyto-management of silver-contaminated soils through sunflower; physiological and biochemical response. Environ. Sci. Pollut. Res. 25 (25), 25390–25400. doi:10.1007/s11356-018-2508-y.

Fellet, G., Marmiroli, M., Marchiol, L., 2014. Elements uptake by metal accumulator species grown on mine tailings amended with three types of biochar. Sci. Tot. Environ. 468–469, 598–608. doi:10.1016/j.scitotenv.2013.08.072.

Ferreras, L., Gomez, E., Toresani, S., Firpo, I., Rotondo, R., 2005. Effect of organic amendments on some physical, chemical and biological properties in a horticultural soil. Bioresour. Technol. 97, 635–640. doi:10.1016/j.biortech.2005.03.018.

Forján, Rubén, Rodríguez-Vila, Alfonso, Cerqueira, Beatriz, Covelo, Emma F., et al., 2017. Comparison of the effects of compost versus compost and biochar on the recovery of a mine soil by improving the nutrient content. J. Geochem. Explor. 183, 46–57. doi:10.1016/j.gexplo.2017.09.013.

Forján, R., Rodríguez-Vila, A., Cerqueira, B., Covelo, E.F., 2018. Comparative effect of compost and technosol enhanced with biochar on the fertility of a degraded soil. Environ. Monit. Assess. 190, 610. doi:10.1007/s10661-018-6997-4.

Fowles, M., 2007. Black carbon sequestration as an alternative to bioenergy. Biomass Bioenergy 31 (6), 426–432. doi:10.1016/j.biombioe.2007.01.012.

Fuhrer, T., Zamboni, N., 2015. High-throughput discovery metabolomics. Curr. Opin. Biotechnol. 31, 73–78. doi:10.1016/j.copbio.2014.08.006.

Gadd, G.M., 2000. Bioremedial potential of microbial mechanisms of metal mobilization and immobilization. Curr. Opin. Biotechnol. 11, 271–279. doi:10.1016/s0958-1669(00)00095-1.

Gandolfi, I., Sicolo, M., Franzetti, A., Fontanarosa, E., Santagostino, A., Bestetti, G., 2010. Influence of compost amendment on microbial community and ecotoxicity of hydrocarbon-contaminated soils. Bioresour. Technol. 101, 568–575. doi:10.1016/j.biortech.2009.08.095.

Gell, K., van Groenigen, J., Cayuela, M.L., 2011. Residues of bioenergy production chains as soil amendments: immediate and temporal phytotoxicity. J. Hazard. Mater. 186, 2017–2025. doi:10.1016/j.jhazmat.2010.12.105.

Gomez-Eyles, J.L., Sizmur, T., Collins, C.D., Hodson, M.E., 2011. Effects of biochar and the earthworm *Eisenia fetida* on the bioavailability of polycyclic aromatic hydrocarbons and potentially toxic elements. Environ. Pollut. 159, 616–622. doi:10.1016/j.envpol.2010.09.037.

Gopinath, K.A., Saha, Supradip M.B.L., Pande, H., et al., 2008. Influence of organic amendments on growth, yield and quality of wheat and on soil properties during transition to organic production. Nutr. Cycl. Agroecosys. 82, 51–60. doi:10.1007/s10705-008-9168-0.

Grenni, P., Barra Caracciolo, A., Rodríguez-Cruz, M.S., Sánchez-Martín, M.J., 2009. Changes in the microbial activity in a soil amended with oak and pine residues and treated with linuron herbicide. Appl. Soil Ecol. 41, 2–7. doi:10.1016/j.apsoil.2008.07.006.

Gu, P., Zhang, Y., Xie, H., Wei, J., Zhang, X., Huang, X., Wang, J., Lou, X., 2020. Effect of cornstalk biochar on phytoremediation of Cd-contaminated soil *by Beta vulgaris* var. cicla. Ecotoxicol. Environ. Saf. 205, 111144. doi:10.1016/j.ecoenv.2020.111144.

Guidi, W., Kadri, H., Labrecque, M., 2012. Establishment techniques to using willow for phytoremediation on a former oil refinery in southern Quebec: achievements and constraints. Chemistry Ecol. 28, 49–64. doi:10.1080/02757540.2011.627857.

Gul, I., Manzoor, M., Kallerhoff, J., Arshad, M., 2020. Enhanced phytoremediation of lead by soil applied organic and inorganic amendments: Pb phytoavailability, accumulation and metal recovery. Chemosphere 258, 127405. doi:10.1016/j.chemosphere.2020.127405.

Hartley, W., Dickinson, N.M., Riby, P., Lepp, N.W., 2009. Arsenic mobility in brownfield soils amended with green waste compost or biochar and planted with *Miscanthus*. Environ. Pollut. 157, 2654–2662. doi:10.1016/j.envpol.2009.05.011.

Han, T., Zhao, Z., Bartlam, M., Wang, Y., 2016. Combination of biochar amendment and phytoremediation for hydrocarbon removal in petroleum-contaminated soil. Environ. Sci. Pollut. Res. 23, 21219–21228. doi:10.1007/s11356-016-7236-6.

Hartl, W., Putz, B., Erhart, E., 2003. Influence of rates and timing of biowaste compost application on rye yield and soil nitrate levels. Eur. J. Soil Biol. 39, 129–139. doi:10.1016/S1164-5563(03)00028-1.

Hilber, I., Blum, F., Leifeld, J., Schmidt, H-P., Bucheli, T.D., 2012. Quantitative determination of PAHs in biochar: a prerequisite to ensure its quality and safe application. J. Agric. Food Chem. 60, 3042–3050. doi:10.1021/jf205278v.

Hossain, M.K., Strezov, V., Chan, K.Y., Nelson, P.F., 2010. Agronomic properties of wastewater sludge biochar and bioavailability of metals in production of cherry tomato (*Lycopersicon esculentum*). Chemosphere 78, 1167–1171. doi:10.1016/j.chemosphere.2010.01.009.

Houben, D., Evrard, L, Sonnet, P., 2013. Beneficial effects of biochar application to contaminated soils on the bioavailability of Cd, Pb and Zn and the biomass production of rapeseed (*Brassica napus* L.). Biomass Bioenergy 57, 196–204. doi:10.1016/j.biombioe.2013.07.019.

Huang, M., Zhu, Y., Li, Z., et al., 2016. Compost as a soil amendment to remediate heavy metal-contaminated agricultural soil: mechanisms, efficacy, problems, and strategies. Water Air Soil Pollut. 227, 359. doi:10.1007/s11270-016-3068-8.

Hussain, F., Hussain, I., Khan, A.H.A, et al., 2018. Combined application of biochar, compost, and bacterial consortia with Italian ryegrass enhanced phytoremediation of petroleum hydrocarbon contaminated soil. Environ. Exp. Bot. 153, 80–88. doi:10.1016/j.envexpbot.2018.05.012.

Jabeen, N., Abbas, Z., Iqbal, M., et al., 2016. Glycinebetaine mediates chromium tolerance in mung bean through lowering of Cr uptake and improved antioxidant system. Arch. Agron. Soil Sci. 62, 648–662. doi:10.1080/03650340.2015.1082032.

Jefferey, S. et al. (2011) 'A quantitative review of the effects of biochar application to soils on crop productivity using meta-analysis'. Agric. Ecosyst. Environ. 144: 175–187, doi:10.1016/j.agee.2011.08.015.

Jin, Y.Y., Luan, Y.N., Ning, Y.C., Wang, L.Y., 2018. Effects and mechanisms of microbial remediation of heavy metals in soil: a critical review. Appl. Sci. 8, 1336. doi:10.3390/app8081336.

Johnson, S.E, Loeppert, R.H, 2006. Role of organic acids in phosphate mobilization from iron oxide. Soil Sci. Soc. Am. J. 70, 222–234. doi:10.2136/sssaj2005.0012.

Kalina, J., 2011. Integrated biomass gasification combined cycle distributed generation plant with reciprocating gas engine and ORC. Appl. Therm. Eng. 31, 2829–2840. doi:10.1016/j.applthermaleng.2011.05.008.

Karami, N., Clemente, R., Moreno-Jiménez, E., Lepp, N.W., Beesley, L., 2011. Efficiency of green waste compost and biochar soil amendments for reducing lead and copper mobility and uptake to ryegrass. J. Hazard. Mater. 191 (1-3), 41–48. doi:10.1016/j.jhazmat.2011.04.025.

Khan, M.S., Zaidi, A., Wani, P.A., et al., 2009. Role of plant growth promoting rhizobacteria in the remediation of metal contaminated soils. Environ. Chem. Lett. 7, 1–19. doi:10.1007/s10311-008-0155-0.

Kidd, P., Barceló, J., Bernal, M.P., Navari-Izzo, F., et al., 2009. Trace element behaviour at the root–soil interface: Implications in phytoremediation. Environ. Exp. Bot. 67, 243–259. doi:10.1016/j.envexpbot.2009.06.013.

Kloss, S., Zehetner, F., Dellantonio, A., Hamid, R., Ottner, F., Liedtke, V., Schwanninger, M., Gerzabek, M.H., Soja, G., 2012. Characterization of slow pyrolysis biochars: effects of feedstocks and pyrolysis temperature on biochar properties. J. Environ. Qual. 41. doi:10.2134/jeq2011.0070.

Kolb, S.E., Fermanich, K.J., Dornbush, M.E., 2009. Effect of charcoal quantity on microbial biomass and activity in temperate soils. Soil Sci. Soc. Am. J. 73, 1173–1181. doi:10.2136/sssaj2008.0232.

Kookana, R.S., Sarmah, A.K., Van Zwieten, L., Krull, E., Singh, B., 2011. Biochar application to soil: agronomic and environmental benefits and unintended consequences. Adv. Agron. 112, 103–143. doi:10.1016/B978-0-12-385538-1.00003-2.

Kumar, S., Chaudhuri, S., Maiti, S.K., 2013. Soil dehydrogenase enzyme activity in natural and mine soil - a review. Middle East J. Sci. Res. 13 (7), 898–906. doi:10.5829/idosi.mejsr.2013.13.7.2801.

Kuppusamy, S., Thavamani, P., Megharaj, M., Venkateswarlu, K, Naidu, R., 2016. Agronomic and remedial benefits and risks of applying biochar to soil: Current knowledge and future research directions. Environ. Int. 87, 1–12 doi:10.1016/j.envint.2015.10.018.

Kwiatkowska-Malin, J., 2018. Functions of organic matter in polluted soils: The effect of organic amendments on phytoavailability of heavy metals. Appl. Soil Ecol. 123, 542–545. doi:10.1016/j.apsoil.2017.06.021.

Laird, D.A., Fleming, P., Davis, D.D., Horton, R., Wang, B., Karlen, D.L., 2010. Impact of biochar amendments on the quality of a typical Midwestern agricultural soil. Geoderma 158 (3–4), 443–449. doi:10.1016/j.geoderma.2010.05.013.

Lehman, J., Rillig, M.C., Thies, J., Masiello, C.A., Hockaday, W.C., Crowley, D, 2011. Biochar effects on soil biota - a review. Soil Biol. Biochem. 43, 1812–1836. doi:10.1016/j.soilbio.2011.04.022.

Lebrun, M., Macri, C., Miard, F., Hattab-Hambli, N., Motelica-Heino, M., Morabito, D., Bourgerie, S., 2016. Effect of biochar amendments on As and Pb mobility and phytoavailability in contaminated mine technosols phytoremediated by *Salix*. J. Geochem. Explor. 182, 149–156. doi:10.1016/j.gexplo.2016.11.016.

Li, M., Ren, L., Zhang, J., Luo, L., Qin, P., Zhou, Y., Huang, C., Tang, J., Huang, H., Chen, A., 2019. Population characteristics and influential factors of nitrogen cycling functional genes in heavy metal contaminated soil remediated by biochar and compost. Sci. Total Environ. 651, 2166–2174. doi:10.1016/j.scitotenv.2018.10.152.

Lima, I.M., Boykin, D.L., Thomas Klasson, K., Uchimiya, M., 2014. Influence of post-treatment strategies on the properties of activated chars from broiler manure. Chemosphere 95, 96–104. doi:10.1016/j.chemosphere.2013.08.027.

Lou, L., Wu, B., Wang, L., Luo, L., Xu, X., Hou, J., Xun, B., Hu, B., Chen, Y., 2011. Sorption and ecotoxicity of pentachlorophenol polluted sediment amended with rice-straw derived biochar. Bioresour. Technol. 102, 4036–4041. doi:10.1016/j.biortech.2010.12.010.

Luna, L., Miralles, I., Andrenelli, M.C., Gispert, M., Pellegrini, S., Vignozzi, N., Solé-Benet, A., 2016. Restoration techniques affect soil organic carbon, glomalin and aggregate stability in degraded soils of a semiarid Mediterranean region. Catena 143, 256–264. doi:10.1016/j.catena.2016.04.013.

Ma, Y., Prasad, M.N., Rajkumar, M., Freitas, H., 2011. Plant growth promoting rhizobacteria and endophytes accelerate phytoremediation of metalliferous soils. Biotechnol. Adv. 29, 248–258. doi:10.1016/j.biotechadv.2010.12.001.

Manzano, R, Silvetti, M., Garau, G., Deiana, S., Castaldi, P., 2016. Influence of iron-rich water treatment residues and compost on the mobility of metal(loid)s in mine soils. Geoderma 283, 1–9. doi:10.1016/j.geoderma.2016.07.024.

Martin, S.M., Kookana, R.S., Van Zwieten, L., Krull, E., 2012. Marked changes in herbicide sorption–desorption upon ageing of biochars in soil. J. Hazard. Mater. 231–232, 70–78. doi:10.1016/j.jhazmat.2012.06.040.

Masiello, C.A., Chen, Y., Gao, X., Liu, S., Cheng, H.Y., Bennett, M.R., Rudgers, J.A., Wagner, D.S., Zygourakis, K., Silberg, J.J., 2013. Biochar and microbial signaling: production conditions determine effects on microbial communication. Environ. Sci. Technol. 7 (20), 11496–11503. doi:10.1021/es401458s.

Menon, V., Rao, M., 2012. Trends in bioconversion of lignocellulose: biofuels, platform chemicals & biorefinery concept. Prog. Energy Combust. Sci. 38, 522–550. doi:10.1016/j.pecs.2012.02.002.

Mercado-Blanco, J., Abrantes, I., Barra Caracciolo, A., Bevivino, A., Ciancio, A., Grenni, P., Hrynkiewicz, K., Kredics, L., Proença, D.N., 2018. Belowground microbiota and the health of tree crops. Front. Microbiol. 9, 1006. doi:10.3389/fmicb.2018.01006.

Miyasaka, S.C., Hollyer, J.R., Kodani, L.S., 2001. Mulch and compost effects on yield and corm rots of taro. Field Crops Res. 71, 101–112. doi:10.1016/S0378-4290(01)00154-X.

Moradi, R., Pourghasemian, N., Naghizadeh, M., 2019. Effect of beeswax waste biochar on growth, physiology and cadmium uptake in saffron. J. Clean. Prod. 229, 1251–1261. doi:10.1016/j.jclepro.2019.05.047.

Mukherjee, A., Zimmerman, A.R., 2013. Organic carbon and nutrient release from a range of laboratory-produced biochars and biochar-soil mixtures. Geoderma 193–194, 122–130. doi:10.1016/j.geoderma.2012.10.002.

Neumann, G., Römheld, V., 2007. The release of root exudates as affected by the plant physiological status. In: Pinton, R., Varanini, Nannipieri P. (Eds.), The Rhizosphere, Biochemistry and Organic Substances at the Soil-Plant Interface2nd edn. CRC Press Taylor & Francis, pp. 23–72. doi:10.1093/aob/mcp166.

Nichols, G.J., Cripps, J.A., Collinson, M.E., Scott, A.C., 2000. Experiments in waterlogging and sedimentology of charcoal: Results and implications. Palaeogeogr. Palaeoclimatol. Palaeoecol. 164 (1–4), 43–56. doi:10.1016/S0031-0182(00)00174-7.

Nigussie, A., Kissi, E, Misganaw, M., Ambaw, G., 2012. Effect of biochar application on soil properties and nutrient uptake of lettuces (*lactuca sativa*) grown in chromium polluted soils. Am. Eurasian J. Agric. Environ. Sci. 12 (3), 369–376.

Nzihou, A., Stanmore, B., 2013. The fate of heavy metals during combustion and gasification of contaminated biomass—A brief review. J. Hazard. Mater. 256-257, 56–66. doi:10.1016/j.jhazmat.2013.02.050.

Norini, M.P., Thouin, H., Miard, F., Battaglia-Brunet, F., Gautret, P., Guégan, R, Le Forestier, L., Morabito, D., Bourgerie, S., Motelica-Heino, M., 2019. Mobility of Pb, Zn, Ba, As and Cd toward soil pore water and plants (willow and ryegrass) from a mine soil amended with biochar. J. Environ. Manag. 232, 117–130. doi:10.1016/j.jenvman.2018.11.021.

Mukherjee, A., Zimmerman, A., 2013. Organic carbon and nutrient release from a range of laboratory-produced biochars and biochar–soil mixtures. Geoderma 193-194, 122–130. doi:10.1016/j.geoderma.2012.10.002.

Novak, J.M., Ippolito, J.A., Ducey, T.F., Watts, D.W., Spokas, K.A., Trippe, K.M., Sigua, G.C., Johnson, M.G., 2018. Remediation of an acidic mine spoil: miscanthus biochar and lime amendment affects metal availability, plant growth, and soil enzyme activity. Chemosphere 205, 709–718. doi:10.1016/j.chemosphere.2018.04.107.

Oburger, E., Jones, D.L., Wenzel, W.W., 2011. Phosphorus saturation and pH differentially regulate the efficiency of organic acid anion-mediated P solubilization mechanisms in soil. Plant Soil 341, 363–382. doi:10.1007/s11104-010-0650-5.

Oburger, E., Dell'mour, M., Hann, S., Wieshammer, G., Puschenreiter, M., Wenzel, W.W., 2013. Evaluation of a novel tool for sampling root exudates from soil-grown plants compared to conventional techniques. Environ. Exp. Bot. 87, 235–247. doi:10.1016/j.envexpbot.2012.11.007.

Oburger, E., Jones, D.L., 2018. Sampling root exudates – mission impossible? Rhizosphere 6, 116–133. doi:10.1016/j.rhisph.2018.06.004.

Ogundiran, M.B., Mekwunyei, N.S., Adejumo, S.A., 2019. Compost and biochar assisted phytoremediation potentials of *Moringa oleifera* for remediation of lead contaminated soil. J. Environ. Chem. Eng. 6, 2206–2213. doi:10.1016/j.jece.2018.03.025.

Palansooriya, K.N., Shaheen, S.M., Chen, S.S., Tsang, D.C.W., Hashimoto, Y., Hou, D., Bolan, N.S., Rinklebe, J., Ok, Y.S, 2020. Soil amendments for immobilization of potentially toxic elements in contaminated soils: a critical review. Environ. Int. 134, 105046. doi:10.1016/j.envint.2019.105046.

Pantigoso, H.A., He, Y., DiLegge, M.J., Vivanco, J.M., 2021. Methods for root exudate collection and analysis. Method. Mol. Biol. 2232, 291–303. doi:10.1007/978-1-0716-1040-4_22.

Park, J.H., Choppala, G.K., Bolan, N.S., et al., 2011. Biochar reduces the bioavailability and phytotoxicity of heavy metals. Plant Soil 348, 439–451. doi:10.1007/s11104-011-0948-y.

Parraga-Aguado, I., González-Alcaraz, M.N., Schulin, R., Conesa, H.M., 2015. The potential use of *Piptatherum miliaceum* for the phytomanagement of mine tailings in semiarid areas: role of soil fertility and plant competition. J. Environ. Manag. 158, 74–84. doi:10.1016/j.jenvman.2015.04.041.

Pausch, J., Kuzyakov, Y., 2017. Carbon input by roots into the soil: quantification of rhizodeposition from root to ecosystem scale. Glob. Change Biol. 24, 1–12. doi:10.1111/gcb.13850.

Paz-Ferreiro, J., Lu, H., Fu, S., Méndez, A., Gascó, G., 2014. Use of phytoremediation and biochar to remediate heavy metal polluted soils: a review. Solid Earth 5, 65–75. doi:10.5194/se-5-65-2014.

Pérez, R., Tapia, Y., Antilén, M., Casanova, M., Vidal, C., Santander, C., Aponte, H., Cornejo, P., 2021. Interactive effect of compost application and inoculation with the fungus *Claroideoglomus claroideum* in *Oenothera picensis* plants growing in mine tailings. Ecotoxicol. Environ. Saf. 208, 111495. doi:10.1016/j.ecoenv.2020.111495.

Philippot, L., Raaijmakers, J.M., Lemanceau, P., van der Putten, W.H., 2013. Going back to the roots: the microbial ecology of the rhizosphere. Nat. Rev. Microbiol. 11, 789–799. doi:10.1038/nrmicro3109.

Puga A.P., Abreu C.A., Melo L.C., Paz-Ferreiro J., Beesley L. (2015) Cadmium, lead, and zinc mobility and plant uptake in a mine soil amended with sugarcane straw biochar. Environ. Sci. Pollut. Res. 22, 17606–17614, doi: 10.1007/s11356-015-4977-6

Qin, G., Gong, D., Fan, M.-Y., 2013. Bioremediation of petroleum-contaminated soil by biostimulation amended with biochar. Int. Biodeterior. Biodegradation 85, 150–155. doi:10.1016/j.ibiod.2013.07.004.

Rajkumar, M., Ae, N., Prasad, M.N., Freitas, H., 2010. Potential of siderophore-producing bacteria for improving heavy metal phytoextraction. Trend. Biotechnol. 28, 142–149. doi:10.1016/j.tibtech.2009.12.002.

Ramanjaneyulu, A.V., Neelima, T.L., Madhavi, A., Ramprakash, T., 2017. Phytoremediation: an overview. In: Humberto, R.M., Ashok, G.R., Thakur, K., Sarkar, N.C. (Eds.), Applied Botany. *American Accademy Press*, pp. 42–84.

Rees, F., Germain, C., Sterckeman, T., et al., 2015. Plant growth and metal uptake by a non-hyperaccumulating species (*Lolium perenne*) and a Cd-Zn hyperaccumulator (*Noccaea caerulescens*) in contaminated soils amended with biochar. Plant Soil 395, 57–73. doi:10.1007/s11104-015-2384-x.

Rhodes, A.H., Carlin, A., Semple, K.T., 2008. Impact of black carbon in the extraction and mineralization of phenanthrene in soil. Environ. Sci. Technol. 42, 740–745.

Robichaud, K., Stewart, K., Labrecque, K., Hijri, M., Cherewyk, J., Amyot, M., 2019. An ecological microsystem to treat waste oil contaminated soil: Using phytoremediation assisted by fungi and local compost, on a mixed contaminant site, in a cold climate. Sci. Tot. Environ. 672, 732–742. doi:10.1016/j.scitotenv.2019.03.447.

Rodríguez-Garrido, B., Balseiro-Romero, M., Kidd, P.S., Monterroso, C., 2020. Effect of plant root exudates on the desorption of hexachlorocyclohexane isomers from contaminated soils. Chemosphere 241, 124920. doi:10.1016/j.chemosphere.2019.124920.

Rodríguez-Vila, A., Asensio, V., Forján, R., Covelo, E.F., 2016. Carbon fractionation in a mine soil amended with compost and biochar and vegetated with Brassica juncea L. J. Geochem. Explor. 169, 137–143. doi:10.1016/j.gexplo.2016.07.021.

Rugova, A., Puschenreiter, M., Koellensperger, G., Hann, S., 2017. Elucidating rhizosphere processes by mass spectrometry - a review. Anal. Chim. Acta 956, 1–13. doi:10.1016/j.aca.2016.12.044.

Sajjad, A., Jabeen, F., Farid, M., Fatima, Q., Akbar, A., Ali, Q., Hussain, I., Iftikhar, U., Farid, S., Ishaq, H.K., 2020. Biochar: a sustainable product for remediation of contaminated soils. In: Hasanuzzaman, M. (Ed.), Plant Ecophysiology and Adaptation under Climate Change: Mechanisms and Perspectives II. Springer, Singapore. doi:10.1007/978-981-15-2172-0_30.

Sarmah, A.K., Srinivasan, P., Smernik, R.J., Manley-Harris, M., Antal, Jr., M.J., Downie A., Van Zwieten, L, 2010. Retention capacity of biochar-amended New Zealand dairy farm soil for an estrogenic steroid hormone and its primary metabolite. Aust. J. Soil Res. 48, 648–658. doi:10.1071/SR10013.

Sarwar, N., Imran, M., Shaheen, M.R., Ishaque, W., Kamran, M.A., Matloob, A., Rehim, A., Hussain, S., 2017. Phytoremediation strategies for soils contaminated with heavy metals: modifications and future perspectives. Chemosphere 171, 710–721. doi:10.1016/j.chemosphere.2016.12.116.

Sasse, J., Martinoia, E., Northen, T., 2018. Feed your friends: do plant exudates shape the root microbiome? Trends Plant Sci. 23 (1), 25–41. doi:10.1016/j.tplants.2017.09.003.

Saum, L., Jiménez, M.B., Crowley, D., 2018. Influence of biochar and compost on phytoremediation of oil-contaminated soil. Int. J. Phytoremediation 20 (1), 54–60. doi:10.1080/15226514.2017.1337063.

Schmidt, HP., 2012. 55 uses of biochar. Ithaka J., 286–289 1/2012.

Sigua, G.C., Novak, J.M., Watts, D.W., Ippolito, J.A., Ducey, T.F., Johnson, M.G., Spokas, K.A., 2019. Phytostabilization of Zn and Cd in mine soil using corn in combination with biochars and manure-based compost. Environments 6, 69. doi:10.3390/environments6060069.

Singh, B., Singh, B.P., Cowie, A.L., 2010. Characterisation and evaluation of biochars for their application as a soil amendment. Aust. J. Soil Res. 48, 516–525.

Singh, C., Tiwari, S., Singh, J.S., 2020. Biochar: A Sustainable Tool in Soil Pollutant Bioremediation. Bioremediation of Industrial Waste for Environmental Safety. Springer, Singapore, pp. 475–494. https://link.springer.com/chapter/10.1007/978-981-13-3426-9_19#citeas.

Sizmur, T., Quillam, R., Puga, A.P., Moreno-Jiménez, E., Beesley, L., Gomez-Eyles, J.L., 2016. Application of biochar for soil remediation. In: Guo, M., He, Z., Uchimiya, S.M. (Eds.). Agricultural and Environmental Applications of Biochar: Advances and Barriers, 63. SSSA Special Publication, pp. 295–324. doi:10.2136/sssaspecpub63.2014.0046.5.

Sohi, S.P., Krull, E., Lopez-Capel, E., Bol, R., 2010. A review of biochars and its use and function in soil. In: Sparks, D.L. (Ed.), Advances in Agronomy. Elsevier, pp. 47–82. doi:10.1016/S0065-2113(10)05002-9.

Tejada, M., Garcia, C., Gonzalez, J.L., Hernandez, M.T., 2006. Organic amendment based on fresh and composted beet vinasse: influence on soil properties and wheat yield. Soil Sci. Soc. Am. J. 70, 900–908. doi:10.2136/sssaj2005.0271.

Uchimiya, M., Lima, I.M., Klasson, K.T., Chang, S., Wartelle, L.H., Rodgers, J.E., 2010a. Immobilization of heavy metal ions (CuII, CdII, NiII, and PbII) by broiler litter-derived biochars in water and soil. J. Agric. Food Chem. 58, 5538–5544.

Uchimiya, M., Lima, I.M., Klasson, K.T., Wartelle, L.H, 2010b. Contaminant immobilization and nutrient release by biochar soil amendment: Roles of natural organic matter. Chemosphere 80 (8), 935–940. doi:10.1016/j.chemosphere.2010.05.020.

Uren, NC., 2007. Types, amounts and possible functions of compounds released into the rhizosphere by soil-grown plants. In: Pinton, R., Varanini, Z., Nannipieri, P (Eds.), The Rhizosphere. Biochemistry and Organic Substances at the Soil-Plant Interface 2nd edn. CRC Press Taylor & Francis Group, Boca Raton, pp. 1–21. doi:10.1093/aob/mcp166. ISBN 978-0-8493-3855-7-7.

Uroz, S., Calvaruso, C., Turpault, M.-P., Frey-Klett, P, 2009. Mineral weathering by bacteria: ecology, actors and mechanisms. Trends Microbiol 17 (8), 378–387. doi:10.1016/j.tim.2009.05.004.

Vaajasaari, K., Joutti, A., 2006. Field-scale assessment of phytotreatment of soil contaminated with weathered hydrocarbons and heavy metals. J. Soils Sediments 6 (3), 128–136. doi:10.1065/jss2006.07.170.

Valentinuzzi, F., Cesco, S., Tomasi, N., Mimmo, T, 2015. Influence of different trap solutions on the determination of root exudates in *Lupinus albus* L. Biol. Fertil. Soils 51, 757–765. doi:10.1007/s00374-015-1015-2.

van Dam, N.M., Bouwmeester, H.J., 2016. Metabolomics in the rhizosphere: tapping into belowground chemical communication. Trends Plant Sci. 21 (3), 256–265. doi:10.1016/j.tplants.2016.01.008.

Van Zwieten, L., Kimber, S., Morris, S., Chan, K.Y., Downie, A., Rust, J., Joseph, S., Cowie, A., 2010. Effects of biochar from slow pyrolysis of papermill waste on agronomic performance and soil fertility. Plant Soil 327, 235–246. doi:10.1007/s11104-009-0050-x.

Venegas, A., Rigol, A., Vidal, M., 2016. Changes in heavy metal extractability from contaminated soils remediated with organic waste or biochar. Geoderma 279, 132–140. doi:10.1016/j.geoderma.2016.06.010.

Vergani, L., Mapelli, F., Marasco, R., Crotti, E., Fusi, M., Di Guardo, A., Armiraglio, S., Daffonchio, D., Borin, S., 2017. Bacteria associated to plants naturally selected in a historical pcb polluted soil show potential to sustain natural attenuation. Front. Microbiol. 8, 1385. doi:10.3389/fmicb.2017.01385.

Verheijen, F.G.A., Jeffery, S., Bastos, A.C., van der Velde, M., Diafas, I, 2010. Biochar Application to Soils - A Critical Scientific Review of Effects on Soil Properties, Processes and Functions. EUR 24099 EN, Office for the Official Publications of the European Communities, Luxembourg, p. 149.

Walker, D.J., Clemente, R., Bernal, M.P., 2004. Contrasting effects of manure and compost on soil pH, heavy metal availability and growth of *Chenopodium album* L. in a soil contaminated by pyritic mine waste. Chemosphere 57, 215–224. doi:10.1016/j.chemosphere.2004.05.020.

Wang, H., Lin, K., Hou, Z., Richardson, B., Gan, J., 2010. Sorption of the herbicide terbuthylazine in two New Zealand forest soils amended with biosolids and biochars. J. Soils Sediments 10 (2), 283–289. doi:10.1007/s11368-009-0111-z.

Wei, L., Shutao, W., Jin, Z., Tong, X., 2014. Biochar influences the microbial community structure during tomato stalk composting with chicken manure. Bioresour. Technol. 154, 148–154. doi:10.1016/j.biortech.2013.12.022.

Wenzel, W.W., 2009. Rhizosphere processes and management in plant-assisted bioremediation (phytoremediation) of soils. Plant Soil 321 (1), 385–408. doi:10.1007/s11104-008-9686-1.

Wyszkowski, M., Ziółkowska, A., 2009. Role of compost, bentonite and calcium oxide in restricting the effect of soil contamination with petrol and diesel oil on plants. Chemosphere 74, 860–865. doi:10.1016/j.chemosphere.2008.10.035.

Wu, H., Zeng, G., liang, J., Chen, J., Xu, J., Dai, J., Li, X., Chen, M., Xu, P., Zhou, Y., Li, F., Hu, L., Wan, J., 2016. Responses of bacterial community and functional marker genes of nitrogen cycling to biochar, compost and combined amendments in soil. Appl. Microbiol. Biotechnol. 100 (19), 8583–8591. doi:10.1007/s00253-016-7614-5.

Wu, B., Wang, Z., Zhao, Y., Gu, Y., Wang, Y., Yu, J., Xu, H., 2019. The performance of biochar-microbe multiple biochemical material on bioremediation and soil micro-ecology in the cadmium aged soil. Sci. Total Environ. 686, 719–728. doi:10.1016/j.scitotenv.2019.06.041.

Xin, J., Liu, X., Liu, W., Zheng, X., 2014. Effects of biochar–BDE-47 interactions on BDE-47 bioaccessibility and biodegradation by Pseudomonas putida TZ-1. Ecotoxicol Environ. Saf. 106, 27–32. doi:10.1016/j.ecoenv.2014.04.036.

Zhang, X., Wang, H., He, L., Lu, K., Sarmah, A., Li, J., Bolan, N.S., Pei, J., Huang, H., 2013. Using biochar for remediation of soils contaminated with heavy metals and organic pollutants. Environ. Sci. Pollut. Res. 20 (12), 8472–8483. doi:10.1007/s11356-013-1659-0.

Zhang, L., Sun, X.-Y., Tian, T., Gong, X.-Q, 2014. Biochar and humic acid amendments improve the quality of composted green waste as a growth medium for the ornamental plant *Calathea insignis*. Sci. Hortic 176, 70–78. doi:10.1016/j.scienta.2014.06.021.

Zhang, R.-H., Li, Z.-G., Liu, X.-D., Wang, B.-C., Zhou, G.-L., Haung, X.-X., Lin, C.-F., Wang, A.-H., Brooks, M., 2017. Immobilization and bioavailability of heavy metals in greenhouse soils amended with rice straw-derived biochar. Ecol. Eng. 98, 183–188. doi:10.1016/j.ecoleng.2016.10.057.

Zhang, M., Wang, J., Bai, S.H., Zhang, Y., Teng, Y., Xu, Z., 2019. Assisted phytoremediation of a co-contaminated soil with biochar amendment: contaminant removals and bacterial community properties. Geoderma 348, 115–123. doi:10.1016/j.geoderma.2019.04.031.

Zheng, R.-L., Cai, C., Liang, J.-H., Huang, Q., Chen, Z., Haung, Y.-Z., Arp, H.P.H., Sun, G.-X, 2012. The effects of biochars from rice residue on the formation of iron plaque and the accumulation of Cd, Zn, Pb, As in rice (*Oryza sativa* L.) seedlings. Chemosphere 89 (7), 856–862. doi:10.1016/j.chemosphere.2012.05.008.

Zhou, R., liu, X., Luo, L., Zhou, Y., Wei, J., Chen, A., Tang, L., Wu, H., Deng, Y., Zhang, F., Wang, Y., 2017. Remediation of Cu, Pb, Zn and Cd-contaminated agricultural soil using a combined red mud and compost amendment. Int. Biodeter. Biodegradation 118, 73–81. doi:10.1016/j.ibiod.2017.01.023.

Zimmerman, J.R., Ghosh, U., Milward, R.N., Bridges, T.S., Luthy, R.G., 2004. Addition of carbon sorbents to reduce PCB and PAH bioavailability in marine sediments: physicochemical tests. Environ. Sci. Technol. 38 (20), 5458–5464. doi:10.1021/es034992v.

Zubillaga, M.S., Bressan, E., Lavado, R.S., et al., 2012. Effects of phytoremediation and application of organic amendment on the mobility of heavy metals in a polluted soil profile. Int. J. Phytoremediation 14 (3), 212–220. doi:10.1080/15226514.2011.587848.

CHAPTER

3

Arbuscular mycorrhizal fungi-assisted phytoremediation: Concepts, challenges, and future perspectives

Anissa Lounès-Hadj Sahraoui[a], Maryline Calonne-Salmon[b], Sonia Labidi[c], Hacène Meglouli[d], Joël Fontaine[a]

[a]*Univ. Littoral Côte d'Opale, UR 4492, UCEIV, Unit of environmental chemistry and interactions with living organisms, SFR Condorcet FR CNRS 3417, Dunkerque, France*

[b]*Earth and Life Institute, Applied Microbiology, Mycology, Université catholique de Louvain, Louvain-la-Neuve, Belgium*

[c]*Université de Carthage, Institut National Agronomique de Tunisie, Laboratoire des Sciences Horticoles, LR13AGR01, Tunis, Mahrajène, Tunisia*

[d]*Institut de Recherche en Biologie Végétale (IRBV), de l'Université de Montréal, Canada*

3.1 Introduction

Soil is the source of terrestrial life due to the multiple ecological and agricultural services that offers. Assessing and maintaining soil quality and security is crucial for global environmental sustainability. Unfortunately, the industrial and agricultural revolutions in America and Europe the latest centuries and recently in the Asian continent, caused an elevation in soil pollution mainly by trace elements (TEs) and persistent organic pollutants (POPs) such as pesticides, chlorophenols, polycyclic aromatic hydrocarbons (PAHs), polychlorinated biphenyls (PCBs), and dioxins/furans (Lombi et al., 1998).

So far, the most widely used soil remediation methods are physicochemical ones, which are most of the time efficient and rapid. But they present several inconveniences as they are costly in energy, irreversibly changing soil structure, decreasing soil microbial activity and depleting available minerals to plants because of aggressive solvents and high temperatures application (Gan et al., 2009). That's why from the last decades, an increasing interest for biological remediation techniques has gained interest over the conventional physicochemical methods. Amongst the gentle remediation options developed, plant remediation techniques, so-named phytoremediation, are innovative and sustainable solutions to restore ecologically polluted soils. Plants can act by direct or indirect means to remove pollutants from the soil either by metabolisation (phyto/rhizodegradation) or by extraction (phytoextraction) or by immobilization of the pollutants in the soil (phytostabilization). They can drive and shape the selection of microbes by secreting root exudates (Bakker et al., 2012; Huang et al., 2014). Therefore, the selection of appropriate plants, which are adapted to the edaphic site conditions and which are able to enhance the soil remediation, is crucial for efficient soil requalification (Inui et al., 2008; Van Aken et al., 2011). The different soil microbial communities are modulated by both plant species and their development

Assisted Phytoremediation. DOI: https://doi.org/10.1016/B978-0-12-822893-7.00008-2

stages (Cadillo-Quiroz et al., 2010; Chaparro et al., 2013; Huang et al., 2014). Many studies showed that successful phytoremediation is highly linked to the suitable plant microbiome (Feng et al., 2017; Fester et al., 2014; Hassan et al., 2014; Thijs et al., 2016).

However, many inconvenients limit these phytotechnologies application, such as the low produced biomass, the slow plant growth, the low bioavailability of the pollutants in the soil or also their phytotoxicity. Thus, to encourage their use, innovative practices that enhance the phytoremediation efficiency are needed. One strategy could be the improvement of their plant tolerance and biomass production through their symbiosis with beneficial soil microorganisms such as arbuscular mycorrhizal fungi (AMF). The presence and survival of AMF in polluted soils have been described in several studies (Lenoir et al., 2016a). These symbiotic fungi are characterized by the development of an extra-radical mycelium able to colonize the environment around the host plant roots. This extraradical mycelium has a fundamental role in plant nutrition by facilitating nutrient and water uptake from the soil. AMF not only improve mineral uptake of plants but they can reduce contaminant toxicity and increase tolerance to environmental stress in the associated hosts (Lenoir et al., 2016b; Smith and Read, 2008). Numerous responses of AMF to mitigate plant stress in the presence of pollutants have been reported (Plouznikoff et al., 2016). Furthermore, many studies illustrated the role of AMF in plant installation on polluted soils, their effect on the fate of some organic pollutants in the plant mycorrhizosphere (Binet et al., 2000; Gao et al., 2011; Lenoir et al., 2016b; Nwoko, 2014; Zhou et al., 2013) and their potential to remediate TEs-polluted soils (Göhre and Paszkowski, 2006). Mycorrhizal colonization can affect the composition of the microbial communities in both the rhizosphere and hyphosphere. This effect may be explained by AMF modulation of plant metabolisms (e.g. carbohydrate metabolism) with significant modifications of root exudation with increasing secretions of nitrogen and phenolic compounds or decreasing secretions of total sugars and phosphorous (Jones et al., 2004). Moreover, AMF themselves may exude substances inducing a selective pressure on the microbial community. The capacity of AMF extraradical hyphae to explore a large surface area and to provide soil with photosynthetically derived carbon to the soil offer attracting niches for bacterial communities (Toljander et al., 2007). Thus, AMF could be considered as an interesting possibility to assist phytoremediation, so named AMF-assisted phytoremediation (or AMF-aided phytoremediation), and to improve phytoremediation and/or phytomanagement efficiencies.

Through this chapter, we tried to present a description of the AMF diversity in POPs and TEs-polluted soils and summarize current knowledge of AMF contribution in phytoremediation and phytomanagement of polluted soils and the impact of mycorrhizal inoculation on soil functionalization. In conclusion, recommendations to promote AMF-assisted phytotechnologies adoption in polluted soil phytomanagement programs will be suggested.

3.2 Arbuscular mycorrhizal fungi diversity in contaminated soils

Diversity of AMF has been extensively studied over the last 20 years in POPs- and TEs-polluted soils, where different plant species were observed to grow, suspecting that AMF may help host plant to adapt to pollutants' induced stresses (Hassan et al., 2014). Several studies have mentioned the presence of AMF associated with numerous plant species growing in hydrocarbons (Cabello, 1997; de la Providencia et al., 2015; Garcés-Ruiz et al., 2017, 2019; Hassan et al., 2014; Iffis et al., 2016; Villacrés et al., 2014) and TEs-aged polluted soils (Deram et al., 2011; Regvar et al., 2010; Ruotsalainen et al., 2007). Their presence has been demonstrated and their identification was based either on morphological criteria of isolated AMF spores or on the use of different molecular tools. Table 3.1 summarizes the AMF taxa isolated

Table 3.1 Identified arbuscular mycorrhizal fungi in aged hydrocarbon- and trace element-contaminated soils.

Pollutant	Samples: soil/host plant	Identified AMF[a]	Identification methodology	Reference
Hydrocarbons	Soil and root system from *Cynodon dactylon*	*R. aggregatus*	Morphological identification of spores extracted from trap cultures	Cabello (1997)
	Soil and root system from *Solidago* sp. and *Dactylis glomerata*	*F. mosseae*		
	Soils	*Sep. deserticola*, *F. geosporus*, *R. intraradices/irregularis*		Cabello (1999)
	Rhizosphere and roots of *Echinochloa polystachya*, *Citrus aurantifolia*, *Citrus aurantium*	• From *E. polystachya*: *G. ambisporum*, *Scl. sinuosa*, *Ac. laevis*, *Am. gerdemannii* • From citrus trees: *De. heterogama*, *G. ambisporum*, *Ac. scrobiculata*, *G. citricola* (probably a *Rhizophagus* sp.)	Morphological identification of spores extracted from trap cultures	Franco-Ramírez et al. (2007)
	Soil	*Sep. constrictum* and *F. mosseae*, *Acaulospora* and *Archeospora*, *Archeosporaceae*, *Claroideoglomeraceae*	454-pyrosequencing 18S on isolated spores	Huang et al. (2007)
	Rhizosphere soil of introduced willow cultivars	*Diversisporaceae*, *Gigasporaceae*, *Glomeraceae* (*Funneliformis*, *Septoglomus*, *Glomus* and *Rhizophagus*), *Paraglomeraceae*	rDNA region	Hassan et al. (2014)
	Roots of *Solidago rugosa*	*Claroideoglomeraceae*, *Diversisporaceae*, *Glomeraceae* and *Archeosporaceae*	454-pyrosequencing 18S rDNA region	Iffis et al. (2014)
	Soil	Predominant genera *Glomus* and *Acaulospora*, *Entrophospora*	Morphological identification of spores extracted from trap cultures	Villacrés et al. (2014)
	Rhizosphere and roots of *Eleocharis obtuse* and *Panicum capillare*	*Claroideoglomus*, *Diversispora*, *Rhizophagus*, *Paraglomus*	High-throughput PCR, cloning and sequencing 18S rDNA	De la Providencia et al. (2015)
	Solidago canadensis, *Populus balsamifera* and *Lycopus europaeus*	Mainly *Diversispora*, *Glomus*, *Acaulospora*, *Claroideoglomus*, *Rhizophagus*, *Eutrophosphora*, *Funneliformis*, unclassified *Archeosporaceae*	454-pyrosequencing 18S rDNA region 454-pyrosequencing ITN dataset	Iffis et al. (2016)
	Soil	*F. mosseae*	Morphological identification of spores extracted from aged soil	Lenoir et al. (2016c)
	Roots from *Carludovica palmate*, *Costuss caber* and *Euterpe precatoria*	*R. prolifer*, *Rhizophagus* sp., *Acaulospora* sp., *Ac. longula*, *Ac. kentinensis*, *Glomus* sp., *Archeospora* sp.	High-throughput PCR, cloning, RFLP from partial SSU, the whole ITS and partial LSU rDNA region	Garcés-Ruiz et al. (2017)

(continued)

Table 3.1 Identified arbuscular mycorrhizal fungi in aged hydrocarbon- and trace element-contaminated soils. ***Continued***

Pollutant	Samples: soil/host plant	Identified AMF[a]	Identification methodology	Reference
	Roots of *Piper* sp., *Costus* sp., *Calathea*, *Anthurium* sp., *Euterpe precatoria*, *Urerabaccifera*, *Ureracaracasana*, *Stigmatopteris*, *Geonomamacrostachys*, *Poaceae*, *Cyathea* sp., *Caladium* sp., *Cordia*, *Cyclanthusbipartitus*, *Faramea* sp., *Paullinia* sp.	*Acaulospora* sp., *Ac. scrobiculata*, *Ac. kentinensis*, *Archeospora* sp., *Glomus* sp., *G. macrocarpum*-like *Rhizophagus* sp., *R. proliferus*, *Rhizophagus/Dominikia* sp., *Sclerocystis* sp., *Kamienskia* sp., unclassified AMF	amplification of the partial SSU, the complete ITS region and partial LSU rRNA gene, using the SSUmAf–LSUmAr or SSUmCf- LSUmBr primers pairs, nested PCR, 454 pyrosequencing	Garcés-Ruiz et al. (2019)
	Roots of *Poa trivialis*, *Phragmites australis*	*F. caledonium*, *Rhizophagus sp.*, *F. mosseae*, *C. luteum*	PCR-DGGE analysis, a three-step nested-PCR on the AMF 18 S rDNA gene, re-amplification with primers NS31/Glo 1 before sequencing	Malicka et al. (2020)
Trace elements	Roots of wild thyme (*Thymus polytrichus*)	*Glomus* sp., *Acaulospora* sp., *F. mosseae*, *R. fasciculatum*	Primary PCR using the primer NS31 and the general fungal primer AM1 to isolate a section of the 18S SSU rDNA. Secondary PCR: amplification using Pfu DNA polymerase using both T3 and T7 primers	Whitfield et al. (2004)
	Roots of *Solidago gigantea*	*Glomus* sp., *C. lamellosum*, *C. claroideum*, *Gi. margarita*, *Gi. decipiens*, *Scutellospora* sp., *Gi gigantea*, *Gi. rosea*, *C. etunicatum*, *Di. spurca*, *Sep. constrictum*, *Sep. viscosum*, *R. intraradices*, *R. fasciculatum*, *F. mosseae*	Amplification of partial ribosomal SSU DNA fragments with RedTaq DNA polymerase and two different sets of primers: NS31/AM1; NS5/ITS4 and ARCH1311/NS8	Vallino et al. (2006)
	Roots of *Verbena officinalis*, *Melilotus alba* and *Senecio inaequidens*	*Glomus* sp., *R. fasciculatum*, *R. intraradices*, *Sep. constrictum*, *Diversispora* sp., *Di. spurca*, *G. proliferum*	Amplification of the partial fungal SSU by NS31/AM1 primers	Bedini et al. (2010)
	Roots and rhizospheric soil of *Phytolacca americana*, *Rehmannia glutinosa*, *Perilla frutescens*, *Litsea cubeba* and *Dysphania ambrosioides*	*Glomus* sp., *R. intraradices*, *C. claroideum*, *R. irregularis*, *A. colombiana*, *Scl. sinuosum*, *C. lamellosum*	Nested PCR: Amplification of the SSU rRNA gene fragments with specific primers AML1/AML2 and the universal eukaryotic primers NS31-GC/G101	Long et al. (2010)

Pollutant	Samples: soil/host plant	Identified AMF[a]	Identification methodology	Reference
	Roots of *Salvia nemorosa*, *S. officinalis*, *Nonnea persica*, *Plantago ovata*	*Glomus* sp. *F. mosseae*, *R. intraradices*, *C. claroideum*, *Di. versiformis*, *Acaulospora* sp.	Nested PCR: Amplification of the ITS region of ribosomal DNA with the primer pair LSU-Glom1/SSU and ITS5/ITS4	Zarei et al. (2010)
	Roots and rhizospheric soil of plantain (*Plantago major* L.)	*Glomus* sp., *F. mosseae*, *R. irregularis*, *C. etunicatum*, *C. lamellosum*, *S. viscosum*	PCR amplifications on the DNA using primer pair AML1 and AML2 to amplify a 790-bp 18S rRNA gene fragment. Nested PCR to amplify 18S rRNA gene fragments. The first PCR using the primer pair NS1 and NS41 and the second PCR using AM1, AM2, AM3 and NS31-GC primers	Hassan et al. (2011)
	Roots of *Phytolacca americana*	*Sep. viscosum*, *Glomus* sp., *Sep. constrictum*, *R. intraradices*	Nested PCR: first amplification using 18S rDNA primers GeoA2 and Geo11. Second amplification: primers AM1/NS31-GC; Third amplification: primers NS31-GC and Glol	Wei et al. (2014)
	Rhizospheric soil of *Cynodon dactylon*, *Boehmeria nivea*, Miscanthus anderss; *Phytolacca americana*, *Lophatherum sinense*	*Sep. viscosum*, *Glomus* sp., *Sep. constrictum*, *R. intraradices*, *R. irregularis*		Wei et al. (2015)
	Rhizospheric soil of *Eleocharis filiculmis* Kunth., *Cyperus sesquiflorus* (Tor.) Mattf and Kuek, *Brachiaria decumbens* (Stapf) Prain, *Vetiveria zizanioides* (L.) Nash, *Pteridium aquilinum* (L.) Kuhn, *Pteris vittata* L., *Anacardium occidentale* L.	*Ac. colombiana*, *Ac. delicata*, *Ac. koskei*, *Ac. mellea*, *Ac. morrowiae*, *Ac. spinosa*, *Acaulospora* sp., *R. clarus*, *R. intraradices*, *R. fasciculatus*, *F. mosseae*, *F. geosporum*, *Sep. constrictum*, *G. macrocarpum*, *G. microcarpum*, *G. glomerulatum*, *Glomus* sp., *C. etunicatum*, *C. lamellosum*, *Gi. margarita*, *Cetraspora gilmorei*, *De. taheterogama*, *De. tarubra*, *Ra. fulgida*, *Scu. pernambucana*, *Scu. Dipurpurescens*, *Scutellospora* sp., *Paraglomus occultum*, *Am. appendicular*, *Am. brasiliensis*	Morphological identification of spores extracted from natural soil	Schneider et al. (2016)

(continued)

Table 3.1 Identified arbuscular mycorrhizal fungi in aged hydrocarbon- and trace element-contaminated soils. ***Continued***

Pollutant	Samples: soil/host plant	Identified AMF[a]	Identification methodology	Reference
Trace elements	*Digitaria violascens* Link, *Eclipta prostrate*, *Veronica didyma Tenore*, *Digitaria ciliaris* (Retz.) Koel, *Herba seu* Radix Amaranthi, *Eleusine indica* (L.) Gaertn, *Solanum nigrum* L., *Miscanthus floridulu* (Labnll.) Warb, *Allium senescens*, *Clinopodium chinense* (Benth.) O.Ktze, *Poa annua* L., *Paspalum paspaloides* (Michx.) Scribn.	*Rhizophagus* sp., *Glomus* sp., *F. mosseae*, *Acaulospora*, *Diversispora*, *C. claroideum*, *Scutellopora*, *Gi. margarita*, *Ambispora* sp., *Paraglomus* sp., *Archaeospora* sp.	Amplification of the SSU rRNA gene fragments by nested PCR with the primer pair of AML1/AML2. The primers used for the second PCR: the primer pairsNS31/AM1	Sun et al. (2016)
	Soil from rehabilitation plantation with grass, canga, Cerrado, native forest, and eucalyptus	*Dominikia* sp., *G. magnicaule*, *G. microaggregatum*, *Glomus* sp., *Sep. viscosum*, *R. clarus*, *R. diaphanus*, *R. fasciculatus*, *Gigaspora* sp., *Cetraspora pellucida*, *De. biornata*, *De. heterogama*, *Diversispora* sp., *Redeckera fulvum*, *Ac. alpina*, *Ac. colombiana*, *Ac. herrerae*, *Ac. lacunosa*, *Ac. mellea*, *Ac. morrowiae*, *Ac. nivalis*, *Ac. scrobiculata*, *Acaulospora* sp., *Ac. spinosa*, *Ac. tuberculata*, *Ac. walkeri*, *Archaeospora trappei*, *Am. leptoticha*, *Am. delicata*, *Am. foveata*, *Am. rehmii*, *Am. spinosissima*, *Paraglomus occultum*, *Corymbiglomus tortuosum*, *De. scutata*, *Gi. albida*, *Gi. gigantea*, *C. etunicatum*, *G. aggregatum*, *G. glomerulatum*, *G. invermaium*, *G. microcarpum*, *G. spinuliferum*, *Scl. coremioides*, *Scl. taiwanensis*, *Scu. sinuosa*, *Scu. pernambucana*, *Septoglomus* sp., *Entrophospora infrequens*	Morphological identification of spores extracted from natural soils and trap cultures	Teixeira et al. (2017)

Pollutant	Samples: soil/host plant	Identified AMF[a]	Identification methodology	Reference
	Soil from a plantation of a mix of grasses species, including *Melinis minutiflora* P. Beauv., *Urochloa brizantha* (Hochst. ex A. Rich.) R.D.Webster and *Panicum maximum* Hochst. ex A. Rich.	Morphological identification: *Archaeospora trappei*, *Am. leptoticha*, *Di. spurca*, *Ac. baetica*, *Ac. colombiana*, *Ac. koskei*, *Ac. mellea*, *Ac. morrowiae*, *Ac. punctata*, *Ac. scrobiculata*, *Ac. spinosissima*, *Ac. tuberculata*, *Ac. walkeri*, *Cetraspora pellucida*, *De. biornata*, *De. savannicola*, *Gi. gigantea*, *Gigaspora* sp., *Ra. coralloidea*, *Ra. fulgida*, *Ra. verrucosa*, *Scutellospora* sp., *C. etunicatum*, *F. mosseae*, *G. ambisporum*, *G. microaggregatum*, *G. spinuliferum*, *R. clarus*, *Glomus* sp., *Sep. viscosum*, *Entrophospora infrequens*., *De. heterogama*, *Gi. albida*, *Gi. decipiens*, *Gi. margarita*, *Ac. spinose* Molecular identification: Glomerales, *Paraglomus*, *Ambispora*, *R. intraradices*, Glomeromycota et *Glomus* sp.	Morphological identification of spores extracted from natural soils and trap cultures + A two-steps nested PCR was used to amplify the 25S rRNA of AMF., The primers used for the first PCR: ITS1/NDL22 LR1/FLR2	Vieira et al. (2018)
Trace elements	Rhizospheric soil of *Bidens pilosa* L., *Tagetes minuta* L., and *Sorghum halepense* (L.)	*Gigasporaceae*, *Diversisporaceae*, *Archaeosporaceae*, *Claroideoglomeraceae*, *Paraglomeraceae*, *Acaulosporaceae*, *Ambisporaceae*, *Claroideoglomeraceae*	Amplification of the 18S rDNA region with primers specific to AMF: AMV4.5NF/AMDGR	Faggioli et al. (2019)
	Rhizospheric soil of *Atriplex halimus* L., *A. canescens* (Pursh) Nutt., *Suaeda fruticosa* (syn. S.vera Forssk. ex J.F. Gmel.), *Marrubium vulgare* L. and *Dittrichia viscosa* (L.) Greuter (syn. *Inula viscosa* (L.) Aiton)	*S. constrictum*, *Sclerocystis* sp., *R. microaggregatum*, *Rhizophagus* sp., *G. maculosum*, *G. nanolumen*, *F. mosseae*, *F. geosporum*, *F. caledonium*, *F. coronatum*, *C. lamellosum*, *C. claroideum*, *Tricispora nevadensis*, *Di. tortuosa*, *Am. reticulate*, *Archaeospora myriocarpa*, *Ac. spinose*, *Ac. rehmii*, *Ac. morrowae*, *Ac. excavata*, *Ac. tortuosa*, *Ac. melea*, *Ac. leavis*, *Ac. cavernata*, *Ac. thomii*	Morphological identification of spores extracted from natural soil	Sidhoum and Fortas (2019)

[a]*Arbuscular mycorrhizal fungi are named according to the current taxonomy (amf-phylogeny.com/amphylo_taxonomy.html). Ac., Acaulospora; Am., Ambispora; AMF, arbuscular mycorrhizal fungi; C., Claroideoglomus; De., Denticutata; Di., Fiversispora; F., Funneliformis; G., Glomus; Gi., Gigaspora; R., Rhizophagus; Ra., Racocetra; Scl., Sclerocystis; Scu., Scutellospora; Sep., Septoglomus.*

and/or identified from different aged contaminated sites. AMF as *Funneliformis mosseae*, *Rhizophagus intraradices*, and *Glomus* sp. were numerously found in soils of different continents (Europe, Asia, and America) polluted by TEs (Ban et al., 2015; Hassan et al., 2011; Krishnamoorthy et al., 2015; Weissenhorn et al., 1994). For TEs-polluted soils and regarding the listed references, the most frequent species of AMF identified in contaminated sites belong principally to *Glomeraceae* family followed by *Claroideoglomeraceae*, *Acaulosporaceae*, *Diversisporaceae*, and *Gigasporaceae*. It seems that AMF species belonging to *Glomeraceae* (i.e., *Septoglomus constrictum*, *F. mosseae*, *R. intraradices*, *Sep. viscosum*) and *Claroideoglomeraceae* (*Claroideoglomus claroideum*, *C. lamellosum*, *C. etunicatum*) are well adapted to high levels of TEs in soils compared to the other ones. In fact, the tolerance of AMF to TEs stress differs between and within taxonomic groups of *Glomeromycota* communities and their diversity are mainly affected by TEs levels in the soils than the plant host species (Faggioli et al., 2019; Long et al., 2010; Sun et al., 2016). Soil polluted with hydrocarbons also contain mainly AMF from the *Glomeraceae* (i.e., *Rhizophagus*, *Funneliformis*, *Glomus* and *Septoglomus* genera), followed by *Acaulosporaceae*, *Claroideoglomeraceae*, *Archeosporaceae*, and *Diversisporaceae* (Table 3.1). *Gigasporaceae* are fewly represented. It is worth to note that these species are similar to those found in TEs-polluted soils.

The presence of these AMF families in TEs- and POPs-polluted soils could be explained by their life history strategy and ecological and functional traits. Indeed, according to the Grimes's C-S-R (competitor, stress tolerator, ruderal) framework, *Glomeraceae* and *Acaulosporaceae* are supposed to be ruderals, stress-tolerators (Chagnon et al., 2013; van der Heyde et al., 2017) and relatively tolerant to disturbances. The constant occurrence of *Glomeraceae* species in areas contaminated with pollutants could be explained by the fact that *Glomeraceae* is the largest AMF group and can be adapted to wide-ranging ecosystems (Schwarzott et al., 2001; Smith and Read, 2008). *Glomeraceae* possess high rates of turnover and have host-generalist tendencies that fit with populations able to quickly recover from most disturbance events (van der Heyde et al., 2017).

3.3 Mechanisms involved in mycorrhizal plant tolerance to soil pollutants

The improvement of growth and tolerance of mycorrhized plants on polluted soils was largely exposed in scientific litterature. Several responses of AMF to mitigate plant stress in the presence of pollutants have been reported, as listed below.

Increased mineral and water nutrition: The main benefit brought by AMF to their host is an improved mineral intake, as inorganic phosphorus (Pi). Indeed, Pi is an essential mineral element for plant growth, which is poorly available for plants because of its low mobility in the soil, leading to the decrease of crop production (Holford, 1997). Thanks to the larger volume of soil explored by their extraradical mycelium, AMF have access to immobile nutrients and non-accessible micro-sites for the host roots, as fine soil pores. In harsh conditions as in presence of pollutants, literature numerously related a better nutrient uptake as well as nitrogen and Pi accumulation in roots and aerial parts of mycorrhizal plants as compared to non-colonized ones (Calonne-Salmon et al., 2018; Joner and Leyval, 2001; Lenoir et al., 2016b; Liu and Dalpé, 2009; Lu and Lu, 2015; Nwoko, 2014; Rabie, 2004; Tang et al., 2009; Yu et al., 2011), even if the improved growth of mycorrhizal plants was not always demonstrated.

In addition, an osmotic stress could be induced in plants grown on polluted soils, similarly to plants grown in drought conditions. This is due to the chemico-physical properties of pollutants such as POPs (hydrophobic and lipophilic), which decrease soil water availability and root gas exchange (Ko and Day, 2004; Merkl et al., 2005; Robertson et al., 2007). A mitigated water deficit stress, a better osmotic potential and a better recovery can be observed in plants colonized by AMF (Le Pioufle et al., 2019; Ruiz-Lozano et al., 2008; Wu, Q.S. et al., 2008) *via* different mechanisms such as a better water-use efficiency, an osmotic potential adjustment, an increased accumulation of proline, glycine, carbohydrates and minerals ions (K and Cl) (reviews of Porcel et al., 2012 and Chitarra et al., 2016).

Attenuation of oxidative stress: Several studies described a better plant tolerance to the oxidative stress-induced by pollutants when colonized by AMF. Indeed, reactive oxygen species (ROS) are generally overproduced in plants submitted to abiotic environmental stresses. The huge increase of ROS production in cells can cause protein and lipid oxidation as well as DNA breakage, which have a harmful effect on cell homeostasis (Miller et al., 2008).

Oxidative stresses-induced by POPs and TEs was mainly demonstrated to be lowered in plants colonized by AMF as compared to non-colonized ones through a reduced accumulation of malondialdehyde (a lipid peroxidation biomarker) and 8-hydroxy-2′-desoxyguanosine (a DNA alteration biomarker) in plants, in concomitance with an enhanced antioxidant enzymatic (superoxide dismutase - SOD, catalase - CAT, peroxidase - POD, ascorbate peroxidase - APX) and non-enzymatic (glutathione/glutathione disulfide ratio) scavenging systems (Debiane et al., 2008, 2009; Driai et al., 2015; Firmin et al., 2015; Lenoir et al., 2017; Xun et al., 2015).

Recently, Lenoir et al. (2017) demonstrated that non-mycorrhizal roots of *Medicago truncatula* L. grown in the presence of benzo[a]pyrene presented an upregulation by 14-, 11-, and 3-folds of genes involved in antioxidant system (*MtSOD*, *MtPOX*, and *MtAPX*), as compared to the non-polluted control. To the contrary, in spite of a lower H_2O_2 accumulation observed in roots colonized by AMF as compared to non-colonized roots, the expressions of these three genes were downregulated in roots colonized by *R. irregularis* exposed to the PAHs, as compared also to their non-polluted control. The authors suggested that the fungus accumulates benzo[a]pyrene in lipid bodies of AMF spores and hyphae, making it less available to the plant, proven by the lower PAH accumulation in roots colonized by AMF than in non-colonized ones.

Lenoir et al. (2017) also demonstrated that two DNA repair genes (*MtTFIIS* and *MtTdp1α*) were downregulated in roots colonized by *R. irregularis* in concomitance with the former results obtained by Debiane et al. (2009), who reported a lower 8-hydroxy-2′-desoxyguanosine accumulation in mycorrhizal chicory roots as compared to non-colonized ones under benzo[a]pyrene exposure.

Accumulation of specific molecules: To limit the damages induced by the accumulation of TEs, an increase of the synthesis of phytochelatins (oligomers of glutathione) could be measured in AMF structures, and these molecules have the capacity to detoxify intracellular TEs (Garg and Aggarwal, 2011; Garg and Chandel, 2012). Metallothionein, an important group of proteins with polypeptides and cysteine residues are also associated with the retention of TEs in mycorrhizal plants (Cabral et al., 2015). The upregulation (from 2- to 9-fold) of gene expression encoding metallothioneins in mycorrhizal plants grown on TE-polluted soil in comparison with non-mycorrhizal ones was confirmed by Cicatelli et al. (2010). To the contrary, another study proved a down-regulation of the expression of two metal chelators (*MsPCS1* (encoding for phytochelatin synthase) and *MsMT2* (encoding for metallothionein)) and two metal transporter genes (*MsIRT1* and *MsNramp1*) in the roots of *M. sativa* colonized by *R. irregularis* as compared to non-colonized plants (Motaharpoor et al., 2019). The authors suggested a Cd sequestration within hyphae of the fungal symbiont, making its concentration at an insufficient level to induce the expression of these genes in root cells. This controversy effect is probably linked to the specific association plant/AMF, the culture conditions and the considered TE.

Arbuscular mycorrhizal fungal colonization in plants exposed to pollutant stress can modulate the pattern of protein expression. For example, annexin, a protein involved in the secretion and maturation of newly synthetized components of the membrane and wall of cells, was found in higher quantity in plants colonized by AMF as compared to the non-colonized ones, suggesting an increase in membrane lipid production in plants associated to AMF (Aloui et al., 2009; Repetto et al., 2003). In the same vein, the concentration of polyunsaturated fatty acids (C18:1, C18:2, and C18:3) dropped in non-mycorrhizal roots grown in presence of benzo[a]pyrene, while it remained similar in roots colonized by AMF (Debiane et al., 2012). Authors thus suggested a repairing of the membrane optimal lipid properties. In the same study, authors observed differences in the root sterol composition and hypothesized that these modifications inhibited PAHs transport along root tissues, protecting the host plant against this harmful pollutant (Debiane et al., 2012). Other studies reported a higher proline accumulation in plants colonized by AMF than in non-colonized ones under petroleum pollution (Alarcón et al., 2008; Tang et al., 2009; Xun et al., 2015). Protection conferred by proline in stressed plants is suggested to be linked with a mediation of the osmotic adjustment and the protection of the subcellular structures *via* a lowered water potential (Ashraf and Foolad, 2007; Xun et al., 2015).

Protection of photosynthesis activity: Accumulated pollutants in the aerial part of the plants can damage the photosynthetic apparatus. However, numerous studies showed that their negative effects on photosynthesis could be counteracted in mycorrhizal plants (Nwoko, 2014; Tang et al., 2009). Indeed, an increase in chlorophyll content in plants colonized by AMF under petroleum and a co-contamination Cd and decabromodiphenyl ether (BDE-209) stress was demonstrated in comparison to non-colonized ones (Abdel Latef, 2013; Li et al., 2019). Yang et al. (2015) also found that plants associated to AMF had much higher photosynthetic pigment contents as chlorophyll compared with those of non-colonized plants with increasing Pb levels. Likewise, Wang et al. (2017) recorded higher net photosynthetic rate in mycorrhizal plants in presence of high Cd concentration, which may be attributed to a greater leaf area in AMF symbiosis than in non-mycorrhizal treatment under the same Cd stress. This positive effect of the mycorrhizal colonization may be linked to an improvement in Mg^{2+} transfer, which contribute to the process of photosynthesis to the host plants *via* the AMF hyphae (Abdel Latef, 2013; Giri et al., 2003), a lower pollutant translocation and thus accumulation of pollutants in aerial part decreasing the induced-stress in this plant part (Malekzadeh et al., 2012) or an alleviated chloroplasts damage (Li et al., 2019).

Pollutant bioaccumulation: For the organic pollution, more than 56% of the articles in the literature showed a higher POPs accumulation in roots colonized by AMF as compared to non-colonized ones, whereas 17 and 26% of the studies showed less and equal POPs accumulations respectively (data not shown). As suggested by Lenoir et al. (2016b), this larger bioaccumulation of POPs in mycorrhizal roots suggests increased transport of POPs from extraradical hyphae to roots, as demonstrated by Gao et al. (2010) with two PAHs, phenanthrene and fluorene. This larger accumulation could also be related to a larger bioavailability of POPs in the mycorrhizosphere, probably related to the nature of the root and fungal exudates. Once accumulated in roots, it was observed that lipid bodies of root cells were able to store POPs as PAHs (Verdin et al., 2006), thus reducing the transport in the root cortex and transport in the aerial parts (Rajtor and Piotrowska-Seget, 2016; Wu, N. et al., 2009). Indeed, 29% of studies show an accumulation of POPs inferior in shoot of plants associated to AMF in comparison with non-colonized ones (Table 3.2). It is well known that mycorrhizal roots have also the capacity

Table 3.2 Table summarizing some studies reporting AMF inoculation effects on plant growth and persistent organic pollutants (POPs) dissipation/degradation in contaminated soils.

Persistent organic pollutant	concentration (mg/kg)	host plant	Inoculum AMF species[a]	culture conditions	culture duration (days)	Dissipation/degradation improvement in comparison with NM plants (%)	Lower transfer in aerial parts of M plants	Shoot growth improvement in M plant	reference
Acenaphtene	35	Alfalfa (*Medicago sativa L.*)	*F. mosseae*	Spiked soil/greenhouse	55	+6.8			Zeng et al. (2010)
			C. etunicatum			+9.8			
Anthracene	5000	Ryegrass (*Lolium perenne* L., var. Barclay)	*F. mosseae* BEG69	Sterile spiked soil/growth chamber	40	=		N	Binet et al. (2000)
	50	Jute (*Corchotus capsulari*)	1:1 mixture of *F. mosseae* and *R. intraradices* (BIORIZE)	Spiked soil/greenhouse	35	~+24		Y	Cheung et al. (2008)
	100					~+12.8		Y	
	150					=		Y	
	10.7	Transformed carrot (*Daucus carota* L.)	*R. custos* (O10MYCO-HSP)	Minimal (M) medium/*in vitro*	49	~+50			Aranda et al. (2013)
	21.4					~+31			
	42.8					~−89			
	30	Transformed chicory (*Cichorium intybus* L.) root	*R. irregularis* DAOM 197198		42	+117			Verdin et al. (2006)
	140					+125			
Phenanthrene	500	Ryegrass (*Lolium perenne* L. cv. Barclay)	*F. mosseae* BEG69	Spiked sand/growth chamber	42	= / −			Corgié et al. (2006)

(continued)

Table 3.2 Table summarizing some studies reporting AMF inoculation effects on plant growth and persistent organic pollutants (POPs dissipation/degradation in contaminated soils. ***Continued***

Persistent organic pollutant	concentration (mg/kg)	host plant	Inoculum AMF species[a]	culture conditions	culture duration (days)	Dissipation/ degradation improvement in comparison with NM plants (%)	Lower transfer in aerial parts of M plants	Shoot growth improvement in M plant	reference
	2.44	Alfalfa (*Medicago sativa* L.)	*C. etunicatum* (BGC USA01)	Sterile spiked soil/ greenhouse	60	~+5.1 in the rhizosphere, = in the bulk soil		N	Wu, N. et al. (2008a)
	4.05					~+6.3 in the rhizosphere, ~+14.9 in the bulk soil		N	
	9.46					~+1.2 in the rhizosphere ~+9.3 in the bulk soil		N	
	103	Alfalfa (*Medicago sativa* L.)	*F. mosseae* (BGC GD01A)*C. etunicatum* (BGC HUN02C)	Sterile spiked mix soil and sand/ greenhouse	30–70	~+3.1 after 30 days, ~+0.9 after 45 days, = after 60 and 70 days	N Y		Gao et al. (2011)
			F. mosseae (BGC GD01A) + *C. etunicatum* (BGC HUN02C)			~+4.4 after 30 days, ~+1.1 after 45 days, = after 60 and 70 days	Y		
	10.7	Transformed carrot (*Daucus carota* L.) root	*R. custos* (O10MYCO-HSP)	Minimal (M) medium/*in vitro*	49	=			Aranda et al. (2013)
	21.4					~−70			
	42.8					~−79			
	5	Alfalfa (*Medicago sativa* L.)	*C. etunicatum* (BGC USA01)	Sterile spiked mix soil and sand / greenhouse	60	~+3.3	N	Y	Wu et al. (2019)

Persistent organic pollutant	concentration (mg/kg)	host plant	Inoculum AMF species[a]	culture conditions	culture duration (days)	Dissipation/ degradation improvement in comparison with NM plants (%)	Lower transfer in aerial parts of M plants	Shoot growth improvement in M plant	reference
Pyrene	74	Alfalfa (*Medicago sativa* L.)	*F. mosseae* (BGC GD01A)	Sterile spiked mix soil and sand/ greenhouse	30–70	from ~+2.2 to ~+62.3			Gao et al. (2011)
			F. mosseae (BGC GD01A) + *C. etunicatum* (BGC BGC HUN02C)			from ~+8.1 to ~+82			
	500	Alfalfa (*Medicago sativa* cv. Europe)	*R. intraradices* (from Institüt für Pflanzenkultur)	Heated spiked mix soil and sand/greenhouse	42	=			Zhou et al. (2013)
		Tall fescue (*Festuca arundinacea* L. cv. Barianne)				=			
		Ryegrass (*Lolium multiflorum* cv. Barclay)				=			
		Celery (*Apium graveolens*)				=			
Fluoranthene	100	Tall fescue (*Festuca arundinacea* L.)	Mix of 4 species (*R. intraradices*, *F. mosseae*, R. *aggregatus*, *C. etunicatum*	Spiked soil in greenhouse	90	=		Y	Rostami and Rostami (2019)
	200					=		Y	
	300					=		Y	
Dibenzothiophene	11.06	Transformed carrot (*Daucus carota* L.) root	*R. custos* (O10MYCO-HSP)	Minimal (M) medium/*in vitro*	49	=			Aranda et al. (2013)
	22.1					~+550			
	44.2					~+980			
Benzo[a] pyrene	1	Alfalfa (*Medicago sativa* L.)	*F. caledonium*	Sterile spiked soil	30 – 90	+13.1 after 90 days			Liu et al. (2004)
	10					+11.5 after 90 days			
	100					+7.5 after 90 days			

(continued)

Table 3.2 Table summarizing some studies reporting AMF inoculation effects on plant growth and persistent organic pollutants (POPs dissipation/degradation in contaminated soils. ***Continued***

Persistent organic pollutant	concentration (mg/kg)	host plant	Inoculum AMF species[a]	culture conditions	culture duration (days)	Dissipation/ degradation improvement in comparison with NM plants (%)	Lower transfer in aerial parts of M plants	Shoot growth improvement in M plant	reference
	25	*Echinochloa polystachya*	*Gi. margarita*	Sterile spiked river sand/growth chamber	70	~−16.4		N	Alarcón et al. (2006)
	50					~−56		N	
	75					~−22.2		N	
	100					=		N	
Anthracene + phenanthrene	3500	Leek (*Allium porrum* L. cv. Musselburgh)	*R. irregularis* DAOM197198	Sterile spiked mix soil and substrate/ greenhouse	84	~+16.5		Y	Liu and Dalpé (2009)
	7000					~+12.5			
	10500					~+39.5			
	3500		*G. versiforme* DAOM 196672			~+26		Y	
	7000					~+25			
	10500					~+39.5			
Phenanthrene + pyrene	50 + 50	Ryegrass (*Lolium multiflorum* L.)	*F. mosseae* (Biorize)	Steam-sterilized spiked soil/ greenhouse	60		N	Y	Wu, F-Y. et al. (2014)
	100 + 100						Y	N	
	200 + 200						N	Y	
	12 + 7.4	Maize (*Zea mays* L.)	*F. mosseae*	Spiked soil/ greenhouse	60	=	N	N	Wu et al. (2011)
	39.01 + 44.05	Ryegrass (*Lolium multiflorum* L.)	*F. mosseae* (Biorize)	Steam-sterilized spiked soil/ greenhouse	60	PHE: ~+1.6	N	N	Yu et al. (2011)
						PYR: =	N		
	69.39 + 83.69					PHE: ~+3.3	Y	Y	
						PYR: =	N		
	158.42 + 189.73					PHE: =	N	Y	
						PYR: =	N		
	43.52 + 49.66	*Sesbania cannabina* (Retz.) Pers.	*F. mosseae* (BGC NM03D)	Spiked soil/ in greenhouse	60	PHE: ~+47.3	N	Y	Ren et al. (2017)
						PYR: ~+25.5	Y		
	71.24 + 93.43					PHE: =	N	Y	
						PYR: =	N		
	174.56 + 192.17					PHE: ~+13,9	Y	Y	
						PYR: =	N		

Persistent organic pollutant	concentration (mg/kg)	host plant	Inoculum AMF species[a]	culture conditions	culture duration (days)	Dissipation/ degradation improvement in comparison with NM plants (%)	Lower transfer in aerial parts of M plants	Shoot growth improvement in M plant	reference
Anthracene + chrysene + dibenz(a,h) anthracene	500 + 500 + 50	Ryegrass (*Lolium perenne* cv. Barclay) and white clover (*Trifolium repens* cv. Grasslands huia)	*F. mosseae* BEG69	Sterile spiked soil/ growth chamber	56	= for the 3 PAHs			Joner et al. (2001)
					112	anthracene: = Chrysene: +17.5 DBA: +110			
Phenanthrene + pyrene + dibenzo(a,h) anthracene	500 + 500 + 50	Alfalfa (*Medicago sativa* cv. Europe)	*R. intraradices* (from Institüt für Pflanzenkultur)	Heated spiked mix soil and sand/greenhouse	42	=		Y	Zhou et al. (2013)
		Tall fescue (*Festuca arundinacea* L. cv. Barianne)				=		N	
		Ryegrass (*Lolium multiflorum* cv. Barclay)				=		Y	
		Celery (*Apium graveolens*)				=		N	
Anthracene + pyrene + chrysene + benz(a,h) anthracene	500 + 500 + 500 + 50	Wheat (*Triticum aestivum* cv. Sakha8)	*F. mosseae*	Sterile spiked soil/ growth chamber	90	=		Y	Rabie (2004)
		Mungbean (*Vigna radiata* V 2010)				=		Y	
		Wheat (*Triticum aestivum* cv. Sakha8)			65	+35.8	N	N	Rabie (2005)
		Mungbean (*Vigna radiata* V 2010)				+58.5	N	N	
		Eggplant. (*Solanium melongena* L.)				=	N	N	

(continued)

Table 3.2 Table summarizing some studies reporting AMF inoculation effects on plant growth and persistent organic pollutants (POPs dissipation/degradation in contaminated soils. ***Continued***

Persistent organic pollutant	concentration (mg/kg)	host plant	Inoculum AMF species[a]	culture conditions	culture duration (days)	Dissipation/ degradation improvement in comparison with NM plants (%)	Lower transfer in aerial parts of M plants	Shoot growth improvement in M plant	reference
8 PAHs	200 (each)	Ryegrass (*Lolium perenne* L. cv. Barclay)	*F. mosseae* BEG69	Sterile spiked soil/ growth chamber	40	=	Y	N	Binet et al. (2000)
10 PAHs	405	Ryegrass (*Lolium perenne* L., cv. Barclay) and white clover (*Trifolium repens* L., cv. Grasslands Huia)	*F. mosseae* BEG69	Sterile aged polluted soils/growth chamber	90	=		N	Joner and Leyval (2003a)
					180	= near the roots, +40.8 in the rhizoplane soil		N	
	2030				90	=		N	
					180	between +52.5 to +150 near the roots, +56 in the rhizoplane soil		Y	
PAHs (15 analyzed)	12.85	Alfalfa (*Medicago sativa* L.)	*F. caledonium*	Aged polluted soil/ greenhouse	90	+110		N	Zhang et al. (2010)
16 US EPA priority PAH	620	Tall fescue (*Festuca arundinacea* L.)	*F. caledonium*		120	~+62.5	N	Y	Lu and Lu (2015)
PAHs	Total PAHs: 210.2; available PAHs: 78.2	Wheat (*Triticum aestivum*)	*R. irregularis* DAOM197198	Sterile aged polluted soil/microcosm	112	total: +16.3 bioavailable: +4.4		Y	Lenoir et al. (2016c)
				Non-sterile aged polluted soil/ microcosm		total: +17.9 bioavailable: +12.9 +47.1 (degradation capacity)	Y	N	
Arabian medium crude oil (ACO)	6000	*Lolium multiflorum* Lam. cv. Passerel Plus	*R. intraradices* (Mycorise ® ASP)	Sterile spiked mix soil and sand/greenhouse	80	+25		N	Alarcón et al. (2008)

Persistent organic pollutant	concentration (mg/kg)	host plant	Inoculum AMF species[a]	culture conditions	culture duration (days)	Dissipation/ degradation improvement in comparison with NM plants (%)	Lower transfer in aerial parts of M plants	Shoot growth improvement in M plant	reference
Diesel	7500	*Melilotus albus* Medik	*C. claroideum, R. diaphanus* and *G. albidum*	Sterile spiked river sand/growth chamber	40 + 60 in presence of diesel	~+60.9		N	Hernández-Ortega et al. (2012)
Crude oil (Nigerian bonny light)	20000 – 40000 - 80000	African bean (*Phaseolus vulgaris* L.)	*F. mosseae*	Spiked soil/screen house	70	residual TPH in soils: -		N at 20000; Y at 40000 and 80000	Nwoko et al. (2013)
						residual TPH in soils: -		Y	
						residual TPH in soils: -		Y	
Crude oil (Nigerian bonny light)	20000	Bean (*Phaseolus vulgaris* L.)				residual TPH in soils: =		N	Nwoko (2014)
	40000					residual TPH in soils: 47.7 lower than in NM treatment		Y	
	80000					residual TPH in soils: 35.5 lower than in NM treatment		Y	
Petroleum (saturated hydrocarbon: 56.1 %, aromatic hydrocarbons: 26.6 %, asphalt: 1.6 %, and non-hydrocarbon compounds: 15.7 %	5000	*A. sativa* cv. Baiyan N°7	*R. intraradices*	Sterile spiked soil/ greenhouse	60	~+29.2		Y	Xun et al. (2015)
	10000					~+13.3		Y	

(continued)

Table 3.2 Table summarizing some studies reporting AMF inoculation effects on plant growth and persistent organic pollutants (POPs dissipation/degradation in contaminated soils. *Continued*

Persistent organic pollutant	concentration (mg/kg)	host plant	Inoculum AMF species[a]	culture conditions	culture duration (days)	Dissipation/ degradation improvement in comparison with NM plants (%)	Lower transfer in aerial parts of M plants	Shoot growth improvement in M plant	reference
Decabromodiphenyl ether (BDE-209)	3.584	Ryegrass (*Lolium multiflorum* L.)	*F. mosseae* (BGC GD01A)	Sterile spiked soil/ growth chamber	60	from ~+24.2 to ~+37.3	N		Wang et al. (2011)
Decabromodiphenyl ether (BDE-209)	4.98	Black nightshade (*Solanum nigrum* L.)	*F. mosseae* (Mycagro)	Sterile spiked soil	35	=	N	Y	Li et al. (2018)
			R. irregularis (Mycagro)			=	N	N	
Decabromodiphenyl ether (BDE-209)	2.2	Amaranth (K11)	*R. irregularis* (Mycagro)	Sterile spiked soil/ greenhouse	65	~+75	N	Y	Li et al. (2019)
PCBs	0.054, 0.245	Alfalfa (*Medicago sativa* L.)	*F. caledonium* 90036	Aged polluted soil		=			Teng et al. (2008)
21 PCB congeners	0.556 – 0.575			Excaved 20-year aged polluted soil/field	180	+48	N	N	Teng et al. (2010)
	0.475	Ryegrass	*F. caledonium*	Aged polluted soil/ greenhouse	180	~+76.5	N	N	Lu et al. (2014)
PCBs	181.14	Maize (*Zea mays* L.)	*F. mosseae* M47V	Aged polluted soil	90	=			Wang, C-D. et al. (2018)

Persistent organic pollutant	concentration (mg/kg)	host plant	Inoculum AMF species[a]	culture conditions	culture duration (days)	Dissipation/ degradation improvement in comparison with NM plants (%)	Lower transfer in aerial parts of M plants	Shoot growth improvement in M plant	reference
Aroclor 1242 (mix of PCBs congeners)	25	Pumpkin (*Cucurbita pepo* L. cv black beauty)	*Ac. leavis* 90034	Spiked soil/ greenhouse	60	~+65 in the bulk soil, '~+43 in the rhizosphere		Y	Qin et al. (2014)
			F. caledonium 90036			= in bulk soil and rhizosphere		N	
			F. mosseae M47V			~+100 in the bulk soil, '~+33 in the rhizosphere		Y	
	15		*F. mosseae* M47V	Controlled climate chamber	40	~from +100 to +220 times			Qin et al. (2016)
17 PCDD/F congeners	2.08 10^{-4} = $1.6.10^{-5}$ mg toxic equivalency quantity/kg	Alfalfa (*Medicago sativa* L.)	*F. mosseae* (Mycagro) + *Glomus* sp. (Agrauxine)	Sterile aged polluted soil/semi-controlled growth room	168	=		Y	Meglouli et al. (2018)
		Tall fescue (*Festuca arundinacea* L.)				=		N	
		Alfalfa (*Medicago sativa* L.)	indigenous inoculum (*Scu. constrictum*, *C. lamellosum*, *F. geosporum*, *F. mosseae*)			=		Y	
		Tall fescue (*Festuca arundinacea* L.)				=		N	
PCDD/F	$2.6.10^{-4}$ = 2.6 10^{-5} mg toxic equivalency quantity/kg	Alfalfa (*Medicago sativa* L.)	*F. mosseae* (MycAgro)	Sterile aged polluted spiked soil / semi-controlled growth room	180	=		Y	Meglouli et al. (2019)

(continued)

Table 3.2 Table summarizing some studies reporting AMF inoculation effects on plant growth and persistent organic pollutants (POPs dissipation/degradation in contaminated soils. ***Continued***

Persistent organic pollutant	concentration (mg/kg)	host plant	Inoculum AMF species[a]	culture conditions	culture duration (days)	Dissipation/degradation improvement in comparison with NM plants (%)	Lower transfer in aerial parts of M plants	Shoot growth improvement in M plant	reference
p, p,-DDE	from 4.87 10^{-5} to 2.25 10^{-4}	Pumpkin (*Curcubita pepo* spp, *pepo* (3 different cultivars)	BioVam	Aged polluted soil/ field	62			N	White et al. (2006)
			Myco-Vam (*R. intraradices, G. aggregatum, F. mosseae*)					N	
			INVAM (*R. intraradices,* F. *mosseae, C. etunicatum, R. clarus*)					N	
DDT	2.5	Alfalfa (*Medicago sativa* L.)	*C. etunicatum* (BGC USA01)	Sterile spiked soil/ greenhouse	60	= in the rhizosphere ~−30 in the bulk soil	Y	N	Wu, N. et al. (2008b)
	5					~+19 in the rhizosphere ~−44 in the bulk soil	Y	Y	
	10					= in the rhizosphere ~−58 in the bulk soil	Y	N	
	0.00173	Pumpkin (*Cucurbita pepo* ssp pepo cv. Howden)	*R. intraradices* (Myke Bulb)	Aged polluted soil/ greenhouse		=	N	N	Whitfield Åslund et al. (2010)

[a]*Arbuscular mycorrhizal fungi are named according to the current taxonomy (amf-phylogeny.com/amphylo_taxonomy.html). Ac., Acaulospora; Am., Ambispora; AMF, arbuscular mycorrhizal fungi; C., Claroideoglomus; De., Denticutata; Di., Diversispora; F., Funneliformis; G., Glomus; Gi., Gigaspora; R., Rhizophagus; Ra., Racocetra; Scl., Sclerocystis; Scu., Scutellospora; Sep., Septoglomus; M, mycorrhizal; NM, non-mycorrhizal; Y, yes; N, no.*

to retain TEs, comprising Fe, Ni, Cd and Zn (Janoušková et al., 2006; Kaldorf et al., 1999; Orlowska et al., 2008; Turnau et al., 1993).

Several studies also demonstrated an adsorption and accumulation of TEs as well as POPs in extraradical mycelium and spores of AMF (Aranda et al., 2013; Ferrol et al., 2009; Gaspar et al., 2002; Gil-Cardeza et al., 2017; González-Guerrero et al., 2008; Verdin et al., 2006; Wu, S. et al., 2015, 2016). Whereas the accumulation and translocation of POPs was fewly studied, a large number of works focused on the TEs accumulation and translocation from the extraradical mycelium to the intraradical mycelium then possibly to the plant. Indeed, only the work of Verdin et al. (2006) proved the accumulation of anthracene in lipid bodies of the hyphae and spores of *R. irregularis*.

Concerning TEs, it is assumed that the cell wall of AMF contains free amino acids, hydroxyl, carboxyl, and other groups representing locations for TEs' adsorption (Demirbas, 2008; Zhou, 1999). Then, TEs are absorbed through the plasma membrane of the AMF extraradical mycelium. Tamayo et al. (2014) and Tisserant et al. (2013) identified, in *R. irregularis*, the presence of genes implicated in the transport of proteins mediating Cu, Zn, Fe, and Mn uptake from the soil solution. Among them, could be find members of the CTR family of Cu transporters (*RiCTR1* and *RiCTR3*), the ZIP (zinc–iron permease or *ZRT-IRT-like protein*) family (*RiZRT1*), the NRAMP (natural resistance-associated macrophage proteins) family (*RiSMF1*), and the iron permease RiFTR1 (Ferrol et al., 2016). The absence of specific transporters of TEs in AMF could be explained by their toxic effects. Thus, TEs transport is probably ensured by the same transport systems for the mineral elements such as Cu, Zn, Fe, and Mn. Moreover, González-Chávez et al. (2011) demonstrated that phosphate transporters are involved in the uptake of the metalloid arsenate. After absorption, TEs, as Zn, Cu and Cd are preferentially stored in fungal vacuoles, as observed in the vacuoles of *R. irregularis* extraradical mycelium exposed to these metals (González-Guerrero et al., 2008; Yao et al., 2014). Handa et al. (2015) reported the induction of Zn transporters and of Fe transport-related proteins (a vacuolar Fe transporter and a ferric-chelate reductase) in mycorrhizal roots of sorghum. This suggestion was confirmed by Chen B. et al. (2018) who demonstrated a negligible accumulation of Cd in roots without fungal structures exposed to TEs, indicating a low translocation from roots colonized by AMF toward roots without AMF structures. These results provide direct evidence for the intraradical immobilization of Cd absorbed by AMF, which may largely promote the increased tolerance of plants to Cd. For example, on a contaminated soil with Cd, Motaharpoor et al. (2019) recorded a drop in the Cd translocation from the roots to the shoots of *M. sativa* colonized by *R. irregularis*. They attributed this decrease to the immobilization of Cd within AMF vacuoles, which possibly decreased Cd availability in root cells and subsequently Cd translocation to shoots . This result is related with those of Gil-Cardeza et al. (2017), who quantified the Cr transport from the extraradical mycelium to the shoot and demonstrated an accumulation of this element in the roots (probably in the intraradical structures), whereas the shoots contained similar concentrations. Nonetheless, while in some studies, it was reported that AMF are able to develop mechanisms helping metal accumulation in cells of roots (Redon et al., 2009) and preventing metal translocation to the shoots (Göhre and Paszkowski, 2006; Joner and Leyval, 1997; Motaharpoor et al., 2019); in other cases, an increased metal translocation from roots to shoots has been observed in plants colonized by AMF (Göhre and Paszkowski, 2006).

The mechanisms cited before point out an intracellular compartmentalization performed by AMF in order to (1) protect themselves against pollutants negative damages (Ferrol et al., 2009), (2) avoid translocation to the plant, constituting, hence, another protection mechanism to pollutant toxicity, and (3) avoid leaching into the soil and water compartments.

3.4 AMF-assisted phytoremediation of polluted soils

Three main phytotechnologies are currently used to restore polluted soils. Whereas, phyto/rhizodegradation is applied in POPs-polluted soils, phytostabilization and phytoextraction are applied in the case of TEs-polluted soils (Fig. 3.1).

Phyto/rhizodegradation

Phyto/rhizodegradation is a phytotechnology involving plants and their associate's rhizosphere microorganisms to modify the structure and toxicity of pollutants, reducing them in non/less-toxic molecules or to completely mineralize organic molecules into innocuous inorganic end products (Schwitzguébel, 2017). AMF-assisted phyto/rhizodegradation has become of interest the last 20 years due to the well-known benefits brought by these obligatory symbionts in term of increased plant tolerance to abiotic stresses and improved POPs dissipation. Only some plant species can tolerate and grow in soils containing high concentrations of POPs (Glick 2010; Lenoir et al., 2016b). Indeed, soils contaminated by toxicants generally present a poor soils structure, a low water-holding capacity, a lack of organic matter, a nutrient deficiency, as well as a modified or low microbial diversity, as compared to a non-contaminated soil. Therefore, AMF can not only improve plant establishment and growth on POPs-polluted soils, but also improve phytostimulation of the microbiome as well as enhance phyto/rhizodegradation (Lenoir et al., 2016b; Plouznikoff et al., 2016; Wang et al., 2019).

Table 3.2 summarizes the studies describing the contribution of AMF inoculation of different plant species in POPs-contaminated soils phytoremediation. As seen on the Fig. 3.2, based on the works reported on the Table 3.2, more than 50% of the studies showed an improved POPs dissipation/degradation by plants colonized by AMF as compared to non-colonized ones. The most used AMF species in these studies are *F. mosseae* followed by *R. irregularis*. It is important to note that most of the published research has been concerned with dissipation of a part or the entire bioavailable extractable POPs, which concerns the decrease in the extractable concentration of POPs in soils, i.e. degradation/metabolization and/or transformation and/or making pollutants unavailable only (volatilization, sorption...). Whereas dissipation has been studied in a number of papers, the degradation of POPs has been fewly described in the literature, and their degradation metabolites are not easy to detect. Nevertheless, the positive effects of AMF inoculation on POPs phytoremediation are generally observed for PAHs and crude oil/petroleum dissipation (which represent 70% of the listed articles) and polybromodiphenylethers (PBDE) (6.8% of the listed articles). Whereas neutral or negative effects are observed with PCBs (11% of the listed articles), dioxins/furans (4.5% of the listed articles), and pesticides (6.8% of the listed articles) dissipation. These man-made compounds are possibly more harmful for AMF, resulting in an inhibited effect on the pollutant dissipation in comparison to non-mycorrhizal plants. Until now, only few studies were conducted with excavated aged-polluted soils which were potted in greenhouse (around 18% of the listed articles), and less were conducted *in situ* (field plots – 4.5% of the listed articles). Knowing that the environmental pressures (rain, temperature, drought conditions...) play a role on the plants and rhizosphere activities as well as pollutant behavior in the soil, the efficacy of *in situ* phytoremediation of POPs by mycorrhizal plants is not currently well-known. In addition, long-term studies are uncommon; the longest studies were monitored after 180 days of plant growth with their AMF associates, in presence of the POPs (Table 3.2). Thus, a good feedback on the efficacy of the phytoremediation of aged POPs-polluted soils by mycorrhizal plants in the long-term is still missing.

As seen on the Fig. 3.2 and Table 3.2, the effect of the fungal plant colonization by AMF on POPs dissipation/degradation, and on the plant accumulation remains unclear. Various factors can explain

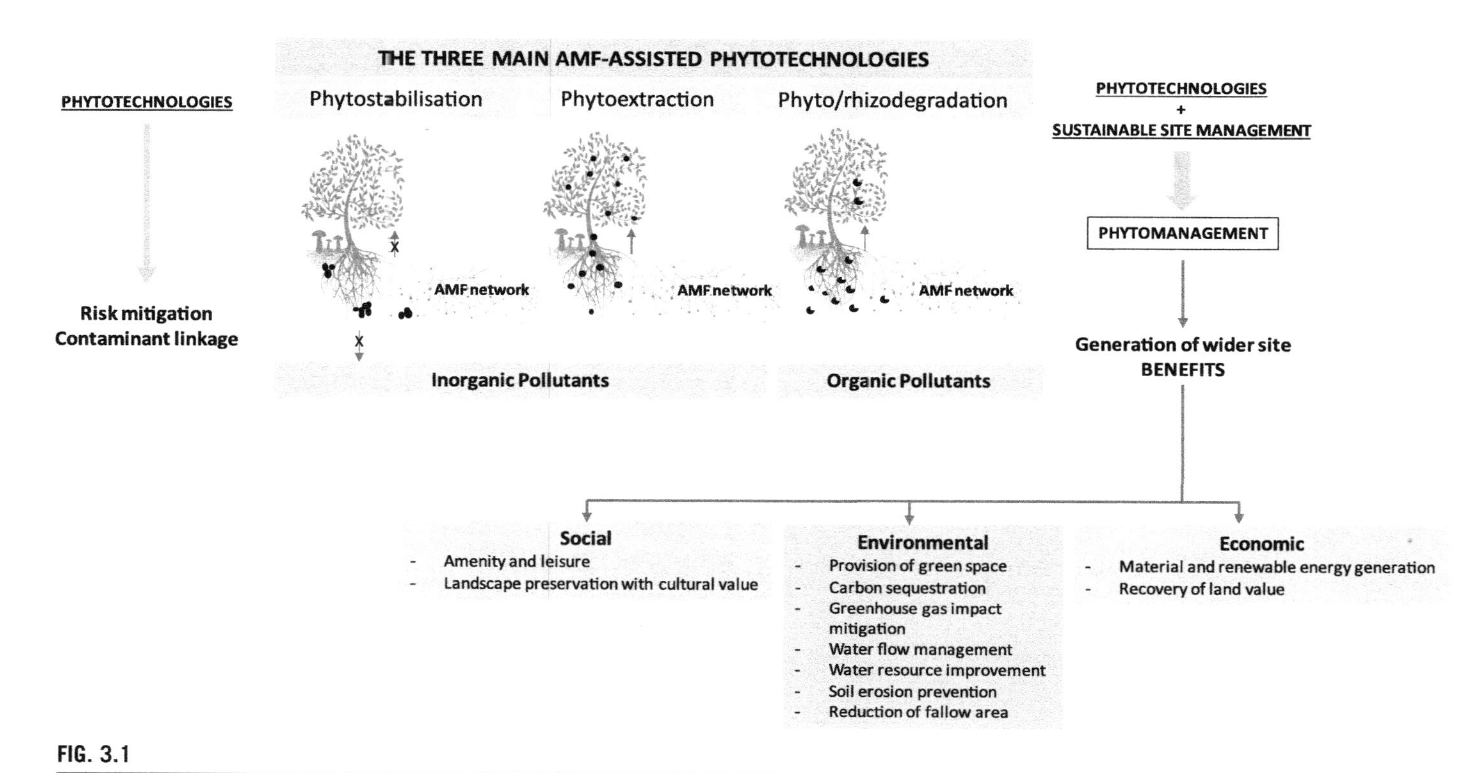

FIG. 3.1

The three main AMF-assisted phytotechnologies applied to clean-up contaminated soils and phytomanagement benefits (adapted from Burges et al., 2018).

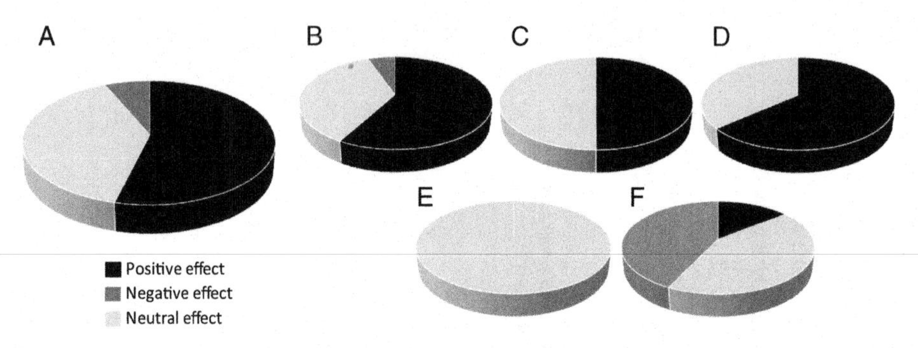

FIG. 3.2

Pie charts on the positive (black), neutral (light grey) or negative (black grey) effects of AMF inoculation on the pollutant dissipation in comparison with non-mycorrhizal plants. Effects of AMF inoculation on dissipation of A: all POPs (134 studies); B: PAHs and crude oil/petroleum (110 studies); C: PBDE (4 studies); D: PCBs (11 studies), E: dioxins/furans (2 studies) and F: pesticides (DDT / *p,p*-DDE) (7 studies). Charts were generated from the Table 3.2.

these disparities: (1) the plant/AMF combination (Liu and Dalpé, 2009; Rabie, 2005), (2) the nature, the concentration and the bioavaibility of POPs in soil (Alarcón et al., 2006; Aranda et al., 2013; Cheung et al., 2008; Ren et al., 2017; Wu, N. et al., 2008a; Yu et al., 2011), (3) the duration of the culture as well as the culture conditions (Gao et al., 2011; Joner and Leyval, 2001, 2003a; Liu et al., 2004), (4) the analysis of pollutants in the rhizosphere or the bulk soil (Joner and Leyval, 2003a; Wu, N. et al., 2008a), and (5) the ability of the polluted soil microbiome to metabolise POPs (Chiapusio et al., 2007; Corgié et al., 2006; Joner and Leyval, 2003b; Wang et al., 2011).

3.5 Phytoextraction and phytostabilization

In the literature, numerous scientific papers studied the beneficial effect of plant inoculation with AMF in the phytoremediation of polluted soils either with only one TE or with a mixture of TEs (Table 3.3). Phytoextraction and phytostabilisation are the two main phytotechnologies used to control TEs-contaminated soils. Phytoextraction consists in contaminants assimilation by roots from the soil and their transfer and accumulation in shoots (Miransari, 2011; Yoon et al., 2006). Phytostabilization (or phytoimmobilization) can be defined as the ability of plants to immobilize the contaminants in soils *via* sorption by roots, precipitation, complexation or metal valence reduction in the rhizosphere (Ghosh and Singh, 2005; Lux et al., 2011; Violante et al., 2010). Over 60% of the cited references in the literature focused on the phytostabilization (Table 3.3). Forest (*Acacia*, *Populus*, *Enterolobium*…), pastoral (*Atriplex*, *Plantago*, *Miscanthus*, *Chrysopogon*…), and forage (*Sorghum*, *Lotus*, *Zea mays*…) species are the most integrated plant species in the requalification of TEs-contaminated soils (De Fátima Pedroso et al., 2018; Lingua et al., 2008; Wang et al., 2018). Inoculation with suitable AMF to restore TEs polluted soils, seems to be an effective solution to overcome unfavorable conditions such as low nutrient availability, poor soil structure, high salinity and low water retention (Hildebrandt et al., 2007). TEs are absorbed through the fungal hyphae and then transported to the plant. Mycorrhizal plants can demonstrate an enhancement in TEs uptake and root-to-shoot transport (phytoextraction). As well, it was proved that AMF contribute to TEs immobilization in the soil (phytostabilization). A

Table 3.3 Table summarizing some studies reporting AMF inoculation effects on plant growth and trace elements (TEs) phytostabilisation/phytoextraction in contaminated soils.

TE Concentration (mg/Kg)	Host plant	Inoculum AMF species[a]	Culture conditions	Culture duration (days)	Shoot growth improvement in M plant	Phytotechnology	Reference
Total As: 21300, extractable As: 35.6	Ribwort plantain (*Plantago lanceolata* L.)	*R. intraradices (BIORIZE)*; *F. geosporum* UNIJAG, PL 14-1; *R. clarus* UNIJAG, PL 13-1; or *Glomus sp.* UNIJAG, PL 21-1	Aged contaminated soil/growth chamber	112	Y	Phytostabilisation	Orłowska et al. (2012)
As: 50	Maize (*Zea mays* L.)	*G. versiforme*	Sterile spiked soil/ greenhouse	65	Y		Wang, S. et al. (2018)
As: 100, 200	Chinese brake (*Pteris vittata* L.), Bermuda grass (*Cynodon dactylon* (L.) Pers.)	*F. mosseae, R. intraradices,* indigenous inoculum	Sterile spiked soil/ greenhouse	56	Y		Leung et al. (2013)
Cd: 50	Trefoil (*Lotus japonicus* L.)	*R. irregularis* DAOM197198	Sterile spiked soil/ greenhouse	60	Y		Chen, B. et al. (2018)
Cd : 250, 700	Fenugreek (*Trigonella Foenum grecum* L.)	*G. monosporum*; *R. clarus*; *Gi. nigra*; *Ac. laevis*	Sterile spiked soil/ greenhouse	40	Y		Abdelhameed and Metwally (2019)
Cd: 14.8	Black nightshade (*Solanum nigrum* L.)	*F. mosseae* (Mycagro)	Sterile spiked soil	35	Y	Phytoextraction	Li et al. (2018)
		R. intraradices (Mycagro)			N		
Cr : 200	Ribwort plantain (*Plantago lanceolata* L.)	*R. intraradices* BEG72	Sterile spiked soil/ greenhouse	98	N	Phytostabilisation /phytoextraction	Nogales et al. (2012)
		indigenous AMF from the contaminated site A			Y		
		indigenous AMF from the contaminated site B			N		
Cr : 400		*R. intraradices* BEG72			Y		
		indigenous AMF from the contaminated site A			N		
		indigenous AMF from the contaminated site B			Y		

(continued)

Table 3.3 Table summarizing some studies reporting AMF inoculation effects on plant growth and trace elements (TEs) phytostabilisation/phytoextraction in contaminated soils. ***Continued***

TE Concentration (mg/Kg)	Host plant	Inoculum AMF species[a]	Culture conditions	Culture duration (days)	Shoot growth improvement in M plant	Phytotechnology	Reference
Cr: 5, 10, 20	Dandelion (*Taraxacum platypecidum* Diels.), Bermudagrass (*Cynodon dactylon*[Linn.] Pers.)	*R. irregularis* BGC AH01	Sterile spiked soil/growth chamber	80	Y	Phytostabilisation	Wu, S-L. et al., 2014
Cr: 40, 80, 160	Mesquite (*Prosopi sjuliflora-velutina*)	*G. deserticola*	Sterile spiked soil/growth chamber	30	N		Arias et al. (2010)
Cu: 1, 5	Common reed (*Phragmites australis* Adans.)	*R.* irregularis	Cu solution/ hydroponic experiment	21	Y		Wu et al. (2020)
Cu: 200, 400	Spotty dotty (*Dysosma versipellis* (Hance) M. Cheng)	*F. mosseae* BGCGZ01A	Sterile spiked soil/ greenhouse	120	Y		Luo et al. (2020)
Cu: 100	Jack bean (Canavalia ensiformis (L.) DC)	*R.* clarus	Sterile spiked soil/ greenhouse	45	Y		Santana et al. (2018)
Fe: 135	Pear millet (*Pennisetum glaucum* (L.) R. Br.), Great millet (*Sorghum bicolor* (L.) Moench)	*Glomus sp., Acaulospora sp, Scutellospora sp.*	Sterile contaminated soil and sand/greenhouse	30, 45	Y	Phytoextraction	Mishra et al. (2016)
Fe: 2500			Sterile spiked contaminated soil and sand/greenhouse		Y		
Hg: 10	Lettuce (*Lactuca sativa* L.)	*R. irregularis*; *F. mosseae*	Sterile spiked soil/ greenhouse	63	Y	Phytostabilisation	Cozzolino et al. (2016)

TE Concentration (mg/Kg)	Host plant	Inoculum AMF species[a]	Culture conditions	Culture duration (days)	Shoot growth improvement in M plant	Phytotechnology	Reference
Hg: 2, 4, 6	Vetiver grass (Chrysopogon zizanoides Nash)	*Glomus sp*	Sterile spiked soil/ greenhouse	30	Y	Phytoextraction	Bretaña et al. (2019)
Mn: 100, 200, 400	White leadtree (*Leucaena leucocephala*) (Lam.) de Wit	*Diversispora sp.*; *C. etunicatum*; *Ac. scrobiculata*	Sterile spiked soil/ greenhouse	60	Y	Phytostabilisation	Garcia et al. (2020)
Mn: 426	*Mimosa caesalpiniaefolia* Benth.	*R. clarus; C. etunicatum*	Sterile aged contaminated soil/ greenhouse	60	Y		Garcia et al. (2018)
Mn: 50, 100	Cow pea (*Vigna unguiculata* (L.) Walp.)	*Scu. reticulata, G. pansihalos*	Spiked soil/ greenhouse	60	-		Alori and Fawole (2012)
Ni: 20, 40	Great millet (*Sorghum vulgare*)	*C. etunicatum isolates* (SFONL and SBH56)	Sterile sand receiving Ni solution	150	Y		Amir et al. (2013)
Ni: 30, 60 Ni: 91.4	*Alphitonia neocaledonica* Endl., *Cloezia artensis* Brongn. & Gris		Sterile aged contaminated soi / greenhouse	365	Y		
Total Ni: 650, available Ni: 157	*Berkheya coddii* Ehrh.	*R. intraradices* (BIORIZE)	Sterile aged contaminated soil/ growth chamber	105	Y	Phytoextraction	Orłowska et al. (2011)
		Mixture of *Di. aurantia, F. mosseae, F. coronatum* and *R. irregularis* from *B. coddii rhizosphere*			Y		
		Mixture of *F. coronatum, F. mosseae, R. intraradices, Di. spurca* and *R. irregularis* from *Senecio coronatus rhizosphere*			Y		
Ni: 1000, 3000, 5000	Soybean (*Glycine max* (L.) Merrill), Lentil (*Lens culinaris* Medic)	*F. mosseae*	Sterile sand receiving Ni solution/ shaded glasshouse	105	Y	Phytoextraction	Jamal et al. (2002)

(continued)

Table 3.3 Table summarizing some studies reporting AMF inoculation effects on plant growth and trace elements (TEs) phytostabilisation/phytoextraction in contaminated soils. ***Continued***

TE Concentration (mg/Kg)	Host plant	Inoculum AMF species[a]	Culture conditions	Culture duration (days)	Shoot growth improvement in M plant	Phytotechnology	Reference
Pb: 250, 500, 1000	Cornflower (*Centaurea cyanus* L.)	*R. intraradices*, *F. mosseae*, *R. fasciculatus*	Sterile spiked soil/ greenhouse	120	Y	Phytostabilisation	Karimi et al. (2018)
Pb: 50, 200, 400, 800	Vetiver (*Chrysopogon zizanioides* (L.) Roberty)	*R. intraradices*; *G. versiforme*	Spiked soil/ greenhouse	120	Y	Phytoextraction	Bahraminia et al. (2016)
Zn: 100, 1000	Silver grass (*Miscanthus sacchariflorus* Maxim. Franch.)	*Gi. margarita*	Spiked soil/ greenhouse	110	Y	Phytostabilisation	Sarkar et al. (2017)
Zn: 25, 50, 100	Green gram (*Vigna radiata* (L.) R. Wilczek)	*Glomus* sp.	Sterile soil receiving Zn solution/ greenhouse	110	Y		Shivakumar et al. (2011)
Zn: 300	Silver poplar (*Populus alba* L.), Black poplar (*Populus nigra* L.)	*F. mosseae* BEG 12; *R. intraradices* BB-E	Spiked soil/ greenhouse	180	N	Phytoextraction	Lingua et al. (2008)
Pb = 150, 300 Cd = 40, 80	Pot marigold (*Calendula officinalis* L.)	*F. mosseae*, *R. intraradices*	Spiked soil/ greenhouse	140	Y		Tabrizi et al. (2015)
Zn = 7200, Cu = 1140, Pb = 480, Cd = 72	Black wattle (*Acacia mangium* Willd.)	*G. macrocarpum, Paraglomus occultum, Glomus sp.*	Aged contaminated soil/greenhouse	120	N	Phytostabilisation	de Fátima Pedroso et al. (2018)
	Great millet (*Sorghum bicolor* (L.) Moench)			100	Y		
	Mauritius grass (*Urochloa brizantha* Hochst. ex A.Rich.)			180	Y		

TE Concentration (mg/Kg)	Host plant	Inoculum AMF species[a]	Culture conditions	Culture duration (days)	Shoot growth improvement in M plant	Phytotechnology	Reference
Zn = 953, Pb = 746, Cd = 13	*Miscanthus* × *gi-ganteus*	*R. irregularis* DAOM 197198	Aged contaminated soil/field	365	-		Firmin et al. (2015)
- **Tailings**: Mn = 13390, Zn = 1080, Fe = 125000 - **Top soil**: Mn = 462, Zn = 40, Fe = 25580	Crested wheat grass (*Agropyron cristatum* (L.) Gaertn), Dahurian wild rye (*Elymus dahuricus* Turcz. ex Griseb.) Aw	*F. mosseae* BGCNM04A *G. versiforme* BGC NM04B *F. mosseae* BGCNM04A *G. versiforme* BGC NM04B	Sterile aged contaminated/greenhouse	90	N Y N Y	Phytostabilisation /phytoextraction	Guo et al. (2013)
Cu = 167.0, Zn = 733.9, Pb = 157.4, Cd = 2.1	Maize (*Zea mays* L.)	*F. caledonium* 90036, *Gi. margarita* ZJ37, *Gi. decipens* ZJ38, *Scu. gilmori* ZJ39, *Acaulospora spp.*, *Glomus sp.*	Aged contaminated soil sunlit/greenhouse	70	N	Phytoextraction	Wang et al. (2007)
Cr = 14855, Cd = 19.5, Ni = 13.4, Pb = 24	Maize (*Zea mays* L.)	*R. fasciculatus*, *R. intraradices*, *F. mosseae*, *R. aggregatus*	Sterile contaminated tannery sludge/greenhouse	90	Y		Singh et al. (2019)

[a]*Arbuscular mycorrhizal fungi are named according to the current taxonomy (amf-phylogeny.com/amphylo_taxonomy.html).Ac., Acaulospora; Am., Ambispora; AMF, arbuscular mycorrhizal fungi; C., Claroideoglomus; De., Denticutata; Di., Diversispora; F., Funneliformis; G., Glomus; Gi., Gigaspora; M, mycorrhizal; NM, non-mycorrhizal; R., Rhizophagus; Ra., Racocetra; Scl., Sclerocystis; Scu., Scutellospora; Sep.: Septoglomus; Y, yes; N, no.*

positive effect of the plant colonization by AMF on plant biomass was observed in 80% of the cases reported in the literature (Table 3.3). The most used AMF species to inoculate plants are *F. mosseae* (35% of the trials) followed by *R. irregularis* (32% of the trials). Mycorrhizal inoculums consisting of a single AMF species are more prevalent than those based on a mixture of species (Table 3.3). The sterilized spiked soil was the largest type of substrate used in greenhouse experiments to regulate TEs concentrations, showing a lack of knowledge of the phytostabilization or phytoextraction efficacy in field culture conditions and with aged-polluted soils. The duration of the trials was also relatively short, from 21 days to one year, again showing the lack of long-term perspective of these phytomanagement techniques of TE-polluted soils as for POPs.

3.6 Mechanisms involved in soil phytoremediation by mycorrhizal plants

Phytoremediation processes are a set of practices aimed at removing and accumulating pollutants (bioaccumulation), changing the pollutant availability in the soil (glomalin production) or rendering harmless environmental contaminants in the soil (TEs transformation or POPs metabolization) (Cristaldi et al., 2017). These mechanisms are discussed below.

Glomalin production: AMF possess the ability to improve the structure of the soil matrix *via* the proliferation of extraradical mycelium in the soils, acting on the aggregation process (Rillig et al., 2010). The soil structure is also influenced by the extraradical mycelium production of glycoproteins as glomalin-related soil protein (GRSP), classified as probable heat-shock proteins (Purin and Rillig, 2007). Chen et al. (2019) mentioned that GRSP produced by the AMF extraradical mycelium is commonly found in various types of soil environments as total GRSP (T-GRSP) and easily extracted GRSP (EE-GRSP) (Schindler et al., 2007; Treseder and Turner, 2007). This stable form of organic matter possesses various roles in soils: (1) improvement of soil organic carbon storage (Kumar et al., 2018; Rillig, 2004), (2) complexation with minerals, organic matter and facilitation of the aggregation of soil particles (Adame et al., 2012; Cornejo et al., 2008; Kohler et al., 2010; Zhang et al., 2017), (3) improvement of the plant resistance against drought stress (Chi et al., 2018), and (4) chelation/sequestration of minerals and TEs (Cd, Cr, Cu, Pb, and Zn) (Cornejo et al., 2008; Gil-Cardeza et al., 2014; González-Chávez et al., 2004; Maleksadeh et al., 2016; Seguel et al., 2016).

A recent study demonstrated a correlation between the colonization of *Sophora viciifolia* and density of hyphal length in link with GRSP production, soil organic matter and soil organic carbon in a polluted soil contaminated with Pb and Zn (Yang et al., 2017). These findings suggest the crucial role played by AMF in accumulation of GRSP, soil organic matter and carbon, which influences the formation and distribution of aggregate and particle-size in TEs-contaminated soils. Similar studies with POPs also demonstrated a larger accumulation of GRSP in the mycorrhizosphere of plants correlated with a higher microbial abundance in the mycorrhizosphere, which enhanced PAHs availability in soil solution, related to a larger root adsorption and accumulation (Chen, S. et al., 2018; Gao et al., 2017a, b). An enhanced microbial degradation in link with an increased bioavailability could be expected.

TE transformation: In the literature, few studies treated the AMF effect in the transformation of toxic TEs to fewer toxic forms. Wu, S. et al. (2015) investigated the uptake, translocation, and transformation of Cr (VI) by *R. irregularis* and they found that AMF can immobilize a part of Cr element *via* reduction of Cr(VI) to Cr(III), forming Cr(III)–phosphate analogues on the fungal surface. In the same study, results showed that the extraradical mycelium of AMF can actively take up Cr (either in the form of Cr(VI) or Cr(III)] and transport Cr [potentially in the form of Cr(III)-histidine analogues) to roots colonized by AMF but immobilize most of the less toxic Cr(III) in the fungal structures. Karandashov and Bucher (2005) also indicated that AMF are well known for improving plant phosphorus acquisition and that phosphate groups are negatively charged which can easily precipitate Cr(III) (Joner et al., 2000). In another work, Wu, S. et al. (2016) examined if the P fertilization of a soil contaminated with Cr(VI)- has the same positive effect of AMF symbiosis on plant Cr tolerance. The authors showed that P amendment in Cr(VI) contaminated soil did not enhance plant Cr tolerance as AMF symbiosis. As well, they demonstrate the role of extra-radical mycorrhizal hyphae and mycorrhizal roots in the immobilization of Cr, which improves plant tolerance to soil contamination with this TE. Zhang et al. (2015) found that AMF also decreased As concentration and increased P uptake by *M. truncatula* under As stress. They observed that AMF increased the portion of arsenite in the total As amont compared to the other forms of As. The authors specified that AMF are able to alleviate the stress induced by As contamination through the methylation of inorganic As into less toxic organic compounds: dimethylarsinic acid and by converting arsenate to arsenite.

POPs metabolization: AMF influence POPs metabolization indirectly through (1) changes in roots exudates that result in a release in soil of catalases, peroxidases, laccases, nitroreductases, and dehalogenases, which promote POPs metabolization/degradation (Alagić et al., 2015; Dorantes et al., 2012) and (2) a modification of the rhizosphere microbiome in terms of quality, abundance and overall rhizosphere bacterial and fungal activities, accelerating the POPs degradation in the mycorrhizosphere (Khan, 2006; Qin et al., 2016). Indeed, the mycorrhizal colonization of the roots demonstrated changes in the root exudates content and quantity, generally increasing sugars and phenolic compounds and decreasing production of organic acids (Bouwmeester et al., 2007; Gomez-Roldan et al., 2007; Hage-Ahmed et al., 2013). The chemical structures of these phenolic compounds is chemically similar to those of POPs such as PAHs (Leigh et al., 2002) and therefore microbial populations in the rhizosphere contribute to the POPs metabolization *via* co-metabolism, using the organic pollutants as carbon and energy substrates for their own growth (Gerhardt et al., 2009; Wojtera-Kwiczor et al., 2014). In addition, exudates produced by the extraradical mycelium could also influence directly bacterial and fungal composition and activity (Qin et al., 2016; Zhang et al., 2018).

The first demonstration of a link between POPs degradation and mycorrhiza associated microbiome was made by Joner and Leyval (2001). They showed that phospholipid fatty acid (PLFA) profiles indicated that the microflora associated to the mycorrhiza correlated for the observed reductions of PAHs concentrations in the presence of mycorrhiza. Since then, a higher fungal count and an increase of bacterial and archaeal numbers, richness, and diversities in PAHs, PBDEs, dioxins/furans and PCBs-polluted soils were determined in link with the POPs dissipation in the mycorrhizosphere (Meglouli et al., 2018; Qin et al., 2014; Teng et al., 2010; Wang et al., 2011). As well, in the presence of PAHs and petroleum, an increase in the population of specific PAHs- and hydrocarbon-degrading bacteria was measured in the mycorrhizosphere in comparison with the rhizosphere of non-colonized ones (Corgié et al., 2006; Małachowska-Jutsz and Kalka, 2010; Małachowska-Jutsz et al., 2011). For example, Qin et al. (2014) found that the association of zucchini with AMF as *Acaulospora laevis* or *F. mosseae* increased the dissipation of PCBs, both in bulk and rhizosphere soils. These authors showed a significant participation of Actinobacteria in PCBs dissipation, demonstrating that the mycorrhizal mycelium could influence the bacterial community and PCB congener profile compositions. Meglouli et al. (2018) showed that the mycorrhizal inoculation of alfalfa and tall fescue by an indigenous inoculum from a dioxins/furans-polluted soil and by a commercial inoculum increased bacterial OTU numbers, richness, and diversities in a dioxins/furans-polluted soil. The relative abundance of some microbial classes, such as Planctomycetia, Solibacteres, and Pezizomycetes, was negatively linked to the concentration of toxic equivalent dioxins/furans, demonstrating their potential involvement in the dioxins/furans dissipation and/or tolerance to the pollutants toxicity (Meglouli et al., 2018).

Therefore, an increased enzymatic activity in relation with an improved POPs dissipation was generally measured in the mycorrhizosphere. Activity of dehydrogenase generally increased in the mycorrhizosphere of various contaminated soils with PAHs, PBDEs and dioxins/furans, in link with a larger concentration of bacteria and a higher POPs dissipation (Alarcón et al., 2006; Li et al., 2018; Liu et al., 2004; Rostami and Rostami, 2019). According to the results obtained by Meglouli et al. (2018, 2019), the activity of dehydrogenase seems to be modulated by the combination plant-AMF grown in presence of dioxins/furans. A higher fluorescein-di-acetate hydrolase activity (considered as a suitable indicator of POPs degradation) was also measured in the alfalfa mycorrhizosphere, whereas it remained stable in the tall fescue mycorrhizosphere (Meglouli et al., 2018). Likewise, soil polyphenol oxidase activity in the mycorrhizosphere increased in comparison to the rhizosphere of non-mycorrhizal plants, in presence of PAHs and participates in PAHs degradation in soil (Durán and Esposito, 2000; Lu and Lu, 2015; Ren et al., 2017). However, in presence of PBDEs and Cd, this activity remained unchanged (Li et al., 2018).

Dioxygenases BphA (encoded by the *bphA* gene) and BphC (encoded by the *bphC* gene) are also implied in the catalysis of PCBs-transformation and belong to some PCBs degraders such as *Acinetobacter*, *Arthrobacter*, *Corynebacterium*, *Pseudomonas*, *Rhodococcus* and *Sphingomonas* (Lu et al., 2014; Pieper and Seeger, 2008; Qin et al., 2014, 2016). Likewise, in PAHs degraders as *Mycobacterium* species, the identified *nid*A gene encodes the large subunits of pyrene dioxygenase and is in charge of the initial dioxygenase step of the PAHs (as pyrene and phenanthrene) metabolization pathways (DeBruyn et al., 2007; Ren et al., 2017; Stingley et al., 2004). In presence of PCBs and PAHs respectively, a higher abundance of *bphA* and *bphC* genes and *nid*A population was quantified in mycorrhizal roots rhizosphere than in non-mycorrhizal roots rhizosphere (Qin et al., 2014, 2016; Ren et al., 2017), generally in link with a larger POPs dissipation. In addition, the abundance of *bphA* and *bphC* genes was found to be correlated with the length of the extraradical AMF in soil. The Burkholderiales were more abundant at the vicinity of the abundant extraradical mycelium of the AMF, suggesting a beneficial effect of the mycorrhizal hyphal exudates on the growth of this group of PCBs-degrading bacteria (Qin et al., 2014, 2016). On the other hand, the abundance of *bphA* and *bphC genes* or *nidA* gene expression remained unchanged in absence or presence of extraradical mycelium in presence of various POPs (Lu et al., 2014).

3.7 Contribution of mycorrhizal inoculation in polluted soil functionalization and in plant biomass valorization

Polluted soils often represent an unfavorable environment for living organisms due to the toxicity of pollutants, the loss of soil structure or the lack of organic matter and nutrients. Therefore, the presence of pollutants in soils and their artificialization can have great effects on microbial activity and microbial community structure, thus altering the soil respiration, nutrient cycles and soil productivity (Morgado et al., 2018). Therefore, strategies such as the use of AMF-assisted phytoremediation to improve the quality and the fertility of polluted soils are necessary to ensure the restoration and refunctionalization of these soils. These phytotechnologies also have subsidiary positive impacts on the surrounding environment, providing ecosystem services with perceptible value for public health and social welfare (e.g., plant, bacterial and fungal biodiversity; nutrients storage and supply; water maintenance, erosion prevention; reduced greenhouse gas emissions; etc.) (Holzman, 2012; Mench et al., 2018). Indeed, the objectives of these remediation processes are not only to reduce soil toxicity by removing or immobilizing pollutants, but also to restore the soil's ability to perform its functions (Hernández-Allica et al., 2006). To estimate the contribution of this phytotechnologies on the health or the functioning of these soils, it is crucial to use various and complementary bioindicators based either on the evaluation of biodiversity or the measure of functional parameters (Bispo et al., 2009). Ubiquitous in the soil and reacting quickly to physical and chemical changes, bacteria and fungi are considered effective bioindicators (Kennedy and Smith, 1995). The biomass, abundance and diversity of these microorganisms can be rapidly assessed using specific biomarkers such PLFA or molecular markers and or measurements of soil respiration, global microbial activity (dehydrogenase activity, Fluorescein diacetate (FDA) hydrolysis, ...) or specific activities involved in C, N, P, and S cycles (Floch et al., 2009; Willers et al., 2015). Likewise, the abundance and diversity of soil micro- and meso-fauna, including nematodes, protozoa and microarthropods can also be measured through direct observation and identified with morphological criteria. These organisms are known to affect the health, structure and fertility of soils, organic matter decomposition, and crop growth (Miller et al, 2017; Pernin et al., 2006). A great number of studies have indicated that the AMF have significant role in ecological restoration of polluted soil (Asmelash et al., 2016). Extramatrical hyphae of AMF form a large network outside of the root, which represent between 20 to 30% of soil microbial biomass and as much as 15% of soil organic C pool (Leake et al., 2004). These extramatrical hyphae grow into the soil to create the skeletal structure that releases glomalin, which binds soil particles together in soil aggregates (Al-Karaki, 2013). This soil aggregation may be of interest when the human exposure pathways are related to the inhalation or ingestion of soil particles (Gray et al., 2006). Moreover, this hyphal network and the promoting effects of AMF on the root system development protect the soil from erosion by wind and water (Gutjahr and Paszkowski, 2013). AMF provide also a critical link between soil, plants and other microorganisms (Hrynkiewicz and Baum, 2012). They can for example positively affect the soil mesofauna population and their activities (Bonkowski et al., 2000). Indeed, many studies showed that these organisms such as collembola play an important role in building up soil from mine residues containing high Cd concentrations (González-Chávez et al., 2009). Moreover, as described in the former section, AMF inoculation influences also bacterial, fungal and archaeal populations in the rhizosphere (Meglouli et al., 2018; Qin et al. 2016; Solís-Domínguez et al. 2011) whether directly through excretion of fungal molecules used as substrates (i.e. carbohydrates, etc.) for microbial growth (Duponnois et al. 2008; Qin et al. 2016; Toljander et al. 2007; Zhang et al., 2018) or indirectly through changes in the quantity and composition of roots exudates (Camprubi et al., 1995; Rodríguez-Caballero et al., 2017). Changes in activity and composition of microbial communities by introduced AMF in polluted soil have been demonstrated through many studies as noted in the previous section. By studying the soil bacterial community-level physiological profiles (Biolog Ecoplates), Wu, J-T. et al. (2014) showed a higher metabolic diversity in mycorrhizal rhizospheric microbial community compared to the non-mycorrhizal one and this diversity can depend of the AMF species. They demonstrated that the rhizosphere inoculated with *F. mosseae* showed a higher microbial metabolic diversity than the one inoculated with *R. irregularis* and concluded that the influence of AMF on rhizospheric microbial communities is specific to the species. Changes in composition of microbial communities especially in terms of abundance, richness and diversity have also been described in dioxins/furans polluted soil by

Meglouli et al. (2018, 2019) using Illumina sequencing of the16S and ITS1 rRNA genes. Meglouli et al. (2018) examined the effects of two mycorrhizal inoculums (indigenous and commercial inoculums) in association with alfalfa and tall fescue on the bacterial, fungal, and archaeal communities in dioxins/furans polluted soil and they showed again, that some bacterial species respond differently depending on the AMF inoculum type suggesting a high degree of specificity between bacteria and AMF species. Likewise, Chen, X.W. et al. (2019) investigated the effect of two AMF species (*F. mosseae* and *R. irregularis*) in the rhizospheric bacterial community associated to rice grown in Cd-contaminated soils in pot experiments. They demonstrated a higher relative abundance of Actinobacteria by 15% (mostly from genus Arthrobacter) in the mycorrhizosphere of *R. irregularis*, whereas it only increased by 1–2% in the mycorrhizosphere of the other treatments. Since soil microbial populations are essential for maintaining soil fertility, the recovery of the composition and the activity of microbial communities can be a key factor to the sustainability of polluted ecosystems (Raveau et al., 2020; Raveau et al., 2021a; Kohler et al., 2016).

On the other hand, many studies have showed that AMF promote the survival of their host plants growing in contaminated soil and improve biomass production and quality. Indeed, these symbiotic fungi enhance the plant nutrition and protect them against pollutant toxicity or pathogenic fungi (Ciadamidaro et al., 2017; Debiane et al., 2012; Firmin et al., 2015; Phanthavongsa et al., 2017). These mycorrhizal fungi can therefore be used as a "biofertilizer" and improve the revegetation of polluted sites and the phytoremediation process (Wu, S-L. et al., 2014). Conversely, AMF could also be utilised in the improvement of metal uptake by hyperaccumulating plants and the phytomining technologies. However, hyperaccumulating plants are generally non-mycorrhizal species and the studies on the role of mycorrhizal symbiosis in metal uptake by hyperaccumulating plants are few (Orłowska et al., 2011). The increase of biomass production and, in some cases, the improvement of biomass quality by limiting the transfer of toxic elements make AMF valuable tools in the strategy for the requalification of polluted soils and the valorization of the plant biomass. During the last few years, an approach that combines phytotechnologies and biomass valorization, known as phytomanagement, is being developed. This concept allows reducing the costs of soil requalification with the production of valuable biomass produced on these polluted lands and its valorization in non-food sectors (Phanthavongsa et al., 2017). Indeed, plant biomass represents a versatile energy source that can substitute fossil energy and a renewable raw materials which can be used in different industrial processes (Scarlat et al., 2015). In their review, Evangelou et al. (2015) described the best valorization options for the biomass produced in polluted sites (bioenergy, wood, biochar, biofortified products). Unfortunately, except for the description of improved biomass, very few studies have demonstrated, at a field scale, the contribution of mycorrhizal inoculation to valorize the biomass produced on contaminated soil. However, it is well known that AMF can facilitate the valorization of biomass through other factors such as the improvement of quality and the number of fibers in fiber plants (Rydlová et al., 2011) or for example the content and composition of secondary metabolites such as essential oils in plants (Khaosaad et al., 2006; Pistelli, et al., 2017).

3.8 Challenges and future perspectives

In conclusion, AMF-assisted phytoremediation is an ecological technology well-adapted to control and manage polluted soils. Its success depends on the ability of plants to colonize harsh environments, which is partly related to the symbiosis that could form with rhizosphere beneficial microorganisms.

Among them, AMF are described as the main drivers of plant growth and tolerance to abiotic and biotic stresses (Ciadamidaro et al., 2017; Miransari, 2017; Otero-Blanca et al., 2018; Phanthavongsa et al., 2017). However, currently, AMF-assisted phytoremediation faces limitations in its use on a commercial and large field scale. Indeed, the success of field inoculation is still unpredictable due to several physicochemical and biological parameters such as the multiple response of plant species to the same AMF species or the competition with the indigenous microbiome (Berruti et al., 2016; Gerhardt et al., 2017). Moreover, the phytotechnologies employment, *via* phyto/rhizodegradation, phytostabilisation or phytoextraction, requires selection of effective plant-AMF associationships. Thus, to improve AMF-assisted phytoremediation efficiency, several avenues can be considered:

1. **Inoculum composition, quality and adaptation**: Some AMF species isolated from contaminated soils besides being better adapted to the edapho climatic conditions of the area, also develop certain adaptation mechanisms to tolerate pollutants, which make them as potential candidates and more suitable for phytoremediation applications (Garg and Chandel, 2010; Plouznikoff et al., 2016; Redon et al., 2009; Wu, Q-S. 2017). Indeed, the risks of failure may be mitigated by considering indigenous strains (Wubs et al., 2016). Moreover, to enhance inoculum quality and efficiency, AMF can be used in combination with (1) other beneficial rhizosphere microorganisms such as PGPR (Plant Growth Promoting Rhizobia) and endophytes, and with (2) eco-friendly and nontoxic organic amendments (Alarcón et al., 2008; Dong et al., 2014; Yu et al., 2011).
2. **Crop rotation:** Crop rotations should be more considered in phytoremediation strategies. This practise contribute to a constant production of biomass and to a reduction of plant diseases, pests populations and weed buildup (Banuelos, 1999). However, crop rotation can negatively affect the communities of microorganisms having a beneficial effect on plant growth including AMF. Therefore, some precautions have to be taken when crop rotations include non-host plants or plants which develop little mycorrhiza colonization (Ocampo and Hayman, 1981).
3. **Co-cropping**: Recent approaches highlight the importance of plant associations (co-cropping) to optimize the efficiency of phytomanagement applications. Some studies point out the importance of assemblages between crops and the spontaneous species (Parraga-Aguado et al., 2013) to favor the ecological rehabilitation of contaminated lands (Boisson et al., 2016). Nevertheless, the biodiversity of this spontaneous vegetation is generally restrained by the soil properties, limiting the panel of species able to develop and to compete with the tolerant ones (Macnair, 1987).
4. **From phytotechnologies to phytomanagement:** To enhance the efficiency, the attractivity and the commercial feasibility of AMF-assisted phytotechnologies, it will be interesting to take advantages from the the potential of AMF in both soil remediation and biomass valorization through integrated approaches. Recent innovative approaches, known as Gentle Remediation Options (GROs), combine plant technologies with sustainable and cost-effective site use, giving rise to the concept of phytomanagement (Cundy et al., 2013). While phytoremediation aims to mitigate risks, phytomanagement allows integrated site management, in which, in addition to take into account risks, the economic, social, ecological and environmental benefits to humankind are considered (Fig. 3.1). Thus, phytomanagement is a technique consisting in the use of plant species to rehabilitate and revalorize polluted environments. For example, by using

fast-growing plant species with high biomass, this phytotechnology allows the production of renewable biomass for the bio-based-economy such as biofuel/energy production from energy crops (willow, poplar, miscanthus), production of plant fibers for material application, production of active essential oils from medicinal and aromatic plants (Pandey et al., 2015; Pandey and Bauddh, 2018; Pandey and Pablo Souza-Alonso, 2019; Pandey and Bajpai, 2019; Perlein et al., 2021; Raveau et al., 2021a; Pandey et al., 2022) or rare and precious metals recuperation (Au and Ni) and eco-calatyser (Zn) using phytomining (Muszyńska and Hanus-Fajerska, 2015). However, the quality of the products, mainly their toxic metals concentrations, must be controlled.

5. **Evaluation of ecosystem services rendered by phytomanagement:** Until now, the cost-effectiveness of phytomanagement projects was assessed using a tool which integrates only economic data and the profitability of these projects was generally lacking (Bert et al., 2017; Kidd et al., 2015; Mench et al., 2010). That's why, one of the current challenges consists in identifying, quantifying and assigning a monetary value to the ecosystem services rendered by phytotechnologies. Indeed, some remediating plant species can contribute to the resilience of the ecosystem by promoting biodiversity (Parraga-Aguado et al., 2014) and positively influence other important ecosystem services by enhancing carbon sequestration and reducing soil erosion (Ciadamidaro et al., 2017; Mench et al., 2010; Robinson et al., 2009; Pandey, 2020a, b).
6. **Long-term and large field scale trials**: To convince the decision-makers and the general public of the economic feasability of phytotechnologies, other trials at a large field scale must be maintained on contaminated soils with the support of the remediation industries.
7. **Performance tests**:The requalification of polluted sites does not only respond to a need for rehabilitation to meet health and environmental standards, but also calls for the reappropriation of these degraded areas, which will depend on the ecological and economic viability of the proposed phytomanagement method. Performance evaluation (*via* toxicity and ecotoxicity tests on the soil before and after phytoremediation), the existence of a real market for by-products and the economic feasibility would be a crucial criterion for the adoption and application of phytotechnologies in polluted soil phytomanagement.
8. Finally, **a decision support tool** should be developed to examine the economic and legislative applicability of phytoremediation for each polluted site.The assessment tool should take into account the influence of phytomanagement on all the components of rehabilitated ecosystems (Burges et al., 2018).

Acknowledgments

This work was supported by l'Agence De l'Environnement et de la Maîtrise de l'Energie (ADEME, Angers, France) whithin PHYTEO and DEPHYTOP projects and has also been carried out in the framework of the Alibiotech project which is financed by European Union, French State and the French Region of Hauts de France.

References

Abdel Latef, A.A., 2013. Growth and some physiological activities of pepper (*Capsicum annuum* L.) in response to cadmium stress and mycorrhizal symbiosis. J. Agr. Sci. Tech. 15, 1437–1448.

Abdelhameed, R.E., Metwally, R.A., 2019. Alleviation of cadmium stress by arbuscular mycorrhizal symbiosis. Int. J Phytoremed. 21 (7), 663–671.
Adame, M.F., Wright, S.F., Grinham, A., Lobb, K., Reymond, C.E., Lovelock, C.E., 2012. Terrestrial–marine connectivity: patterns of terrestrial soil carbon deposition in coastal sediments determined by analysis of glomalin related soil protein. Limnol. Oceanogr. 57 (5), 1492–1502.
Alagić, S.Č., Maluckov, B.S., Radojičić, V.B., 2015. How can plants manage polycyclic aromatic hydrocarbons? May these effects represent a useful tool for an effective soil remediation? A review. Clean Technol. Environ. Policy 17 (3), 597–614.
Alarcón, A., Delgadillo-Martínez, J., Franco-Ramírez, A., Davies Jr, F.T., Ferrera-Cerrato, R., 2006. Influence of two polycyclic aromatic hydrocarbons on spore germination, and phytoremediation potential of *Gigaspora margarita-Echynochloa polystachya* symbiosis in benzo[a]pyrene-polluted substrate. Rev. Int. Contam. Ambient. 22 (1), 39–47.
Alarcón, A., Davies Jr, F.T., Autenrieth, R.L., Zuberer, D.A., 2008. Arbuscular mycorrhiza and petroleum-degrading microorganisms enhance phytoremediation of petroleum-contaminated soil. Int. J. Phytoremed. 10 (4), 251–263.
Al-Karaki, G.N., 2013. The role of mycorrhiza in the reclamation of degraded lands in arid environments. In: Shahid, S.A., Taha, F.K., Abdelfattah, M.A. (Eds.), Developments in Soil Classification, Land Use Planning and Policy Implications. Springer, Dordrecht, pp. 823–836.
Alori, E.T., Fawole, O.B., 2012. Phytoremediation of soils contaminated with aluminium and manganese by two arbuscular mycorrhizal fungi. J. Agric. Sci. 4 (8), 246–252.
Aloui, A., Recorbet, G., Gollotte, A., Robert, F., Valot, B., Gianinazzi-Pearson, V., Aschi-Smiti, S., Dumas-Gaudot, E., 2009. On the mechanisms of cadmium stress alleviation in *Medicago truncatula* by arbuscular mycorrhizal symbiosis: a root proteomic study. Proteomics 9 (2), 420–433.
Amir, H., Lagrange, A., Hassaïne, N., Cavaloc, Y., 2013. Arbuscular mycorrhizal fungi from New Caledonian ultramafic soils improve tolerance to nickel of endemic plant species. Mycorrhiza 23 (7), 585–595.
Aranda, E., Scervino, J.M., Godoy, P., Reina, R., Ocampo, J.A., Wittich, R.M., García-Romera, I., 2013. Role of arbuscular mycorrhizal fungus *Rhizophagus custos* in the dissipation of PAHs under root-organ culture conditions. Environ. Pollut. 181, 182–189.
Arias, J.A., Peralta-Videa, J.R., Ellzey, J.T., Ren, M., Viveros, M.N., Gardea-Torresdey, J.L., 2010. Effects of *Glomus deserticola* inoculation on Prosopis: enhancing chromium and lead uptake and translocation as confirmed by X-ray mapping, ICP-OES and TEM techniques. Environ. Exp. Bot. 68 (2), 139–148.
Ashraf, M., Foolad, M.R., 2007. Roles of glycine betaine and proline in improving plant abiotic stress resistance. Environ. Exp. Bot. 59 (2), 206–216.
Asmelash, F., Bekele, T., Birhane, E., 2016. The potential role of arbuscular mycorrhizal fungi in the restoration of degraded lands. Front. Microbiol. 7, 1095 10.3389/fmicb.2016.01095.
Bahraminia, M., Zarei, M., Ronaghi, A., Ghasemi-Fasaei, R., 2016. Effectiveness of arbuscular mycorrhizal fungi in phytoremediation of lead-contaminated soil by vetiver grass. Int. J. Phytoremed. 18 (7), 730–737.
Bakker, M.G., Manter, D.K., Sheflin, A.M., Weir, T.L., Vivanco, J.M., 2012. Harnessing the rhizosphere microbiome through plant breeding and agricultural management. Plant Soil 360 (1–2), 1–13.
Ban, Y., Xu, Z., Zhang, H., Chen, H., Tang, M., 2015. Soil chemistry properties, translocation of heavy metals, and mycorrhizal fungi associated with six plant species growing on lead-zinc mine tailings. Ann. Microbiol. 65 (1), 503–515.
Banuelos, G.S., 1999. Factors influencing field phytoremediation of selenium laden soils. In: Terry, N., Banuelos, G.S. (Eds.), Phytoremediation of Contaminated Soil and Water. CRC Press, New York, pp. 41–59.
Bedini, S., Turrini, A., Rigo, C., Argese, E., Giovannetti, M., 2010. Molecular characterization and glomalin production of arbuscular mycorrhizal fungi colonizing a heavy metal polluted ash disposal island, downtown Venice. Soil Biol. Biochem. 42 (5), 758–765.

Bert, V., Neub, S., Zdanevitch, I., Friesl-Hanl, W., Collet, S., Gaucher, R., Puschenreiter, M., Müller, I., Kumpiene, J., 2017. How to manage plant biomass originated from phytotechnologies? Gathering perceptions from end-users. Int. J. Phytoremed. 19 (10), 947–954.

Berruti, A., Lumini, E., Balestrini, R., Bianciotto, V., 2016. Arbuscular mycorrhizal fungi as natural biofertilizers: let's benefit from past successes. Front. Microbiol 6, 1559.

Binet, P., Portal, J.M., Leyval, C., 2000. Fate of polycyclic aromatic hydrocarbons (PAH) in the rhizosphere and mycorrhizosphere of ryegrass. Plant Soil 227 (1–2), 207–213.

Bispo, A., Cluzeau, D., Creamer, R., Dombos, M., Graefe, U., Krogh, P.H., Sousa, J.P., Peres, G., Rutgers, M., Winding, A., Römbke, J., 2009. Indicators for monitoring soil biodiversity. Integr. Environ. Assess. Manage. 5 (4), 717–720.

Boisson, S., Le Stradic, S., Collignon, J., Séleck, M., Malaisse, F., Shutcha, M.N., Faucon, M-P., Mahy, G., 2016. Potential of copper-tolerant grasses to implement phytostabilisation strategies on polluted soils in South DR Congo. Environ. Sci. Pollut. Res. 23 (14), 13693–13705.

Bonkowski, M., Cheng, W., Griffiths, B.S., Alphei, J., Scheu, S., 2000. Microbial-faunal interactions in the rhizosphere and effects on plant growth. Eur. J. Soil Biol. 36 (3–4), 135–147.

Bouwmeester, H.J., Roux, C., Lopez-Raez, J.A., Becard, G., 2007. Rhizosphere communication of plants, parasitic plants and AM fungi. Trends Plant Sci 12 (5), 224–230.

Bretaña, B.L., Salcedo, S., Casim, L., Manceras, R., 2019. Growth performance and inorganic mercury uptake of Vetiver (*Chrysopogon zizanoides* Nash) inoculated with arbuscular mycorrhiza fungi (AMF): its implication to phytoremediation. J. Agr. Res. Dev. Ext. Technol. 1 (1), 39–47.

Burges, A., Alkorta, I., Epelde, L., Garbisu, C., 2018. From phytoremediation of soil contaminants to phytomanagement of ecosystem services in metal contaminated sites. Int. J. Phytoremed. 20 (4), 384–397.

Cabello, M.N., 1997. Hydrocarbon pollution: its effect on native arbuscular mycorrhizal fungi (AMF). FEMS Microbiol. Ecol. 22 (3), 233–236.

Cabello, M.N., 1999. Effectiveness of indigenous arbuscular mycorrhizal fungi (AMF) isolated from hydrocarbon polluted soils. J. Basic Microbiol. 39 (2), 89–95.

Cabral, L., Fonsêca Sousa Soares, C.R., Giachini, A.J., Siqueira, J.O., 2015. Arbuscular mycorrhizal fungi in phytoremediation of contaminated areas by trace elements: mechanisms and major benefits of their applications. World J. Microbiol. Bioethanol. 31 (11), 1655–1664.

Cadillo-Quiroz, H., Yavitt, J.B., Zinder, S.H., Thies, J.E., 2010. Diversity and community structure of archaea inhabiting the rhizoplane of two contrasting plants from an acidic bog. Microb. Ecol. 59 (4), 757–767.

Calonne-Salmon, M., Plouznikoff, K., Declerck, S., 2018. The arbuscular mycorrhizal fungus *Rhizophagus irregularis* MUCL 41833 increases the phosphorus uptake and biomass of *Medicago truncatula*, a benzo[a] pyrene-tolerant plant species. Mycorrhiza 28 (8), 761–771.

Camprubi, A., Calvet, C., Estaun, V., 1995. Growth enhancement of *Citrus reshni* after inoculation with *Glomus intraradices* and *Trichoderma aureoviride* and associated effects on microbial populations and enzyme activity in potting mixes. Plant Soil 173 (2), 233–238.

Chagnon, P.L., Bradley, R.L., Maherali, H., Klironomos, J.N., 2013. A trait-based framework to understand life history of mycorrhizal fungi. Trends Plant Sci 18 (9), 484–491.

Chaparro, J.M., Badri, D.V., Bakker, M.G., Sugiyama, A., Manter, D.K., Vivanco, J.M., 2013. Root exudation of phytochemicals in Arabidopsis follows specific patterns that are developmentally programmed and correlate with soil microbial functions. PloS One 8 (2), e55731.

Chen, B., Nayuki, K., Kuga, Y., Zhang, X., Wu, S., Ohtomo, R., 2018. Uptake and intraradical immobilization of cadmium by arbuscular mycorrhizal fungi as revealed by a stable isotope tracer and synchrotron radiation μX-ray fluorescence analysis. Microbes Environ 33 (3), 257–263.

Chen, S., Wang, J., Waigi, M.G., Gao, Y., 2018. Glomalin-related soil protein influences the accumulation of polycyclic aromatic hydrocarbons by plant roots. Sci. Total Environ. 644, 465–473.

Chen, S., Sheng, X., Qin, C., Waigi, M.G., Gao, Y., 2019. Glomalin-related soil protein enhances the sorption of polycyclic aromatic hydrocarbons on cation-modified montmorillonite. Environ. Int. 132, 105093.

Chen, X.W., Wu, L., Luo, N., Mo, C.H., Wong, M.H., Li, H., 2019. Arbuscular mycorrhizal fungi and the associated bacterial community influence the uptake of cadmium in rice. Geoderma 337, 749–757.

Cheung, K-C., Zhang, J-Y., Deng, H-H., Ou, Y-K., Leung, H-M., Wu, S-C., Wong, M-H., 2008. Interaction of higher plant (jute), electrofused bacteria and mycorrhiza on anthracene biodegradation. Biores. Technol. 99 (7), 2148–2155.

Chi, G.G., Srivastava, A.K., Wu, Q.S., 2018. Exogenous easily extractable glomalin-related soil protein improves drought tolerance of trifoliate orange. Arch. Agron. Soil Sci. 64 (10), 1341–1350.

Chiapuso, G., Pujol, S., Toussaint, M.L., Badot, P.M., Binet, P., 2007. Phenanthrene toxicity and dissipation in rhizosphere of grassland plants (*Lolium perenne* L. and *Trifolium partense* L.) in three spiked soils. Plant Soil 294 (1), 103–112.

Chitarra, W., Pagliarani, C., Maserti, B., Lumini, E., Siciliano, I., Cascone, P., Schubert, A., Gambino, G., Balestrini, R., Guerrieri, E., 2016. Insights on the impact of arbuscular mycorrhizal symbiosis on tomato tolerance to water stress. Plant Physiol 171 (2), 1009–1023.

Ciadamidaro, L., Girardclos, O., Bert, V., Zappelini, C., Yung, L., Foulon, J., Papin, A., Roy, S., Blaudez, D., Chalot, M., 2017. Poplar biomass production at phytomanagement sites is significantly enhanced by mycorrhizal inoculation. Environ. Exp. Bot. 139, 48–56.

Cicatelli, A., Lingua, G., Todeschini, V., Biondi, S., Torrigiani, P., Castiglione, S., 2010. Arbuscular mycorrhizal fungi restore normal growth in a white poplar clone grown on heavy metal-contaminated soil, and this is associated with upregulation of foliar metallothionein and polyamine biosynthetic gene expression. Ann. Bot. 106 (5), 791–802.

Corgié, S.C., Fons, F., Beguiristain, T., Leyval, C., 2006. Biodegradation of phenanthrene, spatial distribution of bacterial populations and dioxygenase expression in the mycorrhizosphere of *Lolium perenne* inoculated with *Glomus mosseae*. Mycorrhiza 16 (3), 207–212.

Cornejo, P., Meier, S., Borie, G., Rillig, M.C., Borie, F., 2008. Glomalin-related soil protein in a Mediterranean ecosystem affected by a copper smelter and its contribution to Cu and Zn sequestration. Sci. Total Environ. 406 (1–2), 154–160.

Cozzolino, V., De Martino, A., Nebbioso, A., Di Meo, V., Salluzzo, A., Piccolo, A., 2016. Plant tolerance to mercury in a contaminated soil is enhanced by the combined effects of humic matter addition and inoculation with arbuscular mycorrhizal fungi. Environ. Sci. Pollut. Res. 23 (11), 11312–11322.

Cristaldi, A., Conti, G.O., Jho, E.H., Zuccarello, P., Grasso, A., Copat, C., Ferrante, M., 2017. Phytoremediation of contaminated soils by heavy metals and PAHs. A brief review. Environ. Technol. Inno. 8, 309–326.

Cundy, A.B., Bardos, R.P., Church, A., Puschenreiter, M., Friesl-Hanl, W., Muller, I., Neu, S., Mench, M., Witters, N., Vangronsveld, J., 2013. Developing principles of sustainability and stakeholder engagement for "gentle" remediation approaches: The European context. J. Environ. Manag. 129, 283–291.

do Fátima Pedroso, D., Barbosa, M.V., dos Santos, J.V., Pinto, F.A., Siqueira, J.O., Carneiro, M.A.C., 2018. Arbuscular mycorrhizal fungi favor the initial growth of *Acacia mangium*, *Sorghum bicolor*, and *Urochloa brizantha* in soil contaminated with Zn, Cu, Pb, and Cd. Bull. Environ. Cont. Tox. 101 (3), 386–391.

de la Providencia, I.E., Stefani, F.O., Labridy, M., St-Arnaud, M., Hijri, M., 2015. Arbuscular mycorrhizal fungal diversity associated with Eleocharis obtusa and Panicum capillare growing in an extreme petroleum hydrocarbon-polluted sedimentation basin. FEMS Microbiol. Lett. 362 (12), fnv081 10.1093/femsle/fnv081.

Debiane, D., Garçon, G., Verdin, A., Fontaine, J., Durand, R., Grandmougin-Ferjani, A., Shirali, P., Lounès-Hadj Sahraoui, A., 2008. *In vitro* evaluation of the oxidative stress and genotoxic potentials of anthracene on mycorrhizal chicory roots. Environ. Exp. Bot. 64 (2), 120–127.

Debiane, D., Garçon, G., Verdin, A., Fontaine, J., Durand, R., Shirali, P., Grandmougin-Ferjani, A., Lounès-Hadj Sahraoui, A., 2009. Mycorrhization alleviates benzo[a]pyrene-induced oxidative stress in an *in vitro* chicory root model. Phytochemistry 70 (11–12), 1421–1427.

Debiane, D., Calonne, M., Fontaine, J., Laruelle, F., Grandmougin-Ferjani, A., Lounès-Hadj Sahraoui, A., 2012. Benzo[a]pyrene induced lipid changes in the monoxenic arbuscular mycorrhizal chicory roots. J. Hazard. Mater. 209, 18–26.

DeBruyn, J.M., Chewning, C.S., Sayler, G.S., 2007. Comparative quantitative prevalence of Mycobacteria and functionally abundant nidA, nahAc, and nagAc dioxygenase genes in coal tar contaminated sediments. Environ. Sci. Technol. 41 (15), 5426–5432.

Demirbas, A., 2008. Heavy metal adsorption onto agro-based waste materials: a review. J. Hazard. Mater. 157 (2–3), 220–229.

Deram, A., Languereau, F., Haluwyn, C.V., 2011. Mycorrhizal and endophytic fungal colonization in *Arrhenatherum elatius* L. roots according to the soil contamination in heavy metals. Soil Sediment Contamin 20 (1), 114–127.

Dong, R., Gu, L., Guo, C., Xun, F., Liu, J., 2014. Effect of PGPR *Serratia marcescens* BC-3 and AMF *Glomus intraradices* on phytoremediation of petroleum contaminated soil. Ecotoxicology 23 (4), 674–680.

Dorantes, A.R., Zúñiga, L.A.G., 2012. Phenoloxidases activity in root system and their importance in the phytoremediation of organic contaminants. J. Environ. Chem. Ecotoxicol. 4 (3), 35–40.

Driai, S., Verdin, A., Laruelle, F., Beddiar, A., Lounès-Hadj Sahraoui, A., 2015. Is the arbuscular mycorrhizal fungus *Rhizophagus irregularis* able to fulfil its life cycle in the presence of diesel pollution? Int. Biodeter. Biodegr. 105, 58–65.

Duponnois, R., Galiana, A., Prin, Y., 2008. The mycorrhizosphere effect: a multitrophic interaction complex improves mycorrhizal symbiosis and plant growth. In: Siddiqui, Z.A., Sayeed Akhtar, M., Futai, K. (Eds.), Mycorrhizae: Sustainable Agriculture and Forestry. Springer, Dordrecht, pp. 227–240.

Durán, N., Esposito, E., 2000. Potential applications of oxidative enzymes and phenoloxidase-like compounds in wastewater and soil treatment: a review. Appl. Catal. B: Environ. 28 (2), 83–99.

Evangelou, M.W., Papazoglou, E.G., Robinson, B.H., Schulin, R., 2015. Phytomanagement: phytoremediation and the production of biomass for economic revenue on contaminated land. In: Ansari, A.A., Singh Gill, S., Gill, R., Lanza, G.R., Newman, L. (Eds.). Phytoremediation – Management of Environmental Contaminants, 1. Springer, Cham, pp. 115–132.

Faggioli, V., Menoyo, E., Geml, J., Kemppainen, M., Pardo, A., Salazar, M.J., Becerra, A.G., 2019. Soil lead pollution modifies the structure of arbuscular mycorrhizal fungal communities. Mycorrhiza 29 (4), 363–373.

Feng, N-X., Yu, J., Zhao, H-M., Cheng, Y-T., Mo, C-H., Cai, Q-Y., Li, Y.-W., Li, H., Wong, M-H., 2017. Efficient phytoremediation of organic contaminants in soils using plant–endophyte partnerships. Sci. Total Environ. 583, 352–368.

Ferrol, N., González-Guerrero, M., Valderas, A., Benabdellah, K., Azcón-Aguilar, C., 2009. Survival strategies of arbuscular mycorrhizal fungi in Cu-polluted environments. Phytochem. Rev. 8 (3), 551.

Ferrol, N., Tamayo, E., Vargas, P., 2016. The heavy metal paradox in arbuscular mycorrhizas: from mechanisms to biotechnological applications. J. Exp. Bot. 67 (22), 6253–6265.

Fester, T., Giebler, J., Wick, L.Y., Schlosser, D., Kästner, M., 2014. Plant–microbe interactions as drivers of ecosystem functions relevant for the biodegradation of organic contaminants. Curr. Opin. Biotechnol. 27, 168–175.

Firmin, S., Labidi, S., Fontaine, J., Laruelle, F., Tisserant, B., Nsanganwimana, F., Pourrut, B., Dalpé, Y., Grandmougin-Ferjani, A., Douay, F., Shirali, P., Lounès-Hadj Sahraoui, A., 2015. Arbuscular mycorrhizal fungal inoculation protects *Miscanthus* × *giganteus* against trace element toxicity in a highly metal-contaminated site. Sci. Total Environ. 527–528, 91–99.

Floch, C., Capowiez, Y., Criquet, S., 2009. Enzyme activities in apple orchard agroecosystems: how are they affected by management strategy and soil properties. Soil Biol. Biochem. 41 (1), 61–68.

Franco-Ramírez, A., Ferrera-Cerrato, R., Varela-Fregoso, L., Pérez-Moreno, J., Alarcón, A., 2007. Arbuscular mycorrhizal fungi in chronically petroleum-contaminated soils in Mexico and the effects of petroleum hydrocarbons on spore germination. J. Basic Microbiol. 47 (5), 378–383.

Gan, S., Lau, E.V., Ng, H.K., 2009. Remediation of soils contaminated with polycyclic aromatic hydrocarbons (PAHs). J. Hazard. Mater. 172 (2–3), 532–549.

Gao, Y., Cheng, Z., Ling, W., Huang, J., 2010. Arbuscular mycorrhizal fungal hyphae contribute to the uptake of polycyclic aromatic hydrocarbons by plant roots. Bioresour. Technol. 101 (18), 6895–6901.

Gao, Y., Li, Q., Ling, W., Zhu, X., 2011. Arbuscular mycorrhizal phytoremediation of soils contaminated with phenanthrene and pyrene. J. Hazard. Mater. 185 (2–3), 703–709.

Gao, Y., Zong, J., Que, H., Zhou, Z., Xiao, M., Chen, S., 2017a. Inoculation with arbuscular mycorrhizal fungi increases glomalin-related soil protein content and PAH removal in soils planted with Medicago sativa L. Soil Biol. Biochem. 115, 148–151.

Gao, Y., Zhou, Z., Ling, W., Hu, X., Chen, S., 2017b. Glomalin-related soil protein enhances the availability of polycyclic aromatic hydrocarbons in soil. Soil Biol. Biochem. 107, 129–132.

Garcés-Ruiz, M., Senés-Guerrero, C., Declerck, S., Cranenbrouck, S., 2017. Arbuscular mycorrhizal fungal community composition in *Carludovica palmata*, *Costus scaber* and *Euterpe precatoria* from weathered oil ponds in the Ecuadorian Amazon. Front. Microbiol. 8, 2134 10.3389/fmicb.2017.02134.

Garcés-Ruiz, M., Senés-Guerrero, C., Declerck, S., Cranenbrouck, S., 2019. Community composition of arbuscular mycorrhizal fungi associated with native plants growing in a petroleum-polluted soil of the Amazon region of Ecuador. Microbiology Open 8 (4), e00703 10.1002/mbo3.703.

Garcia, K.G.V., Freire Gomes, V.F., Mendes Filho, P.F., Martins, C.M., Tupinanbá da Silva Júnior, J.M., Medeiros Cunha, C.S., Pinheiro, J.I., 2018. Arbuscular mycorrhizal fungi in the phytostabilization of soil degraded by manganese mining. J. Agric. Sci. 10 (12), 192–202.

Garcia, K.G.V., Mendes Filho, P.F., Pinheiro, J.I., do Carmo, J.F., de Araújo Pereira, A.P., Martins, C.M., Pedroza de Abreu, M.G., de Souza Oliveira Filho, J., 2020. Attenuation of manganese-induced toxicity in *Leucaena leucocephala* colonized by arbuscular mycorrhizae. Water Air Soil Pollut 231 (1), 22 10.1007/s11270-019-4381-9.

Garg, N., Aggarwal, N., 2011. Effects of interactions between cadmium and lead on growth, nitrogen fixation, phytochelatin, and glutathione production in mycorrhizal *Cajanus cajan* (L.) Millsp. J. Plant Growth Regul. 30 (3), 286–300.

Garg, N., Chandel, S., 2010. Arbuscular mycorrhizal networks: process and functions. A review. Agron. Sustain. Dev. 30 (3), 581–599.

Garg, N., Chandel, S., 2012. Role of arbuscular mycorrhizal (AM) fungi on growth, cadmium uptake, osmolyte, and phytochelatin synthesis in *Cajanus cajan* (L.) Millsp. under NaCl and Cd stresses. J. Plant Growth Regul. 31 (3), 292–308.

Gaspar, M., Cabello, M., Cazau, M., Pollero, R., 2002. Effect of phenanthrene and *Rhodotorula glutinis* on arbuscular mycorrhizal fungus colonization of maize roots. Mycorrhiza 12 (2), 55–59.

Gerhardt, K.E., Huang, X.D., Glick, B.R., Greenberg, B.M., 2009. Phytoremediation and rhizoremediation of organic soil contaminants: potential and challenges. Plant Sci 176, 20–30.

Gerhardt, K.E., Gerwing, P.D., Greenberg, B.M., 2017. Opinion: Taking phytoremediation from proven technology to accepted practice. Plant Sci 256, 170–185.

Ghosh, M., Singh, S.P., 2005. A review on phytoremediation of heavy metals and utilization of it's by products. Asian J Energy Environ 6 (4), 214–231.

Gil-Cardeza, M.L., Ferri, A., Cornejo, P., Gomez, E., 2014. Distribution of chromium species in a Cr-polluted soil: presence of Cr (III) in glomalin related protein fraction. Sci. Total Environ. 493, 828–833.

Gil-Cardeza, M.L., Calonne-Salmon, M., Gómez, E., Declerck, S., 2017. Short-term chromium (VI) exposure increases phosphorus uptake by the extraradical mycelium of the arbuscular mycorrhizal fungus *Rhizophagus irregularis* MUCL 41833. Chemosphere 187, 27–34.

Giri, B., Kapoor, R., Mukerji, K.G., 2003. Influence of arbuscular mycorrhizal fungi and salinity on growth, biomass, and mineral nutrition of *Acacia auriculiformis*. Biol. Fertil. Soils 38 (3), 170–175.

Glick, B.R., 2010. Using soil bacteria to facilitate phytoremediation. Biotechnol. Adv 28 (3), 367–374.

Göhre, V., Paszkowski, U., 2006. Contribution of the arbuscular mycorrhizal symbiosis to heavy metal phytoremediation. Planta 223 (6), 1115–1122.

Gomez-Roldan, V., Roux, C., Girard, D., Bécard, G., Puech, V., 2007. Strigolactones: promising plant signals. Plant Signal. Behav. 2 (3), 163–164.

González-Chávez, M.C., Carrillo-Gonzalez, R., Wright, S.F., Nichols, K.A., 2004. The role of glomalin, a protein produced by arbuscular mycorrhizal fungi, in sequestering potentially toxic elements. Environ. Pollut. 130 (3), 317–323.

González-Chávez, M.C., Carrillo-Gonzalez, R., Gutierrez-Castorena, M.C., 2009. Natural attenuation in a slag heap contaminated with cadmium: the role of plants and arbuscular mycorrhizal fungi. J. Hazard. Mater. 161 (2–3), 1288–1298.

González-Chávez, M.C., del Pilar Ortega-Larrocea, M., Carrillo-González, R., López-Meyer, M., Xoconostle-Cázares, B., Gomez, S.K., Harrison, M.J., Figueroa-López, A.M., Maldonado-Mendoza, I.E., 2011. Arsenate induces the expression of fungal genes involved in As transport in arbuscular mycorrhiza. Fungal Biol 115 (12), 1197–1209.

Gonzalez-Guerrero, M., Melville, L.H., Ferrol, N., Lott, J.N., Azcon-Aguilar, C., Peterson, R.L., 2008. Ultrastructural localization of heavy metals in the extraradical mycelium and spores of the arbuscular mycorrhizal fungus *Glomus intraradices*. Can. J. Microbiol. 54 (2), 103–110.

Gray, C.W., Dunham, S.J., Dennis, P.G., Zhao, F.J., McGrath, S.P., 2006. Field evaluation of *in situ* remediation of a heavy metal contaminated soil using lime and red-mud. Environ. Pollut. 142 (3), 530–539.

Guo, W., Zhao, R., Yang, H., Zhao, J., Zhang, J., 2013. Using native plants to evaluate the effect of arbuscular mycorrhizal fungi on revegetation of iron tailings in grasslands. Biol. Fertil. Soils 49 (6), 617–626.

Gutjahr, C., Paszkowski, U., 2013. Multiple control levels of root system remodeling in arbuscular mycorrhiza symbiosis. Front. Plant Sci. 4 (204). doi:10.3389/fpls.2013.00204.

Hage-Ahmed, K., Moyses, A., Voglgruber, A., Hadacek, F., Steinkellner, S., 2013. Alterations in root exudation of intercropped tomato mediated by the arbuscular mycorrhizal fungus *Glomus mosseae* and the soilborne pathogen *Fusarium oxysporum* f. sp. lycopersici. J. Phytopathol. 161 (11–12), 763–773.

Handa, Y., Nishide, H., Takeda, N., Suzuki, Y., Kawaguchi, M., Saito, K., 2015. RNA-seq transcriptional profiling of an arbuscular mycorrhiza provides insights into regulated and coordinated gene expression in *Lotus japonicus* and *Rhizophagus irregularis*. Plant Cell Physiol 56 (8), 1490–1511.

Hassan, S.E.D., Boon, E., St-Arnaud, M., Hijri, M., 2011. Molecular biodiversity of arbuscular mycorrhizal fungi in trace metal-polluted soils. Mol. Ecol. 20 (16), 3469–3483.

Hassan, S.E.D., Bell, T.H., Stefani, F.O., Denis, D., Hijri, M., St-Arnaud, M., 2014. Contrasting the community structure of arbuscular mycorrhizal fungi from hydrocarbon-contaminated and uncontaminated soils following willow (*Salix* spp. L.) planting. PLoS One 9 (7), e102838. doi:10.1371/journal.pone.0102838.

Hernández-Allica, J., Becerril, J.M., Zárate, O., Garbisu, C., 2006. Assessment of the efficiency of a metal phytoextraction process with biological indicators of soil health. Plant Soil 281 (1-2), 147–158.

Hernández-Ortega, H.A., Alarcón, A., Ferrera-Cerrato, R., Zavaleta-Mancera, H.A., López-Delgado, H.A., Mendoza-López, M.R., 2012. Arbuscular mycorrhizal fungi on growth, nutrient status, and total antioxidant activity of *Melilotus albus* during phytoremediation of a diesel-contaminated substrate. J. Environ. Manage. 95, S319–S324.

Hildebrandt, U., Regvar, M., Bothe, H., 2007. Arbuscular mycorrhiza and heavy metal tolerance. Phytochemistry 68 (1), 139–146.

Holford, I.C.R., 1997. Soil phosphorus: its measurement, and its uptake by plants. Soil Res 35 (2), 227–240.

Holzman, D.C., 2012. Accounting for nature's benefits: the dollar value of ecosystem services. Environ. Health Perspect. 120 (4) 10.1289/ehp.120-a152.

Hrynkiewicz, K., Baum, C., 2012. The potential of rhizosphere microorganisms to promote the plant growth in disturbed soils. In: Malik, A., Grohmann, E. (Eds.), Environmental Protection Strategies for Sustainable Development. Springer, Dordrecht, pp. 35–64.

Huang, J-C., Tang, M., Niu, Z-C., Zhang, R-Q., 2007. Arbuscular mycorrhizal fungi in petroleum contaminated soil in Suining area of Sichuan Province. Chin. J. Ecol. 14 http://en.cnki.com.cn/Article_en/CJFDTotal-STXZ200709014.htm.

Huang, X.F., Chaparro, J.M., Reardon, K.F., Zhang, R., Shen, Q., Vivanco, J.M., 2014. Rhizosphere interactions: root exudates, microbes, and microbial communities. Botany 92 (4), 267–275.

Iffis, B., St-Arnaud, M., Hijri, M., 2014. Bacteria associated with arbuscular mycorrhizal fungi within roots of plants growing in a soil highly contaminated with aliphatic and aromatic petroleum hydrocarbons. FEMS Microbiol. Lett. 358 (1), 44–54.

Iffis, B., St-Arnaud, M., Hijri, M., 2016. Petroleum hydrocarbon contamination, plant identity and arbuscular mycorrhizal fungal (AMF) community determine assemblages of the AMF spore-associated microbes. Environ. Microbiol. 18 (8), 2689–2704.

Inui, H., Wakai, T., Gion, K., Kim, Y.S., Eun, H., 2008. Differential uptake for dioxin-like compounds by zucchini subspecies. Chemosphere 73, 1602–1607.

Jamal, A., Ayub, N., Usman, M., Khan, A.G., 2002. Arbuscular mycorrhizal fungi enhance Zn and Ni uptake from contaminated soil by soybean and lentil. Int. J. Phytoremed. 4, 203–221.

Janoušková, M., Pavlíková, D., Vosátka, M., 2006. Potential contribution of arbuscular mycorrhiza to cadmium immobilisation in soil. Chemosphere 65 (11), 1959–1965.

Joner, E.J., Briones, R., Leyval, C., 2000. Metal-binding capacity of arbuscular mycorrhizal mycelium. Plant Soil 226 (2), 227–234.

Joner, E.J., Johansen, A., Loibner, A.P., de la Cruz, M.A., Szolar, O.H., Portal, J.M., Leyval, C., 2001. Rhizosphere effects on microbial community structure and dissipation and toxicity of polycyclic aromatic hydrocarbons (PAHs) in spiked soil. Environ. Sci. Technol. 35 (13), 2773–2777.

Joner, E.J., Leyval, C., 1997. Uptake of ^{109}Cd by roots and hyphae of a *Glomus mosseae/Trifolium subterraneum* mycorrhiza from soil amended with high and low concentrations of cadmium. New Phytol 135 (2), 353–360.

Joner, E.J., Leyval, C., 2001. Influence of arbuscular mycorrhiza on clover and ryegrass grown together in a soil spiked with polycyclic aromatic hydrocarbons. Mycorrhiza 10 (4), 155–159.

Joner, E.J., Leyval, C., 2003a. Rhizosphere gradients of polycyclic aromatic hydrocarbon (PAH) dissipation in two industrial soils and the impact of arbuscular mycorrhiza. Environ. Sci. Technol. 37 (11), 2371–2375.

Joner, E.J., Leyval, C., 2003b. Phytoremediation of organic pollutants using mycorrhizal plants: a new aspect of rhizosphere interactions. Agronomie 23 (5–6), 495–502.

Jones, D.L., Hodge, A., Kuzyakov, Y., 2004. Plant and mycorrhizal regulation of rhizodeposition. New Phytol 163 (3), 459–480.

Kaldorf, M., Kuhn, A.J., Schröder, W.H., Hildebrandt, U., Bothe, H., 1999. Selective element deposits in maize colonized by a heavy metal tolerance conferring arbuscular mycorrhizal fungus. J. Plant Physiol. 154 (5-6), 718–728.

Karandashov, V., Bucher, M., 2005. Symbiotic phosphate transport in arbuscular mycorrhizas. Trends Plant Sci 10 (1), 22 29.

Karimi, A., Khodaverdiloo, H., Rasouli-Sadaghiani, M.H., 2018. Microbial-enhanced phytoremediation of lead contaminated calcareous soil by *Centaurea cyanus* L. Clean-Soil Air Water 46 (2), 1700665 10.1002/clen.201700665.

Kennedy, A.C., Smith, K.L., 1995. Soil microbial diversity and the sustainability of agricultural soils. Plant Soil 170 (1), 75–86.

Khan, A.G., 2006. Mycorrhizoremediation - an enhanced form of phytoremediation. J. Zhejiang Univ. Sci. B 7 (7), 503–514.

Khaosaad, T., Vierheilig, H., Nell, M., Zitterl-Eglseer, K., Novak, J., 2006. Arbuscular mycorrhiza alter the concentration of essential oils in oregano (*Origanum* sp., Lamiaceae). Mycorrhiza 16 (6), 443–446.

Kidd, P., Mench, M., Alvarez-Lopez, V., Bert, V., Dimitriou, I., Friesl-Hanl, W., Herzig, R., Janssen, J.O., Kolbas, A., Müller, I., Neu, S., Renella, G., Ruttens, A., Vangronsveld, J., Puschenreiter, M., 2015. Agronomic practices for improving gentle remediation of trace element-contaminated soils. Int. J. Phytoremed. 17 (11), 1005–1037.

Ko, J.Y., Day, J.W., 2004. A review of ecological impacts of oil and gas development on coastal ecosystems in the Mississippi Delta. Ocean Coastal Manage 47 (11-12), 597–623.

Kohler, J., Caravaca, F., Roldán, A., 2010. An AM fungus and a PGPR intensify the adverse effects of salinity on the stability of rhizosphere soil aggregates of *Lactuca sativa*. Soil Biol. Biochem. 42 (3), 429–434.

Kohler, J., Caravaca, F., Azcón, R., Díaz, G., Roldán, A., 2016. Suitability of the microbial community composition and function in a semiarid mine soil for assessing phytomanagement practices based on mycorrhizal inoculation and amendment addition. J. Environ.Manage. 169, 236–246.

Krishnamoorthy, R., Kim, C.G., Subramanian, P., Kim, K.Y., Selvakumar, G., Sa, T.M., 2015. Arbuscular mycorrhizal fungi community structure, abundance and species richness changes in soil by different levels of heavy metal and metalloid concentration. PLoS One 10 (6), e0128784 10.1371/journal.pone.0128784.

Kumar, S., Meena, R.S., Lal, R., Yadav, G.S., Mitran, T., Meena, B.L., Dotaniya, M.L., El-Sabagh, A., 2018. Role of legumes in soil carbon sequestration. In: Meena, R.S., Das, A., Singh Yadav, G., Lal, R. (Eds.), Legumes for Soil Health and Sustainable Management. Springer, Singapore, pp. 109–138.

Le Pioufle, O., Ganoudi, M., Calonne-Salmon, M., Ben Dhaou, F., Declerck, S., 2019. *Rhizophagus irregularis* MUCL 41833 improves phosphorus uptake and water use efficiency in maize plants during recovery from drought stress. Front. Plant Sci. 10, 897.

Leake, J., Johnson, D., Donnelly, D., Muckle, G., Boddy, L., Read, D., 2004. Networks of power and influence: the role of mycorrhizal mycelium in controlling plant communities and agroecosystem functioning. Can. J. Bot. 82 (8), 1016–1045.

Leigh, M.B., Fletcher, J.S., Fu, X., Schmitz, F.J., 2002. Root turnover: an important source of microbial substrates in rhizosphere remediation of recalcitrant contaminants. Environ. Sci. Technol. 36 (7), 1579–1583.

Lenoir, I., Fontaine, J., Lounès-Hadj Sahraoui, A., 2016a. Arbuscular mycorrhizal fungal responses to abiotic stresses: a review. Phytochemistry 123, 4–15.

Lenoir, I., Lounès-Hadj Sahraoui, A., Fontaine, J., 2016b. Arbuscular mycorrhizal fungal-assisted phytoremediation of soil contaminated with persistent organic pollutants: a review. Eur. J. Soil Sci. 67 (5), 624–640.

Lenoir, I., Lounès-Hadj Sahraoui, A., Laruelle, F., Dalpé, Y., Fontaine, J., 2016c. Arbuscular mycorrhizal wheat inoculation promotes alkane and polycyclic aromatic hydrocarbon biodegradation: Microcosm experiment on aged-contaminated soil. Environ. Pollut. 213, 549–560.

Lenoir, I., Fontaine, J., Tisserant, B., Laruelle, F., Lounès-Hadj Sahraoui, A., 2017. Beneficial contribution of the arbuscular mycorrhizal fungus, *Rhizophagus irregularis*, in the protection of *Medicago truncatula* roots against benzo[a]pyrene toxicity. Mycorrhiza 27 (5), 465–476.

Leung, H.M., Leung, A.O.W., Ye, Z.H., Cheung, K.C., Yung, K.K.L., 2013. Mixed arbuscular mycorrhizal (AM) fungal application to improve growth and arsenic accumulation of *Pteris vittata* (As hyperaccumulator) grown in As-contaminated soil. Chemosphere 92 (10), 1367–1374.

Li, H., Li, X., Xiang, L., Zhao, H-M., Li, Y-W., Cai, Q-Y., Zhu, L., Mo, C-H., Wong, M-H., 2018. Phytoremediation of soil co-contaminated with Cd and BDE-209 using hyperaccumulator enhanced by AM fungi and surfactant. Sci. Total Environ. 613–614, 447–455.

Li, X., Chen, A-Y., Yu, L-E., Chen, X-X., Xiang, L., Zhao, H-M., Mo, C-H., Li, Y-W., Cai, Q-Y., Wong, M-H., Li, H., 2019. Effects of β-cyclodextrin on phytoremediation of soil co-contaminated with Cd and BDE-209 by arbuscular mycorrhizal amaranth. Chemosphere 220, 910–920.

Lingua, G., Franchin, C., Todeschini, V., Castiglione, S., Biondi, S., Burlando, B., Berta, G., 2008. Arbuscular mycorrhizal fungi differentially affect the response to high zinc concentrations of two registered poplar clones. Environ. Pollut. 153 (1), 137–147.

Liu, A., Dalpé, Y., 2009. Reduction in soil polycyclic aromatic hydrocarbons by arbuscular mycorrhizal leek plants. Int. J. Phytoremed. 11 (1), 39–52.

Liu, S-L., Luo, Y-M., Cao, Z-H., Wu, L-H., Ding, K-Q., Christie, P., 2004. Degradation of benzo[a]pyrene in soil with arbuscular mycorrhizal alfalfa. Environ. Geochem. Health 26 (2), 285–293.

Lombi, E., Wenzel, W.W., Adriano, D.C., 1998. Soil contamination, risk reduction and remediation. Land Contam. Reclamat. 6 (4), 183–197.

Long, L.K., Yao, Q., Guo, J., Yang, R.H., Huang, Y.H., Zhu, H.H., 2010. Molecular community analysis of arbuscular mycorrhizal fungi associated with five selected plant species from heavy metal polluted soils. Eur. J. Soil Biol. 46 (5), 288–294.

Lu, Y-F., Lu, M., Peng, F., Wan, Y., Liao, M-H., 2014. Remediation of polychlorinated biphenyl-contaminated soil by using a combination of ryegrass, arbuscular mycorrhizal fungi and earthworms. Chemosphere 106, 44–50.

Lu, Y-F., Lu, M., 2015. Remediation of PAH-contaminated soil by the combination of tall fescue, arbuscular mycorrhizal fungus and epigeic earthworms. J. Hazard. Mater. 285, 535–541.

Luo, J., Li, X., Jin, Y., Traore, I., Dong, L., Yang, G., Wang, Y., 2020. Effects of arbuscular mycorrhizal fungi *Glomus mosseae* on the growth and medicinal components of *Dysosma versipellis* under copper stress. Bull. Environ. Contam. Toxicol., 1–7 10.1007/s00128-019-02780-1.

Lux, A., Martinka, M., Vaculík, M., White, P.J., 2011. Root responses to cadmium in the rhizosphere: a review. J. Exp. Bot. 62 (1), 21–37.

Macnair, M.R., 1987. Heavy metal tolerance in plants: a model evolutionary system. Trends Ecol. Evol. 2 (12), 354–359.

Małachowska-Jutsz, A., Kalka, J., 2010. Influence of mycorrhizal fungi on remediation of soil contaminated by petroleum hydrocarbons. Fresenius Environ. Bull. 19 (12b), 3217–3223.

Małachowska-Jutsz, A., Rudek, J., Janosz, W., 2011. The effect of ribwort (*Plantago lanceolata*) and its myrorrhizas on the growth of microflora in soil contaminated with used engine oil. Arch. Environ. Prot. 37 (1), 99–113.

Malekzadeh, P., Farshian, S.H., Ordubadi, B., 2012. Interaction of arbuscular mycorrhizal fungus (*Glomus intraradices* and *Glomus etunicatum*) with tomato plants grown under copper toxicity. Afr. J. Biotechnol. 11 (46), 10555–10567.

Malekzadeh, E., Aliasgharzad, N., Majidi, J., Abdolalizadeh, J., Aghebati-Maleki, L., 2016. Contribution of glomalin to Pb sequestration by arbuscular mycorrhizal fungus in a sand culture system with clover plant. Eur. J. Soil Biol. 74, 45–51.

Malicka, M., Magurno, F., Piotrowska-Seget, Z., Chmura, D., 2020. Arbuscular mycorrhizal and microbial profiles of an aged phenol–polynuclear aromatic hydrocarbon-contaminated soil. Ecotox. Environ. Safe. 192, 110299 10.1016/j.ecoenv.2020.110299.

Meglouli, H., Lounès-Hadj Sahraoui, A., Magnin-Robert, M., Tisserant, B., Hijri, M., Fontaine, J., 2018. Arbuscular mycorrhizal inoculum sources influence bacterial, archaeal, and fungal communities' structures of historically dioxin/furan-contaminated soil but not the pollutant dissipation rate. Mycorrhiza 28 (7), 635–650.

Meglouli, H., Fontaine, J., Verdin, A., Magnin-Robert, M., Tisserant, B., Hijri, M., Lounès-Hadj Sahraoui, A., 2019. Aided phytoremediation to clean up dioxins/furans-aged contaminated soil: correlation between microbial communities and pollutant dissipation. Microorganisms, 7 (11), 523. doi: 10.3390/microorganisms7110523.

Mench, M., Lepp, N., Bert, V., Schwitzguébel, J.-P., Gawronski, S.W., Schröder, P., Vangronsveld, J., 2010. Successes and limitations of phytotechnologies at field scale: outcomes, assessment and outlook from COST Action 859. J. Soil Sediment 10 (6), 1039–1070.

Mench, M.J., Dellise, M., Bes, C.M., Marchand, L., Kolbas, A., Le Coustumer, P., Oustrière, N., 2018. Phytomanagement and remediation of cu-contaminated soils by high yielding crops at a former wood preservation site: Sunflower biomass and ionome. Front. Ecol. Evol. 6, 123 10.3389/fevo.2018.00123.

Merkl, N., Schultze-Kraft, R., Infante, C., 2005. Phytoremediation in the tropics–influence of heavy crude oil on root morphological characteristics of graminoids. Environ. Pollut. 138 (1), 86–91.

Miller, G., Shulaev, V., Mittler, R., 2008. Reactive oxygen signaling and abiotic stress. Physiol. Plant. 133 (3), 481–489.

Miller, B.P., Sinclair, E.A., Menz, M.H., Elliott, C.P., Bunn, E., Commander, L.E., Dalziell, E., David, E., Davis, B., Erickson, T.E., Golos, P.J., Krauss, S.L., Lewandrowki, W., Mayence, C.E., Merino-Martín, L., Merritt, D.J., Nevill, P.G., Phillips, R.D., Ritchie, A.L., Ruoss, S., Stevens, J.C., 2017. A framework for the practical science necessary to restore sustainable, resilient, and biodiverse ecosystems. Restor. Ecol. 25 (4), 605–617.

Miransari, M., 2011. Hyperaccumulators, arbuscular mycorrhizal fungi and stress of heavy metals. Biotechnol. Adv. 29 (6), 645–653.

Miransari, M., 2017. Arbuscular mycorrhizas and stress tolerance of plants. In: Wu, Q-S. (Ed.), Arbuscular Mycorrhizal Fungi and Heavy Metal Tolerance in Plants. Springer, Singapore, pp. 147–161.

Mishra, V., Gupta, A., Kaur, P., Singh, S., Singh, N., Gehlot, P., Singh, J., 2016. Synergistic effects of arbuscular mycorrhizal fungi and plant growth promoting rhizobacteria in bioremediation of iron contaminated soils. Int. J. Phytoremed. 18 (7), 697–703.

Morgado, R.G., Loureiro, S., González-Alcaraz, M.N., 2018. Changes in soil ecosystem structure and functions due to soil contamination. In: Duarte, A.C., Cachada, A., Rocha-Santos, T. (Eds.), Soil Pollution – From Monitoring to Remediation. Academic Press, London, United Kingdom, pp. 59–87.

Motaharpoor, Z., Taheri, H., Nadian, H., 2019. *Rhizophagus irregularis* modulates cadmium uptake, metal transporter, and chelator gene expression in *Medicago sativa*. Mycorrhiza 29 (4), 389–395.

Muszynska, E., Hanus-Fajerska, E., 2015. Why are heavy metal hyperaccumulating plants so amazing? BioTechnologia. J. Biotechnol. Comput. Biol. Bionanotechnol. 96 (4), 265–271.

Nogales, A., Cortés, A., Velianos, K., Camprubí, A., Estaún, V., Calvet, C., 2012. *Plantago lanceolata* growth and Cr uptake after mycorrhizal inoculation in a Cr amended substrate. Agr. Food Sci. 21 (1), 72–79.

Nwoko, C.O., Okeke, P.N., Ogbonna, P.C., 2013. Influence of soil particle size and arbuscular mycorrhizal fungi (AMF) in the performance of *Phaseolus vulgaris* grown under crude oil contaminated soil. Univers. J. Environ. Res. Technol. 3 (2), 300–310.

Nwoko, C.O., 2014. Effect of arbuscular mycorrhizal (AM) fungi on the physiological performance of *Phaseolus vulgaris* grown under crude oil contaminated soil. J. Geosci. Environ. Protect. 2, 9–14.

Ocampo, J.A., Hayman, D.S., 1981. Influence of plant interactions on vesicular-arbuscular mycorrhizal infections. II. Crop rotations and residual effects of non-host plants. New Phytol 87 (2), 333–343.

Orłowska, E., Mesjasz-Przybyłowicz, J., Przybyłowicz, W., Turnau, K., 2008. Nuclear microprobe studies of elemental distribution in mycorrhizal and non-mycorrhizal roots of Ni-hyperaccumulator *Berkheya coddii*. X-Ray Spectrom. Int. J. 37 (2), 129–132.

Orłowska, E., Przybyłowicz, W., Orlowski, D., Turnau, K., Mesjasz-Przybyłowicz, J., 2011. The effect of mycorrhiza on the growth and elemental composition of Ni-hyperaccumulating plant *Berkheya coddii* Roessler. Environ. Pollut. 159 (12), 3730–3738.

Orłowska, E., Godzik, B., Turnau, K., 2012. Effect of different arbuscular mycorrhizal fungal isolates on growth and arsenic accumulation in *Plantago lanceolata* L. Environ. Pollut. 168, 121–130.

Otero-Blanca, A., Folch-Mallol, J.L., Lira-Ruan, V., Carbente, M.D.R.S., Batista-García, R.A., 2018. Phytoremediation and fungi: an underexplored binomial. In: Prasad, R., Aranda, E. (Eds.), Approaches in Bioremediation. Springer, Cham, pp. 79–95.

Pandey, V.C., 2020a. Phytomanagement of fly ash. Elsevier, Amsterdam, Netherlands, pp. 1–352. doi:10.1016/C2018-0-01318-3. ISBN: 9780128185445.

Pandey, V.C., 2020b. Fly ash ecosystem servicesPhytomanagement of fly ash. Elsevier, Amsterdam, pp. 257–288. doi:10.1016/B978-0-12-818544-5.00009-2.

Pandey, V.C., Bajpai, O., 2019. Phytoremediation: From Theory towards Practice. In: Pandey, V.C., Bauddh, K. (Eds.), Phytomanagement of polluted sites. Elsevier, Amsterdam, Netherlands, pp. 1–49. doi:10.1016/B978-0-12-813912-7.00001-6.

Pandey, V.C., Bauddh, K., 2018. Phytomanagement of polluted sites. Elsevier, Amsterdam, Netherlands, pp. 1–626. doi:10.1016/C2017-0-00586-4. ISBN: 9780128139127.

Pandey, V.C., Mahajan, P., Saikia, P., Praveen, A., 2022. Fibre crop-based Phytoremediation: Socio-economic and Environmental Sustainability. Elsevier, Amsterdam, Netherlands. ISBN: 9780128242469. In press.
Pandey, V.C., Pandey, D.N., Singh, N., 2015. Sustainable phytoremediation based on naturally colonizing and economically valuable plants. J. Clean. Prod. 86, 37–39. doi:10.1016/j.jclepro.2014.08.030.
Pandey, V.C., Souza-Alonso, P., 2019. Market opportunities in sustainable phytoremediation. In: Pandey, V.C., Bauddh, K. (Eds.), Phytomanagement of polluted sites. Elsevier, Amsterdam, Netherlands, pp. 51–82. doi:10.1016/B978-0-12-813912-7.00002-8.
Parraga-Aguado, I., Gonzalez-Alcaraz, M.N., Alvarez-Rogel, J., Jimenez-Carceles, F.J., Conesa, H.M., 2013. The importance of edaphic niches and pioneer plant species succession for the phytomanagement of mine tailings. Environ. Pollut. 176, 134–143.
Parraga-Aguado, I., Querejeta, J-I., González-Alcaraz, M-N., Jiménez-Cárceles, F.J., Conesa, H.M., 2014. Usefulness of pioneer vegetation for the phytomanagement of met-al(loid)s enriched tailings: grasses vs. shrubs vs. trees. J. Environ. Manage. 133, 51–58.
Perlein, A., Zdanevitch, I., Gaucher, R., Robinson, B., Papin, A., Lounès-Hadj Sahraoui, A., Bert, V., 2021. Phytomanagement of a metal(loid)-contaminated agricultural site using aromatic and medicinal plants to produce essential oils: analysis of the metal(loid) fate in the value chain. Environ. Sci. Pollut. Res. doi.org/10.1007/s11356-021-15045-4.
Pernin, C., Joffre, R., Le Petit, J., Torre, F., 2006. Sewage sludge effects on mesofauna and cork oak (*Quercus suber* L.) leaves decomposition in a Mediterranean forest firebreak. J. Environ. Qual. 35 (6), 2283–2292.
Phanthavongsa, P., Chalot, M., Papin, A., Lacercat-Didier, L., Roy, S., Blaudez, D., Bert, V., 2017. Effect of mycorrhizal inoculation on metal accumulation by poplar leaves at phytomanaged sites. Environ. Exp. Bot. 143, 72–81.
Pieper, D.H., Seeger, M., 2008. Bacterial metabolism of polychlorinated biphenyls. J. Mol. Microbiol. Biotechnol. 15 (2-3), 121–138.
Pistelli, L., Ulivieri, V., Giovanelli, S., Avio, L., Giovannetti, M., Pistelli, L., 2017. Arbuscular mycorrhizal fungi alter the content and composition of secondary metabolites in *Bituminaria bituminosa* L. Plant Biol 19 (6), 926–933.
Plouznikoff, K., Declerck, S., Calonne-Salmon, M., 2016. Mitigating abiotic stresses in crop plants by arbuscular mycorrhizal fungi. In: Vos, M.F., Kazan, K. (Eds.), Belowground Defence Strategies in Plants. Springer, Cham, pp. 341–400.
Porcel, R., Aroca, R., Ruiz-Lozano, J.M., 2012. Salinity stress alleviation using arbuscular mycorrhizal fungi. A review. Agron. Sustain. Dev. 32 (1), 181–200.
Purin, S., Rillig, M.C., 2007. The arbuscular mycorrhizal fungal protein glomalin: limitations, progress, and a new hypothesis for its function. Pedobiologia 51 (2), 123–130.
Qin, H., Brookes, P.C., Xu, J., Feng, Y., 2014. Bacterial degradation of Aroclor 1242 in the mycorrhizosphere soils of zucchini (*Cucurbita pepo* L.) inoculated with arbuscular mycorrhizal fungi. Environ. Sci. Pollut. Res. 21 (22), 12790 12799.
Qin, H., Brookes, P.C., Xu, J., 2016. Arbuscular mycorrhizal fungal hyphae alter soil bacterial community and enhance polychlorinated biphenyls dissipation. Front. Microbiol. 7, 939 10.3389/fmicb.2016.00939.
Rabie, G.H., 2004. Using wheat-mungbean plant system and arbuscular mycorrhiza to enhance *in-situ* bioremediation. J Food Agric. Environ. 2 (2), 381–390.
Rabie, G.H., 2005. Role of arbuscular mycorrhizal fungi in phytoremediation of soil rhizosphere spiked with poly aromatic hydrocarbons. Mycobiology 33 (1), 41–50.
Rajtor, M., Piotrowska-Seget, Z., 2016. Prospects for arbuscular mycorrhizal fungi (AMF) to assist in phytoremediation of soil hydrocarbon contaminants. Chemosphere 162, 105–116.
Raveau, R., Fontaine, J., Bert, V., Perlein, A., Tisserant, B., Ferrant, P., Lounès-Hadj Sahraoui, A., 2021a. In situ cultivation of aromatic plant species for the phytomanagement of an aged-trace element polluted soil:

plant biomass improvement options and techno-economic assessment of the essential oil production channel. Science of The Total Environment 147944.

Raveau, R., Fontaine, J., Hijri, M., Lounès-Hadj Sahraoui, A., 2020. The Aromatic Plant Clary Sage Shaped Bacterial Communities in the Roots and in the Trace Element-Contaminated Soil More Than Mycorrhizal Inoculation – A Two-Year Monitoring Field Trial. Front. Microbiol. 11, 1–18. doi:10.3389/fmicb.2020.586050.

Raveau, R., Lounès-Hadj Sahraoui, A., Hijri, M., Fontaine, J., 2021b. Clary sage cultivation and mycorrhizal inoculation influence the rhizosphere fungal community of an aged trace-element polluted soil. Microorganisms 9, 1333. doi:10.3390/microorganisms9061333.

Redon, P.O., Béguiristain, T., Leyval, C., 2009. Differential effects of AM fungal isolates on *Medicago truncatula* growth and metal uptake in a multimetallic (Cd, Zn, Pb) contaminated agricultural soil. Mycorrhiza 19 (3), 187–195.

Regvar, M., Likar, M., Piltaver, A., Kugonič, N., Smith, J.E., 2010. Fungal community structure under goat willows (*Salix caprea* L.) growing at metal polluted site: the potential of screening in a model phytostabilisation study. Plant Soil 330 (1–2), 345–356.

Ren, C-G., Kong, C-C., Bian, B., Liu, W., Li, Y., Luo, Y-M., Xie, Z-H., 2017. Enhanced phytoremediation of soils contaminated with PAHs by arbuscular mycorrhiza and rhizobium. Int. J. Phytoremed. 19 (9), 789–797.

Repetto, O., Bestel-Corre, G., Dumas-Gaudot, E., Berta, G., Gianinazzi-Pearson, V., Gianinazzi, S., 2003. Targeted proteomics to identify cadmium-induced protein modifications in *Glomus mosseae*-inoculated pea roots. New Phytol 157 (3), 555–567.

Rillig, M.C., 2004. Arbuscular mycorrhizae, glomalin, and soil aggregation. Can. J. Soil Sci. 84 (4), 355–363.

Rillig, M.C., Mardatin, N.F., Leifheit, E.F., Antunes, P.M., 2010. Mycelium of arbuscular mycorrhizal fungi increases soil water repellency and is sufficient to maintain water-stable soil aggregates. Soil Biol. Biochem. 42 (7), 1189–1191.

Robertson, S.J., McGill, W.B., Massicotte, H., Rutherford, P.M., 2007. Petroleum hydrocarbon contamination in boreal forest soils: a mycorrhizal ecosystems perspective. Biol. Rev 82 (2), 213–240.

Robinson, B.H., Banuelos, G., Conesa, H.M., Evangelou, M.W.H., Schulin, R., 2009. The phytomanagement of trace elements in soil. Crit. Rev. Plant Sci. 28 (4), 240–266.

Rodríguez-Caballero, G., Caravaca, F., Fernández-González, A.J., Alguacil, M.M., Fernández-López, M., Roldán, A., 2017. Arbuscular mycorrhizal fungi inoculation mediated changes in rhizosphere bacterial community structure while promoting revegetation in a semiarid ecosystem. Sci. Total Environ. 584, 838–848.

Rostami, M., Rostami, S., 2019. Effect of salicylic acid and mycorrhizal symbiosis on improvement of fluoranthene phytoremediation using tall fescue (*Festuca arundinacea* Schreb). Chemosphere 232, 70–75.

Ruiz-Lozano, J.M., Porcel, R., Aroca, R., 2008. Evaluation of the possible participation of drought-induced genes in the enhanced tolerance of arbuscular mycorrhizal plants to water deficit. In: Varma, A. (Ed.), Mycorrhiza. Springer, Berlin, Heidelberg, pp. 185–205.

Ruotsalainen, A.L., Markkola, A., Kozlov, M.V., 2007. Root fungal colonisation in *Deschampsia flexuosa*: Effects of pollution and neighbouring trees. Environ. Pollut. 147 (3), 723–728.

Rydlová, J., Püschel, D., Sudová, R., Gryndler, M., Mikanová, O., Vosátka, M., 2011. Interaction of arbuscular mycorrhizal fungi and rhizobia: effects on flax yield in spoil-bank clay. J. Plant Nutr. Soil Sci. 174 (1), 128–134.

Santana, N.A., Rabuscke, C.M., Soares, V.B., Soriani, H.H., Nicoloso, F.T., Jacques, R.J.S., 2018. Vermicompost dose and mycorrhization determine the efficiency of copper phytoremediation by *Canavalia ensiformis*. Environ. Sci. Pollut. Res. 25 (13), 12663–12677.

Sarkar, A., Asaeda, T., Wang, Q., Kaneko, Y., Rashid, M.H., 2017. Response of *Miscanthus sacchariflorus* to zinc stress mediated by arbuscular mycorrhizal fungi. Flora 234, 60–68.

Scarlat, N., Dallemand, J.F., Monforti-Ferrario, F., Nita, V., 2015. The role of biomass and bioenergy in a future bioeconomy: policies and facts. Environ. Dev. 15, 3–34.

Schindler, F.V., Mercer, E.J., Rice, J.A., 2007. Chemical characteristics of glomalin-related soil protein (GRSP) extracted from soils of varying organic matter content. Soil Biol. Biochem. 39 (1), 320–329.

Schneider, J., Bundschuh, J., do Nascimento, C.W.A., 2016. Arbuscular mycorrhizal fungi-assisted phytoremediation of a lead-contaminated site. Sci. Total Environ. 572, 86–97.

Schwarzott, D., Walker, C., Schüßler, A., 2001. Glomus, the largest genus of the arbuscular mycorrhizal fungi (Glomales), is nonmonophyletic. Mol. Phylogenet. Evol. 21 (2), 190–197.

Schwitzguébel, J-P., 2017. Phytoremediation of soils contaminated by organic compounds: hype, hope and facts. J. Soils Sediment. 17 (5), 1492–1502.

Seguel, A., Cumming, J., Cornejo, P., Borie, F., 2016. Aluminum tolerance of wheat cultivars and relation to arbuscular mycorrhizal colonization in a non-limed and limed. Andisol. Appl. Soil Ecol. 108, 228–237.

Shivakumar, C.K., Hemavani, C., Thippe Swamy, B., Krishnappa, M., 2011. Effect of inoculation with arbuscular mycorrhizal fungi on green gram grown in soil containing heavy metal zinc. J. Exp. Sci. 2 (10) http://updatepublishing.com/journal/index.php/jes/article/view/1887.

Sidhoum, W., Fortas, Z., 2019. The beneficial role of indigenous arbuscular mycorrhizal fungi in phytoremediation of wetland plants and tolerance to metal stress. Arch. Environ. Prot. 45 (1), 103–114.

Singh, G., Pankaj, U., Chand, S., Verma, R.K., 2019. Arbuscular mycorrhizal fungi-assisted phytoextraction of toxic metals by *Zea mays* L. from tannery sludge. Soil Sediment Contam.: Int. J. 28 (8), 729–746.

Smith, S.E., Read, D.J., 2008. Mycorrhizal Symbiosis, 3rd edn. Academic press, New York.

Solís-Domínguez, F.A., Valentín-Vargas, A., Chorover, J., Maier, R.M., 2011. Effect of arbuscular mycorrhizal fungi on plant biomass and the rhizosphere microbial community structure of mesquite grown in acidic lead/zinc mine tailings. Sci. Total Environ. 409 (6), 1009–1016.

Stingley, R.L., Khan, A.A., Cerniglia, C.E., 2004. Molecular characterization of a phenanthrene degradation pathway in *Mycobacterium vanbaalenii* PYR-1. Biochem. Biophys. Res. Commun. 322 (1), 133–146.

Sun, Y., Zhang, X., Wu, Z., Hu, Y., Wu, S., Chen, B., 2016. The molecular diversity of arbuscular mycorrhizal fungi in the arsenic mining impacted sites in Hunan Province of. China. J. Environ. Sci. 39, 110–118.

Tabrizi, L., Mohammadi, S., Delshad, M., Moteshare Zadeh, B., 2015. Effect of arbuscular mycorrhizal fungi on yield and phytoremediation performance of pot marigold (*Calendula officinalis* L.) under heavy metals stress. Int. J. Phytoremed. 17 (12), 1244–1252.

Tamayo, E., Gómez-Gallego, T., Azcón-Aguilar, C., Ferrol, N., 2014. Genome-wide analysis of copper, iron and zinc transporters in the arbuscular mycorrhizal fungus *Rhizophagus irregularis*. Front. Plant Sci 5, 547 10.3389/fpls.2014.00547.

Tang, M., Chen, H., Huang, J.C., Tian, Z-Q., 2009. AM fungi effects on the growth and physiology of *Zea mays* seedlings under diesel stress. Soil Biol. Biochem. 41 (5), 936–940.

Teixeira, A.F.D.S., Kemmelmeier, K., Marascalchi, M.N., Stürmer, S.L., Carneiro, M.A.C., Moreira, F.M.D.S., 2017. Arbuscular mycorrhizal fungal communities in an iron mining area and its surroundings: Inoculum potential, density, and diversity of spores related to soil properties. Ciênc. Agrotec. 41 (5), 511–525.

Teng, Y., Luo, Y-M., Gao, J., Li, Z-G., 2008. Combined remediation effects of arbuscular mycorrhizal fungi legumes-rhizobium symbiosis on PCBs contaminated soils. Huan jing ke xue = Huanjing kexue 29 (10), 2925–2930.

Teng, Y., Luo, Y.-M., Sun, X., Tu, C., Xu, L., Liu, W., Li, Z-G., Christie, P., 2010. Influence of arbuscular mycorrhiza and rhizobium on phytoremediation by alfalfa of an agricultural soil contaminated with weathered PCBs: a field study. Int. J. Phytoremed. 12 (5), 516–533.

Thijs, S., Sillen, W., Rineau, F., Weyens, N., Vangronsveld, J., 2016. Towards an enhanced understanding of plant–microbiome interactions to improve phytoremediation: engineering the metaorganism. Front. Microbiol. 7, 341 10.3389/fmicb.2016.00341.

Tisserant, E., Malbreil, M., Kuo, A., Kohler, A., Symeonidi, A., Balestrini, R., et al., 2013. Genome of an arbuscular mycorrhizal fungus provides insight into the oldest plant symbiosis. PNAS 110 (50), 20117–20122.

Toljander, J.F., Lindahl, B.D., Paul, L.R., Elfstrand, M., Finlay, R.D., 2007. Influence of arbuscular mycorrhizal mycelial exudates on soil bacterial growth and community structure. FEMS Microbiol. Ecol. 61 (2), 295–304.

Treseder, K.K., Turner, K.M., Mack, M.C., 2007. Mycorrhizal responses to nitrogen fertilization in boreal ecosystems: potential consequences for soil carbon storage. Global Change Biol 13 (1), 78–88.

Turnau, K., Kottke, I., Oberwinkler, F., 1993. Element localization in mycorrhizal roots of *Pteridium aquilinum* (L.) Kuhn collected from experimental plots treated with cadmium dust. New Phytol 123 (2), 313–324.

Vallino, M., Massa, N., Lumini, E., Bianciotto, V., Berta, G., Bonfante, P., 2006. Assessment of arbuscular mycorrhizal fungal diversity in roots of Solidago gigantea growing in a polluted soil in Northern Italy. Environ. Microbiol. 8 (6), 971–983.

Van Aken, B., Tehrani, R., Schnoor, J.L., 2011. Endophyte-assisted phytoremediation of explosives in poplar trees by Methylobacterium populi BJ001 T. In: Pirttilä, A.M., Franck, A.C. (Eds.), Endophytes of Forest Trees. Springer, Dordrecht, pp. 217–234.

van der Heyde, M., Ohsowski, B., Abbott, L.K., Hart, M., 2017. Arbuscular mycorrhizal fungus responses to disturbance are context-dependent. Mycorrhiza 27 (5), 431–440.

Verdin, A., Lounès-Hadj Sahraoui, A., Fontaine, J., Grandmougin-Ferjani, A., Durand, R., 2006. Effects of anthracene on development of an arbuscular mycorrhizal fungus and contribution of the symbiotic association to pollutant dissipation. Mycorrhiza 16 (6), 397–405.

Vieira, C.K., Marascalchi, M.N., Rodrigues, A.V., de Armas, R.D., Stürmer, S.L., 2018. Morphological and molecular diversity of arbuscular mycorrhizal fungi in revegetated iron-mining site has the same magnitude of adjacent pristine ecosystems. J. Environ. Sci. 67, 330–343.

Villacrés, B.H., Medina, M., Cumbal, L., Villarroel, A., 2014. Implementación de un banco de Hongos Micorrícicos Arbusculares, aislados de suelos del área de influencia de EP PETROCUADOR y su efecto en el desarrollo de plantas de Maíz (*Zea mays*) en condiciones de estrés por cadmio, en La Joya de los Sachas. Provin. Congr. Cienc. Technol. 9 (1), 24–35.

Violante, A., Cozzolino, V., Perelomov, L., Caporale, A.G., Pigna, M., 2010. Mobility and bioavailability of heavy metals and metalloids in soil environments. J. Soil Sci. Plant Nutr. 10 (3), 268–292.

Wang, C-D., Shan, M-J., Lu, K-P., Wang, H-L., Zhao, WY., Huang, SG., Yu, Y., Maimuli, W., Qin, H., 2018. Combined remediation effects of arbuscular mycorrhizal fungi and dead pig biochar on PCBs contaminated soils. Acta Sci. Circumst. 38 (10), 4157–4164.

Wang, F-Y., Lin, X-G., Yin, R., 2007. Effect of arbuscular mycorrhizal fungal inoculation on heavy metal accumulation of maize grown in a naturally contaminated soil. Int. J. Phytoremed. 9 (4), 345–353.

Wang, F-Y., Adams, C.A., Yang, W., Sun, Y., Shi, Z., 2019. Benefits of arbuscular mycorrhizal fungi in reducing organic contaminant residues in crops: implications for cleaner agricultural production. Crit. Rev. Env. Sci. Technol. doi:10.1080/10643389.2019.1665945.

Wang, L., Huang, X., Ma, F., Ho, S-H., Wu, J., Zhu, S., 2017. Role of *Rhizophagus irregularis* in alleviating cadmium toxicity via improving the growth, micro- and macroelements uptake in *Phragmites australis*. Environ. Sci. Pollut. Res. 4, 3593–3607.

Wang, S., Zhang, S., Huang, H., Christie, P., 2011. Behavior of decabromodiphenyl ether (BDE-209) in soil: effects of rhizosphere and mycorrhizal colonization of ryegrass roots. Environ. Pollut. 159 (3), 749–753.

Wang, S., Pan, S., Shah, G.M., Zhang, Z., Yang, L., Yang, S., 2018. Enhancement in arsenic remediation by maize (*Zea mays* L.) using EDTA in combination with arbuscular mycorrhizal fungi. Appl. Ecol. Environ. Res. 16 (5), 5987–5999.

Wei, Y., Hou, H., Li, J., ShangGuan, Y., Xu, Y., Zhang, J., Zhao, L., Wang, W., 2014. Molecular diversity of arbuscular mycorrhizal fungi associated with an Mn hyperaccumulator - *Phytolacca americana*, in Mn mining area. Appl. Soil Ecol. 82, 11–17.

Wei, Y., Chen, Z., Wu, F., Hou, H., Li, J., Shangguan, Y., Zhang, J., Li, F., Zeng, Q., 2015. Molecular diversity of arbuscular mycorrhizal fungi at a large-scale antimony mining area in southern. China. J. Environ. Sci. 29, 18–26.

Weissenhorn, I., Glashoff, A., Leyval, C., Berthelin, J., 1994. Differential tolerance to Cd and Zn of arbuscular mycorrhizal (AM) fungal spores isolated from heavy metal-polluted and unpolluted soils. Plant Soil 167 (2), 189–196.

White, J.C., Ross, D.W., Gent, M.P., Eitzer, B.D., Mattina, M.I., 2006. Effect of mycorrhizal fungi on the phytoextraction of weathered p, p-DDE by *Cucurbita pepo*. J. Hazard. Mater. 137 (3), 1750–1757.

Whitfield, L., Richards, A.J., Rimmer, D.L., 2004. Relationships between soil heavy metal concentration and mycorrhizal colonisation in *Thymus polytrichus* in northern England. Mycorrhiza 14 (1), 55–62.

Whitfield Åslund, M.L., Lunney, A.I., Rutter, A., Zeeb, B.A., 2010. Effects of amendments on the uptake and distribution of DDT in *Cucurbita pepo* ssp *pepo* plants. Environ. Pollut. 158 (2), 508–513.

Willers, C., Jansen van Rensburg, P.J., Claassens, S., 2015. Phospholipid fatty acid profiling of microbial communities–a review of interpretations and recent applications. J. Appl. Microbiol. 119 (5), 1207–1218.

Wojtera-Kwiczor, J., Żukowska, W., Graj, W., Małecka, A., Piechalak, A., Ciszewska, L., Chrzanowski, Ł., Lisiecki, P., Komorowicz, I., Barałkiewicz, D., Voss, I., Scheibe, R., Tomaszewska, B., 2014. Rhizoremediation of diesel-contaminated soil with two rapeseed varieties and petroleum degraders reveals different responses of the plant defense mechanisms. Int. J. Phytoremed. 16 (7-8), 770–789.

Wu, F-Y., Yu, X-Z., Wu, S-C., Lin, X-G., Wong, M-H., 2011. Phenanthrene and pyrene uptake by arbuscular mycorrhizal maize and their dissipation in soil. J. Hazard. Mater. 187 (1-3), 341–347.

Wu, J.-T., Wang, L., Zhao, L., Huang, X.-C., Ma, F., 2020. Arbuscular mycorrhizal fungi effect growth and photosynthesis of Phragmites australis (Cav.) Trin ex. Steudel under copper stress. Plant Biol. 22 (1), 62–69.

Wu, N., Zhang, S., Huang, H., Christie, P., 2008a. Enhanced dissipation of phenanthrene in spiked soil by arbuscular mycorrhizal alfalfa combined with a non-ionic surfactant amendment. Sci. Total Environ. 394 (2–3), 230–236.

Wu, N., Zhang, S., Huang, H., Shan, X., Christie, P., Wang, Y., 2008b. DDT uptake by arbuscular mycorrhizal alfalfa and depletion in soil as influenced by soil application of a non-ionic surfactant. Environ. Pollut. 151 (3), 569–575.

Wu, N., Huang, H., Zhang, S., Zhu, Y.G., Christie, P., Zhang, Y., 2009. Phenanthrene uptake by *Medicago sativa* L. under the influence of an arbuscular mycorrhizal fungus. Environ. Pollut. 157 (5), 1613–1618.

Wu, Q-S., Xia, R-X., Zou, Y-N., 2008. Improved soil structure and citrus growth after inoculation with three arbuscular mycorrhizal fungi under drought stress. Eur. J. Soil Biol. 44 (1), 122–128.

Wu, Q-S., 2017. Arbuscular Mycorrhizas and Stress Tolerance of Plants. Springer Nature, Singapore.

Wu, S-L., Chen, B-D., Sun, Y-Q., Ren, B-H., Zhang, X., Wang, Y-S., 2014. Chromium resistance of dandelion (*Taraxacum platypecidum* Diels.) and bermudagrass (*Cynodon dactylon* [Linn.] Pers.) is enhanced by arbuscular mycorrhiza in Cr(VI)-contaminated soils. Environ. Toxicol. Chem. 33 (9), 2105–2113.

Wu, S., Zhang, X., Sun, Y., Wu, Z., Li, T., Hu, Y., Su, D., Lv, J., Li, G., Zhang, Z., Zheng, L., Zhang, J., Chen, B., 2015. Transformation and immobilization of chromium by arbuscular mycorrhizal fungi as revealed by SEM–EDS, TEM–EDS, and XAFS. Environ. Sci. Technol. 49 (24), 14036–14047.

Wu, S., Zhang, X., Chen, B., Wu, Z., Li, T., Hu, Y., Sun, Y., Wang, Y., 2016. Chromium immobilization by extraradical mycelium of arbuscular mycorrhiza contributes to plant chromium tolerance. Environ. Exp. Bot. 122, 10–18.

Wubs, E.J., Van der Putten, W.H., Bosch, M., Bezemer, T.M., 2016. Soil inoculation steers restoration of terrestrial ecosystems. Nat. Plants 2 (8), 1–5.

Xun, F., Xie, B., Liu, S., Guo, C., 2015. Effect of plant growth-promoting bacteria (PGPR) and arbuscular mycorrhizal fungi (AMF) inoculation on oats in saline-alkali soil contaminated by petroleum to enhance phytoremediation. Environ. Sci. Pollut. Res. 22 (1), 598–608.

Yang, Y., Han, X., Liang, Y., Ghosh, A., Chen, J., Tang, M., 2015. The combined effects of arbuscular mycorrhizal fungi (AMF) and lead (Pb) stress on Pb accumulation, plant growth parameters, photosynthesis, and antioxidant enzymes in Robinia pseudoacacia L. PloS One 10 (12), e0145726. doi:10.1371/journal.pone.0145726.

Yang, Y., He, C., Huang, L., Ban, Y., Tang, M., 2017. The effects of arbuscular mycorrhizal fungi on glomalin-related soil protein distribution, aggregate stability and their relationships with soil properties at different soil depths in lead-zinc contaminated area. PloS One 12 (8), e0182264. doi:10.1371/journal.pone.0182264.

Yao, Q., Yang, R., Long, L., Zhu, H., 2014. Phosphate application enhances the resistance of arbuscular mycorrhizae in clover plants to cadmium via polyphosphate accumulation in fungal hyphae. Environ. Exp. Bot. 108, 63–70.
Yoon, J., Cao, X., Zhou, Q., Ma, L.Q., 2006. Accumulation of Pb, Cu, and Zn in native plants growing on a contaminated Florida site. Sci. Total Environ. 368 (2–3), 456–464.
Yu, X-Z., Wu, S-C., Wu, F-Y., Wong, M-H., 2011. Enhanced dissipation of PAHs from soil using mycorrhizal ryegrass and PAH-degrading bacteria. J. Hazard. Mater. 186 (2–3), 1206–1217.
Zarei, M., Hempel, S., Wubet, T., Schäfer, T., Savaghebi, G., Jouzani, G.S., Nekouei, M.K., Buscot, F., 2010. Molecular diversity of arbuscular mycorrhizal fungi in relation to soil chemical properties and heavy metal contamination. Environ. Pollut. 158 (8), 2757–2765.
Zeng, Y-C., Li, Q-L., Gao, Y-Z., Ling, W-T., Xiao, M., 2010. Effects of arbuscular mycorrhiza (AM) on residual and forms of polycyclic aromatic hydrocarbons in soils. Soils 42 (1), 106–110.
Zhang, J., Yin, R., Lin, X., Liu, W., Chen, R., Li, X., 2010. Interactive effect of biosurfactant and microorganism to enhance phytoremediation for removal of aged polycyclic aromatic hydrocarbons from contaminated soils. J. Health Sci. 56 (3), 257–266.
Zhang, L., Feng, G., Declerck, S., 2018. Signal beyond nutrient, fructose, exuded by an arbuscular mycorrhizal fungus triggers phytate mineralization by a phosphate solubilizing bacterium. ISME J 12 (10), 2339–2351.
Zhang, X., Ren, B-H., Wu, S-L., Sun, Y-Q., Lin, G., Chen, B-D., 2015. Arbuscular mycorrhizal symbiosis influences arsenic accumulation and speciation in *Medicago truncatula* L. in arsenic-contaminated soil. Chemosphere 119, 224–230.
Zhang, Z., Wang, Q., Wang, H., Nie, S., Liang, Z., 2017. Effects of soil salinity on the content, composition, and ion binding capacity of glomalin-related soil protein (GRSP). Sci. Total Environ. 581, 657–665.
Zhou, J.L., 1999. Zn biosorption by *Rhizopus arrhizus* and other fungi. Appl. Microbiol. Biotechnol. 51 (5), 686–693.
Zhou, X-B., Zhou, J., Xiang, X., Cébron, A., Béguiristain, T., Leyval, C., 2013. Impact of four plant species and arbuscular mycorrhizal (AM) fungi on polycyclic aromatic hydrocarbon (PAH) dissipation in spiked soil. Pol. J. Environ. Stud. 22 (4), 1239–1245.

Further Reading

Raveau, R., Fontaine, J., Lounès-Hadj Sahraoui, A., 2020a. Essential oils as potential alternative biocontrol products against plant pathogens and weeds: A review. foods 9 (3), 365.

CHAPTER

Biochar assisted phytoremediation for metal(loid) contaminated soils

4

Manhattan Lebrun, Romain Nandillon, Florie Miard, Sylvain Bourgerie, Domenico Morabito

University of Orleans, Orléans, France

4.1 Introduction

Europe has 2.5 million contaminated sites. In particular, soil pollution by metals and metalloids (referred to as metal(loid)s hereafter) is an important issue, making up 35% of European contaminated soils (Van Liedekerke et al., 2014, Paya Perez and Rodriguez Eugenio 2018). The sources of such pollution are waste disposal, industrial and commercial activities, military activity, storage, transport and spills on land, as well as nuclear operations (Panagos et al., 2013). Due to the negative effects of metal(loid) polluted soils on both the environment and human health, such sites need to be remediated. One way to do that is through phytoremediation, which is defined as the use of plants and associated microorganisms to reduce the toxic effects of polluted soils (Cristaldi et al., 2017).

Compared to conventional physical and chemical techniques, phytoremediation is an environmentally friendly and low-cost remediation technique, which is also accepted by the public (Pandey et al., 2015; Pandey and Bajpai, 2019). Phytoremediation is divided into two main processes.

The first one is phytoextraction. This process uses (hyper)accumulating plants that take up metal(loid)s via their roots and translocate them into their aerial and harvestable tissues. In general, concentrations in the upper parts are higher than in the roots. This process is a real decontamination, as it removes pollution from the soil; although it requires the harvest and proper disposal of the aerial tissues and is also limited to the root depth (Gomes et al., 2016; Khalid et al., 2016). The other process is phytostabilization, in which metal(loid)s are immobilized at the plant root zone, by absorption into the roots, adsorption onto the root surface and precipitation with root exudates. This technique is not a real decontamination of the soil, but rather an immobilization of the pollutants and stabilization of the soil in order to reduce the spread of contamination through wind erosion and water leaching (Liu et al., 2018; Khalid et al., 2016).

In terms of the remediation of former mining and industrial soils which are highly contaminated by metal(loid)s, extracting the entire contamination using phytoextraction is not feasible, as contamination is often present in high concentrations, and at varying depths in the soil. Therefore, phytostabilization is often preferred.

However, contaminated soils often have a poor fertility, with an extreme pH, and low nutrient and organic matter contents, which together with the elevated metal(loid) concentrations, hinder plant establishment and growth. Therefore, the application of an amendment can be required. Amending

Assisted Phytoremediation. DOI: https://doi.org/10.1016/B978-0-12-822893-7.00010-0

polluted soils will (1) supply nutrients and organic matter to both plants and microorganisms, (2) improve soil physico-chemical properties, and (3) immobilize metal(loid)s. One of the possible amendments which has been gathering attention over the last decade is biochar. The use of biochar will not only immobilize metal(loid)s, and ameliorate soil conditions and microorganism development, but will also sequester carbon and reduce greenhouse gas emissions. This technique is called biochar assisted phytostabilization, and is represented in Fig. 4.1: before amendment application, the soil is bare from vegetation and thus subjected to water leaching and wind erosion, endangering the proximate area and population. After biochar is added to the soil, soil conditions are improved and plants can develop, which limits pollution transfer and thus protects the surrounding environment.

This chapter will focus on the use of biochar in the phytostabilization of metal(loid) contaminated soils. First, we will define what biochar is and how it is obtained, then we will describe the main properties of biochar finishing with a discussion addressing the effects of biochar on the soil, metal(loid)s, microorganisms, and plant development.

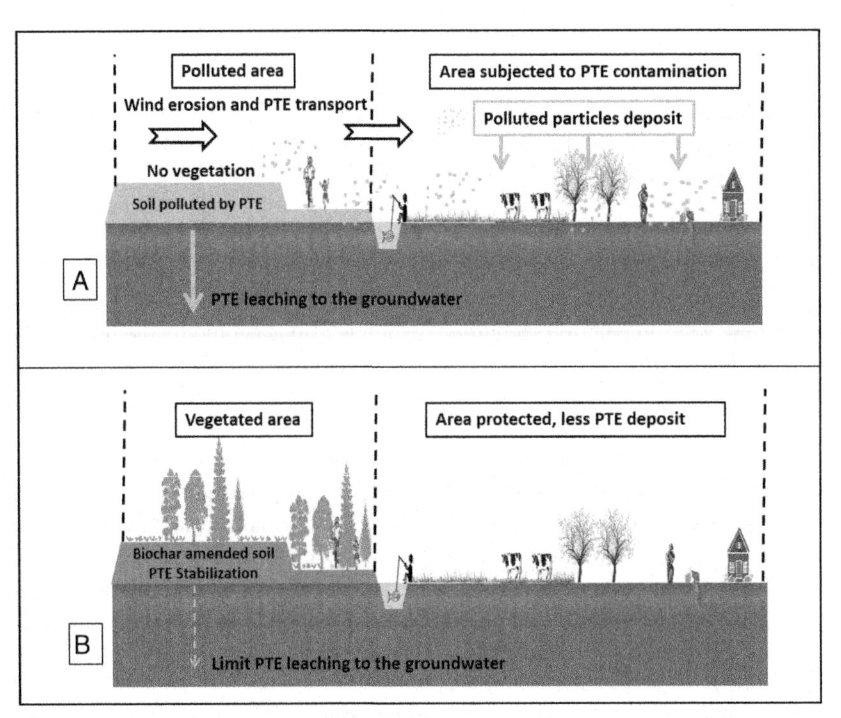

FIG. 4.1 Representation of a biochar assisted phytostabilization process in the goal of stabilizing a site contaminated by potentially toxic elements (PTE) such as metal(loid)s.

(A) Before amendment, the soil is highly contaminated by metal(loid)s and baren of vegetation, therefore it is subjected to wind erosion and water leaching, which contaminates the surrounding environment. In addition to this direct exposure of pollution, the metal(loid)s can enter the food chain (fish, animals, crops) and thus affect human health indirectly. (B) After biochar amendment, soil properties are improved and metal(loid)s immobilized, which allow the establishment of plants (trees and annual crops). Such plant cover will reduce the leaching and erosion of the contaminants, which will protect the surrounding area and thus the population.

4.2 What is biochar?

The origin of biochar goes back to ancient times, during which the Amerindian population made dark earth called *terra preta* (Forján et al., 2017). After its discovery, this *Terra preta* attracted attention of researchers due to its high fertility and elevated carbon content. Since discovering this fertile material of anthropogenic origin, people have been trying to understand its formation process and thus recreate it, leading to the emergence of biochar (Bezerra et al., 2016).

Nowadays, biochar is made through pyrolysis. Pyrolysis is defined as the thermal decomposition of organic materials under limited oxygen conditions, at temperatures generally between 200 and 1000°C (Janus et al., 2015; Yu et al., 2019). Other processes may also be used, such as hydrothermal carbonization, gasification, torrefaction and microwave-assisted pyrolysis. For instance, in hydrothermal carbonization, the material is heated at high temperatures (between 150 and 250°C) in the presence of water and in a confined environment, to give a product generally referred to as hydrochar (Yuan et al., 2017; Yaashikaa et al., 2020). Torrefaction is an emerging process similar to the pyrolysis, carried out under limited oxygen conditions, and with a slower heating rate. This process removes the moisture, oxygen and carbon dioxide from the original biomass (Yuan et al., 2017; Yaashikaa et al., 2020). In microwave-assisted pyrolysis, the residence time is reduced compared to conventional fast and slow pyrolysis and the process requires less energy and is thus less expensive. Moreover, the biochar produced is of higher quality, with a higher surface area and pore volume (Motasemiand Afzal 2013; Li et al., 2016).

Almost any organic material can be pyrolyzed, including crop residues, forestry residues, urban wastes, by-products of the agro-processing industry, animal manure and sewage sludge (Kuppusamy et al., 2016). The type of feedstock that will be processed in the pyrolyzer will greatly affect the properties of the biochar produced.

After pyrolysis, three products are obtained: bio-oil, non-condensable gas, and biochar, which is a solid (Cha et al., 2016; Janus et al., 2015). The proportion of these three products will depend on the pyrolysis process. Indeed, there are three pyrolysis processes: slow pyrolysis, fast pyrolysis and gasification (Qambrani et al., 2017). The slow pyrolysis is characterized by a long residence time (minutes to days), a temperature below 600°C, and a slow heating rate. It produces the same amount of solid, gas and liquid end-products (Qambrani et al., 2017; Qian et al., 2015). For fast pyrolysis, the residence time is shorter (a few seconds), and the heating rate faster. It principally produces bio-oil (around 75%) and lower amounts of gas and biochar (10% to 20%) (Qambrani et al., 2017; Qian et al., 2015). Finally, gasification uses very high temperatures of between 800 and 1000°C, and only yield 10% of biochar and 5% of bio-oil with the organic material principally being broken down into H, CO and CO_2 (Qambrani et al., 2017; Qian et al., 2015). Around 41 million tonnes of biochar are produced annually using these three pyrolysis conditions (Kuppusamy et al., 2016).

Biochar is relatively stable in the environment, having a residence time in the soil of several thousand years (Kuppusamy et al., 2016). It has many applications but its main four areas of use are: (1) waste management, (2) energy production, (3) environmental management, such as reduction of soil and water pollution and increases in soil fertility, and (4) climate change mitigation (Forján et al., 2017).

4.3 What are the properties of biochar?

Biochar is a heterogeneous product and its properties depend on the feedstock used to produce it as well as the pyrolysis parameters, such as heating rate, temperature and residence time (Janus et al., 2015; Shaaban et al., 2018). However, biochars are usually characterized by an alkaline pH, a high surface area and elevated carbon contents, while the H, N and O contents are lower. Table 4.1 list

Table 4.1 Physico-chemical porperties of biochar made from diverse feedstocks and harboring different particle sizes.

Code	Feedstock	Particle size (mm)	pH	EC (μS cm^{-1})	WHC (%)	Specific area ($m^2 g^{-1}$)	Total pore volume (cm3 g^{-1})	Mean pore diameter (nm)	CEC (cmol + kg^{-1})	C%	H%	N%
HW1	Hardwood	<0.1	9.53 ± 0.01	669 ± 9	160 ± 1	223	0.1204	2.16	4.56	80.19 ± 0.84	3.27 ± 0.21	2.63 ± 0.76
HW2	Hardwood	0.2–0.4	8.98 ± 0.02	432 ± 2	183 ± 3	43.91	0.0364	3.32	2.46	96.37 ± 3.97	2.11 ± 0.05	1.19 ± 0.11
HW3	Hardwood	0.5–1	8.46 ± 0.01	302 ± 1	212 ± 4	4.38	0.01	9.13	< 1.05	78.72 ± 1.13	1.74 ± 0.07	2.38 ± 0.78
HW4	Hardwood	1–2.5	8.81 ± 0.02	348 ± 7	200 ± 3	15.53	0.0161	4.16	< 1.03	ND	ND	ND
PW1	Pinewood	<0.1	9.60 ± 0.02	407 ± 2	108 ± 12	234.82	0.1221	2.0797	8.96	ND	ND	ND
PW2	Pinewood	0.2–0.4	8.25 ± 0.01	496 ± 2	308 ± 4	21.245	0.01711	3.2215	2.78	ND	ND	ND
LW1	Lightwood	<0.1	9.66 ± 0.01	1047 ± 2	194 ± 3	255.35	0.1751	2.64	11.7	ND	ND	ND
LW2	Lightwood	0.2–0.4	9.56 ± 0.01	849 ± 4	194 ± 2	251.42	0.1486	2.3639	NA	ND	ND	ND
BS2	Bark-sap	0.5–1	8.1 ± 0	162 ± 2	ND	1.5506	0.0093393	24.093	ND	81.80 ± 3.65	3.61 ± 0.29	1.07 ± 0.06
BA	Bamboo	ND	10.6 ± 0.1	2629 ± 175	ND	14.516	0.018552	5.1122	ND	73.45 ± 2.07	2.30 ± 0.08	2.14 ± 0.15
Ba1	Bark oak	0.2–0.4	9.01 ± 0.2	814 ± 10	ND	99.07	0.0973	3.9289	ND	78.32 ± 9.25	1.73 ± 0.10	0.71 ± 0.12
Ba2	Bark oak	0.5–1	8.88 ± 0.02	729 ± 7	ND	103.45	0.1001	3.8721	ND	69.06 ± 2.32	1.85 ± 0.08	1.00 ± 0.25
Ba3	Bark oak	1–2.5	8.64 ± 0.07	633 ± 34	ND	95.72	0.1022	4.2706	ND	60.99 ± 6.28	1.59 ± 0.22	0.88 ± 0.32
Sa1	Sapwood OAK	0.2–0.4	9.50 ± 0.08	658 ± 58	ND	352.39	0.1617	1.8356	ND	84.90 ± 0.65	2.75 ± 0.16	1.95 ± 0.06
Sa2	Sapwood oak	0.5–1	9.27 ± 0.04	681 ± 9	ND	294.97	0.1558	2.1125	ND	89.70 ± 2.00	2.49 ± 0.03	0.43 ± 0.12
Sa3	Sapwood oak	1–2.5	9.31 ± 0.08	511 ± 51	ND	324.25	0.1588	1.9585	ND	79.63 ± 10.13	2.52 ± 0.10	0.32 ± 0.11
He1	Heartwood oak	0.2–0.4	8.56 ± 0.06	208 ± 7	ND	409.24	0.1854	1.8125	ND	88.52 ± 1.41	2.37 ± 0.11	0.41 ± 0.04
He2	Heartwood oak	0.5–1	8.34 ± 0.11	173 ± 5	ND	403.53	0.1814	1.7983	ND	88.58 ± 2.64	2.31 ± 0.07	0.24 ± 0.03
He3	Heartwood oak	1–2.5	7.95 ± 0.04	191 ± 7	ND	394.48	0.1793	1.8179	ND	70.80 ± 14.44	2.28 ± 0.02	0.18 ± 0.11
VT	Pinewood	ND	8.2	900	85	ND	ND	ND	ND	ND	ND	ND

CEC, cation exchange capacity; EC, electrical conductivity; ND, not determined; WHC, water holding capacity.
Reproduced with permission from Lebrun et al. (2017, 2018b, c, 2020a, 2020b).

the physico-chemical properties of different biochars used during previous studies on metal(loid) contaminated soils: a former mine technosol highly contaminated with As and Pb taken from Pontgibaud (Auvergne-Rhône-Alpes, France), a former fertilizer production site located in Issoudun (Centre-Val-de-Loire, France) contaminated with As, Cu, Pb, and Zn, the mining site of La Petite Faye (Massif Central, France) highly contaminated with As, Sb and Pb, and the former smelting site of Mortagne-du-Nord (Haut-de-France, France) contaminated with Cd, Pb, and Zn.

As can be seen from Table 4.1, biochar is generally alkaline (Chen et al., 2019); all the biochars listed in Table 4.1 had a pH between 7.95 and 10.6, for a heartwood biochar (1–2.5 mm particle size, 500°C pyrolysis temperature) and a bamboo biochar (500°C pyrolysis temperature), respectively. Such alkalinity is due to the decomposition of organic substances, and ash formed during the pyrolysis (Chen et al., 2019; Huang et al., 2017b), as well as the presence of organic functional groups, carbonates and inorganic alkalis on the surface of biochar (Lee et al., 2013). Biochar's alkaline pH is an important property, especially when it is applied to acidic mine soils. However, some biochars showed a neutral or acidic pH. For instance, Zhang et al. (2016) used a biochar made from wildfire with a pH of 3.1, and sewage sludge biochars produced at 300 and 400°C had a pH of 5.32 and 4.87, respectively (Hossain et al., 2011) and a peanut hull biochar had a pH of 6.9 (Zhou et al., 2013).

Biochar pH is influenced by both the feedstock and the pyrolysis temperature. In the study of Brennan et al. (2014), both olive tree pruning biochar and pine woodchip biochar were produced at 450°C, but presented different pH values, alkaline for the olive tree pruning biochar (9.34) and around neutral for the pine woodchip biochar (7.52). Similarly, four biochars were produced at 650°C from different feedstocks: chicken manure, rice husk, sawdust and sugarcane straw. They had pH at 9.96, 8.72, 7.48 and 9.17, respectively (Higashikawa et al., 2016). Finally, when pyrolyzed at 500°C, chicken manure produced a very alkaline biochar (pH 9.1), whereas *Hibiscus cannabinus* and sewage sludge produced neutral biochars, with pH values of 7.5 and 7.1, respectively (Huang et al., 2018).

On a smaller scale, even the type of tissue used can lead to a different pH. For instance, biochars made from different tissues of oak showed different pH values: the biochars made from the bark tissue had a pH of 8.84 on average, those from sapwood 9.36 on average, and those from heartwood 8.28 on average (Table 4.1) (Lebrun et al., 2020a). Regarding the temperature effect, it is generally known that increasing the pyrolysis temperature leads to more alkaline biochars. For instance, sewage sludge pyrolyzed at 300°C and 400°C produced acidic biochars (pH 5.32 and pH 4.87, respectively), whereas when pyrolyzed at 500°C, biochar pH was neutral (pH 7.27) (Hossain et al., 2011). Similarly, in the study of Al-Wabel et al. (2017), date palm pyrolyzed at increasing temperatures led to the production of biochars with increasing pH: 8.32, 9.59, and 11.5 at 300°C, 500°C and 700°C, respectively.

Finally, in addition to these two parameters, biochar particle size can affect biochar parameters. This is illustrated in Table 4.1. In general, the biochar pH decreased with increasing particle size. Hardwood biochars produced by the same pyrolysis process had a pH of 9.53, 8.98, 8.46 and 8.81, after sieving to obtain particle sizes of less than 0.1 mm, between 0.2 and 0.4 mm, between 0.5 and 1 mm and between 1 and 2.5 mm, respectively (Table 4.1) (Lebrun et al., 2018b). Similarly, pinewood biochars harboring particle sizes of < 0.1 mm and 0.2-0.4 mm had pH of 9.60 and 8.25, respectively (Lebrun et al., 2018c), whereas the bark biochars had a pH of 9.01, 8.88 and 8.64, for particle sizes of 0.2-0.4 mm, 0.5-1 mm and 1-2.5 mm, respectively (Lebrun et al., 2020a). A higher pH with small particle sizes may be related to the ash content of the biochar. Guo and Chen (2014) observed that biochar ash content increased with decreasing particle size, and a greater ash content induces a higher biochar pH (Rehrah et al., 2016; Oliveira et al., 2017; Tan et al., 2017b).

Another important parameter is the electrical conductivity (EC) of biochar. The biochars listed in Table 4.1 had EC values between 191 and 2629 $\mu S\ cm^{-1}$. Much the same as for the pH, biochar EC was found to increase with increasing pyrolysis temperature (Cantrell et al., 2012; Al-Wabel et al., 2017), although, as observed by Higashikawa et al. (2016) and Hossain et al. (2011), this relation is not as strong as for pH. However, biochar EC is greatly influenced by the feedstock used (Cantrell et al., 2012; Higashikawa et al., 2016) as well as the organ tissue used for biochar production (Table 4.1) (Lebrun et al., 2020a), and generally decreases when the particle size increases (Table 4.1) (Lebrun et al., 2018b, c, 2020a).

Biochar water holding capacity is high due to its porous structure (Forján et al., 2017; Zhang et al., 2016). Biochar can retain several times its weight in water, as shown in Table 4.1. Biochars in Table 4.1 had a WHC between 108% (pinewood biochar, <0.1 mm) and 308% (pinewood biochar, 0.2–0.4 mm). Similarly, Mary et al. (2016) characterized three biochars made from pea pods, cauliflower leaves, and orange peel and measured WHCs of 200%, 200%, and 132%, respectively.

The principal element in the composition of biochar is carbon, accounting for at least 60% (Chen et al., 2019), and mainly aromatic, in addition to smaller amounts of hydrogen, nitrogen, and oxygen (Chen et al., 2019). Biochar also contains small amounts of phosphorous, calcium, aluminum, potassium, copper, iron, zinc and manganese (Yuan et al., 2018; Yu et al., 2019), which can be important for plant nutrition. The elemental composition of a biochar is also influenced by its feedstock (Brennan et al., 2014; Chen et al., 2010; Kwak et al., 2019) and the temperature of pyrolysis (Al-Wabel et al., 2017; Hass et al., 2012; Jin et al., 2014; Kwak et al., 2019).

Another property which characterizes biochar is its surface area, which varies from a few to hundreds of square meters per gram of biochar (Chen et al., 2019) (Table 4.1). As observed for the other parameters, biochar surface area depends on pyrolysis temperature. For instance, biochars made from dairy manure had a surface area of 1.64 $m^2\ g^{-1}$ when produced at 350°C, whereas its surface area was higher at 700°C (186.5 $m^2\ g^{-1}$) (Cantrell et al., 2012). Similarly, Jindo et al. (2014) observed that the surface area of the biochars increased with increasing pyrolysis temperature, for all the feedstock used that is, apple tree branch, oak tree, rice husk, and rice straw. Moreover, they also showed that surface area was highly dependent on the biochar feedstock: the biochars made from oak trees presented the lowest surface area, followed by apple tree branch biochars, rice straw biochars, and rice husk biochars, which had the highest surface area (Jindo et al., 2014). Such feedstock and temperature effects are related to the composition of the original feedstock, in terms of cellulose, hemicellulose and lignin proportions (Zhao et al., 2013; Gai et al., 2014), which decompose at different temperatures. Moreover, a higher pyrolysis temperature removes more volatile material which results in a higher micropore volume and thus a higher surface area (Ahmad et al., 2012).

The large surface area is accompanied by a total pore volume of between 0.01 and 0.15 cm^3 per gram and a mean pore diameter of several nanometers (Table 4.1). Finally, the surface of biochar presents a large number of functional groups, such as carbonyl, carboxyl and hydroxyl (Chen et al., 2019), which are mainly negatively charged. Such oxygen-containing functional groups are important for the biochar to be able to sorb cation metals.

Overall, it can be concluded that different biochars have some common properties. It is generally alkaline, with a high carbon content and a large surface area, and has oxygen-containing functional groups on its surface. Such properties depend on the feedstock used, the pyrolysis temperature, and the biochar's particle size. These properties make biochar a good soil amendment for the stabilization of contaminated soils, improving soil conditions, immobilizing metal(loid)s, ameliorating microbial activity and plant growth. This allows phytomanagement strategies to be implemented in order to recycle abandoned industrial sites economically and make them less polluting.

4.4 The effects of biochar application on soil and soil pore water properties

After its application to the soil, biochar modifies several soil properties, such as pH, electrical conductivity, organic carbon (OC) and matter (OM) contents, water holding capacity, cation exchange capacity and nutrient content and availability.

4.4.1 Soil pH

Soil pH is an important parameter. It influences many soil properties and affects nutrient and metal(loid) behavior. For instance, cations are more available at acidic pH whereas anion mobility is higher at alkaline pH.

Biochar has been shown in many studies to induce a pH increase after being applied to contaminated soils. For instance, Kelly et al. (2014) observed that soil pH increased following the addition of a beetle-killed lodge pine biochar to a soil which was multi-contaminated due to the from mine extraction of gold, silver, lead and zinc, while Huang et al. (2018) found that the addition of chicken manure biochar to a Pb/Zn tailing, contaminated with high concentrations in Pb, Zn, Cu, Cd and As, increased soil pH.Mokarram-Kashtiban et al. (2019) applied hornbeam biochar to an uncontaminated soil spiked with Cd, Cu and Pb, which induced an increase in soil pH. Finally, a pinewood biochar increased the pH of a former mine soil contaminated by As and Pb (Lebrun et al., 2017).

Several mechanisms can explain such an increase in soil pH following a biochar amendment (Hmid et al., 2015; Xu et al., 2017; Meng et al., 2018; Dai et al., 2018). Biochar is an alkaline material, and therefore its incorporation into the soil induces a liming effect, especially when the soil is very acidic and has a low buffering capacity. In addition, biochar has a high ash content, composed of alkaline mineral elements such as carbonates and oxides. These elements are released into the soil sand can react with H^+ ions and thus reduce the amount of free H^+ ions, thereby increasing the soil pH. Finally, the functional groups on the surface of the biochar are mainly negatively charged (carbonyls, phenols, carboxyls, pyrones), and can thus consume protons.

However, the extent of the biochar-induced pH increase depends on the initial soil pH. For example, when the same pinewood biochar was used on different soils, as shown in Fig. 4.2, it induced a 2.2 and 2.9 unit pH increase when applied at 2% and 5%, respectively, to a former mine soil, whereas when applied at the same rate to the former industrial site of Mortagne-du-Nord it only led to a 0.4 and 0.5 unit increase in pH and had no effect on the alkaline former industrial site of Issoudun (Lebrun et al., 2017; Lomaglio et al., 2018; Lebrun et al., 2018a). The three soils had different initial pHs: the former mine was highly acidic, whereas the industrial site of Mortagne-du-Nord had a slightly acidic pH, and that of Issoudun was already alkaline, which could explain the differences observed in these studies.

The type of biochar applied to the soil can also influence how it will affect soil pH. For instance, several biochars were applied to the same former mine site of Pontgibaud. They were produced from different biomasses, *i.e.* hardwood, pinewood, lightwood, oak bark, oak sapwood, oak heartwood. These biochars generally increased soil pH (Fig. 4.2). However, of these biochars, the one produced

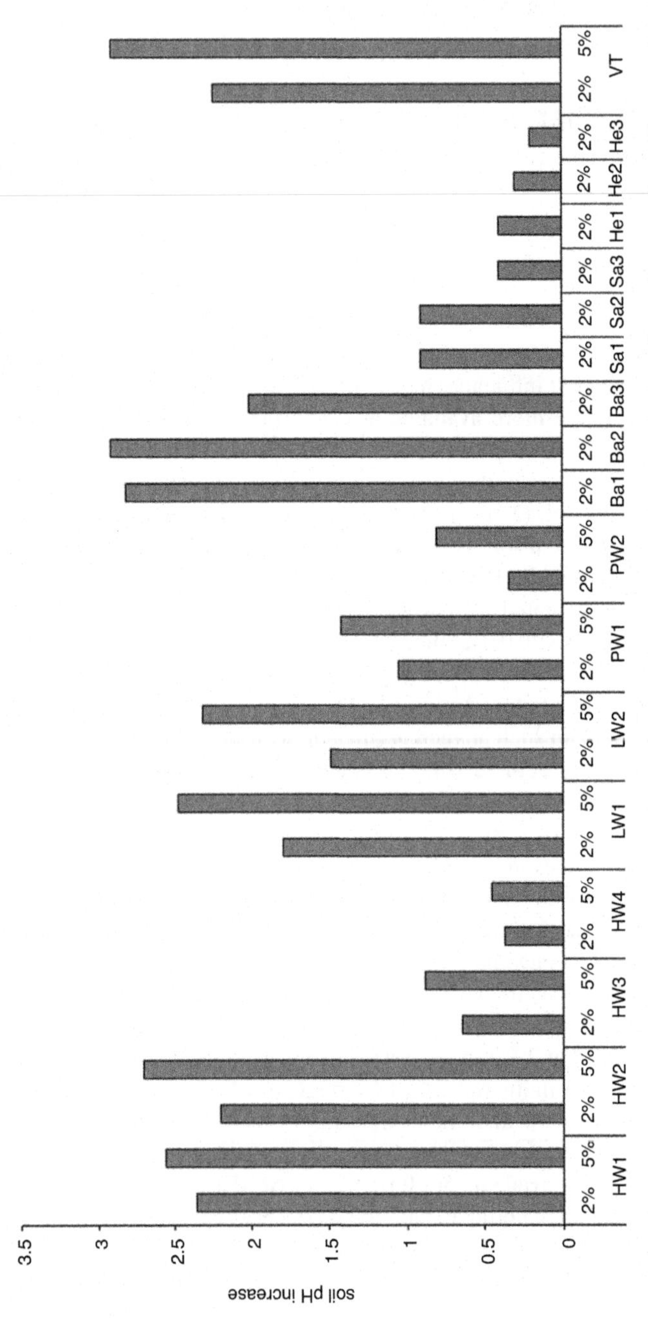

FIG. 4.2 Extent of soil pore water pH increase after the addition of different biochars to the same former mine soil contaminated by arsenic and lead and having an acidic pH (4.6 on average).

The different biochars were applied either at 2% or 5% and the soil pore waters were sampled after a few days of maturation, before any vegetalization was performed. The code meaning is given in Table 4.1. *Reproduced with permission from (Lebrun et al. 2017, Lebrun et al. 2018b, c, 2020a).*

from oak bark induced the highest pH increase (more than 2 units (Lebrun et al., 2020a)) whereas the lowest increase was measured for the heartwood and hardwood biochars (Lebrun et al., 2018b, 2020a). Moreover, when biochar has a low particle size, it induces a higher increase in soil pH than a coarser biochar, as demonstrated in Lebrun et al. (2018b, c). Al-Wabel et al. (2017) observed that, depending on the temperature at which the biochar was produced, it had different effects on soil pH when applied to gold mine soil polluted with Cd, Cu, Pb, Zn, Mn, and Fe: at 300°C, date palm biochar decreased soil pH, whereas at 500°C and 700°C, it increased soil pH.

Finally, the increase in soil pH tends to be higher when biochar is added at a higher dose, as can be seen in Fig. 4.2.

4.4.2 Soil electrical conductivity

The electrical conductivity is generally low on polluted soils (Namgay et al., 2010; Olmo et al., 2014; Lebrun et al., 2017). When biochar is applied to a contaminated soil, soil electrical conductivity was observed to increase. For instance, Hossain et al. (2010) observed a six-fold EC increase following biochar application to a non-contaminated soil, whereas Lebrun et al. (2017) measured a two-fold increase in soil pore water (SPW) EC when biochar was applied to a former silver-lead extraction mine. Such an increase in soil electrical conductivity following biochar application can be explained by the elevated electrical conductivity of the biochar itself (Table 4.1), as well as the presence of soluble salts and cations on the biochar surface that are dissolved into the soil solution (Hmid et al., 2015; Nigussie et al., 2012).

As was found for the pH, the extent by which biochar increases soil electrical conductivity depends on the type and rate of the biochar applied. For instance, the pinewood VT biochar (Table 4.1) induced a two-fold increase in electrical conductivity when applied to the former Pontgibaud mine technosol and a 1.5-fold rise when applied to the La Petite Faye technosol, whereas it had no effect on the industrial site of Issoudun (Lebrun et al., 2017; Lomaglio et al., 2017; Lebrun et al., 2018a), which could be related to the initial soil EC. On the Pontgibaud technosol, the addition of lightwood biochar increased SPW electrical conductivity by two to fivefold when applied at 2% and 5%, whereas the pinewood biochar PW (Table 4.1) only increased soil electrical conductivity by 1.4-fold on average, and only when applied at 5% (Lebrun et al., 2018c).

Finally, biochar particle size is also an important parameter influencing soil electrical conductivity increase. For instance, four hardwood biochars with increasing particle size were applied to the former mine site of Pontgibaud at 2% and 5%. The two finest biochars increased SPW electrical conductivity by 3.3 and 1.5 fold at 2% and 1.9 and 2.7 fold at 5%, whereas the coarser biochars had no effect, whatever the application rate. These two coarse biochars only increased soil electrical conductivity after 46 days of maturation (Lebrun et al., 2018b), which could be related to their lower electrical conductivity compared to the other two fine biochars as well as a lower surface area, resulting in lower exchange.

4.4.3 Soil organic carbon and organic matter contents

The soil organic matter is defined as the sum of the carbon containing compounds. Organic matter also contains mineral elements that are required for plant growth. To be considered fertile, a soil should contain between 2% and 8% organic matter (Pettit 2004). However, most contaminated soils have organic matter levels below this range. Organic matter has other functions within the soil. It aerates the soil and helps retain water and nutrients. It also provides substrates for soil microorganisms (Agegnehu et al., 2015). Moreover, a soil with a higher organic matter content has a higher capacity to retain metals (Forján et al., 2016).

Due to its elevated organic carbon and organic matter levels, biochar is known to increase the organic carbon and organic matter contents of contaminated soils (Janus et al., 2015). For instance, the addition of lightwood biochar (0.2–0.4 mm particle size) increased soil organic matter (SOM) content from 1.36% to 2.77% and 4.94% when applied at 2% and 5% to the Pontgibaud mine soil, respectively (Lebrun et al., 2018c). On the same soil, pinewood biochar (0.2–0.4 mm particle size) led to a high increase in OM content, up to 2.88% and 5.08%, respectively (Lebrun et al., 2018c). However, another pinewood biochar, VT (Table 4.1), did not affect the dissolved organic carbon content of two contaminated soils (the smelting site of Mortagne-du-Nord (Lomaglio et al., 2018) and the gold mine of La Petite Faye (Lomaglio et al., 2017)), while it decreased the content when applied to the silver-lead mine of Pontgibaud (Lebrun et al., 2017). This was probably due to an improvement in the microbial activity, inducing organic carbon degradation (Hass et al., 2012). In 2018, Li et al. (2018), applied a rice straw biochar to a Cd and Pb contaminated soil and measured an increase in SOM content, which was higher at 5% (2.3-fold) than at 2.5% (1.7-fold). Similarly, the application of wheat straw biochar to a Cd contaminated soil had no effect on soil organic carbon (SOC) content when applied at 10 $t.ha^{-1}$, whereas it increased SOC content at 20 and 40 t ha^{-1} (Cui et al., 2012).

4.4.4 Soil water holding capacity

Plants take up the water, and associated nutrients from the soil that they need for their growth, and therefore the ability of a soil to maintain an efficient water holding capacity is important for plant growth. However, many contaminated soils present a low water holding capacity, due to the fact that they result from the crushing of ores and thus do not have a sufficient aggregation structure. When added to the soil, biochar can increase the soil water holding capacity, due to its high sorption capacity, its porous structure, and the increase in soil aggregation (Obia et al., 2016). For instance, lightwood and pinewood biochars applied at 5% increased the water holding capacity of a former mine technosol (Lebrun et al., 2018c). Karhu et al. (2011) observed an increase in the WHC of a non-contaminated soil following the addition of a birch biochar, while Fellet et al. (2011) found that the water content of a former mine contaminated with Pb and Zn increased following the addition of 10% prune residue biochar.Molnar et al. (2016) measured the WHC of an acidic sandy agricultural soil to be 25%, which increased to 29.7% following application of grain husk and paper fiber sludge biochar. However, such an increase is not always observed. For instance, among the four particle sizes of hardwood biochars tested, the finest one had no effect, whereas the other three increased the soil water holding capacity but only when applied at 5% (Lebrun et al., 2018b). This observation can probably be related to the pore diameter of the biochar: larger pores can retain more water.

However, in some studies, biochar did not affect soil WHC. For instance, Jeffery et al. (2015) studied the structure and effects of different biochars and based on their pore structure, biochars should be able to retain water and thus increase soil WHC. But they did not observe these beneficial effects, similarly to the study of Gray et al. (2014). The non-effect of biochar on soil WHC has been linked to its hydrophobicity. The hydrophobicity level of a biochar will depend on its composition: biochars with low hydrogen and oxygen contents generally have a high hydrophobicity, and the presence of a high number of aromatic compounds on the surface of biochar increases its hydrophobic character (Batista et al., 2018). This hydrophobicity creates negative capillary forces and thus prevents the water from infiltrating the biochar pores (Jeffery et al., 2015; Gray et al., 2014).

4.4.5 Soil cation exchange capacity

The cation exchange capacity of a soil corresponds to the total capacity of a material to hold exchangeable cations (Dai et al., 2017), which gives information on the soil sorption capacity towards metal(loid)s.

Biochar is known to have a high cation exchange capacity, as well as a high surface area and negative charges on its surface, which are properties that make biochar capable of increasing soil cation exchange capacity (Yuan et al., 2018). For instance, the addition of a baby corn peel biochar increased the CEC of an acid Arenosol (Rafael et al., 2019), cow manure biochar also increased the CEC of a sandy soil (Uzoma et al., 2011), and the amendment of a Cr contaminated soil with maize stalk biochar increased the soil CEC (Nigussie et al., 2012). The application of a woody biochar to a serpentine soil increased soil CEC. The CEC increase was greater when the application rate was increased (Herath et al., 2014).

The first mechanism by which biochar can increase soil CEC is through the increase in soil pH, as the CEC has been shown to be proportional to the pH (El Naggar et al., 2018). Other mechanisms have been proposed to explain the effect of biochar on soil CEC (Hailegraw et al., 2019): the high surface area, biochar's negative surface charge, the release of carbonates, the oxygenated functional groups on the surface of the biochar, and the adsorption of oxides by the biochar.

However, this beneficial effect is not always observed. For instance, the application of hardwood biochar to a former mine soil did not affect the soil CEC (Nandillon et al., 2019a). This could be due to the relatively low CEC of the applied biochar, as biochars produced at low temperatures generally have a low CEC (Janus et al., 2018).

4.4.6 Soil nutrients

Nutrients, such as nitrogen, phosphorus and potassium, are necessary for plant growth. Unfortunately, contaminated soils generally have a low nutrient content, which are also poorly available (Nandillon et al., 2019a).

Previous studies demonstrated biochar's ability to increase nutrient content and availability in contaminated soils. Nandillon et al. (2019a) observed that hardwood biochar increased available P and available K contents by 3 and 2.7 times, respectively when added at 5% to a former mine soil. Olmo et al. (2014) measured a 1.6-fold increase in soil N and P contents following the addition of an olive tree pruning biochar amendment on a field experiment. Finally, Houben et al. (2013) applied a *Miscanthus straw* biochar to a multi-contaminated soil and measured an increase in available of Ca, K, Mg and P contents. Biochar can ameliorate nutrient content and availability either directly or indirectly. Biochar also contains nutrients which are added to the soil directly (Ahmad et al., 2017; Nigussie et al., 2012; Arienzo et al., 2009). Biochar can also indirectly influence nutrient availability by modifying soil pH. For instance, the availability of phosphorus is dependent on soil pH: phosphorus can form complexes with Al and Fe when soil pH is acidic, while its availability is optimal at neutral pH (El Naggar et al., 2020). Finally, biochar is a source of organic carbon, which will favor the development of microorganisms involved in the mineralization of soil organic matter. This organic matter will be a source of nutrients available for the development of plants.

However, these nutrients released into the soil pore water can be leached and thus lost. Moreover, due to its elevated sorption capacity, biochar can also decrease nutrient availability. Indeed, several studies analyzed the sorption capacities of biochar and observed that biochar can sorb nitrate, ammonium and phosphate (Bakshi et al., 2014; Gai et al., 2014; Gao et al., 2015; Sarkhot et al., 2013; Yang et al., 2017; Zeng et al., 2013; Zhao et al., 2018), making them less leachable but also less available for plants.

4.4.7 Other soil properties

Although they are less studied, biochar can affect other soil properties, such as bulk density, soil aggregate, and total porosity. Indeed, the application of biochar was shown to decrease bulk density and increase soil aggregate and porosity in the studies of Burrell et al. (2016), Devereux et al. (2012), Masulili et al. (2010), and Obia et al. (2016). Several mechanisms can explain the observed decrease in bulk density. Firstly, as biochar has a low density, its addition to a soil with a higher bulk density induces a dilution effect, reducing soil bulk density (Burrell et al., 2016; Blanco-Canqui 2017). Secondly, the addition of biochar supports microbial growth and activity, which can improve soil aeration (Burrell et al., 2016). Similarly, the support of microbial activity can induce the formation of aggregates, increasing soil aggregation (Burrell et al., 2016). Finally, biochar can increase soil porosity due to its induced reduction of soil bulk density and increased soil aggregation, and also because it interacts with mineral soil particles and has a highly porous structure (Blanco-Canqui, 2017).

This section demonstrated the overall positive effect of biochar on the soil properties. Moreover, it also showed that the extent of the effects was dependent on the original soil, the biochar feedstock, and the biochar particle size as well as the biochar application rate. Therefore, it is important to choose the right biochar for each soil.

4.5 The effect of biochar on metals and metalloids

In general, polluted soils are contaminated by more than one element, and often by both metals and metalloids. The application of biochar to soil had different effects on the behavior of pollutants, depending on the element.

Biochar is very effective for the sorption of cations such as lead. Under acidic conditions, which is usually the case for polluted soils, Pb is very mobile and available but the application of biochar can reduce Pb mobility. For instance, when a pinewood biochar was added to the former Ag-Pb mine soil of Pontgibaud (France), Lebrun et al. (2017) measured a 69% and 97% decrease in soil pore water Pb concentrationat rates of 2% and 5%, respectively. This showed that with more biochar, more Pb was immobilized. Similarly, SPW Pb concentration decreased by 86% and 69% on average following the addition of two pinewood biochars and two lightwood biochars to the same soil (Lebrun et al., 2018c). Biochar feedstock is thus an important parameter in the Pb immobilization process. Finally, hardwood biochars with four different particle sizes were tested on this soil. Among these, the two finest biochars decreased Pb mobility just after their application whereas the two coarser biochars led to the same immobilization only after 46 days of maturation (Lebrun et al., 2018b). For the study of Lu et al. (2017), two bamboo biochars and two rice straw biochars with either a fine or a coarse particle size, were applied to a contaminated soil in proximity to a copper smelter, at a 1% or 5% rate. They observed that all biochars decreased the concentration in $CaCl_2$-extractable Pb, with the best treatment

Table 4.2 Lead (Pb) and arsenic (As) sorption (%) capacities of the different biochar. Sorption test were performed in batch using 15 mg of biochar mixed in 15 mL of As or Pb solution. Solutions were mixed for 16 h before filtration, acidification and analyses. A control containing no biochar was performed and the percentage of sorption was calculated by comparing the concentration in the control and in the test. HW1 = hardwood biochar with particle size <0.1 mm; HW2 = hardwood biochar with particle size 0.2–0.4 mm; HW3 = hardwood biochar with particle size 0.5–1 mm; HW4 = hardwood biochar with particle size 1–2.5 mm; HW2-$FeCl_3$ = hardwood biochar with particle size 0.2–0.4 mm, functionalized with $FeCl_3$. Letters indicate significant difference ($p < 0.05$) ($n = 5$).

	HW1 < 0.1 mm	HW2 0.2–0.4 mm	HW3 0.5–1 mm	HW4 1–2.5 mm	HW2-$FeCl_3$ 0.2–0.4 mm
Sorption Pb (%)	85.68 ab	85.52 ab	60.38 ab	40.11 b	99.79 a
Sorption As (%)	3.42 a	3.20 a	3.20 a	4.78 a	48.17 b

Reproduced with permission from Lebrun, M., Miard, F., Renouard, S., Nandillon, R., Scippa, G. S., Morabito, D., Bourgerie, S: Effect of Fe-functionalized biochar on toxicity of a technosol contaminated by Pb and As: sorption and phytotoxicity tests. Environ Sci Pollut Res. 25:33678-33690, 2018.

being 5% rice straw biochar. In this same study, the application of a higher dose was more efficient in immobilizing Pb.

As has been observed for soil nutrients, this Pb immobilization can be due to the direct and indirect effects of biochar.

Biochar was shown to have a sorption capacity towards Pb, as shown in Table 4.2, adapted from the study of Lebrun et al. (2018d). Indeed, the surface of biochar is composed of functional groups with negative charges that can interact with the positively charged Pb ions through electrostatic interaction. In addition, the Fourier Transform Infra-Red Spectrometry analysis of the biochars before and after sorption showed that most of the shifts occurred in the regions corresponding to the OH groups, testifying that sorption of Pb occurred through its interaction with oxygen functional groups (Lebrun et al., 2018d).

Fine biochars showed better sorption efficiency than coarse biochars, which could be explained by the biochars' physical properties: the higher surface area and pore diameter of fine biochars facilitate the transfer of pollutants and also provide more sorption sites for cations; and their higher cation exchange capacity tends to enhance ionic exchange (Lebrun et al., 2018b; Liang et al., 2016).

Biochar can also indirectly induce an immobilization of Pb by modifying other parameters. For instance, the addition of biochar to soil can increase soil pH and it is known that at alkaline pH values, Pb can form hydroxide or carbonate precipitates (Lebrun et al., 2018c; Ahmad et al., 2017; Dai et al., 2018). Finally, by increasing organic carbon content in the soil, biochar promotes microbial activity. Microorganisms can either sequester metal(loid)s or transform them into a less toxic and/or mobile forms (Dai et al., 2018).

On the contrary, biochar is generally not efficient in the sorption of anions such as arsenic. Biochar had no effect on soil pore water As concentration as has been demonstrated in several studies (Lebrun et al., 2018b, c) and sometimes even caused it to increase (Lomaglio et al., 2017). For instance, the application of a water hyacinth biochar to a multi-contaminated soil induced an increase in As bioavailability (Yin et al., 2016). The amendment of a Pb/Zn tailing pond with sewage sludge biochar also increased the extractability of As, as demonstrated in the study of Huang et al. (2018). This non-effect of biochar was confirmed by the sorption tests that showed less than 4% sorption of As onto hardwood biochars (Lebrun et al., 2018d), which can be attributed to a repulsion between the negative charges of the biochars and the negative As ions (Lebrun et al., 2018d). Moreover, biochar can increase As mobility due to the increase in soil pH it induces (Lomaglio et al., 2017).

However, some studies still showed that biochar application was capable of decreasing As mobility. For instance, three sapwood biochars were able to decrease the SPW As concentration of a mine soil (Lebrun et al., 2020a); and a willow biochar reduced the concentration of water soluble As (Gregory et al., 2015).

It can thus be concluded that biochar is very efficient for metal cations, whereas its effect on metal anions varies depending on biochar type and soil conditions.

4.6 The effect of biochar on the soil microbial community

Microorganisms are ubiquitous in soils and sensitive to soil conditions. In particular, the microbial activity can be inhibited by the presence of high concentrations of metal(loid)s in the soil. Boshoff et al. (2014) showed a correlation between the decrease in feeding activity, metabolic activity and functional richness of the soil microorganisms, and an increase in the concentrations of As, Cu and Pb in the soil. Similarly, Xie et al. (2016) observed that the total bioactivity, the richness, as well as the diversity of the soil microorganisms decreased with increasing metal(loid) concentrations. Finally, in samples taken from a metal-polluted soil, the alpha-diversity, which corresponds to the number of species a community contains, was lower than in the samples collected from an unpolluted soil (Chen et al., 2018).

During a phytoremediation process, assisted or not, the increase in the bacterial population and its corresponding activity indicated that soil fertility was ameliorated and the environmental risk of the contaminated soil reduced (Garau et al., 2017).

The application of biochar was shown to be beneficial for soil microorganisms, especially bacteria. Biochar can affect the growth and activity of the microbial community, and can also induce changes in the community structure.

For instance, the application of two different biochars (soybean stover biochar and pine needle biochar) to an agricultural soil adjacent to a mine induced an increase in the soil microbial population that included Gram-positive and Gram-negative bacteria, fungi and actinomycetes (Ahmad et al., 2016). The addition of a corn straw biochar to a Pb contaminated soil increased the soil microbial diversity and the biomass of Gram negative bacteria (Nie et al., 2018). Similarly, the amendment of a gold mine contaminated soil using date palm biochar increased both the soil microbial biomass and the soil microbial activity (Al Wabel et al., 2017). A Cu smelter soil was amended with either chicken manure biochar or oat hull biochar: both biochars increased the microbial activity of the soil, with the chicken manure biochar having a better effect than the oat hull biochar (Moore et al., 2017). The application of eucalyptus biochar and poultry litter biochar to a Cd-contaminated soil near a field landfill induced an increase in the soil basal respiration, the soil metabolic quotient and the soil enzyme activity (Lu et al., 2015). Finally, a biochar made from *Salicornia bigelovii* increased the soil microbial biomass carbon content and the enzyme activities of a non-contaminated soil (Al Marzooqi and Youssef, 2017). Moreover, Chen et al. (2016) measured an increase in the general use of carbon substrates, demonstrating the amelioration in microbial activity following biochar application. However, they also showed a change in the substrate use: in the biochar amended soils, it was observed that a selection of substrates selection including polymers, phenolic compounds and amines were consumed by the microbial communities was observed. Similarly, Tian et al. (2016) measured a higher utilization of the amino acids and amines, which could be a way for microorganisms to compensate for the high C/N ratio after the biochar amendment of a rice paddy soil.

The positive effect of biochar on the soil microbial growth and activity can be explained by direct and indirect mechanisms. Firstly, biochar has a porous structure that can serve as a habitat for soil microorganisms, and protect them from predation (Nie et al., 2018; Lu et al., 2015). Secondly, the labile fraction of biochar could support microbial growth (Purakayastha et al., 2015). Thirdly, biochar also supplies nutrients and carbon for the growth of microorganisms (Nie et al., 2018; Xu et al., 2018). In addition to these direct effects, biochar can indirectly ameliorate microbial growth by improving soil properties (Gul et al., 2015). For instance, the increase in pH following biochar application can ameliorate microbial growth, as slightly alkaline or a neutral soil pH can favor the growth of microorganisms (Huang et al., 2017a). In addition, as demonstrated in a previous section, biochar amendments increase SOM content, water retention and the physical structure of the soil, resulting in better microbial growth (Khadeem and Raiesi 2017; Nie et al; 2018). Finally, biochar immobilizes metal(loid)s, which reduces the toxic stress on all soil constituents, including microorganisms (Park et al., 2011; Moore et al., 2017).

In addition to the beneficial effect of biochar on the growth and activity of the soil microbial community, biochar can also affect its composition (Chen et al., 2016), in terms of diversity, and also by stimulating certain organisms and repressing others (Xu et al., 2017). For instance, Ahmad et al. (2016) observed that a biochar amendment to an agricultural soil near an abandoned mine increased the soil bacterial diversity; and in a deeper analysis, they found that the *Actinobacteria* proportion was enhanced whereas those of *Acidobacteria* and *Chloroflexi* decreased. Huang et al. (2017a) did not find a modification in the diversity of the soil microbial community after biochar was mixed with contaminated river sediments. However, the microbial community composition was changed. In their study, Wang et al. (2019) found that certain bacterial taxa, *i.e. Proteobacteria*, *OD1* and *Verrucomicrobia*, were consistently enriched following a biochar amendment to a contaminated paddy field. Similarly, the abundance of *Actinobacteria*, *Firmicutes*, *Proteobacteria*, *Planctomycetes* and *Cyanobacteria* increased with an increasing supply of biochar to a contaminated industrial site, while that of *Gemmatimonadetes* decreased (Xu et al., 2017). The increase in *Proteobacteria* abundance can be explained by the fact that with biochar, the soil is richer in nutrients (Xu et al., 2017). *Acidobacteria* are generally found in acidic environments, and therefore the increase in soil pH induced by biochar can lead to a decrease in *Acidobacteria* abundance. However, this is not always observed. *Acidobacteria* can degrade organic matter, and thus their abundance can be increased due to the supply of organic matter in biochar (Wang et al., 2019).

The effect of biochar on the structure of the microbial community in the soil can be caused by modifications to the soil physico-chemical properties induced by biochar, such as pH, OM content, nutrient content, WHC, metal(loid) toxicity (Wang et al., 2019; Xu et al., 2017).

4.7 The effects of biochar on plants

As stated in the introduction, polluted soils have properties that make the establishment of a plant cover difficult, which has been demonstrated by previous studies. Indeed, *Salix purpurea*, *Salix alba*, *Salix viminalis* and *Trifolium repens* had a reduced growth on a former mine soil highly contaminated by As and Pb (Lebrun et al., 2017, 2018a, 2019; Nandillon et al., 2019b), whereas *Agrostis capillaris* was not able to grow when the mine technosol was not amended (Nandillon et al., 2019a). This is due to the low fertility of the contaminated soils as well as the elevated pollution levels, which induce toxicity to plants. The addition of biochar can allow plant establishment and highly increase plant growth.

For instance, three *Salix* species (*Salix purpurea*, *Salix alba* and *Salix viminalis*) were grown on Pontgibaud's former mine technosol. Their growth was very low on this substrate but their dry weight production was increased by the addition of 2% and 5% of pinewood biochar (VT Green) (Lebrun et al., 2017) (Fig. 4.3). Yu et al. (2017) also observed that a corn straw biochar improved rice growth on an As contaminated soil. Brennan et al. (2014) applied a biochar obtained from olive tree prunings to a multi-contaminated mine soil and measured an increase in the root length and surface area, leaf surface area, and shoot and root biomass of maize plants.

Salix viminalis and *Populus euramericana* dry weight increased following the addition of lightwood and pinewood biochars to the same technosol (Lebrun et al., 2018c) (Fig. 4.4). *Trifolium repens* and *Agrostis capillaris* growth was also improved by the addition of hardwood biochar (Nandillon et al., 2019 a, b). Tobacco plants also produced bigger and a larger number of leaves following the amendment of a multi-contaminated mine soil by a biochar made from tobacco stems (Zhang et al., 2019).

In addition, as was observed for most of the soil properties, biochar type, particle size and application rate influence the extent of the improvement of plant growth. For instance, four hardwood biochars with different particle sizes were applied to the Pontgibaud mine technosol, and they all improved *Salix viminalis* growth when applied at 2%, except for the coarsest one, which only led to an increase when applied at 5% (Lebrun et al., 2018b). Similarly, Huang et al. (2018) applied three biochars (from *Hibiscus cannabinus*, sewage sludge and chicken manure), to a multi-contaminated tailing at increasing doses. All biochars improved *Cassia alata* shoot and root biomass but the extent of this increase depended on the biochar type and the application rate.

This amelioration of plant growth can be primarily attributed to the effect of biochar on the soil properties. Indeed, as shown in the previous section, biochar application to soil usually increases soil pH, water content, and OM content. Biochar also contains nutrients and can thus improve soil nutrient content and availability. Finally, biochar decreased metal(loid) mobility and availability, thus reducing their toxicity towards plants.

Biochar can also affect the uptake of metal(loid)s by plants. However, it is not possible to draw a definitive conclusion on the effect of biochar on the uptake of pollutants by plants, as the effect differed depending on biochar type, application rate, soil type and species. For instance, the application of hardwood biochar to Pontgibaud decreased the aerial concentration of As in *Salix viminalis*, while Pb aerial concentration increased (Lebrun et al., 2018b). Similarly, As and Pb concentrations decreased in *Salix viminalis* and *Populus euramericana* after the application of lightwood or pinewood biochars (Lebrun et al., 2018c). However, the addition of another pinewood biochar to a former industrial site had no effect on the As and Pb concentrations in *Salix viminalis* and *Salix alba* (Lebrun et al., 2018a). In white willows, the concentrations of Cd, Cu and Pb decreased when hornbeam biochar was applied to a contaminated spiked soil (Mokarram-Kashtiban et al., 2019). Finally, rice As concentrations decreased when an As contaminated soil was amended with corn straw biochar (Yu et al., 2017).

In general, the decrease observed can be explained not only by a dilution effect, as plants produced more biomass, but also, and more importantly, by the decrease in soil metal(loid) availability (Lebrun et al., 2018b). However, the increase could be attributed to the bigger leaves, which induce more transpiration and thus more uptake of water from the soil. Metal(loid) concentrations are thus higher in the roots and may also be more transported in the aerial parts, although translocation can be impaired

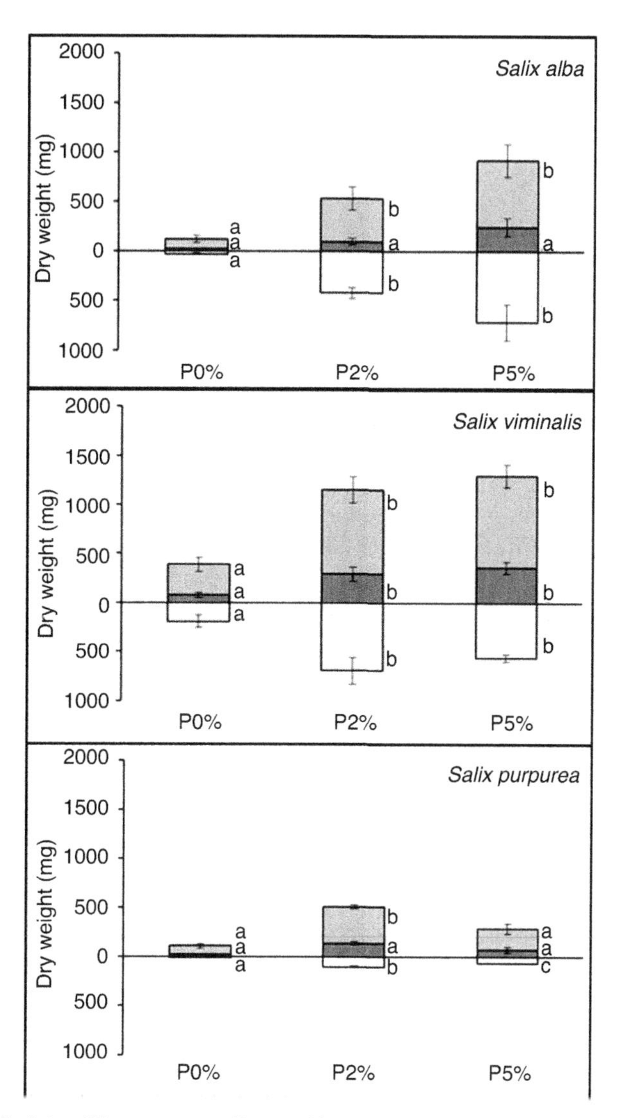

FIG. 4.3 Dry weight (mg) of the different organs (leaves (light grey), stems (dark grey), roots (white)) of *Salix alba*, *Salix viminalis*, *Salix purpurea*.

The plants, coming from non-rooted cuttings, were grown under greenhouse conditions for 63 days on the As and Pb contaminated Pontgibaud technosol (P) amended with 0%, 2% or 5% of pinewood biochar (VT Green, Table 4.1). Letters indicate a significant difference ($p < 0.05$) ($n = 6$). *Reproduced with permission from Lebrun et al. (2017).*

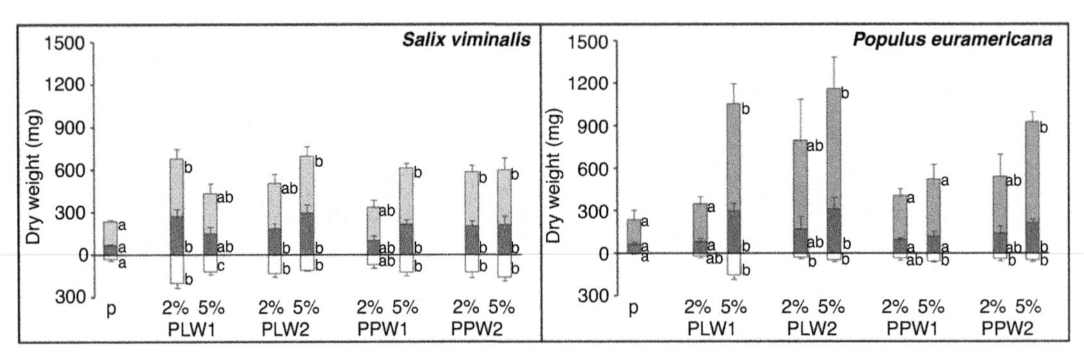

FIG. 4.4 Dry weight (mg) of the organs of *Salix viminalis* and *Populus euramericana*.

The plants, coming from non-rooted cuttings, were grown under greenhouse conditions for 45 days on the As and Pb contaminated Pontgibaud technosol (P) amended with biochars from 2 different feedstocks (LW = lightwood; PW = pinewood) and presenting two particle sizes (1 = inferior to 0.1 mm; 2 = between 0.2 and 0.4 mm). Biochars were added at two rates (2% and 5%). Letters indicate significant difference between the different biochar rates ($p < 0.05$) ($n = 5$). *Reproduced with permission from Lebrun et al. (2018c).*

by the Casparian barrier in the roots. Finally, inside the plants, the translocation pattern was affected following biochar amendment. For instance, Lebrun et al. (2018b) observed that inside *Salix viminalis* plant organs, As and Pb translocation differed depending on the biochar particle size: no As or Pb was detectable with the two fine biochars (<0.1 mm and 0.2–0.4 mm), whereas As and Pb were not detectable in the stems with the two coarse biochars (0.5–1 mm and 1–2.5 mm). Such results show that the growth and metal(loid) uptake of the plant that is, its physiology is also affected by biochar.

Indeed, some studies demonstrated that biochar soil amendment was able to affect plant physiology, notably by reducing the oxidative stress. Metal(loid)s are toxic to plants, as they induce the formation of reactive oxygen species, which are free radicals that can damage the plant at molecular level though DNA mutation and protein oxidation for instance (Ali et al., 2006). Abbas et al. (2018) measured a decrease in oxygen peroxide content as well as a reduction in peroxidase activity, and an increase in superoxide dismutase and catalase activities in wheat plants, showing that biochar reduced the stress in the plants. Similarly, the addition of tea waste biochar to a Cd contaminated soil reduced the oxidative damage in ramie seedlings (Gong et al., 2019). Finally, in a recent study, Lebrun et al. (2020c) studied the response of *Salix viminalis* roots to a biochar amendment. They showed that the biochar amendment modified the phenylpropanoid flux which enhanced the defense and the morphogenic response. In addition, in this amended condition, the oxidative stress and protein misfolding and aggregation in the roots were lower compared to the non-amended condition.

The most obvious reason for such a reduction in oxidative stress in the plants is the immobilization of metal(loid)s by biochar, and the subsequently reduced metal(loid) uptake by plants (Ali et al., 2017). However, the lower oxidative stress in plants was not always correlated with a lower metal(loid) concentration in plants. Indeed, Gong et al. (2019) observed a reduction in oxidative damage markers in ramie seedlings grown on Cd contaminated sediments amended with tea waste biochar. This reduction was related to a reduction in Cd accumulation in the plants. However, the lowest Cd accumulation was not found in plants with the lowest oxidative damage. Thus, they concluded that the reduced Cd accumulation in plants was only partly responsible for the alleviation of the Cd induced oxidative stress.

The reduction in H_2O_2 in the observed plants could be explained by the activation of H_2O_2 by biochar and the formation of superoxide ions on the biochar surface (Gong et al., 2019; Fang et al., 2014). In addition, the functional groups and free radicals on the biochar's surface could have participated to the scavenging of the reactive oxygen species which were produced in response to metal(loid) stress (Gong et al., 2019). Finally, as shown previously, biochar can affect metal(loid) behavior and in particular their speciation, which depends on the soil parameters, can be modified. The toxicity of an element and its uptake by the plant will be affected by its speciation. It is particularly true for arsenic: the form AsIII is more toxic than the form AsV. Also, the absorption of the arsenate ions is three times greater than that of the arsenite ions (INERIS, 2010). Moreover, arsenate ions can compete with phosphate and sulfate ions for sorption sites on the soil and the plant cells, modelling their sorption. Therefore, biochar can reduce metal(loid) toxicity towards plants by inducing a change in their speciation into a less toxic form.

In conclusion, biochar amendments allow plant establishment and increase plant growth by ameliorating soil conditions and immobilizing metal(loid)s. However, its effect on the metal(loid) uptake by plants diverges depending on several factors, *i.e.* soil type, biochar type, biochar particle size, biochar dose application and plant type. Finally, although studies on this detail are still scarce, biochar can help to reduce the oxidative stress induced by metal(loid)s on the plants.

4.8 The importance of biochar dose

As shown in the previous section, biochar is beneficial for the soil, the plant and the microorganisms. However, these effects are not only dependent on biochar and soil properties but also on the dose of biochar applied to the soil.

For instance, Lebrun et al. (2021) tested five doses, from 1% to 5%, of the same biochar on the Pontgibaud technosol. They observed that with increasing biochar doses, soil pore water pH and EC increased while the decrease in Pb concentration was the same whatever the dose applied. Regarding the effect on the plant, flax plants produced higher root biomass with increasing biochar doses, whereas the aerial biomass was higher with more biochar up to 3% and adding a higher concentration of biochar had no further effect. Finally, the lowest translocation of As and Pb towards aerial tissues was observed at 3% biochar. They concluded that for this soil, 3% of hardwood biochar was the optimal dose for the phytostabilization of As and Pb using flax. In 2017, Yue et al. (2017), added a biochar made from municipal sewage sludge to an urban soil with application rates from 0% to 50%. They found that with increasing biochar doses, soil pH, EC, SOC and N contents and available P and K concentrations increased, and the DW of turf grass also increased. However, they also observed that with higher doses of biochar, the total contents in Zn, Cu, Cr, Pb, As, and Cd of the soil increased, which was a negative effect cause by the biochar dose. The study of Herath et al. (2014) evaluated the effect of biochar dose on the soil properties, and plant and microbial growth, when added to a serpentine soil. They measured higher soil pH, EC, TOC and CEC values with higher biochar doses. The biomass of the tomato plants was increased with increasing application rate. However, the effect of the biochar on microbial growth depended on the dose: 1% had no effect, higher biochar doses, up to 2.5%, induced an increase in the number of colony forming units (CFU), while 5% biochar led to a decrease in CFU number. Finalsly, Lu et al. (2017) added three doses, 0%, 1% and 5%, of a rice straw biochar and a bamboo biochar to an abandoned paddy field. They observed that $CaCl_2$-extractable concentrations of Cd, Cu, Pb, and Zn decreased with increasing biochar doses, whereas DTPA extractable concentrations decreased with biochar application, although adding 1% biochar led to a greater decrease than for 5%.

This section shows that it is important to determine the appropriate dose of biochar to apply to a soil. Indeed, adding more biochar will induce a higher cost to the remediation process without necessarily bringing increased benefits and could even have negative effects.

4.9 Improving biochar effects: functionalization and combination with other amendments

The previous sections of this chapter demonstrated that biochar generally has a positive effect on soil properties, metal(loid) immobilization, microbial activity, and plant growth. However, such positive effects were not always found, for instance regarding arsenic. One way to overcome this is to modify and functionalize biochar.

Biochar functionalization consists of a treatment that will modify its surface. There are two main functionalizations: physical functionalization, using magnetization, steam or gas, and chemical functionalization, using amines, methanol, mineral and acid/base treatments (Rajapashka et al., 2016; Tan et al., 2017a). These functionalizations can improve the surface of the biochar and create more sites for metal(loid) sorption. For instance, Lebrun et al. (2018d) performed a functionalization on a hardwood biochar using $FeCl_3$ and observed that the sorption capacity of this biochar towards As was greatly improved, passing from less than 4% to 50% of sorption, without affecting Pb sorption (Table 4.2). A bamboo biochar was functionalized using chitosan, which increased its pH and N, H and O contents, while decreasing its surface area and C level. This functionalization increased the removal capacity of Cd, Cu and Pb (Zhou et al., 2013). Similarly, the functionalization of a canola straw biochar using steam increased its pH and surface area as well as its adsorption capacity towards Pb (Kwak et al., 2019). In the soil, the addition of a corn straw biochar functionalized with ferromanganese oxide led to a reduction in As accumulation in rice, and an increase in amino acid contents in the rice, compared to the non-functionalized biochar (Lin et al., 2017). However, not all functionalizations are beneficial. For instance, Wu et al. (2016) performed several functionalizations on a coconut biochar and observed that Pb sorption was either improved or reduced, depending on the activation. Therefore, it is crucial to select the right functionalization. Moreover, until now most studies have been performed in batch a using sorption test, with few studies moving to the soil. It is thus necessary to verify the preliminary sorption test on the soil, as the beneficial effects can be lost once the biochar is added to the soil, as was observed in the study of Lebrun et al. (2018d). Finally, this functionalization has a cost and can be difficult to do on a large industrial scale, as high amounts (several tonnes per hectare) can be required for field application.

Another way to further improve the effect of biochar is to combine it with one or more other amendments.

One possible amendment is compost, which is already highly used in agriculture. Compost comes from the microbial degradation of organic wastes. It is rich in humic substances, microorganisms, and plant nutrients (Huang et al., 2016). Biochar is known to increase nutrient content and availability, but it is not *per se* considered as a fertilizer, especially if it is made from wood and not crops. Therefore, combining the liming effect and immobilization of metal(loid)s induced by biochar with the high nutrient and organic matter contents of compost can be beneficial for contaminated soils. Moreover, when both biochar and compost are applied together, they can interact. The biochar properties and surface can be modified due to humus oxidation and the microorganisms of the compost, whereas biochar can improve the humification and quality of compost (Liang et al., 2017). For instance, Lebrun et al. (2019) observed that both biochar and compost were able to increase soil WHC, OM content, pH and EC as well as decreasing Pb mobility. However, results were improved when they were applied together rather than alone. They concluded that the combination of biochar and compost was a good option for the phytostabilization of a former mine technosol when applied with *Salix viminalis* plants. Similarly, in two other studies, Nandillon et al. (2019c, 2019d) observed that the combination of biochar and compost was better at improving some soil properties compared to when applied alone. Finally, Yun et al. (2017) found that only the combination of biochar and compost was able to improve water melon yield.

However, neither biochar nor compost were efficient for anions such as arsenic.

For the remediation of arsenic, iron-based amendments, such as iron grit and iron sulfate, can be combined with biochar. These amendmentspossess a high affinity for arsenic due to their high Fe content. The positive effects of these combinations were demonstrated in the studies of Lebrun et al. (2019) and Nandillon et al. (2019a, b, c). Similarly, Fresno et al. (2020) did not observe a beneficial effect on As when using organic amendment although they found that the combination of iron sulfate with biochar improved soil properties, rye growth, and nutrients and decreased the accumulation of As in rye shoots and roots. Finally, the additional immobilization of arsenic by iron-based amendments, combined with the improvements induced by biochar, can lead to a further increase in plant growth, as shown in Fig. 4.5 (Nandillon et al., 2019a).

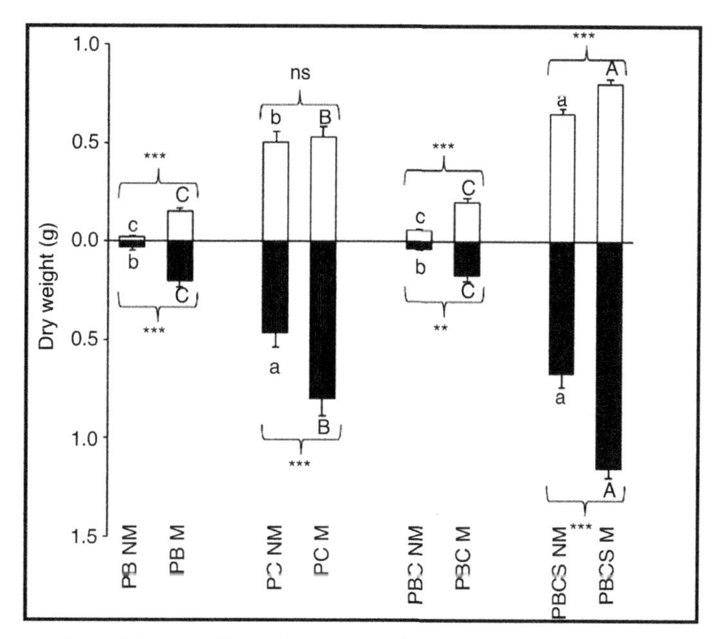

FIG. 4.5 *Agrostis* organ dry weight (g) collected at the end of the experiment period (76 days); white column: aerial parts, black column: roots.

Seeds used were collected on the mine site of Pontgibaud (M: metallicolous) or were of commerical origin (NM: non-metallicolous). Seeds were sown and grown on the different substrates for 76 days under greenhouse conditions. (P): Pontgibaud soil, (PB): P + biochar (B) 2%, (PC): P + compost (C) 5%, (PBC): P + B 2% + C 5%, (PBCS): P + B 2% + C 5% + Iron sulfate 0.15%. Results are expressed as the mean value ± standard error (n = 5). No growth was observed on the non-amended Pontgivaud technosol, threfore it is not represented. Letters lowercase indicate a significant difference ($p < 0.05$) for NM, letters uppercase a significant difference ($p < 0.05$) for M, *** significant difference ($p < 0.001$), ** significant difference ($p < 0.01$) and ns no significant difference between N and NM. *Reproduced with permission from Nandillon et al. (2019a).*

4.10 Conclusions and perspectives

Soils polluted by metal(loid)s cause significant big health and environmental issues worldwide. Therefore, the phytoremediation of those sites has been the focus of much research over the last few decades. Phytoremediation uses plants and microorganisms to reduce the toxic effects of contaminants. This technique is environmentally friendly, and also cost-effective. However, due to the poor fertility of contaminated soils, amendments need to be applied to improve soil conditions and thus allow plant growth. As such, biochar, the product of the pyrolysis of biomass under low oxygen conditions, has been shown to ameliorate soil properties, such as pH, organic matter and nutrient contents, immobilize metal(loid)s, and increase the growth of many different plant species.

However, biochar can also negatively affect the soil, especially in the case of soils contaminated by anionic elements. Biochar can also sorb the nutrients, which is detrimental for microorganisms and plants. To overcome this, biochar can be modified, for instance by fixing iron on its surface which in turn interacts with arsenic. Moreover, biochar can be applied in combinations with other amendments to further improve its beneficial effects: compost for agronomic purposes, and iron-based amendments for anion immobilization for instance. Finally, biochar can also be associated with metal(loid) tolerant microorganisms, which can affect metal(loid) behavior and ameliorate plant growth.

However, as the market for biochar is not fully developed, its cost is still high, and is further increased by its functionalization, which increases the price of the phytoremediation process.

Finally, long-term field experiments need to be performed on a more regular basis in order to evaluate how biochar is ageing in the soils as well as to determine the most efficient biochar application rate (usually between 1% and 5%) in order to have a positive effect at a lowest cost possible. Validation of the use of biochar on soils presenting different properties and types of contaminations will allow the development of a real market for biochar and could thus reduce its cost, especially if wastes or by-products are used to produce it.

References

Abbas, T., Rizwan, M., Ali, S., Adrees, M., Mahmood, A., Rehman, M.Z., Ibrahim, M., Arsad, M., Qayyum, M.F., 2018. Biochar application increased the growth and yield and reduced cadmium in drought stressed wheat grown in an aged contaminated soil. Ecotoxicol. Environ. Saf. 148, 825–833.

Agegnehu, G., Bass, A., Nelson, P., Muirhead, B., Wright, G., Bird, M., 2015. Biochar and biochar-compost as soil amendments: Effects on peanut yield, soil properties and greenhouse gas emissions in tropical North Queensland, Australia. Agric. Ecosyst. Environ. 213, 72–85.

Ahmad, M., Lee, S.S., Dou, X., Mohan, D., Sung, J.K., Yang, J.E., Ok, Y.S., 2012. Effects of pyrolysis temperature on soybean stover-and peanut shell-derived biochar properties and TCE adsorption in water. Bioresour. Technol. 118, 536–544.

Ahmad, M., Ok, Y., Kim, B., Ahn, J., Lee, Y., Zhang, M., Moon, D., Al-Wabel, M., Lee, S., 2016. Impact of soybean stover- and pine needle-derived biochars on Pb and As mobility, microbial community, and carbon stability in a contaminated agricultural soil. J. Environ. Manag. 166, 131–139.

Ahmad, M., Lee, S.S., Lee, S.E., Al-Wabel, M.I., Tsang, D.C., Ok, Y.S., 2017. Biochar-induced changes in soil properties affected immobilization/mobilization of metals/metalloids in contaminated soils. J. Soils Sediments 17 (3), 717–730.

Ali, A., Guo, D., Zhang, Y., Sun, X., Jiang, S., Guo, Z., Huang, H., Liang, W., Li, R., Zhang, Z., 2017. Using bamboo biochar with compost for the stabilization and phytotoxicity reduction of heavy metals in mine-contaminated soils of China. Sci. Rep. 7 (1), 1–12.

Ali, M.B., Singh, N., Shohael, A.M., Hahn, E.J., Paek, K.Y., 2006. Phenolics metabolism and lignin synthesis in root suspension cultures of *Panax ginseng* in response to copper stress. Plant Sci. 171 (1), 147–154.

Al-Wabel, M.I., Usman, A.R.A., Al-Farraj, A.S., Ok, Y.S., Abduljabbar, A., Al-Faraj, A.I., Sallam, A.S., 2017. Date palm waste biochars alter a soil respiration, microbial biomass carbon, and heavy metal mobility in contaminated mined soil. Environ. Geochem. Health 41, 1705–1722.

Arienzo, M., Christen, E.W., Quayle, W., Kumar, A., 2009. A review of the fate of potassium in the soil–plant system after land application of wastewaters. J. Hazard. Mater. 164 (2-3), 415–422.

Bakshi, S., He, Z.L., Harris, W.G., 2014. Biochar amendment affects leaching potential of copper and nutrient release behavior in contaminated sandy soils. J. Environ. Qual. 43 (6), 1894–1902.

Batista, E.M., Shultz, J., Matos, T.T., Fornari, M.R., Ferreira, T.M., Szpoganicz, B., de Freitas, R.A., Mangrich, A.S., 2018. Effect of surface and porosity of biochar on water holding capacity aiming indirectly at preservation of the Amazon biome. Sci. Rep. 8 (1), 1–9.

Bezerra, J., Turnhout, E., Vasquez, I.M., Rittl, T.F., Arts, B., Kuyper, T.W., 2016. The promises of the Amazonian soil: shifts in discourses of Terra Preta and biochar. J. Environ. Pol. Plann. 21 (5), 623–635.

Blanco-Canqui, H., 2017. Biochar and soil physical properties. Soil Sci. Soc. Am. J. 81 (4), 687–711.

Boshoff, M., De Jonge, M., Dardenne, F., Blust, R., Bervoets, L., 2014. The impact of metal pollution on soil faunal and microbial activity in two grassland ecosystems. Environ. Res. 134, 169–180.

Brennan, A., Jiménez, E., Puschenreiter, M., Alburquerque, J., Switzer, C., 2014. Effects of biochar amendment on root traits and contaminant availability of maize plants in a copper and arsenic impacted soil. Plant Soil 379 (1-2), 351–360.

Burrell, L.D., Zehetner, F., Rampazzo, N., Wimmer, B., Soja, G., 2016. Long-term effects of biochar on soil physical properties. Geoderma 282, 96–102.

Cantrell, K., Hunt, P., Uchimiya, M., Novak, J., Ro, K., 2012. Impact of pyrolysis temperature and manure source on physicochemical characteristics of biochar. Bioresour. Technol. 107, 419–428.

Cha, J.S., Park, S.H., Jung, S.C., Ryu, C., Jeon, J.K., Shin, M.C., Park, Y.K., 2016. Production and utilization of biochar: a review. J. Ind. Eng. Chem. 40, 1–15.

Chen, D., Liu, X., Bian, R., Cheng, K., Zhang, X., Zheng, J., Joseph, S., Crowley, D., Pan, G., Li, L., 2018. Effects of biochar on availability and plant uptake of heavy metals–a meta-analysis. J. Environ. Manag. 222, 76–85.

Chen, J., Sun, X., Li, L., Liu, X., Zhang, B., Zheng, J., Pan, G., 2016. Change in active microbial community structure, abundance and carbon cycling in an acid rice paddy soil with the addition of biochar. Eur. Soil Sci. 67 (6), 857–867.

Chen, W., Meng, J., Han, X., Lan, Y., Zhang, W., 2019. Past, present and future of biochar. Biochar 1 (1), 75–87.

Chen, Y., Shinogi, Y., Taira, M., 2010. Influence of biochar use on sugarcane growth, soil parameters, and groundwater quality. Aust. J. Soil Res. 48 (7), 526–530.

Cristaldi, A., Conti, G.O., Jho, E.H., Zuccarello, P., Grasso, A., Copat, C., Ferrante, M., 2017. Phytoremediation of contaminated soils by heavy metals and PAHs. A brief review. Environ. Technol. Innov. 8, 309–326.

Cui, L., Pan, G., Li, L., Yan, J., Zhang, A., Bian, R., Chang, A., 2012. The reduction of wheat Cd uptake in contaminated soil via biochar amendment: a two-year field experiment. BioResources 7 (4), 5666–5676.

Dai, Z., Zhang, X., Tang, C., Muhammad, N., Wu, J., Brookes, P.C., Xu, J., 2017. Potential role of biochars in decreasing soil acidification-A critical review. Sci. Total Environ. 581, 601–611.

Dai, S., Li, H., Yang, Z., Dai, M., Dong, X., Ge, X., Sun, M., Shi, L., 2018. Effects of biochar amendments on speciation and bioavailability of heavy metals in coal-mine-contaminated soil. Hum. Ecol. Risk Assess. 24 (7), 1887–1900.

Devereux, R.C., Sturrock, C.J., Mooney, S.J., 2012. The effects of biochar on soil physical properties and winter wheat growth. Earth Environ. Sci. Trans. R. Soc. Edinb. 103 (1), 13–18.

El-Naggar, A., Lee, S.S., Awad, Y.M., Yang, X., Ryu, C., Rizwan, M., Rinklebe, J., Tsang, D.C.W., Ok, Y.S., 2018. Influence of soil properties and feedstocks on biochar potential for carbon mineralization and improvement of infertile soils. Geoderma 332, 100–108.

El-Naggar, A., Lee, M.H., Hur, J., Lee, Y.H., Igalavithana, A.D., Shaheen, S.M., Ryu, C., Rinklebe, J., Tsang, D.C.W., Ok, Y.S., 2020. Biochar-induced metal immobilization and soil biogeochemical process: an integrated mechanistic approach. Sci. Total Environ. 698, 134112.

Fang, G., Gao, J., Liu, C., Dionysiou, D.D., Wang, Y., Zhou, D., 2014. Key role of persistent free radicals in hydrogen peroxide activation by biochar: implications to organic contaminant degradation. Environ. Sci. Technol. 48 (3), 1902–1910.

Fellet, G., Marchiol, L., Delle Vedove, G., Peressotti, A., 2011. Application of biochar on mine tailings: Effects and perspectives for land reclamation. Chemosphere 83 (9), 1262–1267.

Forján, R., Asensio, V., Guedes, R.S., Rodríguez-Vila, A., Covelo, E.F., Marcet, P., 2017. Remediation of soils polluted with inorganic contaminants: role of organic amendmentsEnhancing Cleanup of Environmental Pollutants. Springer International Publishing, pp. 313–337.

Forján, R., Asensio, V., Rodríguez-Vila, A., Covelo, E., 2016. Contribution of waste and biochar amendment to the sorption of metals in a copper mine tailing. Catena 137, 120–125.

Fresno, T., Peñalosa, J.M., Flagmeier, M., Moreno-Jiménez, E., 2020. Aided phytostabilisation over two years using iron sulphate and organic amendments: Effects on soil quality and rye production. Chemosphere 240, 124827.

Gai, X., Wang, H., Liu, J., Zhai, L., Liu, S., Ren, T., Liu, H., 2014. Effects of feedstock and pyrolysis temperature on biochar adsorption of ammonium and nitrate. PloS One 9 (12).

Gao, F., Xue, Y., Deng, P., Cheng, X., Yang, K., 2015. Removal of aqueous ammonium by biochars derived from agricultural residuals at different pyrolysis temperatures. Chem. Speciat. Bioavailab. 27 (2), 92–97.

Garau, G., Silvetti, M., Vasileiadis, S., Donner, E., Diquattro, S., Deiana, S., Lombi, E., Castaldi, P., 2017. Use of municipal solid wastes for chemical and microbiological recovery of soils contaminated with metal(loid)s. Soil Biol. Biochem. 111, 25–35.

Gomes, M., Hauser-Davis, R., de Souza, A., Vitória, A., 2016. Metal phytoremediation: General strategies, genetically modified plants and applications in metal nanoparticle contamination. Ecotoxicol. Environ. Saf. 134, 133–147.

Gong, X., Huang, D., Liu, Y., Zeng, G., Chen, S., Wang, R., Xu, P., Cheng, M., Zhang, C., Xue, W., 2019. Biochar facilitated the phytoremediation of cadmium contaminated sediments: metal behavior, plant toxicity, and microbial activity. Sci. Total Environ. 666, 1126–1133.

Gray, M., Johnson, M.G., Dragila, M.I., Kleber, M., 2014. Water uptake in biochars: The roles of porosity and hydrophobicity. Biomass Bioenergy 61, 196–205.

Gregory, S., Anderson, C., Camps-Arbestain, M., Biggs, P., Ganley, A., O'Sullivan, J., McManus, M., 2015. Biochar in co-contaminated soil manipulates arsenic solubility and microbiological community structure, and promotes organochlorine degradation. PLOS ONE 10 (4), e0125393.

Gul, S., Whalen, J., Thomas, B., Sachdeva, V., Deng, H., 2015. Physico-chemical properties and microbial responses in biochar-amended soils: mechanisms and future directions. Agric. Ecosyst. Environ. 206, 46–59.

Guo, J., Chen, B., 2014. Insights on the molecular mechanism for the recalcitrance of biochars: interactive effects of carbon and silicon components. Environ. Sci. Technol. 48 (16), 9103–9112.

Hailegnaw, N.S., Mercl, F., Pračke, K., Száková, J., Tlustoš, P., 2019. Mutual relationships of biochar and soil pH, CEC, and exchangeable base cations in a model laboratory experiment. J. Soils Sediments 19, 2405–2416.

Hass, A., Gonzalez, J., Lima, I., Godwin, H., Halvorson, J., Boyer, D., 2012. Chicken manure biochar as liming and nutrient source for acid appalachian soil. J. Environ. Qual. 41 (4), 1096–1106.

Herath, I., Kumarathilaka, P., Navaratne, A., Rajakaruna, N., Vithanage, M., 2014. Immobilization and phytotoxicity reduction of heavy metals in serpentine soil using biochar. J. Soils Sediments 15 (1), 126–138.

Higashikawa, F.S., Conz, R.F., Colzato, M., Cerri, C.E.P., Alleoni, L.R.F., 2016. Effects of feedstock type and slow pyrolysis temperature in the production of biochars on the removal of cadmium and nickel from water. J. Clean. Prod. 137, 965–972.

Hmid, A., Al Chami, Z., Sillen, W., De Vocht, A., Vangronsveld, J., 2015. Olive mill waste biochar: a promising soil amendment for metal immobilization in contaminated soils. Environ. Sci. Pollut. Res. 22 (2), 1444–1456.

Hossain, M., Strezov, V., Chan, K., Ziolkowski, A., Nelson, P., 2011. Influence of pyrolysis temperature on production and nutrient properties of wastewater sludge biochar. J. Environ. Manage. 92 (1), 223–228.

Hossain, M., Strezov, V., Yin Chan, K., Nelson, P., 2010. Agronomic properties of wastewater sludge biochar and bioavailability of metals in production of cherry tomato (*Lycopersicon esculentum*). Chemosphere 78 (9), 1167–1171.

Houben, D., Evrard, L., Sonnet, P., 2013. Beneficial effects of biochar application to contaminated soils on the bioavailability of Cd, Pb and Zn and the biomass production of rapeseed (*Brassica napus* L.). Biomass Bioenergy 57, 196–204.

Huang, D., Liu, L., Zeng, G., Xu, P., Huang, C., Deng, L., Wang, R., Wan, J., 2017a. The effects of rice straw biochar on indigenous microbial community and enzymes activity in heavy metal-contaminated sediment. Chemosphere 174, 545–553.

Huang, H., Yao, W., Li, R., Ali, A., Du, J., Guo, D., Xiao, R., Guo, Z., Zhang, Z., Awasthi, M.K., 2017b. Effect of pyrolysis temperature on chemical form, behavior and environmental risk of Zn, Pb and Cd in biochar produced from phytoremediation residue. Bioresour. Technol. 249, 487–493.

Huang, L., Li, Y., Zhao, M., Chao, Y., Qiu, R., Yang, Y., Wang, S., 2018. Potential of *Cassia alata* L. Coupled with biochar for heavy metal stabilization in multi-metal mine tailings. Int. J. Environ. Res. Public Health 15 (3), 494.

Huang, M., Zhu, Y., Li, Z., Huang, B., Luo, N., Liu, C., Zeng, G., 2016. Compost as a soil amendment to remediate heavy metal-contaminated agricultural soil: mechanisms, efficacy, problems, and strategies. Water Air Soil Pollut. 227 (10), 359.

INERIS, 2010. Arsenic et ses dérivés organiques.

Janus, A., Pelfrêne, A., Heymans, S., Deboffe, C., Douay, F., Waterlot, C., 2015. Elaboration, characteristics and advantages of biochars for the management of contaminated soils with a specific overview on *Miscanthus* biochars. J. Environ. Manage. 162, 275–289.

Janus, A., Waterlot, C., Heymans, S., Deboffe, C., Douay, F., Pelfrêne, A., 2018. Do biochars influence the availability and human oral bioaccessibility of Cd, Pb, and Zn in a contaminated slightly alkaline soil? Environ. Monit. Assess. 190 (4), 218.

Jeffery, S., Meinders, M.B., Stoof, C.R., Bezemer, T.M., van de Voorde, T.F., Mommer, L., van Groenigen, J.W., 2015. Biochar application does not improve the soil hydrological function of a sandy soil. Geoderma 251, 47–54.

Jin, H., Capareda, S., Chang, Z., Gao, J., Xu, Y., Zhang, J., 2014. Biochar pyrolytically produced from municipal solid wastes for aqueous As(V) removal: adsorption property and its improvement with KOH activation. Bioresour. Technol. 169, 622–629.

Jindo, K., Mizumoto, H., Sawada, Y., Sanchez-Monedero, M., Sonoki, T., 2014. Physical and chemical characterization of biochars derived from different agricultural residues. Biogeosciences 11 (23), 6613–6621.

Karhu, K., Mattila, T., Bergström, I., Regina, K., 2011. Biochar addition to agricultural soil increased CH4 uptake and water holding capacity–results from a short-term pilot field study. Agricu. Ecosyst. Environ. 140 (1-2), 309–313.

Kelly, C., Peltz, C., Stanton, M., Rutherford, D., Rostad, C., 2014. Biochar application to hardrock mine tailings: soil quality, microbial activity, and toxic element sorption. Appl. Geochem. 43, 35–48.

Khadem, A., Raiesi, F., 2017. Influence of biochar on potential enzyme activities in two calcareous soils of contrasting texture. Geoderma 308, 149–158.

Khalid, S., Shahid, M., Niazi, N., Murtaza, B., Bibi, I., Dumat, C., 2016. A comparison of technologies for remediation of heavy metal contaminated soils. J. Geochem. Explor. 182 (Part B), 247–268.

Kuppusamy, S., Thavamani, P., Megharaj, M., Venkateswarlu, K., Naidu, R., 2016. Agronomic and remedial benefits and risks of applying biochar to soil: Current knowledge and future research directions. Environ. Int. 87, 1–12.

Kwak, J.H., Islam, M.S., Wang, S., Messele, S.A., Naeth, M.A., El-Din, M.G., Chang, S.X., 2019. Biochar properties and lead (II) adsorption capacity depend on feedstock type, pyrolysis temperature, and steam activation. Chemosphere 231, 393–404.

Lebrun, M., De Zio, E., Miard, F., Scippa, G.S., Renzone, G., Scaloni, A., Bourgerie, S., Morabito, D., Trupiano, D., 2020c. Amending an As/Pb contaminated soil with biochar, compost and iron grit: effect on *Salix viminalis* growth, root proteome profiles and metal (loid) accumulation indexes. Chemosphere 244, 125397.

Lebrun, M., Macri, C., Miard, F., Hattab-Hambli, N., Motelica-Heino, M., Morabito, D., Bourgerie, S., 2017. Effect of biochar amendments on As and Pb mobility and phytoavailability in contaminated mine technosols phytoremediated by *Salix*. J. Geochem. Explor. 182, 149–156.

Lebrun, M., Miard, F., Hattab-Hambli, N., Bourgerie, S., Morabito, D., 2018a. Assisted phytoremediation of a multi-contaminated industrial soil using biochar and garden soil amendments associated with *Salix alba* or *Salix viminalis*: abilities to stabilize As, Pb, and Cu. Water Air Soil Pollut. 229 (5), 163.

Lebrun, M., Miard, F., Hattab-Hambli, N., Scippa, G.S., Bourgerie, S., Morabito, D., 2020a. Effect of different tissue biochar amendments on As and Pb stabilization and phytoavailability in a contaminated mine technosol. Sci. Total Environ. 707, 135657.

Lebrun, M., Miard, F., Nandillon, R., Hattab-Hambli, N., Scippa, G.S., Bourgerie, S., Morabito, D., 2018b. Eco-restoration of a mine technosol according to biochar particle size and dose application: study of soil physico-chemical properties and phytostabilization capacities of *Salix viminalis*. J. Soils Sediments 18 (6), 2188–2202.

Lebrun, M., Miard, F., Nandillon, R., Léger, J.C., Hattab-Hambli, N., Scippa, G.S., Bourgerie, S., Morabito, D., 2018c. Assisted phytostabilization of a multicontaminated mine technosol using biochar amendment: Early stage evaluation of biochar feedstock and particle size effects on As and Pb accumulation of two *Salicaceae* species (*Salix viminalis* and *Populus euramericana*). Chemosphere 194, 316–326.

Lebrun, M., Miard, F., Nandillon, R., Scippa, G.S., Bourgerie, S., Morabito, D., 2019. Biochar effect associated with compost and iron to promote Pb and As soil stabilization and *Salix viminalis* L. growth. Chemosphere 222, 810–822.

Lebrun, M., Miard, F., Renouard, S., Nandillon, R., Scippa, G.S., Morabito, D., Bourgerie, S., 2018d. Effect of Fe-functionalized biochar on toxicity of a technosol contaminated by Pb and As: sorption and phytotoxicity tests. Environ. Sci. Pollut. Res. 25 (33), 33678–33690.

Lebrun, M., Miard, F., Scippa, G.S., Hano, C., Morabito, D., Bourgerie, S., 2020b. Effect of biochar and redmud amendment combinations on *Salix triandra* growth, metal(loid) accumulation and oxidative stress response. Ecotoxicol. Environ. Saf. 195, 110466.

Lebrun, M., Miard, F., Nandillon, R., Morabito, D., Bourgerie, S., 2021. Biochar application rate: improving soil fertility and *linum usitatissimum* growth on an arsenic and lead contaminated technosol. Int. J. Environ. Res 15, 125–134.

Lee, Y., Park, J., Ryu, C., Gang, K., Yang, W., Park, Y., Jung, J., Hyun, S., 2013. Comparison of biochar properties from biomass residues produced by slow pyrolysis at 500°C. Bioresour. Technol. 148, 196–201.

Li, G., Khan, S., Ibrahim, M., Sun, T.R., Tang, J.F., Cotner, J.B., Xu, Y.Y., 2018. Biochars induced modification of dissolved organic matter (DOM) in soil and its impact on mobility and bioaccumulation of arsenic and cadmium. J. Hazard. Mater. 348, 100–108.

Li, J., Dai, J., Liu, G., Zhang, H., Gao, Z., Fu, J., He, Y., Huang, Y., 2016. Biochar from microwave pyrolysis of biomass: a review. Biomass Bioenergy 94, 228–244.

Lin, L., Qiu, W., Wang, D., Huang, Q., Song, Z., Chau, H., 2017. Arsenic removal in aqueous solution by a novel Fe-Mn modified biochar composite: Characterization and mechanism. Ecotoxicol. Environ. Saf. 144, 514–521.

Liang, C., Gascó, G., Fu, S., Méndez, A., Paz-Ferreiro, J., 2016. Biochar from pruning residues as a soil amendment: Effects of pyrolysis temperature and particle size. Soil Tillage Res. 164, 3–10.

Liang, J., Yang, Z., Tang, L., Zeng, G., Yu, M., Li, X., Wu, H., Qian, Y., Li, X., Luo, Y., 2017. Changes in heavy metal mobility and availability from contaminated wetland soil remediated with combined biochar-compost. Chemosphere 181, 281–288.

Liu, L., Li, W., Song, W., Guo, M., 2018. Remediation techniques for heavy metal-contaminated soils: Principles and applicability. Sci. Total Environ. 633, 206–219.

Lomaglio, T., Hattab-Hambli, N., Bret, A., Miard, F., Trupiano, D., Scippa, G.S., Motelica-Heino, M, Bourgerie, S., Morabito, D., 2017. Effect of biochar amendments on the mobility and (bio) availability of As, Sb and Pb in a contaminated mine technosol. J. Geochem. Explor. 182 (Part B), 138–148.

Lomaglio, T., Hattab-Hambli, N., Miard, F., Lebrun, M., Nandillon, R., Trupiano, D., Scippa, G.S, Gauthier, A., Motelica-Heino, M., Bourgerie, S., Morabito, D., 2018. Cd, Pb, and Zn mobility and (bio) availability in contaminated soils from a former smelting site amended with biochar. Environ. Sci. Pollut. Res. 25 (26), 25744–25756.

Lu, H., Li, Z., Fu, S., Méndez, A., Gascó, G., Paz-Ferreiro, J., 2015. Combining phytoextraction and biochar addition improves soil biochemical properties in a soil contaminated with Cd. Chemosphere 119, 209–216.

Lu, K., Yang, X., Gielen, G., Bolan, N., Ok, Y., Niazi, N., Xu, S., Yuan, G., Chen, X., Zhang, X., Liu, D., Song, Z., Liu, X., Wang, H., 2017. Effect of bamboo and rice straw biochars on the mobility and redistribution of heavy metals (Cd, Cu, Pb and Zn) in contaminated soil. J. Environ. Manage. 186, 285–292.

Mary, G.S., Sugumaran, P., Niveditha, S., Ramalakshmi, B., Ravichandran, P., Seshadri, S., 2016. Production, characterization and evaluation of biochar from pod (Pisum sativum), leaf (*Brassica oleracea*) and peel (*Citrus sinensis*) wastes. Int. J. Recycl. Org. Waste Agric. 5 (1), 43–53.

Marzooqi, Al, F., Yousef, F., L., 2017. Biological response of a sandy soil treated with biochar derived from a halophyte (*Salicornia bigelovii*). Appl. Soil Ecol. 114, 9–15.

Masulili, A., Utomo, W.H., Syechfani, M.S., 2010. Rice husk biochar for rice based cropping system in acid soil 1. The characteristics of rice husk biochar and its influence on the properties of acid sulfate soils and rice growth in West Kalimantan, Indonesia. J. Agric. Sci. 2 (1), 39–47.

Meng, J., Tao, M., Wang, L., Liu, X., Xu, J., 2018. Changes in heavy metal bioavailability and speciation from a Pb-Zn mining soil amended with biochars from co-pyrolysis of rice straw and swine manure. Sci. Total Environ. 633, 300–307.

Mokarram-Kashtiban, S., Hosseini, S.M., Kouchaksaraei, M.T., Younesi, H., 2019. Biochar improves the morphological, physiological and biochemical properties of white willow seedlings in heavy metal-contaminated soil. Arch. Biol. Sci. 71 (2), 281–291.

Molnár, M., Vaszita, E., Farkas, É., Ujaczki, É., Fekete-Kertész, I., Tolner, M., Klebercz, O., Kirchkeszner, C., Gruiz, K., Uzinger, N., Feigl, V., 2016. Acidic sandy soil improvement with biochar — a microcosm study. Sci. Total Environ. 563-564, 855–865.

Moore, F., González, M.E., Khan, N., Curaqueo, G., Sanchez-Monedero, M., Rilling, J., Morales, E., Panichini, M., Mutis, A., Jorquera, M., Mejias, J., Hirzel, J., Mejias, J., 2017. Copper immobilization by biochar and microbial community abundance in metal-contaminated soils. Sci. Total Environ. 616-617, 960–969.

Motasemi, F., Afzal, M.T., 2013. A review on the microwave-assisted pyrolysis technique. Renew. Sust. Energy Rev. 28, 317–330.

Namgay, T., Singh, B., Singh, B., 2010. Influence of biochar application to soil on the availability of As, Cd, Cu, Pb, and Zn to maize (Zea mays L.). Aust. J. Soil Res. 48 (7), 638–647.

Nandillon, R., Lahwegue, O., Miard, F., Lebrun, M., Gaillard, M., Sabatier, S., Battaglia-Brunet, F., Morabito, D., Bourgerie, S., 2019b. Potential use of biochar, compost and iron grit associated with *Trifolium repens* to stabilize Pb and As on a multi-contaminated technosol. Ecotoxicol. Environ. Saf. 182, 109432.

Nandillon, R., Lebrun, M., Miard, F., Gaillard, M., Sabatier, S., Morabito, D., Bourgerie, S., 2019a. Contrasted tolerance of *Agrostis capillaris* metallicolous and non-metallicolous ecotypes in the context of a mining technosol amended by biochar, compost and iron sulfate. Environ. Geochem. Health 43, 1457–1575.

Nandillon, R., Lebrun, M., Miard, F., Gaillard, M., Sabatier, S., Villar, M., Bourgerie, S., Morabito, D., 2019d. Capability of amendments (biochar, compost and garden soil) added to a mining technosol contaminated by Pb and As to allow poplar seed (Populus nigra L.) germination. Environ. Monit. Assess. 191 (7), 465.

Nandillon, R., Miard, F., Lebrun, M., Gaillard, M., Sabatier, S., Bourgerie, S., Battaglia-Brunet, F., Morabito, D., 2019c. Effect of biochar and amendments on Pb and As phytotoxicity and phytoavailability in a technosol. CLEAN–Soil, Air, Water 47 (3), 1800220.

Nie, C., Yang, X., Niazi, N.K., Xu, X., Wen, Y., Rinklebe, J., Ok, Y.S, Xu, S., Wang, H., 2018. Impact of sugarcane bagasse-derived biochar on heavy metal availability and microbial activity: A field study. Chemosphere 200, 274–282.

Nigussie, A., Kissi, E., Misganaw, M., Ambaw, G., 2012. Effect of biochar application on soil properties and nutrient uptake of lettuces (*Lactuca sativa*) grown in chromium polluted soils. Am. Eurasian J. Agric. Environ. Sci. 12 (3), 369–376.

Obia, A., Mulder, J., Martinsen, V., Cornelissen, G., Børresen, T., 2016. In situ effects of biochar on aggregation, water retention and porosity in light-textured tropical soils. Soil Tillage Res. 155, 35–44.

Oliveira, F.R., Patel, A.K., Jaisi, D.P., Adhikari, S., Lu, H., Khanal, S.K., 2017. Environmental application of biochar: current status and perspectives. Bioresour. Technol. 246, 110–122.

Olmo, M., Alburquerque, J.A., Barrón, V., Del Campillo, M.C., Gallardo, A., Fuentes, M., Villar, R., 2014. Wheat growth and yield responses to biochar addition under Mediterranean climate conditions. Biol. Fertil. Soils 50 (8), 1177–1187.

Panagos, P., Van Liedekerke, M., Yigini, Y., Montanarella, L., 2013. Contaminated sites in Europe: review of the current situation based on data collected through a European network. J. Environ. Public Health 2013, 158764.

Pandey, V.C., Bajpai, O., 2019. Phytoremediation: from theory towards practice. In: Pandey, V.C., Bauddh, K. (Eds.), Phytomanagement of Polluted Sites. Elsevier, Amsterdam, pp. 1–49. 10.1016/B978-0-12-813912-7.00001-6.

Pandey, V.C., Pandey, D.N., Singh, N., 2015. Sustainable phytoremediation based on naturally colonizing and economically valuable plants. J. Clean. Prod. 86, 37–39.

Park, J., Choppala, G., Bolan, N., Chung, J., Chuasavathi, T., 2011. Biochar reduces the bioavailability and phytotoxicity of heavy metals. Plant Soil 348 (1-2), 439–451.

Paya Perez, A., Rodriguez Eugenio, N., 2018. Status of Local Soil Contamination in Europe: Revision of the Indicator "Progress in the Management Contaminated Sites in Europe. EUR 29124 EN, Publications Office of the European Union, Luxembourg, JRC107508. doi:10.2760/093804.

Pettit, R.E., 2004. Organic matter, humus, humate, humic acid, fulvic acid and humin: their importance in soil fertility and plant health. CTI Res., 1–17.

Purakayastha, T., Kumari, S., Pathak, H., 2015. Characterisation, stability, and microbial effects of four biochars produced from crop residues. Geoderma 239-240, 293–303.

Qambrani, N.A., Rahman, M.M., Won, S., Shim, S., Ra, C., 2017. Biochar properties and eco-friendly applications for climate change mitigation, waste management, and wastewater treatment: a review. Renew. Sust. Energy Rev. 79, 255–273.

Qian, K., Kumar, A., Zhang, H., Bellmer, D., Huhnke, R., 2015. Recent advances in utilization of biochar. Renew. Sust. Energy Rev. 42, 1055–1064.

Rafael, R.B.A., Fernández-Marcos, M.L., Cocco, S., Ruello, M.L., Fornasier, F., Corti, G., 2019. Benefits of biochars and npk fertilizers for soil quality and growth of cowpea (*Vigna unguiculata* L. Walp.) in an acid arenosol. Pedosphere 29 (3), 311–333.

Rajapaksha, A., Chen, S., Tsang, D., Zhang, M., Vithanage, M., Mandal, S., Gao, B., Bolan, N., Ok, Y., 2016. Engineered/designer biochar for contaminant removal/immobilization from soil and water: potential and implication of biochar modification. Chemosphere 148, 276–291.

Rehrah, D., Bansode, R., Hassan, O., Ahmedna, M., 2016. Physico-chemical characterization of biochars from solid municipal waste for use in soil amendment. J. Anal. Appl. Pyrol. 118, 42–53.

Sarkhot, D.V., Ghezzehei, T.A., Berhe, A.A., 2013. Effectiveness of biochar for sorption of ammonium and phosphate from dairy effluent. J. Environ. Qual. 42 (5), 1545–1554.

Shaaban, M., Van Zwieten, L., Bashir, S., Younas, A., Núñez-Delgado, A., Chhajro, M.A., Kubar, K.A., Ali, U., Rana, M.S., Mehmood, M.A., Hu, R., 2018. A concise review of biochar application to agricultural soils to improve soil conditions and fight pollution. J. Environ. Manag. 228, 429–440.

Tan, X.F., Liu, S.B., Liu, Y.G., Gu, Y.L., Zeng, G.M., Hu, X.J., Wang, X., Liu, S.H., Jiang, L.H., 2017a. Biochar as potential sustainable precursors for activated carbon production: multiple applications in environmental protection and energy storage. Bioresour. Technol. 227, 359–372.

Tan, Z., Lin, C.S., Ji, X., Rainey, T.J., 2017b. Returning biochar to fields: A review. Appl. Soil Ecol. 116, 1–11.

Tian, J., Wang, J., Dippold, M., Gao, Y., Blagodatskaya, E., Kuzyakov, Y., 2016. Biochar affects soil organic matter cycling and microbial functions but does not alter microbial community structure in a paddy soil. Sci. Total Environ. 556, 89–97.

Uzoma, K., Inoue, M., Andry, H., Fujimaki, H., Zahoor, A., Nishihara, E., 2011. Effect of cow manure biochar on maize productivity under sandy soil condition. Soil Use Manage. 27 (2), 205–212.

Van Liedekerke, M., Prokop, G., Rabl-Berger, S., Kibblewhite, M., Louwagie, G., 2014. Progress in the management of contaminated sites in Europe. *R*eference Report by the Joint Research Centre of the European Commission 1 (1), 4-6.

Wang, R., Wei, S., Jia, P., Liu, T., Hou, D., Xie, R., Lin, Z., Qiao, Y., Chang, X., Lu, L., Tian, S., 2019. Biochar significantly alters rhizobacterial communities and reduces Cd concentration in rice grains grown on Cd-contaminated soils. Sci. Total Environ. 676, 627–638.

Wu, W., Li, J., Niazi, N.K., Müller, K., Chu, Y., Zhang, L., Yuan, G., Lu, K., Song, Z., Wang, H., 2016. Influence of pyrolysis temperature on lead immobilization by chemically modified coconut fiber-derived biochars in aqueous environments. Environ. Sci. Pollut. Res. 23 (22), 22890–22896.

Xie, Y., Fan, J., Zhu, W., Amombo, E., Lou, Y., Chen, L., Fu, J., 2016. Effect of heavy metals pollution on soil microbial diversity and bermudagrass genetic variation. Front. Plant Sci. 7, 755.

Xu, M., Xia, H., Wu, J., Yang, G., Zhang, X., Peng, H., Yu, X., Li, L., Xiao, H., Qi, H., 2017. Shifts in the relative abundance of bacteria after wine-lees-derived biochar intervention in multi metal-contaminated paddy soil. Sci. Total Environ. 599, 1297–1307.

Xu, Y., Seshadri, B., Sarkar, B., Wang, H., Rumpel, C., Sparks, D., Farrell, M., Hall, T., Yang, X., Bolan, N., 2018. Biochar modulates heavy metal toxicity and improves microbial carbon use efficiency in soil. Sci. Total Environ. 621, 148–159.

Yaashikaa, P.R., Kumar, P.S., Varjani, S., Saravanan, A., 2020. A critical review on the biochar production techniques, characterization, stability and applications for circular bioeconomy. Biotechnol. Rep. 28, e00570.

Yang, J., Li, H., Zhang, D., Wu, M., Pan, B., 2017. Limited role of biochars in nitrogen fixation through nitrate adsorption. Sci. Total Environ. 592, 758–765.

Yin, D., Wang, X., Chen, C., Peng, B., Tan, C., Li, H., 2016. Varying effect of biochar on Cd, Pb and As mobility in a multi-metal contaminated paddy soil. Chemosphere 152, 196–206.

Yu, H., Zou, W., Chen, J., Chen, H., Yu, Z., Huang, J., Tang, H., Wei, X., Gao, B., 2019. Biochar amendment improves crop production in problem soils: A review. J. Environ. Manage. 232, 8–21.

Yu, Z., Qiu, W., Wang, F., Lei, M., Wang, D., Song, Z., 2017. Effects of manganese oxide-modified biochar composites on arsenic speciation and accumulation in an indica rice (*Oryza sativa* L.) cultivar. Chemosphere 168, 341–349.

Yuan, Y., Bolan, N., Prévoteau, A., Vithanage, M., Biswas, J.K., Ok, Y.S., Wang, H., 2017. Applications of biochar in redox-mediated reactions. Bioresour. Technol. 246, 271–281.

Yuan, P., Wang, J., Pan, Y., Shen, B., Wu, C., 2018. Review of biochar for the management of contaminated soil: Preparation, application and prospect. Sci. Total Environ. 659, 473–490.

Yue, Y., Cui, L., Lin, Q., Li, G., Zhao, X., 2017. Efficiency of sewage sludge biochar in improving urban soil properties and promoting grass growth. Chemosphere 173, 551–556.

Yun, C., Yan, M., Dejie, G., Qiujun, W., Guangfei, W., 2017. Chemical properties and microbial responses to biochar and compost amendments in the soil under continuous watermelon crop. Plant Soil Environ. 63 (1), 1–7.

Zeng, Z., Li, T.Q., Zhao, F.L., He, Z.L., Zhao, H.P., Yang, X.E., Wang, H., Zhao, J., Rafiq, M.T., 2013. Sorption of ammonium and phosphate from aqueous solution by biochar derived from phytoremediation plants. J. Zhejiang Univ. Sci. B 14 (12), 1152–1161.

Zhang, H., Shao, J., Zhang, S., Zhang, X., Chen, H., 2019. Effect of phosphorus-modified biochars on immobilization of Cu (II), Cd (II), and As (V) in paddy soil. J. Hazard. Mater. 390, 121349.

Zhang, J., Chen, Q., You, C., 2016. Biochar effect on water evaporation and hydraulic conductivity in sandy soil. Pedosphere 26 (2), 265–272.

Zhao, H., Xue, Y., Long, L., Hu, X., 2018. Adsorption of nitrate onto biochar derived from agricultural residuals. Water Sci. Technol. 77 (2), 548–554.

Zhao, L., Cao, X., Mašek, O., Zimmerman, A., 2013. Heterogeneity of biochar properties as a function of feedstock sources and production temperatures. J. Hazard. Mater. 256-257, 1–9.

Zhou, Y., Gao, B., Zimmerman, A.R., Fang, J., Sun, Y., Cao, X., 2013. Sorption of heavy metals on chitosan-modified biochars and its biological effects. Chem. Eng. J. 231, 512–518.

CHAPTER

Chelate-assisted phytoremediation

5

Dragana Ranđelović[a], Ksenija Jakovljević[b], Tijana Zeremski[c]

[a]*Institute for Technology of Nuclear and Other Mineral Raw Materials, Belgrade, Serbia*

[b]*University of Belgrade, Faculty of Biology, Institute of Botany and Botanical Garden, Belgrade, Serbia*

[c]*Institute of Field and Vegetable Crops, National Institute of the Republic of Serbia, Novi Sad, Serbia*

5.1 Introduction

The rapid development, as a imperative of modern society, has resulted in serious environmental pollution, which has been emphasized as one of the biggest problem in recent decades. One of the leading contributors to soil contamination are metals, especially those labelled as heavy metals. They originate from both natural and anthropogenic sources (Kierczak et al., 2008). Some of these elements are essential for the plants (Fe, Co, Mn, Cr, Zn, etc.), and in certain quantities are necessary for their well-being. However, there are also non-essential ones, such as Hg, Pb, As, Cd, Al, Sr, etc. Although necessary for plant growth, elevated quantites of essential elements can cause toxic effects in plants, while non-essential elements are harmful even in a small dose (Desideri et al., 2011). In view of environmental risk assessment, it is particularly important to estimate the potential bioavailability of each metal under the given conditions.

The increasing pollution of soil with heavy metals raised the issue of their removal. One of the most acceptable methods, taking into account economy and environmental issues is phytoremediation. Phytoremediation is a technique of removing (or rendering them harmless) contaminant from the environment (air, water, and soil) using plants and associated microbes (Ali et al., 2013; Pinto et al., 2015; Singh et al., 2003). There are four main phytoremediation techniques: phytofiltration, phytovolatization, phytostabilization, and phytoextraction (Ali et al., 2013). Phytofiltration or rhizofiltration represent removing of pollutants from water by plants, through their adsorbtion or absorbtion in the plant roots (Pinto et al., 2015). Phytovolatization refers to the uptake of pollutants by plants, conversion to the less toxic volatile form and their subsequent release in the atmosphere; and it is particularly suitable for removing of elemental compounds (Cluis, 2004; Singh et al., 2003). For soil remediation, two basic strategies can be applied: phytostabilization and phytoextraction. Phytostabilization refers not to removal, but to stabilization of pollutants using vegetation cover, mostly on moderately contaminated sites (Krämer 2005). Several processes are assumed to underlie this technique: sorption of pollutants by plants roots, complexation, and precipitation (Pinto et al., 2015; Sarwar et al., 2017). The phytostabilization is particularly useful in heavily polluted industrial sites, where some pioneer species can be used as a

Assisted Phytoremediation. DOI: https://doi.org/10.1016/B978-0-12-822893-7.00004-5

stabilizer of technogenic substrates, preventing erosion and leaching of contaminants to adjacent habitats (Jakovljević et al., 2020).

Phytoextraction represents the most efficient phytoremediation technique for removing pollutants from contaminated soil (Ali et al., 2013; Sarwar et al., 2017). It includes the cultivation of metal-accumulating plants at the contaminated site that absorb pollutants from the soil and concentrate them in their shoots. At the end of vegetation season, these plants are being harvested and taken away, thus reducing the concentration of contaminants at a given site (Cluis, 2004). Underlie this process are metal-accumulating plants, able to absorb pollutants in significant amounts, i.e. above the certain hyperaccumulation threshold, whereby no signs of toxicity (Baker and Brooks, 1989).

Besides the significant biomass production and high metal accumulation capacity, to be able to meet phytoremediation requirements, plants should be non-invasive, fast-growing, deep-rooted, easy for mowing, easily propagated, etc. (Pandey et al., 2015; Pandey and Bajpai, 2019; Mikavica et al., 2020; Robinson et al., 2000).

However, plant accumulation capacity, as the most significant factor in phytoextraction, can be improved for both accumulator and non-accumulator species. The most beneficial procedure for it's intensifying is chemically enhanced phytoremediation, i.e. remediation with chelating agents (Farid et al., 2013). Namely, through increasing solubility of metals, chelates enhance their extraction by plant species, so that non-accumulator species can also be used in phytoextraction if produce sufficient above-ground biomass (Sarwar et al., 2017). Various chelates can be used in process of chelate-assisted phytoremediation.

5.2 Chelating agents

By definition chelate is a chemical compound in which a metal ion is bound to an organic compound called chelating agent. Chelating agents (chelators or chelants) are organic molecules (ligands) which can create one or several bonds to a single metal cation. These metal-chelant complexes are usualy water-soluble and stable due to strong coordinate (dative covalent) bonds (Fig. 5.1). Chelating agents have been used for more than 50 years in fertilizers in order to provide plants grown hidroponicaly or in

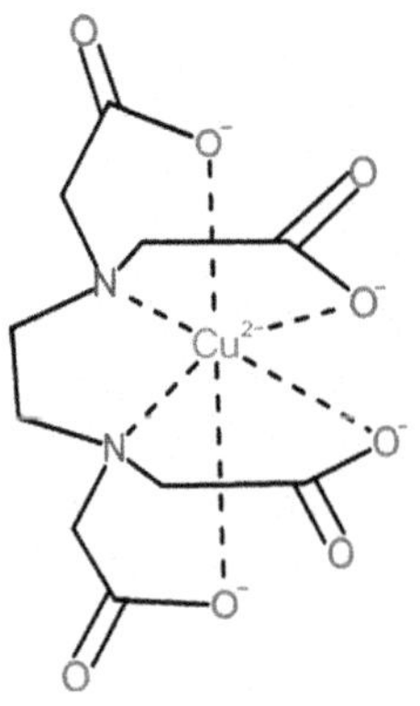

FIG. 5.1

Cu EDTA chelate structure.

soils with micronutrients (Salt et al., 1995). Bonding micronutriens in chelates prevents its precipitation and sorption matrix, thus keeping them available for plant uptake (Salt et al., 1995).

There are several types of chelating agents currently used for chelate-assisted phytoextractions. The usage of each type of chelates has its advantages and disadvantages, depending on targeted metal, soil characteristics, and used plant species. Thus the selection of the most suitable chelant is one of the key steps for successful phytoextraction.

5.2.1 Synthetic/persistent aminopolycarboxylic acids

One of the most efficient chelants from aminopolycarboxylic acids (APCAs) family is ethylene diamine tetraacetic acid (EDTA). Its usage in chelate enhanced phytoextractions started in the late 1980s and early 1990s. Phytoextraction experiments with EDTA showed its ability to efficiently increase metal uptake by plants, especially Pb (Kos et al., 2003; Kos and Leštan, 2004; Luo et al., 2005), but its large-scale field applications were restricted due to low biodegradability of EDTA-metal complexes and prolonged presence in the soil, which increases the risk of heavy metals leaching from soil to water (Grčman et al., 2003; Luo et al., 2005).

Other syntethic APCAs tested and used for chelate-assisted phytoextractions are hydroxylethylene diamine tetraacetic acid (HEDTA), diethylene triamine pentaacetic acid (DTPA), ethylenebis (oxyethylenenitrilo)-tetraacetic acid (EGTA), trans-1,2-cyclohexylene dinitrilo tetraacetic acid (CDTA), ethylenediamine N,N'bis(o-hydroxyphenyl)acetic acid (EDDHA), N (2-hydroxyethyl)iminodiacetic acid (HEIDA), and N,N' di(2-hydroybenzyl) ethylene diamine N,N' diacetic acid (HBED). In most studies, when compared to the EDTA, other syntethic APCAs showed similar or decreased efficiency in increasing metal accumulation in plants (Huang et al., 1997).

5.2.2 Natural/biodegradable aminopolycarboxylic acids

In the last decade, the focus of chelate-assisted phytoextractions has moved towards chelants with higher biodegradability such as ethylene diamine disuccinate (EDDS), nitrilo triacetic acid (NTA), iminodisuccinic acid (IDSA), and methyl glycine diacetic acid (MGDA).

Most of the research has been carried out with EDDS which has three optical isomers: SS-, RS-, and RR-isomer, of which only [S,S']-isomer is biodegradable, and it is produced by microorganisms of which most common is *Amycolatopsis japonicum sp. nov* (Nishikiori et al., 1984; Goodfellow et al., 1997). EDDS used as chelator in phytoextraction increases the uptake of heavy metals similar to EDTA, but its efficiency is different than EDTA. Several authors (Kos and Leštan, 2004; Meers et al., 2005; Luo et al., 2006) found that EDDS is more efficient than EDTA in enhancing the Cu, Ni and Zn uptake by plants, while EDTA has higher efficiency in mobilization od Cd and Pb compared to EDDS. [S,S']-EDDS is a biodegradable chelator and in uncomplexed form is rapidly degraded (Schowanek et al. 1997; Vandevivere et al., 2001). The degradability of metal complexes with [S,S]-EDDS depends on the type of the metal and its properties. EDDS complexes with Ca, Cr, Fe, Pb, Al, Cd, Mg, Na, and Zn are easily degraded in the environment. However, the EDDS complexes with Cu, Ni, Co, and Hg show low biodegradability which can be explained with metal toxicity only in the case of EDTA-Hg complex (Vandevivere et al., 2001).

NTA is chelant primarily used in industry of detergents, but due to its low toxicity and high biodegradability which is correlated to low risk of leaching to groundwater it has been considered to be potentially good candidate for usage in chelate-assisted phytoremediation (Quartacci et al., 2007). Several studies proved the NTA potential in mobilization of heavy metals in soils and enhancing metal uptake by plants thus confirming its potential for usage in phytoextractions (Babaeian et al., 2016).

5.2.3 Natural low molecular weight organic acids

Natural low molecular weight organic acids (NLMWOA) such as gluconic (GA), oxalic (OA), malic (MA), succinic (SA), and citric acid (CA) are produced by plant-associated microbes or exudated by roots. Their role is to interact with metals present in the rhizosphere and by forming complexes to enhance their solubility and mobility (Rajkumar et al., 2012; Ullah et al., 2015b). Thus, NLMWOA can be used as chelating agents in chelate-assisted phytoextractions (Shakoor et al., 2014). One of the biggest advantages of NLMWOA is their very low toxicity since they can easily be degraded to carbon dioxide and water (Huang et al., 1998a), but their efficiency in mobilizing metals depends strongly on the type of metal and is rather low for most common soil pollutants. However, Huang et al. (1998b) found that CA application increased uranium concentrations in *Brassica juncea, B. chinensis, B. narinosa* and amaranth 1000-fold compared to the control as well as compared to other chelates such es EDTA, EDDHA, HEDTA, and DTPA.

5.3 The principle of chelate-assisted phytoremediation

The basic principle of chelate-assisted phytoextraction is illustrated in Fig. 5.1. While hyperaccumulating plants extract heavy metals from soil continuously, plants without hyperaccumulating abilities can extract only small quantities of heavy metals from soils due to its exclusion mechanisms. In chelate-assisted phytoextractions, metals became available for plant uptake after the chelant application to the soils. In contact with soil matrix chelant desorbs metals to mobile forms available to move to the plant roots where they are being uptaken (Fig. 5.2).

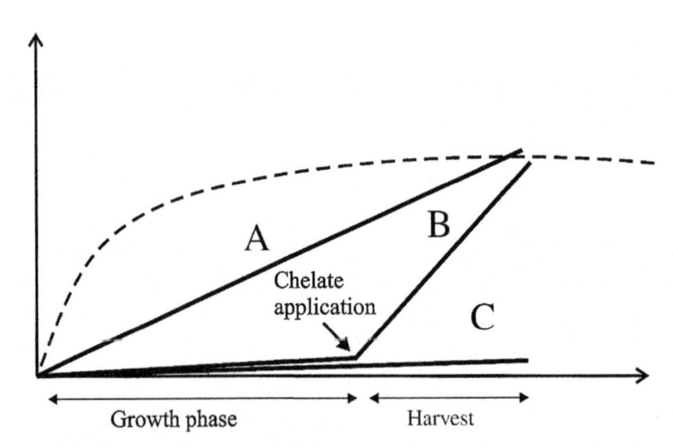

FIG. 5.2 Schematic representation of basic principle for chelate-assisted phytoextraction. Dashed line represents shoot biomass.

Solid line represents metal concentration in shoot of (a) hyperaccumulating plants (b) plants during chelate-assisted phytoremediation (c) non-hyperacumulating plants.

Fig. 5.2: Schematic representation of basic principle for chelate-assisted phytoextraction. Dashed line represents shoot biomass. Solid line represents metal concentration in shoot of (A) hyperaccumulating plants, (B) plants during chelate-assisted phytoremediation, and (C) non-hyperacumulating plants.

Chelates can be added in a single dose after plants develop enough biomass or in a single dose before transplanting, or be gradually added at several lower dosages during the plant growth period (Wu et al., 2007). Different regimes of application are likely to cause diverse effects in soil metal solubility, plant metal uptake, and growth. High concentration of chelates in soil solution often has strong phytotoxic effects on plants since the induction of metal uptake and accumulation of increased levels of toxic metals causes severe plant stress and irreversible damage to the plant (Salt et al., 1998). The time of chelant application is of crucial importance for phytoextraction efficiency as well as for reducing the risk of metal chelates leaching into groundwater (Blaylock et al., 1997; Luo et al., 2007). For that reasons, the application of the chelant is usually done a few days before harvest.

5.4 Metal mobilization in soil

Process of chelate-assisted phytoremediation includes mutual, complex interactions between soil, metal(s), chelate and plant. The main mechanisms included in chelate-assisted phytoremediation are metal solubilization and transport from soil, following the metal uptake and translocation in upper parts of plants.

Upon the addition of chelating agent to soil, metal solubilization starts via dissolution of soil minerals through ligand-exchange reaction and re-mobilization of metals adsorbed on solid surfaces (Nowack, 2002). Metals which are weakly bonded to soil matrix (e.g., in exchangeable and carbonate soil fractions) are solubilized more easily than metals bonded with organic and oxide soil fractions (Ferraro et al., 2015). Metals distribution between different soil fractions and the concentration of metals weakly bonded to soil fractions are one of the key factors affecting the efficiency of chelating agents for solubilization and extraction of metals in soils (Komárek et al., 2008; Tandy, 2004). Metals of anthropogenic origin usually tend to reside in the more mobile fractions, while metals of natural, lithological occurrence are found tightly bound in silicates (Kabala and Singh, 2001), which may additionally justify the usage of chelate for remediation of anthropogenically contaminated sites.

The extraction efficacy of chelate agent also depends on other factors, such as: chelate stability constant, ratio of metal to chelate, basic physical and chemical properties of soil (cation exchange capacity, pH, free carbonate content), soil contamination typology (artificially spiked/contaminated soils or field soils with historical contamination), etc. (Ferraro et al., 2015; Nowack et al., 2006; Ranđelović et al., 2015). Stability constant (log K) describes the stability of metal complex with ligand. Its value expresses the ability of organic ligands to chelate and thus solubilize metals. APCAs are non-selective ligands that can form stabile complexes with number of di- and trivalent cations. Synthetic APCAs generally tend to have higher affinity to chelate metals in comparison to LMWOA (Bian et al., 2018). Stability constants of EDTA complexes with di- and trivalent metal ions are usually higher than constants of EDDS complexes with same ions (Bucheli-Witschel and Egli, 2001).

Although the stability constant values (Ks) can be used to compare different chelants regarding their general efficiency, in real soil environment, where different metals are present in different concentrations, the efficiency of a chelant toward specific metal can be assessed only if the concentrations of competitive metals in soil are considered. For example, stability constants (log K) for Cu-EDTA and

Cu-EDDS complexes are 18.7 and 18.4 respectively (Bucheli-Witschel and Egli, 2001), which makes both chelants potentially good choice for complexing Cu in soil. But, if Ca is present in soil in high concentrations, then EDDS is more efficiant chelant than EDTA since the stability constant (log K) of Ca-EDTA complex is significantly higher than of Ca-EDDS complex (10.6 and 4.2 respectively) and EDTA will be used for complexing Ca instead of Cu.

Chelate-assisted remediation is particularly useful for soils polluted with heavy metals that have low solubility, such as Pb. Synthetic EDTA is one of the most efficient chelates for increasing solubility and availability of Pb from all soil fractions except residual (silica-bound) (Blaylock et al., 1997; Ullah et al., 2015a). Moreover, Barona et al. (2001) and Chrastný et al. (2008) in their experiments found that certain amount of Pb in residual fractions was released upon the addition of EDTA. Consequently, after the EDTA was applied to the soil Pb remains weakly absorbed to the soil components, favoring the further application of (phyto)extraction technologies (Barona et al., 2001). Investigations of Melo et al. (2008) proved that the LMWOA application to the soil increased concentration of Zn, Cu, and Pb in water-soluble and exchangeable soil fractions. Results showed remobilization of Zn from organic fraction, while Pb and Cu were mainly mobilized from reducible fractions. Similarly, the addition of NTA to the soil caused the lowering of Pb fraction bounded to organic matter while increased the concentration of Pb in the soil exchangeable fraction (Ullah et al., 2015a).

5.5 Interfering ions

Since the APCAs are not selective chelants, the complexation of metals other than targeted (such as Ca, Mg, Fe) happens when chelant is applied to the soil (Vogeler and Thayalakumaran, 2005). Therefore, in order to assess the efficiency of the applied chelating agent in solubilization of targeted metal in soil, the affinity of chelant to all the other, non-target metals (e.g., Ca, Fe, and Mg) present in the soil must eventually be considered.

Tandy et al. (2006b) found that in neutral and acidic soils, bounding metals in chelates with EDDS is strongly interfered with Fe. Koopmans et al. (2008) and Hauser et al. (2005) concluded that Al and Fe are the most important competitor ions in acidic soils since free EDDS can dissolve their oxides. Kim et al. (2003) analyzed the efficiency of EDTA for metal mobilization in soil suspensions with pH value below 6 and concluded that Fe is the major interfering ion under acidic conditions.

In calcareous soils, high concentrations of Ca strongly interfere with the process of heavy metals binding with EDTA. Theodoratos et al. (2000) in their study remarked a significant influence of the present Ca on the efficiency of Pb extraction by EDTA and concluded that for efficient removal of Pb from carbonate soils it is necessary to add excess amounts of EDTA. In calcium-rich soils, most of the EDTA is consumed to complex Ca released from $CaCO_3$ (Papassiopi et al., 1999; Sun et al., 2001). Manouchehri et al. (2006) also stated that the $CaCO_3$ content in soil is very important factor influencing the success of heavy metal complexation with EDTA. In their experiments, the efficiency of mobilization with EDTA was increased in carbonate-free soils.

Unfortunately, the presence of interfering ions in the soil is often neglected in phytoremediation studies. Information on concentrations of interfering ions (e.g., Ca, Mg and Fe) should be given together with the concentration of target metal in soil since they are also solubilized by the chelants. With a few notable exceptions (Collins et al., 2002; Lai and Chen, 2004; Shen et al., 2002), the information on the presence of interfering ions in soils is not provided in most studies focused on chelant-enhanced phytoextraction.

5.6 Chelant degradability in soil

Applied chelates should be biodegraded in timely manner and, preferentially, not persist longer than the period during which the plant roots actively uptake elements from the soil solution. Their biodegradability depends on the ligand properties, and the metal ion which is complexed with (Nörtemann, 2005). Generally, APCs that form stronger metal complexes (for example EDTA, EDGA, or DTPA) are relatively resistant to biodegradation, while chelates with low or moderately stability constants (such as EDDS or NTA) are found to degrade much faster (Bolton et al., 1993; Satrudinov et al., 2000). Investigation of Lombi et al. (2001) detected the presence of EDTA-metal complexes in soil even five months after the initial EDTA application. While investigations on biodegradability of EDDS reveal complete degradation of EDDS in soil 22 days after the application (Wang et al., 2012), degradability of uncomplexed EDDS in soils varied from 3.8 to 7.5 days half-life (Meers et al., 2005). Similarly, NTA chelate was found to be readily biodegraded in variety of soils (Tiedje and Mason, 1974). According to Ward (1985) half-life for NTA degradation under aerobic conditions ranged from 87 to 160 h or 3–7 days according to Bucheli-Witschel and Egli (2001). Additionally, natural low molecular organic acids, such as citric acid and oxalic acid were found to be completely biodegraded in soils only 2 weeks after initial application (Meers et al., 2005).

5.7 Chelate uptake by plants

The principles of free metals uptake by plants are described by free-ion activity model (FIAM). The basic assumption of this model is that metals uptake by plants can be provided only by a selective symplastic pathway which requires metals to be in free ionic form (Campbell, 1995). This was demonstrated through a number of various soil and hydroponic experiments where an existence of a significant correlation between the uptake of metals by plants and the concentration of metals in free-ion form was confirmed (Checkai et al., 1987; Sauvé et al., 1996). According to the FIAM, metals bounded in complexes with chelants such as EDDS and EDTA are not in their bioavailable free-ion form and therefore are not available for plant uptake. However, it was experimentally confirmed that chelant application causes increased metal uptake by plants (Kos and Leštan, 2004; Meers et al., 2005; Quartacci et al., 2007). Also, the presence of chelants in plant tissues was confirmed through experiments, and, for example, EDDS and EDTA as complexes or free compounds were detected in different plant tissues (roots, shoots, and xylem sap) (Collins et al., 2002; Tandy et al., 2006a). There are several steps of the metal entrance to the plant roots: metal transport to the root, adsorption on it, and binding to the root epidermis (Degryse et al., 2006; Seregin and Ivanov 2001). EDTA facilitates all these steps and enables the entrance of metal directly to the root. However, there are disagreements regarding the form in which EDTA-complex with metal enters the root – as dissociated or as preserved in the form of complex (Shahid et al., 2014). There are studies supporting both of these statements, reporting the existence of both metal-EDTA complex and free heavy metals ions (Collins et al., 2002; Schaider et al., 2006). On the contrary, according to Wenzel et al. (2003), metal-EDTA complexes are being formed inside the plant roots. Additionally, it was determined that uptake of metal-EDTA complex increases by the existence of root damage, bearing in mind that due to its size, EDTA in complex or alone, cannot pass the plasma membrane (Du et al., 2011). These studies prove that if concentration of chelates in solution is high enough it can be uptaken by the plants.

It has been proposed that chelates are taken up by plants via a fully apoplastic pathway (Collins et al., 2002). Metal-chelant complexes are taking apoplastic pathway passing through spaces between individual cells and cell walls of the roots (Marschner, 1986). However, the apoplastic pathway is not continual. It is interrupted in endodermis of the cortical cells. The Casparian strip whose role in plant roots is to prevent non-selective apoplastic uptake of metals is not an ideal barrier since it is not fully formed at the root tips and has disruptions in places where lateral roots are connected to the main root system (Haynes, 1980). The experiments cofirmed that these disruptions are the points where metals and metal-chelant complexes can enter to xylem without passing through a cell membrane and from there on be transported to the shoots (Haynes 1980; Tandy et al., 2006c). Beside the above described apoplastic pathway common for plant species with the Casparian strips, there are some plant species that have a small nuber of unsuberised cells in the endodermis, called the "passage cells". These cells are responsible for the apoplastic pathway of metals and chelates uptake from surrounding solution into the xylem (Clarkson 1996).

A number of studies confirmed that in the presence of chelators metal uptake is provided by non-selective apoplastic pathway. However, there is a difference between the uptake of low and high concentrations of essential and non-essential metals. Tandy et al. (2006a) analyzed the influence of EDDS on the uptake of Cu, Zn, and Pb in hydroponically grown sunflowers and concluded that at the low metal concentrations in solution and no EDDS added, Cu and Zn (essential metals) entered in to the root cells using primarily the symplastic way, while non-essential metal (Pb) had a very limited uptake due to the high selectivity of cellular uptake mechanisms. When the experiment was repeated but with the addition of EDDS in solution, the uptake of Cu and Zn was decreased, while the uptake of Pb was strongly increased. When metal concentration in solution was high and the EDDS was added, the uptake and translocation of metals to the shoots was significantly increased because the non-selective apoplastic uptake in the presence of chelants exceeds selective symplastic uptake mechanism for both essential and non-essential metals. Aplying chelants to a soil, thus, increases not only the total dissolved metal concentration, but also alters the uptake mechanism from symplastic to apoplastic (Tandy et al., 2006a).

The uptake mechanism of metal complexes with other APCAs is not explained yet, but it is assumed that it is similar to that for EDTA and EDDS. In an experiment conducted by Huang et al. (1997) the leaves of pea and corn plants got purple colour within of 12 h after the EDDHA was applied to the contaminated soil. Since the colour of EDDS solution is purple, they concluded that EDDHA was rapidly uptaken by roots and than translocated to shoots. Lipophilihty of the chelating agent also has the influence on the uptake mechanism and pathway. The symplastc pathway of uptake increases with the increasement of lipophility of chelant-metal complex which was experimentaly confirmed by Wu et al. (1999) who analyzed the influence of lipophilicity of the chelating agent on Pb uptake by *Zea mays.* Uptake and translocation of very hydrophylic compounds such as EDTA by plants is not favoured. That is the reason why highly liphophylic HBED shows better efficiency in increasing Pb root uptake compared to highly hydrophilic EDTA (Wu et al., 1999). Surprisingly, the translocation of HBED-Pb up into the shoots was much lower than EDTA-Pb translocation which the Wu et al. (1999) explained by a significantly higher stability constant of Fe-HBED (log K 37.9) than Pb-HBDE (log K 18.2). Due to its affinity for Fe, the internal replacement of Pb with Fe happens in the plan roots.

Metal translocation from roots to shoots in chelate-assisted phytoextractions differs and depends on the type of plants, chelating agents, and metals. Chen and Cutright (2001) reported that traslocation of Ni and Cd from roots to the shoots of *Helianthus annuus* was increased when HEDTA was applied, while translocation of Cr was decreased in the presence of the same chelating agent. Similar findings were reported by Sekhar et al. (2005) who used different chleating agents for Pb phytoextraction with *Hemidesmus indicus.* They found that EDTA and DTPA application stimulate translocation of Pb from roots to shoots while after the application of CDTA and DTPA the concentration of Pb was lower in the shoots than in the roots.

5.8 Effects of chelates on the plant

In addition to the increased uptake of targeted heavy metals, there are certain effects on the plant itself. Chelate-induced effects on plants depend on the mode of application, dosage and plant tolerance (Wu et al., 2007). Some species are particularly sensitive to chelate applications and pose lower ability to regulate the metal uptake (Souza et al., 2013). Generally, monocotyledons were found to be less sensitive to the chelates than dicotyledons, with lower biomass reductions and less chlorosis (Chen et al., 2004; Luo et al., 2006). Besides chlorosis, necrosis, and curling, leaves of dicotyledons exhibited fast senescence and drying of above-ground tissue (Chen et al., 2004).

The most common consequence of the chelate application is the reduction of plant biomass. It is commonly recorded when chelates are applied in a single higher dose, or gradually added at a couple of lower dosages during the growth period (Wu et al., 2007). In plants with less effective detoxification mechanisms, the increased level of soluble heavy metal may cause the reduction of photosynthesis and biomass, and, eventually, lead to the plant death (Lee and Sung 2014; Prasad and Strzałka 1999). It was found that plant biomass drops to almost half comparing to the untreated plants when applying a dose of 10 mmol kg^{-1} EDTA (Blaylock et al., 1997). Application of different dosages of EDTA in experiments with *Brassica rapa* L. var. *pekinensis* induced the necrotic lesions, especially pronounced in the older leaves, while the highest dose (10 mmol kg^{-1}) caused the strong reduction of plant biomass and senescence of the above-ground tissue (Grčman et al., 2001). Yield reduction, with or without other toxicity symptoms, was confirmed also in *Pisum sativum*, *Helianthus annuus*, *Zea mays*, and *Solidago bicolor* after application of HEDTA and EDTA (Chen and Cutright 2001; Huang et al., 1997). Luo et al. (2005) and Epelde et al. (2008) detected decrease of biomass (along with necrosis and chlorosis) in corn, bean and cardoon plants after EDDS application, in a higher rate than after application of EDTA. A similar was observed in *Nicotiana tabacum* (Evangelou et al., 2007). Application of EDDS also caused the toxic effects, such as necrosis and rapid senescence, in *Amaranthus* sp. *Brassica napus*, *B. rapa*, *Cannabis sativa* (Kos et al., 2003). Rashid et al. (2014) indicated that DTPA induced significant biomass decrease in *Lactuca sativa* in investigations on chelate assisted uptake of metals.

Although organic chelating agents are considered safer in the process of induced phytoextraction, the chelate dosage should be carefully planned. Namely, it was determined that citric acid does not affect the plant growth according to Evangelou et al. (2006), however, according to Turgut et al. (2004) application of CA in higher concentrations had toxic effects on certain plant species. Similarly, Gramss et al. (2004) recorded that a significant concentration of CA, after destroying the physiological barrier responsible for uptake controlling, could allow entry of soil solution rich in metals, causing necrosis in *Brassica chinesis*. Regarding the nitrilotriacetic acid (NTA), one of the natural chelators, according to Wenger et al. (2003), phytotoxic effects were not detected in the experiments with *Nicotiana tabacum*, while Robinson et al. (2000) reported pronounced reduction in biomass and leaves abscission in *Populus* clones. Biomass reduction was also detected in cardon and *Pistacia stratiotes*, although in

Daucus carota adverse effects were detected (Babacian et al., 2016; Epelde et al., 2008; Odjegba and Fasidi, 2004).

Although the chelate residual can express some unwanted post-harvesting effects on plants, those effects can also be beneficial. Chelates eventually increased the solubility of essential elements and enhanced nutrient uptake, which, coupled with detoxifying of heavy metals, enhanced the plant growth and consequently led to increased yields (Najeeb et al., 2009). The biomass increase in *Sedum alfredii* was observed after the application of 200 μM EDTA together with 200 μM Pb compared to the plant treated only with Pb (Tian et al., 2011). Likewise, the biomass of *Tribulus terrestris* greatly increased after the EDTA application, although the number of lateral stems decreased (Markovska et al., 2013). Although often less effective in comparison to synthetic APCAs, the application of organic chelates in assisted phytoremediation is favored due to often positive effects on the plant.

It was observed that humic acids enhanced plant growth, especially in *Brassica* species (Lee and Sung, 2014), mostly through the nutrients providing (Sung et al., 2013). Similarly, it was recorded that CA enhance solubility of metals and uptake of nutrients by *Brassica napus* and *Juncus effusus*, which led to the improved photosynthesis, growth, and biomass characteristics under conditions of stress caused by metals (Ehsan et al., 2014; Najeeb et al., 2009). Evangelou et al. (2006) presented the opposite influence of NLMWOA (citric, oxalic, and tartaric acid) applications on biomass in *Nicotiana tabacum*. Namely, after application of 62.5 mmol kg^{-1} of NLWOA, the increase of yield was detected, while at the high doses they recorded significant biomass decrease.

5.9 Examples of chelate-enhanced phytoremediation studies

The effects of chelates on metal uptake vary with soil type, targeted metal, and plant species (Evangelou et al., 2007). While some studies compared the efficacy of different chelates for increasing metal uptake in selected plants (Table 5.1), others concentrated on specific element or groups of elements important for the environment. Conducted studies differ in terms of experiment duration, pot or field settings, soil spiking or usage of naturally polluted soil and selected plant species. Different regimes of chelate addition are often applied in order to maximize plant metal uptake. In certain cases, the physiological response of plant species is also considered and evaluated, since the application of chelates can cause diverse effects on plant growth and development.

Chemically-assisted metal phytoextraction was thoroughly investigated on different agricultural crop species characterized by high biomass yields, such as *Zea mays*, *Helianthus annuus*, *Brassica juncea*, or *B. napus*. Luo et al. (2005) investigated the effects of EDTA, EDDS, and CA application on the uptake of Cd, Cu, Pb and Zn in *Z. mays*, concluding that application of 5 mmol kg^{-1} EDDS to soil was the most effective, with 45-fold increased concentrations of Cu in corn shoots compared to the control plants. However, treatment negatively affected the plant growth causing 60% decrease of shoot dry matter yields comparing to the control. Treatment with APCAs has significantly increased transfer ratios for Cu, Pb, Zn and Cd in plants, while CA was generally less effective in this sense. Similarly, Huang et al. (1997) investigated the role of synthetic chelates in lead phytoextraction on *Z. mays* L. cv. *Fiesta.* Effectiveness of chelation agents on Pb accumulation in soil was similar to the level of lead desorption in soil, following the series: EDTA > HEDTA > DTPA > EGTA > EDDHA. Addition of EDTA caused 57-fold increase of Pb shoot uptake in corn plants comparing to the control, inducing lead hyperaccumulation. In searching for less toxic EDTA alternatives, Meers et al. (2004) compared several

Table 5.1 Examples of chelate-assisted phytoremediation of various metals with selected plant species.

Chelating agent	Metal	Plant species	Reference
EDTA, EGTA	Cd	*Althaea rosea*	Liu et al. (2009)
NTA, EDDS	Pb, Cu, Zn	*Brassica carinata*	Quartacci et al. (2007)
NTA, DTPA, EDTA	Ni, Co	*Berkheya coddii*	Robinson et al. (1999)
EDTA, EDDS	Zn, Cd	*Chrysanthemum coronaruium*	Luo et al. (2006)
EDTA	Cd, Cu, Fe, Mn, Ni, Pb, Zn	*Helianthus annuus*	Liphadzi and Kirkham, (2006)
EDTA, HEDTA, DTPA, CDTA	Pb	*Hemidesmus indicus*	Sekhar et al. (2005)
EDTA	Cd	*Lemna minor*	Aravind et al. (2015)
EDTA	Cd	*Spirodela polyrhiza*	Aravind et al. (2015)
EDTA	Cd, Zn	*Lolium perenne*	Lambrechts et al. (2011)
OA, CA, EDDS	U	*Macleaya cordata*	Hu et al. (2019)
EDDS, MGDA	Pb, Zn	*Mirabilis jalapa*	Cao et al. (2007)
EDTA, CA	Pb, Zn, Cu, Cd	*Phaseolus vulgaris*	Luo et al. (2005)
EDDTA, HEDTA	Pb	*Pisum sativum*	Huang et al. (1997)
EDDS, CA, OA	As, Pb,Cd	*Pteris vittata*	Liang at al. (2019)
CA, EDDS,EDTA	Pb, Cd	*Ricinus communis*	Zhang et al. (2016)
EDTA, EGTA, HAc	Pb	*Sesbania exaltata*	Miller et al. (2008)
CA, DTPA	Am	*Vetiveria zizanoides*	Singh et al. (2014)
EDTA, EDDS	Pb	*Zea mays*	Komárek et al. (2007)
EDTA, DTPA, NTA, CA, OA, AA, SA	Cu, Zn, Cd, Ni	*Zea mays*	Meers et al. (2004)

synthetic APCAs with biodegradable, LMWOAs for assisted phytoextraction of Cd, Cu, Pb, Ni and Zn in *Z. mays*. While application of EDTA and DTPA increased accumulation values for all elements under study 2–3 times (respectively), significant increase in shoot metal levels was not observed in the treatment with organic acids, which was attributed to their fast degradability in soil and the low dosage applied. Field crop *Helianthus annuus* also demonstrated the ability to uptake and translates heavy metals, even to the level of hyperaccumulation for Cu and Pb (Chauhan and Mathur, 2020; Forte and Mutiti, 2017). Investigations of Chen and Cutright (2001) revealed the limited ability of both EDTA and HEDTA for improving the uptake of Cd, Cr and Ni in plant shoots. However, severe biomass reduction resulted in a final decrease in the total amount of metals removed at the end of the experiment.

Some of the important agricultural crops from Brassicaceae family are well known for their metal accumulation ability (Palmer et al., 2001). Accumulation rate of lead in *Brassica juncea* was tested through the application of several synthetic chelate agents (EDTA, DTPA, CDTA, EGTA), where the metal accumulation in plant shoots markedly increased (even up to several thousand to over ten thousand-fold than the control treatments), as found by Blaylock et al. (1997). The most effective was the concentration of 10 mmol kg^{-1} EDTA, upon whose application plants collected 1.5% Pb in their shoots (calculated to the dry weight). Although the dry yield of *B. juncea* was reduced upon the application of chelates, a very high magnitude of Pb accumulation made this phytoremediation approach potentially useful in practice. Three cultivars of *B. juncea* were also tested on the EDTA - enhanced uptake of Cd and Pb by Lai et al. (2008). As a result of the pot experiment, concentrations of Cd and Pb in shoots were significantly enhanced, approximately 110- to 320-fold and 6.5- to 17-fold, respectively, compared to the control. Similarly, the application of EDTA chelating agent showed enhanced Pb uptake in the experiment with *B. juncea* conducted by Karczewska et al. (2009), but the amounts of Pb leached from soils were higher than those removed by plants. For the phytoremediation of Cu, the same research showed better results in inducing Cu uptake with EDDS than EDTA, followed by increased leaching of Cu from the soil in a similar manner. The efficiency of chelates EDTA and EDDS was tested against the accumulation of Cd, Pb and Zn in *Brassica rapa* (Grčman et al., 2003). Single-dose applications of 10 mmol EDTA and EDDS per kg of soil were the most effective, causing the increase of Pb in the shoots 94.2- and 102.3-fold (respectively) comparing to the control. Addition of 10 mmol of EDTA and EDDS increased the concentration of Cd and Zn in the leaves to the smaller extent: 4.3- and 3.8-fold and 4.7- and 3.5-fold (respectively) compared to the control. Induced phytoextraction of Cu by using EDTA and EDDS in *Brassica napus* showed that EDDS is more efficient amendment for Cu phytoextraction than EDTA (Zeremski-Škorić et al., 2010). However, although increasing doses of EDDS have thereupon increased the shoot concentration of Cu up to 18-fold compared to the control, the accumulated amount of Cu in *B. napus* did not follow the same order of magnitude, as high EDDS concentrations caused significant growth depression (Figs. 5.3–5.6). Conversely, the addition of CA to Cr-contaminated soils with planted *B. napus* has not only increased Cr uptake in plants, but has also minimized Cr stress by significantly increasing plant growth, biomass, and chlorophyll content (Afshan et al., 2015).

Hyperaccumulating species were also subject of experiments with chelate-enhanced phytoextraction, exhibiting different effects and results. For example, *Sedum alfredii* (known as Zn hyperaccumulator) exhibited 1.7-fold Pb uptake after application of 0.1 mmol EDTA in shoots of ecotype collected from Pb/Zn mine area compared to the control, while the addition of EDTA to ecotype of *S. alfredii* in non-accumulating ecotype collected from the suburban garden in China showed a decrease in Pb root content and only slightly increase in Pb shoot content (Liu et al., 2007). Similarly, when 8 mmol kg^{-1}

FIG. 5.3

Brassica napus plants grown on soil polluted by Cu before EDDS application.

FIG. 5.4

Brassica napus plants grown on Cu polluted soil 3 days after EDDS application.

FIG. 5.5

Brassica napus plants grown on Cu polluted soil 14 days after treatment with 8 mmol kg^{-1} EDDS.

of EDTA was applied in pot-culture experiments with *S. alfredii*, concentration in Pb shoots increased 6.06-fold in comparison to the control (Sun et al., 2009). In same experiment, the addition of EDTA and CA increased accumulation of Zn in roots (0.87–2.34 times), stems (2.90–5.85 times) and shoots (3.71–6.06 times) and Cd in roots (0.79–1.68 times), stems (1.77–3.88 times), and shoots (3.37–4.86 times). However, the application of chelates in both experiments caused inhibitory effects on plant growth and biomass. Experiments on chelate-assisted phytoremediation on As-hyperaccumulating fern *Pteris vitatta* by biodegradable EDDS and LMWOAs (CA and OA) showed mainly increased

FIG. 5.6

Brassica napus plants grown on Cu polluted soil 14 days after treatment with 8 mmol kg^{-1} EDTA.

absorption of As, Cd, and Pb in rhizoid and frond (Liang et al., 2019). Content of As increased up to 2.23 times in frond and 1.18 times in rhizoid upon the addition of EDTA, up to 2.17 times in frond and 1.22 times in rhizoid when CA was added, and up to 1.9 times in frond and 1.43 in rhizoid upon the addition of OA. However, the addition of chelating agents EDDS and CA caused a significant decrease in *P. vitatta* rhizoid biomass, while OA showed a positive effect on plant biomass and even had stimulatory effects on the soil microbial structure and diversity. Contrary to these results, the addition of NTA, DTA, and EDTA for enhanced phytoremediation of Ni and Co with *Berkheya coddii* (Ni-hyperaccumulator plant) caused decreased uptake of Ni, while uptake of Co was unaffected (Robinson et al., 1999).

An additional group of plants with potential for use in (enhanced) phytoremediation are ornamental plants. Besides cleaning up the metals from the environment, they have additional advantages such as decorative features, reduced risk of metals entering in the food chain as they are mainly non-edible, the potential for generating income by utilizing their plant parts, and suitability for successful use in urban and suburban areas (Capuana, 2020; Nakbanpote et al., 2016). Application of EDTA and EGTA chelates for enhanced phytoextraction of Cd by *Mirabilis jalapa* caused a significant increase of Cd content in the plant: 149-fold to 189 -fold content of Cd in shoots (respectively), comparing to the control plants (Wang and Liu 2014). The addition of EDDS and MGDA for phytoextraction of Pb and Zn induced significant increase of Pb in leaves of *Mirabilis jalapa*, while Zn accumulation in plant stayed unaffected by the treatments (Cao at al. 2007). Use of biodegradable chelates (EDDS, OA and CA) was also tested for induced phytoremediation of U and Cd in *Zebrina pendula*, where maximum concentration of U and Cd accumulated in the plant (6.60- and 1.72-fold in comparison to the control, respectively) was reached by applying 5 mmol kg^{-1} CA treatment (Chen et al., 2019). The high concentration of EDDS was repressive to the plant growth, while low concentrations of CA and OA enhanced biomass production. Chelate-facilitating Cd uptake with EDTA and EGTA tested on *Calendula officinalis* showed that the application of EGTA significantly increased both total Cd content in the plant (up to 217%) and biomass, while the addition of EDTA caused toxicity to the plants (Liu et al., 2010).

Although woody species have been considered good phytoremediation candidates due to their high biomass production, enhanced growth rate and deep root systems, assisted phytoextraction on woody species has generally been investigated less often than in herbaceous species. However, certain fast growing, high biomass producing tree species such as poplar, willows and eucalyptus have been studied for the efficacy in chelate-assisted phytoremediation. Hu et al. (2014) tested efficiency of different doses of EDTA, EGTA, and CA on Cd uptake by *Populus alba* var. *pyramidalis*. Synthetic chelates showed better results in solubilizing heavy metals than LMWOAs, while addition of chelates generally increased leaf and decreased stem biomass. Addition of 9 mmol kg^{-1} EGTA showed highest leaf-to-soil ratio of 1.18–1.27 in *P. alba*, though for use in remediation practice collection of leaf fall was recommended in order to avoid recycling of bounded Cd. Researches of Komárek et al. (2007) on application of EDTA and EDDS for enhanced remediation of Pb from polluted agricultural soils by *Populus* sp. showed that EDTA addition led to a 12.9–13.7-fold increase of leaf Pb content, confirming high translocation rate and phytoremediation potential. Komárek et al. (2008) conducted research on two-year induced phytoextraction of, Cd, Cu, Pb, and Zn from contaminated agricultural soils on clone *Populus nigra* × *Populus maximoviczii*. Although the application of NH_4Cl proved to be more efficient than EDTA, especially during the phytoextraction of Cd and Zn, authors concluded that this clone is not suitable for chemically enhanced phytoremediation due to generally low metal transfer efficiency coupled with phytotoxic effects. Since fast-growing tree plants have often shown significant uptake and tolerance to Cd in phytoremediation investigations (Luo et al., 2016), Robinson et al. (2000) conducted an investigation on the potential for cadmium accumulation in poplar and willow clones. The addition of 2 g kg^{-1} EDTA and 0.5 g kg^{-1} NTA showed 1.7–1.8-fold increase of Cd content in leaves of poplar clone within 2 weeks of treatment, followed by leaf necrosis and abscission. At the end of the experiment no significant difference in Cd concentration between the control and the treatments was recorded, suggesting that the biomass should be harvested shortly after chelate addition in order to maximize Cd removal.

5.10 Advantages and drawbacks of chelate-assisted phytoremediation

Enhanced phytoremediation is considered to be a promising technology, due to its preserved low cost and shorter time-span needed for remediation compared to traditional phytoremediation process. Chelating agents effectively mobilize metals to soil solution, where they can be readily uptake by the plants. Moreover, chelate-enhanced phytoremediation offers an option for sites where mobility and availability of contaminants are low. Many researches confirmed the efficient metal uptake from soil by selected plant species by using suitable chelating agents (Blaylock et al., 1997; Komárek et al., 2007; Wang and Liu, 2014).

However, several risks connected with the application of this technology arisen with time. Due to the enhanced solubility of metals upon the addition of chelates, leaching of metals and other nutrients has been recorded (Evangelou et al., 2007; Thayalakumaran et al., 2003). If quantity of solubilized metals is larger than the uptake capacity of the plant, mobilized metals may persist in soil longer than the period in which the roots are able to uptake them. Associate drawbacks include the simultaneous extraction and leaching of soil macronutrients important for plant growth (Barona et al., 2001), which in turn may reduce the plant growth.

The excess of mobile metals is prone to leaching in groundwaters or deeper soil layers, which presents an additional environmental hazard and serious drawback for practical application of this technology. Researches considering various technical options, such as the use of horizontal permeable barriers (Kos and Leštan, 2004; Zhao et al., 2011) or polymer-coated chelates designed for slow release of chelant (Shibata et al., 2007) tried to offer viable solutions, yet this problem is still not handled to the level it becomes safe for practical use. Additionally, extended residual times of widely used chelate agents and their complex with metal ions pose a high risk for pollution of the surrounding environment (Tandy et al., 2004). Using biodegradable chelates with short degradation time instead of synthetic, hardly degradable chelating agents become one of the alternatives for minimizing environmental risks (Pinto et al., 2015; Souza et al., 2013).

Additional problem aroused from the persistence of chelates in the soil is their effect on plant growth, which may significantly corroborate the efficiency of phytoextraction. Moreover, upon the addition of chelates, certain negative effects on soil flora and fauna were observed (Romkens et al., 2002). Toxicity of chelates to the biota may be alleviated by the application of smaller doses of chelate, however, this may reduce the extraction efficiency, increase the costs and time needed for a successful remediation. The use of natural organic chelating agents was found to have considerably less toxic effects on plant growth (Evangelou et al., 2006), or even stimulative effects in certain cases (Ehsan et al., 2014). The price of the applied chelate should also be considered, as it increases the overall costs of the remediation process. Therefore, besides the obvious environmental concerns, the addition of chelates should be minimal in order to preserve the economic benefits of the phytoremediation process.

The selection of plant species and suitable agronomic practices that lead to an increase in biomass yield should follow the process of chelate selection and application (Li et al., 2005). Diverse plant species or even the same species grown in different soil, climate and hydrology conditions might exhibit a significant difference in terms of phytoremediation efficiency and tolerance to chelates (Bian et al., 2018). Additionally, majority of researches that stated efficiency of chelate-assisted phytoremediation and high solubilization rate of soil metals in artificial, laboratory conditions, found considerable differences and limitations when applied to the real conditions of multi-contaminated sites (Kim et al., 2003; Nascimento and Xing, 2006).

Although some researchers state that chelate-assisted phytoremediation has reached an impasse (Evangelou et al., 2007; Komárek et al., 2010), to defeat current serious flows, there is a growing need to search for new, effective and easy biodegradable chelating agents. Application technology and dosage of chelate agents is also subject for improvement. Finally, in order to become an acceptable remediation option, enhanced phytoremediation needs to be proven and optimized to the field conditions as safe, cost-effective and environmental-friendly solution.

References

Ali, H., Khan, E., Sajad, M.A., 2013. Phytoremediation of heavy metals—concepts and applications. Chemosphere 91 (7), 869–881.

Afshan, S., Ali, S., Bharwana, S., Rizwan, M., Farid, M., Abbas, F., Ibrahim, M., Mehmood, M., Abbasi, G., 2015. Citric acid enhances the phytoextraction of chromium, plant growth, and photosynthesis by alleviating the oxidative damages in *Brassica napus* L. Environ. Sci. Pollut. Res. 22, 11679–11689.

Aravind, R., Bharti, V.S, Rajkumar, M., Pandey, P.K., Purushothaman, C.S., Shukla, S.P., 2015. Chelating agent mediated enhancement of phytoremediation potential of *Spirodela polyrhiza* and *Lemna minor* for cadmium removal from the water. Int. J. Sci. Res. 5 (3), 2250–3153.

Babaeian, E., Homaee, M., Rahnemaie, R., 2016. Chelate-enhanced phytoextraction and phytostabilization of lead-contaminated soils by carrot (*Daucus carota*). Arch. Agron. Soil Sci 62 (3), 339–358.

Baker, A.J., Brooks, R., 1989. Terrestrial higher plants which hyperaccumulate metallic elements. A review of their distribution, ecology and phytochemistry. Biorecovery 1 (2), 81–126.

Barona, A., Aranguiz, I., Elías, A., 2001. Metal associations in soils before and after EDTA extractive decontamination: implications for the effectiveness of further clean-up procedures. Environ. Pollut. 113 (1), 79–85.

Bian, X., Cui, J., Tang, B., Yang, L., 2018. Chelant-Induced Phytoextraction of Heavy Metals from Contaminated Soils: A Review. Pol. J. Environ. Stud. 27 (6), 2417–2424.

Blaylock, M., Salt, D., Dushenkov, S, Zakharova, O., Gussman, C., Kapulnik, Y., Ensley, B., Salt, D., 1997. Enhanced accumulation of Pb in Indian mustard by soil-applied chelating agents. Environ. Sci. Technol. 31 (3), 860–865.

Bolton, H, Li, S., Workmoan, D, Girvin, D., 1993. Biodegradation of synthetic chelates in subsurface sediments from the Southeast Coastal Plain. J. Environ. Qual. 22 (1), 125–132.

Bucheli-Witschel, M., Egli, T., 2001. Environmental fate and microbial degradation of aminopolycarboxylic acids. FEMS Microbiol. Rev. 25 (1), 69–106.

Campbell, P.G.C., 1995. Interactions between trace metals and aquatic organisms: a critique of the free-ion activity model. In: Tessier, A., Turner, D.R. (Eds.), Metal Speciation and Bioavailability in Aquatic Systems. John Wiley & Sons, New York, USA, pp. 45–102.

Cao, A., Carucci, A., Lai, T., La Colla, P., Tamburini, E., 2007. Effect of biodegradable chelating agents on heavy metals phytoextraction with *Mirabilis jalapa* and on its associated bacteria. Eur. J. Soil Biol. 43 (4), 200–206.

Capuana, M., 2020. A review of the performance of woody and herbaceous ornamental plants for phytoremediation in urban areas. iForest 13 (2), 139–151.

Chauhan, P., Mathur, J., 2020. Phytoremediation efficiency of *Helianthus annuus* L. for reclamation of heavy metals-contaminated industrial soil. Environ. Sci. Pollut. Res. https://doi.org/10.1007/s11356-020-09233-x.

Checkai, R.T., Corey, R.B., Helmke, P.A., 1987. Effects of ionic and complexed metal concentrations on plant uptake of cadmium and micronutrientbmetals from solution. Plant Soil 99 (2), 335–345.

Chen, H., Cutright, T., 2001. EDTA and HEDTA effects on Cd, Cr and Ni uptake by *Helianthus annus*. Chemosphere 45 (1), 21–28.

Chen, L., Wang, D., Long, C., Cui, Z.X., 2019. Effect of biodegradable chelators on induced phytoextraction of uranium- and cadmium- contaminated soil by *Zebrina pendula* Schnizl. NatSci. Rep. 9 (1), 19817.

Chen, Y., Li, X., Shen, Z., 2004. Leaching and uptake of heavy metals by ten different species of plants during an EDTA-assisted phytoextraction process. Chemosphere 57 (3), 187–196.

Chrastný, V., Komárek, M., Jrovcová, E., Štíchová, J., 2008. A critical evaluation of the 0.05 M EDTA extraction of Pb from forest soils. Int. J. Environ. Anal. Chem. 88 (6), 385–396.

Clarkson, D.T., 1996. Root structure and sites of ion uptake. In: Waisel, Y., Eshel, A., Kafkafi, U. (Eds.), Plant Roots: The Hidden Half. Marcel Dekker Inc., New York, USA, pp. 483–510.

Cluis, C., 2004. Junk-greedy greens: phytoremediation as a new option for soil decontamination. BioTeach J. 2 (6), 61–67.

Collins, R.N., Merrington, G., McLaughlin, M.J., Knudsen, C., 2002. Uptake of intact zinc-ethylenediaminetetraacetic acid from soil is dependent on plant species and complex concentration. Envirom. Toxicol. Chem. 21 (9), 1940–1945.

Degryse, F., Smolders, E., Merckx, R., 2006. Labile Cd complexes increase Cd availability to plants. Environ. Sci. Technol. 40 (3), 830–836.

Desideri, D., Meli, M.A., Roselli, C., Feduzi, L., 2011. Polarized X ray fluorescence spectrometer (EDPXRF) for the determination of essential and non essential elements in tea. Microchem. J. 98 (2), 186–189.

Du, R.J., He, E.K., Tang, Y.T., Hu, P.J., Ying, R.R., Morel, J.L., Qiu, R.L., 2011. How phytohormone IAA and chelator EDTA affect lead uptake by Zn/Cd hyperaccumulator Picris divaricata. Int. J. Phytoremediation 13 (10), 1024–1036.

Ehsan, S., Ali, S., Noureen, S., Mahmood, K., Farid, M., Ishaque, W., Shakoor, M.B., Rizwan, M., 2014. Citric acid assisted phytoremediation of cadmium by *Brassica napus* L. Ecotoxicol. Environ. Saf. 106, 164–172.

Epelde, L., Hernández-Allica, J., Becerril, J.M., Blanco, F., Garbisu, C., 2008. Effects of chelates on plants and soil microbial community: comparison of EDTA and EDDS for lead phytoextraction. Sci. Total Environ. 401 (1–3), 21–28.

Evangelou, M.W., Ebel, M., Schaeffer, A., 2006. Evaluation of the effect of small organic acids on phytoextraction of Cu and Pb from soil with tobacco *Nicotiana tabacum*. Chemosphere 63 (6), 996–1004.

Evangelou, M.W., Ebel, M., Schaeffer, A., 2007. Chelate assisted phytoextraction of heavy metals from soil. Effect, mechanism, toxicity, and fate of chelating agents. Chemosphere 68 (6), 989–1003.

Farid, M., Ali, S., Shakoor, M.B., Bharwana, S.A., Rizvi, H., Ehsan, S., Tauqeer, H.M., Iftikhar, U., Hannan, F., 2013. EDTA assisted phytoremediation of cadmium, lead and zinc. Int. J. Agron. Plant Prod. 4 (11), 2833–2846.

Ferraro, A., van Hullebusch, E.D., Huguenot, D., Fabbricino, M., Esposito, G., 2015. Application of an electrochemical treatment for EDDS soil washing solution regeneration and reuse in a multi-step soil washing process: Case of a Cu contaminated soil. J. Environ. Manag. 163, 62–69.

Forte, J., Mutiti, S., 2017. Phytoremediation potential of *Helianthus annuus* and *Hydrangea paniculata* in copper and lead-contaminated Soil. Water Air Soil Pollut 228 (2), 77.

Goodfellow, M., Brown, A.B., Cai, J., Chun, J., Collins, M.D., 1997. Amycolatopsis japonicum sp. nov., an actinomycete producing (S, S)-N, N'-ethylenediaminedisuccinic acid. Syst. Appl. Microbiol. 20 (1), 78–84.

Gramss, G., Voigt, K.D., Bergmann, H., 2004. Plant availability and leaching of (heavy) metals from ammonium-, calcium-, carbohydrate-, and citric acid-treated uranium-mine-dump soil. J. Plant Nutr. Soil Sci. 167 (4), 417–427.

Grčman, H., Velikonja-Bolta, Š., Vodnik, D., Kos, B., Leštan, D., 2001. EDTA enhanced heavy metal phytoextraction: Metal accumulation, leaching and toxicity. Plant Soil 235 (1), 105–114.

Grčman, H., Vodnik, D., Velikonja-Bolta, S., Lesštan, D., 2003. Ethylenediaminedissuccinate as a new chelate for environmentally safe enhanced lead phytoextraction. J. Environ. Qual. 32 (2), 500–506.

Hauser, L., Tandy, S., Schulin, R., Nowack, B., 2005. Column extraction of heavy metals from soils using the biodegradabile chelating agent EDDS. Environ. Sci. Technol. 39 (17), 6819–6824.

Haynes, R., 1980. Ion exchange properties of roots and ionic interactions within the root apoplasm: Their role in ion accumulation by plants. Bot. Rev. 46 (1), 75–99.

Hu, N., Lang, T., Ding, D., Hu, J., Li, C., Zhang, H., Li, G., 2019. Enhancement of repeated applications of chelates on phytoremediation of uranium contaminated soil by *Macleaya cordata*. J. Environ. Radioact. 199, 58–65.

Hu, Y., Nan, Z., Jin, C., Wang, N., Luo, H., 2014. Phytoextraction potential of poplar (*Populus alba* L. Var. *pyramidalis* Bunge) from calcareous agricultural soils contaminated by cadmium. Int. J. Phytoremediation 16 (5), 482–95.

Huang, J.W., Blaylock, M.J., Kapulnik, Y., Ensley, B.D., 1998b. Phytoremediation of uranium- contaminated soils: role of organic acids in triggering uranium hyperaccumulation in plants. Environ. Sci. Technol. 32 (13), 2004–2008.

Huang, J.W., Chen, J., Berti, W.B., Cunningham, S.D., 1997. Phytoremediation of lead-contaminated soils: role of synthetic chelates in lead phytoextraction. Environ. Sci. Technol. 31 (3), 800–805.

Huang, F.C., Brady, P.V., Lindgren, E.R., Guerra, P., 1998a. Biodegradation of uranium-citrate complexes: implications for extraction of uranium from soils. Environ. Sci. Technol. 32 (3), 379–382.

Jakovljević, K., Mišljenović, T., Savović, J., Ranković, D., Ranđelović, D., Mihailović, N., Jovanović, S., 2020. Accumulation of trace elements in *Tussilago farfara* colonizing post-flotation tailing sites in Serbia. Environ. Sci. Pollut. Res. 27 (4), 4089–4103.

Kabala, C., Singh, B.R., 2001. Fractionation and mobility of copper, lead, and zinc in soil profiles in the vicinity of a copper smelter. J. Environ. Qual. 30 (2), 485–492.
Karczewska, A., Gałka, B., Kabała, C., Szopka, K., Kocan, K., Dziamba, K., 2009. Effects of various chelators on the uptake of Cu, Pb, Zn and Fe by maize and indian mustard from silty loam soil polluted by the emissions from copper smelter. Fresenius Environ. Bull. 18 (10), 1967–1974.
Kierczak, J., Neel, C., Aleksander-Kwaterczak, U., Helios-Rybicka, E., Bril, H., Puziewicz, J., 2008. Solid speciation and mobility of potentially toxic elements from natural and contaminated soils: a combined approach. Chemosphere 73 (5), 776–784.
Kim, C., Lee, Y., Ong, S., 2003. Factors affecting EDTA extraction of lead from lead-contaminated soils. Chemosphere 51 (9), 845–853.
Komárek, M., Tlustoš, P., Száková, J., Chrastný, V., 2008. The use of poplar during a two-year induced phytoextraction of metals from contaminated agricultural soils. Environ. Pollut. 151 (1), 27–38.
Komárek, M., Tlustoš, P., Száková, J., Chrastný, V., Ettler, V., 2007. The use of maize and poplar in chelant-enhanced phytoextraction of lead from contaminated agricultural soils. Chemosphere 67 (4), 640–651.
Komárek, M., Vanek, A., Mrnka, L., Sudová, R., Száková, J., Tejnecký, V., Chrastný, V., 2010. Potential and drawbacks of EDDS-enhanced phytoextraction of copper from contaminated soils. Environ. Pollut. 158 (7), 2428–2438.
Koopmans, G., Schenkeveld, W., Song, J., Luo, Y., Japenga, J., Temminghoff, E., 2008. Influence of EDDS on metal speciation in soil extracts: measurement and mechanistic multicomponent modeling. Environ. Sci. Technol. 42 (4), 1123–1130.
Kos, B., Grčman, H., Leštan, D., 2003. Phytoextraction of lead, zinc and cadmium from soil by selected plants. Plant Soil Environ. 49 (12), 548–553.
Kos, B., Leštan, D., 2004. Chelator induced phytoextraction and in situ soil washing of Cu. Environ. Pollut. 134 (2), 333–339.
Krämer, U., 2005. Phytoremediation: novel approaches to cleaning up polluted soils. Curr. Opin. Biotechnol. 16 (2), 133–141.
Lombi, E., Zhao, F.J., Dunham, S.J., McGrath, S.P., 2001. Phytoremediation of heavy-metal contaminated soils: natural hyperaccumulation versus chemically enhanced phytoextraction. J. Environ. Qual. 30 (6), 1919–1926.
Lai, H.Y., Chen, S.W., Chen, Z.S., 2008. Pot experiment to study the uptake of Cd and Pb by three indian mustards (*Brassica juncea*) grown in artificially contaminated soils. Int. J. Phytoremediation 10 (2), 91–105.
Lai, H.Y., Chen, Z.S., 2004. Effects of EDTA on solubility of cadmium, zinc, and lead and their uptake by rainbow pink and vetiver grass. Chemosphere 55 (3), 421–430.
Lambrechts, T., Gustot, Q., Couder, E., Houben, D., Iserentant, A., Lutts, S., 2011. Comparison of EDTA-enhanced phytoextraction and phytostabilisation strategies with *Lolium perenne* on a heavy metal contaminated soil. Chemosphere 85 (8), 1290–1298.
Lee, J., Sung, K., 2014. Effects of chelates on soil microbial properties, plant growth and heavy metal accumulation in plants. Ecol. Eng. 73, 386–394.
Li, T., Yang, X., Jin, X., He, Z., Stoffella, P., Hu, Q., 2005. Root responses and metal accumulation in two contrasting ecotypes of *Sedum alfredii* Hance under lead and zinc toxic stress. J. Environ. Sci. Health A 40 (5), 1081–1096.
Liang, Y., Wang, X., Guo, Z., Xiao, X., Peng, C., Yang, J., Zhou, C., Zeng, P., 2019. Chelator-assisted phytoextraction of arsenic, cadmium and lead by *Pteris vittata* L. and soil microbial community structure response. Int. J. Phytoremediation 21 (10), 1032–1040.
Liphadzi, M.S., Kirkham, M.B., 2006. Availability and plant uptake of heavy metals in EDTA-assisted phytoremediation of soil and composted biosolids. S. Afr. J. Bot. 72, 391–397.
Liu, D., Lia, T.Q., Yang, X.E., Islam, E., Jin, X.F., Mahmood, Q., 2007. Influence of EDTA on lead transportation and accumulation by Sedum alfredii hance. Z. Naturforsch. C 62 (9–10), 717–724.

Liu, J., Zhou, Q., Wang, S., 2010. Evaluation of chemical enhancement on phytoremediation effect of Cd-contaminated soils with *Calendula officinalis* L. Int. J. Phytoremediation 12 (5), 503–515.

Liu, J.N., Zhou, Q.X., Wang, S., Sun, T., 2009. Cadmium tolerance and accumulation of *Althaea rosea* Cav. and its potential as a hyperaccumulator under chemical enhancement. Environ. Monit. Assess. 149, 419–427.

Luo, C., Shen, Z., Li, X., 2005. Enhanced phytoextraction of Cu, Pb, Zn and Cd with EDTA and EDDS. Chemosphere 59 (1), 1–11.

Luo, C., Shen, Z., Lou, L., Li, X., 2006. EDDS and EDTA-enhanced phytoextraction of metals from artificially contaminated sol and residual effects of chelant compounds. Environ. Pollut. 144 (3), 862–871.

Luo, C.L., Shen, Z.G., Li, X.D., 2007. Plant uptake and the leaching of metals during the hot EDDS-enhanced phytoextraction process. Int. J. Phytoremediation 9 (3), 181–196.

Luo, J., Qi, S., Peng, L., Wang, J., 2016. Phytoremediation efficiency OF CD by *Eucalyptus globulus* transplanted from polluted and unpolluted sites. Int. J. Phytoremediation 18 (4), 308–314.

Manouchehri, N., Besancon, S., Bermond, A., 2006. Major and trace metal extraction from soil by EDTA: equilibrium and kinetic studies. Anal. Chim. Acta 559 (1), 105–112.

Markovska, Y., Geneva, M., Petrov, P., Boychinova, M., Lazarova, I., Todorov, I., Stancheva, I., 2013. EDTA reduces heavy metal impacts on *Tribulus terrestris* photosynthesis and antioxidants. Russ. J. Plant Physiol. 60 (5), 623–632.

Marschner, H., 1986. Mineral Nutrition of Higher Plants. Academic Press, San Diego, USA.

Meers, E., Hopgood, M., Lesage, E., Vervaeke, P, Tack, F.M.G., Verloo, M., 2004. Enhanced phytoextraction: in search of EDTA alternatives. Int. J. Phytoremediation 6 (2), 95–109.

Meers, E., Ruttens, A., Hopgood, M.J., Samson, D., Tack, F.M.G., 2005. Comparison of EDTA and EDDS as potential soil amendments for enhanced phytoextraction of heavy metals. Chemosphere 58 (8), 1011–1022.

Melo, E., Nascimento, C., Accioly, A., Santos, A., 2008. Phytoextraction and fractionation of heavy metals in soil after multiple applications of natural chelants. Sci. Agric. 65 (1), 61–68.

Mikavica, I., Ranđelović, D., Djordjevic, V., Gajić, G., Mutić, J., 2020. Orchid species *Anacamptis morio* as a potential bioremediator of As, Cd and Pb. J. Appl. Eng. Sci. 18 (3), 413–421.

Miller, G., Begonia, G., Begonia, M., Ntoni, J., Hundley, O., 2008. Assessment of the efficacy of chelate assisted phytoextraction of lead by coffeeweed (*Sesbania exaltata* Raf.). Int. J. Environ. Res. Public Health 5 (5), 428–435.

Najeeb, U., Xua, L., Ali, S., Jilani, G., Gong, H.J., Shen, W.Q., Zhou, W.J., 2009. Citric acid enhances the phytoextraction of manganese and plant growth by alleviating the ultrastructural damages in *Juncus effusus* L. J. Hazard. Mater. 170 (2–3), 1156–1163.

Nakbanpote, W., Meesungnoen, O., Prasad, M.N.V., 2016. Potential of ornamental plants for phytoremediation of heavy metals and income generation. In: Prasad, M.N.V. (Ed.), Bioremediation and Bioeconomy. Elsevier, Amsterdam, pp. 179–217.

Nascimento, C., Xing, B., 2006. Phytoextraction: a review on enhanced metal availability and plant accumulation. Sci. Agr. 63 (3), 299–311.

Nishikiori, T., Okuyama, A., Naganawa, T., Takita, T., Hamida, M., Takeuchi, T., Aoyagi, T., Umezawa, H., 1984. Production of (S,S)-N,N-ethylenediamine-dissuccinic acid, an inhibitor of phospholipase. J. Antibiot. 37 (4), 426–427.

Nowack, B., 2002. Environmental chemistry of aminopolycarboxylate chelating agents. Environ. Sci. Technol. 36 (19), 4009–4016.

Nowack, B., Schulin, R., Robinson, B., 2006. Critical assessment of chelant-enhanced metal phytoextraction. Environ. Sci. Technol. 40 (17), 5225–5232.

Nörtemann, B., 2005. Biodegradation of chelating agents: EDTA, DTPA, PDTA, NTA, and EDDS. In: Nowack, B., VanBriesen, J.M. (Eds.), Biogeochemistry of Chelating Agents. ACS symposium series 910. American Chemical Society, Washington, DC, pp. 150–170.

Odjegba, V.J., Fasidi, I.O., 2004. Accumulation of trace elements by *Pistia stratiotes*: implications for phytoremediation. Ecotoxicology 13 (7), 637–646.

Palmer, C.E., Warwick, S., Keller, W., 2001. Brassicaceae (Cruciferae) family, plant biotechnology, and phytoremediation. Int. J. Phytoremediation 3 (3), 245–287.

Pandey, V.C., Bajpai, O., 2019. Phytoremediation: From Theory towards Practice. In: Pandey, V.C., Bauddh, K. (Eds.), Phytomanagement of Polluted Sites. Elsevier, Amsterdam, pp. 1–49.

Pandey, V.C., Pandey, D.N., Singh, N., 2015. Sustainable phytoremediation based on naturally colonizing and economically valuable plants. J. Clean Prod. 86, 37–39.

Papassiopi, N., Tamboutis, S., Kontopoulos, A., 1999. Removal of heavy metals from calcareous contaminated soils by EDTA leaching. Water Air Soil Pollut. 109 (1), 1–15.

Pinto, A.P., De Varennes, A., Fonseca, R., Teixeira, D.M., 2015. Phytoremediation of soils contaminated with heavy metals: techniques and strategies. In: Ansari, A., Gill, S., Gill, R., Lanza, G., Newman, L. (Eds.), Phytoremediation. Springer, Cham, pp. 133–155.

Prasad, M.N.V., Strzałka, K., 1999. Impact of heavy metals on photosynthesis. In: Prasad, M.N.V., Hagemeyer, J. (Eds.), Heavy Metal Stress in Plants. Springer, Berlin, Heidelberg, pp. 117–138.

Quartacci, M.F., Irtelli, B., Baker, A.J.M., Navari-Izzo, F., 2007. The use of NTA and EDDS for enhanced phytoextraction of metals from a multiply contaminated soil by *Brassica carinata*. Chemosphere 68 (10), 1920–1928.

Rajkumar, M., Sandhya, S., Prasad, M.N.V., Freitas, H., 2012. Perspectives of plant-associated microbes in heavy metal phytoremediation. Biotechnol. Adv. 30 (6), 1562–1574.

Ranđelović, D., Stanković, S., Mihailović, N., Leštan, D., 2015. Remediation of Cu from copper mine wastes and contaminated soils using (S,S)-Ethylenediaminedisuccinic Acid and acidophilic bacteria. Bioremediation J. 19 (3), 231–238.

Rashid, A., Mahmood, T., Mehmood, F., Khalid, A., Saba, B., Batool, A., Riaz, A., 2014. Phytoaccumulation, competitive adsorption and evaluation of chelators-metal interaction in lettuce plant. Environ. Eng. Manag. J. 13 (10), 2583–2592.

Robinson, B., Brooks, R., Clothier, B., 1999. Soil amendments affecting nickel and cobalt uptake by *Berkheya coddii*: potential use for phytomining and phytoremediation. Ann. Bot. 84 (6), 689–694.

Robinson, B.H., Mills, T.M., Petit, D., Fung, L., Green, S., Clothier, B., 2000. Natural and induced cadmium-accumulation in poplar and willow: Implications for phytoremediation. Plant Soil 227 (1), 301–306.

Römkens, P., Bouwman, L., Japenga, J., Draaisma, C., 2002. Potentials and drawbacks of chelate-enhanced phytoremediation of soils. Environ. Pollut. 116 (1), 109–121.

Salt, D.E., Prince, R.C., Pickering, I.J., 1995. Mechanisms of cadmium mobility and accumulation in Indian mustard. Plant Physiol. 109 (4), 1427–1433.

Salt, D., Smith, R., Raskin, I., 1998. Phytoremediation. Annu. Rev. Plant Physiol. Plant Mol. Biol. 49, 643–668.

Sarwar, N., Imran, M., Shaheen, M.R., Ishaque, W., Kamran, M.A., Matloob, A., Rehim, A., Hussain, S., 2017. Phytoremediation strategies for soils contaminated with heavy metals: modifications and future perspectives. Chemosphere 171, 710–721.

Satroutdinov, A.D., Dedyukhina, E.G., Chistyakova, T.I., Witschel, M., Minkevich, I.G., Eroshin, V.K., Egli, T., 2000. Degradation of metal-EDTA complexes by resting cells of the bacterial strain DSM 9103. Environ. Sci. Technol. 34 (9), 1715–1720.

Sauvé, S., Cook, N., Hendershot, W.H., McBride, M.B., 1996. Linking plant tissue concentrations and soil copper pools in urban contaminated soils. Environ. Pollut. 94 (2), 153–157.

Schaider, L., Parker, D., Sedlak, D., 2006. Uptake of EDTA-complexed Pb, Cd and Fe by solution and sand-cultured *Brassica juncea*. Plant Soil 286 (1), 377–391.

Schowanek, D., Feijtel, T.C.J., Perkins, C.M., Hartman, Federle, T.W., Larson, R.J., 1997. Biodegradation of [S,S], [R,R] and mixed stereoisomers of ethylene diamine disuccinic acid (EDDS), a transition metal chelators. Chemosphere 34 (11), 2375–2391.

Sekhar, K.C., Kamala, C.T., Chary, N.S., Balaram, V., Garcia, G., 2005. Potential of *Hemidesmus indicus* for the phytoextraction of lead from industrially contaminated soils. Chemosphere 58 (4), 507–514.

Seregin, I.V., Ivanov, V.B., 2001. Physiological aspects of cadmium and lead toxic effects on higher plants. Russ. J. Plant Physiol. 48 (4), 523–544.

Shahid, M., Austruy, A., Echevarria, G., Arshad, M., Sanaullah, M., Aslam, M., Nadeem, M., Nasim, W., Dumat, C., 2014. EDTA-enhanced phytoremediation of heavy metals: a review. Soil Sediment Contam 23 (4), 389–416.

Shakoor, M.B., Ali, S., Hameed, A., Farid, M., Hussain, S., Yasmeen, T., Najeeb, U., Bharwana, S.A., Abbasi, G.H., 2014. Citric acid improves lead (Pb) phytoextraction in *Brassica napus* L. by mitigating Pb-induced morphological and biochemical damages. Ecotoxicol. Environ. Saf. 109, 38–47.

Shen, Z.G., Li, X.D., Wang, C.C., Chen, H.M., Chua, H., 2002. Lead phytoextraction from contaminated soils with high biomass plant species. J. Environ. Qual. 31 (6), 1893–1900.

Shibata, M., Konno, T., Akaike, R., Xu, Y., Shen, R., Ma, J.F., 2007. Phytoremediation of Pb contaminated soil with polymer-coated EDTA. Plant Soil 290 (1), 201–208.

Singh, O.V., Labana, S., Pandey, G., Budhiraja, R., Jain, R.K., 2003. Phytoremediation: an overview of metallic ion decontamination from soil. Appl. Microbiol. Biotechnol. 61 (5–6), 405–412.

Singh, S., Kaushik, C.P., Fulzele, D.P., 2014. Remediation of 241Am using *Vetiveria zizanoides* L. nash: influence of chelating agents. Int. J. Life Sci. Pharma Res. 4 (2), 42–48.

Souza, L.A., Piotto, F.A., Nogueirol, R.C., Azevedo, R.A., 2013. Use of non-hyperaccumulator plant species for the phytoextraction of heavy metals using chelating agents. Sci. Agric. 70 (4), 290–295.

Sun, B., Zhao, F.J., Lombi, E., McGrath, S.P., 2001. Leaching of heavy metals from contaminated soils using EDTA. Environ. Pollut. 113 (2), 111–120.

Sun, Y., Zhou, Q., An, J., Liu, W., Liu, R., 2009. Chelator-enhanced phytoextraction of heavy metals from contaminated soil irrigated by industrial wastewater with the hyperaccumulator plant (*Sedum alfredii* Hance). Geoderma 150 (1–2), 106–112.

Sung, K., Kim, K.S., Park, S., 2013. Enhancing degradation of total petroleum hydrocarbons and uptake of heavy metals in a wetland microcosm planted with *Phragmites communis* by humic acids addition. Int. J. Phytoremediation 15 (6), 536–549.

Tandy, S., Ammann, A., Schulin, R., Nowack, B., 2006a. Biodegradation and speciation of residual SS-ethylenediaminedisuccinic acid (EDDS) in soil solution left after soil washing. Environ. Pollut. 142 (2), 191–199.

Tandy, S., Bossart, K., Mueller, R., Ritschel, J., Hauser, L., Schulin, R., Nowack, B., 2004. Extraction of heavy metals from soils using biodegradable chelating agents. Environ. Sci. Technol. 38 (3), 937–944.

Tandy, S., Schulin, R., Nowack, B., 2006b. The influence of EDDS on the uptake of heavy metals in hydroponically grown sunflowers. Chemosphere 62 (9), 1454–1463.

Tandy, S., Schulin, R., Nowack, B., 2006c. Uptake metals during chelant-assisted phytoextraction with EDDs related to the solubilized metal concentration. Environ. Sci. Technol. 40 (8), 2753–2758.

Thayalakumaran, T., Robinson, B., Vogeler, I., Scotter, D., Clothier, B., Percival, H., 2003. Plant uptake and leaching of copper during EDTA-enhanced phytoremediation of repacked and undisturbed soil. Plant Soil 254 (2), 415–423.

Theodoratos, P., Papassiopi, N., Georgoudis, T., Kontopoulos, A., 2000. Selective removal of lead from contaminated soils using EDTA. Water Air Soil Pollut. 122 (3), 351–368.

Tian, S.K., Lu, L.L., Yang, X.E., Huang, H.G., Brown, P., Labavitch, J., Liao, H.B., He, Z.L., 2011. The impact of EDTA on lead distribution and speciation in the accumulator *Sedum alfredii* by synchrotron X-ray investigation. Environ. Pollut. 159 (3), 782–788.

Tiedje, J.M., Mason, B.B., 1974. Biodegradation of nitrilotriacetate (NTA) in soils. Soil Sci. Soc. Am. J. 38, 278–283.

Turgut, C., Katie Pepe, M., Cutright, T., 2004. The effect of EDTA and citric acid on phytoremediation of Cd, Cr, and Ni from soil using *Helianthus annuus*. Environ. Pollut. 131 (1), 147–154.

Ullah, S., Shahid, M., Zia-Ur-Rehman, M., Sabir, M., Ahmad, H.R., 2015. Phytoremediation of Pb-contaminated soils using synthetic chelates. In: Hakeem, K., Sabir, M., Ozturk, M., Mermut, A.R. (Eds.), Soil Remediation and Plants: Prospects and Challenges. Elsevier, Boston, pp. 397–414.

Ullah, A., Heng, S., Munis, M.F.H., Fahad, S., Yang, X., 2015b. Phytoremediation of heavy metals assisted by plant growth promoting (PGP) bacteria: a review. Environ. Exp. Bot. 117, 28–40.

Vandevivere, P.C., Saveyn, H., Verstraete, W., Feijtel, T.C.J., Schowanek, D.R., 2001. Biodegradation of metal-[S,S]-EDDS complexes. Environ. Sci. Technol. 35 (9), 1765–1770.

Vogeler, I., Thayalakumaran, T., 2005. TRansport and reactions of ESTA in soils: experiments and modeling. In: Nowack, B., VanBriesen, J. (Eds.), Biogeochemistry of Chelating Agents. American Chemical Society, Washington, pp. 316–335.

Wang, A., Luo, C., Yang, R., Chen, Y., Shen, Z., Li, X., 2012. Metal leaching along soil profiles after the EDDS application - a field study. Environ. Pollut. 164, 204–210.

Wang, S., Liu, J., 2014. The effectiveness and risk comparison of EDTA with EGTA in enhancing Cd phytoextraction by *Mirabilis jalapa* L. Environ. Monit. Assess. 186 (2), 751–759.

Ward, T.E., 1985. Aerobic and anaerobic biodegradation of nitrilotriacetate in subsurface soils. Ecotoxicol. Environ. Saf. 11 (1), 112–125.

Wenger, K., Gupta, S.K., Furrer, G., Schulin, R., 2003. The role of nitrilotriacetate in copper uptake by tobacco. J. Environ. Qual. 32 (5), 1669–1676.

Wenzel, W.W., Unterbrunner, R., Sommer, P., Sacco, P., 2003. Chelate-assisted phytoextraction using canola (*Brassica napus* L.) in outdoors pot and lysimeter experiments. Plant Soil 249 (1), 83–96.

Wu, J., Hsu, F.C., Cunningham, S.D., 1999. Chelate-assisted Pb phytoextraction: Pb availability, uptake, and translocation constraints. Environ. Sci. Technol. 33 (11), 1898–1904.

Wu, L., Luo, Y., Song, J., 2007. Manipulating soil metal availability using EDTA and low-molwcular –weight organic acids. In: Willey, N. (Ed.), Manipulating soil metal availability using EDTA and low-molwcular –weight organic acids. Phytoremediation. Methods in Biotechnology 23, 291–303.

Zeremski-Škorić, T., Sekulić, P., Maksimović, I., Šeremešić, S., Ninkov, J., Milić, S., Vasin, J., 2010. Chelate-assisted phytoextraction: Effect of EDTA and EDDS on copper uptake by *Brassica napus* L. J. Serb. Chem. Soc. 75 (9), 1279–1289.

Zhang, H., Guo, Q., Yang, J., Ma, J., Chen, G., Chen, T., Zhu, G., Wang, J., Zhang, G., Wang, X., Shao, C., 2016. Comparison of chelates for enhancing *Ricinus communis* L. phytoremediation of Cd and Pb contaminated soil. Ecotoxicol. Environ. Saf. 133, 57–62.

Zhao, S., Lian, F., Duo, L., 2011. EDTA-assisted phytoextraction of heavy metals by turfgrass from municipal solid waste compost using permeable barriers and associated potential leaching risk. Bioresour. Technol. 102 (2), 621–626.

CHAPTER

6 Nanoparticles-assisted phytoremediation: Advances and applications

Omena Bernard Ojuederie[a,b], Adenike Eunice Amoo[b], Shesan John Owonubi[c], Ayansina Segun Ayangbenro[b]

[a]*Department of Biological Sciences, Faculty of Science, Kings University, Odeomu, Osun State, Nigeria*

[b]*Food Security and Safety Niche, Faculty of Natural and Agricultural Sciences, North-West University, Mmabatho, South Africa*

[c]*Department of Chemistry, University of Zululand, KwaDlangezwa, KwaZulu-Natal, South Africa*

6.1 Introduction

There is an increasing burden on the environment due to pollution arising from the accidental or deliberate discharge of toxic wastes into soil and water. These pollutants (organic compounds, heavy metals [HMs] and metalloids, hazardous wastes, and radionuclides) are being released in large quantities into the environment, where they inflict damage on the ecosystem. The release of these wastes has an undesirable effect on human health and has produced toxic effects in plants and microorganisms. Further contamination of soil and water continues to create serious global health concerns (Dixit et al., 2015). The accumulation of toxic wastes in agricultural soil is a threat to food safety and security, posing a health challenge to living organisms in the food web (Tak et al., 2013). To reduce the noxious effects of these pollutants on the environment and health of living organisms, there is an urgent need to reduce and/or remove these pollutants from the environment. To achieve this, several physical and chemical techniques have been employed, but most of these techniques are costly, ineffective at low concentrations of pollutants and are not sustainable and eco-friendly (Dixit et al., 2015). On the other hand, biological techniques are an attractive alternative to physical and chemical techniques. It involves the use of plants and microorganisms or their products to remove hazardous wastes from various environments. The biological technique is cost-effective, sustainable, and eco-friendly, and can re-establish the natural condition of polluted environments (Dixit et al., 2015; Tak et al., 2013).

Phytoremediation is an *in-situ* technique for the degradation, extraction, and/or immobilization of pollutants from a contaminated environment, such as soil or water, using plants (Pulford and Watson, 2003; Gerhardt et al., 2017). This technique employs the abilities of plants to accumulate, degrade, translocate, and selectively uptake pollutants in the environment (Pathak et al., 2020). It renders the pollutants less harmful or harmless to the environment. It is a means of remediation that is primarily driven by the high cost of physical and chemical remediation techniques and the need for an eco-friendly and sustainable means of removal of pollutants (Pathak et al., 2020; Pandey and Bajpai, 2019). Phytoremediation is simple to operate, inexpensive, causes little disruption of the soil structure and

Assisted Phytoremediation. DOI: https://doi.org/10.1016/B978-0-12-822893-7.00011-2

ecosystem with an aesthetic appearance has wide adaptability, and high public acceptance (Ojuederie and Babalola, 2017). Plants can eliminate or degrade organic and inorganic pollutants from environmental media. Hyperaccumulating plants with high accumulating potentials are used in phytoremediation as well as plants which possess lesser extraction abilities than hyperaccumulators but have a high growth rate and biomass yield (Ojuederie and Babalola, 2017; Pandey and Singh, 2019).

The success of using plants as remediation agents depends on the bioavailability of the pollutants, nature of pollutants, the extent of pollution, and the ability of the plants to absorb and accumulate pollutants as biomass (Tak et al., 2013). Phytoremediation has been widely used in different field applications at locations contaminated with pollutants such as chlorinated solvents, crude oil, explosives, HMs, pesticides, and radionuclides (Song et al., 2019; Dubchak and Bondar, 2019). Although the technique is environmentally friendly and cost-effective, the cleanup process takes several years. The process is also limited by pollutant phytotoxicity, poor soil quality, and weather conditions (Song et al., 2019). To optimize and improve phytoremediation, several strategies are being employed. These strategies include the use of rhizospheric microbes, agronomic management practices, soil amendments, genetic engineering, treatment with chemical additives (Gerhardt et al., 2017), and most recently, the use of nanotechnology (Gong et al., 2018). The development of nanotechnology and its synergistic application with biotechnology in remediation processes has opened new pathways for pollutant removal from contaminated media. The first interaction between pollutants and nanomaterials might present a favorable condition for remediation by biotechnology (Gong et al., 2018). Thus, the use of nanomaterials is drawing increasing attention in environmental pollution research.

Nanotechnology involves the production, manipulation, and deployment of nanomaterials, which are defined as particles with sizes between 1 and 100 nm having one or more dimensions (Ghormade et al., 2011; Song et al., 2019; Cai et al., 2019). Nanomaterials possess unique sizes between individual molecules and their corresponding bulk form, which results in unusual physicochemical characteristics (Song et al., 2019). They are excellent absorbents and catalyst as a result of large specific surface areas, which are associated with numerous sorption sites, increased tunable pore size, short intra-particle diffusion, low-temperature modification, and different surface chemistry than other materials (Tang et al., 2014; Cai et al., 2019). These unique properties make nanomaterials to hold a great promise in improving phytoremediation, hence environmental protection. The use of nanomaterials such as graphene oxide and fullerene nanoparticles (NPs) increases the transport of pollutants such as phenanthrene, polychlorinated biphenyl, and naphthol, in soil (Qi et al., 2014; Zhang et al., 2011). This can improve the efficiency of pollutant removal from the contaminated environment by plants and microorganisms (Gong et al., 2018). Nanomaterials can also be used as nutrient carriers and sources of nutrients to plants. This promotes plant growth and development, as well as aiding phytoremediation processes (Gong et al., 2018). The interaction between nanomaterials and living organisms will certainly affect bioremediation efficiency in polluted environments, although it hasn't been concluded if both technologies are combined, it is beneficial to improve the pollutant removal efficiency (Gong et al., 2018).

Nanoparticles have been employed in the remediation of environmental pollutants from air, water, and soil using different methods. The combination of nanotechnology and phytotechnology for the clean-up of environmental pollutants is known as nano-phytoremediation (Srivastav et al., 2018). The efficacy of phytoremediation is augmented by nanotechnology. Previous researches have reported that nano-phytoremediation is more efficient than singly using nanoremediation or phytoremediation (Srivastav et al., 2018; Gong et al., 2017; Zhu et al., 2019). For instance, nano-phytoremediation proved to be more effective in the deterioration and elimination of 2,4,6-trinitrotoluene (TNT) from contaminated soil (Jiamjitrpanich et al., 2013). The stress tolerance of plants can be enhanced using nanoparticles, thus furthering the prospective of phytoremediation (Ma and Wang, 2010; Souri et al., 2017b). The process of phytoremediation using NPs is cheap compared to conventional remediation methods because their large surface area can adsorb contaminants. The maneuverings of phytoremediation are easy, it has extensive adaptability, great community reception and it causes slight destruction to the structure of soil relative to other approaches (Srivastav et al., 2018; Davis et al., 2017; Song et al., 2019). Nevertheless, this green technology is restricted by climatic conditions, phytotoxicity of contaminants, soil properties and it is typically time-consuming (Song et al., 2019). Ideally, NPs utilized for phytoremediation ought to promote plant growth, be non-toxic, intensify the production of phytoenzymes in plants and bind the pollutants while augmenting bioavailability for the plant (Srivastav et al., 2018)

Having seen the prospects of nanomaterials, investigations are ongoing to combine phytoremediation and NPs to deal with different categories of pollutants. Thus, this chapter discusses the synthesis and types of nanomaterials, applications of nanomaterials in phytoremediation of pollutants, as well as enumerate future advancement of nanomaterials-mediated phytoremediation.

6.2 Methods used in phytoremediation

Soil is often contaminated by pollutants generally from industrial processes, the accumulation of which becomes a major threat to public health and the environment globally. Reclaiming the land from these pollutants using plants, require different mechanisms based on the type of pollutant to be removed. Due to the costly approach of reclaiming contaminated environments using chemical and physical techniques, the use of hyperaccumulating plants has gained acceptance as an alternative and cost-effective method of remediation through the process of phytoremediation (Ali et al., 2013). Phytoremediation has developed rapidly due to the array of pollutants to which it could be applied. These include crude oil/hydrocarbon, HMs, explosives, pesticides, chlorinated solvents, and many other contaminants. Plants employ several processes to remove pollutants from contaminated environments. The processes used include phytoextraction, phytodegradation, phytofiltration, phytostabilization, phytostimulation, phytovolatilization, and rhizofiltration (Ojuederie and Babalola, 2017). The selective uptake abilities of the root system of plants coupled with their translocation, bioaccumulation, and contaminant degradation abilities make phytoremediation beneficial to remediation processes (Negri et al., 1996, Tangahu et al., 2011). There are different approaches to phytoremediation (Fig. 6.1)

Phytoextraction is the method of choice for the removal of metals from soil taking advantage of the ability of plants to accumulate metals important to plant growth such as Fe, Mn, Zn, Cu, Mg, Mo, and Ni (Tangahu et al., 2011). Nevertheless, the mechanism used for the uptake of metals also enables plants especially the hyperaccumulating ones, to also take up HMs such as As, Pb, and Hg, which are detrimental to human health. Lead (Pb) accumulation in the soil can be persistent as it is not biodegradable, hence it becomes noxious to biological systems if not removed from the contaminated environment (Ayangbenro and Babalola, 2017). The normal growth and development of plants are affected by mercury uptake, which invariably affects processes such as photosynthesis and oxidative metabolism by disrupting the electron transport in macromolecules such as the chloroplast and mitochondria, as well as reducing water uptake by plants (Tangahu et al., 2011; Sas-Nowosielska et al., 2008).

In humans, it could lead to neurological disorders as well as renal failure (Ayangbenro and Babalola, 2017). Thus, the removal of HMs from the environment is of utmost importance. Several plants

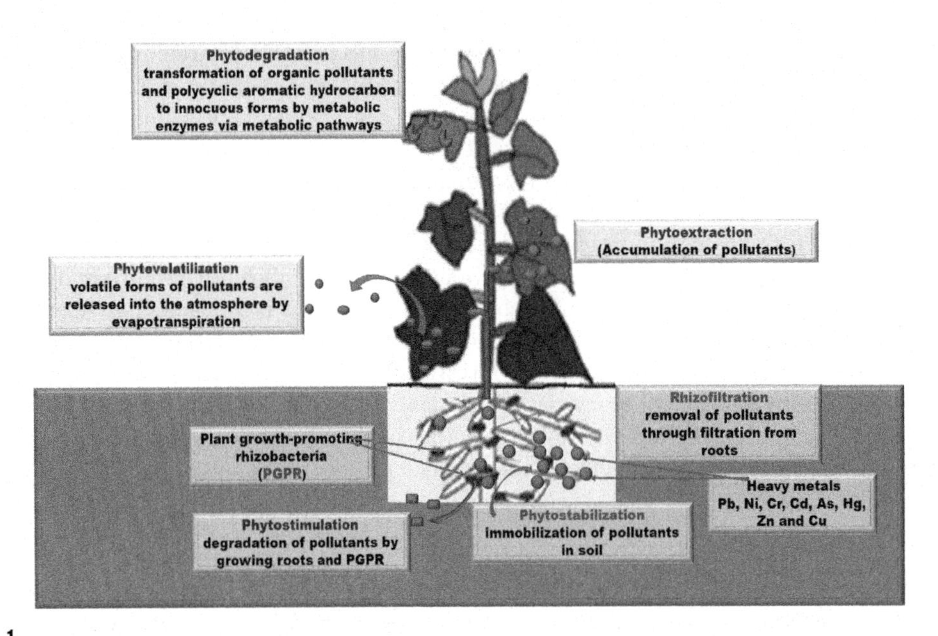

FIG. 6.1

Different methods used in phytoremediation of polluted environments.

have been used for the remediation of HMs due to their ability to grow in the presence of metals and accumulate them through the roots by phytoextraction. Some examples of hyperaccumulating plants used for phytoextraction include Sunflower (*Helianthus annuus* L.) (Laghlimi et al., 2015; Farid et al., 2017), Maize (*Zea mays* L.) (Laghlimi et al., 2015), species within the *Brassica* genus; Rapeseed (*Brassica napus*) and Indian mustard (*Brassica juncea*), which grow rapidly and produces much biomass (Laghlimi et al., 2015; Gomes et al., 2016; Shaheen and Rinklebe, 2015; Kathal et al., 2016), Vetiver grass (*Vetiveria zizanioides*) (Laghlimi et al., 2015), Willow (*Salix viminalis* L.), as well as Poplar trees (Gomes et al., 2016; Kacálková et al., 2015). These plants have a high growth rate, produce more above-ground biomass as well as producing extensive root systems, which are often assisted by the activities of rhizospheric bacteria and symbiotic mycorrhizal fungi.

The HMs are taken up by the roots and translocated to above-ground parts, where they can be continuously harvested and disposed of. The use of non-accumulator woody tree species such as Willow and Populus has been reported to preferentially accumulate metals such as Zn and Cd in their leaves from where they can be harvested (Unterbrunner et al., 2007; Suman et al., 2018). *Tetraena qataranse* has recently been used for the phytoextraction of Cd as revealed by the high translocation factor (TF) (>1.6) reported by Usman et al. (2019), although it also had a low bioconcentration factor (BCF) in its toots (1.5). Despite the usefulness of phytoextraction in the remediation of HMs, its efficiency is hindered by excessive HMs in the contaminated environment, which may reduce plant growth and biomass production. Nevertheless, the success of phytoextraction as an efficient environmental cleanup process, depends on the properties of the metal pollutants as well as the soil they are present in, for instance, the redox state of pollutant and bioavailability in soil (Ali et al., 2013; Pathak et al., 2020)

Phytostabilization involves the conversion of noxious compounds in localized contaminated soils to their innocuous forms thereby preventing the bioavailability of these pollutants in the environment.

This approach has been employed in the removal of inorganic pollutants and metals in sediments. Several plants with the ability to stabilize metals within its roots or around the vicinity of the rhizosphere have gained acceptance as being reliable for decontamination and restoration of the properties of contaminated soils (Kumpiene et al., 2008). The microbiome associated with plant roots in addition to aiding plant growth also promotes metal tolerance ability and decreases the mobilization of metals to the above-ground parts of the plant from the vicinity of the rhizosphere thereby preventing their entrance into the food chain (Ojuederie and Babalola, 2017). Stabilization of pollutants in the rhizosphere of plants reduces the possibility of erosion, leaching, and run-off of pollutants (Tamburini et al., 2017; Pathak et al., 2020)

Metals such as Pb, As, Cd, Cr, Cu, and Zn have been effectively removed from contaminated soils using the phytostabilization approach (Yadav et al., 2018). Recent studies by Usman et al. (2019) revealed the phytostabilization ability of *Tetraena qataranse* in Qatar. The metals Cr, Cu, and Ni were stabilized in the roots of the plant as shown by the high BCF ranging from 1.5 to 12.2 (Usman et al., 2019). This agrees with earlier reports that efficient phytostabilizers can accumulate excess toxic trace metals in their roots while limiting translocation to shoots (Shackira and Puthur, 2019). Thus, plants with BCF greater than one in the roots and TF less than 1 are likely candidates for the phytostabilization of metals in contaminated sites (Shackira and Puthur, 2019). Plants immobilize metals in the rhizosphere soils in several ways by (1) preventing entry of metals via expulsion of metals outside the cell or by a permeability barrier, (2) taking up HM contaminants and binding them to biologically active particles such as extracellular polymers after immobilization, and (3) preventing the absorption of HMs to plant tissues by the roots (Radziemska et al., 2017; Muthusaravanan et al., 2018). Despite the advantages obtained using phytostabilization for remediation of contaminants on site, the method has some limitations. Its effectiveness is hampered by the soil properties (biological, chemical, and physical), which regulates plant growth and metal stabilization in soils. Most often the phytostabilized soils are not suitable for plant growth and would need soil amendments to improve soil properties. This is common in areas highly contaminated with HMs such as mile tailings. It is also time-consuming to select the appropriate plant to use as phytostabilizers. which is expected to have extensive root system with large biomass production and reduced translocation of metals/metalloids to above ground parts of the plant. Moreover, lands reclaimed through phytostabilization will require continuous monitoring to maintain the stabilized condition and if necessary the phytoimmobilization enhanced at intervals by further application of amendments (Bolan et al., 2011).

Phytovolatilization is another mechanism often used in the phytoremediation of organic pollutants and metals. It involves the removal of these contaminants from the environment through the process of volatilization from the leaves of plants. Volatilization of organic contaminants could take place directly in plants by movements of the pollutants from the roots to the above-ground parts of the plant through the stems to leaf surfaces, where the organic compounds get volatilized through the process of evapotranspiration. Alternatively, the volatilization of contaminants is indirectly driven by the activities of plant roots, which elevates volatile pollutant flux from soil and water sub-surfaces (Limmer and Burken, 2016; Yadav et al., 2018). If the contaminants are not volatile as in the case of HMs or metalloids, they are degraded by mechanisms used by plants transforming them into gaseous forms, which are subsequently released into the atmosphere. This has been effectively utilized in the removal of HMs especially mercury (Hg), arsenic (As), and selenium (Se), which are converted to their volatile forms by plants and then released into the atmosphere (Tangahu et al., 2011; Suman et al., 2018). It is important to note that plants used for phytovolatilization should be efficient evapotranspirators. Nevertheless, this technology is limited because the volatilized pollutants easily get redeposited in soil and the pollutants are not completely removed.

Phyto/rhizofiltration is a phytoremediation method in which contaminants are filtered or removed from flowing water surrounding the rhizosphere of plants. Plants used for rhizofiltration possess extensive fibrous root systems through which contaminants can be absorbed efficiently due to their larger surface area. Such plants usually produce more biomass. Rhizofiltration is suitable for the removal of pollutants from water bodies and wetlands with the subsequent harvest of the roots and shoots ones saturated at the end of the process. Several aquatic plants have been utilized for rhizofiltration namely; hydrilla (*Hydrilla verticillata*), water hyacinth (*Eichhornia crassipes*) and water lettuce (*Pistia stratiotes*), sharp dock (*Polygonum amphibium* L.), water dropwort (*Oenathe javanica*) (Haldar and Ghosh, 2020) as well as *Micranthemum umbrosum* (Islam et al., 2015). The aquatic plant *Micranthemum umbrosum* proved to be a good accumulator of As, it accumulated over 1000 mg As g^{-1} in stem and leaf biomasses as well as reduced As concentration in the hydroponic solution by as much as 10-fold, (Islam et al., 2015). After the Chenobyl nuclear disaster in Ukraine in 1986, the environment was highly contaminated with radioactive contaminants. Thirty years after the incidence, efforts to clean up the environment are still ongoing (Beresford et al., 2016). Sunflower (*Helianthus annuus*), a well-known hyperaccumulator of HMs, was found to also have the ability to accumulate radioactive metals, like the cesium-137 and strontium-90 found at Chernobyl (Dushenkov et al., 1999, Dushenkov, 2003). It has been used for decontamination of the environment at Chernobyl via phytoextraction and rhizofiltration from groundwater (Abdullahi, 2015).

The degradation of organic compounds from the environment, the soil, or air within plant tissues is referred to as phytodegradation. This method is useful for the reclamation of land contaminated by organic compounds as well as polycyclic aromatic hydrocarbons (PAHs). Through different metabolic pathways, these organic pollutants are transformed into innocuous forms, some of which are used by the plant for its growth and metabolism. This technique is otherwise called phytotransformation, Phytodegradation is nevertheless, restricted only to the removal of organic pollutants since HMs are non-biodegradable (Ojuederie and Babalola, 2017).

Phytostimulation involves the degradation of pollutants within the vicinity of the rhizosphere of the plant with the aid of microorganisms (Newman and Reynolds, 2004), hence, the technique is sometimes called rhizodegradation. The microbial community within the rhizosphere utilizes the organic carbon released from plant root exudates for their growth and metabolism. Some of these microbes have PAH degrading abilities. These microbes transform the organic compounds to innocuous forms in the soil with the gaseous forms indirectly volatilized. Some bacteria and mycorrhiza fungi act as phytostimulators by enhancing microbiological degradation of petroleum hydrocarbon within the vicinity of the rhizosphere, through their various stimulating activities (Khatoon et al., 2017; Hawrot-Paw et al., 2019).

6.3 Types of nanoparticles

Nanoparticles have been classified according to researchers by their dimensions as 1-D, 2-D, or 3-D NPs (Owonubi et al., 2019), whilst in some other cases according to fabrication source, e.g. organic, inorganic or carbon-based NPs (Ealias and Saravanakumar, 2017). These classifications sometimes overlap, but for this chapter, we will focus on the latter classification (Fig. 6.2). The latter classification is based on the source of the nanoparticle synthesis material, sometimes referred to as the "precursor or starting material."

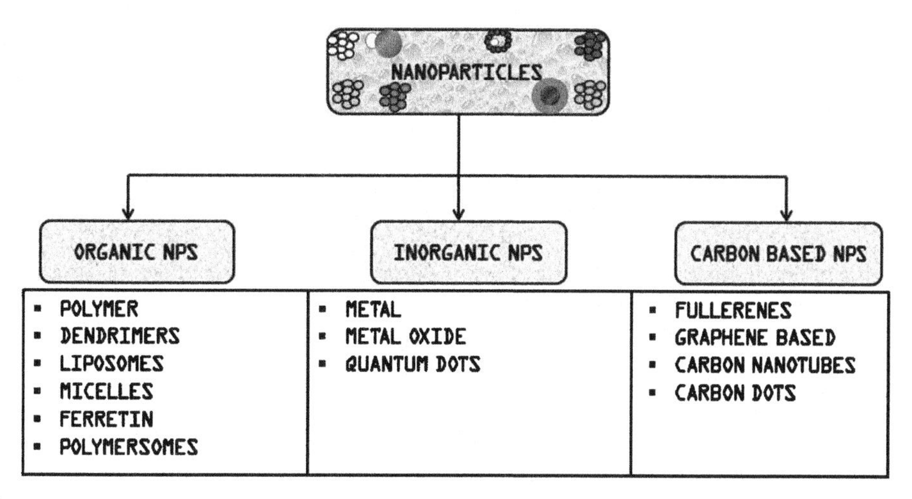

FIG. 6.2

Types of nanoparticles.

6.3.1 Organic nanoparticles

Organic NPs are fabricated from natural or synthetic organic molecules e.g. proteins, viruses, and lipids. These NPs tend to be biodegradable and mostly non-toxic considering their sources and they tend to be used effectively for the encapsulation of materials that are beneficial to drug delivery systems and in general biomedical applications. They are utilized in forms of polymer-drug conjugates, polymer-proteins, liposomes, dendrimers, polymersomes, micelles, and ferritins. These organic NPs tend to be utilized as nanocapsules, which depending on the makeup, could be designed to be sensitive to a certain stimulus, and hence triggered to react in certain ways due to external stimulus such as pH change, heat, light, and electromagnetic radiation. It is important to mention that organic NPs are formed by several organic molecules bound together by chemical binding or self-organization (Wang et al., 2016).

6.3.2 Inorganic nanoparticles

These are NPs that are not made up of carbon materials. Inorganic NPs constitute mostly metal and metal oxide-based nanoparticles. Metal-based NPs are fabricated from metal precursors to their nanometric sizes by either constructive or destructive means and virtually all metals can be fabricated into their nanoparticles. They can be synthesized by photochemical, electrochemical, or simple chemical methods. The chemical method simply involves the reduction by chemical reducing agents of metal-ion precursors in solution. On the other hand, metal oxides can be transformed from bulk into their respective nanometer sizes, metal oxide NPs, to significantly improve their hybrid properties in comparison to the original forms. These modified nanoforms, tend to affect surface characteristics such as high surface area to volume ratio, pore size, surface charge and surface charge density, crystalline, and amorphous structures, shapes like spherical and cylindrical and color, reactivity and sensitivity to environmental factors such as air, moisture, heat, and sunlight (Cartaxo, 2010). These characteristics invariably affect the behavior of the corresponding NPs in comparison to their bulk forms.

6.3.3 Carbon-based nanoparticles

These NPs are made entirely of carbon (Bhaviripudi et al., 2007), with the most common members of this class being fullerenes and carbon nanotubes (CNTs). Graphene and carbon black are at times classified as members of this group when they fall within the nano range. CNTs can be further categorized into multi-walled carbon nanotubes (MWCNTs) and single-walled carbon nanotubes (SWCNTs). The thermal conductivity along the length and non-conduciveness across the tube makes CNTs unique. Fullerenes are the allotropes of carbon having about sixty or more carbon atoms units that are pentagonal and hexagonal in an arrangement. Fullerenes have widespread applications commercially due to their electron affinity, high strength, structure, and electrical conductivity.

6.4 Synthesis of nanoparticles

NPs have been synthesized by various means by researchers in recent times. The choice of synthesis technique to fabricate these NPs is very vital towards attaining NPs with specific properties; biological, optical, magnetic, or electrical properties, that are dependent on their size and dimensions (Rogach et al., 2002). Researchers have resolved that the means of synthesis has undoubtedly been identified to influence NPs crystal structure, morphology, and sizes (Stankic et al., 2016). NPs are widely known to be synthesized by two broad techniques: bottom-up, which involves wet methods or top-down synthesis approach, usually involving physical methods (Fig. 6.3). The bottom-up approach is a building up or constructive approach to NPs synthesis, occurring where there is a building up of NPs from the lowest constituent unit of ordinary matter that comprises a chemical element, atoms, or from molecules or clusters. But, on the other hand, the top-down approach simply involves the breaking down of bulk materials to get the nanoscale materials, for example by mechanical grinding.

Several methods make up these approaches to NPs synthesis including but not limited to: hydrothermal, co-precipitation, sol-gel, sputtering, microwave, microemulsion, laser ablation, ultrasound, spark discharge, and biological synthesis methods. Of these, we shall now briefly discuss some of the more widely used methods.

6.4.1 Co-precipitation method

This synthesis method involves the stages of nucleation, growth, coarsening, and agglomeration processes simultaneously. Co-precipitation method usually involves reactions following the equation:

$$AXb^{+}_{(aq)} + bYa^{-}_{(aq)} \xrightarrow{\Delta} XaYb_{(s)} \tag{6.1}$$

The reactions are formed under high supersaturation conditions and lead to the formation of insoluble species. It begins with the nucleation phase, which is a vital step and it involves the formation of small particles. Secondary processes like aggregation and Ostwald ripening greatly influence the properties of the synthesized NPs. Examples of NPs formed by co-precipitation methods include oxides from the reaction between non-aqueous and aqueous solutions, metal oxides (Owonubi et al., 2020a), metal chalconides from molecular precursors reactions, metals from aqueous solutions by decomposition of metallorganic precursors or electrochemical reduction or reduction of non-aqueous solutions.

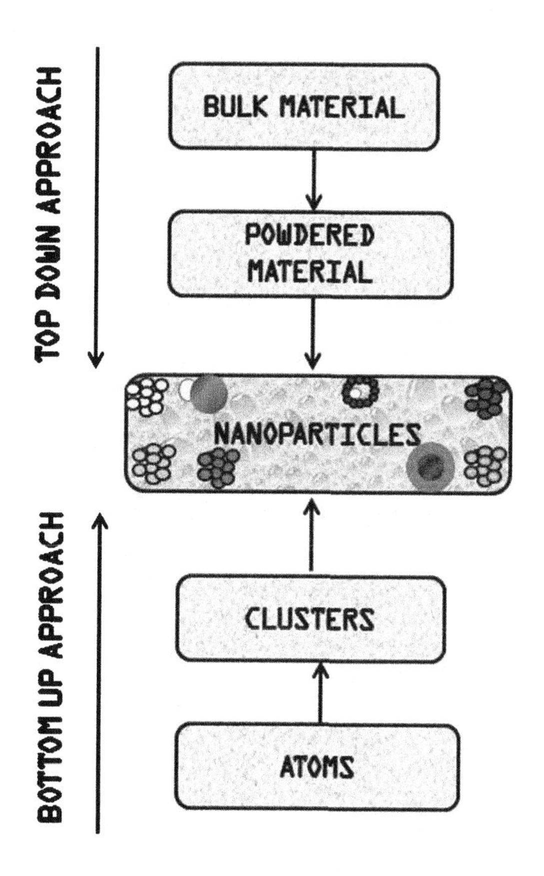

FIG. 6.3

Approaches to nanoparticles synthesis.

6.4.2 Hydrothermal or solvothermal method

The term "hydrothermal" originates from geology and was firstly described by Roderick Murchison a British geologist in the 19th century. He observed changes in the earth's crust and proposed they were as a result of the action of water at higher than normal temperature and pressure leading to minerals and rocks formation. The hydrothermal method involves chemical reactions between substances in an enclosed heated solution, usually under above ambient temperature and pressure (Yang and Park, 2019; Byrappa and Yoshimura, 2012). But the synthesis of single crystals is dependent on the solubility of the minerals under high pressure in a closed chamber (Demazeau, 2008). These crystals are then grown in an autoclave under regulated pressure. This method has been successful in synthesizing several materials for application across disciplines.

6.4.3 Sol-gel method

This method is employed for the production of solid materials from much smaller molecules. It entails a sol, sometimes regarded as the solution, which over time evolves into the formation of a diphasic

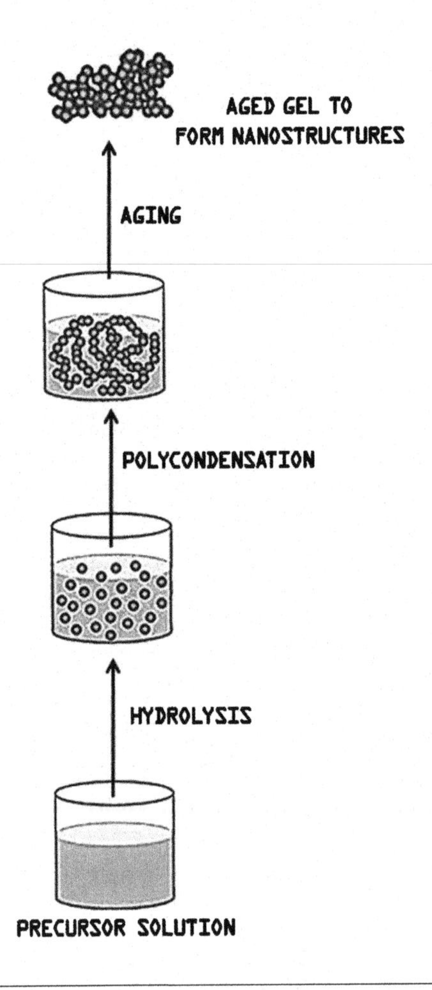

FIG. 6.4

NPs synthesis using the sol-gel approach.

gel-like system that contains a solid and liquid phase. The sol-gel method has been successful in the formation of morphologies ranging from distinct particles to networks of polymers. For example, the synthesis of Metal NPs have employed the sol-gel method by uniformly mixing the precursor starting materials in the liquid phase; employing hydrolysis and polycondensation reactions to encourage the formation of a transparent sol; aging of the prepared sol to allow colloidal particles to aggregate forming gels with 3D networks (Owonubi et al., 2020b), leading to the drying and sintering of the gels to encourage the formation of nanostructures (Fig. 6.4).

6.4.4 Microemulsion method

Although some researchers have a suggestive mechanism for the process, the method is not very well understood. The microemulsion method is one of the most ideal methods for inorganic NP synthesis (Vidal-Vidal et al., 2006; Uskoković and Drofenik, 2005), occurring when the reactants and the microemulsion material are mixed in an enclosed vessel. This ensures the collision of water droplets in the microemulsion as the reactants collide. During the reaction, which occurs at a fast pace, interactions between the reactants occur leading to precipitation within the nanodroplets. This is followed by nucleation growth and coagulation of initial particles, which finally lead to NPs stabilized by surfactants or surrounded by water.

6.4.5 Ultrasound method

The synthesis of NPs using the ultrasound method involves the irradiation of liquids with ultrasonic irradiation forming ultrasonic cavitation. The ultrasonic cavitation encourages unique chemical and physical environments for chemical reactions under extreme conditions. This method encourages the synthesis of NPs with morphologies that can be easily controlled.

6.4.6 Microwave-assisted method

This method has over time become quite common amongst researchers involved in biochemical investigations to nanotechnology. It involves reactions which as a result of the use of microwave reactors are quicker than conventional heating methods. Reactions that are microwave-assisted tend to yield high by-products, with fewer side products. The level of reproducibility with microwave-assisted synthesis is high due to the ability of the reactors to provide excellent control over reaction mixing. This method allows for control over the nucleation and growth phases of the synthesis process providing considering the reaction begins at room temperature. This method encourages selectivity in heating either the precursor molecules or the solvent and thus, can be activated independently and gives room for scalability.

6.4.7 Biomimetic method

This method is sometimes known as the biological approach to NPs synthesis, a process that interconnects biotechnology and nanotechnology. The quest for green chemistry has promoted this method to NP synthesis occurring at much lower temperatures and a slower rate. This results in NPs that are not monodispersed, and it lessens the particle degradation. The benefits and drawbacks of the methods used in NP synthesis are presented (Table 6.1).

It is vital to highlight that in some literature, this classification also categorizes NPs as ceramic-based, polymeric, semi-conductor, or lipid-based centered on the NP make up, but these can easily be fitted into the previously classified groups.

6.5 Applications of nanoparticles in phytoremediation

Environmental remediation and the diminution of pollution using nanomaterials have elicited the increasing attention of scientists with the advent of nanotechnology. Carbon-based and metal-based nanomaterials are the most investigated (Song et al., 2019; Gong et al., 2009; Chen et al., 2017). Nanomaterials have successfully aided phytoremediation as reported in several studies (Rai et al., 2020; Ma and Wang, 2010; Tripathi et al., 2016; Huang et al., 2018) Nano-phytoremediation is

Table 6.1 Benefits and drawbacks of the different methods used in nanoparticles synthesis.

Benefits	Drawbacks
Co-precipitation method	***Co-precipitation***
• Possible green synthesis method, organic solvents are not involved. • Does not require high temperature, thus energy efficient. • NP composition and particle size control efficient. • Preparation by this method is quick and easy. • Allows for modification to NP surface state and homogeneity.	• Reactions not feasible with chemicals with uncharged species. • It could be time-consuming, depending on the reaction. • The reaction can lead to impurities, although trace in the NPs synthesized. • By-products are not always reproducible. • The formation of NPs is highly dependent on the rates of precipitation of the reactants.
Hydrothermal method	Hydrothermal method
The solubility of most materials in solvents is achieved at high pressures and temperatures close to its critical point, thus encouraging this method. The by-product's crystallinity, shape, and size distribution can easily be controlled and adjusted by changing the starting variables. Synthesis of materials with specific phases are easily produced, there is ease with reproducibility of synthesized materials. The method encourages the possibility of solid-state synthesis due to improvements in the chemical activities of reactants. It results in by-products which are thermally decomposed with high vapor pressure and low melting points.	Due to the above ambient temperatures used, observation of the chemical reaction process is not feasible. The safety of most hydrothermal reactions is sometimes difficult to manage. Need for high-end expensive equipment to perform the reactions.
Sol-gel method	Sol-gel method
• Precursor materials are at the molecular level encouraging by-products are homogeneous. • Purity levels of by-products are high. • Ease to control the by-products level of porosity. • Requires low temperature, thus energy efficient. • The method allows for influence over the by-product's chemical makeup, hence suitable for the preparation of materials with multiple components. • The method encourages doping and uniform dispersity of dopants in the by-product.	• The sol-gel method usually takes a long time for the reaction to be concluded. • This method tends to employ the use of organic solvents which may result in toxicity of the by-product.
Microemulsion method	Microemulsion method
• Preparation by this method is quick and easy. • Control over the by-product composition and particle size is easily achieved. • The tendency for synthesized agglomeration of inorganic NPs is low. • Microemulsions are generally thermodynamically stable, this aids in the stability of the NPs synthesized. • NPs formed has a high specific area and crystalline structures.	• The method employs the use of high concentration co (surfactants) which are highly irritable. • Microemulsions stability is dependent on external factors, pH, temperature. • The capacity to solubilize is limited for high melting point substances.

Table 6.1 Benefits and drawbacks of the different methods used in nanoparticles synthesis. *Continued*

Benefits	Drawbacks
Co-precipitation method	***Co-precipitation***
Ultrasound method	Ultrasound method
• Requires low temperature. • Time-saving as the reaction using this method happens at a fast rate. • The method is simple and quick. • The morphologies of NPs synthesized are easily controllable; nanorings, nanobelts, core shelled NPs, etc.	• Materials that are sensitive to heat cannot endure cavitation. • Difficult to scale up. • Although low energy is required, it is an energy-intensive process, with low yield per unit energy utilized.

NPs, nanoparticles.

useful for the elimination of HMs like cadmium (Singh and Lee, 2016) and lead (Liang et al., 2017), metalloids like arsenic (Souri et al., 2017b) and organic pollutants such as trichloroethylene (TCE) (Ma and Wang, 2010). Nanomaterials advance phytoremediation either directly or indirectly. By direct action, the nanomaterials work on the plants and pollutants, while for indirect action, nanomaterials are implicated in the interactions between the plants and pollutants thereby tacitly influencing the ultimate remediation efficacy (Song et al., 2019).

Nanomaterials remove pollutants directly from soils employing adsorption or redox reactions. CNTs for instance have been shown to possess exceptional adsorption ability for different pollutants thereby immobilizing them in the process (Mueller and Nowack, 2010; Song et al., 2019; Kang et al., 2018). The incorporation of the pollutants and CNTs is comparatively stable due to the several connections, which may exist during the adsorption procedure. Through redox reactions, nanomaterials such as nanoscale zerovalent iron (nZVI) are employed as electron contributors for reductive degradation or stabilizing the contaminants (Song et al., 2019). Through indirect action, nanomaterials remove pollutants by intensifying the phytoavailabilty of the pollutants and improving plant growth. The effectiveness of phytoremediation is majorly influenced by the phytoavailabilty of pollutants and plants can take up pollutants just in accessible forms. Other factors that affect the phytoavailability of pollutants are the physico–chemical characteristics of soil and the physiological features of plants. Phytoremediation is usually restricted when there is minimal phytoavailability of pollutants. The efficacy of phytoremediation increases with intensified phytoavailability of pollutants. Nanomaterials can augment the phytoavailability of pollutants by serving as carriers when they gain access into the cell (Song et al., 2019; Zaier et al., 2014; Su et al., 2013). The uptake of pollutants by plants is enhanced by the addition of nanomaterials e.g. fullerene and this is apparent in roots (Ma and Wang, 2010).

The choice of plants for phytoremediation is often informed by their degree of growth and biomass. As a result of low biomass and dawdling growth rate which are caused by constrained tolerance to contaminants and inadequate soil quality, several plants used for phytoremediation function below expectation. Some nanomaterials such as graphene (Zhang et al., 2015), metal-based nanomaterials, Ag nanoparticles (Das et al., 2018), and γ-Fe_2O_3 nanoparticles (Hu et al., 2017) have been used for boosting plant growth using various mechanisms. Some of the mechanisms of action include penetration of the vacuole and deposition in roots (Zhang et al., 2015) and an increase in nitrogen uptake

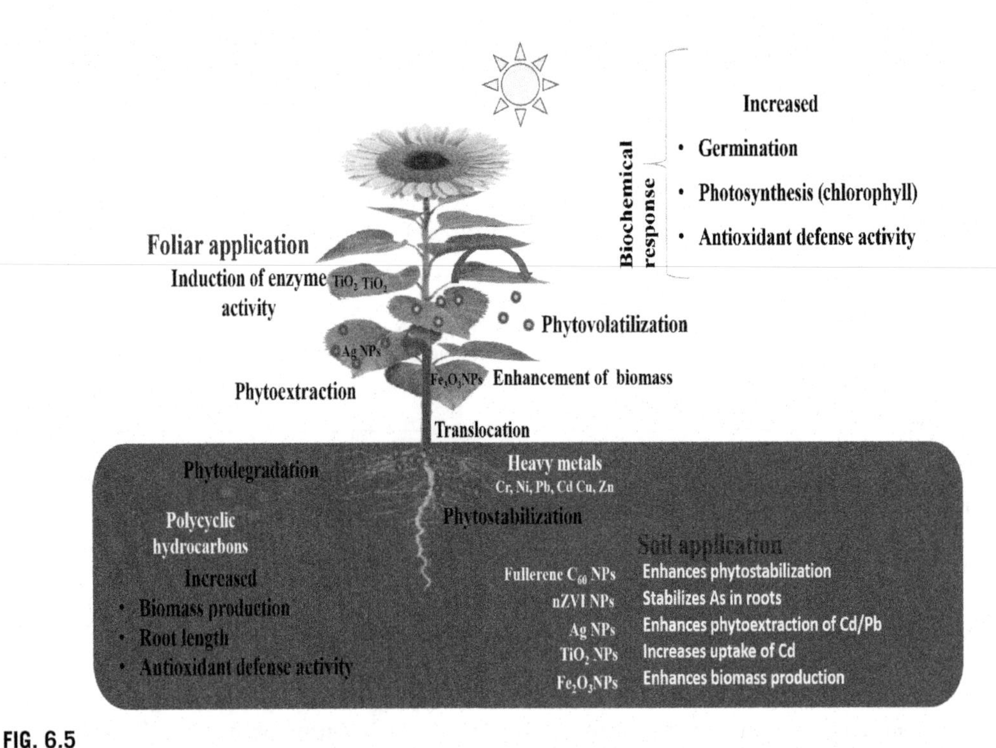

FIG. 6.5

Enhancement of phytoremediation by application of nanoparticles.

(Das et al., 2018). The remediation efficacy of phytoremediation is increased by nanomaterials through plant growth improvement. The efficacy of phytoremediation mediated by NPs is presented in Fig. 6.5.

Nanomaterials can play a role in increasing the resilience of plants to pollutants and improve plant growth in phytoremediation through the augmentation of photosynthesis, regulation of microbial communities in soils, and improvement of abiotic stress. Nanomaterials have been shown to improve the functioning of rhizobacteria under abiotic pressure (Song et al., 2019; Timmusk et al., 2018). Fullerenes have widespread applications commercially due to their electron affinity, high strength, structure, and electrical conductivity. Table 6.2 presents different types of nanoparticles and their applications in phytoremediation. nZVI is known to be highly reactive with good adsorption capacity and reducing power which makes it one of the most engineered nanomaterials (Pasinszki and Krebsz, 2020). The application of nZVI as a remediating agent on contaminated soils could have much impact on plant growth and physiology. It has been used in several studies for the removal of environmental pollutants from wastewaters as well as contaminated agricultural soils. An organic pesticide Endosulfan was remediated from contaminated soil using three hyperaccumulator plants chittaratha (*Alpinia calcarata*), tulsi (*Ocimum sanctum*), and lemongrass (*Cymbopogon citratus*) in an experiment by Pillai and Kottekottil (2016). Nano-phytoremediation using *Alpinia calcarata* within a week removed about 82.2% of endosulfan from the contaminated soil and 94.92% by the third week (Pillai and Kottekottil, 2016). The joint treatment of nZVI and phytoremediation significantly enhanced the removal of the contaminant when compared to using either nZVI or phytoremediation approach alone (17.29% and 51.74% after 7 days) (Pillai and Kottekottil, 2016).

Table 6.2 Application of nanoparticles in phytoremediation.

Nanoparticles	Pollutant	Effect of nanoparticles	Reference
CNTs	Nitrogen removal, organochlorine pesticides, Cadmium	Remediation; promotion of nodulation	(Gong et al., 2019; Yuan et al., 2017; Oloumi et al., 2018; Zhang et al., 2017)
Ferritins	Chromium	Change chromium Cr (VI) into Cr (III)	(Watlington, 2005)
Fullerenes	Trichloroethylene, Nitrogen removal.	Plant uptake of trichloroethylene was increased by ~26% and 82% by the addition of 2 and 15 mg/L of fullerene nanoparticles synthesized through solvent (toluene) exchange; decreased water uptake and nitrogen recovery.	(Ma and Wang, 2010; Yavari et al., 2018)
ZnO NPs	Cadmium and Lead	Accumulation of Cd and Pb in the plant increased enhancing its growth	(Venkatachalam et al., 2017)
Carbon black	Cadmium and Nickel	Modified carbon black significantly reduced the availability of heavy metals in soil and the uptake of Cd and Ni by *Suaeda salsa* by roughly 18 and 10% and improved the growth of plants by alleviating the growth inhibition caused by heavy metals.	(Cheng et al., 2019)
Silicon NPs	Chromium	Accumulation of the concentration of Cr in the root and shoot decreased	(Tripathi et al., 2016)
TiO_2 NPs	Cadmium	Increasing concentration of TiO_2 nanoparticles from 100 to 300 mg/kg caused increased uptake of Cd.	(Singh and Lee, 2016)
Salicylic acid Nanoparticles	Arsenic	Increased accumulation of arsenic was recorded	(Souri et al., 2017a)
Nano zerovalent iron	Endosulfan, trinitrotoluene, arsenic, cadmium, lead, zinc	Removal of the corresponding pollutant	(Pillai and Kottekottil, 2016; Jiamjitrpanich et al., 2012; Vítková et al., 2018)
Ag NPs	Cadmium, Lead and Nickel	Increase in accumulation of Cd, Pb, and Ni in the shoot leading to increased root length	(Khan and Bano, 2016)
Magnetite NPs	Arsenic	Reduction in water-leachable As (V) and the leachability of As (V) remaining in the soil.	(Liang and Zhao, 2014)
Al_2O_3 nanoparticles	Phenanthrene	Reduction in root elongation inhibition in selected plants	(Yang and Watts, 2005)
γ-Fe_2O_3 NPs	Arsenic	50 mg/L γ-Fe_2O_3 NPs significantly enhanced chlorophyll content by 23.2% and root activity by 23.8% but increased MDA formation, and decreased chlorophyll content and root activity at 100 mg/L g-Fe_2O_3	(Hu et al., 2014)

Likewise, evaluation of sunflower and ryegrass sown on As-rich and Zn-rich contaminated soils treated with nZVI led to a reduction in the uptake of As into the roots and shoots of sunflower control plants by 47% and 24% respectively while it significantly reduced CaC12-available fractions of Zn, Pb, and Cd in Zn-rich contaminated soils treated with nZVI (Vítková et al., 2018). Despite the stabilization of the HMs results recorded using nZVI, some As and Zn were taken up by the plants. Likewise, the application of nZVI amended with sodium alginate at lower concentrations (100, 200, and 500 mg kg^{-1}) significantly augmented Pb accumulation in ryegrass (Huang et al., 2016). In a recent pot experiment, nZVI applied as a soil amendment at low concentration (150 mg kg^{-1}), enhanced the phytoextraction ability of White willow (*Salix alba* L.) plants co-inoculated with *Pseudomonas fluorescens* (PGPR) and *Rhizophagus irregularis* (arbuscular mycorrhizal fungus) grown in Pb, Cu, and Cd contaminated soils (Mokarram-Kashtiban et al., 2019). The presence of the rhizospheric microorganisms elevated the BCF of the HMs (Pb > 1, Cu > 2, and Cd 3–7.5) while protecting the plants from the toxicity of the metals, and significantly elongated root length as well as increased leaf area of the treated plants. Moreover, the tolerance index was >1 for most parameters investigated, an indication of the high level of tolerance of the plant in the presence of the rhizospheric microbes and low concentration of the nZVI Nps (Mokarram-Kashtiban et al., 2019). nZVI must have augmented the activity of the rhizospheric microbes by releasing iron which may have prevented HM stress on the Willow plants. However, at higher concentrations, nZVI reduced plant height and leaf area of the plants thereby affecting their growth (Mokarram-Kashtiban et al., 2019). Application of nZVI in nanoparticle-mediated phytoremediation should be done with a low concentration of the nanomaterials as higher concentrations have been reported to be toxic to the plant (Zhu et al., 2019).

Carbon-based nanomaterials such as fullerenes have received widespread applications in phytoremediation compared to graphene or CNTs. Investigation of fullerene C_{60} NPs-mediated phytoremediation of TCE by Ma and Wang (2010) revealed a significant increase in the uptake of TCE on the addition of 2 mg L^{-1} and 15 mg L^{-1} fullerene C_{60} by 26% and 82%, respectively. Likewise, the phytostabilization abilities of *Cucurbita pepo, Zea mays, Solanum lycopersicum,* and *Glycine max* were assessed in 50 g soil contaminated with 2,15 ng g^{-1} weathered chlordane and 118 ng g^{-1} of DDx (DDT + metabolites) and treated with fullerene C_{60} at different concentrations (De La Torre-Roche et al., 2013). The ability of the plants to accumulate the pesticides differed with fullerene C_{60} treated *Glycine max* having a significantly higher DDx content in its roots compared to *Cucurbita pepo.* Moreover, Chordane levels increased by 34.9% in *Solanum lycopersicum* and *Glycine max* while DDx was completed suppressed in *Zea mays* and *Solanum lycopersicum* (De La Torre-Roche et al., 2013; Sanzari et al., 2019).

Several metallic nanoparticles have also been found useful in augmenting phytoremediation. Titanium dioxide (TiO_2) Nps enhanced absorption of cesium (^{133}CS) in soybean roots and environs as well as increased plant protein content (Singh and Lee, 2018). Moreover, the authors reported a higher accumulation of ^{133}CS in the shoot (731.7 μg g^{-1} dw) of the plant than the roots (597.8 μg g^{-1} dw). TiO_2 could also be used for the removal of a range of contaminants from the environment such as organic pesticides, dyes, and noxious compounds from wastewaters (Waghmode et al., 2019). Application of 25 mg L^{-1} zinc oxide (ZnO) Nps to *Leucaena leucocephala* exposed to Cd (50 mg L^{-1}) and Pb (100 mg L^{-1}) reduced oxidative stress produced by the HMs by enhancing antioxidant enzyme activities and decreased the root and shoot lengths tremendously. However, a significant increase was observed in shoot growth by approximately 10.4% compared to the control plants when ZnO was combined with Zn (Venkatachalam et al., 2017). Plant biochemical parameters such as chlorophyll and total soluble proteins were also increased except lipid peroxidation levels in leaves which were reduced (Venkatachalam et al., 2017).

Among the various metallic NPs, silver (Ag) Nps have received much attention due to its catalytic activity, antimicrobial activity as well its high chemical stability (Frattini et al., 2005; Oza et al., 2020). Utilization of AgNps extracted from a medicinal plant was evaluated on four plant species (*Brassica, Ipomea, Camellia,* and *Plantago* in water solution spiked with a neurotoxic and carcinogenetic chemical Fipronil (Romeh, 2018). Interaction of AgNp with *Brassica* resulted in photodegradation of the chemical by 68.8% in contrast to sole treatment with Plantago alone within 6 days (10.14%) (Romeh, 2018). Thus utilization of Ag Nps in association with Brassica plant has many benefits in the removal of Dipronil contaminants in water-logged soils and water (Romeh, 2018). Silver NPs are easily taken up by plants through the xylem and translocated to the above plant parts such as the leaves for harvest or degradation. Despite the benefits derived from the use of these NPs for enhancing plant growth and development, utmost care must be taken to determine the level of phytotoxicity to the plant and the environment. Thus, controlled field trials should be conducted using different concentrations of Nps to ascertain the potential level that may be toxic to the plant or/and the environment. This is relevant because there are beneficial microorganisms in the soil which also play a role in plant biomass production and increased chlorophyll and antioxidant production under biotic and abiotic stresses. Is the composition of these beneficial microbes affected under NP application in the soil for enhanced phytoremediation, and at what level? These are pertinent questions researchers need to take into consideration in the design of a phytoremediation on site program using nanoparticles as a remediation strategy.

6.6 Future perspectives in the utilization of nanoparticle-mediated phytoremediation

Nanoparticle-mediated phytoremediation is a new field of research, which is gradually gaining more awareness as an effective means of cleanup of polluted soil and water due to their large surface areas which makes them highly efficient in absorbing contaminants (Srivastav et al., 2018). However, the effectiveness of NPs in the remediation of contaminated environments is determined by several factors such as the temperature and pH of the soil, particle size, the nature of the nanoparticle and type, as well as the type of plant. Hyperaccumulator plants, bacteria, and fungi are being considered as the most preferred for environmental cleanup. For the effective use of hyperaccumulator plants, such plants must have some basic qualities which include; the ability to grow quickly and produce more biomass, production of an extensive branched root system for increased surface area, low phytotoxic level and possess high contaminant hyperaccumulation capacity. Such plants should also be easily amenable to genetic manipulation and be easily harvested from the sink organ. Genetic modification of hyperaccumulator plants should be considered to improve biomass production, remove mixed contaminants from contaminated sites, as well as augment plant-microbe interactions, which increase microbial rhizosphere diversity. Through plant genetic engineering, specific hyperaccumulator plants could be genetically engineered to better accumulate and degrade toxic contaminants, especially HMs such as arsenic, which is poisonous to living organisms. Due to the fear of contamination of transgenic plants via pollen transfer, chloroplast transformation could be a better alternative in the engineering of hyperaccumulator plants for overexpression of useful genes for bioaccumulation or degradation of contaminants.

The mechanisms by which contaminants are absorbed by such plants via phytoextraction, phytostabilization, and phytodegradation, needs to be well understood. Furthermore, there is a need to study the route of translocation or movement of these NPs to the site of degradation or bioaccumulation in most cases the leaves, to ensure toxic levels of the contaminants are not present and most importantly it would have no noxious effect on the plant and microbes in the soil, which play major roles in plant growth promotion and abiotic stress tolerance. Nanoparticles could gain entrance into plants either through apoplastic means, which is useful for systemic NP delivery by the movement of the particles from the central root cylinder and the vascular tissues, upwards the above plant parts (Sattelmacher, 2009; Larue et al., 2014; Sun et al., 2014; Zhao et al., 2017; Sanzari et al., 2019), or symplastic transport (Roberts and Oparka, 2003). The movement of metals in plants occurs through metal transporters. These metal transporters could also be engineered for enhanced function and their targeting to specific cell types determined to prevent disrupting other cell functions of the plant (Hamon et al., 1997; Yadav et al., 2018). A mechanistic understanding of the metabolic pathways involved in the degradation process will come in handy in the development of plants with enhanced capacity to degrade or volatilize contaminants from the environment. New approaches based on nanoparticle-mediated clustered regularly interspersed palindromic repeats-CRISPR associated proteins (CRISPR-Cas9) technology could be employed to deliver foreign DNA for editing of plant genomes in several biological systems (Glass et al., 2018; Sanzari et al., 2019). However, an efficient delivery system is required to get the desired gene into the nucleus of plant cells. Modulation of influx or efflux of plasma membrane transporters could also be used to augment uptake, accrual, and tolerance of HMs such as arsenic by plants, in addition to regulation of the arsenate reductase activity, and the increase of the number of PCs and glutaredoxins (Srivastav et al., 2018). While trying to enhance phytoremediation abilities of hyperaccumulator plants via genetic engineering and NP-mediated phytoremediation strategies, care must also be taken to evaluate any possible toxicity of such NPs or transgenic plants to the environment when grown on the field or applied on contaminated sites. There is a paucity of information on transgenic plant-mediated phytoremediation field trials which should be considered before wide-scale application on contaminated sites. It is pertinent to note that the success of NP-mediated phytoremediation requires the collaborative effort of researchers in several fields such as nanotechnologists, biochemists, plant breeders, and genetic engineers as well as botanists. More hyperaccumulator plant species also need to be identified and evaluated. Besides, more research should be conducted to determine ways of increasing the bioavailability of metal pollutants in nanoforms for ease of removal by the plants as metallic uptake by plants differs among plant species.

6.7 Conclusion

The activities of man continuously release harmful substances into the environment to the detriment of human health and the environment. The utilization of NPs-mediated phytoremediation has proven to be a more efficient and cost-effective means of removal of organic and inorganic pollutants from the environment compared to the physical or chemical conventional methods. NPs augment the phytoavailability of contaminants when applied as foliar or soil carriers to plants in addition to enhancing the rate of photosynthesis and the composition of the microbiome of the rhizosphere of treated plants, thereby increasing biomass production of plants and the ability to withstand HM stress due to increased activities of antioxidants. With the advancement in technology, highly efficient hyperaccumulators and evapotranspirator plants can be engineered for better efficacy in the remediation process. A better understanding of the interactions taking place within the rhizosphere of plants and the NPs, in addition to studying the metabolic pathways involved in the degradation of noxious pollutants by plants would go a long way in improving the rate at which successful phytoremediation is achieved. Nevertheless, application of NP-assisted phytoremediation on the field needs to be properly monitored to sustain the process, while preventing toxicity to the environment.

References

Abdullahi, M. 2015. Soil contamination, remediation and plants: prospects and challenges. Soil Remediation and Plants; In: Khalid, RH, Muhammad, S., Munir, O., Ahmet, RM, (Eds.), p. 525.

Ali, H., Khan, E., Sajad, M.A., 2013. Phytoremediation of heavy metals—concepts and applications. Chemosphere 91 (7), 869–881.

Ayangbenro, A.S., Babalola, O.O., 2017. A new strategy for heavy metal polluted environments: a review of microbial biosorbents. Int. J. Environ. Res. Public Health 14 (1), 94.

Beresford, N., Fesenko, S., Konoplev, A., Skuterud, L., Smith, J., Voigt, G., 2016. Thirty years after the Chernobyl accident: what lessons have we learnt? J. Environ. Radioact. 157, 77–89.

Bhaviripudi, S., Mile, E., Steiner, S.A., Zare, A.T., Dresselhaus, M.S., Belcher, A.M., Kong, J., 2007. CVD synthesis of single-walled carbon nanotubes from gold nanoparticle catalysts. J. Am. Chem. Soc. 129 (6), 1516–1517.

Bolan, N.S., Park, J.H., Robinson, B., Naidu, R., Huh, K.Y., 2011. Phytostabilization: a green approach to contaminant containment. Adv. Agron. 112, 145–204.

Byrappa, K., Yoshimura, M., 2012. Handbook of Hydrothermal Technology. William Andrew, Elsevier Publishers, p. 800.

Cai, C., Zhao, M., Yu, Z., Rong, H., Zhang, C., 2019. Utilization of nanomaterials for in-situ remediation of heavy metal(loid) contaminated sediments: A review. Science of The Total Environment 662, 205–217.

Cartaxo, A. 2010. Nanoparticles types and properties–understanding these promising devices in the biomedical area. MS thesis, Dept. Biomed. Eng., University of Minho., Braga, Portugal.

Chen, M., Zeng, G., Xu, P., Zhang, Y., Jiang, D., Zhou, S., 2017. Understanding enzymatic degradation of single-walled carbon nanotubes triggered by functionalization using molecular dynamics simulation. Environ. Sci. Nano 4, 720–727.

Cheng, J., Sun, Z., Yu, Y., Li, X., Li, T., 2019. Effects of modified carbon black nanoparticles on plant-microbe remediation of petroleum and heavy metal co-contaminated soils. Int. J. Phytoremediation 21 (7), 634–642.

Das, P., Barua, S., Sarkar, S., Karak, N., Bhattacharyya, P., Raza, N., Kim, K.-H., Bhattacharya, S.S., 2018. Plant extract–mediated green silver nanoparticles: Efficacy as soil conditioner and plant growth promoter. J. Hazard. Mater. 346, 62–72.

Davis, A.S., Prakash, P., Thamaraiselvi, K., 2017. Nanobioremediation technologies for sustainable environmentBioremediation and Sustainable Technologies for Cleaner Environment. Springer, Cham, pp. 13–33.

De La Torre-Roche, R., Hawthorne, J., Deng, Y., Xing, B., Cai, W., Newman, L.A., Wang, Q., Ma, X., Hamdi, H., White, J.C, 2013. Multiwalled carbon nanotubes and C60 fullerenes differentially impact the accumulation of weathered pesticides in four agricultural plants. Environ. Sci. Technol. 47 (21), 12539–12547.

Demazeau, G., 2008. Solvothermal reactions: an original route for the synthesis of novel materials. J. Mater. Sci. 43 (7), 2104–2114.

Dixit, R., Malaviya, D., Pandiyan, K., Singh, U.B., Sahu, A., Shukla, R., Singh, B.P., Rai, J.P., Sharma, P.K., Lade, H., 2015. Bioremediation of heavy metals from soil and aquatic environment: an overview of principles and criteria of fundamental processes. Sustainability 7, 2189–2212.

Dubchak, S., Bondar, O., 2019. Bioremediation and phytoremediation: Best approach for rehabilitation of soils for future use. In: Gupta, D.K., Voronina, A. (Eds.), Remediation Measures for Radioactively Contaminated Areas. Springer, Cham, Switzerland.

Dushenkov, S., 2003. Trends in phytoremediation of radionuclides. Plant Soil 249, 167–175.

Dushenkov, S., Mikheev, A., Prokhnevsky, A., Ruchko, M., Sorochinsky, B., 1999. Phytoremediation of radiocesium-contaminated soil in the vicinity of Chernobyl, UkraineEnviron. Sci. Technol.33, 469–475.

Ealias, A.M., Saravanakumar, M., 2017. A review on the classification, characterisation, synthesis of nanoparticles and their application. IOP Conf. Ser. Mater. Sci. Eng 263 (3), 032019.

Farid, M., Ali, S., Rizwan, M., Ali, Q., Abbas, F., Bukhari, S.A.H., Saeed, R., Wu, L, 2017. Citric acid assisted phytoextraction of chromium by sunflower; morpho-physiological and biochemical alterations in plants. Ecotoxicol. Environ. Saf. 145, 90–102.

Frattini, A., Pellegri, N., Nicastro, D., De Sanctis, O., 2005. Effect of amine groups in the synthesis of Ag nanoparticles using aminosilanes. Mater. Chem. Phys. 94 (1), 148–152.

Gerhardt, K.E., Gerwing, P.D., Greenberg, B.M., 2017. Opinion: taking phytoremediation from proven technology to accepted practice. Plant Sci. 256, 170–185.

Ghormade, V., Deshpande, M.V., Paknikar, K.M., 2011. Perspectives for nano-biotechnology enabled protection and nutrition of plants. Biotechnol. Adv. 29 (6), 792–803.

Glass, Z., Lee, M., Li, Y., Xu, Q., 2018. Engineering the delivery system for CRISPR-based genome editing. Trends Biotechnol. 36 (2), 173–185.

Gomes, M.A.D.C., Hauser-Davis, R.A., Souza, A.N.D., Vitória, A.P, 2016. Metal phytoremediation: General strategies, genetically modified plants and applications in metal nanoparticle contamination. Ecotoxicol. Environ. Saf. 134, 133–147.

Gong, J.-L., Wang, B., Zeng, G.-M., Yang, C.-P., Niu, C.-G., Niu, Q.-Y., Zhou, W.-J., Liang, Y., 2009. Removal of cationic dyes from aqueous solution using magnetic multi-wall carbon nanotube nanocomposite as adsorbent. J. Hazard. Mater. 164 (2–3), 1517–1522.

Gong, X., Huang, D., Liu, Y., Peng, Z., Zeng, G., Xu, P., Cheng, M., Wang, R., Wan, J., 2018. Remediation of contaminated soils by biotechnology with nanomaterials: bio behavior, applications, and perspectives. Crit. Rev. Biotechnol. 38 (3), 455–468.

Gong, X., Huang, D., Liu, Y., Zeng, G., Wang, R., Wan, J., Zhang, C., Cheng, M., Qin, X., Xue, W., 2017. Stabilized nanoscale zerovalent iron mediated cadmium accumulation and oxidative damage of Boehmeria nivea (L.) Gaudich cultivated in cadmium contaminated sediments. Environ. Sci. Technol. 51 (19), 11308–11316.

Gong, X., Huang, D., Liu, Y., Zeng, G., Wang, R., Xu, P., Zhang, C., Cheng, M., Xue, W., Chen, S., 2019. Roles of multiwall carbon nanotubes in phytoremediation: cadmium uptake and oxidative burst in Boehmeria nivea (L.) Gaudich. Environ. Sci. Nano 6 (3), 851–862.

Haldar, S., Ghosh, A., 2020. Microbial and plant-assisted heavy metal remediation in aquatic ecosystems: a comprehensive review. 3 Biotech 10 (5), 1–13.

Hamon, R., Wundke, J., Mclaughlin, M., Naidu, R., 1997. Availability of zinc and cadmium to different plant species. Soil Res. 35, 1267–1278.

Hawrot-Paw, M., Ratomski, P., Mikiciuk, M., Staniewski, J., Koniuszy, A., Ptak, P., Golimowski, W., 2019. Pea cultivar Blauwschokker for the phytostimulation of biodiesel degradation in agricultural soil. Environ. Sci. Pollut. Res. 26 (33), 34594–34602.

Hu, J., Guo, H., Li, J., Gan, Q., Wang, Y., Xing, B., 2017. Comparative impacts of iron oxide nanoparticles and ferric ions on the growth of Citrus maxima. Environ. Pollut. 221, 199–208.

Hu, X., Kang, J., Lu, K., Zhou, R., Mu, L., Zhou, Q., 2014. Graphene oxide amplifies the phytotoxicity of arsenic in wheat. Sci. Rep. 4, 6122.

Huang, D., Qin, X., Peng, Z., Liu, Y., Gong, X., Zeng, G., Huang, C., Cheng, M., Xue, W., Wang, X., Hu, Z., 2018. Nanoscale zero-valent iron assisted phytoremediation of Pb in sediment: impacts on metal accumulation and antioxidative system of Lolium perenne. Ecotoxicol. Environ. Saf. 153, 229–237.

Huang, D., Xue, W., Zeng, G., Wan, J., Chen, G., Huang, C., Zhang, C., Cheng, M., Xu, P., 2016. Immobilization of Cd in river sediments by sodium alginate modified nanoscale zero-valent iron: impact on enzyme activities and microbial community diversity. Water Res. 106, 15–25.

Islam, M.S., Saito, T., Kurasaki, M., 2015. Phytofiltration of arsenic and cadmium by using an aquatic plant, Micranthemum umbrosum: Phytotoxicity, uptake kinetics, and mechanism. Ecotoxicol. Environ. Saf. 112, 193–200.

Jiamjitrpanich, W., Parkpian, P., Polprasert, C., Kosanlavit, R., 2013. Trinitrotoluene and its metabolites in shoots and roots of Panicum maximum in nano-phytoremediation. Int. J. Environ. Sci. Dev. 4, 7.

Jiamjitrpanich, W., Parkpian, P., Polprasert, C., Laurent, F., Kosanlavit, R., 2012. The tolerance efficiency of Panicum maximum and Helianthus annuus in TNT-contaminated soil and nZVI-contaminated soil. J. Environ. Sci. Health A 47 (11), 1506–1513.

Kacálková, L., Tlustoš, P., Száková, J, 2015. Phytoextraction of risk elements by willow and poplar trees. Int. J. Phytoremediation 17 (1–6), 414–421.

Kang, J., Duan, X., Wang, C., Sun, H., Tan, X., Tade, M.O., Wang, S., 2018. Nitrogen-doped bamboo-like carbon nanotubes with Ni encapsulation for persulfate activation to remove emerging contaminants with excellent catalytic stability. Chemical Engineering Journal 332, 398–408.

Kathal, R., Malhotra, P., Kumar, L., Uniyal, P.L., 2016. Phytoextraction of Pb and Ni from the polluted soil by Brassica juncea L. J. Environ. Analy. Toxicol. 6 (5), 1–4.

Khan, N., Bano, A., 2016. Role of plant growth promoting rhizobacteria and Ag-nano particle in the bioremediation of heavy metals and maize growth under municipal wastewater irrigation. Int. J. Phytoremediation 18 (3), 211–221.

Khatoon, H., Pant, A., Rai, J., 2017. Plant adaptation to recalcitrant chemicalsPlant Adaptation Strategies in Changing Environment. Springer, Singapore, pp. 269–290.

Kumpiene, J., Lagerkvist, A., Maurice, C., 2008. Stabilization of As, Cr, Cu, Pb and Zn in soil using amendments–a review. Waste Manag. 28 (1), 215–225.

Laghlimi, M., Baghdad, B., Hadi, H.E., Bouabdli, A., 2015. Phytoremediation mechanisms of heavy metal contaminated soils: a review. Open J Ecol. 5, 375–388.

Larue, C., Castillo-Michel, H., Sobanska, S., Cécillon, L., Bureau, S., Barthès, V., Ouerdane, L., Carrière, M., Sarret, G, 2014. Foliar exposure of the crop Lactuca sativa to silver nanoparticles: evidence for internalization and changes in Ag speciation. J. Hazard. Mater. 264, 98–106.

Liang, J., Yang, Z., Tang, L., Zeng, G., Yu, M., Li, X., Wu, H., Qian, Y., Li, X., Luo, Y., 2017. Changes in heavy metal mobility and availability from contaminated wetland soil remediated with combined biochar-compost. Chemosphere 181, 281–288.

Liang, Q., Zhao, D., 2014. Immobilization of arsenate in a sandy loam soil using starch-stabilized magnetite nanoparticles. J. Hazard. Mater. 271, 16–23.

Limmer, M., Burken, J., 2016. Phytovolatilization of organic contaminants. Environ. Sci. Technol. 50 (13), 6632–6643.

Ma, X., Wang, C., 2010. Fullerene nanoparticles affect the fate and uptake of trichloroethylene in phytoremediation systems. Environ. Eng. Sci. 27 (11), 989–992.

Mokarram-Kashtiban, S., Hosseini, S.M., Kouchaksaraei, M.T., Younesi, H., 2019. The impact of nanoparticles zero-valent iron (nZVI) and rhizosphere microorganisms on the phytoremediation ability of white willow and its response. Environ. Sci. Pollut. Res. 26 (11), 10776–10789.

Mueller, N.C., Nowack, B., 2010. Nanoparticles for remediation: solving big problems with little particles. Elements 6 (6), 395–400.

Muthusaravanan, S., Sivarajasekar, N., Vivek, J.S., Paramasivan, T., Naushad, M., Prakashmaran, J., Gayathri, V., Al-Duaij, O.K., 2018. Phytoremediation of heavy metals: mechanisms, methods and enhancements. Environ. Chem. Lett. 16, 1339–1359.

Negri, M. C., Hinchman, R. R., Gatliff, E. G. 1996. Phytoremediation: Using green plants to clean up contaminated soil, groundwater, and wastewater. (No. ANL/ES/CP-89941; CONF-960804-38). Argonne National Lab., IL Conference, United States.

Newman, L.A., Reynolds, C.M., 2004. Phytodegradation of organic compounds. Curr. Opin. Biotechnol. 15 (3), 225–230.

Ojuederie, O.B., Babalola, O.O., 2017. Microbial and plant-assisted bioremediation of heavy metal polluted environments: a review. Int. J. Environ. Res. Public Health 14 (12), 1504.

Oloumi, H., Mousavi, E.A., Nejad, R.M., 2018. Multi-wall carbon nanotubes effects on plant seedlings growth and cadmium/lead uptake in vitro. Russ. J Plant Physiol. 65 (2), 260–268.

Owonubi, S., Ateba, C., Revaprasadu, N. 2020a. Co-assembled ZnO-Fe2O3x-CuOx nano-oxide materials for antibacterial protection. Phosphorus, Sulfur, Silicon Relat. Elem. 195, (12), 981–987.

Owonubi, S.J., Malima, N.M., Revaprasadu, N., 2020. Metal oxide–based nanocomposites as antimicrobial and biomedical agents. In: KOKKARACHEDU, V., KANIKIREDDY, V., SADIKU, R. (Eds.), Antibiotic Materials in Healthcare. Academic Press. United States, pp. 287–323.

Owonubi, S.J., Mukwevho, E., Revaprasadu, N, 2019. Nanoparticle-based delivery of plant metabolites. In: GOYAL, M.R., SULERIA, H.A.R., AYELESO, A.O., JOEL, T.J., PANDA, S.K. (Eds.), The Therapeutic Properties of Medicinal Plants: Health-Rejuvenating Bioactive Compounds of Native Flora. Apple Academics Press, Boca Raton, p. 139. doi.org/10.1201/9780429265204.

Oza, G., Reyes-Calder6n, A., Mewada, A., Arriaga, L.G., Cabrera, G.B., Luna, D.E., Iqbal, H.M., Sharon, M., Sharma, A, 2020. Plant-based metal and metal alloy nanoparticle synthesis: a comprehensive mechanistic approach. J. Mater. Sci. 55 (4), 1–22.

Pandey, V.C., Bajpai, O., 2019. Phytoremediation: from theory toward practicePhytomanagement of Polluted Sites. Elsevier, Amsterdam, pp. 1–49.

Pandey, V.C., Singh, V., 2019. Exploring the potential and opportunities of current tools for removal of hazardous materials from environments. In: PANDEY, V.C., BAUDDH, K. (Eds.), Phytomanagement of Polluted Sites. Elsevier, Amsterdam, pp. 501–516.

Pasinszki, T., Krebsz, M., 2020. Synthesis and application of zero-valent iron nanoparticles in water treatment, environmental remediation, catalysis, and their biological effects. Nanomaterials 10 (5), 917.

Pathak, S., Agarwal, A.V., Pandey, V.C., 2020. Phytoremediation—a holistic approach for remediation of heavy metals and metalloids. In: PANDEY, V.C., SINGH, V. (Eds.), Bioremediation of Pollutants. Elsevier, Amsterdam, pp. 3–16.

Pillai, H.P., Kottekottil, J., 2016. Nano-phytotechnological remediation of endosulfan using zero valent iron nanoparticles. J. Environ. Prot. 7 (5), 734.

Pulford, I., Watson, C., 2003. Phytoremediation of heavy metal-contaminated land by trees—a review. Environ. Int. 29 (4), 529–540.

Qi, Z., Hou, L., Zhu, D., Ji, R., Chen, W., 2014. Enhanced transport of phenanthrene and 1-naphthol by colloidal graphene oxide nanoparticles in saturated soil. Environ. Sci. Technol. 48 (17), 10136–10144.

Radziemska, M., Vaverková, M.D., Baryła, A, 2017. Phytostabilization—management strategy for stabilizing trace elements in contaminated soils. Int. J Environ. Res. Public Health 14 (9), 958.

Rai, P.K., Kim, K.-H., Lee, S.S., Lee, J.-H., 2020. Molecular mechanisms in phytoremediation of environmental contaminants and prospects of engineered transgenic plants/microbes. Sci. Total Environ. 705, 135858.

Roberts, G.A., Oparka, K.J., 2003. Plasmodesmata and the control of symplastic transport. Plant Cell Environ. 26 (1), 103–124.

Rogach, A.L., Talapin, D.V., Shevchenko, E.V., Kornowski, A., Haase, M., Weller, H., 2002. Organization of matter on different size scales: monodisperse nanocrystals and their superstructures. Adv. Funct. Mater. 12 (10), 653–664.

Romeh, A.A.A, 2018. Green silver nanoparticles for enhancing the phytoremediation of soil and water contaminated by Fipronil and degradation products. Water Air Soil Pollut. 229 (1), 147.

Sanzari, I., Leone, A., Ambrosone, A., 2019. Nanotechnology in plant science: to make a long story short. Front. Bioeng. Biotechnol. 7, 120.

Sas-Nowosielska, A., Galimska-Stypa, R., Kucharski, R., Zielonka, U., Małkowski, E., Gray, L, 2008. Remediation aspect of microbial changes of plant rhizosphere in mercury contaminated soil. Environ. Monit. Assess. 137 (1–3), 101–109.

Sattelmacher B., 2009. The apoplast and its significance for plant mineral nutrition. New Phytologist. 182 (1), 284.
Shackira, A.M., Puthur, J.T., 2019. Phytostabilization of Heavy Metals: Understanding of Principles and Practices. In: Plant-Metal Interactions, Springer, Cham, pp. 263–282.
Shaheen, S.M., Rinklebe, J., 2015. Phytoextraction of potentially toxic elements by Indian mustard, rapeseed, and sunflower from a contaminated riparian soil. Environ. Geochem. Health 37 (6), 953–967.
Singh, J., Lee, B.-K., 2016. Influence of nano-TiO2 particles on the bioaccumulation of Cd in soybean plants (Glycine max): a possible mechanism for the removal of Cd from the contaminated soil. J. Environ. Manag. 170, 88–96.
Singh, J., Lee, B.-K., 2018. Effects of nano-TiO2 particles on bioaccumulation of 133Cs from the contaminated soil by soybean (glycine max). Proc. Saf. Environ. Prot. 116, 301–311.
Song, B., Xu, P., Chen, M., Tang, W., Zeng, G., Gong, J., Zhang, P., Ye, S., 2019. Using nanomaterials to facilitate the phytoremediation of contaminated soil. Crit. Rev. Environ. Sci. Technol. 49 (9), 791–824.
Souri, Z., Karimi, N., Sandalio, L.M., 2017a. Arsenic hyperaccumulation strategies: an overview. Front. Cell Dev. Biol. 5, 67.
Souri, Z., Karimi, N., Sarmadi, M., Rostami, E., 2017b. Salicylic acid nanoparticles (SANPs) improve growth and phytoremediation efficiency of Isatis cappadocica Desv., under As stress. IET Nanobiotechnol. 11, 650–655.
Srivastav, A., Yadav, K.K., Yadav, S., Gupta, N., Singh, J.K., Katiyar, R., Kumar, V., 2018. Nano-phytoremediation of Pollutants from Contaminated Soil Environment: Current Scenario and Future Prospects. Phytoremediation, Springer, Cham, pp. 383–401.
Stankic, S., Suman, S., Haque, F., Vidic, J., 2016. Pure and multi metal oxide nanoparticles: synthesis, antibacterial and cytotoxic properties. J. Nanobiotechnol. 14, 73.
Su, Y., Yan, X., Pu, Y., Xiao, F., Wang, D., Yang, M., 2013. Risks of single-walled carbon nanotubes acting as contaminants-carriers: potential release of phenanthrene in Japanese medaka (Oryzias latipes. Environ. Sci. Technol. 47 (9), 4704–4710.
Suman, J., Uhlik, O., Viktorova, J., Macek, T., 2018. Phytoextraction of heavy metals: a promising tool for clean-up of polluted environment? Front. Plant Sci. 9, 1476.
Sun, D., Hussain, H.I., Yi, Z., Siegele, R., Cresswell, T., Kong, L., Cahill, D.M., 2014. Uptake and cellular distribution, in four plant species, of fluorescently labeled mesoporous silica nanoparticles. Plant Cell Rep. 33 (8), 1389–1402.
Tak, H.I., Ahmad, F., Babalola, O.O., 2013. Advances in the application of plant growth-promoting rhizobacteria in phytoremediation of heavy metals. In: WHITACRE, D.M. (Ed.), Reviews of Environmental Contamination and Toxicology. Springer, New York, pp. 33–52.
Tamburini, E., Sergi, S., Serreli, L., Bacchetta, G., Milia, S., Cappai, G., Carucci, A., 2017. Bioaugmentation-assisted phytostabilisation of abandoned mine sites in south west Sardinia. Bull. Environ. Contam. Toxicol. 98, 310–316.
Tang, W.-W., Zeng, G.-M., Gong, J.-L., Liang, J., Xu, P., Zhang, C., Huang, B.-B., 2014. Impact of humic/fulvic acid on the removal of heavy metals from aqueous solutions using nanomaterials: a review. Sci. Total Environ. 468, 1014–1027.
Tangahu, B.V., Abdullah, S.R.S., Basri, H., Idris, M., Anuar, N., Mukhlisin, M, 2011. A review on heavy metals (As, Pb, and Hg) uptake by plants through phytoremediation. Int. J. Chem. Eng. 1–31. doi:10.1155/2011/939161.
Timmusk, S., Seisenbaeva, G., Behers, L., 2018. Titania (TiO 2) nanoparticles enhance the performance of growth-promoting rhizobacteria. Sci. Rep. 8, 1–13.
Tripathi, D.K., Singh, S., Singh, V.P., Prasad, S.M., Chauhan, D.K., Dubey, N.K., 2016. Silicon nanoparticles more efficiently alleviate arsenate toxicity than silicon in maize cultiver and hybrid differing in arsenate tolerance. Front. Environ. Sci. 4, 46.

Unterbrunner, R., Wieshammer, G., Hollender, U., Felderer, B., Wieshammer-Zivkovic, M., Puschenreiter, M., Wenzel, W.W., 2007. Plant and fertiliser effects on rhizodegradation of crude oil in two soils with different nutrient status. Plant Soil 300 (1), 117–126.

Uskoković, V., Drofenik, M, 2005. Synthesis of materials within reverse micelles. Surf. Rev. Lett. 12 (2), 239–277.

Usman, K., Al-Ghouti, M.A., Abu-Dieyeh, M.H., 2019. The assessment of cadmium, chromium, copper, and nickel tolerance and bioaccumulation by shrub plant Tetraena qataranse. Sci. Rep. 9 (1), 5658.

Venkatachalam, P., Jayaraj, M., Manikandan, R., Geetha, N., Rene, E.R., Sharma, N., Sahi, S., 2017. Zinc oxide nanoparticles (ZnONPs) alleviate heavy metal-induced toxicity in Leucaena leucocephala seedlings: a physiochemical analysis. Plant Physiol. Biochem 110, 59–69.

Vidal-Vidal, J., Rivas, J., López-Quintela, M, 2006. Synthesis of monodisperse maghemite nanoparticles by the microemulsion method. Coll. Surf A: Physicochem. Eng. Asp. 288, 44–51.

Vítková, M., Puschenreiter, M., Komárek, M, 2018. Effect of nano zero-valent iron application on As, Cd, Pb, and Zn availability in the rhizosphere of metal (loid) contaminated soils. Chemosphere 200, 217–226.

Waghmode, M.S., Gunjal, A.B., Mulla, J.A., Patil, N.N., Nawani, N.N., 2019. Studies on the titanium dioxide nanoparticles: Biosynthesis, applications and remediation. SN Appl. Sci., 1 (4), 310.

Wang, D., Wu, Y., Xia, J., 2016. Review on photoacoustic imaging of the brain using nanoprobes. Neurophotonics 3 (1), 010901.

Watlington, K., 2005. Emerging Nanotechnologies for Site Remediation and Wastewater Treatment. Environmental Protection Agency, Washington DC.

Yadav, K.K., Gupta, N., Kumar, A., Reece, L.M., Singh, N., Rezania, S., Khan, S.A., 2018. Mechanistic understanding and holistic approach of phytoremediation: a review on application and future prospects. Ecol. Eng. 120, 274–298.

Yang, G., Park, S.-J., 2019. Conventional and microwave hydrothermal synthesis and application of functional materials: a review. Materials 12 (21), 1177.

Yang, L., Watts, D.J., 2005. Particle surface characteristics may play an important role in phytotoxicity of alumina nanoparticles. Toxicol. Lett. 158 (2), 122–132.

Yavari, S., Malakahmad, A., Sapari, N.B., Yavari, S., 2018. Fullerene C60 for enhancing phytoremediation of urea plant wastewater by timber plants. Environ. Sci. Pollut. Res. 25 (12), 11351–11363.

Yuan, Z., Zhang, Z., Wang, X., Li, L., Cai, K., Han, H., 2017. Novel impacts of functionalized multi-walled carbon nanotubes in plants: promotion of nodulation and nitrogenase activity in the rhizobium-legume system. Nanoscale 9 (28), 9921–9937.

Zaier, H., Ghnaya, T., Ghabriche, R., Chmingui, W., Lakhdar, A., Lutts, S., Abdelly, C., 2014. EDTA-enhanced phytoremediation of lead-contaminated soil by the halophyte Sesuvium portulacastrum. Environ. Sci. Pollut. Res. 21 (12), 7607–7615.

Zhang, J., Gong, J.-L., Zeng, G.-M., Yang, H.-C., Zhang, P., 2017. Carbon nanotube amendment for treating dichlorodiphenyltrichloroethane and hexachlorocyclohexane remaining in dong-ting lake sediment—an implication for in-situ remediation. Sci. Total Environ. 579, 283–291.

Zhang, L., Wang, L., Zhang, P., Kan, A.T., Chen, W., Tomson, M.B., 2011. Facilitated transport of 2, 2′, 5, 5′-polychlorinated biphenyl and phenanthrene by fullerene nanoparticles through sandy soil columns. Environ. Sci. Technol. 45 (4), 1341–1348.

Zhang, M., Gao, B., Chen, J., Li, Y., 2015. Effects of graphene on seed germination and seedling growth. J. Nanopart. Res. 17 (2), 1–8.

Zhao, X., Meng, Z., Wang, Y., Chen, W., Sun, C., Cui, B., Cui, J., Yu, M., Zeng, Z., Guo, S., 2017. Pollen magnetofection for genetic modification with magnetic nanoparticles as gene carriers. Nat. Plants 3 (12), 956–964.

Zhu, Y., Xu, F., Liu, Q., Chen, M., Liu, X., Wang, Y., Sun, Y., Zhang, L., 2019. Nanomaterials and plants: Positive effects, toxicity and the remediation of metal and metalloid pollution in soil. Sci. Total Environ 662, 414–422. doi:10.1016/j.scitotenv.2019.01.234.

CHAPTER

7

Transgenic plant-mediated phytoremediation: Applications, challenges, and prospects

Omena Bernard Ojuederie[a,b], David Okeh Igwe[c,d,f], Jacob Olagbenro Popoola[e]

[a]*Department of Biological Sciences, Biotechnology Unit, Kings University, Odeomu, Osun State, Nigeria*

[b]*Food Security and Safety Niche, Faculty of Natural and Agricultural Sciences, North-West University, Mmabatho, South Africa*

[c]*Department of Biotechnology, Ebonyi State University, Abakaliki, Ebonyi State, Nigeria*

[d]*Section of Plant Pathology, Boyce Thompson Institute for Plant Research, Ithaca, New York, United States*

[e]*Department of Biological Sciences, Covenant University, Ota, Ogun State, Nigeria*

[f]*Plant Pathology and Plant-Microbe Biology, School of Integrated Plant Sciences, Cornell University, Ithaca, NY, USA*

7.1 Introduction

Globally, the environment is constantly altered, contaminated, and degraded due to human activities including unregulated land use, deforestation, pollution/effluents from industries, climate change, and lately from acts of terrorism (Ibañez et al., 2015; Pilon-Smits and Freeman, 2006). Industrialization, natural process, and war releases large amounts of toxic compounds into the biosphere (Daghan, 2019; Eapen and D'souza, 2005; Fahimirad and Hatami, 2017). Other primary anthropogenic sources of heavy metals (HMs) into the biosphere include coal-burning power plants, mines, smelters waste combustion, and emissions from automobiles (Echem and Kabari, 2013; Pathak et al., 2020). The contamination of surface water, agricultural farmlands, industrial effluents, domestic, organic, and inorganic pollutants are increasingly becoming an environmental nuisance and health hazard in several regions of the world (Gu et al., 2015; Heaton et al., 1998; Wang et al., 2015). The combined factors highlighted above alongside the reckless release of HMs and other contaminants require a holistic approach of remediation for a cleaner environment for healthy living. A healthy environment is non-negotiable and desirable particularly in the developing nations where the rate of pollution is on the increase. Thus, the biosphere containing biological organisms (plants, animals, humans, microorganisms among others) requires constant monitoring, regulation, and replenishing approaches such as afforestation and remediation amongst others. Although chemical, thermal, and aeration remediation strategies have been variously used and adopted, they are limited in several ways due to generation of secondary pollutants, low efficiency and high cost associated (Komives and Gullner, 2005; Meagher, 1999; Prasad, 2003). The shortcomings of these strategies and innovations in biotechnology and genetic engineering, led to the advancement in phytoremediation technology (Gunarathne et al., 2019; Meagher, 1999; Nagata et al., 2009; Pandey et. al., 2015; Pandey and Singh, 2019).

Assisted Phytoremediation. DOI: https://doi.org/10.1016/B978-0-12-822893-7.00009-4

In contrast to other clean-up technologies, phytoremediation is beneficial and characterized by low installation and maintenance cost, eco-friendliness, capability of carbon sequestration, and production of biofuel (Makhzoum et al., 2013; Pandey et al., 2015; Van Aken, 2008). Simple phytoremediation strategy is based on diverse detoxification mechanisms involving several approaches such as phytoextraction, phytodegradation, phytovolatilization, enzymatic degradation, and sequestration (Aken et al., 2010; Mena-Benitez et al., 2008; Van Aken, 2008). Some years back, Pandey et al. (2015) provided a novel strategy of sustainable phytoremediation that involves the use of naturally colonizing species to remediate heavy metal contaminated sites. This strategy with no risk of use of main products from the naturally colonizing species, consisted of carbon sequestration, enhancement of substrate quality, pleasant landscape, and biodiversity conservation with the advantage of low-cost inputs and limited maintenance (Pandey et al., 2015). Generally, phytoremediation is seen as an economically viable strategy that can be adopted globally in the remediation and management of contaminated sites because it is eco-friendly to animal communities, can generate revenue for landowners, and provide income for phyto-based industries (Pandey and Bajpai, 2019). Notably, a number of plants with the inherent capacity to tolerate and amass contaminants, metals, and pollutants have been identified and used as phytoremediation agents (Ager et al., 2003; Pandey et al., 2016; Weisman et al., 2010). *Arabidopsis thaliana, Nicotiana tabacum, Vernonia amygadalina, Oryza sativa*, amongst others have been variously used and are being exploited in many phytoremediation trials to take up noxious compounds through their extensive root systems (Bodnar et al., 2012; Echem and Kabari, 2013). More so, plant species such as *Jatropha curcas, Populus species, Ricinus communis*, and *Miscanthus species* are promising energy crops in sustainable phytoremediation and biofuel production (Pandey et al., 2016). Nevertheless, not all conventional plant species have the capacity to adequately clean up or takes up toxic substances. Low removal rates, poor tolerability, and inadequate accumulation ability are major constraints to their effective and efficient use (Cherian et al., 2012; Van Aken, 2008).

7.2 Pros and cons of phytoremediation using genetically engineered plants

The adoption and use of genetically engineered plants as hyperaccumulators for environmental cleanup of polluted sites have been perceived as a cost-effective, eco-friendly, and a better option than other strategies. In retrospect, some of the traditional plant-based techniques are relatively expensive, time-consuming, and environmentally harmful, causing adverse impacts on the ecosystem (Yadav et al., 2018). As such, biotechnological approaches, including genetic engineering/recombinant DNA technology and sophisticated tissue culture techniques of producing transgenic plants offer great hope and a better alternative to other methods. This has revolutionized the field of phytoremediation and resolved most of the limitations associated with the use of natural plants. Transgenic plants with improved sequestration, enhanced phytovolatilization, and rhizofiltration, increased phytostabilization and uptake, and rapid translocation have been produced and are currently being used to remediate polluted environments (Elias et al., 2012; Gunarathne et al., 2019). They are known to be efficient, less expensive, and exhibit features of an ideal hyperaccumulator such as high biomass production and extensive rooting system. Such transgenic plants accumulate toxic substances rapidly and detoxify them for a long period and reduce the treatment cost (Abhilash et al., 2009; Gunarathne et al., 2019). These plants are considered hyperaccumulators, user- and eco-friendly with high capacity to remediate contaminated sites without any known negative side-effects (Alkorta et al., 2004; Azab et al., 2016). Consequently, more than 400 hyperaccumulator plants characterized by high rooting systems that accumulate a high quantity of contaminants are being used and explored in phytoremediation (Abhilash et al., 2009; Chaney et al., 1997; Gunarathne et al., 2019). These plants can absorb metals much higher than the conventional plants attributable to their root-to-shoot transport system with high pollutant tolerance (Gunarathne et al., 2019), and massive biomass (Chaney et al., 1997). They are also resistant to pest and diseases, tolerate negative effects of climate change via their extensive root systems, and thrive in poor soil (Aken et al., 2010). Genetically engineered plants also play a substantial role in biogeochemical prospecting, elemental allelopathy, and resistance against fungal pathogens (Yadav et al., 2018), and perhaps have implications on human health by way of the food chain (Freeman and Salt, 2007; Pilon-Smits and Freeman, 2006). Transgenic plants that are not edible can generate energy via anaerobic digestion, control soil erosion, improve soil, and stabilize water canals (Gunarathne et al., 2019; Krämer and Chardonnens, 2001; Lang et al., 2004). Another major merit of transgenic plants for phytoremediation is their green chemistry which can be explored to generate various high plant-based products (proteins – enzymes, biofuels, sugars amongst others), and phyto-siderophores (PS) which are efficient in metal absorption to replace the use of toxic chemicals (Gomes et al., 2016; Gunarathne et al., 2019). PS such as avenic acid, mugineic acid, nicotinamine are organic substances formed by plants under Zinc or Iron-deficient conditions, which can chelate Zn or Fe and increase their uptake by plants (Kobayashi et al., 2010). Research findings have revealed that PS solubilizes other metals such as Cu, Cd, Ni, Zn in contaminated and uncontaminated substrates (Kobayashi et al., 2010; Reichman and Parker, 2005). While the economic benefits and merits of transgenic plant mediated phytoremediation cannot be overemphasized, their environmental limitations are of great concern. The potential risk of becoming an invasive plant or weed is high, influencing biodiversity by altering the natural gene composition (Gunarathne et al., 2019; Raybould et al., 2012). They could also be harmful to wildlife and, by extension, human consumers while their long-life makes risk assessment in some way, challenging (Davison, 2005). The advantages of transgenic plants for phytoremediation purposes outweigh the few disadvantages which in most cases have not been scientifically proven beyond reasonable doubt.

Lately, the use of transgenic plants in nanoparticle-mediated phytoremediation is increasingly gaining unprecedented attention as a sustainable and cost-effective means of remediating contaminated lands, water bodies, and related polluted sites (Kiiskila et al., 2015). In this review, we discussed the various ways nano-phytoremediation could be enhanced using the modern techniques of genetic engineering and elaborated on the challenges encountered in transgenic plants mediated phytoremediation, in addition to its applications and prospects.

7.3 Transgenic plants mediated phytoremediation

The methods used in producing transgenic plants vary and include but are not limited to *Agrobacterium*-mediated transformation, biolistic bombardment or gene gun, electroporation as well as protoplast fusion. The soil-borne bacterium *Agrobacterium tumefaciens* which naturally causes crown gall disease has been a widely used method for obtaining transgenic plants. The bacterium carries a tumor-inducing (Ti) plasmid. Plasmids are circular independent DNA molecules capable of self-replication and extrachromosomal inheritance. They occur in bacteria cells and are transferred from one bacterial cell to another by conjugation or transformation. Insertion of any piece of foreign genetic material between the left and right 25 bp borders of the plasmid's T- DNA region gets transferred into the plant's chromosomes within the nucleus where it is subsequently integrated. This knowledge has been exploited by genetic engineers to produce transgenic plants. Thus, genetic engineers disarm the Ti plasmid by replacing the genes responsible for the production of hormones in the oncogenic region of the T-DNA region with foreign genetic material with sticky ends, and the addition of genes for antibiotic resistance. Fig. 7.1 illustrates the steps used in *Agrobacterium*-mediated plant transformation. For the efficiency of genetic transformation using *Agrobacterium tumefaciens* or *A. rhizogenes* in the case of root transformation, the left and right 25 base borders must be in place around the T-DNA region as it permits recombination of the genes in the host genome. Transfer of the T-DNA region is also controlled by the virulence genes (*Vir* A - *Vir* H) which consist of the regulatory system for transcriptional activation of the other vir genes involved in directing the T-DNA transfer from the bacterium to the plant cell (Gelvin, 2012). In addition to the 25 bp borders and the virulence genes needed for T-DNA transfer, a third essential component are the chromosomal genes *chv*A, *chv*B, *chv*D, *chv*E, and *psc*A found on the *A. tumefaciens* chromosome. Exopolysaccharides produced by genes *chv*A and *chv*B enables the bacterium to attach to plant cells, while gene *chvE* encodes a glucose/galactose transporter for binding specific sugars that induces vir genes (Gelvin, 2012).

After the transformation of plant cells using *Agrobacterium*, marker genes are required to monitor and detect plant transformation systems to know if the DNA has been successfully transferred into recipient cells. These marker genes: scorable or reporter and selectable marker genes are inserted into the Ti plasmid before the transformation process. Reporter genes are test genes whose expression results in a quantifiable phenotype. Examples include the use of *β glucuronidase* (gus), green fluorescent protein, or firefly *luciferase* gene which are detected by fluorescence or bioluminescence. On the other hand, the selectable marker genes kill untransformed cells while keeping alive the transformed ones on media containing high levels of selection agent. Selection agents are used in agroinfection to identify the transformed plants containing the transgene from the untransformed plants. Some selection agents used are; antibiotics, herbicide resistance genes, antimetabolites, hormone biosynthetic genes, as well as genes conferring resistance to toxic levels of amino acids (Gelvin, 2012). Identified transformed cells are regenerated using tissue culture techniques and later acclimatized to field conditions for trials.

The success of *Agrobacterium*-mediated transformation, however, depends on the use of a robust promoter and an appropriate signal sequence to regulate the preferred subcellular localization (Weerakoon, 2019). Gisbert et al. (2003) successfully transformed the leaves of *Nicotiana glauca* plants by introducing the wheat phytochelatin synthase 1 cDNA (*TaPCS1*) using the *Agrobacterium*-mediated transformation approach. The *TaPCS1* gene was overexpressed in the transformed plants leading to a remarkable increase in the plants' tolerance to Pb and Cd, due to a 160% increase in seedling roots than in the wild type plants, greener leaves, and twice the rate of Pb uptake by the transgenic plants (Gisbert et al., 2003). The *TaPCS1* gene isolated from wheat was also found to confer Cd tolerance to *Saccharomyces cerevisiae* and *Nicotiana glauca* plants (Gisbert et al., 2003). Likewise, a *ThMT3* gene that encodes a type 3-metallothionein was transferred from *Tamarix hispida* into *Salix matsudana* plants to confer greater tolerance to Cu stress (Yang et al., 2015). The gene construct was made by cloning the *ThMT3* gene into a binary vector PROKII regulated by the *Cauliflower mosaic virus* (CaMV) 35S promoter, with a *neomycin phosphotransferase* II (npt II) gene regulated by a *nopaline synthase* (nos) promoter, which was subsequently transferred into *Agrobacterium tumefaciens* strain LBA4404

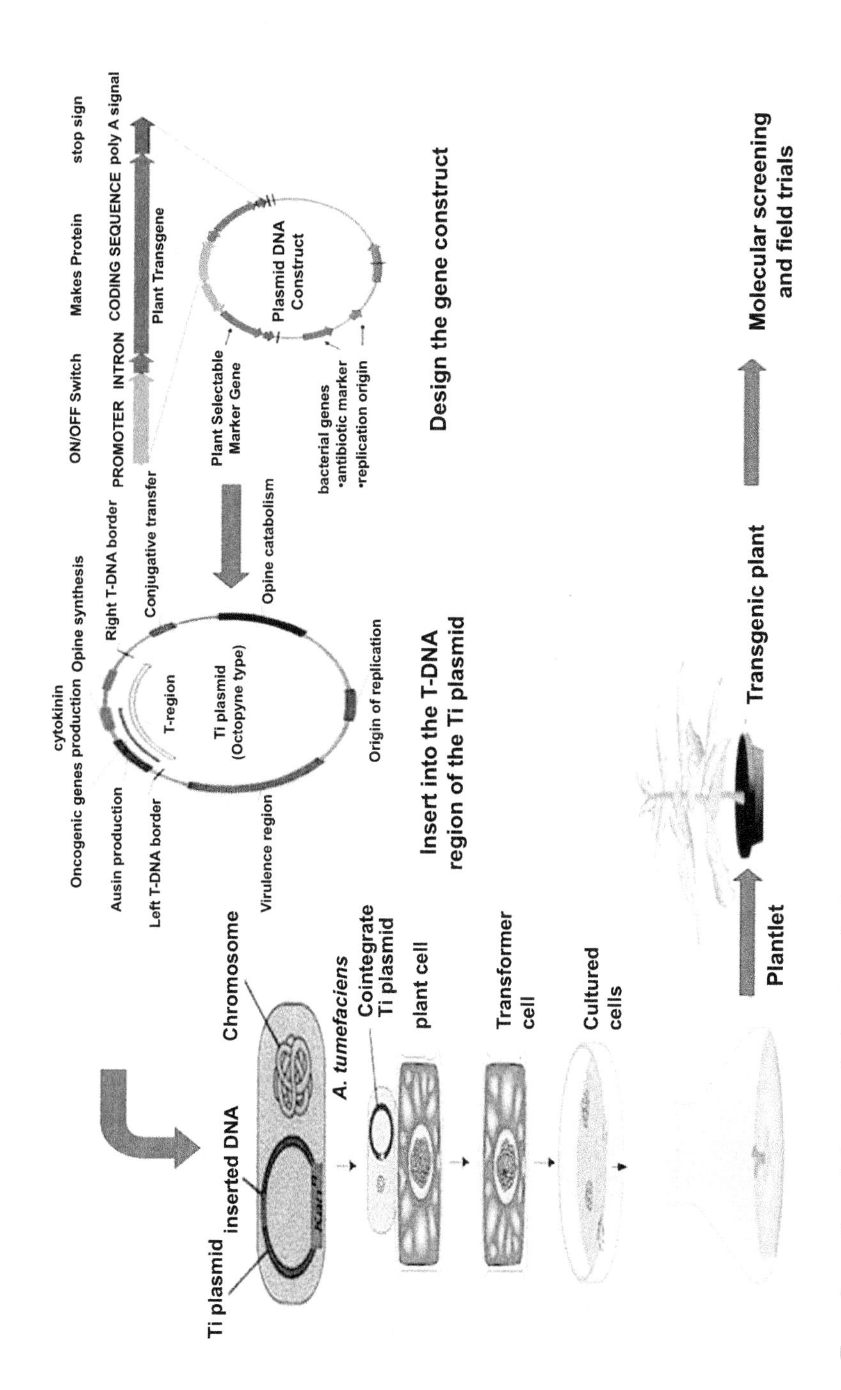

FIG. 7.1

Steps in *Agrobacterium*-mediated plant transformation.

(Chen et al., 2003; Yang et al., 2015). Transgenic lines obtained had improved growth parameters; higher plant height and longer roots compared to the un-transformed plants under Cu stress.

Likewise, the ACC deaminase gene from *Enterobacter cloacae* was transferred into tomato plants regulated by specific promoters; the CaMV 35S promoter which controls constitutive expression, the *Agrobacterium rhizogenes* rolD promoter for specific expression in the roots, or the tobacco pathogenesis-related PRB-1b promoter (Fasani et al., 2018) that enhances the accumulation of several HMs such as Cd, Co, Cu, Ni, Pb, and Zn, and the tolerance abilities of the plants. The transgenic tomato plants had a greater ability to accumulate a higher concentration of the HMs compared with the untransformed plants. Moreover, significant differences were observed in the uptake of the HMs based on the type of promoter used. The greatest tolerance and metal accumulation were observed in plants engineered using the PRB-*1b* promoter with the least tolerance obtained from the construct made with the CaMV 35S promoter (Fasani et al., 2018; Grichko et al., 2000). Despite the benefits of the *Agrobacterium*-mediated plant transformation method, its use is limited because it is host specific and not suitable for high-throughput applications, and foreign DNA integration into the host genome (Demirer et al., 2019c).

7.4 Application of transgenic plants mediated phytoremediation of polluted environments

Phytoremediation is increasingly being used as a cheap and alternative approach for the removal of pollutants from the environments This is because certain plants can naturally take up or sequester HMs from the environment, detoxify them using different metabolic pathways, and release the nontoxic forms of these metals into the environment through various processes such as phytoextraction, phytostabilization, phytodegradation, phytovolatilization as well as phytostimulation or rhizodegradation (Ojuederie and Babalola, 2017). Most of the plants used for phytoremediation are hyperaccumulators. Nonetheless, some of these plants are slow-growing and do not produce sufficient biomass. Moreover, they are selective to the type of metals they can accumulate or detoxify, and the metabolic pathways are not fully understood.

The advent of genetic manipulation or recombinant DNA (rDNA) technology in the early 1970s, broke the barriers of transfer of genes across species and paved the way for the transfer of useful genes from one species to another where it gets expressed if integrated into the genome of the host organism. Since then, several efforts have been made to develop genetically improved plants with phytoremediation abilities (Ibañez et al., 2015; Macek et al., 2008). The first-generation transgenic plants concentrated on innovations that benefitted the farmers such as crop protection against insect pests (Bt. technology) and diseases or viruses or through enhanced tolerance to herbicides. Transgenic tobacco plant with resistance to herbicides was first reported in France and the United States in 1986 and subsequently virus-resistant tobacco in China in 1992 (Agnihotri and Seth, 2019). Incorporating the gene for Bt toxin from the bacterium *Bacillus thuringiensis* (Bt.) into crops, made insect resistance in such a crop possible. This technology has reduced the quantity of pesticides applied to agricultural crops and has been used for cotton, maize, and cowpea.

The rDNA technology later found its usefulness for environmental cleanup of pollutants such as HMs and xenobiotics. The constraints to the utilization of hyperaccumulator plants for phytoremediation could be mitigated with genetic engineering of known hyperaccumulator plants for increased biomass production, enhanced detoxification mechanism, as well as faster growth since most are slow growing. Nevertheless, it has been said that the genetic manipulation of fast-growing high biomass producing plant species offers a better chance of obtaining more tolerance and HM uptake than modifying hyperaccumulator plants (Yan et al., 2020), but this requires having a mechanistic understanding of HM tolerance and uptake in plants which determines the appropriate genes to be selected for genetic improvement.

Genetic modification of plants for phytoremediation over the years have focused on some key genes or traits that are manipulated and introduced into transgenic plants for better tolerance and uptake of pollutants such as HMs by phytoextraction, in addition to increased detoxification of organic pollutants. Advances in genetic engineering and biotechnology have revolutionized the field of phytoremediation with the understanding that plants can be transformed (genetically modified) with enhanced ability to accumulate and degrade toxic substances in their tissues which gets volatilized to the atmosphere (Eapen and D'souza, 2005; Gunarathne et al., 2019). Transgenic plants such as Poplar, Arabidopsis, and Tobacco, have been transformed with genes leading to greater uptake and metabolism of toxic volatile pollutants (Doty et al., 2007; Gullner et al., 2001; Kiiskila et al., 2015). Overexpression of bacterial γ-glutamylcysteine synthetase in the cytosol of the transgenic Poplar plants produced higher glutathione (GSH) concentrations in leaves and thus highly active in the Cd remediation (He et al., 2015). In another study, the T2 generation of the *ScMTII* genes in transgenic and non-transgenic tobacco plants exposed to Cd pollution indicated that 19.8 % more accumulation of Cd was obtained in the *ScMTII* gene bearing transgenic tobacco than the non-transgenic tobacco plant (Daghan et al., 2010). In contrast, transgenic tobacco plants transformed with a P1B-ATPase gene from *Populus tomentosa* (*PtoHMA5*) has been demonstrated to improve Cd transport as a better option to non-transgenic tobacco plants (Wang et al., 2018). Transgenic potato plants have also been demonstrated to display outstanding cross-tolerance for the photosynthesis-inhibiting herbicides atrazine, chlortoluron, and methabenzthiazuron, the lipid biosynthesis-inhibiting herbicides acetochlor and metolachlor, as well as the carotenoid biosynthesis-inhibiting herbicide norflurazon (Inui et al., 2000). In reality, many of these transgenic plants are characterized by extensive root systems and large size, thus having capacity to efficiently remediate areas polluted with several contaminants at a higher frequency and at lower costs (Doty et al., 2007; Gullner et al., 2001; Kiiskila et al., 2015; Van Aken, 2008). The development of transgenic plants for enhanced phytoremediation is based on identification of the precise genes for increased tolerance to HMs or accumulation of more HMs from polluted environments. Application of transgenic plant mediated phytoremediation have focused on engineering fast-growing, high biomass-producing plants to achieve the key properties of enhanced metal tolerance and HM accumulation. Thus, the selection of genes for transgenic plant-mediated phytoremediation should take into consideration the knowledge of metal tolerance and the mechanisms of metal uptake in plants (Yan et al., 2020).

7.5 Application of advanced omic technologies in enhancing phytoremediation

The breakthroughs in genomics, proteomics, transcriptomics, and metagenomics have enhanced the production of transgenic crops for phytoremediation purposes (Kumar et al., 2014). These omic technologies coupled with high-throughput next-generation sequencing platforms, and bioinformatics, have permitted a greater comprehension of the plant system biology as well as gene expressions and the various metabolic systems in plants through metabolomics, thereby providing the foundation for plant synthetic biology (Bell et al., 2014). Transcriptomic studies on hyperaccumulator plants have given new insights to the molecular mechanisms regulating the accumulation of metals and tolerance, in addition to the identification of several genes that are constitutively overexpressed and could be involved in enhancing hyperaccumulation in plants. Utilization of transcriptomics together with next-generation sequencing and bioinformatics technology has rapidly increased the speed of identification of novel genes or proteins expressed under heavy metal stress as well as genes conferring tolerance to xenobiotics and degradation of pollutants.

Genetic engineering of plants for enhanced gene expression in phytoremediation is based on (1) manipulating uptake systems and metal/metalloid transporter genes; (2) augmenting the production of ligands from metals and metalloids; and (3) transforming metals and metalloids to innocuous and volatile forms (Basu et al., 2018; Kotrba et al., 2009; Mosa et al., 2016). Genes that play essential roles in the uptake, translocation, and sequestration of inorganic pollutants such as metal transporters, phytochelatins (PCs) metallothioneins (MTs) as well as metal chelators and enzymes could be easily identified in plants exposed to environmental pollutants. Identified genes can then be introgressed and overexpressed in fast-growing plants with high-biomass or hyperaccumulation which could enhance their functions in phytoremediation.

7.5.1 Engineering metal transporters for improved efficiency in phytoextraction

Metal transporters are indispensable for the uptake and translocation of metals essential to plants as well as HMs during phytoextraction (Fasani et al., 2018). These transporters direct the uptake of metals from the soil to the roots (plasma membrane transporters) where they get translocated to the shoots via the xylem and subsequently detoxified within storage vacuoles. These specialized H+-coupled carrier proteins which are crucial for the accumulation of HMs ions from the soil, reside within the root cell membrane where they transport precise metals across cellular membranes and facilitate influx–efflux of metal translocation from roots to shoots (DalCorso et al., 2019; Yan et al., 2020). However, these transporters are often damaged by the high concentration of HMs in the soil especially mercury which damages transporters of the cell membrane such as aquaporins, thereby obstructing the free flow of nutrients and water (Ruiz and Daniell, 2009; Zhang and Tyerman, 1999).

The use of rDNA technology to modify gene expression activity in addition to localization of HM ion transporters, could be a possible approach for transgenic plant-mediated phytoremediation. This is achievable because these proteins regulate the uptake, distribution, and accumulation of numerous HMs in plants (Agnihotri and Seth, 2019). The road map to achieving the transformation of hyperaccumulating plants for efficient phytoremediation was proposed by Eapen and D'souza (2005). Tonoplast metal transporters belonging to heavy metal *PIB- ATPase*, the Natural resistance-associated macrophage protein (NRAMP), the Cation diffusion facilitator (CDF) (Jan and Parray, 2016; Williams et al., 2000), and the ZIP families are known to sustain the physiological concentrations of heavy metals and could be responsible for responses to metal stress by plants. CPx-type or ATPase metal transporters have been implicated in homeostasis of metal ions and tolerance in addition to been responsible for compartmentalization of HMs in plants. Tolerance to Cd, cobalt, Pb, and Zn was enhanced in *Arabidopsis thaliana* (Morel et al., 2009).

Knowledge of the type of transporters needed for the uptake of different metal contaminants will be instrumental in enhancing the phytoremediation abilities of select hyperaccumulator plants for efficient remediation of pollutants. One way of enhancing metal transporters is to use genetic engineering techniques for the overexpression of metal transporter genes which then augments the abilities of hyperaccumulator plants to sequester such metals from the environment at a higher frequency. Multiple genes regulate HM uptake, translocation, and sequestration. Most of these genes are not obtained from hyperaccumulators but mainly from model plants such as *Arabidopsis thaliana* which are not tolerant or hyperaccumulators or the genes are cloned from some microorganisms with a recognized mechanism of detoxification of heavy metals or tolerance (Verbruggen et al., 2009).

To genetically engineer metal transporters, certain factors must be taken into consideration. These include determination of the plant organs and cell types involved in metal compartmentalization, as well as how the metals are transported and localized in subcellular structures and storage vacuoles. Most of the achievements made from phytoextraction of heavy metal(loid)s and their subsequent accumulation in the above-ground organs of plants have been accomplished through expressing metal ligands or transporters, enzymes involved in sulfur metabolism, enzymes that alter the chemical forms or redox state of metal(loid)s as well as the principal constituents of metabolism (Fasani et al., 2018; Weerakoon, 2019). By manipulating the influx or efflux transporter genes or elevating the levels of chelators, the tolerance abilities of hyperaccumulator plants can be augmented.

The process of phytovolatilization could be enhanced by targeting the genes that encode enzymes capable of converting toxic pollutants such as HMs or xenobiotics into innocuous and volatile forms. Moreover, the increased presence of chelators would enable the export of these toxic pollutants from the cytoplasmic matrix into vacuoles or the cell wall (Mosa et al., 2016; Weerakoon, 2019). Over the years, several researchers have enhanced the phytoremediation abilities of plants through introgression of transporter genes into plants (Table 7.1). Considering the toxic nature of mercury (Hg) to living organisms, its accumulation in the environment is a threat to human health and should be removed from the environment using an eco-friendly technology such as phytoremediation.

The ATP-binding cassette (ABC) family is recognized as a family that has several genes encoding proteins involved in the removal of toxins and ion-regulation (Martinoia et al., 2002; Koźmińska et al., 2018). Transgenic plants with higher Hg tolerance were developed by overexpression of a *Populus trichocarpa* ATP-binding ABC transporter gene *PtABCC1* in *Arabidopsis* and poplar which resulted in 26%–72% or 7%–160% increase of Hg uptake in both plants, respectively (Sun et al., 2018), compared to the wide type plants. The ABC genes are particularly useful in the genetic transformation of plants for phytoremediation when overexpressed as they augment movement of mobile metal ions such as Cd and Cu or otherwise attach to non-mobile ions in plant roots such as Pb (Koźmińska et al., 2018). Apart from the *PtABCC1* gene, other high mercury tolerant genes (*ScYCF1, MerC-H,* and *AtABCC1,2*) have also been identified (Sun et al., 2018), and could be used for the transformation of plants for better remediation purposes.

7.5.2 Engineering metal-binding ligands for enhanced phytoremediation

The genes involved in HM tolerance falls into three categories: metallothioneins (MTs) required for metal hyper-tolerance, phytochelatins (PCs) essential for improved HMs tolerance and sequestration in vacuoles, and glutathione (GSH). In addition to these genes for augmenting phytoremediation, Eapen and D'souza (2005) recommended modification of metabolic pathways as well as oxidative stress mechanisms that increase tolerance to HMs/metalloids, engineering metal chelators and transporters for xylem loading and translocation, and manipulation of plant growth regulator synthesis which is essential to augment plant growth and biomass production (Eapen and D'souza, 2005). These genes

Table 7.1 Engineering of metal transporters for enhanced metal uptake, translocation, and tolerance.

Genes	Source plant/ organism	Engineered plant	Transporter type	Achievements	References
AtPHT1, AtPHT7	Arabidopsis	Tobacco	Pi transporter	Enhanced uptake and accumulation of As	(LeBlanc et al., 2013; Jha, 2020)
PHT1ox, PHT7ox PHT1ox+YCF1 PHT7ox +YCF1	Wild type *Arabidopsis thaliana* and *Saccharomyces cerevisiae*	*Arabidopsis*	Pi transporter + ABC transporter	Increased uptake of As, and enhanced resistance to As	(LeBlanc et al., 2013)
IRT1	*Arabidopsis*	*Arabidopsis*	IRT1 transporter	Accumulation of increased levels of Zn and Cd	(Connolly et al., 2002)
AtABCC1 and AtABCC2,	*Arabidopsis*	*Arabidopsis*	Phytochelatins transporters	Increased tolerance to As	(Song et al., 2010)
YCF1	*Saccharomyces cerevisiae*	*Arabidopsis*	ABC vacuolar transporters.	Improved Pb and Cd tolerance and accumulation	(Song et al., 2003)
Mere	*Escherichia coli*	*Arabidopsis*	Mercury transporter	Enhanced uptake of methylmercury and mercuric ions	(Sone et al., 2013)
PgIREG1	*Psychotria gabriellae*	*Arabidopsis*	Nickel transporter	Enhanced phytoextraction of nickel in leaves of Arabidopsis	(Merlot et al., 2014)
OsMTP1	Oryza sativa	Tobacco	CDF transporter	Increased plant biomass and uptake of Cd with moderate tolerance and uptake of As	(Das et al., 2016)
PtABCC1	Arabidopsis	Populus tomentosa	ABC transporter	Enhanced accumulation of Hg in Arabidopsis and Populus	(Sun et al., 2018; Jha, 2020)

could be isolated from microbes, humans, or plants and introgressed into high biomass producing plants to improve their efficiency in phytoremediation. They act as metal-binding ligands which are effective for metal uptake and detoxification.

Metallothioneins (MT) are free radical scavenging low molecular weight proteins rich in cysteine (about 25–33%) with a high affinity for metal ions (Keeran et al., 2019). The MT gene is activated when a plant is under abiotic stress such as HM contamination, or oxidative stress, and it aids hyperaccumulation of toxic metals at elevated levels in hyperaccumulating plants. The Sulphur molecule present in cysteine binds to HMs when a plant is under stress leading to their detoxification due to the synthesis of a metalloenzymes complex, which activates the transcription genes, and metabolizes metallodrugs (Bratić et al., 2009; Chaudhary et al., 2018; Gautam et al., 2012). MTs have been reported to sequester HMs such as Cd^{2+} and Hg^{+} when exposed to it. Thus, overexpression of MTs could be a profitable means of enhancing the remediation of these toxic metals from the environment and subsequent detoxification when introduced into plants using rDNA techniques. For instance, two MTs *PjMT1* and *PjMT2* obtained from *Prosopis juliflora* were overexpressed in transgenic tobacco plants resulting in increased tolerance and uptake of Cd^{2+} by nine-fold and fivefold respectively, compared with the untransformed plants (Balasundaram et al., 2014). Likewise, overexpression of mammalian MT in tobacco plants led to increased tolerance to Cd^{2+} at the seedling stage (Gunarathne et al., 2019; Pan et al., 1994) and *Arabidopsis thaliana* overexpressing an MT gene *PsMTA* from *Pisum sativum* accumulated higher levels of Cu^{2+} in its roots (Weerakoon, 2019).

Phytochelatins (PCs) are thio-rich small peptides found in the cytoplasm of plant cells that contain sulfhydryl (–SH) groups for binding various metals (Anjum et al., 2015). These PCs are synthesized by phytochelatins synthase (PCS) and c-glutamyl cysteine synthetase (c-GCS) (Hirata et al., 2005; Koźmińska et al., 2018). PCs aid chelation of metal ions by regulating the homeostasis of HMs in plants which increases their availability for detoxification (Yadav et al., 2018). It is believed that mannose plays an essential role in inducing genes involved in the biosynthesis of PCs. Thus manipulating the metabolic pathway involved in mannose metabolism as well as cell metabolism, can induce synthesis of these phytochelatin enzymes which subsequently confers tolerance of plant cells to elevated levels of exposure from excess HM ions (Fasani et al., 2018). Sequestering HM inside vacuoles after complexation with PCs and other small peptides has been regarded as the major method of detoxification of metals in plants (Anjum et al., 2015; Shri et al., 2014,). This principle was used to develop transgenic rice lines with increased As tolerance by expressing a PC synthase, *CdPCS1*, gotten from an aquatic As-accumulator plant, *Ceratophyllum demersum* (Shri et al., 2014). This led to a higher rate of accumulation of As in shoots and roots in transgenic lines than the wild-type plants but limited As obtained in the grains. Thus the elevated level of PCs resulted in the sequesteration and detoxification of more As in the shoots and roots compared with the grains with more accumulation obtained in the roots.

7.5.3 Overexpression of cytochrome P450 enzymes for enhanced phytoremediation

Cytochrome 450 (CYP) enzymes consist of a superfamily of heme proteins involved in the metabolism and detoxification of xenobiotics such as herbicides, pharmaceuticals, petrochemicals, polycyclic aromatic hydrocarbons (PAHs), and industrial pollutants (Kumar et al., 2014). These enzymes abound in different forms in most living organisms with several cytochrome P450 isoforms and a nonspecific NADPH cytochrome P450 oxidoreductase present in microsomes (Abhilash et al., 2009). Active genes from plants, animals, and microbes are being incorporated into plants to tolerate, degrade, and remove toxic substances (Doty, 2008). For instance, augmented metabolism and removal of many organic herbicides and pollutants was obtained from the overexpression of mammalian genes encoding cytochrome P450s (Doty, 2008; Kawahigashi et al., 2008; Rylott et al., 2006).

Extensive research have been conducted on the production of transgenic plants using Cytochrome P450 in several plants including *Nicotiana tabacum* for tolerance to the herbicide phenyl urea and resistance to sulfonylurea (O'keefe et al.1994; Siminszky et al., 1999), oxidation of trichloroethylene (TCE) and ethylene Dibromide (Doty et al., 2000), tolerance of *Arabidopsis thaliana* to phenyl urea herbicide (Siminszky et al., 1999), *Oryza sativa* for enhanced metabolism of atrazine, simazine, metolachlor chlorotoluron and norflurazon (Abhilash et al., 2009; Kawahigashi et al., 2005, 2006, 2007,

2008), hybrid poplar (*Populus tremula* x *Populous alba*) for augmented removal of TCE, vinyl chloride, carbon tetrachloride, benzene and chloroform from hydroponic solution and air (Abhilash et al., 2009; Doty et al., 2007; Umesh and Vishwas, 2017), as well as resistance to a combination of Cd and trichloroethylene in Alfalfa (Umesh and Vishwas, 2017; Zhang and Liu, 2011).

Human CYP1A1, CYP2B6, and CYP2C19 have been identified with the ability to metabolize numerous herbicides (Inui et al., 2000; Kumar et al., 2014), CYP1A1 and CYP1A2 that can metabolize PAHs, and CYP2B11 that metabolizes polychlorinated dibenzo-p-dioxins (Anzenbacher and Anzenbacherova, 2001; Kumar et al., 2014; Sakaki and Munetsuna, 2010), as well as CYP2E1 that can metabolize volatile organic compounds (James et al., 2008; Kumar et al., 2014). Transforming biomass producing plants with these genes will invariably increase the efficiency and success of phytoremediation processes. The P450 enzymes with enhanced metabolism can degrade xenobiotic pollutants. Mammalian cytochrome P450 2EI was overexpressed in transgenic tobacco plants resulting in augmentation of trichloroethylene (TCE) and ethylene dibromide metabolism (Cherian and Oliveira, 2005; Doty et al., 2000). Transgenic rice (*Oryza sativa*) with cytochrome P450 monooxygenases CYP1A1, CYP2B6, and CYP2C19 has been demonstrated to tolerate herbicides/xenobiotic pollutants compared to non-transgenic rice plants (Kawahigashi et al., 2008). Transgenic Poplar (*P. alba* x *P. tremula*) plants were developed and characterized with overexpression of cytochrome P450 2E1 with their extensive root systems having capacity to metabolize and remove toxic substances in the halogen class by (Doty et al., 2007). Notably, these plants with experimental evidence, exhibited increased removal rates of volatile hydrocarbons, such as vinyl chloride, TCE, carbon tetrachloride, chloroform, and benzene (Doty et al., 2007).

7.6 Nanoparticle-mediated plant transformation

Nanotechnology is an advanced technology that studies the application of extremely small nanoparticles (NPs) at nanoscales between 1 and 100 nanometers (nm) in diameter. It is progressively gaining more attention as an innovative method for the transformation of plants and could be extremely useful in agriculture as well as phytoremediation of polluted environments. Several types of NPs have been developed in recent times and include DNA NPs, carbon-based NPs, metal-based NPs such as nano aluminum, nano zinc, nano gold, TiO_2, ZnO, and Al_2O_3 (Demirer et al., 2019a, 2019b, 2019c; Lv et al., 2020; Zhang et al., 2019), as well as composite NPs like chitosan-complexed single-walled carbon nanotubes (Lv et al., 2020). Despite the fact that gene delivery into plants using nanomaterials is relatively new, some achievements have been made. Due to the high aspect ratio, excellent tensile strength, high surface area to volume ratio, and biocompatibility with facile surface modification, carbon nanotubes (CNTs) in recent times have been revealed to pass through extracted chloroplast (Demirer et al., 2018; Wong et al., 2016,) and plant membranes (Giraldo et al., 2014). CNTs could be single-walled carbon nanotubes (SWCNTs) or multiwalled carbon nanotubes (MWCNTs). The SWCNTs are efficient gene-delivery vehicles that can enter the membranes of organelles and possess several benefits in plant genetic transformation (Kwak et al., 2019; Lv et al., 2020). Since chloroplast and mitochondria are maternally inherited, they could be used for plant genetic transformation which overcomes the limitations involved in nuclear transformation using *Agrobacterium*.

Demirer et al. (2018), recently applied carbon nanotubes scaffolds to external plant tissue through CNT-mediated delivery of linear and plasmid DNA, as well as small interfering RNA (siRNA) thereby obtaining efficient transient gene expression and silencing in mature plants (Demirer et al., 2018). The authors succeeded in the delivery of DNA with strong transient expression of green fluorescent protein in mature *Eruca sativa* (arugula-dicot) and *Triticum aestivum* (wheat-monocot) leaves and protoplasts and recorded 95% efficiency in silencing a gene by the delivery of siRNA to mature *Nicotiana benthamiana* leaves. Thus, DNA–CNT conjugates not only permit the DNA to gain access into plant cells but also prevents degradation of polynucleotides (Demirer et al., 2019a, 2019b, 2019c; Lv et al., 2020). Likewise, Chitosan complexed SWCNTs was developed by Kwak et al. (2019) for the specific transfer of foreign DNA to the chloroplasts of various plants including arugula, watercress, tobacco, and spinach (Lv et al., 2020), with high precision without any integration of the transgene in the nuclear genome. These techniques could be used for different plant species both model and non-model plants and thus serve as a useful approach to the engineering of hyperaccumulator plants or fast-growing plants with high biomass production for efficient phytoremediation of polluted soil and water. NPs can outstandingly regulate the molecular and genetic aspects of hyperaccumulators and consequently the phytoremediation process (Rai et al., 2020). The nanomaterials alleviate the stress of high concentrations of pollutants by reducing the phytotoxicity via a reduction in the quantity of available pollutants while increasing the tolerance of the plant. In a recent review by Lv et al. (2020), the authors extensively discussed the different approaches for NP transformation including magnetofection, carbon nanotubes, DNA nanostructures, peptide NPs as well as clay nanosheets which gives different merits in modulation of plant growth. Nanostructures made up of DNA are the basic parts of the genetic material in plants which makes it easy to metabolize, are highly effective in plant genetic transformation (Lv et al., 2020). Besides, nanostructures possess exceptional biological stability and are not degraded by nucleases (Kim et al., 2020; Lv et al., 2020) and also highly biocompatible with the target tissues without triggering any immune response by plants (Li et al.,2019; Zhan et al., 2020)

Nevertheless, NP mediated transformation of plants still has its drawbacks. For instance, the method of delivery of the NPs into plant tissues still remain a major challenge and further research is needed. Moreover, there is possibility that there may be leakage of NPs into the environment which will invariably get into the food chain and negatively affect human health. Thus, the effects of such leakage into the environment should be considered when applying NP mediated transgenic plants for environmental cleanup. With the Np-mediated gene transformation approach, it is difficult to achieve stable genetic transformation leading to low yield of transgenic plants since most of the NP-mediated transformations are transient (Lv et al., 2020).

7.7 Safety issues in the use of transgenic plants for phytoremediation

In phytoremediation processes (phytoextraction, rhizofiltration, phytostabilization, and phytovolatilization), significant progress has been achieved in combating environmental degradation resulting from various pollutants including hazardous metals, metalloids, and organic contaminants (He and Yang, 2007; Koźmińska et al. 2018; Wan et al., 2016,). The processes of rhizofiltration, phytostabilization, phytovolatilization, and phytoextraction, lead to adsorption, precipitation, absorption contaminants, and release of pollutants to bio-remedy the soil and water resources (Glick, 2003; He and Yang, 2007; Rafati et al., 2011). Aside from the enormous success so far recorded in the application of phytoremediation, the utility of these transgenics crops over the natural plants has some limitations, hence, there is a need to enhance the competence of bioconcentration and root-to-shoot transporting efficiency of contaminants by plants for more effective, improved, and environmentally friendly options to address the issues of phytoremediation. The use of genetic manipulation and plant transformation technologies to obtain transgenic plants with well-endowed phytoremediation potential has been introduced (Ellis et al., 2004). In phytoremediation, techniques of extraction, sequestration, and detoxification of environmental contaminants, are favorable, gainful, and environmentally friendly approaches (Cunningham et al., 1995; Singh et al., 2003; Gunarathne et al., 2019). In the process of phytoremediation, plants are capable of absorbing different categories of toxic metals, metalloids, and organic contaminants (Gunarathne et al., 2019), and some of the contaminants are convertible to harmless substances through catabolic and anabolic processes that take place within plants (Meagher, 2000). Some of the suspected risks involved in the process of phytoremediation have been established to have no impact on the plants and surroundings as in the cases of mercury volatilization in crops, emission of Selenium, and gene escape from transgenic crops (Lin et al., 2000; Meagher et al., 2000).

Transgenic plants, genetically modified plants (GMPs), are transformed plants obtained using recombinant DNA technology to facilitate the full expression of a gene of interest that is foreign to the plant or modification of endogenous traits inherent in various genes (Key et al., 2008). In this process, novel genes that are identified from hyperaccumulators are biotechnologically manipulated and passed on to plants that possess fast growth rate, deep rooting systems, tolerance to variable biotic and abiotic stressors (Cherian and Oliveira, 2005; Gullner et al., 2001; Gunarathne et al., 2019). Besides, tissue culture techniques can play a significant role in generating transformed plants with enhanced phytoremediation competence. The development of transgenic crops through phytoremediation began with the aim of achieving crops that are tolerant to heavy metals (Van Aken, 2008). These genetically modified plants have shown some merits in bioremediation over naturally grown crops, but not without some suspected issues that are connected with the surroundings, human health, and animals. Members of the public including researchers consider the use of transgenic plants for bioremediations as major factors that militate against the maintenance of balance in environmental habitation, resulting in disruption of biodiversity occasioned by the invasiveness of the super transgenics crops.

Most of the transgenics possess potentially super traits/genes (inherited from the transgenes) that enable them to axiomatically outgrow plants with non-transgenes, thereby demonstrating great dominance and competition over the natural or unmodified ones (Auer, 2008; Ellstrand and Schierenbeck, 2006). Undoubtedly, these transgenes, when biotechnologically and successfully integrated into plants, could serve as remedial plants in phytoremediation of polluted environments by thriving very well in such contaminated areas, where natural crops cannot grow. However, depriving such environments of the needed biodiversity could further pose more risks. Biodiversity ensures the availability of various species with naturally endowed genetic make-ups (that permit variation or polymorphisms due to the exchange of novel traits/genes) that could withstand adverse effects (contaminants and pollutants) in the surroundings (Culley and Hardiman, 2007; Ellstrand and Schierenbeck, 2006; Gunarathne et al., 2019). Possibly, the exchange of genes between transgenics and natural crops could lead to the evolution of weeds that are resistant to herbicide application, posing a great challenge in weed management strategies (Gunarathne et al., 2019; Simard et al., 2006; Zelaya et al., 2007). Other potentially catastrophic effects linked to destruction of biodiversity through the use of transgenic plants include the extinction of population of non-transgenic ones consisting of wild relatives, deprivation of soil microorganisms (beneficial microbes), scarcity of variety of food supply to animals, and a drastic reduction in insect population due to destabilization of the natural structure of ecosystems (Auer, 2008; Hancock, 2003; Wilkinson et al., 2003). For instance, in phytoextraction, weather plays a vital role in the trace elements to be taken up through chelating agents that are hazardous to soil ecosystems (Wei et al., 2008). In this process, there is concern of the incorporation of the pollutants/contaminants that could be disastrous to environments, humans, and animals after consumption. In the phytoremediation process, some chemicals such as carbon tetrachloride (CCl_4) and trichloroethylene (C_2HCl_3) are not completely eliminated and could induce phytotoxic concentration and unproductivity in the plants involved. However, some vital enzymes, biphenyl dioxygenases, have been integrated into plants to overcome some of these risks (Aken et al. 2010; Sylvestre et al. 2009; Van Aken et al., 2010). Also, deleterious traits inherent in transgenic plants (inherited via gene flow) that control fruit ripening and male sterility can result in total dominance over native plant population and depletion of biodiversity (Hancock, 2003; Hegde et al., 2006). In addition, studies need to be conducted to determine the most suitable Np to apply on specific plant species as their bioavailability may differ.

7.8 Future prospects of genetically modified plants in phytoremediation

Despite the efficiency and effectiveness of transgenic crops generated through a biotechnological approach, to enhance bioremediation, the technique is still faced with some challenges. The technique requires improvement to capture a wider range of pollutants since the current utility is within a subset of pollutants/contaminants when several sites are enveloped with various degrees of pollutants/contaminants. Understanding the detailed biochemical pathways and molecular interplays of plants that partake extensively in the uptake of the pollutants is very vital for the improvement of the technique and to enhance its precision and efficiency. Integrating Clustered regularly interspaced short palindromic repeats (CRISPR), and CRISPR-associated protein-9 nuclease (CRISPR/Cas9) and other associated technologies (gene silencing through RNA interference, RNAi; zinc-finger nucleases, ZFNs; transcription-activator-like effector nucleases, TALENs) through adequate funding will possibly improve efficiency and also address the issue of safety concerns that are linked to gene flow from the transgenic crops to natural or wild species of crops. These three methods (CRISPR/Cas9, TALENS, and ZFNs) have over the years been used as the major method for targeted genetic mutations to enhance crop productivity. (Cantos et al., 2014; Doman et al., 2020: Kazama et al., 2019; Lv et al., 2020). Regulation of gene expression using CRISPR enables genome reprograming of hyperaccumulator plants through transcriptional repression or activation (CRISPRi and CRISPRa) which strongly augments the phytoremediation potential of environmental contaminants (Basharat et al., 2018; Rai et al., 2020). With advanced biotechnological techniques, some imperfections in phytoremediation will be overcome to enhance its uses (Pilon-Smits and Pilon, 2002).

By means of genetically engineered plants, the phytoremediation abilities of high-biomass, fast-growing plants and hyperaccumulators have been improved through the overexpression of genes for metal transporters, phytochelatins, and metallothionein which enhances metal uptake, translocation and sequestration of inorganic pollutants thereby conferring higher tolerance to pollutants. NPs mediated genetic transformation has overcome the barrier of delivery of foreign genetic material into plants as they could be easily integrated into the chloroplast or mitochondria DNA of cells without the need for biolistic bombardment compared to nuclear transformation with *Agrobacterium*. Despite the successes achieved in the use of transgenic plants mediated phytoremediation, the technology is yet to be fully accepted due to regulatory bottlenecks and the fear of the impact such plants would have on the environment. Utmost care must be taken at the end of the cleanup process to properly dispose the wastes obtained. Nevertheless, recent advancement in the gene editing technology using the CRISPR/Cas9 system coupled with nanotechnology would greatly enhance the production and precisions of transgenic plants for efficient remediation of polluted environments.

References

Abhilash, P., Jamil, S., Singh, N., 2009. Transgenic plants for enhanced biodegradation and phytoremediation of organic xenobiotics. Biotechnol. Adv. 27 (4), 474–488.

Ager, F., Ynsa, M., Domınguez-Solıs, J., López-Martın, M., Gotor, C., Romero, L., 2003. Nuclear micro-probe analysis of *Arabidopsis thaliana* leaves. Nucl. Instrum. Methods Phys. Res. B: Beam Interact. Mater. Atom. 210, 401–406.

Agnihotri, A., Seth, C.S., 2019. Transgenic brassicaceae: a promising approach for phytoremediation of heavy metals. In: Prasad, M.N.V (Ed.), Transgenic Plant Technology for Remediation of Toxic Metals and Metalloids. Elsevier, pp. 239–255.

Aken, B.V., Correa, P.A., Schnoor, J.L., 2010. Phytoremediation of polychlorinated biphenyls: new trends and promises. Environ. Sci. Technol. 44 (8), 2767–2776.

Alkorta, I., Hernández-Allica, J., Garbisu, C., 2004. Plants against the global epidemic of arsenic poisoning. Environ. Int. 30 (7), 949–951.

Anjum, N.A., Hasanuzzaman, M., Hossain, M.A., Thangavel, P., Roychoudhury, A., Gill, S.S., et al., 2015. Jacks of metal/metalloid chelation trade in plants—an overview. Front. Plant Sci. 6, 192.

Anzenbacher, P., Anzenbacherova, E., 2001. Cytochromes P450 and metabolism of xenobiotics. Cell. Mol. Life Sci. 58, 737–747.

AUER, C., 2008. Ecological risk assessment and regulation for genetically modified ornamental plants. Critic. Rev. Plant Sci. 27 (4), 255–271.

Azab, E., Hegazy, A.K., El-Sharnouby, M.E., Abd Elsalam, H.E., 2016. Phytoremediation of the organic Xenobiotic simazine by p450-1a2 transgenic *Arabidopsis thaliana* plants. Int. J. Phytoremed. 18 (7), 738–746.

Balasundaram, U., Venkataraman, G., George, S., Parida, A., 2014. Metallothioneins from a hyperaccumulating plant *Prosopis juliflora* show difference in heavy metal accumulation in transgenic tobacco. Int. J. Agric. Environ. Biotechnol. 7 (2), 241–246.

Basharat, Z., Novo, L., Yasmin, A., 2018. Genome editing weds CRISPR: what is in it for phytoremediation? Plants 28 (7), 3. doi:10.3390/plants7030051.

Basu, S., Rabara, R.C., Negi, S., Shukla, P., 2018. Engineering PGPMOs through gene editing and systems biology: a solution for phytoremediation? Trend Biotechnol 36 (5), 499–510.

Bell, T.H., Joly, S., Pitre, F.E., Yergeau, E., 2014. Increasing phytoremediation efficiency and reliability using novel omics approaches. Trends Biotechnol. 32, 271–280.

Bodnar, M., Konieczka, P., Namiesnik, J., 2012. The properties, functions, and use of selenium compounds in living organisms. J. Environ. Sci. Health. Part C 30 (3), 225–252.

Bratić, A.M., Majić, D.B., Samardžić, J.T., Maksimović, V.R., 2009. Functional analysis of the buckwheat metallothionein promoter: tissue specificity pattern and up-regulation under complex stress stimuli. J. Plant Physiol. 166, 996–1000.

Cantos, C., Francisco, P., Trijatmiko, K.R., Slamet-Loedin, I., Chadha-Mohanty, P.K., 2014. Identification of 'safe harbor' loci in indica rice genome by harnessing the property of zinc-finger nucleases to induce DNA damage and repair. Front. Plant Sci. 5, 302.

Chaney, R.L., Malik, M., Li, Y.M., Brown, S.L., Brewer, E.P., Angle, J.S., et al., 2018. Role of phytochelatins (PCs), metallothioneins (MTs), and heavy metal ATPase (HMA) genes in heavy metal tolerance. In: Prasad, R. (Ed.), Mycoremediation and Environmental Sustainability. Fungal Biology. Springer, Cham. http://doi-org-443.webvpn.fjmu.edu.cn/10.1007/978-3-319-77386-5_2.

Chaney, R.L., Malik, M., Li, Y.M., Brown, S.L., Brewer, E.P., Angle, J.S., et al., 1997. Phytoremediation of soil metals. Curr. Opin. Biotechnol. 8 (3), 279-284.

Chen, P.-Y., Wang, C.-K., Soong, S.-C., To, K.-Y., 2003. Complete sequence of the binary vector pBI121 and its application in cloning T-DNA insertion from transgenic plants. Mol. Breed. 11 (4), 287–293.

Cherian, S., Oliveira, M.M., 2005. Transgenic plants in phytoremediation: recent advances and new possibilities. Environ. Sci. Technol. 39 (24), 9377–9390.

Cherian, S., Weyens, N., Lindberg, S., Vangronsveld, J., 2012. Phytoremediation of trace element–contaminated environments and the potential of endophytic bacteria for improving this process. Critic. Rev. Environ. Sci. Technol. 42, 2215–2260.

Chaudhary, K., Agarwal, S., Khan, S., 2018. Role of phytochelatins (PCs), metallothioneins (MTs), and heavy metal ATPase (HMA) genes in heavy metal tolerance. In: Prasad R. (Ed.), Mycoremediation and Environmental Sustainability. Fungal Biology. Springer, Cham. http://doi-org-443.webvpn.fjmu.edu.cn/10.1007/978-3-319-77386-5_2.

Connolly, E.L., Fett, J.P., Guerinot, M.L., 2002. Expression of the IRT1 metal transporter is controlled by metals at the levels of transcript and protein accumulation. Plant Cell 14 (6), 1347–1357.

Cunningham, S.D., Berti, W.R., Huang, J.W.W., 1995. Phytoremediation of contaminated soils. Trend. Biotechnol. 13, 393–397.

Culley, T.M., Hardiman, N.A., 2007. The beginning of a new invasive plant: a history of the ornamental Callery pear in the United States. AIBS Bull. 57 (11), 956964.

Daghan, H., 2019. Transgenic tobacco for phytoremediation of metals and metalloids. In: Prasad M.N.V. (Ed.) Transgenic Plant Technology for Remediation of Toxic Metals and Metalloids. Elsevier, p. 564.

Daghan, H., Arslan, M., Uygur, V., Koleli, N., Eren, A., 2010. The cadmium phytoextraction efficiency of ScMTII gene bearing transgenic tobacco plant. Biotechnol. Biotechnologic. Equip. 24 (3), 1974–1978.

Dalcorso, G., Fasani, E., Manara, A., Visioli, G., Furini, A., 2019. Heavy metal pollutions: State of the art and innovation in phytoremediation. Int. J. Mol. Sci. 20, 3412.

Das, N., Bhattacharya, S., Maiti, M.K., 2016. Enhanced cadmium accumulation and tolerance in transgenic tobacco overexpressing rice metal tolerance protein gene OsMTP1 is promising for phytoremediation. Plant Physiol. Biochem. 105, 297–309.

Davison, J., 2005. Risk mitigation of genetically modified bacteria and plants designed for bioremediation. J. Indust. Microbiol. Biotechnol. 32 (11–12), 639–650.

Demirer, G., Zhang, H., Goh, N., Chang, R., Landry, M., 2019a. Nanotubes effectively deliver siRNA to intact plant cells and protect siRNA against nuclease degradation. https://www.researchgate.net/publication/331462099_Nanotubes_effectively_deliver_siRNA_to_intact_plant_cells_and_protect_siRNA_against_nuclease_degradation.

Demirer, G.S., Zhang, H., Goh, N.S., Gonzalez-Grandio, E., Landry, M.P., 2019b. Carbon nanotube-mediated DNA delivery without transgene integration in intact plants. Nat. Protoc. 14, 2954–2971.

Demirer, G.S., Zhang, H., Matos, J.L., Goh, N.S., Cunningham, F.J., Sung, Y., 2019c. High aspect ratio nanomaterials enable delivery of functional genetic material without DNA integration in mature plants. Nat. Nanotechnol. 14, 456–464.

Demirer, G.S., Chang, R., Zhang, H., Chio, L., Landry, M.P., 2018. Nanoparticle-guided biomolecule delivery for transgene expression and gene silencing in mature plants. Biophy. J. 114, 217a.

Doman, J.L., Raguram, A., Newby, G.A., Liu, D.R., 2020. Evaluation and minimization of Cas9-independent off-target DNA editing by cytosine base editors. Nat. Biotechnol. 38, 620–628.

Doty, S.L., 2008. Enhancing phytoremediation through the use of transgenics and endophytes. New Phytol 179 (2), 318–333.

Doty, S.L., James, C.A., Moore, A.L., Vajzovica., Singleton G.L., Ma, C., Khan, Z., 2007. Enhanced phytoremediation of volatile environmental pollutants with transgenic trees. Proceed. Natl. Acad. Sci. USA 104 (43), 16816–16821.

Doty, S.L., Shang, T.Q., Wilson, A.M., Tangen, J., Westergreen, A.D., Newman, L.A., et al., 2000. Enhanced metabolism of halogenated hydrocarbons in transgenic plants containing mammalian cytochrome P450 2E1. Proceed. Natl. Acad. Sci. USA 97 (12), 6287–6291.

Eapen, S., D'souza, S., 2005. Prospects of genetic engineering of plants for phytoremediation of toxic metals. Biotechnol. Adv. 23 (2), 97–114.

Echem, O.G., Kabari, L., 2013. Heavy metal content in Bitter Leaf (*Vernonia amygdalina*) grown along heavy traffic routes in Port Harcourt. Agric. Chem. 201–210.

Elias, A.A., Busov, V.B., Kosola, K.R., Ma, C., Etherington, E., Shevchenko, O., et al., 2012. Green revolution trees: semidwarfism transgenes modify gibberellins, promote root growth, enhance morphological diversity, and reduce competitiveness in hybrid poplar. Plant Physiol. 160, 1130–1144.

Ellis, D.R, Sors, T.G., Brunk, D.G., Albrecht, C., Orser, C., Lahner, B., 2004. Production of Se-methylselenocysteine in Transgenic Plants Expressing Selenocysteine Methyltransferase. BMC Plant Biol. 4, 1.

Ellstrand, N.C., Schierenbeck, K.A., 2006. Hybridization as a stimulus for the evolution of invasiveness in plants? Euphytica 148 (1-2), 35–46.

Fahimirad, S., Hatami, M., 2017. Heavy metal-mediated changes in growth and phytochemicals of edible and medicinal plants. In: Ghorbanpour, M., Varma, A. (Eds.), Medicinal Plants and Environmental Challenges. Springer International Publishing. https://doi.org/10.1007/978-3-319-68717-9_11.

Fasani, E., Manara, A., Martini, F., Furini, A., Dalcorso, G., 2018. The potential of genetic engineering of plants for the remediation of soils contaminated with heavy metals. Plant Cell Environ 41 (5), 1201–1232.

Freeman, J.L., Salt, D.E., 2007. The metal tolerance profile of Thlaspi goesingense is mimicked in *Arabidopsis thaliana* heterologously expressing serine acetyltransferase. BMC Plant Biol 7, 63.

Gautam, N., Verma, P.K., Verma, S., Tripathi, R.D., Trivedi, P.K., Adhikari, B., Chakrabarty, D., 2012. Genome-wide identification of rice class I metallothionein gene: tissue expression patterns and induction in response to heavy metal stress. Funct. Integr. Genomics 12, 635–647.

Gelvin, S., 2012. Plant Molecular Biology Manual. Springer, Netherlands. https://doi.org/10.1007/978-94-009-0951-9.

Giraldo, J.P., Landry, M.P., Faltermeier, S.M., Mcnicholas, T.P., Iverson, N.M., Boghossian, A.A., 2014. Plant nanobionics approach to augment photosynthesis and biochemical sensing. Nat. Mater. 13 (4), 400–408.

Gisbert, C., Ros, R., De Haro, A., Walker, D.J., Bernal, M.P., Serrano, R., 2003. A plant genetically modified that accumulates Pb is especially promising for phytoremediation. Biochem. Biophy. Res. Commun. 303 (2), 440–445.

Glick, B.R., 2003. Phytoremediation: synergistic use of plants and bacteria to clean up the environment. Biotechnol. Adv. 21 (5), 383–393.

Gomes, M.A., Hauser-Davis, R.A., De Souza, A.N., Vitória, A.P., 2016. Metal phytoremediation: general strategies, genetically modified plants, and applications in metal nanoparticle contamination. Ecotoxicol. Environ. Saf. 134P1, 133–147. doi: 10.1016/j.ecoenv.2016.08.024.

Grichko, V.P., Filby, B., Glick, B.R., 2000. Increased ability of transgenic plants expressing the bacterial enzyme ACC deaminase to accumulate Cd, Co, Cu, Ni, Pb, and Zn. J. Biotechnol. 81 (1), 45–53.

Gu, C.-S., Liu, L.-Q., Deng, Y.-M., Zhu, X.-D., Huang, S.-Z., Lu, X.-Q., 2015. The heterologous expression of the Iris lactea var. chinensis type 2 metallothionein IlMT2b gene enhances copper tolerance in *Arabidopsis thaliana*. Bull. Environ. Contam. Toxicol. 94 (2), 247–253.

Gullner, G., Kömives, T., Rennenberg, H., 2001. Enhanced tolerance of transgenic poplar plants overexpressing γ-glutamylcysteine synthetase towards chloroacetanilide herbicides. J. Exp. Bot. 52 (358), 971–979.

Gunarathne, V., Mayakaduwa, S., Ashiq, A., Weerakoon, S.R., Biswas, J.K., Vithanage, M., 2019. Transgenic plants: Benefits, applications, and potential risks in phytoremediation. In: Prasad, M.N.V (Ed.), Transgenic Plant Technology for Remediation of Toxic Metals and Metalloids. Elsevier, pp. 89–102.

Hancock, J.F., 2003. A framework for assessing the risk of transgenic crops. Bioscience 53 (5), 512–519.

He, J., Li, H., Ma, C., Zhang, Y., Polle, A., Rennenberg, H., et al., 2015. Overexpression of bacterial γ-glutamylcysteine synthetase mediates changes in cadmium influx, allocation, and detoxification in poplar. New Phytol. 205 (1), 240–254.

He, Z.-L., Yang, X.-E., 2007. Role of soil rhizobacteria in phytoremediation of heavy metal contaminated soils. J. Zhejiang Univ. Sci. B. 8 (3), 192–207.

Heaton, A.C., Rugh, C.L., Wang, N.J., Meagher, R.B., 1998. Phytoremediation of mercury-and methylmercury-polluted soils using genetically engineered plants. J. Soil Contam. 7 (4), 497–509.

Hegde, S.G., Nason, J.D., Clegg, J.M., Ellstrand, N.C., 2006. The evolution of California's wild radish has resulted in the extinction of its progenitors. Evolution 60 (6), 1187–1197.

Hirata, K., Tsuji, N., Miyamoto, K., 2005. Biosynthetic regulation of phytochelatins, heavy metal-binding peptides. J. Biosci. Bioeng. 100 (6), 593–599.

Ibañez, S.G., Paisio, C.E., Oller, A.L.W., Talano, M.A., González, P.S., Medina, M.I., et al., 2015. Overview and new insights of genetically engineered plants for improving phytoremediation. In: Ansari, A., Gill, S., Gill, R., Lanza, G., Newman, L. (Eds.), Phytoremediation. Springer, Cham. https://doi.org/10.1007/978-3-319-10395-2_8.

Inui, H., Kodama, T., Ohkawa, Y., Ohkawa, H., 2000. Herbicide metabolism and cross-tolerance in transgenic potato plants co-expressing human CYP1A1, CYP2B6, and CYP2C19. Pest. Biochem. Physiol. 66 (2), 116–129.

James, C.A., Xin, G., Doty, S.L., Strand, S.E., 2008. Degradation of low molecular weight volatile organic compounds by plants genetically modified with mammalian cytochrome P450 2E1. Environ. Sci. Technol. 42 (1), 289–293.

Jan, S., Parray, J.A., 2016. Approaches to heavy metal tolerance in plants. Springer, Singapore doi: 0.1007/978-981-10-1693-6.

Kawahigashi, H., Hirose, S., Ohkawa, H., Ohkawa, Y., 2005. Phytoremediation of metolachlor by transgenic rice plants expressing human CYP2B6. J. Agric. Food Chem. 53, 9155–9160.

Kawahigashi, H., Hirose, S., Ohkawa, H., Ohkawa, Y., 2006. Phytoremediation of herbicide atrazine and metolachlor by transgenic rice plants expressing human CYP1A1, CYP2B6 and CYP2C19. J. Agric. Food Chem. 54, 2985–2991.

Kawahigashi, H., Hirose, S., Ohkawa, H., Ohkawa, Y., 2007. Herbicide resistance of transgenic rice plants expressing human CYP1A1. Biotechnol. Adv. 25, 75–84.

Kawahigashi, H., Hirose, S., Ohkawa, H., Ohkawa, Y., 2008. Transgenic rice plants expressing human P450 genes involved in xenobiotic metabolism for phytoremediation. J. Molecular Microbiol. Biotechnol. 15 (2), 212–219.

Kazama, T., Okuno, M., Watari, Y., Yanase, S., Koizuka, C., Tsuruta, Y., 2019. Curing cytoplasmic male sterility via TALEN-mediated mitochondrial genome editing. Nat. Plants 5, 722–730.

Keeran, N.S., Balasundaram, U., Govindan, G., Parida, A.K., 2019. *Prosopis juliflora*: a potential plant for mining of genes for genetic engineering to enhance phytoremediation of metals. In: Prasad, M.N.V. (Ed.), Transgenic Plant Technology for Remediation of Toxic Metals and Metalloids. Elsevier, pp. 381–393.

Key, S., Ma, J.K.C., Drake, P.M.W., 2008. Genetically modified plants and human health. J. Royal Soc. Med. 101 (6), 290–298.

Kiiskila, J.D., Das, P., Sarkar, D., Datta, R., 2015. Phytoremediation of explosive-contaminated soils. Curr. Pollut. Rep. 1, 23–34.

Kim, C.J., Park, J.E., Hu, X., Albert, S.K., Park, S.J., 2020. Peptide-driven shape control of low-dimensional DNA nanostructures. ACS Nanotechnol. 14 (2), 2276–2284.

Kobayashi, T., Nakanishi, H., Nishizawa, N.K., 2010. Recent insights into iron homeostasis and their application in graminaceous crops. Proceed. Japan Acad. Ser. B Phy. Biol. Sciences 86 (9), 900–913.

Komives, T., Gullner, G., 2005. Phase I xenobiotic metabolic systems in plants. Z Naturforsch C. J. Biosci. 60, (3–4), 179–185.

Kotrba, P., Najmanova, J., Macek, T., Ruml, T., Mackova, M., 2009. Genetically modified plants in phytoremediation of heavy metal and metalloid soil and sediment pollution. Biotechnol. Adv 27, (6), 799–810.

Koźmińska, A., Wiszniewska, A., Hanus-Fajerska, E., Muszyńska, E., 2018. Recent strategies of increasing metal tolerance and phytoremediation potential using genetic transformation of plants. Plant Biotechnol. Rep. 12, 1–14.

Krämer, U., Chardonnens, A.N., 2001. The use of transgenic plants in the bioremediation of soils contaminated with trace elements. Appl. Microbiol. Biotechnol. 55 (6), 661–672.

Kumar, S., Jin, M., Weemhoff, J.L., 2014. Cytochrome P450-mediated phytoremediation using transgenic plants: a need for engineered cytochrome P450 enzymes. J. Pet. Env. Biotechnol. 3 (5), 1000127 doi: 10.4172/2157-7463.1000127.

Kwak, S.Y., Lew, T.T.S., Sweeney, C.J., Koman, V.B., Wong, M.H., Bohmert-Tatarev, K., et al., 2019. Chloroplast-selective gene delivery and expression in planta using chitosan-complexed single-walled carbon nanotube carriers. Nat. Nanotechnol. 14 (5), 447–455.

Lang, M.L., Zhang, Y.X., Chai, T.Y., 2004. Advances in the research of genetic engineering of heavy metal resistance and accumulation in plants. Chinese J. Biotechnol. 20 (2), 157–164.

Leblanc, M.S., Mckinney, E.C., Meagher, R.B., Smith, A.P., 2013. Hijacking membrane transporters for arsenic phytoextraction. J. Biotechnol. 163 (1), 1–9.

Li, M., Liu, J., Deng, M., Ge, Z., Afshan, N., Zuo, X., Li, Q., 2019. Rapid transmembrane transport of DNA nanostructures by chemically anchoring artificial receptors on cell membranes. ChemPlusChem, 84 (4) 323–327.

Liu, M., Deng, J., Ge, M., Afshan, Z., Zuo, X., Li, Q., 2019. Rapid transmembrane transport of DNA nanostructures by chemically anchoring artificial receptors on cell membranes. ChemPlusChem. 84 (4), 323–327.

Lin, Z.Q., Schemenauer, R.S., Cervinka, V., Zayed, A., Lee, A., Terry, N., 2000. Selenium volatilization from the soil-Salicornia bigelovii for treatment system for the remediation of contaminated water and soil in the San Joaquin valley. J. Environ. Quality 29 (4), 1048–1056.

Lv, Z., Jiang, R., Chen, J., Chen, W., 2020. Nanoparticle-mediated gene transformation strategies for plant genetic engineering. Plant J. 104 (4), 880-891. doi:10.1111/tpj.14973.

Macek, T., Kotrba, P., Svatos, A., Novakova, M., Demnerova, K., Mackova, M., 2008. Novel roles for genetically modified plants in environmental protection. Trend. Biotechnol. 26 (3), 146–152.

Mahar, A, Wang, P, Ali, A, Mk, Awasthi, Ah, Lahori, Wang, Q, Li, R., Zhang, Z, 2016. Challenges and opportunities in the phytoremediation of heavy metals contaminated soils: a review. Ecotoxicol. Environ. Saf. 126, 111–121. doi: 10.1016/j.ecoenv.2015.12.023.

Makhzoum, A.B., Sharma, P., Bernards, M.A., Trémouillaux-Guiller, J., 2013. Hairy roots: an ideal platform for transgenic plant production and other promising applications. In: Gang, R.R. (Ed.), Phytochemicals, Plant Growth, and the Environment. Springer Science Plus Business Media, New York. doi:10.1007/978-1-4614-4066-6_6.

Martinoia, E., Klein, M., Geisler, M., Bovet, L., Forestier, C., Kolukisaoglu, U., et al., 2002. Multifunctionality of plant ABC transporters–more than just detoxifiers. Planta 214, 345–355.

Meagher, R.B., 1999. Phytoremediation of Ionic and Methyl Mercury P. University of Georgia, Athens, Georgia(US).

Meagher, R.B., 2000. Phytoremediation of toxic elemental and organic pollutants. Curr. Opin. Plant Biol. 3 (2), 153–162.

Meagher, R.B., Rugh, C.L., Kandasamy, M.K., Gragson, G., Wang, N.J., 2000. Engineered phytoremediation of mercury pollution in soil and water using bacterial genes. In: Terry, N, Banuelos, G (Eds.), Phytoremediation of Contaminated Soil and Water. Boca Raton7 Lewis, pp. 201–221.

Mena-Benitez, G.L., Gandia-Herrero, F., Graham, S., Larson, T.R., Mcqueen-Mason, S.J., French, C.E., et al., 2008. Engineering a catabolic pathway in plants for the degradation of 1, 2-dichloroethane. Plant Physiol. 147, (3), 1192–1198.

Merlot, S., Hannibal, L., Martins, S., Martinelli, L., Amir, H., Lebrun, M., 2014. The metal transporter PgIREG1 from the hyperaccumulator Psychotria gabriellae is a candidate gene for nickel tolerance and accumulation. J. Exp. Bot. 65 (6), 1551–1564.

Morel, M., Crouzet, J., Gravot, A., Auroy, P., Leonhardt, N., Vavasseur, A., et al., 2009. AtHMA3, a P1B-ATPase allowing Cd/Zn/co/Pb vacuolar storage in *Arabidopsis*. Plant Physiol. 149 (2), 894–904.

Mosa, K.A., Saadoun, I., Kumar, K., Helmy, M., Dhankher, O.P., 2016. Potential biotechnological strategies for the cleanup of heavy metals and metalloids. Front. Plant Sci. 7, 303. doi: 10.1016/S0006-291X(03)00349-8.

Nagata, T., Nakamura, A., Akizawa, T., Pan-Hou, H., 2009. Genetic engineering of transgenic tobacco for enhanced uptake and bioaccumulation of mercury. Biol. Pharma. Bull. 32 (9), 1491–1495.

Ojuederie, O.B., Babalola, O.O., 2017. Microbial and plant-assisted bioremediation of heavy metal polluted environments: a review. Int. J. Environ. Res. Public Health 14 (12), 1504.

Pandey, V.C., Pandey, D.N., Singh, N., 2015. Sustainable phytoremediation based on naturally colonizing and economically valuable plants. J. Clean. Prod. 86, 37–39.

Pandey, V.C., Bajpai, O., Singh, N., 2016. Energy crops in sustainable phytoremediation. Renew. Sustain. Ener. Rev. 54, 58–73.

Pandey, V.C., Bajpai, O., 2019. Phytoremediation: From Theory towards Practice. In: Pandey, V.C., Bauddh, K. (Eds.), Phytomanagement of Polluted Sites. Elsevier, Amsterdam, pp. 1–49. https://doi.org/10.1016/B978-0-12-813912-7.00001-6.

Pandey, V.C., Singh, V., 2019. Exploring the potential and opportunities of recent tools for removal of hazardous materials from environments. In: Pandey, V.C., Bauddh, K. (Eds.), Phytomanagement of Polluted Sites. Elsevier, Amsterdam, pp. 501–516. https://doi.org/10.1016/B978-0-12-813912-7.00020-X.

Pathak, S., Agarwal, A.V., Pandey, V.C., 2020. Phytoremediation—a holistic approach for remediation of heavy metals and metalloids. In: Pandey, V.C., Singh, V. (Eds.), Bioremediation of Pollutants. Elsevier, Amsterdam, pp. 3–14. https://doi.org/10.1016/B978-0-12-819025-8.00001-6.

Pan, A., Yang, M., Tie, F., Li, L., Chen, Z., Ru, B., 1994. Expression of mouse metallothionein-I gene confers cadmium resistance in transgenic tobacco plants. Plant Mol. Biol. 24 (2), 341–351.

Pilon-Smits, E, Pilon, M., 2002. Phytoremediation of metal using transgenic plants. Critic. Rev. Plant Sci. 21 (5), 439–456.

Pilon-Smits, E.A., Freeman, J.L., 2006. Environmental cleanup using plants: biotechnological advances and ecological considerations. Front. Ecol. Environ. 4 (4), 203–210.

Prasad, M., 2003. Phytoremediation of metal-polluted ecosystems: hype for commercialization. Russ. J. Plant Physiol. 50, 686–701.

Rafati, M., Khorasani, N., Moattar, F., Shirvany, A., Moraghebi, F., Hosseinzadeh, S., 2011. Phytoremediation potential of *Populus alba* and *Morus alba* for cadmium, chromium, and nickel absorption from polluted soil. Int. J. Environ. Res. 5 (4), 961–970.

Rai, P.K., Kim, K.-H., Lee, S.S., Lee, J.-H., 2020. Molecular mechanisms in phytoremediation of environmental contaminants and prospects of engineered transgenic plants/microbes. Sci. Total Environ. 705, 135858.

Raybould, A., Higgins, L.S., Horak, M.J., Layton, R.J., Storer, N.P., De La Fuente, J.M., et al., 2012. Assessing the ecological risks from the persistence and spread of feral populations of insect-resistant transgenic maize. Trans. Res. 21 (3), 655–664.

Reichman, S., Parker, D., 2005. Metal complexation by phytosiderophores in the rhizosphere. In: Huang, P.M, Gobran, G (Eds.), Biogeochemistry of Trace Elements in the Rhizosphere. Elsevier, pp. 129–156.

Ruiz, O.N., Daniell, H., 2009. Genetic engineering to enhance mercury phytoremediation. Curr. Opin. Biotechnol. 20, (2), 213–219.

Rylott, E.L., Jackson, R.G., Edwards, J., Womack, G.L., Seth-Smith, H.M., Rathbone, D.A., et al., 2006. An explosive-degrading cytochrome P450 activity and its targeted application for the phytoremediation of RDX. Nat. Biotechnol. 24 (2), 216–219.

Sakaki, T., Munetsuna, E., 2010. Enzyme systems for biodegradation of polychlorinated dibenzo-p-dioxins. Appl. Microbiol. Biotechnol. 88 (1), 23–30.

Sheoran, V., Sheoran, A.S., Poonia, P., 2009. Phytomining: a review. Miner. Eng. 22 (12), 1007–1019.

Shri, M., Dave, R., Diwedi, S., Shukla, D., Kesari, R., Tripathi, R.D., et al., 2014. Heterologous expression of *Ceratophyllum demersum* phytochelatin synthase, CdPCS1, in rice leads to lower arsenic accumulation in grain. Sci. Rep. 4, 5784.

Singh, O., Labana, S., Pandey, G., Budhiraja, R., Jain, R.K., 2003. Phytoremediation: an overview of metallic ion decontamination from soil. Appl. Microbiol. Biotechnol. 61 (5/6), 405–412.

Simard, M.-J., Legere, A., Warwick, S.I., 2006. Transgenic *Brassica napus* fields and *Brassica rapa* weeds in Quebec: sympatry and weed crop *in situ* hybridization. Botany 84 (12), 1842–1851.

Siminszky, B., Corbin, F.T., Ward, E.R., Fleischmann, T.J., Dewey, R.E. 1999. Expression of a soybean cytochrome P450 monooxygenase cDNA in yeast and tobacco enhances the metabolism of phenylurea herbicides. Proc. Natl. Acad. Sci. USA. 96, 1750–1755.

Sone, Y., Nakamura, R., Pan-Hou, H., Sato, M.H., Itoh, T., Kiyono, M., 2013. Increase methylmercury accumulation in *Arabidopsis thaliana* expressing bacterial broad-spectrum mercury transporter MerE. AMB Exp. 3 (1), 52.

Song, W.-Y., Park, J., Mendoza-Cózatl, D.G., Suter-Grotemeyer, M., Shim, D., Hörtensteiner, S., et al., 2010. Arsenic tolerance in *Arabidopsis* is mediated by two ABCC-type phytochelatin transporters. Proceed. Natl. Acad. Sci. 107, 21187–21192.

Song, W.-Y., Sohn, E.J., Martinoia, E., Lee, Y.J., Yang, Y.-Y., Jasinski, M., et al., 2003. Engineering tolerance and accumulation of lead and cadmium in transgenic plants. Nat. Biotechnol. 21 (8), 914–919.

Sun, L., Ma, Y., Wang, H., Huang, W., Wang, X., Han, L., Sun, W., Han, E., Wang, B., 2018. Overexpression of PtABCC1 contributes to mercury tolerance and accumulation in Arabidopsis and poplar. Biochem. Biophys. Res. Commun. 497, 997–1002.

Sylvestre, M., Macek, T., Mackova, M., 2009. Transgenic plants to improve rhizoremediation of polychlorinated biphenyls (PCBs). Curr. Opin. Biotechnol. 20 (2), 242–247.

Umesh, B.J., Vishwas A.B., 2017. Transgenic Approaches for Building Plant Armor and Weaponry to Combat Xenobiotic Pollutants: Current Trends and Future Prospects. In: Hashmi M.Z. et al. (eds.), Xenobiotics in the Soil Environment, Soil Biology 49. doi:10.1007/978-3-319-47744-2_14.

Van Aken, B., 2008. Transgenic plants for phytoremediation: helping nature to clean up environmental pollution. Trend. Biotechnol. 26 (5), 225–227.

Van Aken, B., Correa, P.A., Schnoor, J.L., 2010. Phytoremediation of polychlorinated biphenyls: new trends and promises. Environ. Sci. Technol. 44 (8), 2767–2776.

Verbruggen, N., Leduc, D., Vanek, T., 2009. Potential of plant genetic engineering for phytoremediation of toxic trace elements. Phytotechnol. Solut. Sustain. Land Manage., 1–24.

Wan, X., Lei, M., Chen, T., 2016. Cost–benefit calculation of phytoremediation technology for heavy-metal-contaminated soil. Sci. Total Environ. 563, 796–802.

Wang, X., Zhi, J., Liu, X., Zhang, H., Liu, H., Xu, J., 2018. Transgenic tobacco plants expressing a P1B-ATPase gene from *Populus tomentosa* Carr. (PtoHMA5) demonstrate improved cadmium transport. Int. J. Biol. Macromol. 113, 655–661.

Wang, Y., Ren, H., Pan, H., Liu, J., Zhang, L., 2015. Enhanced tolerance and remediation to mixed contaminates of PCBs and 2, 4-DCP by transgenic alfalfa plants expressing the 2, 3-dihydroxybiphenyl-1, 2-dioxygenase. J. Hazard. Mater. 286, 269–275. doi: 10.1016/j.jhazmat.2014.12.04.

Weerakoon, S.R., 2019. Genetic engineering for metal and metalloid detoxification. In: Prasad, M.N.V. (Ed.), Transgenic Plant Technology for Remediation of Toxic Metals and Metalloids. Elsevier, pp. 23–41.

Weisman, D., Alkio, M., Colón-Carmona, A., 2010. Transcriptional responses to polycyclic aromatic hydrocarbon-induced stress in *Arabidopsis thaliana* reveal the involvement of hormone and defense signaling pathways. BMC Plant Biol. 10, 59.

Wei, S., Teixeira Da Silva, J.A., Zhou, Q., 2008. Agro-improving method of phytoextracting heavy metal contaminated soil. J. Hazard. Mater. 150 (3), 662–668.

Williams, L.E., Pittman, J.K., Hall, J., 2000. Emerging mechanisms for heavy metal transport in plants. Biochim. Biophy. Acta (BBA)-Biomemb. 1465 (1–2), 104–126.

Wilkinson, M.J., Sweet, J., Poppy, G.M., 2003. Risk assessment of GM plants: avoiding gridlock? Trend. Plant Sci. 8 (5), 208–212.

Wong, M.H., Misra, R.P., Giraldo, J.P., Kwak, S.Y., Son, Y., Landry, M.P., 2016. Lipid exchange envelope penetration (LEEP) of nanoparticles for plant engineering: a universal localization mechanism. Nanotechnol. Lett. 16 (2), 1161–1172.

Yadav, B.K., Sibel, M.A., Van Bruggen, J.J.A, 2011. Rhizofiltration of a heavy metal (lead) containing wastewater using the wetland plant Carex pendula. Clean Soil Air Water 39 (5), 467–474.

Yadav, K.K., Gupta, N., Kumar, A., Reece, L.M., Singh, N., Rezania, S., et al., 2018. Mechanistic understanding and holistic approach of phytoremediation: a review on application and future prospects. Ecol. Eng. 120 (2018), 274–298.

Yan, A., Wang, Y., Tan, S.N., Yusof, M.L.M., Ghosh, S., Chen, Z., 2020. Phytoremediation: a promising approach for revegetation of heavy metal-polluted land. Front. Plant Sci. 11.

Yang, J., Chen, Z., Wu, S., Cui, Y., Zhang, L., Dong, H., et al., 2015. Overexpression of the *Tamarix hispida* ThMT3 gene increases copper tolerance and adventitious root induction in *Salix matsudana* Koidz. Plant Cell Tiss, Organ Cult. 121, 469–479.

Zelaya, I.A., Owen, M.D.K., Van Gessel, M.J., 2007. Transfer of glyphosate resistance: evidence of hybridization in Conyza (Asteraceae). Am. J. Bot. 94 (4), 660–673.

Zhan, Y., Ma, W., Zhang, Y., Mao, C., Shao, X., Xie, X., et al., 2020. Diversity of DNA nanostructures and applications in oncotherapy. Biotechnol. J. 15 (1), e1900094.

Zhang, H., Demirer, G.S., Zhang, H., Ye, T., Goh, N.S., Aditham, A.J., et al. 2019. DNA nanostructures coordinate gene silencing in mature plants. Proc. Natl. Acad. Sci. USA, 116 (15), 7543–7548.

Zhang, W.-H., Tyerman, S.D., 1999. Inhibition of water channels by HgCl2 in intact wheat root cells. Plant Physiol 120 (3), 849–858.

Zhang, Y., Liu, J., 2011. Transgenic alfalfa plants co-expressing glutathione S-transferase (GST) and human CYP2E1 show enhanced resistance to mixed contaminates of heavy metals and organic pollutants. J. Hazard. Mater. 189, 357–362.

CHAPTER

8 CRISPR-assisted strategies for futuristic phytoremediation

Henny Patel[a], Shreya Shakhreliya[a], Rupesh Maurya[a], Vimal Chandra Pandey[b], Nisarg Gohil[a], Gargi Bhattacharjee[a], Khalid J. Alzahrani[c], Vijai Singh[a]

[a]*Department of Biosciences, School of Science, Indrashil University, Rajpur, Mehsana, Gujarat, India*
[b]*Department of Environmental Science, Babasaheb Bhimrao Ambedkar University, Lucknow, Uttar Pradesh, India*
[c]*Department of Clinical Laboratories Sciences, College of Applied Medical Sciences, Taif University, Saudi Arabia*

8.1 Introduction

In the past few decades, due to urbanization, the accumulation of industrial and domestic waste and their discharge have incessantly contributed to increasing environment pollution. It has adversely impacted all forms of life and has led to cause a number of diseases including respiratory disorders, infant mortality, increased stress, cardiovascular diseases, endothelial dysfunction, cancer, and others (Kelishadi et al., 2009; Kelishadi and Poursafa, 2010). Environmental pollutants in nature are increasing day-by-day and many of these are non-degradable by nature. In order to remove these harmful pollutants, microorganisms and plants having the natural ability to degrade and accumulate pollutants are used. Some of them have the potential to convert toxic complex compounds into simpler non-toxic forms. However, the capability to degrade pollutants is poor as the rate and amount of pollutants released as industrial and domestic wastes is humongous (Ramírez-García et al., 2019; Vishwakarma et al., 2020). Therefore, a pressing need has arisen to engineer or genetically manipulate microorganisms or plants to increase their genetic ability for hyper-accumulation or degradation of pollutants in order to clean environmental pollutants.

Fine-tuning of genetic components including promoters, ribosome binding sites, small regulatory RNA, and others are exchanged with strong genetic components in the genome to increase the efficiency of organisms for improved gene expression that can result in better functional performance (Singh, 2014; Gohil et al., 2017; Patel et al., 2018; Bhattacharjee et al., 2020a). A number of rate-limiting genes that reduce the ability of organisms for degradation of pollutants that can be overexpressed by inserting an additional copy of genes or increased functionality by optimizing gene expression or protein engineering to increase protein functional activity for the higher rate of degradation or accumulation of pollutants.

Recently developed clustered regularly interspaced short palindromic repeats (CRISPR) and CRISPR associated protein (Cas) systems is a breakthrough genome editing technology. It is a simple, rapid, specific and versatile technology that have been used in the broad spectrum of organisms including bacteria, yeast, mammals, plant, viruses, and many more (Jinek et al., 2012; Jiang et al., 2013; Singh et al., 2017; Singh et al., 2018; Khambhati et al., 2020). Currently, CRISPR-Cas9 technology is less explored

Assisted Phytoremediation. DOI: https://doi.org/10.1016/B978-0-12-822893-7.00006-9

in microorganisms and plants that have a natural ability to degrade or accumulate pollutants. In this chapter, we highlight CRISPR-Cas9 technology used in microorganisms and plants for enhanced bio-remediation in order to clean environmental pollution at a certain scale.

8.2 Basic of CRISPR biology

CRISPR-Cas systems are classified into two classes. The classification is based on effector proteins that stimulate the immune system for cleaving foreign nucleic acids. Class 1 CRISPR-Cas systems include types I, III and IV, and the system works with an effector module consisting of a different subunit complex involving 4 - 7 cas proteins located in an irregular stoichiometry, whereas class 2 systems include types II, V and VI, which use only single multidomain effector protein. It is further subdivided into type I A-G, type III Cas10, IIIA-D and type IV Csf1, IVA-C. On the other hand, Class 2 comprises type II cas9, IIA-C, type V cas12, VA-U and type VI cas13, VIA-D (Singh, 2020).

Type I includes protein cas3 helicase/nuclease, multi-subunit cas6e/cas6f complex linked with processed crRNA and forms a ribonucleoprotein complex that cleaves the invasive DNA. Type II is composed of the key Cas9 protein which has two active domains, RuvC-like and HNH domains that aids its endonuclease activity. It forms a complex of Cas9 and crRNA-trans-activating crRNA (tracrRNA) that binds to the target DNA sequence in presence of protospacer adjacent motif (PAM). The single guide RNA (sgRNA) is composed of a 20–25 nucleotide region for target-DNA binding crRNA and a 42-nucleotide long hairpin tracrRNA for the Cas9 binding and it directs the Cas9 nuclease to the target site Type III involves a number of RAMP proteins (Repair Associated Mysterious Proteins), Cas10 and Cas6 which help in the processing of crRNA and specific DNA cleavage and recognition of foreign DNA does not require a PAM sequence and cuts are made randomly (Jinek et al., 2012; Singh et al., 2017).

8.3 Molecular mechanism of the CRISPR-Cas9 system

A genome editing tool provides researchers the ability to alter desired gene sequences by inserting and deleting sequences in the genome of any organism. It has become a promising genome editing platform technology in the biological, biomedical and therapeutics field. In 2012, Jennifer Doudna's group developed the CRISPR-Cas9 system as a genome editing technology (Jinek et al., 2012). Later, it was extended in a wide range of organisms including bacteria, yeast, fungi, *Drosophila*, zebrafish, and mammals (Jinek et al., 2012; Jakočiūnas et al., 2015; Singh et al., 2017; Singh et al., 2018). CRISPR-Cas9 requires expression of Cas9 along with sgRNA in cells that can bind onto the desired DNA sequences in presence of PAM. As shown in Fig. 8.1, the CRISPR-Cas9 system can form a complex and bind to desired sequences that can subsequently generate a double-stranded break (DSB). The DSB is then repaired by either non-homologous end joining (NHEJ) or homology-directed repair (HDR) in presence of the provided homologous DNA sequence. This creates an insertion/deletion (indels) or repairing of a defective gene in order to restore or rescue gene function (Singh et al., 2017; Bhattacharjee et al., 2020b). CRISPR-Cas9 system has been applied to the number of microorganisms and plants but it is still less explored in plants that have the potential to degrade environmental pollution through bioremediation.

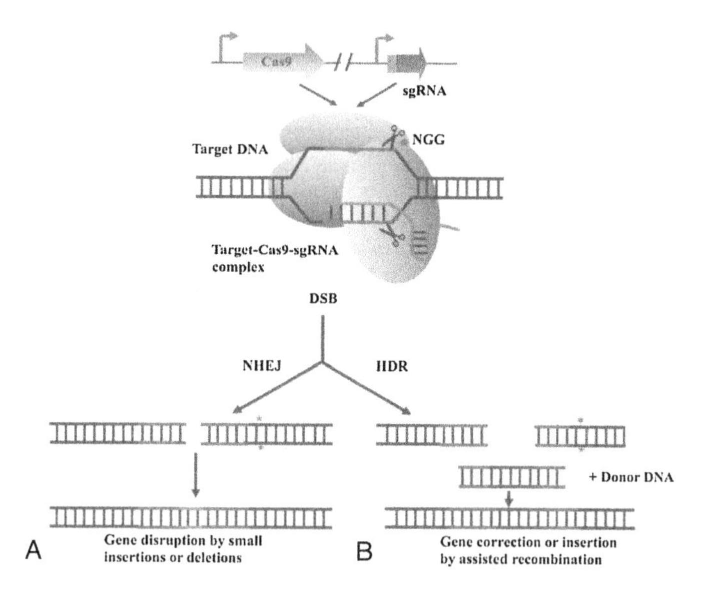

FIG. 8.1

Molecular mechanism of CRISPR-Cas9 system. Expression of Cas9 protein along with sgRNA that can bind onto desired target DNA and generate a double stranded break which is repaired by NHEJ or HDR mechanism. It allows incorporation of indels and also correction of defective gene repair. Figure reproduced with permission from Singh et al. (2017), Elsevier.

8.4 Phytoremediation for removal of pollutants

Phytoremediation is a way to remove environmental pollutants by using plants (Pandey et al., 2015, 2016; Pandey and Bajpai, 2019). Some of the plants have an inherent biosynthetic pathway for degradation or accumulation of complex pollutants which allows them to convert it into simple or non-toxic chemicals (Pandey and Bajpai, 2019). Industrial development is chiefly responsible for organic contaminants (Dsikowitzky and Schwarzbauer, 2014; Patel et al., 2017). Major emerging contaminants are generated from pharmaceuticals, nanomaterials, disinfection byproducts, dioxane, algal toxins, etc (Richardson and Kimura, 2017). Safe disposal of industrial waste is a major ecological challenge. Environmental contaminants leave toxic effects and cause diseases, for example, Minamata disease which is caused due to methylmercury poisoning (Harada, 1995) and thalassogenic diseases (Shuval, 2003). Effluents can also be a source of human enteric viruses (Okoh et al., 2010). A phytoremediation is a significant tool based on bioaccumulation of toxic non-degradable pollutant which is cost-effective and environment friendly (Rungwa et al., 2013; Ali et al., 2013; Pirzadah et al., 2015; Malik et al., 2017; Sumiahadi and Acar, 2018; Pandey and Bajpai, 2019). Plants used for bioremediation have been listed in Table 8.1.

In phytoremediation, plants utilize their import pumps to accumulate pollutants. They are used for remediating contaminated soil, sediments, surface water, and groundwater. Water, solutes, and organic contents present in water are pumped by plants as part of their physiological processes. This phenomenon of pumping can be exploited for stabilizing, removing, or breaking down. Several plants

Table 8.1 Summary of plant used in phytoremediation.

Plants name	Role of plants	Reference
Daucus carota (Carrot)	Absorbs Phenolic compound	De Araujo et al. (2002)
Brassica juncea (mustard)	Removal of metal contaminants from soil (Zn, Cd). Absorb from the root of plants.	Singh and Fulekar (2012)
Eichhornia crassipes	Ability to take up Ag, Cd, Cr, Cu, Hg, Ni, Pb, and Zn.	Odjegba and Fasidi (2007)
Thlaspi caerulescens	Metal tolerant hyperaccumulator plant. Removes Zn and Cd contaminants from soil	Zhao et al. (2003)
Potamogeton crispus, Myriophyllum spicatum	Atrazine uptake	Li et al. (2019)
Helianthus annuus	High accumulation of Pb	Seth et al. (2011)
Salvinia (water fern)	Cu hyper-accumulation in freshwater ecosystem	Al-Hamdani and Blair (2004).
Alternanthera sessilis	Remediation of Pb, Zn, Fe, Mg	Mazumdar and Das (2015)
Mosses and lichens	Uptake of airborne pollutants	Thomas et al. (1984)

growing in degraded areas are suggested for land rehabilitation (Gudin and Syratt, 1975) through the uptake of heavy metals and metalloids (Tangahu et al., 2011), remediation of petroleum hydrocarbon-contaminated soil (Kaimi et al., 2007; Ndimele, 2010; Brandt et al., 2006; Sun et al., 2004), remediation of heavy metal-enriched fly ash (Pandey, 2012; Maiti and Pandey, 2021), organic pollutants, radionuclides, and many more.

Several ways for remediation through plants are hyper-accumulation of heavy metals, metalloids and other pollutants (Pollard et al., 2014; Pandey and Bajpai, 2019; Pandey and Singh, 2019; Pathak et al., 2020) through phytodegradation (phytotransformation) (Newman and Reynolds, 2004; Wang et al., 2008), phytostabilization (phyto-immobilization) (Alkorta and Garbisu, 2010), phytovolatilization (Limmer and Burken, 2016), phytoextraction (Shen et al., 2002; McGrath and Zhao, 2003), phytofiltration (Pratas et al., 2014), rhizo-degradation, and phytostimulation (Memarian and Ramamurthy, 2012). While the bioremediation of contaminated aquatic ecosystem that can be achieved by utilization of wetland plants (Rai, 2008) such as free-floating wetland species including *Eichhornia crassipes* (Carrion et al., 2012; Pandey, 2016), lemna minor (Bokhari et al., 2016), *Spirodela polyrhiza* (Rahman et al., 2007). *Pistia stratiotes* (Khan et al., 2014), submerged aquatic species such as *Hydrilla verticillata* (Chang et al., 2020), *Najas indica* (Singh et al., 2010), and emergent species including *Typha domingensis* (Hegazy et al., 2011), *Cana indica* (Cheng et al., 2007), *Lucena lucocephala* (Bento et al., 2012), Azolla spp (Pandey, 2012), and bioremediation of polycyclic aromatic compounds remediation by poplar (Wittig et al., 2003).

Constructed wetlands are a complex natural treatment system. Several ecosystem services work by treating eutrophic water (Yang et al., 2008) and landscape-scale water treatment by utilizing and absorbing pollutants (Harrington and McInnes, 2009). Wetlands are also the most productive ecosystem services which provide habitat for a large variety of flora and fauna. Several other ecosystem services include carbon sequestration, flood control, managing water quality, maintaining hydrological and hydrogeochemical cycle (Boyer and Polasky, 2004; Tatu and Anderson, 2017). Constructed wetlands are an efficient system for phytoremediation (Samecka-Cymerman et al., 2004). Effluent from dairy (Dipu et al., 2011), urban waste (Zhang et al., 2007), domestic wastewater (Akinbile et al., 2016; Elfanssi et al., 2018), salt phytoremediation (Shelef et al., 2012), diesel (Al-Baldawi et al., 2014), mixture of contaminants (Guittonny-Philippe et al., 2015), industrial wastewater (Worku et al., 2018; Bhattacharjee et al., 2020c) is significantly remediated through constructed wetlands.

Heavy metals such as chromium (Mant et al., 2006), mercury led by *Typha* (Gomes et al., 2014) and *Scirpus grossus* (Tangahu et al., 2013), cadmium by *Ipomoea aquatica* (Bhaduri et al., 2012) and *Nymphaea aurora* (Schor-Fumbarov et al., 2003) copper by using *Eichhornia crassipus*, *Pistia spp.*, *Echinodorus species*, and *Nymphaea species* (Lu et al., 2018), zinc by using *Azolla pinnata* (Akinbile et al., 2016), and several other plants have been explored for their remediation potential and merging different aquatic plant qualities for metal uptake (Panfili et al., 2017). Suspension cells of eucalyptus were used for biotransformation of bisphenol degradation (Hamada et al., 2002). *Nopalea cochenillifera* was used for dye degradation (Adki et al., 2012), *Evolvulus alsinoides* for detoxification of azo dyes (Shanmugam et al., 2018), and trichloroethylene was transformed by poplar suspension culture (Shang and Gordon, 2002).

8.5 Phytoremediation by enriching microbes-plant interaction

A wide range of studies have exhibited *Arbuscular mycorrhizal* fungi as beneficial for plant growth regulations (Begum et al., 2019; Barea and Azcón-Aguilar, 1982), nutrition (Smith and Smith, 2012), and conservation (Bothe et al., 2010) either by increasing or by decreasing acquisition of minerals (Clark and Zeto, 2000; George 2000). Further, it also has significant role in heavy metal detoxification and decreasing the bioaccumulation of heavy metals (Orłowska et al., 2005; Bhaduri and Fulekar, 2012). Effective results can also be achieved by selecting plant and fungi partners accurately (see review Deng and Cao, 2017). So far it is proven that plants have the tremendous ability for phytoremediation. It can be turned into a cost-efficient technology by introducing new approaches including engineered plant species for phytoremediation. It requires more research on large scale remediation models as well as responses in different climatic conditions and soils.

8.6 Phytoremediation using engineered plant

Phytoremediation is productive in relation to its cost and eco-friendly manner (Salt et al., 1998). Transgenic plants have been engineered to remove pollutants (Van Aken, 2008), herbicides (Kawahigashi, 2009), and toxic metals (Eapen and D'souza, 2005). It has enhanced plant ability including high tolerance, accumulation, and mobilization of heavy metals and toxic contaminants that can be further manipulated by genetic engineering of plants for significant remediation (Pilon-Smits et al., 2002; Hsieh et al., 2009). Transgenic strategies also involve the expression of genes taking part in metabolism, uptake, and localization (Koźmińska et a., 2018). Expression of genes for enhanced production of proteins or peptides in microorganisms and plants for improving heavy metal accumulation and/or tolerance possess great potential (Mejáre and Bülow et al., 2001). Plants have several mechanisms such as cellular and molecular mechanisms for detoxification contamination. Cell wall composition, rhizosphere, plasma membrane properties and integrity, and enzymatic transformation (De Mello-Farias et al., 2011) are being explored for improving the phytoremediation potential of plants. Several metal chelators, transporters, metallothionein (MT), and phytochelatin (PC) genes have been expressed into plants for better metal tolerance, accumulation, and sequestration (Eapen and D'souza, 2005). As depicted in Fig. 8.2, engineered plants are used for higher uptake, translocation, degradation, and accumulation of toxic pollutants (Yan et al., 2020).

8.6.1 Phytochelatins

Phytochelatins are known for their metal-binding properties (Vögeli-Lang et al., 1996). They are synthesized by γ-Glu-Cys dipeptidyl transpeptidase called phytochelatin synthase. Phytochelatins keep plant proteins safe from the harmful effects of heavy metal (Kneer and Zenk, 1992). Phytochelatins form a phytochelatin metal complex and accumulate in different plant parts such as leaves, stems, and many. In the case of Pb accumulation, phytochelatins bind with Pb and form PCn and Pb–PCn complexes within the tissue (Andra et al., 2009). Phytochelatin have also been used for remediation of arsenic (Schmöger et al., 2000) and cadmium (Szalai, 2002). Genes expressing phytochelatin synthase play a significant role in the field of phytoremediation (Cobbett, 2000). It has been observed that, over-expression of wheat *TaPCS1* enhances tolerance against lead (Pb) and cadmium (Gisbert et al., 2003), while *AtPCS1 gene* enhanced cadmium accumulation (Pomponi et al., 2006; Cahoon et al., 2015). *PCSAt* gene expressed in *Mesorhizobium huakuii* subsp. *rengei* B3 bacterial strain has been shown to increase the ability of cells to bind to Cd^{2+}. *M. huakuii* subsp. *rengei* B3 upon establishing a symbiotic relationship with *Astragalus sinicus*, the symbionts enhanced Cd^{2+} accumulation in nodules (Sriprang et al., 2003).

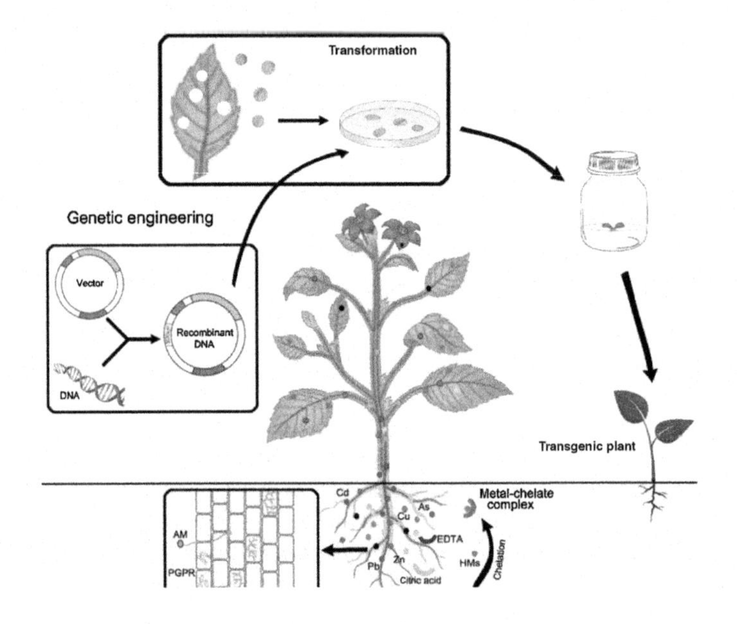

FIG. 8.2

Engineered plant for uptake, translocation and accumulation of pollutants. Adopted from Yan et al. (2020).

8.6.2 Proteins

MTs are cysteine-rich, low molecular weight proteins (Grennan, 2011) that are classified based on metal binding abilities, encoded genes, and amino acids. MTs show significance in homeostasis, protection against oxidative stress, and buffering against toxic heavy metals (Shabb et al., 2017; Smith and Nordberg, 2015). Chelator genes *ppk* and *merP* gene expression in plants via the nuclear genome shows resistance up to 10 μm of mercury (Nagata et al., 2006). Mercury remediation is achieved by the expression of mouse *Mt1* genes in the chloroplast of tobacco plants. Transgenic plants by chloroplast engineering aid more mercury accumulation (Ruiz et al., 2011). The first legume- rhizobium system was developed based on two constructed agrobacterium stains, 3841-PsMT1 and 3841-PsMT2. Pea MT in genes *PsMT1* and *PsMT2,* are expressed to control cadmium accumulation. Strain 3841-PsMT1 improved the biomass with wild type pea positively and decreased the cadmium content in the shoot. It was also found that nodules were able to maintain organization under the exposure to Cd. Thereafter, the 3841-PsMT1 strain became important for inoculating pea plants to increase biomass and also decrease cadmium concentration in plant tissues. Other strain 3841-PsMT2 with the pea Cd-tolerant mutant that has accumulated the highest Cd concentration than any other variant (Tsyganov et al., 2020).

8.6.3 Transporters

Vacuoles in plants are responsible for protoplasmic volume and storage. Overexpression of vacuolar transporters causes efficient phytoremediation by plants (Mendoza-Cózatl et al., 2011). Vacuolar transporters play a significant role in the accumulation of heavy metals. Various vacuolar transporters have been reported for sequestration of heavy metals such as YCF1, the overexpression of which increases tolerance towards Cd and Pb (Tong et al., 2004). P-type heavy metal ATPase transporter acts significantly in the accumulation of heavy metals. Overexpressing AtHMA4 in *A. thaliana* enhanced the efficiency of P1B-ATPase transporter for zinc and cadmium accumulation (Verret et al., 2004; Siemianowski et al., 2011). Expression of AtHMA4 enhanced Zn accumulation in aerial parts of tomato plants (Kendziorek et al., 2014) while *ZntA* transgenics uptake metal ions (Lee et al., 2003).

Expression of modified *merA* gene is a chief target in yellow poplar for phytoremediation of mercury (Rugh et al., 1998). Bacterial mercury reductase gene, *merA* possess the capacity to convert Hg(II) into less toxic Hg(O), whereas, bacterial organo-mercurial lyase gene *merB* is capable of converting methylmercury into sulfhydryl-bound Hg(II) which has subsequently taken up by plant roots (Heaton et al., 1998). Expressing important genes organomercurial lyase (*mer*B) and mercuric reductase (*merA*) mercuric ion binding protein (MerP) originating from transposon Tn*MERI1* of *Bacillus megaterium* strain MB1 in transgenic *Arabidopsis* showed higher tolerance and accumulation of heavy metals (Hsieh et al., 2009).

8.7 Biofortification and phytoremediation

Approaches such as the biofortification of metals in crops are not explored much (Zhao and McGrath, 2009). But it can be done by combining the biofortification and phytoremediation to tackle malnutrition as well as environmental remediation regarding to mineral nutrients namely, Fe, I, Cu, Zn, and Se. Therefore, linking both approaches and its implementation is a current need for environmental and societal sustainability (Pandey, 2013). The use of the remediation process to incorporate nutrients in crops such as selenium (Schiavon and Pilon-Smits, 2017) and zinc (Rouached, 2013) has been attempted. Bañuelos et al. (2015) explored selenium accumulation in *Stanleya pinnata* and used the dried powder of *S. pinnata* for selenium enriching soil. Selenium biofortification in wheat by using YAM2 bacterium (Yasin et al., 2015) have been reported. Recently, biofortification of selenium in *Cannabis sativa* has been reported (Stonehouse et al., 2020). Overall, above-engineered plants have shown the potential to degrade and hyperaccumulation of environmental pollutants for cleaning and management of pollutions. Due to some environmental biosafety and regulatory issues, engineered plants cannot be grown in fields for controlling pollutants. A pressing need has arisen to fine tune or genome edit small sequence of the genome that has less regulatory issues and recently developed CRISPR-Cas9 technology can be a plausible solution for addressing such issues.

8.8 CRISPR-Cas9 system for genome editing towards bioremediation

Recent development of CRISPR-Cas9 technology for altering gene function for better adaptation and flexibility allows optimal growth and development of plant during dynamic changing environmental conditions (DalCorso et al., 2019; Bhattacharjee et al., 2020d; Gohil et al., 2020). For reserving the farming land, many types of research are going on for removing toxicity from soil and water by the genome editing of high heavy metal tolerant plants (hyperaccumulators) in order to increase the tolerance strength of species. Phytoremediation is a remediation process in which *in-situ* and *ex-situ* applications are governed by the characteristics of the soil and water at the site, nutrient sustainability, meteorology, hydrology, feasible ecosystems, and contaminant characteristics. Plants play a significant role in the removal of toxic contaminants (Lasat, 2002; Seth et al., 2012; Salt et al., 1995). Processes including uptake, translocation and transformation can help in the removal of contaminants from soil and water (Kvesitadze et al., 2015). Fig. 8.3 shows phytoremediation using the CRISPR-Cas9 system in plants for increasing transport, accumulation of toxic, enhanced tolerance, and detoxification of toxic chemicals (Basharat et al., 2018).

CRISPR-Cas9 system has been used to modify the genome of a number of plants (Shan et al., 2013). The ease-of-use, affordability, and efficiency of CRISPR could soon be employed to enhance and improve the qualities of plants and bacteria involved in phytotechnologies, such as phytomining, phytostabilization, phytoextraction, phytovolatilization and bio-energy generation. CRISPR systems have already been attempted to edit a number of plants, including energy crops which are known for their ability to sustain, tolerate, immobilize and stabilize inorganic and organic pollutants. Moreover, quite a few genes playing role in increased metal tolerance, metal uptake and hyperaccumulation have already been investigated. Thus, the CRISPR-mediated genome reprogramming of plants, including its use in gene expression regulation through transcriptional repression or activation (CRISPRi and CRISPRa), could be of huge importance for phytoremediation. Plants with significant potential of phytoremediation including *Arabidopsis thaliana* have been sequenced. About 700 genes in *Arabidopsis* translate to form protein which can act on environmental pollutants (Briskine et al., 2017; Mandáková et al., 2015; Cobbett and Meagher, 2002).

CRISPR-Cas9 has the potential for developing improved plant lines with significant traits or deletion of unwanted traits to facilitate crop improvement and fruit quality improvement by chromosomal mutagenesis through NHEJ of DSBs generated by the CRISPR–Cas9 system (Yin et al., 2017; Wang et al., 2019). *Pseudomonas* sp. and *Acinetobacter* sp. facilitates plants in improving biomass (Tak

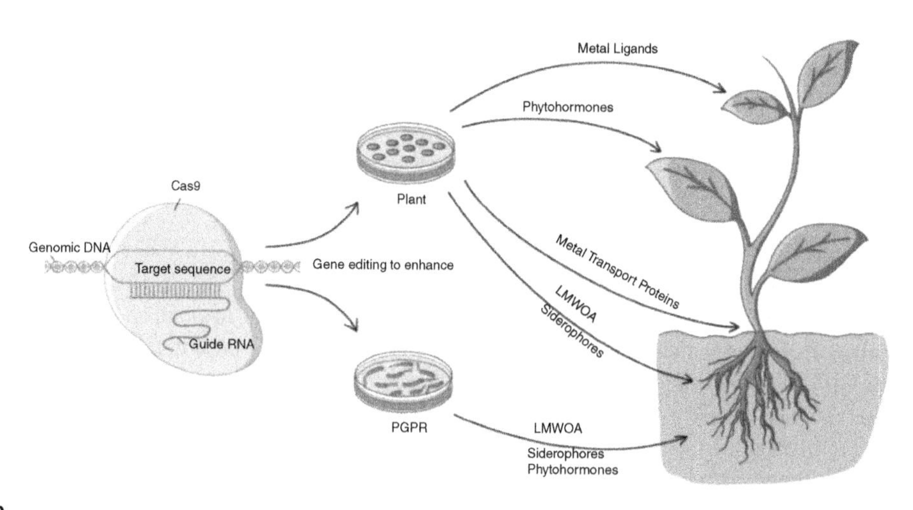

FIG. 8.3

Phytoremediation using CRISPR-Cas9 system. Genome editing of plants are shown to increase transport, accumulation of toxic, enhanced tolerance and detoxification of toxic chemicals. Adopted from Basharat et al. (2018).

et al., 2013; Gurska et al., 2009; Hansda et al., 2014). The CRISPR-Cas9 tool enhances the synthesis of a number of plant growth-promoting bacteria that are important to enhance the amount of biomass produced. Currently, the CRISPR-Cas systems are less explored in plants and microbes that have the natural ability to degrade or accumulate pollutants. However, fine-tuning of those microbes and plants genome allows higher rate of pollutants degradation. In near future, CRISPR-Cas systems could be expended in plants and microbes for genome editing towards the higher rate of degradation or accumulation of pollutants for better cleaning of environmental pollutions.

8.9 CRISPR-Cas9 technology and climate resilient phytoremediation

Areas selected for remediation are made without considering climate change. They may face modification in sea level, high degree of rainfall, extreme weather, forest fire, and winds. Contaminated site shows presence of toxicity due to metals. Climate change can affect contaminated site in many ways such as metal mobilization and risk associated with metals, soil, water distribution (Jarsjö et al., 2020; Augustsson et al., 2011), contamination mobilization due to uneasiness in aquifers (Libera et al., 2019), increase in greenhouse gases (Hou et al., 2018), phytoextraction (Srivastava et al., 2012), and deposition of heavy metals (Zhang et al., 2018). Climate resilience is recommended to be incorporated during planning, designing (Arisz and Burrell, 2006), during course of action, functioning, and maintenance of the remediation strategy for better environmental economic and social impacts (Scott et al., 2016).

Temperature has effects on physiology and morphology. Due to global temperature rise, there is an urgent requirement of production of heat tolerant varieties for better exploitation in phytoremediation programs. Many hyper-accumulators do not perform under high temperatures. Genome editing can be easily implemented by incorporating artificial nucleases. Bioengineering of tolerance pathways (Agarwal et al., 2013; Ganjewala et al., 2019), antioxidant enzymes (Jobby et al., 2016), osmolyte accumulation (Sharma et al., 2019), hyper-accumulators, and transporters and their overexpression (Luo et al., 2020) will be significant futuristic phytoremediation strategy.

8.10 Conclusion and future remarks

Bioremediation is currently used for the degradation and accumulation of environmental pollutants. Natural ability of microorganisms and plants show bioremediation of complex hazardous molecules/toxic metals into a simple and non-toxic form. However, its efficacy is quite less as compared to the enormous amount of the pollutants that are generated every day from industry and as domestic wastes. Therefore, engineering of microorganisms and plants are required that have strong potential and are powerful than natural organisms for the degradation and accumulation of pollutants at a much higher rate. Recently developed CRISPR-Cas9 system has been used for improving the strength of microorganisms and plants for rapid and efficient bioremediation of pollutants. It is currently less explored in plants and microorganisms for bioremediation. However, in the near future, the CRISPR-Cas9 system can be expanded in more organisms and also extended to the field level applications for better control and management of pollutants.

Acknowledgement

H.P., S.S., R.M., N.G., G.B. and V.S. thank Indrashil University, Rajpur, Mehsana for the infrastructure facility. N.G. acknowledges the Indian Council of Medical Research, Government of India for financial support as Senior Research Fellowship (File No. 5/3/8/63/ITR-F/2020). The financial assistance from Gujarat State Biotechnology Mission (GSBTM Project ID: 5LY45F), Gujarat, India to G.B. and V.S. is duly acknowledged.

References

Adki, V.S., Jadhav, J.P., Bapat, V.A., 2012. Exploring the phytoremediation potential of cactus (*Nopalea cochenillifera* Salm. Dyck.) cell cultures for textile dye degradation. Int. J. Phytoremediation 14 (6), 554–569.

Agarwal, P.K., Shukla, P.S., Gupta, K., Jha, B., 2013. Bioengineering for salinity tolerance in plants: state of the art. Mol. Biotechnol. 54 (1), 102–123.

Akinbile, C.O., Ogunrinde, T.A., Chebt Man, H., Aziz, H.A., 2016. Phytoremediation of domestic wastewaters in free water surface constructed wetlands using *Azolla pinnata*. Int. J. Phytoremediation 18 (1), 54–61.

Al-Baldawi, I.A.W., Abdullah, S.R.S., Hasan, H.A., Suja, F., Anuar, N., Mushrifah, I., 2014. Optimized conditions for phytoremediation of diesel by *Scirpus grossus* in horizontal subsurface flow constructed wetlands (HSFCWs) using response surface methodology. J. Environ. Manage. 140, 152–159.

Al-Hamdani, S.H., Blair, S.L., 2004. Influence of copper on selected physiological responses in Salvinia minima and its potential use in copper remediation. American Fern J. 94 (1), 47–56.

Ali, H., Khan, E., Sajad, M.A., 2013. Phytoremediation of heavy metals—concepts and applications. Chemosphere 91 (7), 869–881.

Alkorta, I., Becerril, J.M., Garbisu, C., 2010. Phytostabilization of metal contaminated soils. Rev. Environ. Health 25 (2), 135–146.

Andra, S.S., Datta, R., Sarkar, D., Makris, K.C., Mullens, C.P., Sahi, S.V., et al., 2009. Induction of lead-binding phytochelatins in vetiver grass [*Vetiveria zizanioides* (L.)]. J. Environ. Qual. 38 (3), 868–877.

Arisz, H., Burrell, B.C., 2006. Urban drainage infrastructure planning and design considering climate change, 2006 IEEE EIC Climate Change Conference. IEEE, pp. 1–9.

Augustsson, A., Filipsson, M., Öberg, T., Bergbäck, B., 2011. Climate change—an uncertainty factor in risk analysis of contaminated land. Sci. Total Environ. 409 (22), 4693–4700.

Bañuelos, G.S., Arroyo, I., Pickering, I.J., Yang, S.I., Freeman, J.L., 2015. Selenium biofortification of broccoli and carrots grown in soil amended with Se-enriched hyperaccumulator *Stanleya pinnata*. Food Chem 166, 603–608.

Barea, J.M., Azcón-Aguilar, C., 1982. Production of plant growth-regulating substances by the vesicular-arbuscular mycorrhizal fungus *Glomus mosseae*. Appl. Environ. Microbiol. 43 (4), 810–813.

Basharat, Z., Novo, L.A.B., Yasmin, A., 2018. Genome editing weds CRISPR: what is in it for phytoremediation? Plants (Basel) 7 (3), 51.

Begum, N., Qin, C., Ahanger, M.A., Raza, S., Khan, M.I., Ahmed, N., et al., 2019. Role of arbuscular mycorrhizal fungi in plant growth regulation: implications in abiotic stress tolerance. Front. Plant Sci. 10, 1068.

Bento, R.A., Saggin-Júnior, O.J., Pitard, R.M., Straliotto, R., da Silva, E.M.R., de Lucena Tavares, S.R., et al., 2012. Selection of leguminous trees associated with symbiont microorganisms for phytoremediation of petroleum-contaminated soil. Water Air Soil Pollut. 223, 5659–5671.

Bhaduri, A.M., Fulekar, M.H., 2012. Assessment of arbuscular mycorrhizal fungi on the phytoremediation potential of *Ipomoea aquatica* on cadmium uptake. 3 Biotech 2, 193–198.

Bhattacharjee, G., Gohil, N., Singh, V., 2020. An introduction to design of microbial strain using synthetic biology toolboxes for production of biomolecules. In: Singh, V., Singh, A.K, Bhargava, P., Joshi, M., Joshi, C. (Eds.), Engineering of Microbial Biosynthetic Pathways. Springer, Singapore, pp. 1–10.

Bhattacharjee, G., Gohil, N., Singh, V., 2020. Synthetic biology approaches for bioremediation. In: Pandey, V.C., Singh, V. (Eds.), Bioremediation of Pollutants. Academic press, Amsterdam, pp. 303–312.

Bhattacharjee, G., Gohil, N., Vaidh, S., Joshi, K., Vishwakarma, G.S., Singh, V., 2020. Microbial bioremediation of industrial effluents and pesticides. In: Pandey, V.C., Singh, V. (Eds.), Bioremediation of Pollutants. Academic press, Amsterdam, pp. 287–302.

Bhattacharjee, G., Mani, I., Gohil, N., Khambhati, K., Braddick, D., Panchasara, H., et al., 2020. CRISPR technology for genome editing. In: Faintuch, J., Faintuch, S. (Eds.), Precision Medicine for Investigators, Practitioners and Providers. Academic Press, London, pp. 59–69.

Bokhari, S.H., Ahmad, I., Mahmood-Ul-Hassan, M., Mohammad, A., 2016. Phytoremediation potential of *Lemna minor* L. for heavy metals. Int. J. Phytoremediation 18 (1), 25–32.

Bothe, H., Turnau, K., Regvar, M., 2010. The potential role of arbuscular mycorrhizal fungi in protecting endangered plants and habitats. Mycorrhiza 20 (7), 445–457.

Brandt, R., Merkl, N., Schultze-Kraft, R., Infante, C., Broll, G., 2006. Potential of vetiver (*Vetiveria zizanioides* (L.) Nash) for phytoremediation of petroleum hydrocarbon-contaminated soils in Venezuela. Int. J. Phytoremediation 8 (4), 273–284.

Briskine, R.V., Paape, T., Shimizu-Inatsugi, R., Nishiyama, T., Akama, S., Sese, J., et al., 2017. Genome assembly and annotation of *Arabidopsis halleri*, a model for heavy metal hyperaccumulation and evolutionary ecology. Mol. Ecol. Resour. 17 (5), 1025–1036.

Cahoon, R.E., Lutke, W.K., Cameron, J.C., Chen, S., Lee, S.G., Rivard, R.S., et al., 2015. Adaptive engineering of phytochelatin-based heavy metal tolerance. J. Biol. Chem. 290 (28), 17321–17330.

Carrion, C., Ponce-de Leon, C., Cram, S., Sommer, I., Hernandez, M., Vanegas, C., 2012. Potential use of water hyacinth (*Eichhornia crassipes*) in Xochimilco for metal phytoremediation. Agrociencia 46 (6), 609–620.

Chang, G., Yue, B., Gao, T., Yan, W., Pan, G., 2020. Phytoremediation of phenol by *Hydrilla verticillata* (Lf) Royle and associated effects on physiological parameters. J. Hazard. Mater. 388, 121569.

Cheng, S., Xiao, J., Xiao, H., Zhang, L., Wu, Z., 2007. Phytoremediation of triazophos by *Canna indica* Linn. in a hydroponic system. Int. J. Phytoremediation 9 (6), 453–463.

Clark, R.Á., Zeto, S.K., 2000. Mineral acquisition by arbuscular mycorrhizal plants. J. Plant Nutr. 23 (7), 867–902.

Cobbett, C.S., 2000. Phytochelatin biosynthesis and function in heavy-metal detoxification. Curr. Opin. Plant Biol. 3 (3), 211–216.

Cobbett, C.S., Meagher, R.B., 2002. Arabidopsis and the genetic potential for the phytoremediation of toxic elemental and organic pollutants. Arabidopsis Book 1, e0032.

Czakó, M., Feng, X., He, Y., Liang, D., Márton, L., 2005. Genetic modification of wetland grasses for phytoremediation. Z. Naturforsch C. J. Biosci. 60 (3-4), 285–291.

DalCorso, G., Fasani, E., Manara, A., Visioli, G., Furini, A., 2019. Heavy metal pollutions: state of the art and innovation in phytoremediation. Int. J. Mol. Sci. 20 (14), 3412.

De Araujo, B.S., Charlwood, B.V., Pletsch, M., 2002. Tolerance and metabolism of phenol and chloroderivatives by hairy root cultures of *Daucus carota* L. Environ. Pollut. 117 (2), 329–335.

de Mello-Farias, P.C., Chaves, A.L.S., Lencina, C.L., 2011. Transgenic plants for enhanced phytoremediation-physiological studies. In: Alvarez, M. (Ed.), Genetic Transformation. InTech, Rijeka, pp. 305–328.

Deng, Z., Cao, L., 2017. Fungal endophytes and their interactions with plants in phytoremediation: a review. Chemosphere 168, 1100–1106.

Dipu, S., Kumar, A.A., Thanga, V.S.G., 2011. Phytoremediation of dairy effluent by constructed wetland technology. Environmentalist 31, 263–278.

Dsikowitzky, L., Schwarzbauer, J., 2014. Industrial organic contaminants: identification, toxicity and fate in the environment. Environ. Chem. Lett. 12, 371–386.

Eapen, S., D'souza, S.F., 2005. Prospects of genetic engineering of plants for phytoremediation of toxic metals. Biotechnol. Adv. 23 (2), 97–114.

Elfanssi, S., Ouazzani, N., Latrach, L., Hejjaj, A., Mandi, L., 2018. Phytoremediation of domestic wastewater using a hybrid constructed wetland in mountainous rural area. Int. J. Phytoremediation 20 (1), 75–87.

Ganjewala, D., Kaur, G., Srivastava, N., 2019. Metabolic engineering of stress protectant secondary metabolites to confer abiotic stress tolerance in plantsMolecular Approaches in Plant Biology and Environmental Challenges. Springer, Singapore, pp. 207–227.

George, E., 2000. Nutrient uptake. In: Kapulnik, Y., DoudsJr, D.D. (Eds.), Arbuscular Mycorrhizas: Physiology and Function. Springer, Dordrecht, pp. 307–343.

Gisbert, C., Ros, R., De Haro, A., Walker, D.J., Bernal, M.P., Serrano, R., et al., 2003. A plant genetically modified that accumulates Pb is especially promising for phytoremediation. Biochem. Biophys. Res. Commun. 303 (2), 440–445.

Gohil, N., Bhattacharjee, G., Singh, V., 2020. Genetic engineering approaches for detecting environmental pollutants. In: Pandey, V.C., Singh, V. (Eds.), Bioremediation of Pollutants. Academic press, Amsterdam, pp. 387–401.

Gohil, N., Panchasara, H., Patel, S., Ramírez-García, R., Singh, V., 2017. Book review: recent advances in yeast metabolic engineering. Front. Bioeng. Biotechnol. 5, 71.

Gomes, M.V.T., de Souza, R.R., Teles, V.S., Mendes, É.A., 2014. Phytoremediation of water contaminated with mercury using *Typha domingensis* in constructed wetland. Chemosphere 103, 228–233.

Grennan, A.K., 2011. Metallothioneins, a diverse protein family. Plant Physiol. 155 (4), 1750–1751.

Gudin, C., Syratt, W.J., 1975. Biological aspects of land rehabilitation following hydrocarbon contamination. Environ. Pollut. 8 (2), 107–112.

Guittonny-Philippe, A., Petit, M.E., Masotti, V., Monnier, Y., Malleret, L., Coulomb, B., et al., 2015. Selection of wild macrophytes for use in constructed wetlands for phytoremediation of contaminant mixtures. J. Environ. Manage. 147, 108–123.

Gurska, J., Wang, W., Gerhardt, K.E., Khalid, A.M., Isherwood, D.M., Huang, X.D., et al., 2009. Three year field test of a plant growth promoting rhizobacteria enhanced phytoremediation system at a land farm for treatment of hydrocarbon waste. Environ. Sci. Technol. 43 (12), 4472–4479.

Hamada, H., Tomi, R., Asada, Y., Furuya, T., 2002. Phytoremediation of bisphenol A by cultured suspension cells of *Eucalyptus perriniana*-regioselective hydroxylation and glycosylation. Tetrahedron Lett 43 (22), 4087–4089.

Hansda, A., Kumar, V., Anshumali, A., Usmani, Z., 2014. Phytoremediation of heavy metals contaminated soil using plant growth promoting rhizobacteria (PGPR): A current perspective. Recent Res. Sci. Technol. 6 (1), 131–134.

Harada, M., 1995. Minamata disease: methylmercury poisoning in Japan caused by environmental pollution. Crit. Rev. Toxicol. 25 (1), 1–24.

Harrington, R., McInnes, R., 2009. Integrated constructed wetlands (ICW) for livestock wastewater management. Bioresour. Technol. 100 (22), 5498–5505.

Heaton, A.C., Rugh, C.L., Wang, N.J., Meagher, R.B., 1998. Phytoremediation of mercury-and methylmercury-polluted soils using genetically engineered plants. J. Soil Contam. 7 (4), 497–509.

Hegazy, A.K., Abdel-Ghani, N.T., El-Chaghaby, G.A., 2011. Phytoremediation of industrial wastewater potentiality by *Typha domingensis*. Int. J. Environ. Sci. Technol. 8, 639–648.

Hou, D., Song, Y., Zhang, J., Hou, M., O'Connor, D., Harclerode, M., 2018. Climate change mitigation potential of contaminated land redevelopment: A city-level assessment method. J. Clean. Prod. 171, 1396–1406.

Hsieh, J.L., Chen, C.Y., Chiu, M.H., Chein, M.F., Chang, J.S., Endo, G., et al., 2009. Expressing a bacterial mercuric ion binding protein in plant for phytoremediation of heavy metals. J. Hazard. Mater. 161 (2–3), 920–925.

Jakočiūnas, T., Bonde, I., Herrgård, M., Harrison, S.J., Kristensen, M., Pedersen, L.E., et al., 2015. Multiplex metabolic pathway engineering using CRISPR/Cas9 in *Saccharomyces cerevisiae*. Metab. Eng. 28, 213–222.

Jarsjö, J., Andersson-Sköld, Y., Fröberg, M., Pietroń, J., Borgström, R., Löv, Å., Kleja, D.B., 2020. Projecting impacts of climate change on metal mobilization at contaminated sites: controls by the groundwater level. Sci. Total Environ. 712, 135560.

Jiang, W., Bikard, D., Cox, D., Zhang, F., Marraffini, L.A., 2013. RNA-guided editing of bacterial genomes using CRISPR-Cas systems. Nat. Biotechnol. 31 (3), 233–239.

Jinek, M., Chylinski, K., Fonfara, I., Hauer, M., Doudna, J.A., Charpentier, E., 2012. A programmable dual-RNA-guided DNA endonuclease in adaptive bacterial immunity. Science 337 (6096), 816–821.

Jobby, R., Shah, K., Shah, R., Jha, P., Desai, N., 2016. Differential expression of antioxidant enzymes under arsenic stress in Enterobacter sp. Environ. Prog. Sustain. Energy 35 (6), 1642–1645.

Kaimi, E., Mukaidani, T., Tamaki, M., 2007. Screening of twelve plant species for phytoremediation of petroleum hydrocarbon-contaminated soil. Plant Prod. Sci. 10 (2), 211–218.

Kawahigashi, H., 2009. Transgenic plants for phytoremediation of herbicides. Curr. Opin. Biotechnol. 20 (2), 225–230.

Kelishadi, R., Mirghaffari, N., Poursafa, P., Gidding, S.S., 2009. Lifestyle and environmental factors associated with inflammation, oxidative stress and insulin resistance in children. Atherosclerosis 203 (1), 311–319.

Kelishadi, R., Poursafa, P., 2010. Air pollution and non-respiratory health hazards for children. Arch. Med. Sci. 6 (4), 483–495.

Kendziorek, M., Barabasz, A., Rudzka, J., Tracz, K., Mills, R.F., Williams, L.E., et al., 2014. Approach to engineer tomato by expression of *AtHMA4* to enhance Zn in the aerial parts. J. Plant Physiol. 171 (15), 1413–1422.

Khambhati, K., Gohil, N., Bhattacharjee, G., Singh, V., 2020. Development and challenges of using CRISPR-Cas9 system in mammalians. In: Singh, V., Dhar, P.K. (Eds.), Genome Engineering via CRISPR-Cas9 System. Academic Press, London, pp. 83–93.

Khan, M.A., Marwat, K.B., Gul, B., Wahid, F., Khan, H., Hashim, S., 2014. Pistia stratiotes L. (Araceae): Phytochemistry, use in medicines, phytoremediation, biogas and management options. Pakistan J. Bot. 46 (3), 851–860.

Kneer, R., Zenk, M.H., 1992. Phytochelatins keep safe plant enzymes from heavy metal poisoning. Phytochemistry 31, 2663–2667.

Koźmińska, A., Wiszniewska, A., Hanus-Fajerska, E., Muszyńsk, E., 2018. Recent strategies of increasing metal tolerance and phytoremediation potential using genetic transformation of plants. Plant Biotechnol. Rep. 12 (1), 1–14. doi:10.1007/s11816-017-0467-2 2018.

Kvesitadze, G., Khatisashvili, G., Sadunishvili, T., Kvesitadze, E., 2015. Plants for remediation: uptake, translocation and transformation of organic pollutants. In: Öztürk, M., Ashraf, M., Aksoy, A., Ahmad, M.S.A., Hakeem, K.R. (Eds.), Plants, Pollutants and Remediation. Springer, Dordrecht, pp. 241–308.

Lasat, M.M., 2002. Phytoextraction of toxic metals: a review of biological mechanisms. J. Environ. Qual. 31 (1), 109–120.

Lee, J., Bae, H., Jeong, J., Lee, J.Y., Yang, Y.Y., Hwang, I., et al., 2003. Functional expression of a bacterial heavy metal transporter in Arabidopsis enhances resistance to and decreases uptake of heavy metals. Plant Physiol. 133 (2), 589–596.

Li, H., Qu, M., Lu, X., Chen, Z., Guan, S., Du, H., et al., 2019. Evaluation of the potential of *Potamogeton crispus* and *Myriophyllum* spicatum on phytoremediation of atrazine. Int. J. Environ. Anal. Chem. 99 (3), 243–257.

Libera, A., de Barros, F.P., Faybishenko, B., Eddy-Dilek, C., Denham, M., Lipnikov, K., Wainwright, H., 2019. Climate change impact on residual contaminants under sustainable remediation. J. Contam. Hydrol. 226, 103518.

Limmer, M., Burken, J., 2016. Phytovolatilization of organic contaminants. Environ. Sci. Technol. 50 (13), 6632–6643.

Lu, D., Huang, Q., Deng, C., Zheng, Y., 2018. Phytoremediation of copper pollution by eight aquatic plants. Polish J. Environ. Stud. 27 (1), 175–181.

Luo, J.S., Xiao, Y., Yao, J., Wu, Z., Yang, Y., Ismail, A.M., Zhang, Z., 2020. Overexpression of a defensin-like gene CAL2 enhances cadmium accumulation in plants. Front. Plant Sci. 11, 217.

Maiti, D., Pandey, V.C., 2021. Metal remediation potential of naturally occurring plants growing on barren fly ash dumps. Environ. Geochem. Health 43 (4), 1415–1426. doi:10.1007/s10653-020-00679-z.

Malik, Z.H., Ravindran, K.C., Sathiyaraj, G., 2017. Phytoremediation: a novel strategy and eco-friendly green technology for removal of toxic metals. Int. J. Agric. Environ. Res. 3, 1–18.

Mandáková, T., Singh, V., Krämer, U., Lysak, M.A., 2015. Genome structure of the heavy metal hyperaccumulator *Noccaea caerulescens* and its stability on metalliferous and nonmetalliferous soils. Plant Physiol 169 (1), 674–689.

Mant, C., Costa, S., Williams, J., Tambourgi, E., 2006. Phytoremediation of chromium by model constructed wetland. Bioresour. Technol. 97 (15), 1767–1772.

Mazumdar, K., Das, S., 2015. Phytoremediation of Pb, Zn, Fe, and Mg with 25 wetland plant species from a paper mill contaminated site in North East India. Environ. Sci. Pollut. Res. 22 (1), 701–710.

McGrath, S.P., Zhao, F.J., 2003. Phytoextraction of metals and metalloids from contaminated soils. Curr. Opin. Biotechnol. 14 (3), 277–282.

Mejáre, M., Bülow, L., 2001. Metal-binding proteins and peptides in bioremediation and phytoremediation of heavy metals. Trends Biotechnol 19 (2), 67–73.

Memarian, R., Ramamurthy, A.S., 2012. Effects of surfactants on rhizodegradation of oil in a contaminated soil. J. Environ. Sci. Health 47 (10), 1486–1490.

Mendoza-Cózatl, D.G., Jobe, T.O., Hauser, F., Schroeder, J.I., 2011. Long-distance transport, vacuolar sequestration, tolerance, and transcriptional responses induced by cadmium and arsenic. Curr. Opin. Plant Biol. 14 (5), 554–562.

Nagata, T., Kiyono, M., Pan-Hou, H., 2006. Engineering expression of bacterial polyphosphate kinase in tobacco for mercury remediation. Appl. Microbiol. Biotechnol. 72 (4), 777–782.

Ndimele, P.E., 2010. A review on the phytoremediation of petroleum hydrocarbon. Pakistan J. Biol. Sci. 13 (15), 715–722.

Newman, L.A., Reynolds, C.M., 2004. Phytodegradation of organic compounds. Curr. Opin. Biotechnol. 15 (3), 225–230.

Odjegba, V.J., Fasidi, I.O., 2007. Phytoremediation of heavy metals by *Eichhornia crassipes*. Environmentalist 27, 349–355.

Okoh, A.I., Sibanda, T., Gusha, S.S., 2010. Inadequately treated wastewater as a source of human enteric viruses in the environment. Int. J. Environ. Res. Public Health 7 (6), 2620–2637.

Orłowska, E., Ryszka, P., Jurkiewicz, A., Turnau, K., 2005. Effectiveness of arbuscular mycorrhizal fungal (AMF) strains in colonisation of plants involved in phytostabilisation of zinc wastes. Geoderma 129 (1–2), 92–98.

Pandey, VC, 2012. Phytoremediation of heavy metals from fly ash pond by *Azolla caroliniana*. Ecotoxicol. Environ. Saf. 82, 8–12.

Pandey, VC, 2013. Book review—phytoremediation and biofortification: two sides of one coin. In: Xuebin, Yin, Linxi, Yuan (Eds.). Environmental Engineering and Management Journal, 12. Springer, Netherlands, pp. 851–852.

Pandey, VC, Pandey, DN, Singh, N, 2015. Sustainable phytoremediation based on naturally colonizing and economically valuable plants. J. Clean. Prod. 86, 37–39.

Pandey, VC, Bajpai, O, 2019. Phytoremediation: from theory towards practice. In: Pandey, V.C., Bauddh, K. (Eds.), Phytomanagement of Polluted Sites. Elsevier, Amsterdam, pp. 1–49. https://doi.org/10.1016/B978-0-12-813912-7.00001-6.

Pandey, VC, Bajpai, O, Singh, N, 2016. Energy crops in sustainable phytoremediation. Renew. Sustain. Energy Rev. 54, 58–73.

Pandey, VC, Singh, V, 2019. Exploring the potential and opportunities of recent tools for removal of hazardous materials from environments. In: Pandey, V.C., Bauddh, K. (Eds.), Phytomanagement of Polluted Sites. Elsevier, Amsterdam, pp. 501–516. https://doi.org/10.1016/B978-0-12-813912-7.00020-X.

Pandey, V.C., 2016. Phytoremediation efficiency of *Eichhornia crassipes* in fly ash pond. Int. J. Phytoremediation 18 (5), 450–452.

Panfili, I., Bartucca, M.L., Ballerini, E., Del Buono, D., 2017. Combination of aquatic species and safeners improves the remediation of copper polluted water. Sci. Total Environ. 601, 1263–1270.

Patel, S., Maurya, R., Solanki, H., 2017. Phytoremediation of treated industrial effluent collected from Ahmedabad mega pipe line. J. Ind. Pollut. Control 33 (2), 1202–1208.

Patel, S., Panchasara, H., Braddick, D., Gohil, N., Singh, V., 2018. Synthetic small RNAs: current status, challenges, and opportunities. J. Cell. Biochem. 119 (12), 9619–9639.

Pathak, S., Agarwal, A.V., Pandey, V.C., 2020. Phytoremediation—a holistic approach for remediation of heavy metals and metalloids. In: Pandey, V.C., Singh, V. (Eds.), Bioremediation of Pollutants. Elsevier, Amsterdam, pp. 3–14. https://doi.org/10.1016/B978-0-12-819025-8.00001-6.

Pilon-Smits, E., Pilon, M., 2002. Phytoremediation of metals using transgenic plants. Crit. Rev. Plant Sci. 21 (5), 439–456.

Pirzadah, T.B., Malik, B., Tahir, I., Kumar, M., Varma, A., Rehman, R.U., 2015. Phytoremediation: an eco-friendly green technology for pollution prevention, control and remediation. Soil Rem Plants Prospect. Chall 5, 107–126.

Pollard, A.J., Reeves, R.D., Baker, A.J., 2014. Facultative hyperaccumulation of heavy metals and metalloids. Plant Sci. 217, 8–17.

Pomponi, M., Censi, V., Di Girolamo, V., De Paolis, A., Di Toppi, L.S., Aromolo, R., et al., 2006. Overexpression of Arabidopsis phytochelatin synthase in tobacco plants enhances Cd 2+ tolerance and accumulation but not translocation to the shoot. Planta 223 (2), 180–190.

Pratas, J., Paulo, C., Favas, P.J., Venkatachalam, P., 2014. Potential of aquatic plants for phytofiltration of uranium-contaminated waters in laboratory conditions. Ecol. Eng. 69, 170–176.

Rahman, M.A., Hasegawa, H., Ueda, K., Maki, T., Okumura, C., Rahman, M.M., 2007. Arsenic accumulation in duckweed (*Spirodela polyrhiza* L.): a good option for phytoremediation. Chemosphere 69 (3), 493–499.

Rai, P.K., 2008. Heavy metal pollution in aquatic ecosystems and its phytoremediation using wetland plants: an eco-sustainable approach. Int. J. Phytoremediation 10 (2), 133–160.

Ramírez-García, R., Gohil, N., Singh, V., 2019. Recent advances, challenges, and opportunities in bioremediation of hazardous materials. In: Pandey, V.C., Bauddh, K. (Eds.), Phytomanagement of Polluted Sites. Academic Press, Amsterdam, pp. 517–568.

Richardson, S.D., Kimura, S.Y., 2017. Emerging environmental contaminants: challenges facing our next generation and potential engineering solutions. Environ. Technol. Innovation 8, 40–56.

Rouached, H., 2013. Recent developments in plant zinc homeostasis and the path toward improved biofortification and phytoremediation programs. Plant Signaling Behav. 8 (1), e22681.
Rugh, C.L., Senecoff, J.F., Meagher, R.B., Merkle, S.A., 1998. Development of transgenic yellow poplar for mercury phytoremediation. Nat. Biotechnol. 16, 925–928.
Ruiz, O.N., Alvarez, D., Torres, C., Roman, L., Daniell, H., 2011. Metallothionein expression in chloroplasts enhances mercury accumulation and phytoremediation capability. Plant Biotechnol. J. 9 (5), 609–617.
Rungwa, S., Arpa, G., Sakulas, H., Harakuwe, A., Timi, D., 2013. Phytoremediation–an eco-friendly and sustainable method of heavy metal removal from closed mine environments in Papua New Guinea. Procedia Earth Planet. Sci. 6, 269–277.
Salt, D.E., Blaylock, M., Kumar, N.P., Dushenkov, V., Ensley, B.D., Chet, I., et al., 1995. Phytoremediation: a novel strategy for the removal of toxic metals from the environment using plants. Biotechnology 13 (5), 468–474.
Salt, D.E., Smith, R.D., Raskin, I., 1998. Phytoremediation. Annu. Rev. Plant Biol. 49, 643–668.
Samecka-Cymerman, A., Stepien, D., Kempers, A.J., 2004. Efficiency in removing pollutants by constructed wetland purification systems in Poland. J. Toxicol. Environ. Health 67 (4), 265–275.
Schiavon, M., Pilon-Smits, E.A., 2017. Selenium biofortification and phytoremediation phytotechnologies: a review. J. Env. Qual. 46 (1), 10–19.
Schmöger, M.E., Oven, M., Grill, E., 2000. Detoxification of arsenic by phytochelatins in plants. Plant Physiol 122 (3), 793–802.
Schor-Fumbarov, T., Keilin, Z., Tel-Or, E., 2003. Characterization of cadmium uptake by the water lily Nymphaea aurora. Int. J. Phytoremediation 5 (2), 169–179.
Scott, M., Lennon, M., Haase, D., Kazmierczak, A., Clabby, G., Beatley, T., 2016. Nature-based solutions for the contemporary city/Re-naturing the city/Reflections on urban landscapes, ecosystems services and nature-based solutions in cities/Multifunctional green infrastructure and climate change adaptation: brownfield greening as an adaptation strategy for vulnerable communities?/Delivering green infrastructure through planning: insights from practice in Fingal, Ireland/Planning for biophilic cities: from theory to practice. Plan. Theory Pract 17 (2), 267–300.
Seth, C.S., Misra, V., Singh, R.R., Zolla, L., 2011. EDTA-enhanced lead phytoremediation in sunflower (*Helianthus annuus* L.) hydroponic culture. Plant Soil 347, 231.
Seth, C.S., Remans, T., Keunen, E., Jozefczak, M., Gielen, H., Opdenakker, K., et al., 2012. Phytoextraction of toxic metals: a central role for glutathione. Plant Cell Environ. 35 (2), 334–346.
Shabb, J.B., Muhonen, W.W., Mehus, A.A., 2017. Quantitation of human metallothionein isoforms in cells, tissues, and cerebrospinal fluid by mass spectrometry. Methods Enzymol. 586, 413–431.
Shan, Q., Wang, Y., Li, J., Zhang, Y., Chen, K., Liang, Z., et al., 2013. Targeted genome modification of crop plants using a CRISPR-Cas system. Nat. Biotechnol. 31 (8), 686–688.
Shang, T.Q., Gordon, M.P., 2002. Transformation of [14C] trichloroethylene by poplar suspension cells. Chemosphere 47 (9), 957–962.
Shanmugam, L., Naikawadi, V.B., Ahire, M.L., Nikam, T.D., 2018. Decolorization and detoxification of reactive azo dyes using cell cultures of a memory tonic herb *Evolvulus alsinoides* L. J. Environ. Chem. Eng. 6 (5), 6479–6488.
Sharma, A., Shahzad, B., Kumar, V., Kohli, S.K., Sidhu, G.P.S., Bali, A.S., Zheng, B, 2019. Phytohormones regulate accumulation of osmolytes under abiotic stress. Biomolecules 9 (7), 285.
Shelef, O., Gross, A., Rachmilevitch, S., 2012. The use of *Bassia indica* for salt phytoremediation in constructed wetlands. Water Res. 46 (13), 3967–3976.
Shen, Z.G., Li, X.D., Wang, C.C., Chen, H.M., Chua, H., 2002. Lead phytoextraction from contaminated soil with high-biomass plant species. J. Environ. Qual. 31 (6), 1893–1900.
Shuval, H., 2003. Estimating the global burden of thalassogenic diseases: human infectious diseases caused by wastewater pollution of the marine environment. J. Water Health 1 (2), 53–64.
Siemianowski, O., Mills, R.F., Williams, L.E., Antosiewicz, D.M., 2011. Expression of the P1B-type ATPase AtHMA4 in tobacco modifies Zn and Cd root to shoot partitioning and metal tolerance a. Plant Biotechnol. J. 9 (1), 64–74.

Singh, 2020. An introduction to genome editing CRISPR-Cas systems. In: Singh, V., Dhar, P.K. (Eds.), Genome Engineering via CRISPR-Cas9 System. Academic Press, London, pp. 1–13.

Singh, A., Fulekar, M.H., 2012. Phytoremediation of heavy metals by Brassica juncea in aquatic and terrestrial environment. In: Anjum, N.A., Ahmad, I., Pereira, M.E., Duarte, A.C., Umar, S. (Eds.), The Plant Family *Brassicaceae*. Springer, Dordrecht, pp. 153–169.

Singh, R., Tripathi, R.D., Dwivedi, S., Kumar, A., Trivedi, P.K., Chakrabarty, D., 2010. Lead bioaccumulation potential of an aquatic macrophyte *Najas indica* are related to antioxidant system. Bioresour. Technol. 101 (9), 3025–3032.

Singh, V., 2014. Recent advancements in synthetic biology: current status and challenges. Gene 535 (1), 1–11.

Singh, V., Braddick, D., Dhar, P.K., 2017. Exploring the potential of genome editing CRISPR-Cas9 technology. Gene 599, 1–18.

Singh, V., Gohil, N., Ramirez Garcia, R., Braddick, D., Fofié, C.K., 2018. Recent advances in CRISPR-Cas9 genome editing technology for biological and biomedical investigations. J. Cell. Biochem. 119 (1), 81–94.

Smith, D.R., Nordberg, M., 2015. General chemistry, sampling, analytical methods, and speciation. In: Nordberg, G.F., Fowler, B.A., Nordberg, M. (Eds.), Handbook on the Toxicology of Metals. Academic Press, London, pp. 15–44.

Smith, S.E., Smith, F.A., 2012. Fresh perspectives on the roles of arbuscularmycorrhizal fungi in plant nutrition and growth. Mycologia 104 (1), 1–13.

Sriprang, R., Hayashi, M., Ono, H., Takagi, M., Hirata, K., Murooka, Y., 2003. Enhanced accumulation of Cd2+ by a *Mesorhizobium* sp. transformed with a gene from Arabidopsis thaliana coding for phytochelatin synthase. Appl. Environ. Microbiol. 69 (3), 1791–1796.

Srivastava, J., Chandra, H., Nautiyal, A.R., Kalra, S.J., 2012. Response of C3 and C4 plant systems exposed to heavy metals for phytoextraction at elevated atmospheric CO2 and at elevated temperatureEnvironmental Contamination. Intech Open Publisher, Croatia, pp. 3–16.

Stonehouse, G.C., McCarron, B.J., Guignardi, Z.S., El Mehdawi, A.F., Lima, L.W., Fakra, S.C., et al., 2020. Selenium metabolism in hemp (Cannabis sativa L.)-potential for phytoremediation and biofortification. Environ. Sci. Technol. 54 (7), 4221–4230.

Sumiahadi, A., Acar, R., 2018. A review of phytoremediation technology: heavy metals uptake by plants, IOP Conference Series: Earth and Environmental Science, 142, 012023.

Sun, W.H., Lo, J.B., Robert, F.M., Ray, C., Tang, C.S., 2004. Phytoremediation of petroleum hydrocarbons in tropical coastal soils I. Selection of promising woody plants. Environ. Sci. Pollut. Res. 11 (4), 260–266.

Szalai, G., 2002. Effect of Cd treatment on phytochelatin synthesis in maize. Acta Biol. Szeged. 46 (3–4), 121–122.

Tak, H.I., Ahmad, F., Babalola, O.O., 2013. Advances in the application of plant growth-promoting rhizobacteria in phytoremediation of heavy metals. Rev. Environ. Contam. Toxicol. 223, 33–52.

Tangahu, B.V., Abdullah, S.R.S., Basri, H., Idris, M., Anuar, N., Mukhlisin, M., 2013. Phytoremediation of wastewater containing lead (Pb) in pilot reed bed using Scirpusgrossus. Int. J. Phytoremediation 15 (7), 663–676.

Tangahu, B.V., Sheikh Abdullah, S.R., Basri, H., Idris, M., Anuar, N., Mukhlisin, M., 2011. A review on heavy metals (As, Pb, and Hg) uptake by plants through phytoremediation. Int. J. Chem. Eng. 2011, 939161.

Tatu, K.S., Anderson, J.T., 2017. An introduction to wetland science and South Asian wetlands. In: Prusty, B.A.K., Chandra, R., Azeez, P.A. (Eds.), Wetland Science. Springer, New Delhi, pp. 3–30.

Thomas, W., Rühling, Å., Simon, H., 1984. Accumulation of airborne pollutants (PAH, chlorinated hydrocarbons, heavy metals) in various plant species and humus. Environ. Pollut. Ser. A, Ecol. Biol. 36, 295–310.

Tong, Y.P., Kneer, R., Zhu, Y.G., 2004. Vacuolar compartmentalization: a second-generation approach to engineering plants for phytoremediation. Trends Plant Sci. 9 (4), 7–9.

Tsyganov, V.E., Tsyganova, A.V., Gorshkov, A.P., Seliverstova, E.V., Kim, V.E., Chizhevskaya, E.P., et al., 2020. Efficacy of a plant-microbe system: *Pisum sativum* (L.) cadmium-tolerant mutant and rhizobium leguminosarum setrains, xpressing Pea Metallothionein genes *PsMT1* and *PsMT2*, for cadmium phytoremediation. Front. Microbiol. 11, 15.

Van Aken, B., 2008. Transgenic plants for phytoremediation: helping nature to clean up environmental pollution. Trends Biotechnol. 26 (5), 225–227.

Verret, F., Gravot, A., Auroy, P., Leonhardt, N., David, P., Nussaume, L., et al., 2004. Overexpression of AtHMA4 enhances root-to-shoot translocation of zinc and cadmium and plant metal tolerance. FEBS Lett. 576 (3), 306–312.

Vishwakarma, G.S., Bhattacharjee, G., Gohil, N., Singh, V., 2020. Current status, challenges and future of bioremediation. In: Pandey, V.C., Singh, V. (Eds.), Bioremediation of Pollutants. Academic press, Amsterdam, pp. 403–415.

Vögeli-Lange, R., Wagner, G.J., 1996. Relationship between cadmium, glutathione and cadmium-binding peptides (phytochelatins) in leaves of intact tobacco seedlings. Plant Sci. 114 (1), 11–18.

Wang, T., Zhang, H., Zhu, H., 2019. CRISPR technology is revolutionizing the improvement of tomato and other fruit crops. Hortic. Res. 6, 77.

Wang, X., Wu, N., Guo, J., Chu, X., Tian, J., Yao, B., et al., 2008. Phytodegradation of organophosphorus compounds by transgenic plants expressing a bacterial organophosphorus hydrolase. Biochem. Biophys. Res. Commun. 365 (3), 453–458.

Wittig, R., Ballach, H.J., Kuhn, A., 2003. Exposure of the roots of *Populus nigra* L. cv. *Loenen* to PAHs and its effect on growth and water balance. Environ. Sci. Pollut. Res. 10 (4), 235.

Worku, A., Tefera, N., Kloos, H., Benor, S., 2018. Constructed wetlands for phytoremediation of industrial wastewater in Addis Ababa. Ethiopia. Nanotechnol. Environ. Eng. 3, 9.

Yan, A., Wang, Y., Tan, S.N., MohdYusof, M.L., Ghosh, S., Chen, Z., 2020. Phytoremediation: a promising approach for revegetation of heavy metal-polluted land. Front. Plant Sci. 11, 359.

Yang, X.E., Wu, X., Hao, H.L., He, Z.L., 2008. Mechanisms and assessment of water eutrophication. J. Zhejiang Univ. Sci. B 9 (3), 197–209.

Yasin, M., El-Mehdawi, A.F., Anwar, A., Pilon-Smits, E.A., Faisal, M., 2015. Microbial-enhanced selenium and iron biofortification of wheat (*Triticum aestivum* L.)-applications in phytoremediation and biofortification. Int. J. Phytoremediation 17 (1–6), 341–347.

Yin, K., Gao, C., Qiu, J.L., 2017. Progress and prospects in plant genome editing. Nat. Plants 3, 17107.

Zhang, H., Huo, S., Yeager, K.M., Xi, B., Zhang, J., He, Z., Wu, F., 2018. Accumulation of arsenic, mercury and heavy metals in lacustrine sediment in relation to eutrophication: Impacts of sources and climate change. Ecol. Indic. 93, 771–780.

Zhang, X.B., Peng, L.I.U., Yang, Y.S., Chen, W.R., 2007. Phytoremediation of urban wastewater by model wetlands with ornamental hydrophytes. J. Environ. Sci. 19 (8), 902–909.

Zhao, F.J., Lombi, E., McGrath, S.P., 2003. Assessing the potential for zinc and cadmium phytoremediation with the hyperaccumulator *Thlaspi caerulescens*. Plant Soil 249, 37–43.

Zhao, F.J., McGrath, S.P., 2009. Biofortification and phytoremediation. Curr. Opin. Plant Biol. 12 (3), 373–380.

CHAPTER

Approaches for assisted phytoremediation of arsenic contaminated sites

9

Ankita Gupta[a], Arnab Majumdar[b], Sudhakar Srivastava[a]
[a]*Plant Stress Biology Laboratory, Institute of Environment and Sustainable Development, Banaras Hindu University, Varanasi, India*
[b]*Department of Earth Sciences, Indian Institute of Science Education and Research (IISER) Kolkata, Mohanpur, West Bengal, India*

9.1 Introduction

Arsenic (As) is a toxic metalloid, which is present as a contaminant in soil and groundwater in many regions of the world, especially in the South and Southeast Asian countries. The contamination of As is majorly due to natural biogeochemical processes and also due to anthropogenic activities (Pandey et al., 2011; Srivastava et al., 2012; Zhang et al., 2013; Podgorski et al.,2017). The areas having the highest As contamination include the Ganga, Brahmaputra, and Meghna river basins in India and Bangladesh (Mazumder and Dasgupta, 2011; Rahman et al., 2015; Verma et al., 2015). The groundwater in the As contaminated regions is used during crop cultivation, thereby leading to the transfer of this toxic metalloid into the humans via As-contaminated crops (Sarkar and Paul, 2016). Among the staple crops, rice is known to accumulate the highest amount of As in its edible parts (Williams et al., 2007). This is possible due to the availability of As in the form of arsenite (As[III]) in the rice fields, where the environment is largely anaerobic (Hu et al., 2013). Also, rice contains highly expressed silicic acid transporters through which arsenite gets efficiently transported into the plant parts (Ma et al., 2007). The other staple crops such as wheat and maize are grown aerobically and, in such conditions, arsenate (As[V]) is present predominantly. As exists in several organic forms also such as monomethylarsonic acid, dimethylarsinic acid, trimethylarsine oxide (TMAO), and arsenobetaine (Awasthi et al., 2017; Majumdar and Bose, 2017). When humans are exposed to the As contaminated crops for a long period of time, it can lead to dysfunction of various organs and tissues in the human body, such as kidney, skin, gastrointestinal tract, and nervous system (Zavala et al., 2008; Islam et al., 2017). It can also cause cancers; the condition referred to as "arsenicosis" (Banerjee et al., 2013; Abdul et al., 2015). Due to its poisonous nature and carcinogenicity, efficient methods are needed to be developed for the remediation of As from the groundwater and the soil. Further, such variety of rice needs to be developed in which As accumulation is minimum (Tripathi et al., 2007).

Many methods have been developed for As remediation; these include various physical, chemical, and biological methods (Wan et al., 2020). However, the efficiency of various methods varies and depends on several factors of the site and As concentration. The physico-chemical methods include chemical precipitation, ion chelation, size exclusion methods, adsorption, immobilization, coagulation, and filtration (Cegłowski and Schroeder, 2015; Naseri et al., 2017; Duarte et al., 2009; Mahimairaja

Assisted Phytoremediation. DOI: https://doi.org/10.1016/B978-0-12-822893-7.00007-0

et al., 2005). These methods are efficient and used to remove As from contaminated sites. But these methods are cost-intensive and are also unable to remove As when it is present at very low concentrations (Hammaini et al., 2003). Further, physical and chemical treatment methods are not eco-friendly, cause damage to the fertile soil and changes in the microflora of the soil. The biological method for As remediation involves the use of living microorganisms and plants and even the dead biomass of various organisms. The use of microorganisms is known as bioremediation while the use of plants is called as phytoremediation. However, it is obvious that during phytoremediation, a number of rhizospheric and endophytic microorganisms naturally assist plants in the process of remediation. Phytoremediation is a cost-effective, eco-friendly, and sustainable method, which can be used to remove even very low concentrations of As (Majumdar et al., 2018; Jasrotia et al., 2017). Phytoremediation is made up of two words; "phyto" means plants and "remediation," which indicates clean-up of a contaminated site (Pilon-Smits, 2005). A successful plant for application in phytoremediation should possess some basic qualities like rapid growth, high biomass, tolerance to As and ability to take up and accumulate it in high quantities (Tripathi et al., 2008; de Souza et al., 2019; Pandey and Singh, 2015; Pandey and Bajpai, 2019). However, very few plant species are naturally present in the environment, which have high As accumulation capacity and optimum growth habits for successful removal of As in a timely manner. The lack of a desirable attribute of plants can lead to extended and very long time taking phytoremediation projects (Ernst, 2005; Pandey and Bajpai, 2019). Therefore, plants may sometime need to be assisted by various methods so as to increase their efficiency of As removal. These include the use of plant growth promoting microorganisms (PGPMs), fungi, biochar, etc. (Mesa et al., 2017; Franchi et al., 2017). Further, genetic engineering has been used to increase the As accumulation capacity and As tolerance of the plants to achieve the maximum removal of As in the minimum possible time (Nahar et al., 2017). This chapter presents the chemical, biological, and genetic engineering methods used to augment the phytoremediation capacity of plants for As removal from the contaminated sites.

9.2 Methods of phytoremediation

9.2.1 Phytoextraction

Phytoextraction is a method of phytoremediation in which plants uptake the contaminant from the medium via their roots and the contaminants are subsequently translocated for accumulation and storage in the shoot part of the plant (Yoon et al., 2006; Rafati et al., 2011). In this way, the metal present in soil is extracted with the help of plants and concentrated in specific harvestable tissues and can therefore be easily removed from the site. This indicates that higher the biomass of plants greater will be the metal removal. Apart from biomass, the plant should have a rapid growth rate so as to harvest the biomass at short intervals (Li et al., 2018). However, in case of rapidly growing and high biomass producing plants like Indian mustard, the cropping time is limited to a particular season. Hence, such crops can be grown only once in a year offsetting any benefits of metal removal from the site as it can again build up in the soil during rest of the year. To solve such situation, wetland plants like cat tails (*Typha latifolia*) can be used to take up As and other soil toxic elements (Sarkar et al., 2017). The use of perennial trees is hindered by the fact that they grow slowly and shed only a few leaves per year. However, if a plant can hyperaccumulate a metal beyond normal range, it can remove high amount of metal in one crop. The hyperaccumulator plant in case of As should accumulate metal in more than 1000 mg kg^{-1} dry weight (0.1% of biomass) in harvestable tissue (Souri et al., 2017). Only a few plants have been found to hyperaccumulate As, such as *Pteris vittata, P. cretica, Pityrogramma calomelanos,* and *Isatis capadoica* (Niazi et al., 2012; Karimi and Souri, 2015). *P. vittata* is the first identified As hyperaccumulator that can contain up to 2.2% of As in its biomass (Vetterlein et al., 2009; Xie et al., 2009).The prolific As accumulation ability of *P. vittata* comes from the fact that it can effectively transport As to the leaves and store it in vacuoles. The vacuolar sequestration of As in *P. vittata* occurs in ionic form as arsenite (As[III]) through As compounds resistance 3 (ACR3) transporter (Indriolo et al., 2010). This is in contrast to angiosperms that require arsenite to be complexed with thiols before they can be sequestered to vacuoles (Song et al., 2014). Nonetheless, some complexation of As with various ligands such as reduced glutathione (GSH) and phytochelatins (PCs) occurs in *P. vittata* also (Vetterlein et al., 2009). In addition, *P. vittata* shows high tolerance to As toxicity via the stimulation of various antioxidants, hormones, and changes in metabolic processes (Zemanova et al., 2019). The use of a plant for phytoextraction must also ensure that they should not be edible for the herbivores so as to avoid the entry of the toxic heavy metal into the food web (Mejare and Bulow, 2001; Pilon-Smits, 2005; Ali et al., 2013). It is important to note that hyperaccumulation of a metal can itself protect a plant from attack by insects and make them less tasty to animals (Behmer et al., 2005).

9.2.2 Phytostabilization

Phytostabilization is a method of phytoremediation in which the heavy metal contaminants are immobilized in the medium itself with the help of plants. The bioavailability of As depends on a number of factors, including pH and redox state (Eh), level of iron hydroxides and oxyhydroxides, level of other mineral elements like manganese and phosphate, organic matter content, etc. (Majumdar et al., 2020).The plants are known to excrete a number of chemicals to the soil through roots that mainly include various organic acids (citric acid [CA], malic acid) (Jones and Darrah, 1994; Jones, 1998). The secretion of acids alters the pH of soil and affects chemical and biochemical reactions. In this way, the growth of some plants can stabilize As in soil and restrict its access to crop plants in subsequent season. There are certain metallicolous plants, which are able to grow in highly As contaminated sites without its much accumulation in their tissues. Such plants like *Holcus lanatus, Silene vulgaris,* and *Agrostis tenuis* can tolerate excessively high As concentrations and still complete their life cycle. These plants accumulate As in high concentrations but at a slow rate and thus avoid toxic response (Hartley-Whitaker et al., 2001). Some other plants can uptake high quantities of As and accumulate it into the roots, basically precipitating As within the root zone. Thus, these plants decrease the transfer of As to shoots and also to the external medium (soil). In this way, As remains restricted in soil. But this solution is not a permanent solution for As removal from the contaminated site because it is only held in a place and not removed permanently from the contaminated site (Vangronsveld et al., 2009; Praveen et al., 2019). However, the use of such plants in between cropping seasons may restrict As in soil and may decrease As accumulation in crop plants at least in one or two subsequent seasons.

9.2.3 Phytovolatilization

As is known to exist in several inorganic and organic forms. TMAO and trimethyl arsine are the volatile species of As (Frankenberger and Arshad, 2002). Apart from As, mercury, and selenium also have volatile species. Phytovolatilization is a method of phytoremediation in which the plants uptake and transpire the contaminants into the atmosphere by converting the metals into lesser toxic organic and volatile forms and release the metal into the atmosphere through the leaves (Limmer and Burken,

2016). However, in case of As, the plant enzymes and processes involved in the conversion of As to volatile species are not discovered yet. It is known that the presence of organic As species in plants is due to direct uptake from soil where organic As species are formed in microbes and released to the soil (Wang et al., 2014). Hence, till date, it is known that it is the microbes in soil which actually convert As to organic and volatile species. Therefore, phytovolatilization strategy for As can only be achieved via co-inoculation of microbes or via transgenic development (Guarino et al., 2020). Further, this method may not be feasible for use in the fields as it releases As vapors into the atmosphere, which can further be transferred in the people working on and near the contaminated fields (Padmavathiamma and Li, 2007).

9.3 Assisted phytoremediation

Phytoremediation is a very effective way of clean-up of As contaminated fields, but it still lacks in many ways. Therefore, the present research aims to further augment phytoremediation efficiency by providing additional assistance to plants. To augment phytoremediation capacity of plants, chemicals, microbes, and genetic engineering may be used.

9.3.1 Chemical-assisted phytoremediation

In this, the plants used for phytoremediation purposes are assisted with various types of chemicals so as to increase the efficiency of the plants. The heavy metal concentration is not always very high in the soil. So to increase the availability of the metals to the plants, certain chemicals (chelating agents) are used which increase the phytoextraction capacity of the plants. Some of the chelating agents used are ethylenediamine tetraacetic acid (EDTA), diethylenetriamine pentaacetic acid, ethyleneglycol tetraacetic acid, and sodium dodecyl sulfate (Guo et al., 2014; Xiang et al., 2020). These chemicals increase the efficiency of phytoremediation, but they have several disadvantages as well. They tend to change the soil microflora, soil enzyme activities and can cause harm to the environment. EDTA for example is not only toxic but also a non-biodegradable compound which can persist in the environment for several months and can possibly get transferred in the food chain (Hasan et al., 2019). So instead of these harmful chemicals, organic chelating agents such as acetic acid, oxalic acid (OA), malic acid, and CA can be applied to increase the accumulation capacity of the plants used for phytoaccumulation purposes (Souza et al., 2013). Sometimes, the chelating agents can also enhance the bioavailability of other non-desirable elements as well. The increased available level of elements may not be fully absorbed by the plants. Such excess freely mobile metal becomes prone to reach to groundwater (Wang and Mulligan, 2006). Another approach of chemical assisted phytoremediation can be the use of plant growth stimulating chemicals and fertilizers (Chen and Cutright, 2002). In this way, if the plant growth and biomass generation would be high, the eventual removal in a particular time frame would also be high.

Biochar is a carbonaceous porous substance which is obtained through pyrolysis of plant and animal wastes under minimal oxygen supply (Paz-Ferreiro et al., 2014). It has a high surface area due to which the metal particles can easily adsorb onto the biochar particles, thereby immobilizing the toxic heavy metals. It has recently gathered attention due to its carbon sequestration potential, ability to improve the soil fertility, and bioenergy production. Simiele et al. (2020) amended former tin mine soil, contaminated with As, with biochar and iron sulfate and studied the effects of soil properties and the metal uptake capacity of three Salicaceae species. It was seen that the physico-chemical properties of the soil improved, and the uptake of metal also increased by the plant root, thereby increasing the phytostabilization capacity of the plant species. As the biochar application has many advantages, it can be used along with other techniques to increase the rate of bioremediation of heavy metals. Liang et al. (2019) conducted a pot experiment with multi-metal contaminated soil for assessing phytoremediation potential of *P. vittata* plants. They also assessed the improvement of phytoremediation in presence of biodegradable chelators, ethylenediamine disuccinic acid, CA, and OA. The addition of 2.5 mmol kg^{-1} OA was found to be beneficial in improving biomass of rhizoid and shoots, as well as As accumulation as compared to control. Soil microbial diversity was also found to improve upon OA addition. Jeong et al. (2014) demonstrated that siderophores from *Pseudomonas aeruginosa* complex As in solution. The pot experiments with the supplementation of siderophores showed 3.7-fold more As accumulation in plants than control *P. cretica* plants (Jeong et al., 2014).

Phosphate is another preferred choice by many researchers for enhancement of phytoremediation. Studies suggests that the application of phosphorus in the form of phosphate fertilizer (triple superphosphate [$Ca(H_2PO_4)_2.2H_2O$]) can effectively enhance plant's growth and also helps to make arsenate more available to the plant (Cao and Ma, 2004). Reports on using potassium dihydrogen phosphate (KH_2PO_4) assisting *B. juncea* and *B. napus* can enhance the efficacy of phytoremediation many folds with increase in plant's biomass (Niazi et al., 2017). A combination of bi-ammonium phosphate [$(NH_4)_2HPO_4$] and *L. albus* were found to be efficient in As removal from the soil microcosm (Tassi et al., 2004). In another study, *B. juncea, L. albus*, and *Helianthus annuus* were assisted by EDTA and phosphate (in the form of dipotassium hydrogen phosphate [K_2HPO_4]). The study showed that phosphate ions triggered the release of arsenate ions in the soil and made it bioavailable to the plant root system (Franchi et al., 2019). Barbafieri et al. (2017) assessed the effect of phosphate addition on the performance of selected plants with respect to growth and As removal in a set of experiments. It was found that addition of phosphate significantly improved the ability of plants (*L. albus, Helianthus annuus, B. juncea,* and *P. vittata*) to extract As from soil.

9.3.2 Microbe-assisted phytoremediation

Numerous microorganisms reside in soil everywhere in the world; be it even the contaminated soil. In As contaminated fields also, As-tolerant PGPMs have been found that can be used to increase the As accumulation capacity of the plants. PGPMs are cost-effective, safe, and sustainable method, and can help in decreasing the use of chemical fertilizers.

The interaction between the microorganisms and the plants in the rhizosphere is very important for proper growth of the plants as this interaction affects the uptake and transport of the nutrients from the soil into the plants (Ullah et al., 2015). The root exudates consist of organic metabolites such as phenolics, sterols, fatty acids, nucleotides, amino acids, sugars, and vitamins, which cause the interactions between the microbes and the plant roots. PGPMs help in the growth of plants and also reduce As stress (Ma et al., 2016). The most important factors which are essential for plant growth are phosphate, nitrogen, soil organic carbon, and PGPMs enhance the availability of N, P, and K nutrition for proper growth of the plant species (Duarah et al., 2011). They also cause improvement in the soil organic matter, which leads to enhancement of soil fertility (Khan et al., 2010; Rashid et al., 2016). The other factor needed for the plant growth is balanced phytohormone metabolism and response. Auxin, ethylene, gibberellin, and cytokinins are some of the important phytohormones, which are needed for proper plant growth. PGPMs are known to produce indole-3-acetic acid (IAA) and have 1-aminocyclopropane-1-carboxylic acid (ACC) deaminase activity (Etesami et al., 2014). They also regulate the level of abscisic acid and ethylene (Tsukanovaet al., 2017). Thus, PGPMs are known to regulate the level of various phytohormones in plants during As stress so as to sustain and improve plant's growth and As tolerance.

Further, microbes are known to metabolize inorganic As to organic less toxic As species, which are more mobile in the plants and reach at a greater speed in shoots of plants. Siderophores are iron-chelating ligands, which are produced in conditions when iron is present in very low concentrations. They enhance iron-uptake in the form of ferric ions (Fe^{3+}). There are three families of siderophore namely, hydroxamates, catecholates, and carboxylates. Fe forms iron plaque on rice roots under anaerobic conditions due to which the bioavailability of As decreases as it can bind strongly on the root surface itself (Upadhyay et al., 2018).

9.3.2.1 Types of microorganisms

PGPMs mainly consists of three broad categories of microorganisms that is, bacteria, fungi, and algae. These microorganisms play a vital role in As cycling in the environment; therefore, these microorganisms can be utilized to enhance the phytoremediation potential of the plants in a sustainable way (Table 9.1).

Table 9.1 Summary of examples of PGPMs used to reduce arsenic stress and arsenic accumulation in plants.

Plant growth promoting microorganisms (PGPMs)	Targeted plant species	Resulted effect due to the use of PGPMs	References
Bacteria			
Pseudomonas sp., Delftia sp., Bacillus sp., Variovorax sp., Pseudoxanthomonas sp.	*Pteris vittata*	As was removed from the medium along with the increased biomass of the plant.	Lampis et al. (2015)
Delftia, Comamonas, Streptomyces	*Pteris vittata*	As removal increased from 7% to 15% after bacterial inoculation. The biomass of the plant was increased by 50%.	Yang et al. (2012)
Ensifer adhaerens, Rhizobium herbae, Variovorax paradoxus, Phyllobacterium myrsinacearum	*Betula celtiberica*	The consortia of the bacteria used improved the accumulation of As in the plant.	Mesa et al. (2017)
Kocuria flava, Bacillus vietnamensis	*Oryza sativa*	The As(III) uptake was decreased and the growth of the plant was increased in As-amended soil.	Mallick et al. (2018)
Serratia liquefaciens	*Oryza sativa*	Decreased As accumulation was seen in the rice grains due to decreased translocation of As from the roots to the grains.	Cheng et al. (2020)
Pseudomonas vancouverensis	*Pteris vittata, Pteris multifida*	The capacity of As removal was enhanced due to biotransformation of the rhizospheric As by the bacteria.	Yang et al. (2020)
Fungi			
Rhizoglomus intraradices, Glomus etunicatum	*Triticum aestivum*	Enhanced growth was seen in the fungal inoculated plants in comparison with the uninoculated plants.	Sharma et al. (2017)

Table 9.1 Summary of examples of PGPMs used to reduce arsenic stress and arsenic accumulation in plants. ***Continued***

Plant growth promoting microorganisms (PGPMs)	Targeted plant species	Resulted effect due to the use of PGPMs	References
Glomus geosporum, Glomus versiforme, Glomus mosseae	Isolated from *Pteris vittata,* used for *Oryza sativa*	As accumulation in the rice grains was reduced, thereby improvement in the quality of rice grains was seen.	Wu et al. (2015), Chen et al. (2006), Chan et al. (2013)
Glomus mosseae, Glomus intraradices, Glomus etunicatum	*Pityrogramma calomelanos, Tagetes erecta, Melastoma malabathricum*	The phytoavailability of As was increased, helping in the removal of As in the soil, along with the increased plant biomass.	Jankong et al. (2007)
Piriformospora indica	*Oryza sativa*	Decrease in the bioavailability of free As in the soil was seen. Also, the fungi modulated the antioxidative system of the plants to decrease As stress.	Mohd et al. (2017)
Trichoderma sp.	*Helianthus annuus*	Biomass production and As accumulation in plants was increased due to siderophore production, IAA, ACC-deaminase activity, and phosphate solubilization.	Govarthanan et al. (2018)
Humicola sp.	*Bacopa monnieri*	As content was reduced due to volatilization of As from the leaves.	Tripathi et al. (2020)
Algae			
Chlorodesmis sp., Cladophora sp.	*Eichhornia crassipes*	*Cladophora* reduced As concentration from 6 to 0.1 mg L^{-1} and *Chlorodesmis* reduced As concentration by 40%–50% in 10 days.	Jasrotia et al. (2015)
Chlorella vulgaris, Nannochloropsis sp.	*Oryza sativa*	As accumulation was reduced in the roots and shoots of the plants after algal inoculation in the rice.	Upadhyay et al. (2016)
Anabaena sp.	*Oryza sativa*	Improvement in the growth of rice plant was seen along with the decreased accumulation of As.	Ranjan et al. (2018)
Pseudomonas putida and *Chlorella vulgaris*	*Oryza sativa*	As concentration was reduced in the root and shoot in the inoculated plants. Also enhanced plant growth was seen after inoculation with the algal consortium.	Awasthi et al. (2018)

9.3.2.1.1 Bacteria

Plant growth promoting bacteria (PGPB) helps in proper plant growth through reduction of metal stress in the rhizospheric zone of plants (de Andrade et al., 2019). PGPB improve the growth of hyperaccumulator plants which have less biomass and cannot have proper growth. PGPB also increase the phytoremediation capacity by increasing plant tolerance. PGPB produce siderophores in nutrient deficient conditions and thereby increase As solubility and iron supply (Khanna et al., 2019). There are two types of PGPB: endophytic ones, which colonize the internal plant tissues, and rhizospheric bacteria residing in the rhizosphere of plants (Afzal et al., 2019; Soldan et al., 2019; Ullah et al., 2015). PGPR secrete ACC-deaminase, which acts on and degrades ACC, a precursor of ethylene, and thus reduce ethylene levels and promote growth of the plant (Ma et al., 2016; Alka et al., 2020). The important PGPB consists of *Frankia, Rhizobium, Klebsiella, Pseudomonas, Bacillus,* and *Arthrobacter* (Checcucci et al., 2017).

Lampis et al. (2015) used *Pseudomonas* sp., *Delftia* sp., *Bacillus* sp., *Variovorax* sp., and *Pseudoxanthomonas* sp., in As-contaminated medium, and found the biomass of *P. vittata* to be increased along with As removal from the medium due to production of IAA and siderophores by the bacterial strains. Similarly, Yang et al. (2012) inoculated *P. vittata* plants with *Delftia, Comamonas,* and *Streptomyces* and found about 50% increase in biomass of plants after 4 months. The As removal increased from 7% without bacterial supplementation to 15% with bacterial strains. An increase in As accumulation and biomass in Indian mustard plants was observed upon application of *Staphylococcus aureus* isolated from As contaminated soil (Srivastava et al., 2013). *Acinetobacter*, which causes As(III) oxidation and *Comamonas, Flavobacterium, Pseudomonas,* and *Staphylococcus,* which cause As(V) reduction and As(III) oxidation were isolated from the rhizosphere of *P. vittata* (Wang et al., 2012; Singh et al., 2015). These microbes can act as resource material to be utilized with other high biomass plants for enhancing phytoremediation efficiency. The possibility of combined chemical (thiosulfate) and microbial (consortium of bacteria) assisted phytoremediation of As by *Brassica juncea* and *Lupinus albus* was tested by Franchi et al. (2017). Thiosulfate addition was found to significantly increase the mobilization of As and increase its bioavailability to plants. The addition of microbial consortium along with thiosulfate increased accumulation up to 85% for As by plants. Thiosulfate breaks into sulfur and sulfate, which in turn can compete with As ions in soils to get associated with Fe-oxyanions and release As in the bioavailable form for uptake by the plants (Fukushi and Sverjensky, 2007; Petruzzelli et al., 2014). Mesa et al. (2017) identified autochthonous As tolerant rhizospheric and endophytic bacteria from Betula celtiberica. The microbiome of plants was found to be dominated by Flavobacteriales, Burkholderiales, and Pseudomonades with predominance of Pseudomonas and Flavobacterium genera. Out of several bacteria, four bacteria were tested at field-scale experiment. It was found that the siderophore and IAA producing endophytic bacterial consortium (*Ensifer adhaerens, Rhizobium herbae, Variovorax paradoxus,* and *Phyllobacterium myrsinacearum*) enhanced As accumulation in Betula celtiberica plants. This report suggested a prominent relation of endophytic bacteria responsible for the As tolerance activity as well as growth promotion to the associated plant.

Kocuria flava and *Bacillus vietnamensis* were used in As-amended soil and it was seen that As(III) uptake was significantly reduced and the plant (*Oryza sativa*) growth was improved. It was possible due to reduced bioavailability of As (Mallick et al., 2018).From aqueous system, As removal by *Echinodorus cordifolius* associated with an endophytic species *Arthrobacter creatinolyticus* has been assessed. The plants were shown to have great potential to remove considerable amount of As from water (Prum et al., 2018). Mukherjee et al. (2018) isolated endophytic bacteria having tolerance to As from *Lantana camara* and *Solanum nigrum* possessing varying As tolerance and identified several potential species including *Kocuria* sp., *Enterobacter* sp., and *Kosakonia* sp. The application of microbial consortium improved plant growth and increased accumulation of As along with its root to shoot translocation. The effects were attributable to holistic biochemical and molecular changes including the altered expression of aquaporins and a MRP transporter. Cheng et al. (2020) isolated *Serratia liquefaciens* F2, an As-resistant facultative endophytic bacterial strain and applied to As contaminated medium. It was found that rice plants inoculated with strain F2 showed lower As content in the rice grain reduced. The strain F2 thus reduced the available As in the soil and reduced the translocation of As from the roots to the rice grains. Such microbes can be used in the strategy of phytostabilization to restrict As in root zone and allowing good plant growth. Strains of *Stenotrophomonas maltophilia* and *Agrobacterium* sp. were used as a consortium with *Arundo donax* to assess the As removal via phytovolatilization. About 75% of the As from the experimental system was removed with this system (Guarino et al., 2020). Among several rhizospheric microbes analyzed from the rice plant rhizosphere, some species of *Bacillus* and *Pseudomonas* were found to be efficient in As removal from the rhizospheric soil with secretion of plant growth enhancing substances (Xiao et al., 2020). In a recent study, *Pseudomonas vancouverensis* strain m318 has been found to effectively biotransform rhizospheric As and increase the As removal capacity of the associated ferns *P. vittata* and *P. multifida* (Yang et al., 2020). *Actinomycetales*, *Betaproteobacteria*, and *Bacilli* were assessed in association with *B. juncea* and *L. albus* to remove soil As that showed a potential removal efficacy (Alka et al., 2020).

9.3.2.1.2 Fungi

The fungi used for enhancing phytoremediation basically consist of two types, arbuscular mycorrhizal fungi and rhizospheric fungi. The AM fungi help in enhancing phytoremediation by increasing the absorption area of roots and increasing the contaminant uptake capacity of the plant (Gamalero et al., 2009; Cabral et al., 2015). The AM fungi can also produce organic acids, glycoproteins and cyclosporin to form metal complexes which affect the availability of the metal particles for translocation (Khan, 2005). Sharma et al. (2017) used two fungi species, *Rhizoglomus intraradices* and *Glomus etunicatum* to decrease the As stress in *Triticum aestivum.* The plants inoculated with mycorrhiza showed better growth than the plants not inoculated with the fungal strains. The mycorrhizal inoculation helped in fighting the phosphorus deficiency caused due to As, thereby maintaining the P:As ratio (He and Lilleskov, 2014). An inoculum of AM fungi (*Glomus* spp. and *Acaulospora* spp.) from As-contaminated soils increased As accumulation in shoots of maize plants (Bai et al., 2008). Total As accumulations was also found to be increased in *Glomus mosseae*-inoculated white clover and ryegrass plants (Dong et al., 2008). Three AM fungi, *Glomus geosporum, Glomus versiforme,* and *Glomus mosseae* isolated from *Pteris vittata* were used for rice plant in As-contaminated fields. Rice plants showed improved grain quality due to less As accumulation in the rice grains through enhancement of P:As ratios (Wu et al., 2015; Chen et al., 2006; Chan et al., 2013). *Glomus mosseae, G. intraradices,* and *G. etunicatum* were reported to enhance associated plants' biomass, *Pityrogramma calomelanos*, *Tagetes erecta,* and *Melastoma malabathricum,* and also removed As from the soil by making it more phytoavailable (Jankonget al., 2007).

Piriformospora indica when inoculated along with rice plants decreased the As bioavailability in soil and also attuned antioxidant responses of plants to reduce As stress (Mohd et al., 2017). As per hypothesis, the As tolerance can be induced by the AM fungi by providing high phosphorus availability to the plant's root system (He and Lilleskov, 2014). Several species of *Glomus sp.* (*Glomus etunicatum, Glomus constrictum, and Glomus mosseae*) have been observed to do so in association with maize (Xia et al., 2007; Yu et al., 2010). *Trichoderma* sp. MG is one of the filamentous fungi that can used effectively with *Helianthus annuus* for the enhancement of phytoremediation perspective as reported by Zeng et al. (2010) and Govarthanan et al. (2018). Govarthanan et al. (2018) studied the impact of *Trichoderma* sp. inoculation to *Helianthus annus* growing in As-amended soil. A significant increase in biomass production and As accumulation in plants was seen in As-amended soil due to siderophore production, IAA, ACC-deaminase activity, and phosphate solubilization. In an indirect way, filamentous fungi can change soil As to methylated species and subsequently these organic forms of As can be taken up by plants. *Scrophulariopsis brevicaulis, Aspergillus niger*, and *Gliocladium roseum* are some of the filamentous fungi that can transform inorganic As to methylated forms (Dabrowska et al., 2012) and reports suggests that several plant species can take up organic form of As with variable preference (Raab et al., 2007). Tripathi et al. (2020) isolated 45 fungal strains from As contaminated sites of West Bengal, India. Out of these, some of the highly tolerant fungal isolates could tolerate up to 5000 mg L^{-1} As(V) and accumulated As in range of 0.146 to 11.36 g kg^{-1} when subjected to As stress of 10 mg L^{-1} for 21 d. As volatilization (0.05 to 53.39 mg kg^{-1}) by fungal strains was also noticed. The highest As volatilizing strain 2WS1 was identified as *Humicola* sp. Tripathi et al. (2020) also conducted a pot experiment with best fungal strains to evaluate effects on *Bacopa monnieri*. The lowest As content in leaf tissues was seen in presence of *Humicola* strain presumably due to As volatilization from leaves. Therefore, fungi can also find application for supplementation to plants for both phytoextraction and phytostabilization angles.

9.3.2.1.3 Algae

Algal use for assisting phytoremediation of metal contaminants is an eco-friendly and sustainable process. There are various functional groups present on the cell walls of algae such as sulphuryl, hydroxyl, imidazole, carboxyl, phosphoryl, and amine, which help in the biosorption of heavy metals from the contaminated surrounding into the algae (Kaplan, 2013). During the process of As complexation, production of PCs is very important which help in metal complexation and also in averting the oxidative stress (Munoz et al., 2014). Due to the high biomass production, *Spirogyra hyaline* has been suggested for phytoremediation of multiple toxic elements, including As (Kumar and Oommen, 2012). Considering the relation of phosphorus to As, *Chlamydomonas reinhardtii* and *Scenedesmus obliquus* were investigated under phosphorus supplemented and deprived conditions where these two alga were found to be efficient in As uptake and removal under phosphorus supplemented condition (Wang et al., 2013). *Chara vulgaris* was tested for the As removal from naturally contaminated water bodies and was found to be effective by reduction of 66.25% As from the water (Fereshteh et al., 2007). In another recent study, *Chara vulgaris* has been used in combination of water hyacinth and vetiver grass to remove As from contaminated soils (Taleei et al., 2019).In a report, four taxa, chlorophyta, cyanophyta, euglenophyta, and heterokontophyta were found to be efficient in As phytoremediation with high bioaccumulation rate (Mitra et al., 2012). In another report by Yin et al., (2012), *Synechocysis* sp. PCC6803, a single-cell algae, was shown to have potential for phytoremediation with ability of rapid As oxidation from As(III) to As(V) and subsequent accumulation intracellularly. Jasrotia et al. (2015) tested phytoremediation efficiency of *Eichhornia crassipes* and two algae (*Chlorodesmis* sp. and *Cladophora* sp.). They found that *Cladophora* could survive up to 6 mg L^{-1} As and reduce As concentration of medium from 6 to 0.1 mg L^{-1} in 10 d. *Chlorodesmis* could reduce As concentration by 40%–50% in 10 d. A consortium of *Pseudomonas putida, Chlorella vulgaris* and *Nannochloropsis* sp. when inoculated with rice (*Oryza sativa*) showed reduction in As concentration in the plant, thereby improving the plant growth (Upadhyay et al., 2016; Awasthi et al., 2018). *Anabaena,* when grown along with rice plants, decreased the As(III) and As(V) accumulation and also improved the growth of the plant by improving the activities of nitrogen metabolism genes. The activity of antioxidant enzymes reduced the expression of As transporter genes (Ranjan et al., 2018). The algal-assisted phytoremediation is a very effective process (Mitra et al., 2017), but further research needs to be done in this prospect.

9.3.3 Genetic engineering-assisted phytoremediation

Genetic engineering has become an integral part of scientific research, due to which we have been able to solve many of our agricultural problems. It has been also applied to solve the problem of environmental pollution, especially in soil and water. As contamination is one such problem and the most appropriate way of solving this problem is phytoremediation. Plants used in phytoremediation are efficient enough, but their efficiency can be increased with the help of genetic engineering (Zhu and Rosen, 2009). There are various genes present in plants which are responsible for As uptake, translocation, and detoxification and also for regulation of various responses of plants (Chen et al., 2013; Chauhan et al., 2020) (Fig. 9.1). These genes can be utilized and transgenic plants can be used to reduce As contamination from the environment (Tong et al., 2004).

9.3.3.1 Transporters involved in arsenic uptake and translocation

There are various types of transporters present in the plant tissues that are responsible for the uptake of As by the root tissues, translocation of As from the root to the shoot, transfer of As from one cell to another, and in compartmentalization to vacuole. These transporters are aquaglyceroporins for As(III) and phosphate transporters (PHTs) for As(V), which regulates the concentration of As in the plant

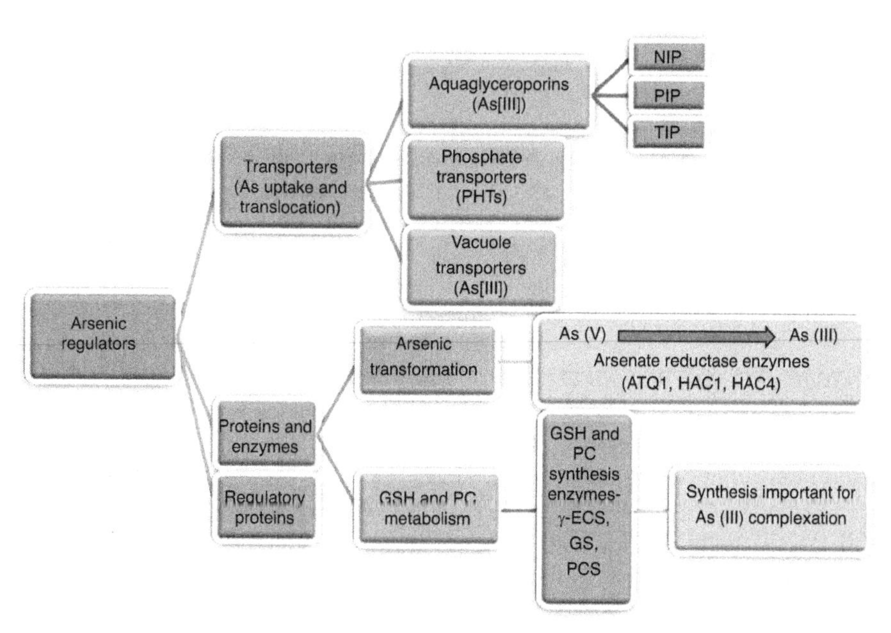

FIG. 9.1 Summary of transporters, proteins and enzymes responsible for the regulation of arsenic accumulation in the plants. The genes involved can be utilised to make transgenic plants to ameliorate arsenic-stress growing in the arsenic-contaminated regions.

ATQ1, arsenate tolerance QTL1; GS, glutathione synthetase; GSH, glutathione, HAC1, high arsenic content 1; HAC4, high arsenic content 4; NIP, nodulin 26-like intrinsic protein, PC, phytochelatin; PCS, phytochelatin synthase; PIP, plasma membrane intrinsic protein; TIP, tonoplast intrinsic protein; γ-ECS, γ-glutamylcysteine synthetase.

tissues and organs (Awasthi et al., 2017; Majumdar et al., 2020). As(III) is transported via aquaglyceroporins and the major classes of aquaglyceroporins are nodulin 26-like intrinsic protein (NIP), plasma membrane intrinsic proteins (PIP) and tonoplast intrinsic proteins (TIP) (Zhao et al., 2010; Chen et al., 2017; Majumdar and Bose, 2018). *PvTIP4;1*, a TIP class transporter obtained from *P. vittata. PvTIP4;1* was expressed in *Arabidopsis thaliana* and showed an increase in As accumulation in the transgenic plant in comparison to wild type, but with a decrease in root length and shoot biomass (He et al., 2016). *OsPIP2;4, OsPI2;6,* and *OsPIP2;7*, PIP class transporters from rice, were overexpressed in rice. The transgenic plant showed improved As tolerance in comparison with wild type. One more transporter, under the family of natural resistance-associated macrophage proteins (*OsNRAMP1*) from rice, is involved in the transport of As(III). *OsNRAMP1* was expressed in *A. thaliana* and it resulted in higher biomass in the transgenic plant in comparison with the wild type with increased As(III) tolerance (Tiwari et al., 2014). PHTs are the type of transporters which are involved in the uptake of As(V) (Ai et al., 2009; Kumari et al., 2018). *OsPHR2* is a phosphate starvation response 2 gene found in rice plants. This gene along with *OsPT8* (phosphate transporter 8) was over-expressed in *O. sativa* and the transgenic plants showed higher phosphate and As(V) uptake as well as translocation (Wu et al., 2011).

Vacuole transporters are another kind of transporters which are very important for constraining the movement of As inside the plant tissues by performing the role of a mediator that helps in the transport of As(III) or As(III)-PC complexes into the vacuoles. Complexes of As with PCs and GSH can be pumped and sequestered to the vacuoles by a group of proteins from the ATP-binding cassette (ABC) superfamily (Tommasini et al., 1998). *PvACR3;1*is a vacuolar transporter which is derived from *P. vittata* and transports As(III) into the vacuoles (Indriolo et al., 2010). When this transporter gene was expressed in *A. thaliana* and *Nicotiana tabacum,* it was seen that the translocation of As(III) to the shoots was reduced and higher retention of As(III) was seen in the roots of the transgenic plants in comparison with the wild type plants (Chen et al., 2017). ABC family transporter, *OsABCC1* and a vacuolar transporter, yeast cadmium factor 1 (*ScYCF1*) derived from *O. sativa* and *S. cerevisiae,* respectively were co-overexpressed in rice plant and the plants showed reduction in As concentration in rice grains (Deng et al., 2018). Therefore, tissue specific expression of important transporter genes can be utilized to develop transgenic plants with increased accumulation of As.

9.3.3.2 Proteins and enzymes involved in arsenic transformation and glutathione and phytochelatin metabolism

As transformation involves the processes which enable the availability of As in such a way that it is ready for either efflux, thiol complexation, or vacuolar transport (Upadhyay et al., 2018). The reduction of As(V) to As(III) occurs enzymatically by arsenate reductase enzymes (Mukhopadhyay and Rosen, 2002). As(III) is important for complexation purposes and for vacuolar transport. The major arsenate reductase enzymes are arsenate tolerance QTL1 (ATQ1) and high arsenic content 1 (HAC1), and HAC4 (Sanchez-Bermejo et al., 2014; Chao et al., 2014; Shi et al., 2016; Xu et al., 2017).In *O. sativa, OsHAC1;1* and *OsHAC1;2* are the two genes, which regulate the As(V) reduction into As(III). The overexpression of these two genes resulted in reduced accumulation of As in the shoots because of the increased efflux of As(III) into the external medium due to enhanced reduction of As(V) to As(III) (Shi et al., 2016). *OsHAC4* overexpression reduced As concentration in the rice plants and increased As tolerance whereas its knock-out mutant showed opposite results (Xu et al., 2017). In a study by Dhankher et al. (2002), *Escherichia coli*arsenate reductase gene (*arsC*) and γ-glutamylcysteine synthetase (γ-ECS) were overexpressed in *A. thaliana,* and this resulted in the coupled As(V) to As(III) reduction followed by complexation with stimulated GSH biosynthesis and thus a higher As accumulation in plants. *PvGrx5* is a gene regulating the activity of glutaredoxin in *P. vittata*. When this gene was expressed in *A. thaliana*, it was seen that the transgenic plants showed improvement in As(V) reduction and cellular As(III) levels were also changed along with increase in As tolerance (Sundaram et al., 2009).

As(III) can make complexes with thiol (-SH) containing molecules such as PCs, GSH, and cysteine (Kumar et al., 2019). The tripeptide GSH gets converted into PCs with the help of the enzyme, phytochelatin synthase (PCS) (Schmöger et al., 2000). GSH and PCs are very important for As(III) complexation process and the enzymes which are important for their synthesis are γ-glutamylcysteine synthetase (γ-ECS), glutathione synthetase (GS), and PCS (Tripathi et al., 2007). Therefore the expression of these enzymes has been studied to increase the As accumulation capacity of the plants and to ameliorate As stress (Kumari et al., 2018). *GSH1* gene, responsible for the production of γ-glutamylcysteine synthetase enzyme in *S. cerevisiae*, when expressed in *A. thaliana* along with *AsPCS1* (isolated from *Allium sativum*) showed more tolerance toward As in comparison to the wild type plants. The dual gene transgenic plant was more As tolerant than the single gene transformants (Guo et al., 2008). *CdPCS1*, a PCS producing gene in *C. demersum*, when overexpressed in *Oryza sativa*, showed improved PCS activity leading to more sequestration of As(III) in root and shoot and reduced As content in rice grains (Shri et al., 2014).

9.3.3.3 Regulatory proteins

Other than the enzymes, transporters and proteins, there are various regulatory proteins, hormones, transcription factors etc., which are involved in the response mechanism of the plant to As stress (Di and Tamas, 2007). *MIR528* is a miRNA528 gene which is expressed in *O. sativa*. This gene was overexpressed in the rice plant, and the transgenic plants were found to be more sensitive to As(III) in comparison to the wild type plants (Liu et al., 2015). WRKY6 is regulatory transcription factor involved in suppression of *PHT1;1* expression in response to As(V) in *A. thaliana* plants. It was found that WRKY6 overexpression in *Arabidopsis* enhanced tolerance to As(V) and root length was markedly increased (Castrillo et al., 2013).

9.4 Conclusions and future perspectives

As contamination in the environment is a worldwide problem that needs to be tackled efficiently to prevent the ill effects of As in the environment, especially in the soil due to which the soil quality gets degraded and the soil microflora gets changed. Also, As can cause various health effects in humans. Therefore, various methods have been researched for As remediation. The conventional physical and chemical methods remove As from the environment, but they have their own disadvantages such as high cost, quality reduction of the soil and water, and are not sustainable methods. Phytoremediation is the most eco-friendly, cost-effective, and sustainable approach toward As bioremediation, but the efficiency of the phytoremediation methods alone is not sufficient to ameliorate As stress in the environment. Therefore, phytoremediation needs to be assisted with various ways including chemicals, microorganisms, and via transgenic plant development. Further research should be done toward the use of consortia of bacteria-fungi, bacteria-algae, fungi-algae, bacteria-fungi-algae, and to see whether the consortia based approach is more efficient than the use of single type of microorganism. The use of transgenic plants is also a successful and reliable option to reduce As stress in the environment due to its high specificity and sensitivity. Future research needs to focus on integration of various suggested approaches into one platform to achieve greater As phytoremediation in lesser time than that achieved via a single approach.

References

Abdul, K.S.M., Jayasinghe, S.S., Chandana, E.P., Jayasumana, C., De Silva, P.M.C., 2015. Arsenic and human health effects: a review. Environ. Toxicol. Pharmacol. 40 (3), 828–846.

Afzal, I., Shinwari, Z.K., Sikandar, S., Shahzad, S., 2019. Plant beneficial endophytic bacteria: Mechanisms, diversity, host range and genetic determinants. Microbiol. Res. 221, 36–49.

Ai, P., Sun, S., Zhao, J., Fan, X., Xin, W., Guo, Q., Yu, L., Shen, Q., Wu, P., Miller, A.J., Xu, G., 2009. Two rice phosphate transporters, OsPht1; 2 and OsPht1; 6, have different functions and kinetic properties in uptake and translocation. Plant J. 57 (5), 798–809.

Ali, H., Khan, E., Sajad, M.A., 2013. Phytoremediation of heavy metals—concepts and applications. Chemosphere 91 (7), 869–881.

Alka, S., Shahir, S., Ibrahim, N., Chai, T.T., Bahari, Z.M., Manan, F.A., 2020. The role of plant growth promoting bacteria on arsenic removal: a review of existing perspectives. Environ. Technol. Innov. 17, 100602.

Awasthi, S., Chauhan, R., Dwivedi, S., Srivastava, S., Srivastava, S., Tripathi, R.D., 2018. A consortium of alga (Chlorella vulgaris) and bacterium (Pseudomonas putida) for amelioration of arsenic toxicity in rice: a promising and feasible approach. Environ. Exp. Bot. 150, 115–126.

Awasthi, S., Chauhan, R., Srivastava, S., Tripathi, R.D., 2017. The journey of arsenic from soil to grain in rice. Front. Plant Sci. 8, 1007.

Bai, J.F., Lin, X.G., Yin, R., Zhang, H.Y., Wang, J.H., Chen, X.M., Luo, Y.M., 2008. The influence of arbuscular mycorrhizal fungi on As and P uptake by maize (*Zea mays* L.) from As-contaminated soils. Appl. Soil Ecol. 3, 137–145.

Banerjee, M., Banerjee, N., Bhattacharjee, P., Mondal, D., Lythgoe, P.R., Martínez, M., Pan, J., Polya, D.A., Giri, A.K., 2013. High arsenic in rice is associated with elevated genotoxic effects in humans. Sci. Rep. 3, 2195.

Barbafieri, M., Pedron, F., Petruzzelli, G., Rosellini, I., Franchi, E., Bagatin, R., Vocciante, M., 2017. Assisted phytoremediation of a multi-contaminated soil: Investigation on arsenic and lead combined mobilization and removal. J. Environ. Manage. 203, 316–329.

Behmer, S.T., Lloyd, C.M., Raubenheimer, D., Clark, S.J., Knight, J., Leighton, R.S., Harper, F.A., Smith, J.A.C., 2005. Metal hyperaccumulation in plants: mechanisms of defence against insect herbivores. Funct. Ecol. 19 (1), 55–66.

Cabral, L., Soares, C.R.F.S., Giachini, A.J., Siqueira, J.O., 2015. Arbuscular mycorrhizal fungi in phytoremediation of contaminated areas by trace elements: mechanisms and major benefits of their applications. World J. Microbiol. Biotechnol. 31 (11), 1655–1664.

Cao, X., Ma, L.Q., 2004. Effects of compost and phosphate on plant arsenic accumulation from soils near pressure-treated wood. Environ. Pollut. 132 (3), 435–442.

Castrillo, G., Sánchez-Bermejo, E., de Lorenzo, L., Crevillén, P., Fraile-Escanciano, A., Mohan, T.C., et al., 2013. WRKY6 transcription factor restricts arsenate uptake and transposon activation in Arabidopsis. Plant Cell 25, 2944–2957.

Cegłowski, M., Schroeder, G., 2015. Removal of heavy metal ions with the use of chelating polymers obtained by grafting pyridine–pyrazole ligands onto polymethylhydrosiloxane. Chem. Eng. J. 259, 885–893.

Chan, W.F., Li, H., Wu, F.Y., Wu, S.C., Wong, M.H., 2013. Arsenic uptake in upland rice inoculated with a combination or single arbuscular mycorrhizal fungi. J. Hazard. Mater. 262, 1116–1122.

Chao, D.Y., Chen, Y., Chen, J., Shi, S., Chen, Z., Wang, C., Danku, J.M., Zhao, F.J., Salt, D.E., 2014. Genome-wide association mapping identifies a new arsenate reductase enzyme critical for limiting arsenic accumulation in plants. PLoS Biol. 12 (12), e1002009.

Chauhan, R, Awasthi, S, Indoliya, Y, Chauhan, AS, Mishra, S, Agrawal, L, Srivastava, S, Dwivedi, S, Singh, PC, Mallick, S, Chauhan, PS, Pande, V, Chakrabarty, D, Tripathi, RD, 2020. Transcriptome and proteome analyses reveal selenium mediated amelioration of arsenic toxicity in rice (*Oryza sativa* L.). J. Hazard. Mater. 390, 122122.

Checcucci, A., DiCenzo, G.C., Bazzicalupo, M., Mengoni, A., 2017. Trade, diplomacy, and warfare: the quest for elite rhizobia inoculant strains. Front. Microbiol. 8, 2207.

Chen, B.D., Zhu, Y.G., Smith, F.A., 2006. Effects of arbuscular mycorrhizal inoculation on uranium and arsenic accumulation by Chinese brake fern (Pteris vittata L.) from a uranium mining-impacted soil. Chemosphere 62 (9), 1464–1473.

Chen, H., Cutright, T.J., 2002. The interactive effects of chelator, fertilizer, and rhizobacteria for enhancing phytoremediation of heavy metal contaminated soil. J. Soils Sediments 2 (4), 203–210.

Chen, Y., Hua, C.Y., Jia, M.R., Fu, J.W., Liu, X., Han, Y.H., Liu, Y., Rathinasabapathi, B., Cao, Y., Ma, L.Q., 2017. Heterologous expression of Pteris vittata arsenite antiporter PvACR3; 1 reduces arsenic accumulation in plant shoots. Environ. Sci. Technol. 51 (18), 10387–10395.

Chen, Y., Sun, S.K., Tang, Z., Liu, G., Moore, K.L., Maathuis, F.J., Miller, A.J., McGrath, S.P., Zhao, F.J., 2017. The Nodulin 26-like intrinsic membrane protein OsNIP3; 2 is involved in arsenite uptake by lateral roots in rice. J. Exp. Bot. 68 (11), 3007–3016.

Chen, Y., Xu, W., Shen, H., Yan, H., Xu, W., He, Z., Ma, M., 2013. Engineering arsenic tolerance and hyperaccumulation in plants for phytoremediation by a PvACR3 transgenic approach. Environ. Sci. Technol. 47 (16), 9355–9362.

Cheng, C., Nie, Z.W., He, L.Y., Sheng, X.F., 2020. Rice-derived facultative endophytic Serratia liquefaciens F2 decreases rice grain arsenic accumulation in arsenic-polluted soil. Environ. Pollut. 259, 113832.

Dabrowska, B.B., Vithanage, M., Gunaratna, K.R., Mukherjee, A.B., Bhattacharya, P., 2012. Bioremediation of arsenic in contaminated terrestrial and aquatic environmentsEnvironmental Chemistry for a Sustainable World. Springer, Dordrecht, pp. 475–509.

de Andrade, F.M., de Assis Pereira, T., Souza, T.P., Guimarães, P.H.S., Martins, A.D., Schwan, R.F., Pasqual, M., Dória, J., 2019. Beneficial effects of inoculation of growth-promoting bacteria in strawberry. Microbiol. Res. 223, 120–128.

de Souza, T.D., Borges, A.C., Braga, A.F., Veloso, R.W., de Matos, A.T., 2019. Phytoremediation of arsenic-contaminated water by Lemna Valdiviana: an optimization study. Chemosphere 234, 402–408.

Deng, F., Yamaji, N., Ma, J.F., Lee, S.K., Jeon, J.S., Martinoia, E., Lee, Y., Song, W.Y., 2018. Engineering rice with lower grain arsenic. Plant Biotechnol. J. 16 (10), 1691–1699.

Di, Y., Tamás, M.J., 2007. Regulation of the arsenic-responsive transcription factor Yap8p involves the ubiquitin-proteasome pathway. J. Cell Sci. 120 (2), 256–264.

Dhankher, O.P., Li, Y., Rosen, B.P., Shi, J., Salt, D., Senecoff, J.F., Sashti, N.A., Meagher, R.B., 2002. Engineering tolerance and hyperaccumulation of arsenic in plants by combining arsenate reductase and γ-glutamylcysteine synthetase expression. Nat. Biotechnol. 20, 1140–1145.

Dong, Y, Zhu, Y.G., Smith, F.A., Wang, Y.S., Chen, B.D., 2008. Arbuscular mycorrhiza enhanced arsenic resistance of both white clover (*Trifoliumrepens*Linn.) and ryegrass (*Lolium perenne*L.) plants in an arsenic-contaminated soil. Environ. Pollut. 15, 174–181.

Duarah, I., Deka, M., Saikia, N., Boruah, H.D., 2011. Phosphate solubilizers enhance NPK fertilizer use efficiency in rice and legume cultivation. 3 Biotech 1 (4), 227–238.

Duarte, A.A., Cardoso, S.J., Alçada, A.J., 2009. Emerging and innovative techniques for arsenic removal applied to a small water supply system. Sustainability 1 (4), 1288–1304.

Ernst WHO, 2005. Phytoextraction of mine wastes: Opinion and impossibilities. Chem Erde-Geochem 65, 29–42.

Etesami, H., Hosseini, H.M., Alikhani, H.A., Mohammadi, L., 2014. Bacterial biosynthesis of 1-aminocyclopropane-1-carboxylate (ACC) deaminase and indole-3-acetic acid (IAA) as endophytic preferential selection traits by rice plant seedlings. J. Plant Growth Regul. 33 (3), 654–670.

Fereshteh, G., Yassaman, B., Reza, A.M.M., Zavar, A., Hossein, M., 2007. Phytoremediation of arsenic by macroalga: Implication in natural contaminated water, Northeast Iran. J. Appl. Sci. 7 (12), 1614–1619.

Franchi, E., Cosmina, P., Pedron, F., Rosellini, I., Barbafieri, M., Petruzzelli, G., Vocciante, M., 2019. Improved arsenic phytoextraction by combined use of mobilizing chemicals and autochthonous soil bacteria. Sci. Total Environ. 655, 328–336.

Franchi, E., Rolli, E., Marasco, R., Agazzi, G., Borin, S., Cosmina, P., Pedron, F., Rosellini, I., Barbafieri, M., Petruzzelli, G., 2017. Phytoremediation of a multi contaminated soil: mercury and arsenic phytoextraction assisted by mobilizing agent and plant growth promoting bacteria. J. Soils and Sediments 17, 1224–1236.

Frankenberger, W.T. and Arshad, M., 2002. Volatilization of arsenic in environmental chemistry of arsenic, 363–380.

Fukushi, K., Sverjensky, D.A., 2007. A surface complexation model for sulfate and selenate on iron oxides consistent with spectroscopic and theoretical molecular evidence. Geochim. Cosmochim. Acta 71 (1), 1–24.

Gamalero, E., Lingua, G., Berta, G., Glick, B.R., 2009. Beneficial role of plant growth promoting bacteria and arbuscular mycorrhizal fungi on plant responses to heavy metal stress. Can. J. Microbiol. 55 (5), 501–514.

Govarthanan, M., Mythili, R., Selvankumar, T., Kamala-Kannan, S., Kim, H., 2018. Myco-phytoremediation of arsenic- and lead-contaminated soils by Helianthus annuus and wood rot fungi, Trichoderma sp. isolated from decayed wood. Ecotoxicol. Environ. Saf. 151, 279–284.

Guarino, F., Miranda, A., Cicatelli, A., Castiglione, S., 2020. Arsenic phytovolatilization and epigenetic modifications in Arundo donax L. assisted by a PGPR consortium. Chemosphere, 126310.

Guo, J., Dai, X., Xu, W., Ma, M., 2008. Overexpressing GSH1 and AsPCS1 simultaneously increases the tolerance and accumulation of cadmium and arsenic in Arabidopsis thaliana. Chemosphere 72 (7), 1020–1026.

Guo, J., Feng, R., Ding, Y., Wang, R., 2014. Applying carbon dioxide, plant growth-promoting rhizobacterium and EDTA can enhance the phytoremediation efficiency of ryegrass in a soil polluted with zinc, arsenic, cadmium and lead. J. Environ. Manage. 141, 1–8.

Hammaini, A., González, F., Ballester, A., Blázquez, M.L., Munoz, J.A., 2003. Simultaneous uptake of metals by activated sludge. Miner. Eng. 16 (8), 723–729.

Hartley-Whitaker, J., Ainsworth, G., Vooijs, R., ten Bookum, W., Schat, H., Meharg, A.A., 2001. Phytochelatins are involved in differential arsenate tolerance in *Holcus lanatus*. Plant Physiol. 126, 299–306.

Hasan, M., Uddin, M., Ara-Sharmeen, I., Alharby, F, Alzahrani, H., Hakeem, Y.K.R. and Zhang, L., 2019. Assisting phytoremediation of heavy metals using chemical amendments. Plants 8 (9), 295.

He, X., Lilleskov, E., 2014. Arsenic uptake and phytoremediation potential by arbuscular mycorrhizal fungiMycorrhizal Fungi: Use in Sustainable Agriculture and Land Restoration. Springer, Berlin, Heidelberg, pp. 259–275.

He, Z., Yan, H., Chen, Y., Shen, H., Xu, W., Zhang, H., Shi, L., Zhu, Y.G., Ma, M., 2016. An aquaporin Pv TIP 4; 1 from Pteris vittata may mediate arsenite uptake. New Phytol. 209 (2), 746–761.

Hu, P., Huang, J., Ouyang, Y., Wu, L., Song, J., Wang, S., Li, Z., Han, C., Zhou, L., Huang, Y., Luo, Y., 2013. Water management affects arsenic and cadmium accumulation in different rice cultivars. Environ. Geochem. Health 35 (6), 767–778.

Indriolo, E., Na, G., Ellis, D., Salt, D.E., Banks, J.A., 2010. A vacuolar arsenite transporter necessary for arsenic tolerance in the arsenic hyperaccumulating fern Pteris vittata is missing in flowering plants. Plant Cell 22 (6), 2045–2057.

Islam, S., Rahman, M.M., Rahman, M.A., Naidu, R., 2017. Inorganic arsenic in rice and rice-based diets: health risk assessment. Food Control 82, 196–202.

Jankong, P., Visoottiviseth, P., Khokiattiwong, S., 2007. Enhanced phytoremediation of arsenic contaminated land. Chemosphere 68 (10), 1906–1912.

Jasrotia, S., Kansal, A., Mehra, A., 2017. Performance of aquatic plant species for phytoremediation of arsenic-contaminated water. Appl. Water Sci. 7 (2), 889–896.

Jeong, S., Hee Sun Moon, H.S., Nam, K., 2014. Enhanced uptake and translocation of arsenic in cretan brake fern (*Pteris cretica* L.) through siderophore arsenic complex formation with an aid of rhizospheric bacterial activity. J. Hazard. Mater. 280, 536–543.

Jesrotia, S., Kansal, A., Mehra, A., 2015. Performance of aquatic plant species for phytoremediation of arsenic-contaminated water. Appl. Water Sci. 7, 889–896.

Jones, D.L., Darrah, P.R., 1994. Role of root derived organic acids in the mobilization of nutrients from the rhizosphere. Plant Soil 166 (2), 247–257.

Jones, D.L., 1998. Organic acids in the rhizosphere–a critical review. Plant Soil 205 (1), 25–44.

Kaplan, D., 2013. Absorption and adsorption of heavy metals by microalgae. Handbook of microalgal culture. Appl. Phycol. Biotechnol. 2, 602–611.

Karimi, N., Souri, Z., 2015. Effect of phosphorus on arsenic accumulation and detoxification in arsenic hyperaccumulator, Isatis cappadocica. J. Plant Growth Regul. 34 (1), 88–95.

Khan, A.G., 2005. Role of soil microbes in the rhizospheres of plants growing on trace metal contaminated soils in phytoremediation. J. Trace Elem. Med. Biol. 18 (4), 355–364.

Khan, M.S., Zaidi, A., Ahemad, M., Oves, M., Wani, P.A., 2010. Plant growth promotion by phosphate solubilizing fungi–current perspective. Arch. Agron. Soil Sci. 56 (1), 73–98.

Khanna, K., Jamwal, V.L., Gandhi, S.G., Ohri, P., Bhardwaj, R., 2019. Metal resistant PGPR lowered Cd uptake and expression of metal transporter genes with improved growth and photosynthetic pigments in Lycopersiconesculentum under metal toxicity. Sci. Rep. 9 (1), 1–14.

Kumar, J.N., Oommen, C., 2012. Removal of heavy metals by biosorption using freshwater alga Spirogyra hyalina. J. Environ. Biol. 33 (1), 27.

Kumar, K., Gupta, D., Mosa, K.A., Ramamoorthy, K., Sharma, P., 2019. Arsenic transport, metabolism, and possible mitigation strategies in plantsPlant-Metal Interactions. Springer, Cham, pp. 141–168.

Kumari, P., Rastogi, A., Shukla, A., Srivastava, S., Yadav, S., 2018. Prospects of genetic engineering utilizing potential genes for regulating arsenic accumulation in plants. Chemosphere 211, 397–406.

Lampis, S., Santi, C., Ciurli, A., Andreolli, M., Vallini, G., 2015. Promotion of arsenic phytoextraction efficiency in the fern Pteris vittata by the inoculation of As-resistant bacteria: a soil bioremediation perspective. Front. Plant Sci. 6, 80.

Li, B., Gu, B., Yang, Z., Zhang, T., 2018. The role of submerged macrophytes in phytoremediation of arsenic from contaminated water: A case study on Vallisnerianatans (Lour.) Hara. Ecotoxicol. Environ. Saf. 165, 224–231.

Liang, Y., Wang, X., Guo, Z., Xiao, X., Peng, C., Yang, J., Zhou, C., Zeng, P., 2019. Chelator-assisted phytoextraction of arsenic, cadmium and lead by *Pteris vittata* L. and soil microbial community structure response. Int. J. Phytoremediation 21, 1032–1040.

Limmer, M., Burken, J., 2016. Phytovolatilization of organic contaminants. Environ. Sci. Technol. 50 (13), 6632–6643.

Liu, Q., Hu, H., Zhu, L., Li, R., Feng, Y., Zhang, L., Yang, Y., Liu, X., Zhang, H., 2015. Involvement of miR528 in the regulation of arsenite tolerance in rice (Oryza sativa L.). J. Agric. Food Chem. 63 (40), 8849–8861.

Ma, J.F., Yamaji, N., Mitani, N., Tamai, K., Konishi, S., Fujiwara, T., Katsuhara, M., Yano, M., 2007. An efflux transporter of silicon in rice. Nature 448 (7150), 209–212.

Ma, Y., Oliveira, R.S., Freitas, H., Zhang, C., 2016. Biochemical and molecular mechanisms of plant-microbe-metal interactions: relevance for phytoremediation. Front. Plant Sci. 7, 918.

Ma, Y., Rajkumar, M., Zhang, C., Freitas, H., 2016. Beneficial role of bacterial endophytes in heavy metal phytoremediation. J. Environ. Manage. 174, 14–25.

Mahimairaja, S., Bolan, N.S., Adriano, D.C., Robinson, B., 2005. Arsenic contamination and its risk management in complex environmental settings. Adv. Agron. 86, 1–82.

Majumdar, A., Bose, S., 2017. Toxicogenesis and metabolism of arsenic in rice and wheat plants with probable mitigation strategiesArsenic: Risks of Exposure, Behavior in the Environment and Toxicology. Nova Science Publishers, New York, USA, pp. 149–166.

Majumdar, A., Bose, S., 2018. A glimpse on uptake kinetics and molecular responses of arsenic tolerance in Rice plantsMechanisms of Arsenic Toxicity and Tolerance In Plants by Mirza Hasanuzzaman, Kamrun Nahar, Masayuki Fujita. Springer, Singapore, pp. 299–315.

Majumdar, A., Barla, A., Upadhyay, M.K., Ghosh, D., Chaudhuri, P., Srivastava, S., Bose, S., 2018. Vermiremediation of metal (loid) s via Eichornia crassipes phytomass extraction: a sustainable technique for plant amelioration. J. Environ. Manage. 220, 118–125.

Majumdar, A., Kumar, J.S., Bose, S., 2020. Agricultural water management practices and environmental influences on arsenic dynamics in rice fieldArsenic in Drinking Water and Food by Sudhakar Srivastava. Springer, Singapore, pp. 425–443.

Mallick, I., Bhattacharyya, C., Mukherji, S., Dey, D., Sarkar, S.C., Mukhopadhyay, U.K., Ghosh, A., 2018. Effective rhizoinoculation and biofilm formation by arsenic immobilizing halophilic plant growth promoting bacteria (PGPB) isolated from mangrove rhizosphere: a step towards arsenic rhizoremediation. Sci. Total Environ. 610, 1239–1250.

Mazumder, D.G., Dasgupta, U.B., 2011. Chronic arsenic toxicity: studies in West Bengal, India. Kaohsiung J. Med. Sci. 27 (9), 360–370.

Mejáre, M., Bülow, L., 2001. Metal-binding proteins and peptides in bioremediation and phytoremediation of heavy metals. Trends Biotechnol. 19 (2), 67–73.

Mesa, V., Navazas, A., González-Gil, R., González, A., Weyens, N., Lauga, B., Gallego, J.L.R., Sánchez, J., Peláez, A.I., 2017. Use of endophytic and rhizosphere bacteria to improve phytoremediation of arsenic-contaminated industrial soils by autochthonous Betula celtiberica. Appl. Environ. Microbiol. 83 (8), e03411–e03416.

Mitra, N., Rezvan, Z., Ahmad, M.S., Hosein, M.G.M., 2012. Studies of water arsenic and boron pollutants and algae phytoremediation in three springs. Iran Int. J. Ecosyst. 2 (3), 32–37.

Mitra, A., Chatterjee, S., Gupta, D.K., 2017. Uptake, transport, and remediation of arsenic by algae and higher plantsArsenic Contamination in the Environment. Springer, Cham, pp. 145–169.

Mohd, S., Shukla, J., Kushwaha, A.S., Mandrah, K., Shankar, J., Arjaria, N., Saxena, P.N., Narayan, R., Roy, S.K., Kumar, M., 2017. Endophytic fungi Piriformospora indica mediated protection of host from arsenic toxicity. Front. Microbiol. 8, 754.

Mukherjee, G., Saha, C., Naskar, N., Mukherjee, A., Mukherjee, A., Lahiri, S., Majumder, A.L., Seal, A., 2018. An endophytic bacterial consortium modulates multiple strategies to improve arsenic phytoremediation efficiency in *Solanum nigrum*. Sci. Rep. 8, 6979.

Mukhopadhyay, R., Rosen, B.P., 2002. Arsenate reductases in prokaryotes and eukaryotesEnviron. Health Perspect.110, 745–748.

Munoz, L.P., Purchase, D., Jones, H., Feldmann, J., Garelick, H., 2014. Enhanced determination of As–phytochelatin complexes in Chlorella vulgaris using focused sonication for extraction of water-soluble species. Anal. Methods 6 (3), 791–797.

Nahar, N., Rahman, A., Nawani, N.N., Ghosh, S., Mandal, A., 2017. Phytoremediation of arsenic from the contaminated soil using transgenic tobacco plants expressing ACR2 gene of Arabidopsis thaliana. J. Plant Physiol. 218, 121–126.

Naseri, E., Ndé-Tchoupé, A.I., Mwakabona, H.T., Nanseu-Njiki, C.P., Noubactep, C., Njau, K.N., Wydra, K.D., 2017. Making Fe-based filters a universal solution for safe drinking water provision. Sustainability 9 (7), 1224.

Niazi, N.K., Bibi, I., Fatimah, A., Shahid, M., Javed, M.T., Wang, H., Ok, Y.S., Bashir, S., Murtaza, B., Saqib, Z.A., Shakoor, M.B., 2017. Phosphate-assisted phytoremediation of arsenic by Brassica napus and Brassica juncea: morphological and physiological response. Int. J. Phytoremediation 19 (7), 670–678.

Niazi, N.K., Singh, B., Van Zwieten, L., Kachenko, A.G., 2012. Phytoremediation of an arsenic-contaminated site using Pteris vittata L. and Pityrogramma calomelanos var. austroamericana: a long-term study. Environ. Sci. Pollut. Res. 19 (8), 3506–3515.

Padmavathiamma, P.K., Li, L.Y., 2007. Phytoremediation technology: hyper-accumulation metals in plants. Water Air. Soil Pollut. 184 (1-4), 105–126.

Pandey, V.C., Bajpai, O., 2019. Phytoremediation: From Theory towards Practice. In: Pandey, V.C., Bauddh, K. (Eds.), Phytomanagement of Polluted Sites. Elsevier, Amsterdam, pp. 1–49. https://doi.org/10.1016/B978-0-12-813912-7.00001-6.

Pandey, V.C., Singh, J.S., Singh, R.P., Singh, N., Yunus, M., 2011. Arsenic hazards in coal fly ash and its fate in Indian scenario. Resour. Conserv. Recycl. 55 (9-10), 819–835.

Pandey, V.C., Singh, N., 2015. Aromatic plants versus arsenic hazards in soils. J. Geochem. Explor. 157, 77–80.

Paz-Ferreiro, J., Lu, H., Fu, S., Méndez, A., Gascó, G., 2014. Use of phytoremediation and biochar to remediate heavy metal polluted soils: a review. Solid Earth 5 (1), 65.

Petruzzelli, G., Pedron, F., Rosellini, I., 2014. Effects of thiosulfate on the adsorption of arsenate on hematite with a view to phytoextraction. Res. J. Environ. Earth Sci. 6 (6), 326–332.

Pilon-Smits, E., 2005. Phytoremediation. Annu. Rev. Plant Biol. 56, 15–39.

Podgorski, J.E., SAMAS, E., Khanam, T., Ullah, R., Shen, H., Berg, M., 2017. Extensive arsenic contamination in high-pH unconfined aquifers in the Indus Valley. Sci. Adv. 3, e1700935.

Praveen, A., Pandey, V.C., Mehrotra, S., Singh, N., 2019. Arsenic accumulation in Canna: effect on antioxidative defense system. Appl. Geochem. 108, 104360.

Prum, C., Dolphen, R., Thiravetyan, P., 2018. Enhancing arsenic removal from arsenic-contaminated water by Echinodorus cordifolius– endophytic Arthrobacter creatinolyticus interactions. J. Environ. Manage. 213, 11–19.

Raab, A., Williams, P.N., Meharg, A., Feldmann, J., 2007. Uptake and translocation of inorganic and methylated arsenic species by plants. Environ. Chem. 4 (3), 197–203.

Rafati, M., Khorasani, N., Moattar, F., Shirvany, A., Moraghebi, F., Hosseinzadeh, S., 2011. Phytoremediation potential of Populus alba and Morus alba for cadmium, chromuim and nickel absorption from polluted soil. Int. J. Environ. Res. 5 (4), 961–970.

Rahman, M.M., Dong, Z., Naidu, R., 2015. Concentrations of arsenic and other elements in groundwater of Bangladesh and West Bengal, India: potential cancer risk. Chemosphere 139, 54–64.

Ranjan, R., Kumar, N., Dubey, A.K., Gautam, A., Pandey, S.N., Mallick, S., 2018. Diminution of arsenic accumulation in rice seedlings co-cultured with Anabaena sp.: modulation in the expression of lower silicon transporters, two nitrogen dependent genes and lowering of antioxidants activity. Ecotoxicol Environ. Saf. 151, 109–117.

Rashid, M.I., Mujawar, L.H., Shahzad, T., Almeelbi, T., Ismail, I.M., Oves, M., 2016. Bacteria and fungi can contribute to nutrients bioavailability and aggregate formation in degraded soils. Microbiol. Res. 183, 26–41.

Sánchez-Bermejo, E., Castrillo, G., Del Llano, B., Navarro, C., Zarco-Fernández, S., Martinez-Herrera, D.J., Leo-del Puerto, Y., Muñoz, R., Cámara, C., Paz-Ares, J., Alonso-Blanco, C., 2014. Natural variation in arsenate tolerance identifies an arsenate reductase in Arabidopsis thaliana. Nat. Commun. 5 (1), 1–9.

Sarkar, A., Paul, B., 2016. The global menace of arsenic and its conventional remediation-a critical review. Chemosphere 158, 37–49.

Sarkar, S.R., Majumdar, A., Barla, A., Pradhan, N., Singh, S., Ojha, N., Bose, S., 2017. A conjugative study of Typha latifolia for expunge of phyto-available heavy metals in fly ash ameliorated soil. Geoderma 305, 354–362.

Schmöger, M.E., Oven, M., Grill, E., 2000. Detoxification of arsenic by phytochelatins in plants. Plant Physiol. 122 (3), 793–802.

Sharma, S., Anand, G., Singh, N., Kapoor, R., 2017. Arbuscular mycorrhiza augments arsenic tolerance in wheat (Triticum aestivum L.) by strengthening antioxidant defense system and thiol metabolism. Front. Plant Sci. 8, 906.

Shi, S., Wang, T., Chen, Z., Tang, Z., Wu, Z., Salt, D.E., Chao, D.Y., Zhao, F.J., 2016. OsHAC1; 1 and OsHAC1; 2 function as arsenate reductases and regulate arsenic accumulation. Plant Physiol. 172 (3), 1708–1719.

Shri, M., Dave, R., Diwedi, S., Shukla, D., Kesari, R., Tripathi, R.D., Trivedi, P.K., Chakrabarty, D., 2014. Heterologous expression of Ceratophyllum demersum phytochelatin synthase, CdPCS1, in rice leads to lower arsenic accumulation in grain. Sci. Rep. 4, 5784.

Simiele, M., Lebrun, M., Miard, F., Trupiano, D., Poupart, P., Forestier, O., Scippa, G.S., Bourgerie, S., Morabito, D., 2020. Assisted phytoremediation of a former mine soil using biochar and iron sulphate: effects on As soil immobilization and accumulation in three Salicaceae species. Sci. Total Environ. 710, 136203.

Singh, S., Shrivastava, A., Barla, A., Bose, S., 2015. Isolation of arsenic-resistant bacteria from Bengal delta sediments and their efficacy in arsenic removal from soil in association with Pteris vittata. Geomicrobiol. J. 32 (8), 712–723.

Soldan, R., Mapelli, F., Crotti, E., Schnell, S., Daffonchio, D., Marasco, R., Fusi, M., Borin, S., Cardinale, M., 2019. Bacterial endophytes of mangrove propagules elicit early establishment of the natural host and promote growth of cereal crops under salt stress. Microbiol. Res. 223, 33–43.

Song, W.Y., Yamaki, T., Yamaji, N., Ko, D., Jung, K.H., Fujii-Kashino, M., An, G., Martinoia, E., Lee, Y., Ma, F., 2014. A rice ABC transporter, OsABCC1, reduces arsenic accumulation in the grain. Proc. Natl. Acad. Sci. USA 111, 15699–15704.

Souri, Z., Karimi, N., Sandalio, L.M., 2017. Arsenic hyperaccumulation strategies: an overview. Front. Cell Dev. Biol. 5, 67.

Souza, L.A., Piotto, F.A., Nogueirol, R.C., Azevedo, R.A., 2013. Use of non-hyperaccumulator plant species for the phytoextraction of heavy metals using chelating agents. Sci. Agric. 70 (4), 290–295.

Srivastava, S., Suprasanna, P., D'souza, S.F., 2012. Mechanisms of arsenic tolerance and detoxification in plants and their application in transgenic technology: a critical appraisal. Int. J. Phytoremediation 14 (5), 506–517.

Srivastava, S., Verma, P.C., Chaudhry, V., Singh, N., Abhilash, P.C., Kumar, K.V., Sharma, N., Singh, N., 2013. Influence of inoculation of arsenic-resistant staphylococcus arlettae on growth and arsenic uptake in *Brassica Juncea* (L.) Czern. Var. R-46. J. Hazard. Mater. 262, 1039–1047.

Sundaram, S., Wu, S., Ma, L.Q., Rathinasabapathi, B., 2009. Expression of a Pteris vittata glutaredoxin PvGRX5 in transgenic Arabidopsis thaliana increases plant arsenic tolerance and decreases arsenic accumulation in the leaves. Plant Cell Environ. 32 (7), 851–858.

Taleei, MM, Karbalaei, G.N., Jozi, S.A., 2019. Arsenic removal of contaminated soils by phytoremediatoin of vetiver grass, chara algae and water hyacinth. Bull. Environ. Contam. Toxicol. 102, 134–139.

Tassi, E., Pedron, F., Barbafieri, M., Petruzzelli, G., 2004. Phosphate-assisted phytoextraction in As-contaminated soil. Eng. Life Sci. 4 (4), 341–346.

Tiwari, M., Sharma, D., Dwivedi, S., Singh, M., Tripathi, R.D., Trivedi, P.K., 2014. Expression in Arabidopsis and cellular localization reveal involvement of rice NRAMP, OsNRAMP1, in arsenic transport and tolerance. Plant Cell Environ. 37, 140–152.

Tommasini, R., Vogt, E., Fromenteau, M., Hörtensteiner, S., Matile, P., Amrhein, N., Martinoia, E., 1998. An ABC-transporter of Arabidopsis thaliana has both glutathione-conjugate and chlorophyll catabolite transport activity. Plant J. 13 (6), 773–780.

Tong, Y.P., Kneer, R., Zhu, Y.G., 2004. Vacuolar compartmentalization: a second-generation approach to engineering plants for phytoremediation. Trends Plant Sci. 9 (1), 7–9.

Tripathi, R.D., Srivastava, S., Mishra, S., Dwivedi, S., 2008. Strategies for phytoremediation of environmental contamination. In: Bose, B., Hemantaranjan, A. (Eds.), Development in physiology, biochemistry and molecular biology of plants. Vol.2, New India Publishing Agency, New Delhi, India, pp. 175-220.

Tripathi, R.D., Srivastava, S., Mishra, S., Singh, N., Tuli, R., Gupta, D.K., Maathuis, F.J., 2007. Arsenic hazards: strategies for tolerance and remediation by plants. Trends Biotechnol. 25 (4), 158–165.

Tripathi, P., Khare, P., Barnawal, D., Shanker, K., Srivastava, P.K., Tripathi, R.D., Kalra, A., 2020. Bioremediation of arsenic by soil methylating fungi: Role of Humicola sp.strain 2WS1 in amelioration of arsenic phytotoxicity in Bacopa monnieri L. Sci. Total Environ. 716, 136758.

Tsukanova, K.A., Meyer, J.J.M., Bibikova, T.N., 2017. Effect of plant growth-promoting Rhizobacteria on plant hormone homeostasis. S. Afr. J. Bot. 113, 91–102.

Ullah, A., Heng, S., Munis, M.F.H., Fahad, S., Yang, X., 2015. Phytoremediation of heavy metals assisted by plant growth promoting (PGP) bacteria: a review. Environ. Exp. Bot. 117, 28–40.

Upadhyay, A.K., Singh, N.K., Singh, R., Rai, U.N., 2016. Amelioration of arsenic toxicity in rice: comparative effect of inoculation of Chlorella vulgaris and Nannochloropsis sp. on growth, biochemical changes and arsenic uptake. Ecotoxicol. Environ. Saf. 124, 68–73.

Upadhyay, M.K., Yadav, P., Shukla, A., Srivastava, S., 2018. Utilizing the potential of microorganisms for managing arsenic contamination: a feasible and sustainable approach. Front. Environ. Sci. 6, 24.

Vangronsveld, J., Herzig, R., Weyens, N., Boulet, J., Adriaensen, K., Ruttens, A., Thewys, T., Vassilev, A., Meers, E., Nehnevajova, E., van der Lelie, D., 2009. Phytoremediation of contaminated soils and groundwater: lessons from the field. Environ. Sci. Pollut. Res. 16 (7), 765–794.

Verma, S., Mukherjee, A., Choudhury, R., Mahanta, C., 2015. Brahmaputra river basin groundwater: solute distribution, chemical evolution and arsenic occurrences in different geomorphic settings. J. Hydrol. Reg. Stud. 4, 131–153.

Vetterlein, D., Wesenberg, D., Nathan, P., Bräutigam, A., Schierhorn, A., Mattusch, J., Jahn, R., 2009. Pteris vittata–revisited: uptake of As and its speciation, impact of P, role of phytochelatins and S. Environ. Pollut. 157 (11), 3016–3024.

Wan, X., Lei, M., Chen, T., 2020. Review on remediation technologies for arsenic-contaminated soil. Front. Environ. Sci. Eng. 14, 24.

Wang, N.X., Li, Y., Deng, X.H., Miao, A.J., Ji, R., Yang, L.Y., 2013. Toxicity and bioaccumulation kinetics of arsenate in two freshwater green algae under different phosphate regimes. Water Res. 47 (7), 2497–2506.

Wang, P., Sun, G., Jia, Y., Meharg, A.A., Zhu, Y., 2014. A review on completing arsenic biogeochemical cycle: microbial volatilization of arsines in environment. J. Environ. Sci. 26 (2), 371–381.

Wang, S., Mulligan, C.N., 2006. Effect of natural organic matter on arsenic release from soils and sediments into groundwater. Environ. Geochem. Health 28 (3), 197–214.

Wang, X., Rathinasabapathi, B., Oliveira, L.M.D., Guilherme, L.R., Ma, L.Q., 2012. Bacteria-mediated arsenic oxidation and reduction in the growth media of arsenic hyperaccumulator Pteris vittata. Environ. Sci. Technol. 46 (20), 11259–11266.

Williams, P.N., Villada, A., Deacon, C., Raab, A., Figuerola, J., Green, A.J., Feldmann, J., Meharg, A.A., 2007. Greatly enhanced arsenic shoot assimilation in rice leads to elevated grain levels compared to wheat and barley. Environ. Sci. Technol. 41 (19), 6854–6859.

Wu, F., Hu, J., Wu, S., Wong, M.H., 2015. Grain yield and arsenic uptake of upland rice inoculated with arbuscular mycorrhizal fungi in As spiked soils. Environ. Sci. Pollut. Res. 22, 8919–8926.

Wu, Z., Ren, H., McGrath, S.P., Wu, P., Zhao, F.J., 2011. Investigating the contribution of the phosphate transport pathway to arsenic accumulation in rice. Plant Physiol. 157 (1), 498–508.

Xia, Y.S., Chen, B.D., Christie, P., Wang, Y.S., Li, X.L., 2007. Arsenic uptake by arbuscular mycorrhizal maize (*Zea mays* L.) grown in an arsenic-contaminated soil with added phosphorus. J. Environ. Sci. 19 (10), 1245–1251.

Xiang, D., Liao, S., Tu, S., Zhu, D., Xie, T., Wang, G., 2020. Surfactants enhanced soil arsenic phytoextraction efficiency by Pteris vittata L. Bull. Environ. Contam. Toxicol. 104, 259–264.

Xiao, A.W., Li, Z., Li, W.C., Ye, Z.H., 2020. The effect of plant growth-promoting rhizobacteria (PGPR) on arsenicaccumulation and the growth of rice plants (Oryza sativa L.). Chemosphere 242, 125136.

Xie, Q.E., Yan, X.L., Liao, X.Y., Li, X., 2009. The arsenic hyperaccumulator fern Pteris vittata L. Environ. Sci. Technol. 43 (22), 8488–8495.

Xu, J., Shi, S., Wang, L., Tang, Z., Lv, T., Zhu, X., Ding, X., Wang, Y., Zhao, F.J., Wu, Z., 2017. OsHAC4 is critical for arsenate tolerance and regulates arsenic accumulation in rice. New Phytol. 215 (3), 1090–1101.

Yang, Q., Tu, S., Wang, G., Liao, X., Yan, X, 2012. Effectiveness of applying arsenate reducing bacteria to enhance arsenic removal from polluted soils by Pteris vittata L. Int. J. Phytoremediation 14, 89–99.

Yang, C., Ho, Y.N., Makita, R., Inoue, C., Chien, M.F., 2020. A multifunctional rhizobacterial strain with wide application in different ferns facilitates arsenic phytoremediation. Sci. Total Environ. 712, 134504.

Yin, X.X., Wang, L.H., Bai, R., Huang, H., Sun, G.X., 2012. Accumulation and transformation of arsenic in the blue-green alga Synechocysis sp. PCC6803. Water Air Soil Pollut. 223 (3), 1183–1190.

Yoon, J., Cao, X., Zhou, Q., Ma, L.Q., 2006. Accumulation of Pb, Cu, and Zn in native plants growing on a contaminated Florida site. Sci. Total Environ. 368, 456–464.

Yu, Y., Zhang, S., Huang, H., Wu, N., 2010. Uptake of arsenic by maize inoculated with three different arbuscular mycorrhizal fungi. Commun. Soil Sci. Plant Anal. 41 (6), 735–743.

Zavala, Y.J., Gerads, R., Gürleyük, H., Duxbury, J.M., 2008. Arsenic in rice: II. Arsenic speciation in USA grain and implications for human health. Environ. Sci. Technol. 42 (10), 3861–3866.

Zemanová, V., Pavlíková, D., Dobrev, P.I., Motyka, V., Pavlík, M., 2019. Endogenous phytohormone profiles in Pteris fern species differing in arsenic accumulating ability. Environ. Exp. Bot. 166, 103822.

Zeng, X., Su, S., Jiang, X., Li, L., Bai, L., Zhang, Y., 2010. Capability of pentavalent arsenic bioaccumulation and biovolatilization of three fungal strains under laboratory conditions. Clean–Soil Air Water, 38 (3), 238–241.

Zhang, Q., Berg, M., Zheng, Q., Sun, G., Rodríguez, L.A.G., Xue, H., Johnson, C., 2013. Groundwater arsenic contamination throughout China. Science 341, 866–868.

Zhao, F.J., McGrath, S.P., Meharg, A.A., 2010. Arsenic as a food chain contaminant: mechanisms of plant uptake and metabolism and mitigation strategies. Annu. Rev. Plant Biol. 61, 535–559.

Zhu, Y.G., Rosen, B.P., 2009. Perspectives for genetic engineering for the phytoremediation of arsenic-contaminated environments: from imagination to reality? Curr. Opin. Biotechnol. 20 (2), 220–224.

CHAPTER

10 Compost-assisted phytoremediation

Janhvi Pandey[a,b], Sougata Sarkar[b,c], Vimal Chandra Pandey[d]

[a]*Academy of Scientific and Innovative Research (AcSIR), India*
[b]*Division of Agronomy and Soil Science, CSIR-Central Institute of Medicinal and Aromatic Plants, Lucknow, India*
[c]*Genetic Resources and Agro-Technology Division, CSIR-Indian Institute of Integrative Medicine, Jammu, India*
[d]*Department of Environmental Science, Babasaheb Bhimrao Ambedkar University, Lucknow, India*

10.1 Introduction

In the modern era, the environment's major problem is the contamination of soil due to various industrial activities. The wastes produced from many industries incorporate a significant amount of toxic elements like heavy metals (HMs). When disposed of either in water streams or landfill sites, these wastes further contaminate these areas to a great extent. Therefore, the hour's need is to reclaim these heavily polluted sites in the most cost-effective, sustainable way possible. It is often said that "the ultimate goal of any soil remediation process must be not only to reduce the concentration of the target contaminants but, most importantly, also to improve soil health" (Epelde et al., 2014).

Due to population explosion, there is an ever-increasing demand for poultry or livestock products, which create enormous amounts of organic residues. Apart from livestock/poultry ventures, many household and industrial activities also produce a large amount of organic waste products. These waste products must be dealt in the most sustainable way possible to be disposed of safely. It has been reported that organic waste products can be composted to form nutrient-rich amendments, which can enhance soil fertility and reclaim contaminated sites (Zhuang et al., 2009). The continuous use of inorganic fertilizers can lead to soil acidification and degradation and are expensive to use (Das et al., 2017). Therefore, composts produced from organic waste residues can be utilized in the place of inorganic fertilizers, as they have been proved to assist in the enhancement of soil fertility and plant production cost-effectively (Kravchenkoet al., 2017).

Phytoremediation processes can be assisted with compost amendments to reduce the long term threat included with compost applications like biosolids or sewage sludge composts (Singh et al., 2020; Sæbø and Ferrini, 2006). The soil fertility of the contaminated sites is alleviated by the utilization of composts in phytoremediation trials. The mechanisms involved are forming stable organic matter-mineral aggregates that improve soil structure, activation of nitrogen cycling microorganisms, and degradation of pollutants by an increment in microbial activity. Composts derived from various waste

Assisted Phytoremediation. DOI: https://doi.org/10.1016/B978-0-12-822893-7.00001-X

products have been reported to contain the property that immobilizes contaminants in soil and enhances plant productivity by providing essential nutrients (Gadepalle et al., 2007). For centuries, the utilization of organic amendments has been proposed to improve soil health, revegetation, and reduce the toxic metal availability to the plants (Abbott et al., 2001). Compost application in HM contaminated soil also impacts the bioavailability and mobility of potentially toxic elements (Fagnano et al., 2011) either due to immobilization or by the generation of insoluble complexes (Achiba et al., 2009). But, before application of these compost amendments on fields, their proper evaluation is needed to ensure metal immobilization to reduce the risk of food chain contamination (Clemente et al., 2015; Puga et al., 2015).

Phytoremediation processes can be augmented to a great extent by the utilization of composts in contaminated soils which also helps in their nutrient enrichment (Pandey and Bajpai, 2018). In the present chapter, an effort has been made to elaborate the significance of compost amendments in assisted phytoremediation strategies. The impact of compost applications on soil fertility, plant productivity, metal mobility, microbial activity, and the mechanisms involved has been discussed in detail. The focus has been given to the mechanisms that undergo in the soil after compost application and their impact on phytoremediation.

10.2 What is compost?

Compost can be defined as a natural material that is enormously rich in organic matter and nutrients. Compost is made by a process in which certain organic residues like municipal waste, crop leftovers, and animal excreta are converted into useful material by aerobic stabilization. This technique promotes the biological degradation and conversion of organic waste material into a humus-rich product. Organic waste materials are microbially degraded to obtain compost due to which it is enriched in humus, nutrients, and microorganisms (Sæbø and Ferrini, 2006).

The process of "composting" can successfully manage an enormous amount of organic waste material, and is a straightforward and sustainable approach (Onwosi et al., 2017). In this technique, a large volume of organic waste is reduced, and its nutrients are recycled (Oliveira et al., 2017). By utilizing the composting process, the readily degradable portions of organic wastes like proteinaceous substances, carbohydrates, etc. are reduced and an enhancement in the humification is witnessed (Baffi et al., 2007).

This process is majorly conducted by various aerobic thermophilic microorganisms and the biostable and hygienic product obtained as an end product can be used as an organic fertilizer. For the microorganisms to flourish and be generally useful in their work, they should be given appropriate supplements and environment (>40°C). The physical characteristics of materials used for compost production influence certain physical properties of the same (Agnew and Leonard, 2003a). In the agriculture sector, modest source of nutrients are organic wastes. The supplement estimation of natural squanders is viewed as moderate, even though the waste material's quality depends mainly on the type of waste material and its handling strategy (Petersen et al., 2003).

Since the last few decades, a surge in waste management practices has enhanced the range and types of composts widely. The properties of all these composts specifically rely on their original material. Some common sources comprise sludge from municipal wastes, agricultural crop residues, and household waste materials (Sæbø and Ferrini, 2006). Due to large amounts of humic substances, the compost possesses metal complexation as characteristic property (Clemente and Bernal, 2006). It can also enhance soil productivity (Fagnano et al., 2011) when applied to the soil as a fertilizer.

10.3 Compost quality evaluation

The compost's quality is a crucial aspect that must be tested before using it as an amendment. Specific characteristics or indices are measured for evaluating the compost maturity. Without attaining full maturity, the compost may have potentially no benefit to the soil or plants (Zaha et al., 2013).Only composted manure is utilized for soil remediation because fresh manure contains elevated ammonia concentrations, hence may prove harmful for plants (Gadepalle et al., 2007). Composted manure also reduces the risk of nitrogen loss due to leaching and surface erosion and eliminates the soil-borne pathogens (Escribano, 2016).

Quality of the compost is evaluated based on specific properties which have been categorized (Zaha et al., 2013) as:

1. Physical properties which includes temperature, odor, etc.
2. Biological properties which can be evaluated by plant bioassay-germination test
3. Chemical properties like pH, electrical conductivity (EC), carbohydrates content, and C/N ratio.
4. Microbiological properties which is evaluated by respiration analysis, etc.

During all the stages in the production of composts, physical properties hold an essential position. The water content in mature compost should not exceed 45%. This water percentage provides optimum conditions during compost processing like sieving, mixing with additives (Agnew and Leonard, 2003b). If the water content is much lower than 45%, then microorganisms will become inactive. In contrast, if the water percentage is very high, it may displace air from pore space creating anaerobic regions. The permeability in compost decreases if the wet bulk density increases, which might result from the compost's matrix compaction. Various mechanical characteristics of the compost like porosity and strength are also influenced by the compost's bulk density (Agnew and Leonard, 2003b).

Apart from this, among the biological properties, the germination test is done to assess the compost's maturity, toxicity, and nutrient content. This test is the most sensitive parameter to evaluate the biological properties of the compost.

10.4 Types of compost

A sustainable approach for low-input intensive agriculture practices is the utilization of organic waste materials as manures, enhancing soil fertility, and maintaining soil health. Various types of organic amendments utilized for the remediation of contaminated soil include manures, biosolids, composts derived from different source materials, sewage sludge, bark, woodchips, and other amendments (Gadepalle et al., 2007).

Different types of composts have been reported in past studies. The sources of compost can be broadly categorized into three categories (Fig. 10.1). These are:

1. Animal/livestock-based composts
2. Plants based compost
3. Composts based on human activities

The compost production based on the wastes produced by human activities can be further divided into two categories as:

1. Household activities
2. Industrial activities

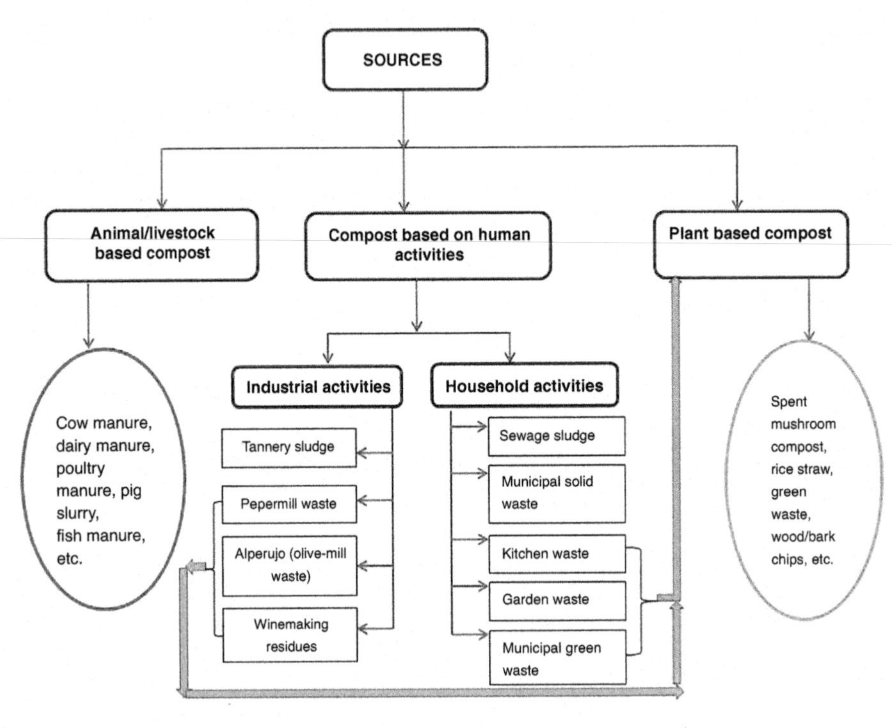

FIG. 10.1

Broad categorization of sources for compost production.

10.5 Impact of compost application on soil systems

For centuries, organic waste has been used in agricultural operations, such as soil amendments (Sims and Pierzynski, 2018). Various organic amendments are frequently utilized in agricultural operations because they have been known to improve soil health properties like water holding capacity, aeration, nutrient availability, and porosity and biological properties (like microbial activity) (Das et al., 2017). During biological remediation processes, these amendments also boost plant growth and development. As far as inorganic mineral fertilizers are concerned, composts offer an additional advantage over them because they improve overall soil health, especially soil framework, thus creating a better environment for developing plant roots (Madejón et al., 2014). For the maintenance of soil quality and proper regulation of essential soil functions, organic matter is crucial (Elsgaard et al., 2001).Manures derived from livestock waste materials have been extensively utilized in the past because they act as sources of nutrients and improve soil fertility.

Compost, when added to the soil as an amendment, has an immediate impact on various physicochemical properties like pH, organic carbon percentage, and metal(loid) solubility (Solís-Dominguez et al., 2012). Compost should be applied to the agricultural fields one month before planting of crops to get appropriately stabilized without posing any nutrient shock to the plants (Fiorentino et al., 2013). The fertility of the soil in which compost is added as an amendment improves because it provides organic matter, for example, fulvic and humic acids, macronutrients like nitrogen, potassium, phosphorus, and various micronutrients (Wu et al., 2017; Gondek et al., 2018; Liang et al., 2018). It has been suggested via several studies that if composts are regularly applied as a soil amendment, they improve the soil fertility (physical, chemical, and biological) to a great extent (Park et al., 2011).

In acid tainted soils, the augmentation of organic amendments increases soil alkalinity (Madejón et al., 2010). It has been reported that augmentation of compost in soil decreases the soil's bulk density, whereas enhances water retention, porosity, and availability of water to the plants is also increased (Agnew and Leonard, 2003a). When added to the soil, organic matter of composts gradually degrades into simpler components, thereby releasing nutrients for plant uptake (Agegnehu et al.,2017). Aoyama et al. (1999) reported that when manure is applied to the soil, there is a surge in mineral-bound organic matter in total fractions that might be attributed to the degradation of the bulk of organic matter mass macroaggregates. It has been reported that long term administration of organic amendments/composts enhances the total carbon content, specifically in macroaggregate fraction of soils (Saroa and Lal, 2003; Mikha and Rice, 2004). Likewise, organic amendments' total nitrogen concentration is shielded against microbial degradation in macroaggregates of soil (Sodhi et al.,2009).

Since the early 1980s, many studies have been conducted to investigate the role of compost addition in enhancing the biodegradation process in polluted soils (Wu et al., 2017). In the contaminated sites, composts' addition is associated with an improvement in soil health by enhancing organic matter and nutrient concentration. Compost addition to these polluted sites also reduces metals' leaching and transform their exchangeable fractions into an organic and Mn, Fe oxyhydroxide fragments (Frutos et al., 2017).The direct impact of organic amendments or composts in the soil might be associated with improved soil structure as these inputs associate physically with the soil particles. Along with it, compost addition also invigorates the soil's biological activity by enhancing microbial activity (Wu et al., 2017).The organic matter concentration and pH are increased in polluted soil when organic amendments are added to it (Kiikkilä et al., 2001). If the soil is acidic, the organic amendment minimizes the impact of soil acidity by elevating Ca^{2+} ions in the soil solution, which dislocates Al^{3+} and H^+; as a result, Al^{3+} ions bind firmly into insoluble Al^{3+} complexes (Ross et al., 2008).

During the cropping season, compost added to the soil leads to its degradation and production of humic acid, which has the immense chelating capacity, thereby posing an affirmative impact on agriculture (Fiorentino et al., 2013).

The cation exchange capacity (CEC) of the soil is also positively enhanced by organic matter. During the phytoremediation process, organic matter added in the form of composts improves aeration in soil and its physical framework, thereby creating favorable conditions for plant development and growth (Clemente et al., 2015). A significant objective during the phytostabilization process is the restoration of soil properties.

A large amount of work has been carried out for the past few decades, which have reported the positive impact of different compost augmentation types on soil health and fertility. It has been reported that composted cattle manure like cow manure, swine manure, etc. improve soil fertility and can be used widely as soil amendments (Das et al., 2017). Medina et al. (2012) reported that spent mushroom compost could be utilized as a suitable amendment in contaminated soils as it enhances the organic matter content as well as macronutrients in the medium like N, K, P, Mg, and Ca, thus improving the soil fertility thereby diminishing the pollutant toxicity. The organic matter released from the food waste compost when it degrades in the soil might establish soluble complexes with soil aggregates, modifying the soil surface electrostatic properties, thereby enhancing soil properties (Tsang et al., 2014).

The addition of rice straw compost to the soil has been reported to enhance stable water aggregates that might have resulted from a higher amount of carbon availability and microbial activity (Sodhi et al., 2009). Thus, rice straw compost counters the negative impacts of intensive tillage practices on soil aggregates in rice-wheat cropping systems (Sodhi et al., 2009).Yang et al. (2005) reported that after applying farmyard manure and wheat straw, the Water stability of soil aggregates (WSA) was significantly greater than 0.25 mm size fraction (Yang et al.,2005).

Soil physical properties suitable for crop growth and development might be indicated by specific parameters like mean weight diameter (MWD) and improved soil particles' aggregation. Several studies have reported that compost like rice straw manure has a positive impact on MWD, when applied to the soil (Tripathy and Singh, 2004; Singh et al., 2007).Rice straw compost elevates the carbon accumulation in soil macroaggregates. This might be attributed to the fact that when manure is applied to the soil, it induces the degradation of organic matter, forming complexes called microaggregates. They further accumulate by humifying organic content to form macroaggregates rich in carbon content, which is shielded against any runoff (Sodhi et al., 2009). In another experiment, it was suggested that cow manure in flooded rice cropping systems was more effective in enhancing soil fertility characteristics than swine compost (Das et al., 2017).

It has also been reported that municipal solid waste (MSW)/biosolid compost encompasses various plant-available nutrients and can be utilized as suitable, cost-effective manure, along with it they also increase the phosphorus absorption capability of the soil (Petersen et al., 2003; Montemurro and Maiorana, 2007; Qazi et al., 2009). By their utilization as soil amendments, they often lower the pH of the rhizospheric region of soil (Santibáñez et al., 2008). This decrement in the pH might be due to certain factors like secretion of root exudates or organic acids like maleic, citric, oxalic acid, etc., cation-anion exchange reactions, or redox-coupled mechanisms in which oxidation states of certain elements like Mn and Fe are altered, and H^+ is either liberated or used (Hinsinger et al., 2003). Apart from essential macro and micronutrients, MSW/biosolid compost also contains various toxic metals that can get accumulated in the soil if the long-term application of compost is made (Zhao et al., 2013). To overcome this problem, Zhaoet al. (2013) suggested that MSW/biosolid compost can be utilized to develop sod after extricating toxic metals present in the same.

Likewise, sewage sludge compost can be used as a soil conditioner because it enhances the soil fertility properties like carbon content, N, P, K, and water retention capacity. It also positively influences soil physical characteristics like bulk density and porosity. It has been suggested that sewage sludge compost can be utilized to revive the fertility of reclaimed soil from landfills; thus, it can act as a cost-effective solution to the modern disposal problems (Song and Lee, 2010).

Recommendations for compost application have also been suggested. For compost to be utilized as fertilizer, it's pH must be between 6 and 9, EC less than 3.5 mS/cm, and C/N ratio 26–35 (Zaha et al., 2013). A standard dose of 20–90 Mg ha^{-1} compost (fresh weight basis) must be applied in the fertility deteriorated soils. Apart from this, in soils where fertility is not compromised, 10–20 Mg ha^{-1} compost (fresh weight basis) can be applied to improve plant growth and enhance potential toxic elements availability (Fiorentino et al., 2013).

10.6 Impact of compost on metal(loids) mobility in soil/plant systems

Due to various industrial operations, a large amount of waste is produced. When discharged, this waste contaminates soil/water/air because it contains a significant concentration of potentially toxic elements.

Thus, it is a major environmental problem that needs to be managed in the most sustainable way possible.

The immobilization of these toxic elements in the soil can be done either by using certain chemicals like dolomites, zeolites, and lime (Garau et al., 2007) or using organic waste amendments like compost. Utilization of chemicals for this purpose is not a sustainable approach because they are not cost-effective to use on large fields along with it, they leave a negative effect on soil health. On the other hand, organic amendments like composts and animal/plant manures are environmentally excellent and cheap (Adejumo et al. 2011).

Two types of processes occur and impact the HM availability in soil by the addition of compost (Garcia-Mina, 2006). Large molecular weight humic substances present in compost form complexes with HM to immobilize them. Apart from this, humic substances, which are of low molecular weight, apparently increase metals' mobility. Hence, critical evaluation of the impact of various types of compost application on the availability of HM is needed. When applied to HM contaminated soils, organic waste amendments not only enhance its fertility properties but also diminish bioavailability and mobility of potentially toxic elements by promoting certain processes like precipitation, complexation, sorption, etc. Thus, these amendments ameliorate the phytoremediation process in a contaminated site (Madejón et al., 2014). This amelioration occurs through various techniques like volatilization, immobilization, and reduction (Park et al., 2011). It has been reported that the application of organic residues like biosolids enhances the leaching of metals in soil (Richards et al., 2000).

The process of phytostabilization/phytoremediation can be boosted by utilizing plant species tolerant to various stresses and amendments that reduce the mobility of metal (loid)s in soil (Bolan et al., 2011). Organic amendments can successfully immobilize trace elements in soil due to their liming effect (Brallier et al., 1996), and also, they are cost-effective (Madejón et al., 2010). These amendments also adjust the soil acidity, thereby reducing the trace element mobility (Brallier et al., 1996), sustaining plant nutritional requirements (Madejón et al., 2014). The dissolved organic carbon in composts forms soluble complexes with metals, thereby enhancing the availability of these elements for plant's growth (Brunetti et al., 2012; Zheljazkov and Warman, 2003).Organic matter, specifically humic substances (containing phenol and carboxylic groups with negatively charged sites) present in the composts/organic waste amendments, forms stable complexes with trace elements, thereby decreasing their mobility and solubility (Walker et al., 2004a). Many studies have reported reducing HM uptake by plants by applying composts in soil (Gadepalle et al., 2007). It has also been reported that compost application diminishes the soil's water-soluble metal concentrations (Solís-Dominguez et al., 2012).

Many factors affect the impact of compost amendment on the availability and mobility of metals in the soil like source material of compost, type of soil, and compost's physico-chemical characteristics (Walker et al., 2004b) like sewage compost, pig slurry compost application has been reported to reduce HM mobility in soil (Pandey and Singh, 2018; Pathak et al., 2020). Keeping track of the trace element availability is vital because alteration can happen in organic matter due to microbial activity (Pardo et al., 2011). While assessing the mobility of pollutants and their plant availability, their propensity for the complexation to soil organic matter is crucial. The octanol partition coefficient well demonstrates this affinity to water (Kow). If log Know > 4, the possibility for plant uptake is low as most of the elements are retained at roots due to complex formations (Duarte-Davidson and Jones, 1996). The augmentation in the pH brought about the immobilization of metals in soils because it expands the variable negative charge locales into organic matter and iron and manganese (Fe/Mn) oxyhydroxides, which increases the ability to complex and adsorb metals, respectively, and conceivable precipitation.

The reduction in metals' mobility in soil enhances the pH because negatively charged sites on organic matter and Fe/Mn oxyhydroxides are increased, thereby increasing the complexation, adsorption, or precipitation of metals (Pérez-Esteban et al., 2014). Reports suggest that by adding compost in the soil, soil's capacity to form complexes with HM, and their realignment enhances, which ultimately promotes their uptake in plants. This might be attributed to the alterations in the functional groups in soil due to compost addition (Sung et al., 2011; Hattab et al., 2015). A study reported that Cr form complexes with organic matter present in organic residue amendment by reducing from Cr^{6+} to Cr^{3+} and finally precipitating to form the chromic hydroxide, which pertains to the soil for a more extended period (Park et al., 2011).

Some contradictory results have also been reported in which increment in metal mobility and plant uptake was observed post-application of composts to the soil. Cao et al. (2003) reported the increase in As adsorption by applying biosolid compost in neutral/acidic soil. This might have occurred due to a large concentration of available phosphorus, which may have been displaced by As from binding sites (Hartley et al., 2010). In acidic soil, As mobility is reduced by forming complexes with organic matter present in soil (Wang and Mulligan, 2009).Reports suggest that the recalcitrant portions (Fe-oxide and residual) of some metals like As gets converted to labile fractions by compost application (Fitz and Wenzel, 2002). A study reported a decrement in Zn, Pb, and Cu mobile fractions but increased in As bioavailable fractions by applying mixed municipal sewage waste compost and garden waste compost in soil (Alvarenga et al., 2009).

A study suggested that some extra benefits can be obtained by co-composting for the reclamation of HM contaminated sites (Tandy et al., 2009).Numerous studies have reported the usage of composted organic residues like poultry manure, cow manure, and MSW manure for reclamation of HM contaminated sites (Cao et al., 2003; Walker et al., 2004a) and some of them are listed in Table 10.1.

Table 10.1 Some examples of impact of compost amendments on metal(loid)s mobility in phytoremediation experiments.

Type of compost	Test plants	Metal contaminants	Results regarding metal mobility in plant/soil	References
Olive-mill waste (alperujo) compost	*B. bituminosa*	Multimetals	• Though solubility of metals was low, yet total Pb and Zn concentration in soil was high. • Study suggests that thresholds for toxicity must be based on the concentration of trace elements which are "available to plants".	(Pardo et al., 2014)
Municipal green waste compost	*Lepidium sativum*	Hg	• Hg translocation and uptake by plants increased by amendment addition. • Increment in compost dose application decreased the extracted Hg concentration.	(Smolinska, 2015)
Dairy manure and green waste compost	-	Multimetals	• Compost treatment had a significant effect on water extractable conc. of metals. • Extractability of Pb followed another path in soil.	(Solís-Dominguez et al., 2012)
Spent mushroom compost (SMC)	*Atriplex halimus*	Cd, Pb, and Cu	• SMC decreases metal mobility in soil. • Metals were immobilized in the roots of test plants and translocation to shoots was reduced significantly.	(Frutos et al., 2017)

(continued)

Table 10.1 Some examples of impact of compost amendments on metal(loid)s mobility in phytoremediation experiments. ***Continued***

Type of compost	Test plants	Metal contaminants	Results regarding metal mobility in plant/soil	References
Manure compost	*Brassica juncea*	Cu and Zn	• Manure compost addition reduced the percent of residual and exchangeable Cu fractions in soil. • Similar trend was observed for Fe-Mn oxide bound Zn, exchangeable and residual Zn fractions.	(Huang et al., 2020)
Biosolid compost	*Daucus carota, Lactuca sativa*	As	• As uptake was reduced in both test plants by compost application. • Exchangeable, water soluble, and carbonate fraction of As in soil was reduced.	(Cao and Ma, 2004)
MSW and biosolid compost	*Pteris vittata*	As	• 50 g/kg of compost application in soil reduced As uptake in test plants as well as decreased its water soluble fraction in soil.	(Cao et al., 2003)
Biosolid compost	*L. rigidum, T. subterraneum*	Multimetals	• Metal concentration in plant tissues increased with enhancement in biosolid compost doses in soil.	(Rate et al., 2004)
Biosolid compost, farmyard manure, fish manure, horse manure, spent mushroom, pig manure, and poultry manure	-	Cr(VI)	• Cr(VI) to Cr(III) reductions increased in soil by amendment application.	(Bolan et al., 2003)
Municipal solid waste compost	Vegetable crops	Multimetals	• Availability of heavy metals increased by compost addition and micronutrient concentration in plants increased.	(Ghaly and Alkoaik, 2010)

10.7 Impact of composts on soil microbial activity

In the metalliferous soil environment, soil microorganisms, along with plants, play a significant role. Amendment of soil with compost enhances the growth of plants and favors the activity of soil microbiota (Ghanem et al., 2013). These two associate together to strengthen the process of phytoremediation, plant resilience toward stress as well increase the production of plant biomass (Mukherjee et al., 2018; Mesa et al., 2017). It has been reported that the application of several organic amendments/composts like biosolids and livestock manures enhance the concentration of total organic carbon in soil aggregates because these amendments either serve as the source of carbon or increase the disintegration of organic matter present in soil (Bolan et al., 2003;Wang et al., 2010).This increased concentration of organic carbon is readily oxidized, and it acts as an energy source for microorganisms present in the soil, which take part in the metal(loid)s reduction (Chiu et al., 2009; Losi and Frankenberger, 1997). In comparison to physicochemical properties, microbial parameters are more progressively utilized as pointers of soil wellbeing, inferable from their affectability, quick reaction, biological pertinence, and ability to provide data that incorporates distinctive natural variables (Galende et al., 2014).

Organic amendments, like compost application, might influence the type of soil microflora and its activity. Microorganisms present in the soil play a significant role in nutrient and carbon pool cycling, thereby affecting soil fertility. Therefore, they are crucial for agricultural ecosystems (Nannipieri et al., 2002).

To better understand the importance of organic waste amendments in agronomic performance and its environmental impact, the complex interrelationship amidst microflora, transitions of carbon pools and nutrients, and crop production in amended soils must be explained in detail (Franco-Otero et al., 2012). During bioremediation in contaminated soils, the compost act as a nutrient source and provide structural support to the microorganisms (Wyszkowski and Ziółkowska, 2009). Organic residues, when freshly applied to the soil, act as sites of nucleation for soil microbial and fungal growth, in this way making complexes of soil particles and residues in the form of microaggregates (Angers and Giroux, 1996). It has been reported that microbial activity is increased by adding organic amendments in soil, which in turn enhances available carbon that leads to the formation of aggregates due to the accumulation of extracellular polysaccharides (Angers and Giroux, 1996).

Petersen et al., 2003 reported that the application of sludge compost enhanced microorganisms' activity, which was proved by increased fluorescein diacetate hydrolysis (FDA) activity, an indicator of hydrolytic activity. Also, no adverse effect was observed on the process of nitrification by the application of sludge in soil. A change in the phospholipid fatty acid configuration of the sludge compost was also observed, which prompted that the succession of microbial flora somehow diverted the microbial community more like that of the soil (Petersen et al., 2003). The significant advantage of utilizing organic waste residues as an amendment is that they influence the soil microfloral activity, which impacts the mobilization of nutrients and plant growth and development (Sun et al., 2015).

The microorganisms present in the soil are mainly sensitive toward macronutrient (N, P, K) changes in the medium (Allison and Martiny, 2008). González et al. (2019) reported that significant DGFE bands were observed when compost was added in the soil, which might be attributed to the alterations in the microbial community that largely depends on the source material of compost. Fiorentino et al. (2013) reported that the compost application to the soil enhanced the community of nitrite-oxidizing microbes in the rhizospheric soil of giant reed plants in a phytoremediation experiment.

Alperujo compost (olive-mill waste) and pig slurry have also been reported to enhance soil microbial biomass carbon and nitrogen compared to the mineral fertilizer application, proving the organic matter's significance in amendments (Pardo et al., 2011). Similarly, maximum microbial enzymatic activities (acid phosphatase, dehydrogenase, β-glucosidase, urease) were reported in the soil amended with sewage sludge (Alvarenga et al., 2009b). It has been reported that the microbial community present in sewage sludge is tolerant towards organic contaminants and can degrade the same when applied discreetly in soil (Petersen et al., 2003). It has been reported that in metal contaminated soils, the application of organic waste residues as fertilizers diminish the toxicity of soil contaminants towards soil microbiota (Kiikkilä et al., 2001).

The reproduction rate of microorganisms is reported to increase with the addition of organic amendments in soil. Various composts like green waste compost, biosolids, spent mushroom compost, were added to the soil in an experiment. After their application, an increment in microbial biomass carbon, nitrogen, and enzymatic activities' was recorded (Park et al., 2011). This might be attributed to the fact that compost addition increases porosity, which in turn enhances oxygen concentration and dispersion, leading to enhanced microbial activity. In another experiment, it was reported that the spore population of vesicular-arbuscular mycorrhiza fungus was increased by adding chicken litter and dairy cow manure compost in the plots compared to the plots amended with inorganic fertilizers (Oyi Dai, 2011). Enhancement in microbial community activity might be one of the reasons behind the significant role of organic waste amendments in soil fertility and crop production (Caravaca et al., 2002).

Wheat straw compost has been reported to enhance the fungal hyphae length, and propagules count while its application reduces the actinomycetes and ammonifers bacterial community. In contrast, the addition of poultry manure favors the aforesaid bacterial community along with fungal populations (Acea and Carballas, 1999). Thus, the type of microbiota in the growing medium largely depends on the compost amendment's source material.

In HM polluted sites, the structure and function of microbiota can be retrieved by adding organic amendments. Along with it, compost addition in metal contaminated soil also enhances the diversification of microbial community (Barbarick et al., 2004), expediting the metal(loid)s reduction. Livestock waste compost application in soil has been suggested to regulate the soil microbiota, thereby improving soil fertility and health (Hartmann et al., 2015).

In another experiment, poultry manure and crushed cotton gin compost were applied to the soil polluted with nickel, and their impact on the microbial community was recorded. Maximum soil enzymatic activity was observed in soil amended with cotton gin compost, which might be attributed to the higher absorption capability of nickel in humic acid (Tejada et al., 2008).

A nine-year long term phytomanagement experiment was conducted in sites polluted with copper to assess the alterations into soil microbial population dynamics and community types (Burges et al., 2020). This evaluation was based on sunflower and tobacco plants-based crop rotation system, and compost along with limestone was applied in the soil on an annual basis. It was reported that at the end of nine years, an alteration in the types of soil microbiota was observed, and microbial communities responsible for nitrogen cycling flourished in the soil. This shift in genetically diverse community structure might be accredited to alter the soil's physicochemical properties in a nine-year-long phytomanagement trial involving compost application. Compost application renewal in the sixth year replenished the nutrients, and organic content reduced the availability of copper and enhanced soil fertility, which increased the activity of soil microorganisms crucial for soil health (Burges et al., 2020).

Zhen et al. (2014) reported that the application of compost and certain bacterial fertilizers significantly increased the activity of the soil microbial community. They suggested from the soil microbial perspective that in cropland systems where soil fertility has been compromised, compost can be applied for speedy reclamation of these sites. In another experiment, it was reported that organic amendments promoted microbial groups like Zygomycota, Proteobacteria, which grow specifically in nutrient-rich growing mediums (Francioli et al., 2016). González et al. (2019) deduced that the plant species is one of the major factors that define the microbial community structure in the rhizospheric soil. This finding was reported based on the experiment conducted to assess the impact of compost application on wheat and barley plants' arsenic resilience. Each plant species showed different microbial community shape.

10.8 Impact of compost on plants

Numerous studies have reported that organic amendments like compost addition ameliorate the eroded, acidic, polluted, or nutrient-deficient soils. By their incorporation, soil porosity increases, leading to better circulation of air, promoting the growth and activity of soil microbiota, thereby helping in revegetation and plant growth (Guidi et al., 2012).

Compared to inorganic mineral fertilizers, compost addition in soil provides an additional advantage because it diminishes the shock experienced by plants in stress conditions. This benefit might be attributed to the improvement of soil texture (aeration, porosity, water holding capacity, etc.) by compost addition, thereby generating a better environment for the development and growth of plant's roots (Madejón et al., 2014). The positive impact of green waste compost application on the growth of *Lepidium sativum* in contaminated soil was reported. This might be accredited to the soil fertility and plant growth, enhancing properties of compost (Smolinska, 2015). Another phytoremediation experiment conducted on Ryegrass (*Lolium* species) reported that compost addition enhanced plant nutrition and increased plants' biomass. Apart from it, a higher concentration of metals was observed in plants' harvested biomass (Karami et al., 2011).

Municipal sewage waste compost application in soils has been reported to enhance plant yield and diminish metal availability (Farrell et al., 2010). When compost produced from the crop residues was applied to Selenium contaminated soils, it was found to decrease Se availability to the *Brassica napus* plants (Ajwa et al., 1998). The soil fertility of a polluted site was reported to increase by the addition of alperujo compost, which promoted the growth and development of *Paulownia fortunei* trees (Madejón et al., 2014). In another experiment, the combined impact of poultry manure compost, farmyard manure, and plant leaves was evaluated on the selenium mobility in contaminated soil. These amendments enhanced Se volatilization and reduced uptake of the same in *Zea mays* and *Vigna unguiculata* (Dhillon et al., 2010).

Likewise, it has been reported that the amendment of biosolid compost to Cr, As, and Cu contaminated soil enhances the productivity of lettuce and carrot three-fivefold in comparison to the same crops grown in unamended soil (Cao et al., 2003). An increase in the plant nutrient content and biomass was also reported in *Rosmarinus officinalis* L. when cultivated in compost amended soil (Cala et al., 2005). An increment in the root yield by 5.6-fold and shoot biomass by 3.6-fold in *Lupinus albus* L. was reported when grown in soil amended by compost comprising sewage sludge (25%), olive (50%), and vegetable waste (25%) (Castaldi et al., 2005). Another finding reported that biosolids' application on topsoil enhanced the dry plant yield of *Trifolium subterraneum* and *Lolium rigidum* rather than when applied on clay/sand (Rate et al., 2004).

Compost application in pollute and degraded soil has also been reported to promote more massive vegetative coverage growth than inorganic fertilizers (Robichaud et al., 2019). It has also been suggested that compost application instigates differential impact on the plants according to their species, like compost amendment in As contaminated soil induced more considerable uptake of this metal in shoot biomass of barley rather than wheat plants. Additionally, higher potassium concentration, Na, and calcium were recorded in barley, whereas Fe content was higher in wheat plants (González et al., 2019).

A combined application of compost and biochar in polluted soil promoted *Moringa oleifera* plants' development and enhanced their capability for Pb phytoextraction (Ogundiran et al., 2018). Similarly, it has been reported that Zn and Pb concentration in leaves of *Festuca arundinacea* and ryegrass is reduced by the application of compost in the soil. Their growth is still improved due to the enormous organic matter content present in compost (Rizzi et al., 2004). Similar findings on Pb and Zn mobility in soil and plant tissues of *Brassica oleracea* and *Spinacea oleracea*, after compost addition in soil, was reported by Pichtel and Bradway (Pichtel and Bradway, 2008). It has also been suggested that compost application and lime can effectively stabilize contaminants present in soil and enhance the biomass of *Chrysanthemum coronarium*to a significant level (Córdova et al., 2011). Smolińska, 2019 in their experiment reported that green waste compost incorporation (at an optimum dose of 3:1) with soil enhances *L. sativum* plants' phytoextraction capability in Hg polluted sites and also increased the production of plants. Additionally, green waste compost has been suggested to promote plant production and diminish Pb uptake in plants' aerial parts (Karami et al., 2011).

Compost application in acidic mine soil has been reported to reduce the acidity and make the growing medium suitable for plants' proper growth and development (Madejón et al., 2006). Compost application in mine soils has been reported to reduce uptake of As, Cu, and Pb by plants. Its application enhances the mobility of copper ions in soil but reduces the same to the plants' bioavailability. This might be attributed to the fact that plants naturally absorb Cu^{2+} free ions, but due to compost addition, the free metal ions mostly make complexes with organic ligands (Tandy et al., 2009).

Another study suggested that application of sewage sludge as an amendment in contaminated soil enhanced the physiological parameters and biomass of three tree species (*Liriodendron tulipifera*, *Quercus acutissima*, and *Betula schmidtii*). In contrast, no significant accumulation of HM was observed in the leaves of these trees (Song and Lee, 2010).

10.9 Augmenting compost impact by mixing with other amendments

The utilization of compost as an organic amendment in various agricultural practices has been reported to enhance soil fertility, crop production as well as assist in phytoremediation of metal-contaminated sites (Clemente et al., 2006; Mench et al., 2010). Along with compost, additional amendments can be used to decrease the leaching of nutrients/metals from soil and increase soil fertility (Phillips, 2002). There are many instances that applying a single amendment is not enough to rectify the soil problems like acidity, HM contamination, etc. and a combination of different amendments can fulfill the purpose. Hence, the utilization of various mixtures of amendments and compost has been suggested for the reclamation of contaminated soils (Park et al., 2011).Many potential amendments have been reported that are beneficial in phytoremediation trials when mixed with various composts types. Biochar has specific characteristics like counteracting soil acidity (due to its alkaline nature), nutrient binding, preventing surface nutrient runoff, and enhanced water retention capacity, which prove beneficial for the soil health and plant production (Fowles, 2007). But to its exorbitant nature, it is not cost-effective to use it at a massive scale. Hence, it has been suggested to use it combined with other economic organic amendments like compost (Rodríguez-Vila et al., 2014). The addition of biochar with vermicompost has also been reported to promote plant growth and also diminishes the negative effect of agriculture on the quality of water near fields (Doan et al., 2015). In the metal-contaminated sites, various inorganic amendments like lime and gypsum, along with compost, have been suggested to reduce the bioavailability of metals (Ruttens et al., 2010) because lime addition increases the soil pH, which immobilizes the metals to a significant extent (Gadepalle et al., 2007). The utilization of organic amendments, along with chelants like EDTA and EDDS has also been recommended to reduce leaching of metals in soil (Belyuan et al., 2018). Likewise, certain Arbuscular mycorrhizal fungi have also been reported to enhance contaminated soil reclamation when applied along with compost (Robichaud et al., 2019). Some examples of the augmentation of compost impact on plants, soil, and phytoremediation by using other amendments are given in Table 10.2.

10.10 Conclusions and future prospects

Compost amendments play a significant role in the reclamation of polluted locales. By compost production, the problem of ever-increasing waste disposal problem can be tackled most sustainably. It can be used as an organic amendment for rectifying soil fertility problems and assisting in phytoremediation. Sustainable development can be achieved by utilizing these recycled waste products as soil amendments, which in turn reclaims contaminated sites.

Table 10.2 Some examples of the augmentation of compost impact on plants, soil, and phytoremediation by the use of other amendments.

Compost types + other amendments	Plant species	Contaminants and type of soil	Results	References
Biosolid compost + Bauxsol	*Bothrichloa insculpt, Eucalyptus paniculata*	Multimetals; acidic soil	• Biomass of both grass and tree species increased by combination of amendments.	(Maddocks et al., 2004)
Olive mill compost + cow manure	*Chenopodium album*	Cu Zn, Fb, Fe; pyritic soil	• Zn, Mn, Fe, and Pb concentration in plant tissues reduced by application of composts. • Zn and Mn concentration in shoots were reduced significantly by cow manure amendment.	(Castaldi et al., 2005)
Compost + zeolite and/or iron oxide	Rye grass (*Lolium perenne*)	As	• Compost (15%) mixed with around 5% zeolite and/or Fe_2O_3 decreases As availability to plants.	(Gadepalle et al., 2008)
GWC + biochar	Ryegrass (*Lolium perenne*)	Cu and Pb	• Compost addition reduced Pb concentration whereas biochar reduced Cu concentration in pore water and in shoots of ryegrass. • Biomass yield enhanced by compost addition alone or along with biochar.	(Karami et al., 2011)
Compost + lime (LCO)	*C. coronarium, L. perenne*	Multimetals; acidic soil	• Cultivated plant species and amendment type significantly interact with each other. • Best results - for plant cover = *C. coronarium/L. perenne* + LCO, biomass = *C. coronarium* + LCO • Cu^{2+} mobilization decreased in the soil solution.	(Córdova et al., 2011)
Compost (garden waste + MSW) + *T. harzianum* A6 inoculants	Giant reed (*Arundo donax*)	Cd and other PTEs	• PTE and nitrogen availability enhanced by compost application. • Only PTE absorption was increased by *Trichodermaharzianum* A6 inoculation. • Both amendments together enhanced Cd mobility in plant leaves.	(Fiorentino et al., 2013)
MSWC + EDTA and $(NH_4)_2SO_4$	Mulberry	Cd, Cr, and Pb	• Total metal concentration enhanced in mulberry plant tissues by amendments.	(Zhao et al., 2013)
Compost + Biochar	*Brassica juncea*	Cu, Co; settling pond soil	• Compost + biochar can reduce soil acidity. • Amendment reduced the pseudo total concentration of Co in soil.	(Rodríguez-Vila et al., 2014)
MMSWC + GWC + NPK + CaO	*Agrostis tenuis*	Mutlimetals; acidic soil	• Increase in biomass yield of plants amended by GWC + NPK + CaO. • Cu and Zn were immobilized but As extractability increased.	(Alvarenga et al., 2014)

COW + AMF inoculation	*Tetraclinis articulata*	Multimetals	• Biomass yield of *T. articulate* increased by application of COW and AMF. • Ni, Pb, and Cr concentration in shoots whereas As and Cr concentration in roots was reduced by COW application.	(Curaqueo et al., 2014)
BMC + vermi-compost + biochar	*Zea mays*	Degraded acrisol	• Plant productivity increased along with significant reduction in negative effects of agricultural practices on environment.	(Doan et al., 2015)
Compost + biochar	*Zea mays*		• Both amendments enhance soil health properties as well as biomass yield of maize.	(Agegnehu et al., 2016)
PMC + RHB + GSB	*Moringa oleifera*	Pb	• *M. oleifera* survivability and growth enhanced. • Test plant tolerated significant Pb concentration without any toxicity symptoms.	(Ogundiran et al., 2018)
FWC + chelant:EDTA + EDDS		Cu, Zn, and Pb	• In EDTA-assisted phytoremediation, FWC can reduce the excessive leaching of Pb and Cu.	(Beiyuan et al., 2018)
GWC + nitrilotriacetic acid (NTA)	*Lepidium sativum*	Hg	• Mobile Hg fraction was reduced in soil. • Total Hg concentration increased in test plants by both amendments together.	(Smolińska, 2019)
Compost + *Trametesversicolor* fungal spawn	*Salix planifolia*, *Salix alaxensis*	PHC and As, Cd, Co, Cr, Cu, Pb, Zn; waste oil pit	• Maximum Zn uptake was recorded in *S. planifolia*.	(Robichaud et al., 2019)
MSWC + nano-silica (NS)	*Secale montanum*	Pb	• Biomass of *Secale* increased by 2% application of MSWC. • NS500 mg kg^{-1} and MSWC (2%) enhances the phytoremediation potential of test plant.	(Moameri and Abbasi Khalaki, 2019)
Compost of WSB and sludge + N	*Brassica rapa*	Zn, Cd	• Bioavailability of Cd and Zn was reduced to a great extent by amendment application.	(Awasthi et al., 2019)
MWC + biochar	*Bromus tomentellus*	Cr, Zn	• Artificial neural network can be used for overpredicting data. • *B. tomentellus* can act as a putative candidate for Cr and Zn contaminated sites.	(Roohi et al., 2020)
Compost + dolomitic limestone	Tobacco, Sunflower	Cu	• Structure and type of soil microbiota altered in the 9th year of the long term phytomanagement experiment, but microbial diversity remained the same.	(Burges et al., 2020)

But, before applying these composts derived from various waste products, it must be ensured that they should not increase the burden of organic/inorganic pollutants in soil. Certain concerns must be addressed. It has been hypothesized that compost application (those derived from biosolids or municipal sewage sludge, etc.) might release toxic metals over time by microbial activity reapplication might enhance metal concentrations in soil with time. There is a need to evaluate each organic waste amendment/compost for its effect on metal mobility before its application. More intensive studies are needed to assess the impact of compost application on different element's mobility in soil or plants as each element has its behavioral pattern in the soil in the presence of large amounts of organic matter. Long-term experiments must be conducted to assess these organic amendments' impact on the mobility of metal contaminants and soil properties.

References

Abbott, D.E., Essington, M.E., Mullen, M.D., Ammons, J.T., 2001. Fly ash and lime-stabilized biosolid mixtures in mine spoil reclamation: simulated weathering. J. Environ. Qual. 30 (2), 608–616.

Acea, M.J., Carballas, T., 1999. Microbial fluctuations after soil heating and organic amendment. Bioresour. Technol. 67 (1), 65–71.

Achiba, W.Ben, Gabteni, N., Lakhdar, A., Laing, G.Du, Verloo, M., Jedidi, N., Gallali, T., 2009. Effects of 5-year application of municipal solid waste compost on the distribution and mobility of heavy metals in a Tunisian calcareous soil. Agric. Ecosyst. Environ. 130 (3–4), 156–163.

Adejumo, S., Togun, A., Adediran, J., Ogundiran, M., 2011. Field assessment of progressive remediation of soil contaminated with lead-acid battery waste in response to compost application (Symposium 3.5.1 Heavy Metal Contaminated Soils, International Symposium: Soil Degradation Control, Remediation, and Reclamation, Tokyo Metropolitan University Symposium Series No.2, 2010). ペドロジスト 54 (3), 182–193.

Agegnehu, G., Srivastava, A.K., Bird, M.I., 2017. The role of biochar and biochar-compost in improving soil quality and crop performance: a review. Appl. Soil Ecol. 119, 156–170.

Agegnehu, G., Bass, A.M., Nelson, P.N., Bird, M.I., 2016. Benefits of biochar, compost and biochar–compost for soil quality, maize yield and greenhouse gas emissions in a tropical agricultural soil. Sci. Total Environ. 543 (Feb 1), 295–306.

Agnew, J.M., Leonard, J.J., 2003a. The physical properties of compost. Compost Sci. Util. 11 (13), 238–264.

Agnew, J.M., Leonard, J.J., 2003b. The physical properties of compost. Compost Sci. Util. 11 (13), 238–264.

Ajwa, H.A., Bañuelos, G.S., Mayland, H.F., 1998. Selenium uptake by plants from soils amended with inorganic and organic materials. J. Environ. Qual. 27 (5), 1218–1227.

Akram Qazi, M., Akram, M., Ahmad, N., Artiola, J.F., Tuller, M., 2009. Economical and environmental implications of solid waste compost applications to agricultural fields in Punjab, Pakistan. Waste Manag 29 (9), 2437–2445.

Allison, S.D., Martiny, J.B.H., 2008. Resistance, resilience, and redundancy in microbial communities. Proc. Natl. Acad. Sci. U. S. A. 105 (Suppl. 1), 11512–11519.

Alvarenga, P., De Varennes, A., Cunha-Queda, A.C., 2014. The effect of compost treatments and a plant cover with Agrostis tenuis on the immobilization/mobilization of trace elements in a mine-contaminated soil. Int. J. Phytoremediation 16 (2), 138–154.

Alvarenga, P., Gonçalves, A.P., Fernandes, R.M., de Varennes, A., Vallini, G., Duarte, E., Cunha-Queda, A.C., 2009a. Organic residues as immobilizing agents in aided phytostabilization: (I) Effects on soil chemical characteristics. Chemosphere 74 (10), 1292–1300.

Alvarenga, P., Palma, P., Gonçalves, A.P., Fernandes, R.M., de Varennes, A., Vallini, G., Duarte, E., Cunha-Queda, A.C., 2009b. Organic residues as immobilizing agents in aided phytostabilization: (II) effects on soil biochemical and ecotoxicological characteristics. Chemosphere 74 (10), 1301–1308.

Angers, D.A., Giroux, M., 1996. Recently deposited organic matter in soil water-stable aggregates. Soil Sci. Soc. Am. J. 60 (5), 1547–1551.

Aoyama, M., Angers, D.A., N'dayegamiye, A., Bissonnette, N., 1999. Protected organic matter in water-stable aggregates as affected by mineral fertilizer and manure applications. Canadian J. Soil Sci. 79 (3), 419–425.

Awasthi, M.K., Wang, Q., Chen, H., Liu, T., Awasthi, S.K., Duan, Y., Varjani, S., Pandey, A., Zhang, Z., 2019. Role of compost biochar amendment on the (im)mobilization of cadmium and zinc for Chinese cabbage (Brassica rapa L.) from contaminated soil. J. Soils Sed. 19 (12), 3883–3897.

Baffi, C., Dell'Abate, M.T., Nassisi, A., Silva, S., Benedetti, A., Genevini, P.L., Adani, F., 2007. Determination of biological stability in compost: a comparison of methodologies. Soil Biol. Biochem. 39 (6), 1284–1293.

Barbarick, K.A., Doxtader, K.G., Redente, E.F., Brobst, R.B., 2004. Biosolids effects on microbial activity in shrubland and grassland soils. Soil Sci. 169 (3), 176–187.

Beiyuan, J., Lau, A.Y.T., Tsang, D.C.W., Zhang, W., Kao, C.M., Baek, K., Ok, Y.S., Li, X.D., 2018. Chelant-enhanced washing of CCA-contaminated soil: coupled with selective dissolution or soil stabilization. Sci. Total Environ. 612 (Jan 15), 1463–1472.

Bolan, N.S., Adriano, D.C., Natesan, R., Koo, B.-J., 2003. Effects of organic amendments on the reduction and phytoavailability of chromate in mineral soil. J. Environ. Qual. 32 (1), 120–128.

Bolan, N.S., Park, J.H., Robinson, B., Naidu, R., Huh, K.Y., 2011. Phytostabilization. A green approach to contaminant containment. Adv. Agron. 112 (Jan 1), 145–204.

Brallier, S., Harrison, R.B., Henry, C.L., Dongsen, X., 1996. Liming effects on availability of Cd, Cu, Ni and Zn in a soil amended with sewage sludge 16 years previously. Water. Air. Soil Pollut. 86 (1–4), 195–206.

Brunetti, G., Farrag, K., Soler-Rovira, P., Ferrara, M., Nigro, F., Senesi, N., 2012. The effect of compost and Bacillus licheniformis on the phytoextraction of Cr, Cu, Pb and Zn by three brassicaceae species from contaminated soils in the Apulia region, Southern Italy. Geoderma 170 (Jan 2015), 322–330.

Burges, A., Fievet, V., Oustriere, N., Epelde, L., Garbisu, C., Becerril, J.M., Mench, M., 2020. Long-term phytomanagement with compost and a sunflower – Tobacco rotation influences the structural microbial diversity of a Cu-contaminated soil. Sci. Total Environ. 700 (Jan 2015), 134529.

Cala, V., Cases, M.A., Walter, I., 2005. Biomass production and heavy metal content of Rosmarinus officinalis grown on organic waste-amended soil. J. Arid Environ. 62 (3), 401–412.

Cao, X., Ma, L.Q., Shiralipour, A., 2003. Effects of compost and phosphate amendments on arsenic mobility in soils and arsenic uptake by the hyperaccumulator, Pteris vittata L. Environ. Pollut. 126 (2), 157–167.

Cao, X., Ma, L.Q., 2004. Effects of compost and phosphate on plant arsenic accumulation from soils near pressure-treated wood. Environ. Pollut. 132 (3), 435–442.

Caravaca, F., Hernández, T., García, C., Roldán, A., 2002. Improvement of rhizosphere aggregate stability of afforested semiarid plant species subjected to mycorrhizal inoculation and compost addition. Geoderma 108 (1–2), 133–144.

Castaldi, P., Santona, L., Melis, P., 2005. Heavy metal immobilization by chemical amendments in a polluted soil and influence on white lupin growth. Chemosphere 60 (3), 365–371.

Chiu, C.C., Cheng, C.J., Lin, T.H., Juang, K.W., Lee, D.Y., 2009. The effectiveness of four organic matter amendments for decreasing resin-extractable Cr (VI) in Cr (VI)-contaminated soils. J. Hazard. Mater. 161 (2–3), 1239–1244.

Clemente, R., Almela, C., Bernal, M.P., 2006. A remediation strategy based on active phytoremediation followed by natural attenuation in a soil contaminated by pyrite waste. Environ. Pollut. 143 (3), 397–406.

Clemente, R., Bernal, M.P., 2006. Fractionation of heavy metals and distribution of organic carbon in two contaminated soils amended with humic acids. Chemosphere 64 (8), 1264–1273.

Clemente, R., Pardo, T., Madejón, P., Madejón, E., Bernal, M.P., 2015. Food byproducts as amendments in trace elements contaminated soils. Food Res. Int. 73 (July 1), 176–189.

Córdova, S., Neaman, A., González, I., Ginocchio, R., Fine, P., 2011. The effect of lime and compost amendments on the potential for the revegetation of metal-polluted, acidic soils. Geoderma 166 (1), 135–144.

Curaqueo, G., Schoebitz, M., Borie, F., Caravaca, F., Roldán, A., 2014. Inoculation with arbuscular mycorrhizal fungi and addition of composted olive-mill waste enhance plant establishment and soil properties in the regeneration of a heavy metal-polluted environment. Environ. Sci. Pollut. Res. 21 (12), 7403–7412.

Das, Suvendu, Jeong, S.T., Das, Subhasis, Kim, P.J., 2017. Composted cattle manure increases microbial activity and soil fertility more than composted swine manure in a submerged rice paddy. Front. Microbiol. 8 (Sep), 1702.

Dhillon, K.S., Dhillon, S.K., Dogra, R., 2010. Selenium accumulation by forage and grain crops and volatilization from seleniferous soils amended with different organic materials. Chemosphere 78 (5), 548–556.

Doan, T.T., Henry-Des-Tureaux, T., Rumpel, C., Janeau, J.L., Jouquet, P., 2015. Impact of compost, vermicompost and biochar on soil fertility, maize yield and soil erosion in Northern Vietnam: a three year mesocosm experiment. Sci. Total Environ. 514 (May 1), 147–154.

Duarte-Davidson, R., Jones, K.C., 1996. Screening the environmental fate of organic contaminants in sewage sludge applied to agricultural soils: II. The potential for transfers to plants and grazing animals. Sci. Total Environ. 185 (1–3), 59–70.

Elsgaard, L., Petersen, S.O., Debosz, K., 2001. Effects and risk assessment of linear alkylbenzene sulfonates in agricultural soil. 2. Effects on soil microbiology as influenced by sewage sludge and incubation time. Environ. Toxicol. Chem. 20 (8), 1664–1672.

Epelde, L., Burges, A., Mijangos, I., Garbisu, C., 2014. Microbial properties and attributes of ecological relevance for soil quality monitoring during a chemical stabilization field study. Appl. Soil Ecol. 75 (Mar 1), 1–12.

Escribano, A.J., 2016. Organic farming: a promising way of food production.

Fagnano, M., Adamo, P., Zampella, M., Fiorentino, N., 2011. Environmental and agronomic impact of fertilization with composted organic fraction from municipal solid waste: a case study in the region of Naples, Italy. Agric. Ecosyst. Environ. 141 (1–2), 100–107.

Farrell, M., Perkins, W.T., Hobbs, P.J., Griffith, G.W., Jones, D.L., 2010. Migration of heavy metals in soil as influenced by compost amendments. Environ. Pollut. 158 (1), 55–64.

Fiorentino, N., Fagnano, M., Adamo, P., Impagliazzo, A., Mori, M., Pepe, O., Ventorino, V., Zoina, A., 2013. Assisted phytoextraction of heavy metals: Compost and Trichoderma effects on giant reed (Arundo donax L.) uptake and soil N-cycle microflora. Ital. J. Agron. 8 (4), 244–254.

Fitz, W.J., Wenzel, W.W., 2002. Arsenic transformations in the soil-rhizosphere-plant system: fundamentals and potential application to phytoremediation. J. Biotechnol. 99 (3), 259–278.

Fowles, M., 2007. Black carbon sequestration as an alternative to bioenergy. Biomass Bioenergy 31 (6), 426–432.

Francioli, D., Schulz, E., Lentendu, G., Wubet, T., Buscot, F., Reitz, T., 2016. Mineral vs. organic amendments: microbial community structure, activity and abundance of agriculturally relevant microbes are driven by long-term fertilization strategies. Front. Microbiol. 7 (Sep), 1446.

Franco-Otero, V.G., Soler-Rovira, P., Hernández, D., López-de-Sá, E.G., Plaza, C., 2012. Short-term effects of organic municipal wastes on wheat yield, microbial biomass, microbial activity, and chemical properties of soil. Biol. Fertil. Soil 48 (2), 205–216.

Frutos, I., García-Delgado, C., Cala, V., Gárate, A., Eymar, E., 2017. The use of spent mushroom compost to enhance the ability of Atriplex halimus to phytoremediate contaminated mine soils. Environ. Technol. 38 (9), 1075–1084.

Gadepalle, V.P., Ouki, S.K., Herwijnen, R. Van, Hutchings, T., 2007. Immobilization of heavy metals in soil using natural and waste materials for vegetation establishment on contaminated sites. Soil Sediment Contam. 16 (2), 233–251.

Gadepalle, V.P., Ouki, S.K., Herwijnen, R. Van, Hutchings, T., 2008. Effects of amended compost on mobility and uptake of arsenic by rye grass in contaminated soil. Chemosphere 72 (7), 1056–1061.

Galende, M.A., Becerril, J.M., Barrutia, O., Artetxe, U., Garbisu, C., Hernández, A., 2014. Field assessment of the effectiveness of organic amendments for aided phytostabilization of a Pb-Zn contaminated mine soil. J. Geochemical Explor. 145 (Oct 1), 181–189.

Garau, G., Castaldi, P., Santona, L., Deiana, P., Melis, P., 2007. Influence of red mud, zeolite and lime on heavy metal immobilization, culturable heterotrophic microbial populations and enzyme activities in a contaminated soil. Geoderma 142 (1–2), 47–57.

Garcia-Mina, J.M., 2006. Stability, solubility and maximum metal binding capacity in metal-humic complexes involving humic substances extracted from peat and organic compost. Org. Geochem. 37 (12), 1960–1972.

Ghaly, A.E., Alkoaik, F.N., 2010. Effect of municipal solid waste compost on the growth and production of vegetable crops. American J. Ag. Bio. Sci. 5 (3), 274–281.

Ghanem, A., D'Orazio, V., Senesi, N., 2013. Effects of compost addition on pyrene removal from soil cultivated with three selected plant species. CLEAN – Soil Air Water 41 (12), 1222–1228.

Gondek, K., Mierzwa-Hersztek, M., Kopeć, M., 2018. Mobility of heavy metals in sandy soil after application of composts produced from maize straw, sewage sludge and biochar. J. Environ. Manage. 210 (Mar 15), 87–95.

González, Á., García-gonzalo, P., Gil-díaz, M.M., Alonso, J., Lobo, M.C., 2019. Compost-assisted phytoremediation of As-polluted soil. J. Soils Sed. 19 (7), 2971–2983.

Guidi, W., Kadri, H., Labrecque, M., 2012. Establishment techniques to using willow for phytoremediation on a former oil refinery in southern Quebec: achievements and constraints. Chem. Ecol. 28 (1), 49–64.

Hartley, W., Dickinson, N.M., Riby, P., Leese, E., Morton, J., Lepp, N.W., 2010. Arsenic mobility and speciation in a contaminated urban soil are affected by different methods of green waste compost application. Environ. Pollut. 158 (12), 3560–3570.

Hartmann, M., Frey, B., Mayer, J., Mäder, P., Widmer, F., 2015. Distinct soil microbial diversity under long-term organic and conventional farming. ISME J. 9 (5), 1177–1194.

Hattab, N., Motelica-Heino, M., Faure, O., Bouchardon, J.L., 2015. Effect of fresh and mature organic amendments on the phytoremediation of technosols contaminated with high concentrations of trace elements. J. Environ. Manage. 159 (Aug 15), 37–47.

Hinsinger, P., Plassard, C., Tang, C., Jaillard, B., 2003. Origins of root-mediated pH changes in the rhizosphere and their responses to environmental constraints: a review. Plant Soil. 248 (1), 43–59.

Huang, H., Luo, L., Huang, L., Zhang, J., Gikas, P., Zhou, Y., 2020. Effect of manure compost on distribution of Cu and Zn in rhizosphere soil and heavy metal accumulation by Brassica juncea. Water Air Soil Pollut. 231 (5), 1–10.

Karami, N., Clemente, R., Moreno-Jiménez, E., Lepp, N.W., Beesley, L., 2011. Efficiency of green waste compost and biochar soil amendments for reducing lead and copper mobility and uptake to ryegrass. J. Hazard. Mater. 191 (July 15), 41–48.

Kiikkilä, O., Perkiömäki, J., Barnette, M., Derome, J., Pennanen, T., Tulisalo, E., Fritze, H., 2001. Situ Bioremediation through mulching of soil polluted by a copper-nickel smelter. J. Environ. Qual. 30 (4), 1134–1143.

Kravchenko, A.N., Snapp, S.S., Robertson, G.P., 2017. Field-scale experiments reveal persistent yield gaps in low-input and organic cropping systems. Proc. Natl. Acad. Sci. U. S. A. 114 (5), 926–931.

Liang, H., Wu, W.L., Zhang, Y.H., Zhou, S.J., Long, C.Y., Wen, J., Wang, B.Y., Liu, Z.T., Zhang, C.Z., Huang, P.P., Liu, N., Deng, X.L., Zou, F., 2018. Levels, temporal trend and health risk assessment of five heavy metals in fresh vegetables marketed in Guangdong Province of China during 2014–2017. Food Control 92 (Oct 1), 107–120.

Losi, M.E., Frankenberger, W.T., 1997. Reduction of selenium oxyanions by Enterobacter cloacae SLD1a-1: isolation and growth of the bacterium and its expulsion of selenium particles. Appl. Environ. Microbiol. 63 (8), 3079–3084.

Madejón, E., De Mora, A.P., Felipe, E., Burgos, P., Cabrera, F., 2006. Soil amendments reduce trace element solubility in a contaminated soil and allow regrowth of natural vegetation. Environ. Pollut. 139 (1), 40–52.

Madejón, P., Pérez-de-Mora, A., Burgos, P., Cabrera, F., Lepp, N.W., Madejón, E., 2010. Do amended, polluted soils require re-treatment for sustainable risk reduction? - Evidence from field experiments. Geoderma 159 (1–2), 174–181.

Madejón, P., Xiong, J., Cabrera, F., Madejón, E., 2014. Quality of trace element contaminated soils amended with compost under fast growing tree Paulownia fortunei plantation. J. Environ. Manage. 144 (Nov 1), 176–185.

Maddocks, G., Lin, C., McConchie, D., 2004. Effects of Bauxsol™ and biosolids on soil conditions of acid-generating mine spoil for plant growth. Environ. Pollut. 127 (2), 157–167.

Medina, E., Paredes, C., Bustamante, M.A., Moral, R., Moreno-Caselles, J., 2012. Relationships between soil physico-chemical, chemical and biological properties in a soil amended with spent mushroom substrate. Geoderma 173–174 (Mar 1), 152–161.

Mench, Michel, Lepp, Nick, Bert, Valérie, Schwitzguébel, Jean-Paul, Gawronski, Stanislaw W, Schröder, Peter, Vangronsveld, Jaco, Norra, S., Mench, M, Lepp, N, Bert, V, Schwitzguébel, J.-P, Gawronski, S W, Schröder, P, Vangronsveld, J, 2010. Successes and limitations of phytotechnologies at field scale: outcomes, assessment and outlook from COST Action 859. J Soils Sediments 10 (6), 1039–1070.

Mesa, V., Navazas, A., González-Gil, R., González, A., Weyens, N., Lauga, B., Gallego, J.L.R., Sánchez, J., Peláez, A.I., 2017. Use of endophytic and rhizosphere bacteria to improve phytoremediation of arsenic-contaminated industrial soils by autochthonous Betula celtiberica. Appl. Environ. Microbiol. 83 (8), e03411–e03416.

Mikha, M.M., Rice, C.W., 2004. Tillage and manure effects on soil and aggregate-associated carbon and nitrogen. Soil Sci. Soc. Am. J. 68 (3), 809–816.

Moameri, M., Khalaki, M.A., 2019. Capability of Secale montanum trusted for phytoremediation of lead and cadmium in soils amended with nano-silica and municipal solid waste compost. Environ. Sci. Pollut. Res. 26 (24), 24315–24322.

Montemurro, F., Maiorana, M., 2007. Nitrogen utilization, yield, quality and soil properties in a sugarbeet crop amended with municipal solid waste compost. Compost Sci. Util. 15 (2), 84–92.

Mukherjee, G., Saha, C., Naskar, N., Mukherjee, Abhishek, Mukherjee, Arghya, Lahiri, S., Majumder, A.L., Seal, A., 2018. An endophytic bacterial consortium modulates multiple strategies to improve arsenic phytoremediation efficacy in Solanum nigrum. Sci. Rep. 8 (1), 1–16.

Nannipieri, P, Kandeler, E, Ruggiero, P., 2002. Enzymes in the Environment: Activity, Ecology, and Applications. Marcel Dekker, New York, pp. 1–34.

Ogundiran, M.B., Mekwunyei, N.S., Adejumo, S.A., 2018. Compost and biochar assisted phytoremediation potentials of Moringa oleifera for remediation of lead contaminated soil. J. Environ. Chem. Eng. 6 (2), 2206–2213.

Oliveira, L.S.B.L., Oliveira, D.S.B.L., Bezerra, B.S., Silva Pereira, B., Battistelle, R.A.G., 2017. Environmental analysis of organic waste treatment focusing on composting scenarios. J. Clean. Prod. 155 (July 1), 229–237.

Onwosi, C.O., Igbokwe, V.C., Odimba, J.N., Eke, I.E., Nwankwoala, M.O., Iroh, I.N., Ezeogu, L.I., 2017. Composting technology in waste stabilization: On the methods, challenges and future prospects. J. Environ. Manage. 190 (Apr 1), 140–157.

Dai, Oyi, 2011. Effect of arbuscular mycorrhizal (AM) inoculation on growth of Chili plant. African J. Microbiol. Res. 5 (28), 5004–5012.

Pandey, V.C., Bajpai, O., 2018. Phytoremediation: from theory toward practice. In Phytomanagement of Polluted Sites: Market Opportunities in Sustainable Phytoremediation. Elsevier Inc., pp. 1–49.

Pandey, V.C., Singh, V., 2018. Exploring the potential and opportunities of current tools for removal of hazardous materials from environments. In Phytomanagement of Polluted Sites: Market Opportunities in Sustainable Phytoremediation. Elsevier Inc., pp. 501–516.

Pardo, T., Martínez-Fernández, D., Clemente, R., Walker, D.J., Bernal, M.P., 2014. The use of olive-mill waste compost to promote the plant vegetation cover in a trace-element-contaminated soil. Environ. Sci. Pollut. Res. 21 (2), 1029–1038.

Pardo, T., Clemente, R., Bernal, M.P., 2011. Effects of compost, pig slurry and lime on trace element solubility and toxicity in two soils differently affected by mining activities. Chemosphere 84 (5), 642–650.

Park, J.H., Lamb, D., Paneerselvam, P., Choppala, G., Bolan, N., Chung, J.W., 2011. Role of organic amendments on enhanced bioremediation of heavy metal (loid) contaminated soils. J. Hazard. Mater. 185 (2–3), 549–574.

Pathak, S., Agarwal, A.V., Pandey, V.C., 2020. Phytoremediation—a holistic approach for remediation of heavy metals and metalloidsBioremediation of Pollutants. Elsevier, pp. 3–16.

Pérez-Esteban, J., Escolástico, C., Masaguer, A., Vargas, C., Moliner, A., 2014. Soluble organic carbon and pH of organic amendments affect metal mobility and chemical speciation in mine soils. Chemosphere 103, 164–171.

Petersen, S.O., Henriksen, K., Mortensen, G.K., Krogh, P.H., Brandt, K.K., Sørensen, J., Madsen, T., Petersen, J., Grøn, C., 2003. Recycling of sewage sludge and household compost to arable land: Fate and effects of organic contaminants, and impact on soil fertility. Soil Tillage Res. 72 (2), 139–152.

Phillips, I.R., 2002. Phosphorus sorption and nitrogen transformation in two soils treated with piggery wastewater. Aust. J. Soil Res. 40 (2), 335–349.

Pichtel, J., Bradway, D.J., 2008. Conventional crops and organic amendments for Pb, Cd and Zn treatment at a severely contaminated site. Bioresour. Technol. 99 (5), 1242–1251.

Puga, A.P., Abreu, C.A., Melo, L.C.A., Beesley, L., 2015. Biochar application to a contaminated soil reduces the availability and plant uptake of zinc, lead and cadmium. J. Environ. Manage. 159 (Aug 15), 86–93.

Rate, A.W., Lee, K.M., French, P.A., 2004. Application of biosolids in mineral sands mine rehabilitation: use of stockpiled topsoil decreases trace element uptake by plants. Bioresour. Technol. 91 (3), 223–231.

Richards, B.K., Steenhuis, T.S., Peverly, J.H., McBride, M.B., 2000. Effect of sludge-processing mode, soil texture and soil pH on metal mobility in undisturbed soil columns under accelerated loading. Environ. Pollut. 109 (2), 327–346.

Rizzi, L., Petruzzelli, G., Poggio, G., Guidi, G.V., 2004. Soil physical changes and plant availability of Zn and Pb in a treatability test of phytostabilization. Chemosphere 57 (9), 1039–1046.

Robichaud, K., Stewart, K., Labrecque, M., Hijri, M., Cherewyk, J., Amyot, M., 2019. An ecological microsystem to treat waste oil contaminated soil: using phytoremediation assisted by fungi and local compost, on a mixed-contaminant site, in a cold climate. Sci. Total Environ. 672 (July 1), 732–742.

Rodríguez-Vila, A., Covelo, E.F., Forján, R., Asensio, V., 2014. Phytoremediating a copper mine soil with Brassica juncea L., compost and biochar. Environ. Sci. Pollut. Res. 21 (19), 11293–11304.

Roohi, R., Jafari, M., Jahantab, E., Aman, M.S., Moameri, M., Zare, S., 2020. Application of artificial neural network model for the identification the effect of municipal waste compost and biochar on phytoremediation of contaminated soils. J. Geochem. Explor. 208 (Jan 1), 106399.

Ross, D.S., Matschonat, G., Skyllberg, U., 2008. Cation exchange in forest soils: the need for a new perspective. Eur. J. Soil Sci. 59 (6), 1141–1159.

Ruttens, A., Adriaensen, K., Meers, E., De Vocht, A., Geebelen, W., Carleer, R., Mench, M., Vangronsveld, J., 2010. Long-term sustainability of metal immobilization by soil amendments: cyclonic ashes versus lime addition. Environ. Pollut. 158 (5), 1428–1434.

Sæbø, A., Ferrini, F., 2006. The use of compost in urban green areas - a review for practical applicationUrban For. Urban Green 4 (3–4), 159–169.

Santibáñez, C., Verdugo, C., Ginocchio, R., 2008. Phytostabilization of copper mine tailings with biosolids: implications for metal uptake and productivity of Lolium perenne. Sci. Total Environ. 395 (1), 1–10.

Saroa, G.S., Lal, R., 2003. Soil restorative effects of mulching on aggregation and carbon sequestration in a Miamian soil in central Ohio. L. Degrad. Dev. 14 (5), 481–493.

Sims, J.T., Pierzynski, G.M., 2018. Assessing the Impacts of Agricultural, Municipal, and Industrial By-Products on Soil Quality. John Wiley & Sons, Ltd, pp. 237–261.

Singh, G., Jalota, S.K., Singh, Y., 2007. Manuring and residue management effects on physical properties of a soil under the rice-wheat system in Punjab. India. Soil Tillage Res. 94 (1), 229–238.

Singh, G., Pankaj, U., Ajayakumar, P.V., Verma, R.K., 2020. Phytoremediation of sewage sludge by Cymbopogon martinii (Roxb.) Wats. var. motia Burk. grown under soil amended with varying levels of sewage sludge. Int. J. Phytoremediation 22 (5), 540–550.

Smolinska, B., 2015. Green waste compost as an amendment during induced phytoextraction of mercury-contaminated soil. Environ. Sci. Pollut. Res. 22 (5), 3528–3537.

Smolińska, B., 2019. The influence of compost and nitrilotriacetic acid on mercury phytoextraction by Lepidium sativum L. J. Chem. Technol. Biotechnol. 95 (4), 950–958.

Sodhi, G.P.S., Beri, V., Benbi, D.K., 2009. Soil aggregation and distribution of carbon and nitrogen in different fractions under long-term application of compost in rice-wheat system. Soil Tillage Res. 103 (2), 412–418.

Solís-Dominguez, F.A., White, S.A., Hutter, T.B., Amistadi, M.K., Root, R.A., Chorover, J., Maier, R.M., 2012. Response of key soil parameters during compost-assisted phytostabilization in extremely acidic tailings: effect of plant species. Environ. Sci. Technol. 46 (2), 1019–1027.

Song, U., Lee, E.J., 2010. Environmental and economical assessment of sewage sludge compost application on soil and plants in a landfill. Resour. Conserv. Recycl. 54 (12), 1109–1116.

Sun, R., Zhang, X.X., Guo, X., Wang, D., Chu, H., 2015. Bacterial diversity in soils subjected to long-term chemical fertilization can be more stably maintained with the addition of livestock manure than wheat straw. Soil Biol. Biochem. 88, 9–18.

Sung, M., Lee, C.Y., Lee, S.Z., 2011. Combined mild soil washing and compost-assisted phytoremediation in treatment of silt loams contaminated with copper, nickel, and chromium. J. Hazard. Mater. 190 (1–3), 744–754.

Tandy, S., Healey, J.R., Nason, M.A., Williamson, J.C., Jones, D.L., 2009. Remediation of metal polluted mine soil with compost: co-composting versus incorporation. Environ. Pollut. 157 (2), 690–697.

Tejada, M., Moreno, J.L., Hernández, M.T., García, C., 2008. Soil amendments with organic wastes reduce the toxicity of nickel to soil enzyme activities. Eur. J. Soil Biol. 44 (1), 129–140.

Tripathy, R., Singh, A.K., 2004. Effect of water and nitrogen management on aggregate size and carbon enrichment of soil in rice-wheat cropping system. J. Plant Nutr. Soil Sci. 167 (2), 216–228.

Tsang, D.C.W., Yip, A.C.K., Olds, W.E., Weber, P.A., 2014. Arsenic and copper stabilisation in a contaminated soil by coal fly ash and green waste compost. Environ. Sci. Pollut. Res. 21 (17), 10194–10204.

Walker, D.J., Clemente, R., Bernal, M.P., 2004a. Contrasting effects of manure and compost on soil pH, heavy metal availability and growth of Chenopodium album L. in a soil contaminated by pyritic mine waste. Chemosphere 57 (3), 215–224.

Walker, D.J., Clemente, R., Bernal, M.P., 2004b. Contrasting effects of manure and compost on soil pH, heavy metal availability and growth of Chenopodium album L. in a soil contaminated by pyritic mine waste. Chemosphere 57 (3), 215–224.

Wang, S., Mulligan, C.N., 2009. Effect of natural organic matter on arsenic mobilization from mine tailings. J. Hazard. Mater. 168 (2–3), 721–726.

Wang, X.S., Chen, L.F., Li, F.Y., Chen, K.L., Wan, W.Y., Tang, Y.J., 2010. Removal of Cr (VI) with wheat-residue derived black carbon: Reaction mechanism and adsorption performance. J. Hazard. Mater. 175, 816–822.

Wu, H., Lai, C., Zeng, G., Liang, J., Chen, J., Xu, J., Dai, J., Li, X., Liu, J., Chen, M., Lu, L., Hu, L., Wan, J., 2017. The interactions of composting and biochar and their implications for soil amendment and pollution remediation: a review. Crit. Rev. Biotechnol. 37 (6), 754–764.

Wyszkowski, M., Ziółkowska, A., 2009. Role of compost, bentonite and calcium oxide in restricting the effect of soil contamination with petrol and diesel oil on plants. Chemosphere 74 (6), 860–865.

Yang, C., Yang, L., Ouyang, Z., 2005. Organic carbon and its fractions in paddy soil as affected by different nutrient and water regimes. Geoderma 124 (1–2), 133–142.

Zaha, C., Dumitrescu, L., Manciulea, I., 2013. Correlations between composting conditions and characteristics of compost as biofertilizer. Bull. Transilv. Univ. Braşov Ser. I Eng. Sci. 2013.

Zhao, S., Shang, X., Duo, L., 2013. Accumulation and spatial distribution of Cd, Cr, and Pb in mulberry from municipal solid waste compost following application of EDTA and $(NH_4)_2SO_4$. Environ. Sci. Pollut. Res. 20 (2), 967–975.

Zheljazkov, V.D., Warman, P.R., 2003. Application of high Cu compost to Swiss chard and basil. Sci. Total Environ. 302 (1–3), 13–26.

Zhen, Z., Liu, H., Wang, N., Guo, L., Meng, J., Ding, N., Wu, G., Jiang, G., 2014. Effects of manure compost application on soil microbial community diversity and soil microenvironments in a temperate cropland in China. PLoS One 9 (10).

Zhuang, P., McBride, M.B., Xia, H., Li, N., Li, Z., 2009. Health risk from heavy metals via consumption of food crops in the vicinity of Dabaoshan mine, South China. Sci. Total Environ. 407 (5), 1551–1561.

CHAPTER

11

Bioremediation of contaminated soil with plant growth rhizobium bacteria

Metin Turan[a], Sanem Argin[b], Parisa Bolouri[a], Tuba Arjumend[c], Nilda Ersoy[d], Ertan Yıldırım[e], Adem Güneş[f], Melek Ekinci[e], Dilara Birinci[a]

[a]*Department of Genetics and Bioengineering, Faculty of Engineering, Yeditepe University, Istanbul, Turkey*

[b]*Department of Agricultural Trade and Management, School of Applied Science, Yeditepe University, Istanbul, Turkey*

[c]*Department of Plant Protection, Faculty of Agriculture, Usak University, Uşak, Turkey*

[d]*Department of Organic Agriculture, Vocational School of Technical Sciences, Akdeniz University, Antalya, Turkey*

[e]*Department of Horticulture, Faculty of Agriculture, Atatürk University, Erzurum, Turkey*

[f]*Department of Soil Science, Faculty of Agriculture, Erciyes University, Kayseri, Turkey*

11.1 Introduction

The rapidly increasing population in the world and the nutritional requirements of this population as well as the technological and industrial developments in meeting this need are the basic components of the problems that exist in the world today. Variations in ecology and climatic conditions in recent years have changed the course of agricultural practices and made excessive use of pesticides and chemical fertilizers inevitable.

Owing to the threats it presents to global health and the functioning of the ecosystem, environmental pollution has become a major concern worldwide. "The most significant toxins in agricultural soils are inorganic materials such as metalloids, chemicals, heavy metals, petroleum hydrocarbons, and organic pollutants, but radionuclides can also pose a potential hazard in certain areas" (FAO, 2015). Heavy metals emerge either from natural or anthropogenic causes, such as soil depletion, emissions from petrochemical and agrochemical industry, fossil fuel burning, mining, and pesticide usage (Zhang et al., 2020). By an anthropocentric perspective, environmental contamination, might have serious threats to socio-economic well-being, people's health and soil production. "The most prominent pollutants in this process are industrial by-products, chemical solvents, dyes, industrial solvents, petroleum hydrocarbons, trichlorethylene, nitro aromatic compounds, polycyclic aromatic hydrocarbons (PAHs), polychlorinated biphenyls, phthalate esters, ethyl benzene, benzene, xylene (BTEX), and toluene (Dangi et al., 2018)". In addition to these, due to the recent developments in technology, overproduction and use of nanomaterials have caused great risks in human health and ecosystems. These chemicals made their way into soils, rivers, seas and even the atmosphere in an uncontrolled manner.

Assisted Phytoremediation. DOI: https://doi.org/10.1016/B978-0-12-822893-7.00013-6

Besides direct exposure, many species can be harmed by the pollutants through the food chain. Since plants and microbes act as sources of energy for other species, the presence of pollutants can also contaminate or damage those other species. This feature may lead to a phenomenon known as biomagnification, where higher degree of absorption than excretion causes a rise in the abundance of a particular material in living organisms. Such mechanism could contribute to higher concentration levels of contaminants higher up in the trophic chain which may ultimately have a negative effect on human health. Bibliometric findings indicate that heavy metal contamination has gained the most interest of scholars (Guo et al., 2014). The major inorganic pollutants including metalloids and metals such as lead, copper, cadmium, mercury, arsenic and selenium are by far the most harmful contaminants in land and groundwater and can quite easily bioaccumulate in tissues and living beings. Such metals cannot be transformed into non-toxic forms spontaneously, but can have significant and enduring accumulation in the soil (Ayangbenro, 2017). In this situation, it is necessary to monitor levels, versatility and chemical configurations to minimize the risk of biota toxicity and entry into the food chain.

Besides heavy metal contaminants, the global emission of nitrogen and phosphorus from fertilizers, waste and drainage of dairy farmings causes eutrophication of rivers and wetlands and, as a result, major disruption to water quality and marine life. Herbicides and pesticides are among the most commonly found organic toxins in soils due to their extensive or inappropriate use in cultivation. The pesticide utilization rate is poor and most leftover pesticides are dispersed in the atmosphere by either precipitation or runoff (Sultana et al., 2005). In the meantime, pesticides may be degraded but commonly spread at comparatively trace amounts due to their overuse and prolonged application (Cycon et al., 2017; Alvarez et al., 2017).

The so-called PAHs, including compounds like those of pyrene and naphthalene, petroleum and petroleum products such as ethylbenzene, xylene, toluene, and benzene, and polychlorinated biphenylsene are other major classes of organic contaminants (Hooper et al., 2009). Organic compounds linked to medicinal substances are emerging pollutants in the water-based settings (Zhang et al., 2012). Co-existance of metallic elements and pesticide contaminants in the soil has exacerbated significant environmental issues and concerns worldwide. The interrelationship of pesticides and heavy metals makes the pollution more complex and can have an effect on the quality and functionality of soil.

Traditional remediation methods involve chemical, physicochemical, thermal and physical treatments. Traditional chemical methods and physicochemical methods that are being used for removal of toxic elements involve ion exchange, precipitation, oxidation-reduction, neutralization, evaporation, filtration, solidification, incineration, reverse electrochemical, osmosis and ion exchange treatments. Burning at high temperatures is known as the thermal method which causes severe air pollution. Physical methods involve solidification of pollutants by encapsulation in concrete, asphalt or plastic. However, these treatments are limited by high manufacturing costs, higher reagent needs, massive amounts, and the production of secondary environmental contaminants. These techniques' limitations become more pronounced in tailings wastewater, polluted groundwater and other industrial wastewater. Moreover, whatever the technology is, there is a serious threat of leakage. For this reason, all the on-going processes to address the challenge of detoxifying the hazardous compounds have a potential harm to the ecosystem. However, since the end products are minimally toxic, in overall the volume of toxicity is reduced at the cost of energy and money by using these methods.

As an alternative to the traditional methods, bioremediation is a cost effective, environmentally friendly, and sustainable technique (Adams et al., 2015) where the pollutants are cleaned from the soil, water and air by the help of microorganisms (plant growth-promoting rhizobacteria, PGPR). These special non-pathogenic microorganisms chemically break down and dispose the waste in the nature, by converting organic substances into harmless compounds such as fatty acids, carbon dioxide, amino acids while providing energy and nutrients for themselves via this process. The function of PGPR in bioremediation is either through its direct effect on the dynamics of plant growth (phytoremediation) or through acidification, chelation, precipitation, immobilization of rhizospheric heavy metals (Singh et al., 2014).

In this chapter, different bioremediation approaches in the removal of pollutants are explained; the importance and role of PGPR in bioremediation and phytoremediation are highlighted and the effects of co-existing contaminants on the bioremediation of pollutants are discussed.

11.2 Bioremediation

In the term bioremediation, "bio" refers to living beings, while "remediate" refers to the act of resolving an issue. Bioremediation can be defined as using biological species to address environmental issues such as polluted soil or freshwater. Bioremediation is a mechanism in which hazardous contaminants are converted into non-toxic substances by living organisms (Asha et al., 2013). The desired consequence of bioremediation is that the contaminated areas will be returned to their original states without causing any other adverse impact on the ecosystem (Adhikari et al., 2004; Jobby et al., 2018). A more detailed description is that, bioremediation is a way of cleaning up the polluted habitats through leveraging the microorganisms' diversified metabolic capacities to transform pollutants into non-harmful substances through mineralizing, converting into microbial biomass or producing water and carbon (IV) oxide (Mentzer and Ebere, 1996; Lovley, 2003).

Bioremediation has three different approaches, which are biostimulation, bioaugmentation and intrinsic bioremediation (Fig. 11.1). Biostimulation involves the addition of adequate nutrients such as inorganic fertilizers (NPK), organic manure and the oxygen source to a contaminated site to

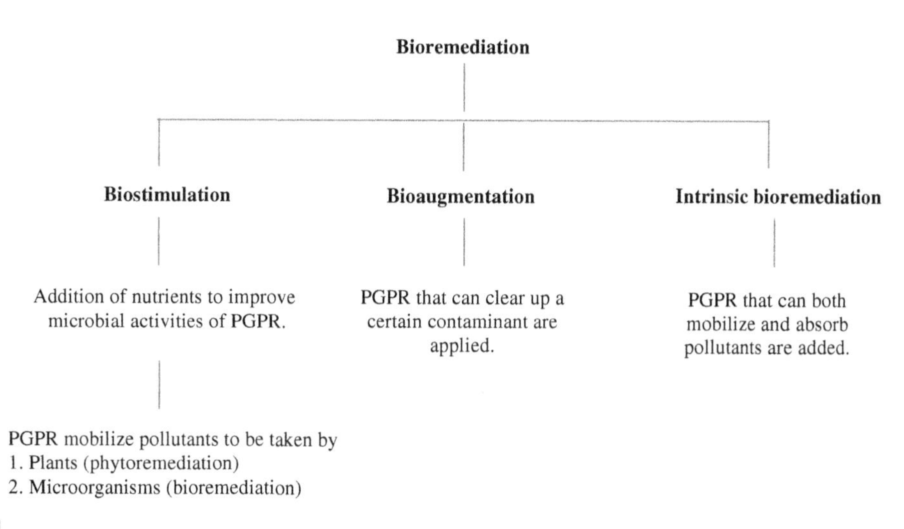

FIG. 11.1

Bioremediation approach.

improve the microbial activities of indigenous PGPR communities. These microorganisms either release essential substances into the soil to mobilize pollutants to be taken by plants (phytoremediation) or by other microorganisms (bioremediation). On the other hand, in bioaugmentation processes microorganisms (PGPR) that can clear up a certain contaminant are applied to the polluted soil and water. In intrinsic bioremediation, which is also called natural attenuation, microorganisms (PGPR) that can both mobilize and absorb pollutants and toxins are added to the contaminated soil and water.

An argument to highlight here is that the term bioremediation may be confused with biodegradation. As a technique, bioremediation can use biodegradation as just one of the pathways engaged or implemented in the bioremediation process. It should be noted that not all contaminants are biodegradable and in bioremediation a fraction of the biodegradable contaminants can be degraded by certain microorganisms. Therefore, studying the biodegrading potential of microorganisms would be worthwhile. Regardless of the reality that microorganisms have been used for the care and conversion of waste materials for at least a century, bioremediation is seen as a revolutionary technique for sustainable decontamination of contaminated ecosystems. Although the time to see the first decontamination effect is longer in bioremediation compared to the traditional methods, the cost is lower and the impact is longer. Moreover, techniques of bioremediation could be implemented together and combined with other therapeutic measures in an interconnected way. Depending on the type of living species associated with the technique, the forms of bioremediation may be classified as phytoremediation, microbial remediation and integrated remediation. Living species, particularly fungi, actinomycetes, plants and bacteria have demonstrated the potential to remediate soil polluted by pesticides and poisonous heavy metals such as Cd^{2+} and Hg^{2+}. Table 11.1 outlines the bioremediation reports of soils co-polluted with pesticides and heavy metals. Microbes are either used to detoxify or convert toxins, whereas plants are used to stabilize or eliminate soil pollutants (Fig. 11.2). Additionally, certain Rhizobacteria contribute significantly in the bioremediation of contaminated soil through heavy metal sequestration and xenobiotic compound degradation (pesticides, for example).

Bioremediation is a natural practice and is thus deemed by the community to be an appropriate process of waste disposal for polluted materials such as soil. Bioremediation has the potential advantages such as being highly effective (especially in big, low-contamination areas), easy-to-avail, eco-friendly, cost effective (Zhang et al., 2020). A diverse variety of bacteria from the rhizosphere (PGPR) such as *Rhizobium, Bradyrhizobium, Mycobcterium, Aerobacter, Azospirillum, Arthrobacter, Pseudomonas, Azotobacter, Micrococcus, Serratia, Clostridium, Cellulomonas, Bacillus, Flavbacterium, Derxia, Klebsiella, Azomonas, Actinomycetes,* etc., have been used for bioremediation (Keister et al., 1999). These PGPR use different pathways to increase the immunity of plants to abiotic stress, including osmoregulation; oligotrophic, endogenous metabolism; adaptation to starving; and effective metabolic functions (Lugtenberg et al., 2009; Egamberdiyeva et al., 2008). Abiotic and biotic components (temperature, pH, bioaccumulation of contaminants, interactions with toxicants, biological competition and biological condition) may all influence the biosorption of pesticides and metallic elements (Zhang et al., 2020). Bacteria with physiological adaptation and genetic capacity for increased resistance to drought, to salinity and to high temperatures, may enhance crop productivity at degraded sites (Yang et al., 2009; Enebel and Babalola, 2018). Similarly, in bioremediation of explosive materials, many factors such as environmental factors, type and number of microorganisms, source and concentration of nutrients and substrates affect the process (Abatenh et al., 2017).

Table 11.1 Bioremediation by microorganisms.

Microorganism used in bioremediation	Usage areas	Mechanism of action	Reference
Rhizobacteria (PGPR) in the rhizo-sphere	The toxic metals in soil (Zn, Pb, Cd)	"*Pseudomonas* strain RJ10 and *Bacillus* strain RJ16 to improve the soil mobility of lead and cadmium and to make it easier for a tomato cultivar with a Cd hyperaccumulator to absorb Cd and Pb during plant growth". Their study reported a 58%–104% and 67%–93% percent rise in the usable types of Cd and Pb in inoculated soil, respectively.	Bhattacharyya and Jha (2012)
Dreissena polymorpha	Wastewater	D. polymorpha is able to sustain itself until 21 days under good physiological environments, but with a high mortality rate (24 percent) at 28 days, a drop in filtration potential (8/15 mussels screened and 17.0 percent filtration rate) and stimulation of the antioxidant mechanisms.	Geba et al. (2021)
Geobacter sulfurreducens and *Shewanella Oneidensis*	Wastewater	Microbial fuel cells (MFCs) are built on the foundation of such bacteria, which are capable of producing an electric charge from organic pollutant degradation. MFCs' output voltage was used to illustrate p-nitrophenol biosensors in industrial wastewater.	Malvankar and Lovley (2014) Chen et al. (2016)
Polyphosphate kinase in *Escherichia coli*	-	A possible phosphate reduction application. Proteins on the luminal side of an artificial microcompartment of bacteria were successfully targeted in a recent study.	Lee et al. (2018)
Xanthobacter autotrophicus	Chlorinated aliphatic compounds	Microbial species present in the rhizosphere of plants are included. It has been shown that metabolites produced into this zone by plant roots can increase the biodegradation of 1,2-dichloroethylene	Mena-Benitez et al. (2008) Fraraccio et al. (2017)
Ideonella sakaiensis 201-F6 enzymes	Degradation of polyethylene terephthalate (PET),	Extracellular MHET operation was successfully passed to the marine microalgae Phaeodactylum tricornutum to use this device in aquatic environments.	Yoshida et al. (2016). Moog et al. (2020)
Streptomyces sp. M7 Actinobacteria	Real co-contaminated soils.	A decrease in the bioaccessibility of lindane (42%) and Cr (VI) (52%) show the positive outcomes for the remediation of Cr(VI) and lindane following Streptomyces sp. supplementation. M7	Polti et al. (2014)
Enzymes that are degrading (e.g., oxygenases, laccases, peroxidases, and dehydrogenases),	Pollution, especially pollution of soil and oceans	In order to form oxygenated intermediates, molecular oxygen is trapped and injected into the PAHs, which are further processed into inert, non-toxic, simpler products where laccases serve a broad and pivotal role depending on the specific characteristics and consequent distinct function.	Zhang et al. (2020)

(*continued*)

Table 11.1 Bioremediation by microorganisms. ***Continued***

Microorganism used in bioremediation	Usage areas	Mechanism of action	Reference
Pseudomonas, Acinetobacter, Rhizobium, Brevundimonas Penicillium canescens and *Aeromonas.*	Environmental bioremediation of heavy metals	The review of the literature shows that the overall percentage of transformation by microorganisms of various heavy metals were 27 % for Cr, 20% for Co, 31% for Cd, and 22% for Pb, comprising approximately 70% of the total material. Besides that, some metals, such as Ni is 07%, Zn 05%, Hg and As are around 18 %.	Ayangbenro and Babalola (2017) Gupta and Singh (2017) Layton et al. (2014)
Thiobacillus ferrooxidans, T. thiooxidans	Industrial waste containing radioactive heavy metal waste is often soluble in water and is readily combined with soil or water.	Catalyzed by microorganism-released enzymes	Cumberland et al. (2016)
Pseudomonas Rhodococcus Alcaligenes, Ralstonia Acinetobacter, Nocardia, Vibrio, and Achromobacter	Petroleum-hydrocarbon degradation	Environmental cleanup has traditionally relied on microorganisms or bioremediators.	Prince (2005) Head et al. (2006) Andreoni et al. (2000) Jayashree et al. (2012) Prakash and Muhammad (2011)
DR1-bf+ biofilm	Bioremediation of uranium	Results have shown that uranium remediation in the presence of Ca^2 by DR1-bf+ biofilm In the first 5 minutes after treatment, it was very rapid, extracting almost 25 percent of the original uranium concentration (1000 mg/L) and achieving a limit of 75 ± 2 percent within 30 minutes of treatment.	Manobala et al. (2019)
Stenotrophomonas maltophilia PAH-degrading strain	Agricultural waste that was powdery and high in potassium, nitrogen and phosphorus	The addition of a 1% soil conditioner will greatly modify soil properties and provide ample N, P, and K for microorganisms to encourage microbial growth and are essential in the bioremediation of oil-contaminated soil.	Liu et al. (2020)
Sulfate reducing bacteria (SRB) and iron-reducing microbes	A microbial-induced mineral transformation scheme for the bioremediation of pyritic waste rocks.	The acidic IRB lowers and increases the pH of Fe3+ to Fe2+ and then there's the SRB, which is less acid-tolerant.	Phyo et al. (2020)
Bacillus firmus B. iodinium Staphylococcus sp. *Streptomyces* sp., *Methylobacterium organophilum, Pseudomonas* sp., *Gemella* sp.,	Pb bioremediation of soil and water	PGPR have potential to accumulate the Pb.	Kumaran et al. (2011) Marzan et al. (2017)

Table 11.1 Bioremediation by microorganisms. ***Continued***

Microorganism used in bioremediation	Usage areas	Mechanism of action	Reference
Gemella sp. *Streptomyces* sp. *Bacillus firmus*, *Staphylococcus* sp., *Arthrobacter* strain D9, *Enterobacter cloacae*, *Micrococcus* sp. *Flavobacterium* sp. *Pseudomonas* sp. *Pseudomonas aeruginosa* (CH07) *A. faecalis* (GP06),	Cu bioremediation of soil and water	PGPR have potential to accumulate the Cu and increasing availabllity for phytoremediation.	Kumar et al. (2011) Bhattacharya and Gupta (2013)
Micrococcus sp., *Pseudomonas* sp., *Acinetobacter* sp.	Ni bioremediation of soil and water	PGPR have potential to accumulate the Ni and increase the availabllity for phytoremediaition	Vijayaraghavan and Yun (2008)
Acinetobacter sp. *Bacillus subtilis* *E. coli*, *Streptomyces sp.* *Pseudomonas aeruginosa*, *Immobilized P. aeruginosa (P bead) Immobilized* *Pseudomonas aeruginosa (B bead), B. subtilis*	Cr bioremediation of soil and water	PGPR have potential to accumulate the Cr.	Nayan et al. (2018) Onta˜non et al. (2018) Benazir et al. (2010)
Vibrio parahaemolyticus, Pseudomonas, *Vibrio fluvialis* *Bacillus licheniformis* *Klebsiella pneumoniae*,	Hg bioremediation of soil and water	PGPR have potential to accumulate the Hg.	Al- Garni et al. (2010), Jafari et al. (2015), Saranya et al. (2017)
Enterobacter cloacae	Co bioremediation of soil and water	PGPR have potential to accumulate the Co.	Jafari et al. (2015)
Brevibacterium epidermidis EZ-K02	Acetophenone, benzoate, gentisate, p-hydroxybenzoate, catechol, cobalt, cadmium and arsenic in waste and soil	In addition to its ability to accumulate these toxic substances, PGPR can break them down into simple organic compounds with its enzymes and hormones.	Ziganshina et al. (2018)

(*continued*)

Table 11.1 Bioremediation by microorganisms. *Continued*

Microorganism used in bioremediation	Usage areas	Mechanism of action	Reference
Pseudomonas, Rhodococcus, Bordetella, Chromobacter and Variovorax, Bacillus subtilis	Hexachlorocyclohexane (HCH), Polychlorinated biphenyls (protocatechuate biphenyl and Benzoate) in waste and soil		Sun et al. (2016)
Actinobacteria, Rhodococcus, Mycobacterium, Rhizobiales, Burkholderiales, Actinomycetales	Hydrocarbons, diesel, oil additives, gasoline, and Polycyclic aromatic hydrocarbons anthracene in waste and soil		Auffret et al. (2015) Cappelletti et al. (2016)

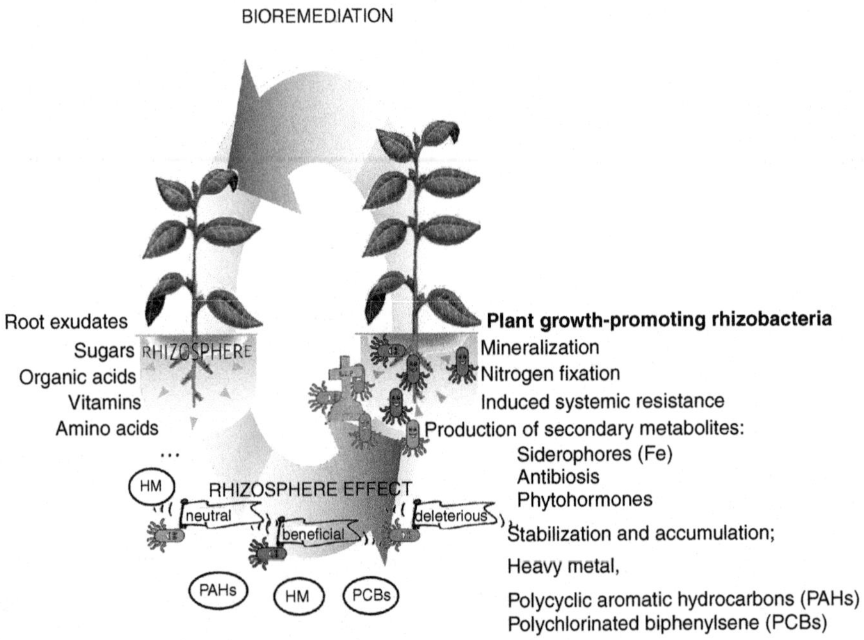

FIG. 11.2

Bioremedial mechanism of PGPR.

11.3 Importance of plant growth promoting rhizobacteria in bioremediaiton and phytoremediation

The function of PGPR in bioremediation (Fig. 11.2) is either through its direct effect on the plant growth (phytoremediation) or through acidification, chelation, precipitation, immobilization of rhizospheric heavy metals (Singh et al., 2014).

Rhizobacteria are soil bacteria capable of colonizing the mycorrhizosphere and stimulating the plant development with a broad range of mechanisms such as organic matter mineralization, soil-borne pathogen biological regulation, biological nitrogen fixation, root growth stimulation, phytohormone and biofertilizer production. A very fascinating aspect of PGPR is its ability to increase the bioavailability of nutrients. Several bacterial strains have been recognized as phosphorus-solubilizing microbes, while some other species are seen to improve the solubility of micronutrients, such as the ones that generate iron chelating compounds (siderophores). Nitrogen (N) in its elemental form is undesirable for plants. Nitrogen can be a source for plants only after many enzymatic processes are performed by the bacteria to convert molecular nitrogen into ammonium and nitrate. Thus, by increasing the bioavailability of nutrients, PGPRs play a beneficial role in biostimulation (Fig. 11.1) since (1) they positively affect phytoremediation by improving the plant tolerance to abiotic stress, and decreased water supply (Parnell et al., 2016) and (2) nitrogen can be utilized more by non-nitrogen-fixing PGPR in the microbial community due to the actions of nitrogen-fixing PGPR.

Many heavy metals and toxic elements have low mobility in nature. For this reason, their removal from soil, water and air by plants can sometimes be a problem due to their low availability. However, beneficial organisms with high biosorption capacities increase the uptake of metals by changing the bioavailability of hyper-accumulator plants. The use of these desirable bacteria increases the phytoremediation potential of nutrient-poor metal contaminated areas. Therefore, in order to ensure adequate level of remediation efficiency, besides the necessary nutrients for plants, microbes with metal tolerance are also added, and hyper accumulator plants and biosoption microorganisms take a joint role in cleaning these soils from harmful toxins. Thus, harmful toxins that cannot be removed from the soil with the help of plants are easily removed with the joint effect of PGPR and the plant. The use of metal tolerant microbes can therefore be crucial to promote detoxification of heavy metal contaminated soil (Ayangberno and Babalola, 2017). Moreover, Rhizobacteria are also vital for improving physical conditions of the soil by (1) increasing the water holding capacity, (2) increasing accumulation of humus, (3) increasing deposition of humus which can contribute to an increase in the buffer capacity, (4) gradual release of nutrients, and (5) enhancing the cation exchange capacity. Bacteria such as *Clostridium sp., Streptomyces sp., Cellulomonas sp., and Bacillus sp.,* aid in the decomposition of organic matter, in which complex organic compounds, such as cellulose, pectin, hemicellulose, lignin are transformed into simpler substances. These processes make the soil ideal for agricultural production (Ramprasad et al., 2014), thus for the optimum growth of plants which positively affect the phytoremediation. In addition, besides the stresses caused by the heavy metals, plants are exposed to many abiotic stress conditions such as salinity, low/high temperatures, high light intensity, alkalinity. PGPR has a very important effect in increasing the survival rates of plants by eliminating the negative outcomes of such stress conditions (Tak et al., 2013) which serves for the efficiency of phytoremediation.

11.4 Mechanisms involved in bioremediation by plant growth promoting rhizobacteria

Pesticide and heavy metal polluted land can be restored using living species, including bacteria, actinomycetes, fungi, and plants. The pollution might affect directly the organisms or the mechanisms of bioremediation are influenced by the co-existing pollutants (Zhang et al., 2020). The simultaneous presence of metals and pesticides can have a beneficial or harmful impact on the growth (microorganisms and plants), depending on habitats, growth phases, pollutant levels and characteristics, as well as soil properties (Lu et al., 2014).

Via a sequence of biochemical processes, heavy metals influence biological functions. Metal cations may link with multiple organic ligands, including enzymes, cell membranes, nucleic acids and microorganism proteins, when taken up by the cell. In certain situations, the binding is preferable. Biological activity can be induced by low metal concentrations of Cu, Fe Mn, Zn (Sandrin et al., 2007). Typically, though, metals could impair the functions of enzymes or proteins by masquerading as catalytic active groups, or modifying structures of proteins (Ye et al., 2017). Toxic metal cations can substitute enzyme cations that are physiologically necessary and make them non-functional (Sandrin et al., 2003). Moreover, oxidative pressure is exerted on microorganisms by metals (Kachur et al., 1998). Tight adherence of metallic ions to groups of sulfhydryl enzymes which are essential to microbial metabolism, is the most prevalent cause of metal toxicity (Olaniran et al., 2009). Toxic metals icluding copper (Cu), zinc (Zn), cromium (Cr), lead (Pb), and mercury (Hg) can impair some or all biological activities, triggering various kind of reactions (Simon et al., 2017).

Despite the likely toxicity, depending on the resistance/tolerance mechanisms, many species may thrive in metal-contaminated environments. However, their biomass, physiology, development, biological activity and mycelium colonization will also be influenced to various degrees. Also, metallic elements can influence macronutrient content such as Na, K, N, P, Ca and Mg. Lower Cu amounts, for instance, enhanced the uptake of macronutrients for *Vigna radiata* L., while elevated heavy metal levels, on the other hand, disrupt macronutrient adsorption (Ca, Mg and Na) (Karaca et al., 2010; Bartlett, 2013).

Biodegradative-bacteria, plant growth promoting rhizobacteria (PGPR), and arbuscular mycorrhizal fungi (AMF) have acquired importance in soil remediation amongst the rhizospheric microbes. Biodegradative-bacteria are the soil bacteria capable of degrading organic contaminants. PGPR produce different compounds or metabolites, such as phytohormones, siderophores, and ammonia, that stimulate plant development, by fixing atmospheric N and turning nonsoluble P into the soluble form, certain PGPR may supply N and P for plants. Furthermore, PGPR excrete some other metabolites, including organic acids, exopolysaccharide (EPS), siderophores, biosurfactants, metal chelating agents, antioxidants and antibiotics, which minimize plant stress induced by pathogens, organic contaminants and phytotoxicity of heavy metals (Hussain et al., 2018). In the meanwhile, such metabolites and compounds are not just useful for plant development, yet often supportive in degrading organic contaminants and altering the existence of heavy metals (Manoj et al., 2020).

Moreover, microorganisms in the rhizosphere may use their own oxidation mechanisms to decompose organic contaminants. Plant-microbe interactions increase the production of rhizosphere catabolic enzymes (Hong et al., 2015). Microorganism-produced enzymes are involved in oxidizing, mineralizing, stabilizing, sequestrating and detoxifiying organic and heavy metal contaminants (Glick, 2010; Gullap et al., 2014). For toxins that microorganisms do not use as sources of carbon and energy, the joint effect of the host plant and microbes promotes biodegradation of the toxins. It is interesting that another significant plant-microbe interaction for the oxidation of organic compounds is the endophytic (bacteria and fungi) relationship. Endophytes deteriorate organic pollutants and promote the endosphere and rhizosphere when treated with heavy metals (Feng et al., 2017). When a consortium of microbial strains is used instead of a single strain culture, bioremediation of heavy metals becomes more effective (Wang and Chen, 2009).

Metals can decrease the enzyme activation in the soil and extend the half-life of organic contaminants. The lower levels of metallic elements will activate the microbes of the rhizosphere and encourage organic pollutant degradation, whereas elevated levels of heavy metals suppress microbial activity and impede organic pollutant degradation (Yang, 2004). High levels of heavy metals can be tolerated by PGPR microorganisms isolated from polluted soils (Pishchik et al., 2002). Such microbes have mechanisms in place to boost and remove or perform metabolic processes from cells to convert metals into non-toxic types (Grobelak et al., 2015). To prolong the lifetime of PGPR introduced into the soil, immobilization techniques are being used. The type of carrier used for immobilization must be chosen carefully (Yang et al., 2009).

In contrast to uninoculated treatment, the application of PGPR to soil treated with waste sludge enhances the bioavailability of organic N and P. The boost observed after microbial inoculation implies that waste sludge and/or potassium phosphate are synergistically correlated with PGPR.

The use of microbial products in precision agriculture can be viewed as a desirable improvement in the soil. PGPR may be a feasible way of improving soil conservation, soil bioremediation, and the number of plant cultivation areas. Heavy metals are believed to dislocate essential components of biological molecules, impairing the molecules' roles and modifying the composition or role of the proteins, enzymes or membrane transporters, hence having toxic effects on plants (Tak et al., 2013). With the current focus on supporting sustainable development and limiting the consumption of natural resources, Rhizobacteria can be considered as a good alternative to synthetic fertilizers to encourage crop productivity and increase the quality of agri-food sludge.

As discussed earlier, traditional methods require electricity and chemical reagents, these methods are usually costly and even though they are successful, secondary noxious byproducts are produced (Selatnia et al., 2004). Using indigenous microorganisms with pathways capable of degrading certain heavy metals or genetically modified microorganisms to handle contaminated habitats by transforming radioactive heavy metals into non-hazardous forms is an important way to remove toxic metal pollutants from the atmosphere and stabilize the ecosystem (Gupta and Diwan, 2016). However, for the bioremediation process to be effective, only microbes with a demonstrated capacity to mitigate and withstand excessive toxicity are to be used.

11.5 The functions of plant growth promoting rhizobacteria in phytoremediation

In phytoremediation, plants are used to remove toxic substances from soil, water and air. The plants used for this purpose can be cultivated plants, as well as hyperaccumulators that have tolerance to high concentrations of non-essential material in their bodies. PGPR have a direct and indirect effect on phytoremediation. They might have a direct impact on the growth of plants and the root development; or an indirect effect, by transforming toxic elements into forms with increased mobility for an easier uptake (Turan and Angın, 2004).

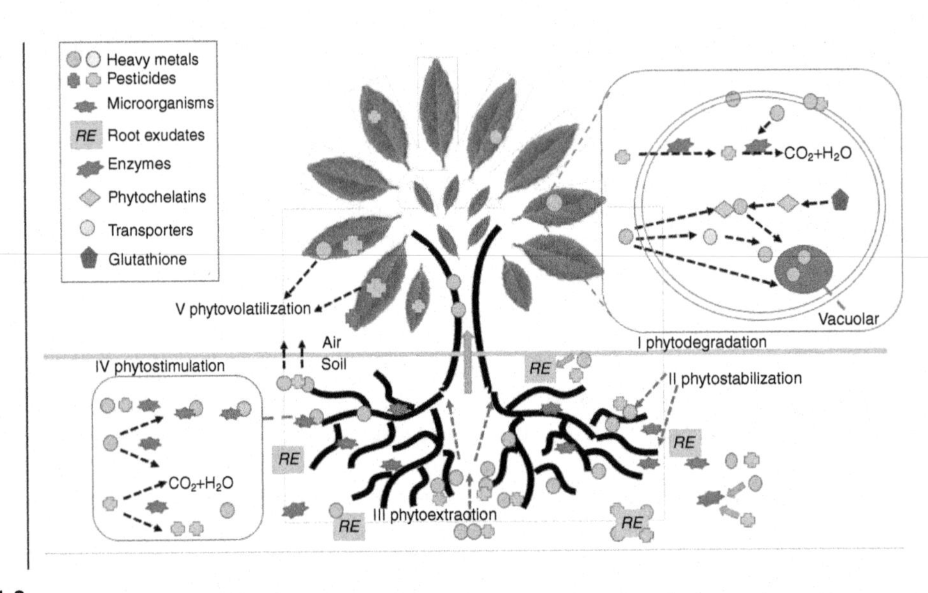

FIG. 11.3

The phyto-remediation mechanisms used for heavy metal and pesticide co-contaminated soil.

Phytoremediation requires the use of biomass and rhizospheric microbiota to cope with the pollution resulted from organic toxins and metals. Phytoremediation is amongst the most low-cost mitigation alternatives, but is sometimes not feasible enough as a self-sufficient tool to the farmers, particularly in economically deprived areas. If the removed contaminants are not adequately disposed, there is a possibility that the remediation process simply turn into a bigger concern (pollution shell game) (Song and Park, 2017). There are several methods used in phytoremediation. Each segment of plants appears to adopt a particular mode of action, including phytodegradation, phytostabilization, phytoextraction, phytostimulation, and phytovolatilization (Fig. 11.3). Phytodegradation is decomposition of pollutants into simpler forms by the enzymes and metabolites secreted from plants roots. Phytostabilization is the use of plants to reduce heavy metal mobility, thus restrict their bioavailability, via absorption and precipitation. Phytoextraction entails a process that is used by plants to accumulate metals from the soil into the plant's roots and shoots. Phytostimulation is mobilizing the heavy metals in rhizosphere by secreting secondary metabolites from their roots. Phytovolatilization is the absorption and emission of volatile compounds containing mercury or arsenic metals into the atmosphere (Jing et al., 2007).

In phytoremediation studies, the focus has been on detecting hyperaccumulators such as *Vittata* (As), *Alyssum* Sp (Ni and Co), *Noccaea caerulescens* (Cd, Ni, Pb, Zn), *Berkheya coddii* (Ni), and roses (*Rosa gallica*) (Cd, Ni, Pb) (Dickinson et al., 2009; Esringu et al., 2015; Mahar et al., 2016). Plants that extract phytonutrients but do not accumulate toxins in the edible part have not been thoroughly examined, however some promising findings have suggested that certain plant species may pass organochlorines and heavy metals to roots, stems, and leaves, despite the fact that edible components have much lower concentrations (Liu et al., 2012; Haller et al., 2017). Harvesting is perhaps amongst the most time-consuming and expensive phytoremediation steps (Song et al., 2016) and the processing of polluted plant matter is the main deterrent to use phytoremediation commercially (Mohanty, 2016).

Certain plants demonstrate remarkable resistance and reclamation capability against co-existence of various pesticides and metallic elements. Via root systems, plants remove toxins, shift from the below ground parts to the shoots, and either absorb or convert them, (heavy metals) in the body parts or destroy or mineralize (pesticides) during metabolic processes. In particular, plant root exudates encourage the growth of rhizosphere organisms and increase the diversity of the bacterial population, all of which foster the degradation of pesticides (Teng et al., 2011; Huang et al., 2011; Montiel-Rozas et al., 2016). Preliminary phytoremediation trials on DDT (dichlorodiphenyltrichloroethane) and Cd co-contaminated soil in China's Zhejiang Province using 23 Ricinus communis genotypes were performed (Huang et al., 2011). Certain genotypes of castor may be considered as potential species of DDT and Cd accumulator plants. Plenty other plants have been shown to be able to fix co-contaminated soils with pesticides and metals, including *Lolium perenne* L., *Medicago sativa* L., *Spinacia oleracea*, *Cucurbita pepo*, *Brassica napus* L., *Vetiveria zizanioides* L., and *Brassica juncea* L. (Lin et al., 2006; Turan and Esringu, 2007; Angın et al., 2008; Zhang et al., 2019).

The remediation capacity of plants is, however, often not as successful as thought. The phyto-remediation capacity of wetland plant species *Phragmites australis* in co-polluted soils with Cd and PCP (pentachlorophenol) was assessed (Hechmi et al., 2014). The bioconcentration factor and translocation factor values of Cd were lower than 1.0 and the presence of Cd greatly affected the dissipation of PCP, which showed that *P. australis* cannot be used for phytoremediation. Likewise, high PCP concentrations (100 mg/kg) prevented PCP degradation and Cd deposition in maize (Hechmi et al., 2013).

In plants with real landfill soils co-contaminated with organic compounds and heavy metals, the growth of *Lupinus luteus* L. was inhibited, endophytes became impossible to colonize, except at the lowest level of contamination (Gutierrez Gines et al., 2014). This phytoremediation insufficiency was largely due to the combined effect of metallic elements and organic contaminants: (1) plant growth was greatly impacted by heavy metals, organic compounds and their associations, therefore, the overall biomass of plants was drastically decreased, (2) the absorption and conversion of contaminants by plants was restricted, (3) the alteration in the microbial population of the rhizosphere and behavior and/or root physiology was not conducive to the dissipation of organic compounds and the absorption of metals, (4) the biota of soils had been synergistically disrupted, (5) colonization by endophytic bacteria was ineffective. In addition, the effectiveness of phytoremediation is often influenced by plant varieties, plant growth phases, physicochemical properties of soil and experimental circumstances. For effective remediation, the biotic and abiotic variables are needed to be taken into consideration.

Widely used PGPR strains for bioremediation (Table 11.1) are *Azospirillum brasilense, Pseudomonas putida, Serratia liquefaciens, Enterobacter cloacae,* and *Pseudomonas aeruginosa* (Bhattacharyya and Jha, 2012). The inclusion of endophytes will greatly improve the effectiveness of phytoextraction (He et al., 2009, 2013). Two *Pseudomonas* sp. strains tolerant to Cd, RJ10 as well as *Bacillus sp.* RJ16 were used to improve the movement of lead and cadmium in soil to make it easier for a tomato cultivar with a Cd hyperaccumulator to absorb Cd and Pb during plant growth. Compared with uninoculated controls, in inoculated soil, usable forms of Pb and Cd and increased by 67%–93% and 58%–104%, respectively. In the tomato plants tested, inoculated plants grown in heavy metal-polluted soil relative to uninoculated plants, the rise in Cd and Pb concentration in the above-ground ranged from 70% to over 110%, respectively (He et al., 2009).

PGPR inoculation also has the ability to boost the efficacy of phytoremediation (He et al., 2013). Plants of the *Brassica napus* genus inoculated with *Rahnella* JN6 strain the stress caused by metal behavior was reduced, due to the secreted ACC deaminase by the bacteria. Plants showed improved shoots and roots length and root biomass. The concentrations and accumulation of Cd, Pb, and Zn in rape plants inoculated with JN6 isolate were significantly higher in over and underground plant parts (shoots and roots tissues) in comparison to Cd, Pb, and Zn-modified soils without inoculation. Such findings imply that the bacterial strains may be used to optimize microbial phytoextraction of heavy metals such as Pb and Cd contaminated lands. However, it's difficult to optimize conditions for microbe inoculation of particular plants since the bacterial consortium's influence is proportional to the inoculum density and the species of the plant, as well as on the stage of plant growth (Karami and Shamsuddin, 2010). PGPR also increases the phytostabilization efficiency. A strain of *Pseudomonas maltophilia*, for example, can convert toxic and mobile Cr^{6+} to non-toxic and immobile Cr^{3+} which also decrease the sensitivity of the plant to other toxic ions such as Hg^{2+}, Pb^{2+}, and Cd^{2+} (Blake et al., 1993; Park et al., 1999).

The productivity of metal phytoextraction can be improved by joint plantation and inoculation with rhizobacteria, primarily by improving the production of dry biomass and the overall survivorship of crops cultivated on infected land. For instance, the addition of K ascorbata SUD165/26 increased the amount of Indian mustard seeds that sprouted in a soil contaminated with nickel and the achievable plant size between 50%–100% (Burd et al., 1998).

Significant biological interactions have arisen in recent years between microbes and native plant species, suggesting that many types of bacteria live in these ecosystems play an important role in allowing these plants to withstand certain environmental factors and even pollutants (Al-Thani and Yasseen, 2018). A significant variety of microbes and plants have already been recognized with their capacity to extract toxins of different forms from contaminated waters and soils and/or destroy them. Thus, phytoremediation approaches to degrade pollutants and remove different forms of contaminants from soil and water, needs to be in harmony with native plants and microorganisms.

11.6 Conclusions and future outlooks

With bioremediation methods, it is possible to convert/degrade/remove all toxic substances released as a result of natural phenomena or human activities in a more environmentally friendly, more economical and more sustainable manner. In order for this technology to be more effective in the near future, it must be integrated with information systems. The diagnosis of the organisms involved, the pollution parameters they affect, the degree of effectiveness, the cost, duration, and ecological definitions should be shared via information systems to create a common network with scientists and sector representatives working on this issue all over the world. The flow of the information through this network would be crucial in solving current problems rapidly and maintaining sustainability.

The presence of co-contaminated heavy metal soils and pesticides poses a threat to natural ecosystems and also has a detrimental effect on human health. In the remediation of such soils, bioremediation has been shown to be successful. This chapter outlined the bioremediation applications, particularly for metal and pesticide co-contaminated soils which is quite challenging. Focus was given to the impact of co-existing contaminants on the bioremediation of pollutants.

The following concerns should be addressed in future studies to facilitate bioremediation of co-contaminated soils: (1) When organic contaminants are included in the remediation goals, remediation plans must be established not only for the parent organic contaminants but also for their metabolites. These metabolites should be taken into account since certain derivatives of such organic contaminants have a prolonged half-life, or are much more hazardous than their parent toxins. (2) In order to prevent a secondary contamination, after-treatments should be planned to handle microorganisms and plants used for bioremediation. The biomass should be disposed as hazardous waste. Besides, more research is needed to understand plant-plant, microbe-microbe and plant-microbe relationships in polluted environments. In this respect, heavy metals may be eliminated from contaminated soils, either by increasing metal-accumulating capacity of the plant or by increasing the plant biomass. Especially in highly contaminated soils where the metal content reaches the plant tolerance level, plant biomass can be increased by inoculating plants with rhizobacteria.

In conclusion, bioremediation is an economical and sustainable approach to solve the pollutant problems and PGPR play an important role in the efficiency of remediation. Since the contamination of soil, particularly with heavy metals and pesticides, has come to a critical level which affects the ecology and human health dramatically, there is a dire need to involve information systems to create data banks that provide a specific solution to a specific problem in order to act rapidly to conserve soil, water and air, in a word our planet.

References

Adams, G.O., Fufeyin, P.T., Okoro, S.E., 2015. Ehinomen I. bioremediation, biostimulation and bioaugmention: a review. Int. J. Environ. Bioremediat. Biodegrad. 3 (1), 28–39.

Abatenh, E., Gizaw, B., Tsegaye, Z., Wassie, M., 2017. The role of microorganisms in bioremediation - a review. Open J. Environ. Biol. 2 (1), 038–046.

Adhikari, T., Manna, M.C., Singh, M.V., Wanjari, R.H., 2004. Bioremediation measure to minimize heavy metals accumulation in soils and crops irrigated with city effluent. J. Food. Agric. Environ. 2 (1), 266–270.

Al-Garni, S.M., Ghanem, K.M., Ibrahim, A.S., 2010. Biosorption of mercury by capsulated and slime layerforming Gram and *bacilli* from an aqueous solution. Afr. J. Biotechnol. 9 (38), 6413–6421.

Al-Thani, R.F., Yasseen, B.T., 2018. Solutes in native plants in the Arabian Gulf region and the role of microorganisms: future research. J. Plant Ecol. 11 (5), 671–684. https://doi.org/10.1093/jpe/rtx066.

Alvarez, A., Saez, J.M., Davila Costa, J.S., Colin, V.L., Fuentes, M.S., Cuozzo, S.A., Benimeli, C.S., Polti, M.A., Amoroso, M.J., 2017. Actinobacteria: current research and perspectives for bioremediation of pesticides and heavy metals. Chemosphere 166, 41–62.

Andreoni, V., Bernasconi, S., Colombo, M.J.B., Cavalca, L., 2000. Detection of genes for alkane and naphthalene catabolism in Rhodococcus sp. strain 1BN. Environ. Microbiol. 2 (5), 572–577.

Angın, İ., Turan, M., Quirine, M.K., Çakıcı, A., 2008. Humic acid addition enhances B and Pb phytoextraction by vetiver grass (*Vetiveria zizanioides* (L). Nash). Water Air Soil Pollut. 188 (1), 335–343.

Asha, L., Sandeep, P., Reddy, S., 2013. Review on bioremediation potential tool for removing. Environ. Pollut., Int. J. Basic Appl. Chem. Sci. 3 (3), 21–33.

Auffret, M.D., Yergeau, E., Labb, D., Fayolle-Guichard, F.C., Greer, W., 2015. Importance of rhodococcus strains in a bacterial consortium degrading a mixture of hydrocarbons, gasoline, and diesel oil additives revealed by metatranscriptomic analysis. Appl. Microbiol. Biotechnol. 99 (5), 2419–2430.

Ayangbenro, A.S., Babalola, O.O., 2017. New strategy for heavy metal polluted environments: a review of microbial biosorbents. Int. J. Environ. Res. Public Health 14, 94.

Bartlett, R., 2013. Interactions of copper and pyrene on phytoremediation potential of Brassica juncea in copper–pyrene co-contaminated soil. Chemosphere 90 (10), 2542–2548.

Benazir, J.F., Suganthi, R., Rajvel, D., Pooja, M.P., Mathithumilan, B., 2010. Bioremediation of chromiumin tannery effluent by microbial consortia. Afr. J. Biotechnol. 9 (21), 3140–3143.

Bhattacharyya, P.N., Jha, D.K., 2012. Plant growth-promoting rhizobacteria (PGPR): emergence in agriculture. World J. Microbiol. Biotech. 28 (4), 1327–1350. https://doi.org/10.1007/s11274-011-0979-9.

Bhattacharya, A, Gupta, A., 2013. Evaluation of Acinetobacter sp. B9 for Cr (VI) resistance and detoxification with potential application in bioremediation of heavy-metals-rich industrial wastewater. Environ. Sci. Pollut. Res. 20 (9), 6628–6637.

Blake, R.C., Choate, D.M., Bardhan, S., Revis, N., Barton, L.L., Zocco, T.G., 1993. Chemical transformation of toxic metals by *a Pseudomonas* strain from a toxic waste site. Environ. Toxicol. Chem. 12 (5), 1365–1376.

Burd, G.I., Dixon, D.G., Glick, B.R., 1998. A plant growth-promoting bacterium that decreases nickel toxicity in seedlings. Appl. Environ. Microbiol. 64 (3), 3663–3668.

Cappelletti, M., Fedi, S., Zampolli, J., A.Di Canito, P., D'Ursi, A., Orro, C., Viti, L., Milanes, D., Zannoni, P., Gennaro, Di., 2016. Phenotype microarray analysis may unravel genetic determinants of the stress response by Rhodococcus aetherivorans BCP1 and Rhodococcus opacus R7. Res. Microbiol. 167 (9–10), 766–773.

Chen, Z., Niu, Y., Zhao, S., Khan, A., Ling, Z., Chen, Y., Liu, P., Li, X., 2016. A novel biosensor for p-nitrophenol based on an aerobic anode microbial fuel cell. Biosens. Bioelectron. 15 (85), 860–868.

Cumberland, SA., Douglas, G., Grice, K., Moreau, JW., 2016. Uranium mobilityin organic matter-rich sediments: a review of geological and geochemical processes. Earth Sci. Rev. 159, 160–185.

Cycon, M., Mrozik, A., Piotrowska-Seget, Z., 2017. Bioaugmentation as a strategy for the remediation of pesticide-polluted soil: a review. Chemosphere 172, 52–71.

Dangy, AK., Sharma, B., Hill, RT., Shukla, P., 2018. Bioremediation through microbes: systems biology and metabolic engineering approach: a review. Critic. Rev. Biotech. 39 (1), 79–98.

Dickinson, N.M., Baker, A.J.M., Doronila, A., Laidlaw, S., Reeves, R.D., 2009. Phytoremediation of inorganics: realism and synergies. Int. J. Phytoremediation 11 (2), 97–114.

Egamberdiyeva, D., Islam, KR., 2008. Salt tolerant rhizobacteria:plant growth promoting traits and physiological characterization within ecologically stressed environment. In: Ahmad, I, Pichtel, J, Hayat, S (Eds.), Plant-Bacteria Interactions: Strategies and Techniques to Promote Plant Growth. Wiley-VCH, Weinheim, pp. 257–281.

Esringu, A., Kulekçi, A.E., Turan, M., Ercişili, S., 2015. Phytoremediation of some heavy metals by different tissues of roses grown in the main intersections in Erzurum city, Turkey. Fresenius. Environ. Bull. 24 (9), 2787–2791.

Enebel, M.C., Babalola, O.O., 2018. The influence of plant growth-promoting rhizobacteria in planttolerance to abiotic stress: a survival strategy. Appl. Microbiol. Biotechnol. 102 (18), 7821–7835.

FAO, ITPS, 2015. Status of the World's Soil Resources (SWSR) – Main Report. Food and Agriculture Organization of the United Nations and Intergovernmental Technical Panel on Soils (Italy Rome). ISBN 978-92-5-109004-6.

Feng, N.X., Yu, J., Zhao, H.M., Cheng, Y.T., Mo, C.H., Cai, Q.Y., Li, Y.W., Li, H., Wong, M.H., 2017. Efficient phytoremediation of organic contaminants in soils using plant-endophyte partnerships. Sci. Total. Environ. 583 (1), 352–368.

Fraraccio, S., Strejcek, M., Dolinova, I., Macek, T., Uhlik, O., 2017. Secondary compound hypothesis revisited: selected plant secondary metabolites promote bacterial degradation of cis-1,2- dichloroethylene (cDCE). Sci. Rep. 7, 8406. doi:10.1038/s41598-017-07760-1.

Geba, E., Rioult, D., Palluel, O., Dedourge-Geffard, O., Betoulle, S., Aubert, D., Bigot-Clivot, A.E., 2021. Resilience of *Dreissena polymorpha* in wastewater effluent: use as a bioremediation tool. J. Environ. Manage. 278, 111513. doi:10.1016/j.jenvman.2020.111513.

Glick, B.R., 2010. Using soil bacteria to facilitate phytoremediation. Biotechnol. Adv. 28 (3), 367–374.

Grobelak, A., Napora, A., Kacprzak, M., 2015. Using plant growth-promoting rhizobacteria (PGPR) to improve plant growth. Ecol. Eng. 84, 22–28. http://dx.doi.org/10.1016/j.ecoleng.2015.07.019.

Guo, K., Liu, Y.F., Zeng, C., Chen, Y.Y., Wei, X.J., 2014. Global research on soilcontamination from 1999 to 2012: a bibliometric analysis. Acta Agric. Scand. B Soil Plant Sci. 64 (5), 377–391.

Gullap, M.K., Dascı, M., Erkovan, H.İ., Koç, A., Turan, M., 2014. Plant growth promoting rhizobacteria (PGPR) and phosphorus fertilizer assisted phytoextraction of toxic heavy metal elements from contaminated soils with meadow. Commun. Soil Sci. Plant Anal. 45 (19), 2593–2606.

Gupta, P., Diwan, B., 2017. Bacterial exopolysaccharide mediated heavy metal removal: a review on biosynthesis, mechanism and remediation strategies. Biotechnol. 13, 58–71.

Gupta, S., Singh, D., et al., 2017. Role of genetically modified microorganisms in heavy metal bioremediation. In: Kumar, R. et al. (Ed.), Advances in Environmental Biotechnology. Springer, Singapore. https://doi.org/10.1007/978-981-10-4041-2_12.

Gutierrez-Gines, M.J., Hernandez, A.J., Perez-Leblic, M.I., Pastor, J., Vangronsveld, J., 2014. Phytoremediation of soils co-contaminated by organic compounds and heavy metals: bioassays with Lupinus luteus L. and associated endophytic bacteria. J. Environ. Manage. 143, 197–207.

Haller, H., Jonsson, A., Lacayo Romero, M., Jarquín Pascua, M., 2018. Bioaccumulation and translocation of field-weathered toxaphene and other persistent organic pollutants in three cultivars of amaranth (A. Cruentus 'R127 M_exico', A. Cruentus 'Don Le_on' Y A. Caudatus 'CAC 48 Perú') e a field study from former cotton fields in Chinandega, Nicaragua. Ecol. Eng. 121, 65–71.

He, L.Y., Chen, Z.J., Ren, G.D., Zhang, Y.F., Qian, M., Sheng, X.F., 2009. Increased cadmium and lead uptake of a cadmium hyperaccumulator tomato by cadmium-resistant bacteria. Ecotoxicol. Environ. Saf. 72 (5), 1343–1348.

He, H.D., Ye, Z.H., Yang, D.J., Yan, J.L., Xiao, L., Zhong, T., 2013. Characterization of endophytic Rahnella sp. JN6 from Polygonum pubescens and its potential in promoting growth and Cd, Pb, Zn uptake by Brassica napus. Chemosphere 90 (6), 1960–1965. https://doi.org/10.1016/j.

Head, I.M, Jones, D.M., Rooling, W.F., 2006. Marine microorganisms make a meal of oil. Nat Rev Microbiol 4, 173–182.

Hechmi, N, Ben Aissa, B., Abdennaceur, H., Jedidi, N, 2013. Phytoremediation potential of maize (*Zea mays* L.) in co-contaminated soils with pentachlorophenol and cadmium. Int. J. Phytorem. 15 (7), 703–713.

Hechmi, N., Ben Aissa, N., Abdenaceur, H., Jedidi, N., 2014. Evaluating the phytoremediation potential of Phragmites australis grown in pentachlorophenol and cadmium co-contaminated soils. Environ. Sci. Pollut. Res. 21 (2), 1304–1313.

Hong, Y., Liao, D., Chen, J., Khan, S., Su, J., Li, H.A., 2015. Comprehensive study of the impact of polycyclic aromatic hydrocarbons (PAHs) contamination on salt marsh plants Spartina alterniflora: implication for plant-microbe interactions in phytoremediation. Environ. Sci. Pollut. Res. 22 (9), 7071–7081.

Hooper, S.W., Pettigrew, C.A., Sayler, G.S., 2009. Ecological fate, effects and prospects for the elimination of environmental polychlorinated biphenyls (PCBs). Environ. Toxicol. Chem. 9 (5), 655–667.

Huang, H., Yu, N., Wang, L., Gupta, D.K, He, Z., Wang, K., Zhu, Z., Yan, X, Li, T., Yang, X.E., 2011. The phytoremediation potential of bioenergy crop Ricinus communis for DDTs and cadmium co-contaminated soil. Bioresour. Technol. 102 (23), 11034–11038.

Hussain, I., Aleti, G., Naidu, R., Puschenreiter, M., Mahmood, Q., Rahman, M.M., Wang, F., Shaheen, S., Syed, J.H., 2018. Reichenauer, Microbe and plant assistedremediation of organic xenobiotics and its enhancement by genetically modified organisms and recombinant technology: a review. Sci. Total Environ. 628–629, 1582–1599.

Jafari, S.A., Cheraghi, S., Mirbakhsh, M., Mirza, R., Maryamabadi, A., 2015. Employing response surface methodology for optimization of mercury bioremediation by *Vibrio parahaemolyticus* PG02 in coastal sediments of Bushehr, Iran. CLEAN Soil, Air, Water 43 (1), 118–126.

Jayashree, R., Nithya, S.E., Rajesh, P.P., Krishnaraju, M., 2012. Biodegradation capability of bacterial species isolated from oil contaminated soil. J. Acad. Ind. 1 (3), 140–143.

Jing, Y., Zhen, H.E., Xiao, Y., 2007. Role of soil rhizobacteria in phytoremediation of heavy metal contaminated soils. J. Zhejiang Univ. Sci. B. 8 (3), 192–207.

Jobby, R., Jha, P., Yadav, A.K., Desai, N., 2018. Biosorption and biotransformation ofhexavalent chromium [Cr(VI)]: a comprehensive review. Chemosphere 207, 255–266.

Kachur, A.V., Koch, C.J., Biaglow, J.E., 1998. Mechanism of copper-catalyzed oxidation of glutathione. Free Radical Res. 28 (3), 259–269.

Karaca, A., Cetin, S.C., Turgay, O.C., Kizilkaya, R., 2010. Effects of heavy metals on soil enzyme activities, soil heavy metals, Springer, Berlin, Heidelberg, pp. 237–262.

Karami, A, Shamsuddin, Z., 2010. Phytoremediation of heavy metals with several efficiency enhancer methods. Afr. J. Biotech. 9 (25), 3689–3698.

Keister, D.L., Cregan, P.B., 1999The Rhizosphere and Plant Growth10. Kluver Academic Publishers, Dordrecht, Netherlands, pp. 315–326.

Kumar, R., Bhatia, D., Singh, R., Rani, S., Bishnoi, N.R., 2011. Sorption of heavymetals fromelectroplating effluent using immobilized biomass Trichoderma viride in a continuous packed-bed column. Int. Biodeter. Biodegr. 65 (8), 1133–1139.

Kumaran, N.S., Sundaramanicam, A., Bragadeeswaran, S., 2011. Adsorption studies on heavymetals by isolated cyanobacterial strain (nostoc sp.) from uppanar estuarine water, southeast coast of India. J. Appl. Sci. Res. 7 (11), 1609–1615.

Layton, A.C., Chauhan, A., Williams, D.E., Mailloux, B., Knappett, P.S, Ferguson, A.S., McKay, L.D., Alam, M.J., Ahmed, K.M., van Geen, A., Sayler, G.S., 2014. Metagenomes of microbial communities in arsenic- and pathogen-contaminated well and surface water from Bangladesh. Genome Announc. 2 (6), e01170–14.

Lee, M.J., Mantell, J., Brown, I.R., Fletcher, J.M., Verkade, P., Pickersgill, R.W., Woolfson, D.N., Frank, S., Warren, M.J., 2018. De novo targeting to the cytoplasmic and luminal side of bacterial microcompartments. Nat. Commum. 9, 3413.

Lin, Q., Wang, Z., Ma, S., Chen, Y., 2006. Evaluation of dissipation mechanisms by Lolium perenne L, and Raphanus sativus for pentachlorophenol (PCP) in copper co-contaminated soil. Sci. Total. Environ. 368 (2–3), 814–822.

Liu, L., Hu, L., Tang, J., Li, Y., Zhang, Q., Chen, X., 2012. Food safety assessment of planting patterns of four vegetable-type crops grown in soil contaminated by electronic waste activities. Environ. Manage 93 (1), 22–30. doi:10.1016/j.jenvman.2011.08.021.

Liu, H., Tan, X., Guo, J., Liang, X., Xie, Q., Chen, S., 2020. Bioremediation of oil-contaminated soil by combination of soil conditioner and microorganism. J. Soils Sediments 20 (4), 2121–2129. https://doi.org/10.1007/s11368-020-02591-6.

Lovley, D.R., 2003. Cleaning up with genomics: applying molecular biology to bioremediation. Nat. Rev. Microbiol. 1 (1), 35–44.

Lu, M., Zhang, Z.Z., Wang, J.X., Zhang, M., Xu, Y.X., Wu, X.J., 2014. Interaction of heavy metals and pyrene on their fates in soil and tall fescue (*Festuca arundinacea*). Environ. Sci. Technol. 48 (2), 1158–1165.

Lugtenberg, K., 2009. Plant growth promoting rhizobacteria. Ann. Rev. Microbiol 63, 549–556.

Mahar, A., Wang, P., Ali, A., Awasthi, M.K., Lahori, A.H., Wang, Q., Li, R., Zhang, Z., 2016. Challenges and opportunities in the phytoremediation of heavy metals contaminated soils: a review. Ecotoxicol. Environ. Saf. 126, 111–121.

Malvankar, N.S., Lovley, D.R., 2014. Microbial nanowires for bioenergy applications. Curr. Opin. Biotechnol. 27, 88–95.

Manobala, T., Sudhır, K., Shukla, S., Rao, S.T., Kumar, D.M., 2019. A new uranium bioremediation approach using radio-tolerant Deinococcus radiodurans biofilm. Indian Academy of Sciences. J. Biosci. 44 (5), 122. doi:10.1007/s12038-019-9942-y.

Manoj, S.R., Karthik, C., Kadirvelu, K., Arulselvi, P.I., Shanmugasundaram, T., Bruno, B., Rajkumar, M., 2020. Understanding the molecular mechanisms for the enhanced phytoremediation of heavy metals through plant growth promoting rhizobacteria: a review. J. Environ. Manage. 254, 109779.

Marzan, L.W., Hossain, M.S., Mina, A. Akter, Y., Chowdhury, A.M.M.A., 2017. Isolation and biochemical characterization of heavy-metal resistant bacteria from tannery effluent in Chittagong city, Bangladesh: Bioremediation viewpoint. Egypt. J. Aquat. Res. 43 (1), 65–74.

Mena-Benitez, G.L., Gandia-Herrero, F., Graham, S., Larson, T.R., McQueen-Mason, S.J., French, C.E., Rylott, E.L, Bruce, N.C., 2008. Engineering a catabolic pathway in plants for the degradation of 1,2-dichloroethane. Plant Physiol. 147 (3), 1192–1198.

Mentzer, E., Ebere, D., 1996. Remediation of hydrocarbon contaminated sites A paper presented at, 8th Biennial International Seminar on the Petroleum Industry and the Nigerian Environment. November, Port Harcourt.

Mohanty, M., 2016. Post-harvest management of phytoremediation technology. J. Environ. Anal. Toxicol. 6 (5), 398.

Montiel-Rozas, M.M., Madejon, E., Madejon, P., 2016. Effect of heavy metals and organic matter on root exudates (low molecular weight organic acids) of herbaceous species: An assessment in sand and soil conditions under different levels of contamination. Environ. Pollut. 216, 273–281.

Moog, D., Schmitt, J., Senger, J., Zarzycki, J., Rexer, K.H., Linne, U., Erb, T.J., Maier, U.G., 2020. Using a marine microalga as a chassis for polyethylene terephthalate (PET) degradation. Microb Cell actories 18 (1), 171.

Nayan, A.K., Panda, S.S., Basu, A., Dhal, N.K., 2018. Enhancement of toxic Cr (VI), Fe, and other heavymetals phytoremediation by the synergistic combination of native Bacillus cereus strain and veltiveria of phytoremediation. J. Phytoremediation 20 (7), 682–691.

Olaniran, A.O., Balgobind, A., Pillay, B., 2009. Impacts of heavy metals on 1,2-dichloroethane biodegradation in co-contaminated soil. J. Environ. Sci. 21 (5), 661–666.

Ontãnon, O.M., Fernandez, M., Agostini, E., Gonz´alez, P.S., 2018. Identification of the main mechanisms involved in the tolerance and bioremediation of Cr(VI) by Bacillus sp. SFC 500-1E. Environ. Sci. Pollut. Res. 25 (16), 16111–16120.

Park, C.H., Keyhan, M., Matin, A., 1999. Purification and characterization of chromate reductase in *Pseudomonas putida*. Abs. Gen. Meet. American Soc. Microbial. 99 (4), 536–548.

Parnell, J.J., Berka, R., Young, H.A., Sturino, J.M., Kang, Y., Barnhart, D.M., DiLeo, M.V., 2016. From the lab to the farm: an industrial perspective of plant beneficial microorganisms. Front. Plant Sci. 7, 1110.

Pishchik, V.N., Vorobyev, N.I., Chernyaeva, I.I., Timofeeva, S.V., Koyhemyakov, A.P., Alexeev, Y.V., Lukin, S.M., 2002. Experimental and mathematical simulation of plant growth promoting rhizobacteria and plant interaction under cadmium stress. Plant Soil 243 (2), 173–186.

Phyo, A.K., Jia, Y., Tan, Q., Sun, H., Liu, Y., Dong, B., Ruan, R., 2020. Article competitive growth of sulfate-reducing bacteria with bioleaching acidophiles for bioremediation of heap bioleaching residue. Int. J. Environ. Res. Public Health 17 (8), 2715.

Polti, M.A., Aparicio, J.D., Benimeli, C.S., Amoroso, M.J., 2014. Simultaneous bioremediation of Cr(VI) and lindane in soil by actinobacteria. Int. Biodeterior. Biodegrad. 88, 48–55.

Prakash, B., Muhammad, I., 2011. Pseudomonas aeruginosa is present in crude oil contaminated sites of Barmer Region (India). J. Bioremed. Biodegrad. 2 (5), 1–3.

Prince, R.C., 2005. The microbiology of marine oil spill bioremediation. In: Ollivier, B, Magot, M (Eds.), Petroleum Microbiology. American Society for Microbiology, Washington, DC, pp. 317–335.

Ramprasad, D., Debasish, S., Sreedhar, B., 2014. Plant growth promoting rhizobacteria – an overview. European. J. Biotechnol. Biosci. 2 (2), 30–34.

Sandrin, T.R., Maier, R.M., 2003. Impact of metals on the biodegradation of organic pollutants. Environ. Health Perspect. 111 (8), 1093–1101.

Saranya, K., Sundaramanickam, A., Shekhar, S., Swaminathan, S., Balasubramanian, T., 2017. Bioremediation of mercury by vibrio fluvialis screened from industrial effluents. BioMed Res. Int. 7, 1–6. https://doi.org/10.1155/2017/6509648.

Sandrin, T.R., Hoffman, D.R., 2007. Bioremediation of organic and metal co-contaminated environments: effects of metal toxicity, speciation, and bioavailability on biodegradation. In: Singh, S.N., Tripathi, R.D. (Eds.), Environmental Bioremediation Technologies. Springer, Berlin, Heidelberg, pp. 1–34.

Singh, R., Singh, P., Sharma, R., 2014. Microorganism as a tool of biore-mediation technology for cleaning environment: a review. Int. Acad. Ecol. Environ. Sci. 4 (1), 1–6.

Simon, M.Z., Sola, D., Visnuk, P., Benimeli, C.S., Polti, M.A., Alvarez, A., 2017. Cr (VI) and lindane removal by Streptomyces M7 is improved by maize root exudates. J. Basic Microbiol. 57 (12), 1037–1044.

Selatnia, A., Boukazoula, A., Kechid, N., Bakhti, M., Chergui, A., Kerchich, Y., 2004. Biosorption of lead (II) from aqueous solution by a bacterial dead streptomyces rimosus biomass. Biochem. Eng. J. 19 (2), 127–135.

Song, U., Kim, D.W., Waldman, B., Lee, E.J., 2016. From phytoaccumulation to postharvest use of water fern for landfill management. J. Environ. Manag. 182, 13–20.

Song, U., Park, H., 2017. Importance of biomass management acts and policies after phytoremediation. J. Ecol. Environ. 41 (1), 13.

Sultana, P., Testuyuki, K., Moloy, B., Nobukazu, N., 2005. Predicting herbicides concentrations in paddy water and runoff to the river basin. J. Environ. Sci. 17 (4), 631–636.

Sun, S, Xie, S., Chen, H., Cheng, C., Qin, X., Dai, S.Y., Zhang, X., Yuan, J.S., 2016. Genomic and molecular mechanisms for efficient biodegradation of aromatic dye. J. Hazard Mater. 302, 286–295.

Tak, H.I., Ahmad, F., Babalola, O.O., 2013. Advances in the application of plant growth-promoting rhizobacteria in phytoremediation of heavy metals. In: Whitacre, D.M. (Ed.). Reviews of Environmental Contamination and Toxicology, 223. Springer, New York, NY, USA, pp. 33–52.

Teng, Y., Shen, Y.Y., Luo, Y.M., Sun, X.H., Sun, M.M., Fu, D.Q., Li, Z.G., Christie, P., 2011. Influence of Rhizobium meliloti on phytoremediation of polycyclic aromatic hydrocarbons by alfalfa in an aged contaminated soil. J. Hazard. Mater. 186 (2-3), 1271–1276.

Turan, M., Angın, I., 2004. Organic chelate assisted phytoextraction of B, Cd, Mo and Pb from contaminated soils using two agricultural crop species. Acta Agric., Scand., Sect. B, Soil and Plant Sci. 54 (4), 221–231.

Turan, M., Esringu, A., 2007. Phytoremediation based on canola (*Brassica napus* L.) and Indian mustard *(Brassica juncea* L.) planted on spiked soil by aliquot amount Cd, Cu, Pb, and Zn. Plant Soil Environ. 53 (1), 7–15.

Vijayaraghavan, K., Yun, Y.S., 2008. Bacterial biosorbents and biosorption. Biotechnol. Adv. 26 (3), 266–291.

Wang, J., Chen, C., 2009. Biosorbents for heavy metals removal and their future. Biotechnol. Adv. 27 (2), 195–226.

Yang, Q., 2004. Phytoremediation in Soil Co-Contaminated with Organic and Heavy Metals Pollutants. Zhejiang University.

Yang, J., Kloepper, J.W., Ryu, C.M., 2009. Rhizosphere bacteria help plants tolerate abiotic stress. Trends Plant Sci. 14 (1), 1–4.

Ye, S., Zeng, G., Wu, H., Zhang, C., Liang, J., Dai, J., Liu, Z., Xiong, W., Wan, J., Xu, P., Cheng, M., 2017. Co-occurrence and interactions of pollutants, and their impacts on soil remediation—A review. Crit. Rev. Environ. Sci. Technol. 47 (16), 1528–1553.

Yoshida, S., Hiraga, K., Takehana, T., Taniguchi, I., Yamaji, H., Maeda, Y., Toyohara, K., Miyamoto, K., Kimura, Y., Oda, K., 2016. A bacterium that degrades and assimilates poly(ethylene terephthalate). Science 351 (6278), 1196–1199.

Zhang, D.Q., Tan, S.K., Gersberg, R.M., Sadreddini, S., Zhu, J., Tuan, N.A., 2012. Removal of pharmaceutical compounds in tropical constructed wetlands. Ecol. Eng. 37 (3), 460–464.

Zhang, M., Wang, J., Hosseini, S., Yaling, B., 2019. Assisted phytoremediation of a co-contaminated soil with biochar amendment_ Contaminant removals and bacterial community properties. Geoderma 348, 115–123.

Zhang, H., Yuan, X., Xiong, T., Wang, H., Jiang, L., 2020. Bioremediation of co-contaminated soil with heavy metals and pesticides:Influence factors, mechanisms and evaluation methods. Chem. Eng. J. 398, 125657.

Ziganshina, E.E., Mohammed, W.S., Doijad, S.P., Elena, I., Shagimardanova, D., Natalia, E., Gogoleva, D., Ziganshin, M., 2018. Draft genome sequence of Brevibacterium epidermidis EZ-K02 isolated from nitrocellulose-contaminated wastewater environments. Data Brief 1 (7), 119–123.

CHAPTER

Phytobial remediation by bacteria and fungi

12

Gordana Gajić, Miroslava Mitrović, Pavle Pavlović
Department of Ecology, Institute of Biological Research "Siniša Stanković", National Institute of Republic of Serbia, University of Belgrade, Belgrade, Serbia

12.1 Introduction to phytobial remediation by bacteria and fungi

Anthropogenic activities such as mining, industrial processes, manufacturing, coal and fossil combustion, vehicular traffic, disposal of domestic and industrial waste, pesticides, and fertilizers release in the environment a huge amount of inorganic and organic pollutants that require urgent attention due to their serious effects on the environment and human beings (Prasad et al., 2006; Gajić et al., 2009; Pandey et al. 2011; Prasad et al., 2018; Gajić and Pavlović, 2018; Gajić et al., 2018; Pandey and Bauddh, 2019; Pandey, 2020; Pandey and Singh, 2020). Contaminants, such as metal(loid)s (As, B, Cd, Co, Cr, Cu, Hg, Mn, Mo, Ni, Pb, Se, Sb, Zn), radionuclides (Cs, Sr, U), chlorinated solvents (TCE, PCE), petroleum hydrocarbons (BTEX), polychlorinated biphenyls (PCBs), polychlorinated naphthalenes (PCNs), perfluorooctanoic acid (PFOA), polycyclic aromatic hydrocarbons (PAH), explosives (TNT, DNT, TNB, RDX, HMX) have negative effects on earth's climate, soil biogeochemical cycling, biodiversity, and food chain (Mc Cutcheon and Schnoor, 2003; Lohman et al., 2007; Teng et al., 2015) (Fig. 12.1A).

Phytoremediation is an environmentally friendly, cost effective and public accepted green technology that uses plants to clean up contaminated sites from pollutants (Salt et al., 1998; Pilon-Smits, 2005; Gajić and Pavlović, 2018; Pandey and Bauddh, 2019; Pandey, 2020; Pandey and Singh, 2020). The main phytoremediation technologies are phytostabilization, phytoextraction, rhizodegradation, phytodegradation, and volatilization and they are useful for revitalization of polluted soils from contaminants to the ecologically safe levels (Pilon-Smits, 2005; Gajić and Pavlović, 2018; Gajić et al., 2018; Pandey and Bauddh, 2019; Pandey and Singh, 2020). Therefore, selected plant species for sustainable phytoremediation possess a high ecological potential to exclude/take up/accumulate/degrade/transform/volatilize inorganic and organic pollutants reducing their mobility/availability/toxicity in the surrounding environment and settlements (Gajić and Pavlović, 2018; Pandey and Bauddh, 2019; Pandey and Singh, 2020). Effective phytoremediation depends on soil/mine spoil/fly ash physico-chemical properties (soil texture, hygroscopic water, pH, electrical conductivity, carbon and nitrogen content, available potassium and phosphorous content, organic matter content), type, total and bioavailable content of pollutants, plants, and underground and aboveground plant biomass (Kostić et al., 2012; Gajić et al., 2013; Pandey and Singh, 2014; Pavlović et al., 2016; Kostić et al., 2018; Pavlović et al., 2019). Successful ecorestoration and revitalization of contaminated lands is achieved by seeding the grass – legume mixtures and planting shrubs/trees and they are creating ecological conditions for the arrival of spontaneous colonizers from the surrounding area (Pavlović et al., 2004; Djurdjević et al., 2006; Kostić et al., 2012;

Assisted Phytoremediation. DOI: https://doi.org/10.1016/B978-0-12-822893-7.00002-1

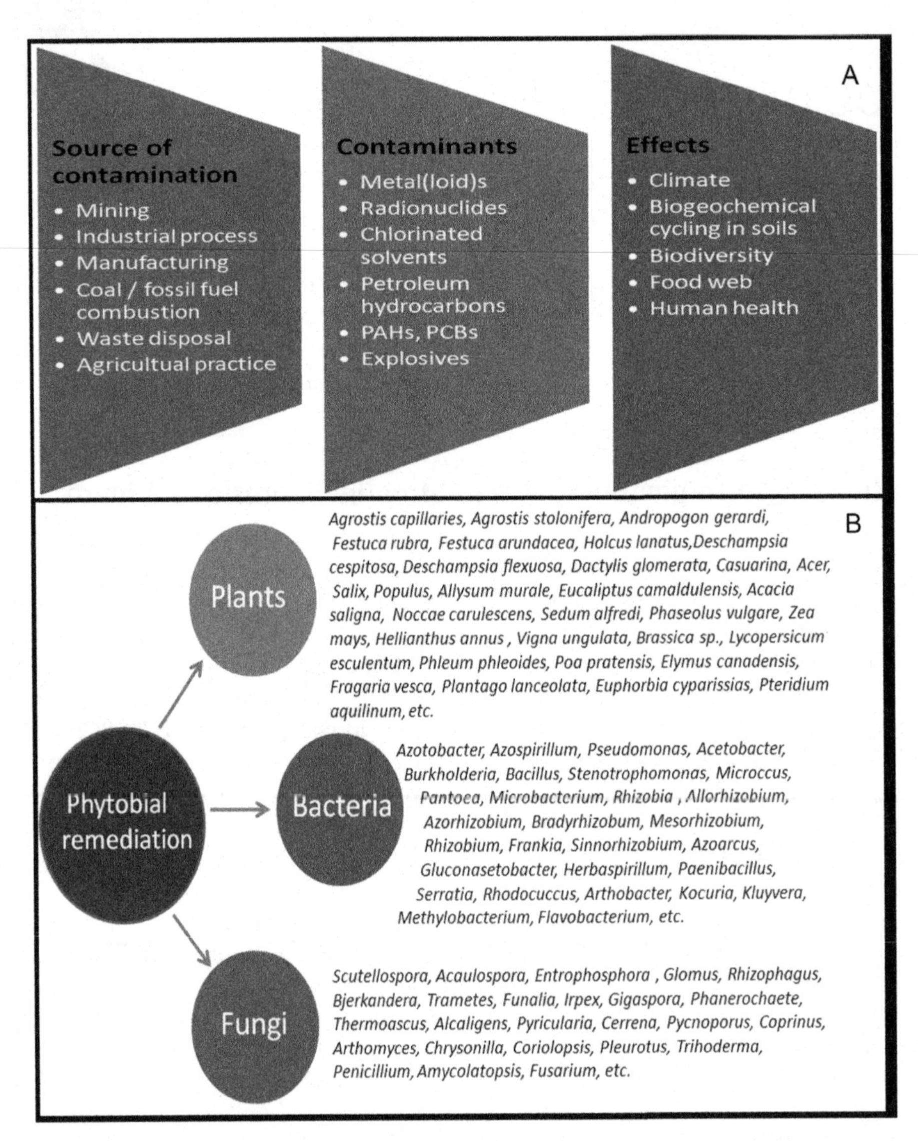

FIG. 12.1

The main sources of contamination, contaminants and their effects on environment and human health (A); phytobial remediation, technology that employed plants, bacteria and fungi (B).

Pandey, 2013; Gajić et al., 2013; Pandey, 2015; Gajić et al., 2016; Gajić et al., 2020a). Established self – sustaining vegetation on contaminated sites stabilizes substrate, prevents wind erosion, improves the physico-chemical characteristics of soils, and provides the organic substances binding the pollutants and reducing the transfer of toxic contaminants in the web's food (Mitrović et al., 2008; Pandey, 2012; Pandey et al., 2012; Maiti, 2013; Pandey et al., 2015a; Pandey et al., 2015b; Pandey et al., 2016; Gajić et al., 2016; Grbović et al., 2019; Grbović et al., 2020). Phytomanagement of polluted sites as part of ecosystem management services can be successful if practitioners and decision – makers use native plants that are resilient to local climate (tolerate drought, high light and temperatures), and show high tolerance to adverse physico-chemical characteristics of substrate and high concentrations of pollutants; they are fast growing and easy to establish and they possess a high ability to improve substrate fertility through nutrient cycling (Djurdjević et al., 2006; Pavlović et al., 2004; Mitrović et al., 2008; Gajić et al., 2019; Gajić et al., 2020a; Gajić et al., 2020b).

Assisted phytoremediation technology that combines bioremediation with phytoremediation in order to reduce the level of pollutants can be called phytobial remediation (Roy et al., 2015; Selvi et al., 2019; Asad et al., 2019). This technology employs microbes (rhizospheric and endophytic bacteria and fungi) with plants increasing their phytoremediation potential in revitalization of an ecosystem on contaminated sites (Roy et al., 2015; Selvi et al., 2019; Asad et al., 2019; Roy and Pandey, 2020) (Fig. 12.1B). Therefore, the exploitation of plant - microbe interactions may represent integrated remediation process toward pollutant removal/recovery in the environment (Selvi et al., 2019). Phytobial remediation is recognized as a clean technology for vast areas of contaminated soil, sediment and groundwater with minimum cost together with high social acceptance (Roy et al., 2015; Selvi et al., 2019). The plants and their associated microorganisms have been applied to remove metal(loid)s, radionucleoids, halogenated compounds, polycyclic aromatic hydrocarbons, petroleum products, and explosives more in the laboratory and greenhouse conditions, and less on the field (Pandey et al., 2014; Mc Cutcheon and Schnoor, 2003; Glick, 2010; Teng et al., 2015; Roy and Pandey, 2020).

Metaorganism presents a plant host and its microorganisms and it has shown to be effective in improving plant nutrition and growth as well as disease resistance (Berendsen et al., 2012; Berg et al., 2014). According to Heijden Van Der et al. (2008) soil microbiome drives biogeochemical processes, regulates plant productivity, and recycles nutrients. Plant microbiome is capable to regulate the expression of some plant characteristics which can promote plant's physiological state and its resilience to stress and all that can potentially hold the key to improving phytoremediation (Mendes et al., 2013). Plants can promote the abundance bacteria and fungi releasing root exudates creating habitats for them and in return, they promote plant growth and capability to reduce pollutant level (Teng et al., 2015; Thijs et al., 2016). Effectiveness of phytobial remediation depends on soil physico-chemical characteristics (texture, pH, nutrient, and organic matter content), bioavailability of contaminants, plant species, diversity, and richness of bacterial and fungal communities (Segura et al., 2009).

12.2 Phytobial remediation by plant growth-promoting bacteria

Plant growth-promoting bacteria (PGPB) presents plant associated microbes that are beneficial for plants because they stimulate plant growth, reduce diseases, decrease metal(loid) toxicity, and degrade organic compounds in order to accelerate phytoremediation (Glick, 2010; Glick, 2012). PGPB can be classified as free – living/endophytic soil bacteria, such as *Azotobacter, Azospirillum, Pseudomonas, Acetobacter, Burkholderia, Bacillus, Stenotrophomonas, Micrococus, Pantoea, Microbacterium, Rhizobia* (Glick, 2003; Santoyo et al., 2016) and symbiotic rhizospheric bacteria, such as *Allorhizobium, Azorhizobium, Bradyrhizobum, Mesorhizobium, Rhizobium, Frankia,* and *Sinnorhizobium* (Vessey, 2003). In the rhizosphere, plants create favorable conditions for bacterial growth and stimulate their action (Glick, 2010) releasing root exudates at a high rate (organic acids, amino acids, sugars) which they can use as a food source (Doornbos et al., 2012). According to Fujishige et al. (2006) when bacterial cells bind to the root, they proliferate by forming microcolonies that are embedded in an extracellular matrix. Survival of microbes under environmental conditions that are not optimal for their growth is linked to "quorum sensing" in which microbes synthetize signaling molecules in rhizosphere where microbial population develops and expands by forming a mantle over the root surface, and working together against the stress (Daniels and Vanderleydan, 2004; Asad et al., 2019). Endophytic bacteria enter through plant's root cracks, root hairs, stomata, lenticels colonizing the internal plant tissue of plant without harming the host (Santoyo et al., 2016)

PGPB can enhance plant growth facilitating plant nutrition acquisition (nitrogen, phosphorus, iron), and regulating plant hormone level (ethylene, indoleacetic acid) (Solano et al., 2008; Glick, 2010; Glick, 2012) (Fig. 12.2). They assist phytoremediation of contaminants through different mechanisms, such as mobilization/immobilization/bioaccumulation/biotransformation/biovolatilization (Gadd, 2010; Ullah et al., 2015; Roy et al., 2015; Selvi et al., 2019; Asad et al., 2019). Therefore, PGPB reduce the level of site contamination affecting soil biogeochemistry and pollutant insolubility and producing organic acids, siderophores as well as various phytohormones and enzymes (Tangahu et al., 2011) (Fig. 12.2). However, the bacterial inoculation depends on total contaminant level in soils that is, extremely high metal concentrations may inhibit the population size of bacteria and together with decrease of activities of soil enzymes may point out that the highest bacterial inoculation is the best solution for slightly and moderately contaminated sites (Wu et al., 2006).

PGPB assisted-phytoremediation of contaminated sites was given in Table 12.1. Application of microbes in nutrient-poor polluted sites may provide plants with essential elements, such as N/P/Fe

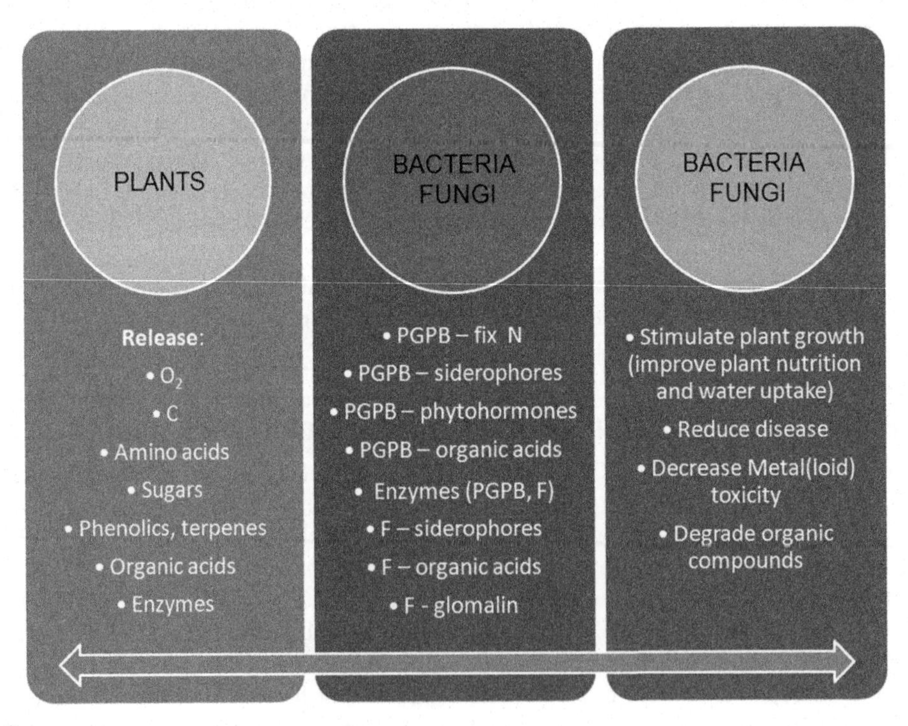

FIG. 12.2

Link between plants – plant growth promoting bacteria (PGPB) – fungi (F) system.

Table 12.1 Plant growth promoting bacteria (PGPB) assisted - phytoremediation of contaminated sites.

PGPB	Contaminants	References
PGPB – fix nitrogen		
Rhizobia, Bacillus, Pseudomonas, Azospirillum, Azoarcus, Burkholderia, Gluconacetobacter, Herbaspirillum, Azotobacter, Paenibacillus, Rhizobium, Bradyrhizobium, Mesorhizobium	Cd, Cr, Cu, Ni, Zn	Barriuso et al. (2008) Wani et al. (2008) Purchase et al. (1997) Khan et al. (2009)
PGPB – organic acid		
Gluconacetobacter diazototrophicus	Zn, Cd	Delvasto et al. (2009) Li et al. (2010)
PGPB – siderophores		
Yersinia pestis	Zn, Ci, Fe, Ni	Perry et al. (2015) Robinson et al. (2018)
Pseudomonas putida KNP9	Cd	Tripathi et al. (2005)
Pseudomonas aeruginosa	Pb, Cr	Braud et al. (2009)
Pseudomonas aeruginosa PAO1	Al, Co, Cu, Ni, Pb, Zn	Braud et al. (2010)
Streptomyces acidiscabies	Cd, Ni	Dimkpa et al. (2008)
Streptomyces tendae		Dimkpa et al. (2009)
Pseudomonas azotoformas	Cd, Pb, Ni, As(III), Zn, Cu, Co	Nair et al. (2007)
Rhodococcus *Arthobacter* *Kocuria*	As (III)	Retamal – Morales et al. (2018)
Kluyvera ascorbata SUD165	Ni, Pb, Zn	Burd et al. (1998) Burd et al. (2000)
PGPB – phytochormones		
Methylobacterium oryzae CBM20 *Burholderia* sp. CBMB40	Ni, Cd	Madhaiyan et al. (2007)
Kluyvera ascorbata SUD165	Ni, Pb, Zn	Burd et al. (1998) Burd et al. (2000)
Phychrobacter sp., SRA1, SRA2 *Bacillus cereus* SRA10 *Bacillus* sp., SRP4 *Bacillus wehenstephanensis* SRP12	Ni	Ma et al. (2009a)
Pseudomonas putida *Flavobacterium sp.* *Pseudomonas aeruginosa*	16 priority PAHs	Huang et al. (2004)

(Gamalero and Glick, 2012). A number of PGPB are active to fix nitrogen and provide it to plants: *Rhizobia, Bacillus, Pseudomonas, Azospirillum, Azoarcus, Burkholderia, Gluconacetobacter, Herbaspirillum, Azotobacter, Paenibacillus* (Barriuso et al., 2008). According to Wani et al. (2008) inoculation with *Rhizobium* protects plants against toxic levels of heavy metals reducing their uptake by roots and shoots. Therefore, nitrogen fixing microbes, such as *Rhizobium leguminosarum bv. Trifolii* show tolerance to Cd, Cu, Ni, and Zn (Purchase et al., 1997), whereas *Rhizobium* sp. RP5 is resistant to Ni and Zn, *Rhizobium* sp. RL9 to Zn, *Bradyrhizobium* sp. RM8 to Ni and Zn, and *Mesorhizobium* sp. RC3 to Cr (Khan et al., 2009). However, there are differences between nitrogen fixing microbes in their tolerance to heavy metals: *Streptomyces* AR16 showed high Zn tolerance whereas *Pseudomonas* PR04 showed low resistance to Zn (Kuffner et al. 2008). Solubilization of inorganic P occurs through the synthesis of organic acids (gluconic acid, citric acid) by phosphate – solubilizing PGPB (Rodrigez and Fraga, 1999; Rodrigez et al., 2004) whereas mineralization of organic P occurs through the synthesis of phosphatases (Rodrigez and Fraga, 1999). In the condition of limited supply of iron, bacteria induce synthesis of siderophores that have high affinity for Fe^{3+} and membrane receptors that are capable to bind iron – siderophores complex thereby allowing Fe uptake by plants (Hider and Kong, 2010; Glick, 2012). Vansuyt et al. (2007) showed that *Arabidopsis thaliana* is capable to take up Fe – pyoverdine complex synthetized by *Pseudomonas fluorescens* C7.

Rhizospheric PGPB mobilize metal(loid)s through chelation, redox transformation, and volatilization (Roy et al., 2015). Bacteria can affect metal mobility, solubility, and uptake by plants altering soil pH, and producing biosurfactants, organic acids, phytosiderophores, and phytohormones (Lasat et al., 2002; Khan et al., 2009; Ma et al., 2011). Therefore, *Microbacterium liquefaciens* and *Microbacterium arabinogalactanolyticum* reduce soil pH and increase Ni uptake by *Alyssum murale* (Abou-Shanab et al., 2003). Furthermore, PGPB produce biosurfactants that induce desorption of metal from soil increasing its solubility (Rajkumar et al., 2012), as *Pseudomonas aeruginosa* for Pb and Cd (Ullah et al., 2015). Guarino and Sciarrillo (2017) noted that *Eucalyptus camaldulensis* and *Acacia saligna* increased their biomass when they were inoculated with rhizospheric bacteria in the condition of high concentrations of As, Cd, Pb, and Zn. However, inoculation of hyperaccumulator plant *Noccae carulescens* with PGPB increased Zn uptake and its accumulation in plant shoots whereas in non-hyperaccumulator plant *Noccae arvense* Zn was not accumulated (Whiting et al. 2001).

Organic acids. Rhizospheric PGPB produce organic acids (oxalic acid, acetic acid, citric acid, succinic acid, formic acid, tartaric acid) that mobilize and solubilize metal(loid)s in the soil (Li et al., 2010; Ullah et al., 2015), binding them in complex metal – ligand. According to Ryan et al. (2001) stability of ligands depends on pH, metal type and organic acid. Therefore, *Gluconacetobacter diazotrophicus* released gluconic acid derivate 5-ketogluconic acid which solubilized Zn compounds (Saravanan et al. 2007) whereas inoculated *Sedum alfredi* with rhizospheric PGPB accumulates Zn and Cd due to organic acids secreted from bacteria compared to plants that are not inoculated (Delvasto et al., 2009; Li et al., 2010).

Siderophores. Siderophores (Greek "iron carrier") are secondary metabolites with high affinity for Fe and they enable its acquisition through specific uptake systems (Wang et al., 2014). The main chemical classes of siderophores are catecholate (enterobactin), hydroxamate (ferrioxamine B), carboxylate (rhizobactin), and phenolate, and there are siderophores with a mixture of functional groups, such as pyoverdine (Cornelis, 2010; Sullivan and Gadd, 2019; Kramer et al., 2020). PGPB, such as *Rhizobium, Pseudomonas, Streptomyces, Azotobacter, Serratia,* and *Bradyrhizobium* are able to produce siderophores (Ali and Vidhale, 2013; Kramer et al., 2020). According to Thiem et al. (2018) PGPB are capable to produce siderophores under stressful environmental conditions, such as lack of nutrients or toxicity. In this context, siderophores can bind heavy metals, such as Zn, Mn, Cu, Ni, Cr, Co, Pb, Mo representing metallophores (Sullivan and Gadd, 2019; Kramer et al., 2020). The metal-siderophore stability determines the availability of metal to microbes and plants, and these siderophores can mobilize or immobilize specific metal, depending on its essentiality or toxicity (Ahmed and Holmstrom, 2014). These metallophores in some cases do not enter the cell, they bind metals without transfering them into the cells (Braud et al., 2010; Schalk et al., 2011). Yersiniabactin is produced from *Yersinia pestis* scavenges Zn, Cu and Ni, and Fe for bacterial metabolism (Perry et al., 2015; Robinson et al., 2018). According to Tripathi et al. (2005) in the presence of Cd, siderophores produced from *Pseudomonas putida* KNP9 can increase biomass of *Phaseolus vulgaris* without any toxicity symptoms. However, some siderophores can enhance phytoextraction in plants and speed up phytoremediation, such as pyochelin produced from *Pseudomonas aeruginosa* that increases availability of Pb and Cr in the rhizosphere of *Zea mays* (Braud et al., 2009). However, *Pseudomonas aeruginosa* strain PAO1 produced pyoverdine that was able to sequester Al, Co, Cu, Ni, Pb, and Zn from the extracellular medium of the bacteria as well as pyochelin that was able to chelate Al, Co, Cu, Ni, Pb, and Zn; thus, decreasing metal diffusion into bacteria, reducing toxic metal accumulation, and contributing to bacterial resistance to heavy metals (Braud et al., 2010). Similarly, siderophores from *Streptomyces acidiscabies* and *Streptomyces tendae* are able to bind Cd and Ni protecting bacterial cells and plants (*Helianthus annuus* and *Vigna ungulata*) from metal toxicity (Dimkpa et al., 2008; Dimkpa et al., 2009). Furthermore, Nair et al. (2007) found that siderophores produced from *Pseudomonas azotoformas* can chelate Cd, Pb, Ni, As(III), Zn, Cu, Co, St, whereas Retamal–Morales et al. (2018) showed that siderophores – like compounds produced from Actinobacteria (*Rhodococcus, Arthobacter*, and *Kocuria*) were able to bind As (III) in arsenic tolerant isolates. In this context, siderophores produced from *Kluyvera ascorbata* SUD165 can protect *Brassica juncea*, *Brassica napus*, and *Lycopersicum esculentum* from Ni, Pb, and Zn toxicity (Burd et al., 1998; Burd et al., 2000). In addition, siderophores can promote plant hormone production in the presence of heavy metals in order to increase the growth of plants (Dimpka et al., 2009). Hence, siderophores from *Streptomyces* stimulate auxin production binding Cd, Cu, Ni, Al, Mn and U (Dimpka et al., 2008; Dimpka et al., 2009).

Phytohormone production. Phytohormones regulate plant growth affecting the plant metabolic activity for their normal functioning and they stimulate protection from abiotic and biotic stress factors (Gamalero and Glick, 2012). PGPB are capable to modify plant hormone levels which enable plants to overcome stress conditions (Verbon and Liberman, 2016). Indole-3-acetic acid (IAA) produced by PGPB enhances plants lateral and adventitious rooting that is important for nutrient uptake and it is involved in cell signaling in the condition of environmental stresses (Spaepen et al., 2007; Gravel et al., 2007) whereas auxin released from PGPB can be significant for plant development in polluted sites (Shin et al., 2012). Ethylene leads to the inhibition of plant growth and causes senescence and abscission, especially when the plants have been exposed to heavy metal contamination (Gamalero and Glick, 2012). PGPB bacteria, such as *Methylobacterium oryzae* CBM20 and *Burholderia* sp. CBMB40 promote growth of *Oryza sativa* reducing ethylene level and decreasing Ni and Cd toxicity (Madhaiyan et al., 2007). PGPB are able to reduce ethylene production in plants thereby enhancing the plant growth by producing 1-aminocyclopropane-1-carboxylate (ACC) deaminase, which hydrolyzes the ethylene precursor ACC to NH_3 and α-ketobutyrate by using N from the rhizosphere (Glick, 2003). PGPB are capable to release up to 1000 times more ACC-deaminase, especially in the stressful condition, such as contamination with heavy metals (Glick, 2003; Farwell et al., 2006) or organic compounds (Saleh et al., 2004; Reed and Glick, 2005). In this context, it was showed that *Kluyvera ascorbata* SUD165 can protect *Brassica juncea, Brassica napus,* and *Lycopersicum esculentum* from Ni, Pb, and Zn toxicity and this is attributed to great bacterial potential for production of ACC deaminase, which in turn reduces the level of ethylene in plants Burd et al., 1998; Burd et al., 2000). Similarly, bacterial ACC deaminase enhances plant growth and Cd accumulation in *Brassica juncea* (Belimov et al., 2005). PGPB, such as *Phychrobacter* sp., SRA1, SRA2, *Bacillus cereus* SRA10, *Bacillus* sp., SRP4, and *Bacillus wehenstephanensis* SRP12 produce ACC deaminase that reduces ethylene level and contributes to Ni resistance of *Alyssum serpyllifolium* and *Phleum phleoides* (Ma et al., 2009a). Inoculation of *Festuca arundinacea* with PAH degrading bacteria (*Pseudomonas putida, Flavobacterium sp., Pseudomonas aeruginosa*) promotes a microbial degradation process through ACC deaminase activity for removal of PAHs compounds that is, the addition of PGPB provides better plant growth by increasing plant tolerance to contaminants in the soil and greatly accelerates the remediation process (Huang et al., 2004). Also, inoculation of *Festuca arundinacea, Poa pratensis,* and *Elymus canadensis* grown in creosote-contaminated soil with PGPB containing ACC deaminase increases plant biomass (root and shoot) and greatly enhances phytoremediation process (Huang et al., 2004).

12.3 Phytobial remediation by mycorrhizal fungi

Mycorrhizal fungi form symbiosis with plants and have a prominent role in productivity, stability and biodiversity of ecosystems (van der Heijden et al., 1998) as well as in phytoremediation of contaminated sites (Gonzáles – Chávez et al., 2006; Turnau et al., 2006). The hyphae of mycorrhizal fungi stabilize the soil and provide water and nutrients for plants, whereas plants provide food for fungi (Sylvia et al., 2005) (Fig. 12.2). There are two types of mycorrhiza: (1) *ectomycorrhiza* in which mycelium forms compacted cover on the root's surface, having a protection role for the plant; mycelium penetrates between root's cortical cells creating Hartig – net presenting the site where the exchange of compounds between the fungus and the plant takes place, and (2) *endomycorrhiza* in which hyphae penetrate inside the live cortical cells, crossing 9AM) cell walls and they continue to spread and develop to plasma membrane of the plant cell; type vesicular – arbuscular mycorrhiza (AM) forms arbuscules, fungal structures within cortical cells of plants; external mycelium of AM functioning as a "bridge" between root and soil (Gonzáles – Chávez et al., 2006; Turnau et al., 2006). AM occurs in 80% of plant species that belong to Bryophyta, Pterydophyta, Gymnospermae, and Angiospermae (Smith and Read, 1997), whereas about 120 species of fungi belong to Glomeromycetes (Schussler et al., 2001).

Arbuscular mycorrhizal fungi (AMF) participate in soil stability, plant nutrient and water uptake, protection from pathogens and resistance to environmental stress (pH, salt stress, drought, high levels of metal(loid)s and organic pollutants in soils) (Gonzáles – Chávez et al., 2006; Turnau et al., 2006) (Fig. 12.2). Active hyphae exudate glomalin (glycoprotein) which can act as a biological cement increasing soil aggregation (Wright and Upadhyaya, 1998), and contribute to carbon sequestration and organic C pool (Rilling et al., 1999). AMF have benefits for plant host, because they improve plant nutrition in nutrient-deficient soils and affect water relations in plants (Smith and Read, 1997) due to the great absorptive soil surface that they occupy (Gonzáles – Chávez et al., 2006). Transpiration rates and larger leaf area of mycorrhizal plants are significantly higher compared to nonmycorrhizal plants (Kothari et al., 1990).

AMF assist in phytoremediation of contaminated sites with metal(loid)s and organic compounds (Gonzáles – Chávez et al., 2006; Turnau et al., 2006; Roy et al., 2015; Selvi et al. 2019). Inoculation of selected plants with best fungal isolates of AMF leads to the successful restoration of contaminated sites because they are well adapted to climate and polluted conditions that is, they can effectively stabilize the soil, take up essential nutrients, accumulate and sequence pollutants (Colpaert, 1998). Grasses, trees, and shrubs colonized by AMF have already been successfully used in phytoremediation management practice: *Agrostis capillaries, Agrostis stolonifera, Andropogon gerardi, Festuca rubra, Festuca arundacea, Holcus lanatus, Deschampsia cespitosa, Deschampsia flexuosa, Dactylis glomerata, Casuarina, Acer, Salix, Populus* (Colpaert, 1998; Schat and Verkleij, 1998; Gonzáles – Chávez et al., 2006). C3 grasses colonized by AMF were noted at the sites that are highly contaminated, such as industrial waste (Daft and Nicolson, 1974), and heavy metals (Khan et al., 2000). AMF-assisted phytoremediation of contaminated sites is given in Table 12.2. Gonzáles – Chávez et al. (2006) found different plant species colonized by external mycelium of *Scutellospora, Acaulospora, Entrophosphora* and *Glomus* growing on the Cd tailings. Mycorrhizal fungi, such as *Glomus mossae*, *Glomus geosporum*, and *Glomus etunicatum* enhance As accumulation in *Plantago lanceolata* growing on As mine waste (Orlowska et al., 2012). Furthermore, *Rhizophagus custos* under root cultures is capable to remove PAHs, especially anthracene (Aranda et al., 2013) whereas plant-associated fungi, such as *Bjerkanderaadusta, Trametes hirsute, Trametes viride, Funalia trogii, Irpex lacteus* could survive in the presence of decolorized textile industry effluents (Tegli et al., 2013). Vinichuk et al. (2013) showed that AMF colonized quinoa plants enhanced ^{137}Cs uptake.

Metal(loid)s can affect the fungal species, richness, abundance and diversity Gadd, 1993; Ross and Kaye, 1994). According to del Val et al. (1999) fungal diversity was higher in sites moderately contaminated with Pb, Cd, Cr, Cu, Ni, Hg and Zn compared to highly contaminated soil. Sambandan et al. (1992) found 15 different fungal species in the soils polluted by Zn, Cu, Pb, Ni, and Cd whereas Turnau et al. (2001) found different AMF species that colonized *Fragaria vesca* on industrial waste, but with different frequency: *Glomus* sp. HM-CL4 (52%), *G. claroideum* and *Glomus* sp. MH-CL5 (32%–34%), *G. intraradices* and *Glomus mosseae* (5%–7%). Generally, the number of spores in contaminated area can be affected by metal(loid)s, but different fungi show various sensitivity and strain special behavior (Pawlowska et al., 2000).

Mycorrhizal fungi enhance the resistance of plants to metal(loid)s by (1) cell wall – binding, (2) limited influx through plasma membrane, (3) sequestration in vacuole and cytoplasm, (4) active efflux of metal(loid)s (Gonzáles – Chávez et al., 2006). Thus, external mycelium and internal hyphae of AMF-colonized roots bind heavy metals to the cell wall (Turnau, 1998; Gonzáles – Chavez et al., 2004). The surface of intraradical mycelium colonized roots of *Euphorbia cyparissias* consists of deposits with high levels of As, Fe, Zn, Cu and Pb (Turnau, 1998) whereas extraradical mycelium of AMF (*Glomus mosseae* BEG-132, *Glomus claroideum* BEG-134) is characterized with high Cu-sorptive capacity (Gonzáles – Chavez et al., 2004). According to Gadd (1993) organic fungal metabolites are efficient metal(loid) chelators, such as oxalic acid, citric acid, siderophores, and riboflavin. Glomalin from *Gigaspora rosea* sequestered Pb, Cu, and Cd reducing their availability and toxicity in soil (Gonzáles – Chavez et al., 2004), whereas hyphae from external mycelium of AMF possessed a high capacity for Cu sequestration. Chelation of heavy metals with metallothionines (MTs) as metal binding polypeptides was noted in ectomycorrhizal and endomycorrhizal fungi (Gonzáles – Chávez et al., 2006). Gene GmarMT1 that encodes a fungal metallothionein which confers resistance to Cu and Cd was identified in *Gigaspora margarita* BEG-34 (Lanfranco et al., 2002). Furthermore, Cu in contaminated sites may induce a high tyrosinase activity in external mycelium of AMF (*Glomus caledonicum* BEG-133, *Glomus mosseae*

Table 12.2 Arbuscular mycorrhizal fungi (AMF) assisted - phytoremediation of contaminated sites.

AMF	Contaminantion	References
Scutellospora *Acaulospora* *Entrophosphora* *Glomus*	Cd-rich slag and tailings	Gonzáles – Chávez et al. (2006)
Glomus mossae *Glomus geosporum* *Glomus etunicatum*	As mine waste	Orlowska et al. (2012)
Bjerkandera adusta *Trametes hirsute* *Trametes viride* *Funalia trogii* *Irpex lacteus*	Textile industry effluents	Tegli et al. (2013)
Glomus sp. HM-CL4 *G. claroideum* *Glomus* sp. MH-CL5 *G. intraradices* *Glomus mosseae*	Industrial waste	Turnau et al. (2001)
Glomus mosseae BEG-132 *Glomus claroideum* BEG-134	Cu	Gonzáles – Chavez et al. (2004)
Gigaspora rosea	Cu, Pb, Cd	Gonzáles – Chávez et al. (2006)
Gigaspora margarita BEG-34	Cu, Cd	Lanfranco et al. (2002)
Glomus caledonicum BEG-133 *Glomus mosseae* BEG-132 *Glomus claroideum* BEG-134	Cu	Gonzáles (2000)
Glomus mosseae BEG132 *Glomus claroideum* BEG-134	Cu/As	Gonzáles – Chavez et al. (2004)

BEG-132, and *Glomus claroideum* BEG-134) (Gonzáles, 2000). Crystal-like aggregates in mucilaginous outer hyphal cell wall of *Glomus mosseae* BEG132, *Glomus claroideum* BEG-134 contained Fe and Cu when grown in Cu/As polluted soils (Gonzáles et al., 2002). In addition, polyphosphate granules vacuole in hyphae of AMF colonized roots of *Pteridium aquilinum* contained Cd more than the cytoplasm of host cell and they can contribute to decreasing transfer of metals to the plant (Turnau et al., 1993).

12.4 Enzymatic degradation of organic compounds

Industrial (metal processing/mining/petrochemical complexes/industry effluents/chemical weapon production/pulp/paper industry/dye industry), and agricultural activities, traffic as well as waste disposal release a number highly toxic/hazardous persistent organic pollutants (POPs) in the environment, such as PAHs, PCBs, pesticides, nitriles, cyanides, phenols, chlorophenols, naphthalene, plastics, and dyes (Mc Cutcheon and Schnoor, 2003). The employability of bacteria, fungi and plants is significant for efficient remediation of organic compounds at the contaminated sites through their biotransformation and biodegradation mediated by microbial and plant enzymes (Mc Cutcheon and Schnoor, 2003; Rao et al., 2010; Karigar and Rao, 2011; Teng et al., 2015) (Fig. 12.2). Pollutant biodegradability, availability, contaminant concentration, pH, temperature and redox conditions are factors that affect the biochemistry and bioactivity of pollutant microbial degradation (Gevao et al., 2000; Luecking et al., 2000; Gianfreda and Nannipieri, 2001). Degradation of pollutants by enzymes produced from bacteria, fungi and plants goes to complete mineralization to CO_2 and H_2O or contaminants are degraded/transformed to intermediates that are less toxic and harmful than parent compounds (Gianfreda and Bollag, 2002; Gianfreda and Rao, 2004; Rao et al., 2010). Contaminant susceptibility and accessibility are influenced by its chemical structure and concentrations, as well as by environmental conditions (Gianfreda and Bollag, 2002; Gianfreda and Rao, 2004; Rao et al., 2010).

The main classes of enzymes with high potential in degradation of organic compounds derived from bacteria, fungi and plants are oxidoreductases, transferases, hydrolases, lyases, isomerases, and ligases (Gianfreda and Rao, 2004; Rao et al., 2010; Karigar and Rao, 2011).

Degradation of organic compounds by enzymes produced from plants/bacteria/fungi is given in Table 12.3. The main *oxidoreductases* are oxygenases, laccases, and peroxidases (Karigar and Rao, 2011). They participate in detoxification of phenolics, chlorinated solvents and halogenated compounds (herbicides, insecticides, fungicides), petroleum hydrocarbons, explosives (Duran and Esposito, 2000; Leung, 2004). According to Karigar and Rao (2011) plant families, such as *Fabaceae, Graminae* and *Solanacea* release oxidoreductases responsible for oxidative degradation of some soil constituents. *Mono-oxigenases* are involved in the processes of desulfurization, dehalogenation, denitrification, ammonification, hydroxylation, biodegradation and biotransformation of aromatic and aliphatic compounds (alkanes, cycloalkanes, alkenes, haloalkenes, aromatic, and heterocyclic hydrocarbons) (Fox et al., 1990; Arora et al., 2010), whereas dioxygenases are involved in oxygenation of aromatic compounds and some aliphatic products (naphthalene) (Dua et al., 2002). Therefore, catechol dioxygenase from *Comamonas testoteroni* and *Pseudomonas pseudoalcaligens* are involved in bioremediation of chlorophenols and PCB whereas phenol oxidase from *Gleophyllum trabeum, Trametes versicolor, Phanerochaete chrysosporium* and *Thermoascus auranticus* are involved in Kraft effluent decontamination and chlorinated compounds degradation (Duran and Esposito, 2000). Dioxygenase from *Alcaligens eutrophs* transformed herbicide 2,4-D (Radnoti de Lipthay et al., 1999) whereas biodegradation of naphthalene is achieved through dioxygenase produced from *Pseudomonas* sp. (Barriault et al., 1999).

Laccases enzymes (p-diphenol:dioxygen oxidoreductase) produced by plants, fungi and bacteria catalyzed the oxidation of reduced phenolic and aromatic compounds, lignins, and aryldiamines (Mai et al., 2000; Rezende et al., 2005). Laccases from plants degrade chlorophenols, whereas laccases from fungi *Trametes hispida* and *Pyricularia oryzae* degrade azo-dyes, laccases from *Trametes versicolor* degrade chlorophenols, and PAHs from *Cerrena unicolor* degrade phenols and 2.4–dichlorophenols, and laccases from *Pycnoporus cinnabarinus* degrade benzopyrene (Duran and Esposito, 2000).

Table 12.3 Degradation of organic compounds by enzymes produced from plants/bacteria/fungi.

Enzymes	Plant/bacteria/fungi	Contaminant	References
Dioxygenase	*Comamonas testoteroni* *Pseudomonas pseudoalcaligens*	Chlorophenols, PCB	Duran and Esposito (2000)
Dioxygenase	*Alcaligens eutrophs*	Herbicide 2,4-D	Radnoti de Lipthay et al. (1999)
Dioxygenase	*Pseudomonas* sp.	Naphthalene	Barriault et al. (1999)
Phenoloxydase	*Gleophyllum trabeum* *Trametes versicolor* *Phanerochaete chrysosporium* *Thermoascus auranticus*	Kraft effluent Chlorinated compounds	Duran and Esposito (2000)
Laccasses	*Plants* *Trametes hispida* *Pyricularia oryzae* *Trametes versicolor* *Cerrena unicolor* *Pycnoporus cinnabarinus*	Chlorophenols Azo-dyes Chlorophenols PAHs Phenols 2.4-dichlorphenols Benzopyrene	Duran and Esposito (2000)
Peroxidases (Horseradish)	*Barley* *Soybean*	Phenols, Chlorophenols Kraft effluents	Duran and Esposito (2000) Karigar and Rao (2011)
Peroxidases (LiP, MnP)	*Phanerochaete chrysosporium* *Chrysonilia sitophyla*	Aromatic phenolics Kraft effluents	Duran and Esposito (2000)
Peroxidases (LiP, MnP)	*Coriolopsis polyzona* *Pleurotus ostreatus* *Trametes versicolor*	PCBs	Novotny et al. (1997)
Peroxidases (LiP, MnP)	*Phanerochaete chrysosporium*	TNT, RDX	Cameron et al. (2000)
Hydrolases (cellulase)	*Trichoderma resei* *Penicilium funiculosum*	Cellulose materials	van Wyk (1999)
Hydrolases (xylanase, β-xylosidase)	*Streptomyces thermoviolaceus*	Kraft pulp	Christov and Prior (1996)

Table 12.3 Degradation of organic compounds by enzymes produced from plants/bacteria/fungi. ***Continued***

Enzymes	Plant/bacteria/fungi	Contaminant	References
Hydrolases (amylase)	*Bacillus lichneiformis*	Starch materials	Copinet et al. (2001)
Proteases (keratinase)	*Chrysosporium keratinophilum*	Keratin	Singh (2002)
Nylon degrading enzyme	*Phanerochaete chrysosporium* *Trametes versicolor* strain IZU-154	Nylon	Deguchi et al. (1997) Deguchi et al. (1998)
Depolymerases Esterases	*Amycolatopsis sp.* *Bacillus sp.*	Plastic (polylactic acid)	Nakamura et al. (2001)
Ccellobiose dehydrogenase	*Phanerochaete chrysosporium*	Plastic (polylactic acid)	Cameron and Aust (1999)
Esterases	*Curvularia senegalensis* *Coryneabacterium*	Plastic (polyurethane)	Nakajima – Kambe et al. (1999
Nitrilase Nitrile hydratases Amidases	*Nocardia* sp. *Rhodococcus* sp. *Fusarium solani*	Nitrile compounds	Banerjee et al. (2002)
Cynidases, Cyanide hydratases Formamide hydrolases Rhodanese	*Trihoderma* sp. *Alcaligenes denitrificans*	Cyanide – wastes	Barclay et al. (1998) Basheer et al. (1993) Lynch (2002)

Peroxidases: Class I enzymes include yeast cytochrome *c* peroxidase, ascorbate peroxidase (APX) from plants and bacterial peroxidases; class II enzymes include fungal peroxidases (lignin peroxidase, LiP and manganese peroxidase, MnP) from *Phanerochaete chrysosporium*, *Coprinus cinereus*, and *Arthromyces ramosus* (ARP) and versatile peroxidases (VP); class III contains plant peroxidases (horseradish, HRP) from barley and soybean (Karigar and Rao, 2011). Therefore, LiP degrades halogenated phenolics, PAH, PCB, MnP degrades lignin and other phenolic compounds, whereas VP degrades methoxybenzenes and phenolic aromatic compounds (Karigar and Rao, 2011). Horseradish peroxidase is involved in degradation of phenols, chlorophenols, and Kraft effluents (Duran and Esposito, 2000). LiP and MnP derived from *Phanerochaete chrysosporium* and *Chrysonilia sitophyla* degrade aromatic and phenolic compounds, Kraft effluents (Duran and Esposito, 2000), LiP, and MnP released from *Coriolopsis polyzona*, *Pleurotus ostreatus*, *Trametes versicolor* degrade PCBs (Novotny et al., 1997), whereas LiP and MnP produced from *Phanerochaete chrysosporium* degrade explosives (TNT and RDX) (Cameron et al., 2000).

Hydrolitic enzymes (hydrolases) include lipase, carbohydratases, proteases, esterases, phosphatases, phytases that are produced from bacteria, fungi, and plants (Karigar and Rao, 2011). *Lipases* can catalyze reactions, such as hydrolysis (triacylglycerol to glycerol and free-fatty acids), interesterification, esterification, alcoholysis and aminolysis of organics, such as oil spill, detergents, pollutants from baking, paper and pulp industry (Joseph et al., 2006; Hermansyah et al., 2007; Sharma et al., 2011). *Carbohydratases* hydrolyze substrate derived from textile manufacturing, detergent production, paper and pulp industry to simple carbohydrates and include enzymes, such as cellulases, hemicellulases, pectinases, amylases and xylanases (Rixon et al., 1992; Adriano-Anaya et al., 2005). Therefore, cellulase from *Trichoderma resei* and *Penicilium funiculosum* hydrolyze cellulose materials (van Wyk, 1999), xylanase and β-xylosidase from *Streptomyces thermoviolaceus* hydrolyze Kraft pulp (Christov and Prior, 1996), amylase from *Bacillus lichneiformis* hydrolyze starch materials (Copinet et al., 2001). *Proteases* hydrolyze the breakdown of proteinaceous substances derived from food, leather, detergent and pharmaceutical industry (Singh, 2002; Beena and Geevarghese, 2010). Thus, keratinase produced from *Chrysosporium keratinophilum* can hydrolyze keratin (Singh, 2002).

Synthetic, man-made materials widely used worldwide include nylon, plastic (polyacrylate, polyurethane), and mixed composites can be degraded by bacteria and fungi (Howard, 2002). Nylon (linear polymer with amide bonds, and not hydrolyzed by proteases) can be degraded by nylon-degrading enzyme in ligninolytic culture of *Phanerochaete chrysosporium* and *Trametes versicolor* strain IZU-154 (Deguchi et al., 1997; Deguchi et al., 1998). Plastic biodegradation is achieved by different enzymes, such as depolymerases and esterases that is, *Amycolatopsis* sp. and *Bacillus* sp. were able to depolymerize the polylactic acid (Nakamura et al., 2001), *Phanerochaete chrysosporium* through cellobiose dehydrogenaze degrades polyacrylate (Cameron and Aust, 1999), *Curvularia senegalensis,* and *Coryneabacterium* degrade polyurethane through esterases (Nakajima – Kambe et al., 1999). Nitrile compounds that are widely used in chemical industries, feedstock solvents, pharmaceutical and drug intermediates, pesticides, etc., can be degraded by nitrile-degrading enzymes (nitrilases, nitrile hydratases, and amidases) released from bacteria and fungi (*Nocardia* sp., *Rhodococcus* sp. and *Fusarium solani*) (Banerjee et al., 2002). Furthermore, cyanides from cyanide – waste can be effectively catabolized cyanides through cyanide-catabolizing enzymes (cyanidases, cyanide hydratases, formamide hydrolases, and rhodanese) from *Trihoderma* sp., *Alcaligenes denitrificans* (Barclay et al., 1998; Basheer et al., 1993; Lynch, 2002).

12.5 Integrated phytobial remediation for functional cleanup environment

12.5.1 Phytobial remediation toward metal(loid)s removal from contaminated sites

Plant associated bacteria and fungi enhance phytoremediation alleviated metal(loid) toxicity altering their availability through acidification, chelation, precipitation, and redox reactions, and increasing plant growth (Hassan et al., 2017). Extracellular mechanisms of bacteria and fungi include metal(loid) chelation and cell wall binding, whereas intracellular mechanisms involve metal(loid) binding to organic acids, peptides, proteins, amino acids, phenols, sulfur compounds (sequestration and compartmentalization), and an increased plant antioxidative system (Bellion et al., 2006; Guarino et al., 2018). It was found that metal(loid)s may increase microbial community diversity through mixing of different microbiomes promoting horizontal gene transferring (HGT) (Rilling et al., 2015) that is, plasmids have resistance genes for different metal(loid)s, such as As, Hg, Cd, Cu, Co, Zn, Pb, Ag, Ni,

Te (Silver and Phung, 2009; Pal et al., 2017). Metal(loid) removal by plant/bacteria/fungi system from contaminated sites is given in Table 12.4.

Bacteria-mediated processes may alleviate metal(loid) toxicity and enhance tolerance to pollutants by increasing plant biomass, metal(loid) uptake, accumulation, and inhibiting of oxidative stress: *Pseudomonas koreensis* AGB-1 increased biomass of *Miscanthus sinensis*, chlorophyll, and protein content, as well as its enzymatic activity in the presence of Cd, Pb, Cu, Zn, and As (Babu et al., 2015); *Bacillus thuringiensis* GDB-1 increased biomass of *Alnus firma*, chlorophyll content and Cd, Pb, Cu, Zn, and As accumulation (Babu et al., 2013); *Pseudomonas* sp. Lk9 enhanced biomass of *Solanum nigrum* and uptake of Cd, Zn, Cu, and Cr (Chen et al., 2014a); *Bacillus* sp. SLS18 increased biomass of *Sorghum bicolor* and Cd and Mn uptake (Luo et al., 2012); *Exiguobacterium* sp. increased growth of *Vigna radiata*, chlorophyll content, and reduced As oxidative stress (Pandey and Bhatt, 2016); *Achromobacter xyloxidans* Ax10 enhanced Cu accumulation and phytoextraction by *Brassica juncea* (Ma et al., 2009a,b); *Kluyvera ascorbata* SUD165/26 reduced Pb, Ni, and Zn toxicity in plants (Burd et al., 2000); *Pseudomonas* sp. PsA4 and *Bacillus* sp. Ba32 protected *Brassica juncea* from Cr toxicity (Rajkumar et al., 2006); *Bacillus subtilis* SJ-101 enhanced growth of *Brassica juncea*, and Ni accumulation (Zaidi et al., 2006); *Pseudomonas asplenii* AC and *P.* AC-1 increased growth of *Phragmites australis* and protected from Cu toxicity (Reed et al., 2005); *Pseudomonas tolaasii* ACC23, *P. fluorescens* ACC9, *Alcaligens* sp. ZN4, and *Mycobacterium* sp. ACC14 may protect *Brassica napus* from Cd toxicity (Dell'Amico et al., 2008). Dolphen and Thiravetyan (2019) showed that *Bacillus pumilus* decreases As accumulation in grains of *Oryza sativa* and a combination of leonardite with bacteria reduces plant oxidative stress and down-regulates transporters Lsi1, Lsi2, OsPT4 increasing Si accumulation in roots and facilitating phytoremediation of As contaminated soils. A number of PGPB were isolated from the rhizosphere, root and shoot of *Prosopis julifora* growing on the tannery effluent contaminated soils (Cd, Co, Cr, Cu, Fe, Mn, Ni, Pb, and Zn) and showed significant activity od ACC-deaminase, P-solubilization, IAA, and siderophore production (Khan et al. 2015). However, *Pantoea stewartii* ASI11, *Enterobacter* sp., *Microbacterium arborescens* HU33 and their consortium (mixture of these three PGPB) showed high seed germination, root and shoot length, and weight of *Prosopis julifora.* Also, *Pseudomonas aeruginosa* PJRS20, *Pantoea stewartii* ASI11, and *Microbacterium arborescens* HU33 showed resistance to excess Cr (3000 mg/L) enabling growth and survival of *Prosopis julifora* in highly polluted sites (Khan et al. 2015). The same authors also showed that inoculation of *Pantoea stewartii* ASI11, *Enterobacter* sp., *Microbacterium arborescens* HU33 to *Lolium multiflorum* sown on the tannery effluent soils improved its growth and they were able to remove Cr enhanced phytoremediation of contaminated soils (Khan et al. 2015). Inoculation of *Micrococcus* sp., *Pseudomonas* sp., and *Arthrobacter* sp. to *Glycine max* promoted root elongation, Cd accumulation in root assisted in Cd phytoremediation efficiency (Rojjanateeranaj et al., 2017). Endophytic bacteria *Serratia marcescens* isolated from the roots of *Hedysarum pallidum* growing on mining antimony (Sb) spoil show that Sb induced oxidative stress (high production of H_2O_2 and malondialdehyde, MDA content), but increased antioxidant enzyme activities, such as catalase (CAT), ascorbat peroxidase (APX), peroxidase (POD), and superoxide dismutase (SOD) indicating efficient plant-bacteria assisted phytoremediation of Sb polluted soils (Kassa-Laouar et al., 2020). Lampis et al. (2009) found that selenite resistant rhizobacteria *Bacillius mycoides* SeITE01 and *Stenotrophomonas maltophilia* SeITE02 can stimulate SO_3^{2-} phytoextraxtion efficiency up to 65% in *Brassica juncea* growing in hydroponic conditions.

Fungi-mediated process may alleviate metal(loid) toxicity and increase the effectiveness of phytoremediation of contaminated sites: inoculation of *Zea mays* with AM fungi reduced Pb stress

Table 12.4 Metal(loid)s removal by plant/bacteria/fungi system from contaminated sites.

Bacteria/fungi	Plants	Contamination	References
BACTERIA			
Pseudomonas koreensis AGB-1	*Miscanthus sinensis*	Cd, Pb, Cu, Zn, As	Babu et al. (2015)
Bacillus thuringiensis GDB-1	*Alnus firma*	Cd, Pb, Cu, Zn, As	Babu et al. (2013)
Pseudomonas sp. Lk9	*Solanum nigrum*	Cd, Zn, Cu, Cr	Chen et al. (2014a)
Bacillus sp. SLS18	*Sorghum bicolor*	Cd, Mn	Luo et al. (2012)
Exiguobacterium sp.	*Vigna radiata*	As	Pandey and Bhatt (2016)
Achromobacter xyloxidans Ax10	*Brassica juncea*	Cu	Ma et al. (2009a,b)
Kluyvera ascorbata SUD165/26	*Brassica juncea* *Brassica napus* *Lycopersicum esculentum*	Ni, Pb, Zn	Burd et al. (2000)
Pseudomonas sp. PsA4 *Bacillus* sp. Ba32	*Brassica juncea*	Cr	Rajkumar et al. (2006)
Bacillus subtilis SJ-101	*Brassica juncea*	Ni	Zaidi et al. (2006)
Pseudomonas asplenii AC	*Phragmites australis*	Cu	Reed et al. (2005)
Pseudomonas tolaasii ACC23 *P. fluorescens* ACC9 *Alcaligens* sp. ZN4 *Mycobacterium* sp. ACC14	*Brassica napus*	Cd	Dell'Amico et al. (2008)
Bacillus pumilus	*Oryza sativa*	As	Dolphen and Thiravetyan (2019)
Pantoea stewartii ASI11 *Enterobacter* sp. *Microbacterium arborescens* HU33 consortium	*Prosopis julifora*	Cd, Co, Cr, Cu, Fe, Mn, Ni, Pb, Zn	Khan et al. (2015)

Table 12.4 Metal(loid)s removal by plant/bacteria/fungi system from contaminated sites. *Continued*

Bacteria/fungi	Plants	Contamination	References
Pantoea stewartii ASI11 *Enterobacter* sp. *Microbacterium arborescens* HU33	*Lolium multiflorum*	Cr	Khan et al. (2015)
Micrococcus sp. *Pseudomonas* sp. *Arthrobacter* sp.	*Glycine max*	Cd	Rojjanateeranaj et al. (2017)
Serratia marcescens	*Hedysarum pallidum*	Sb	Kassa-Laouar et al. (2020)
Bacillius mycoides SeITE01 *Stenotrophomonas maltophilia* SeITE02	*Brassica juncea*	Se	Lampis et al. (2009)
Fungi			
AMF	*Zea mays*	Pb	Zhang et al. (2010)
Rhizophagus irregularis *Funneliformis mossaeae*	*Helianthus annuus*	Cu, Cd, Zn	Hassan et al. (2013)
Glomus intraradices	*Helianthus annuus*	Ni	Ker and Charest (2010)
AMF	*Hordeum vulgare*	As, Al, Cr, Zn, Cu, Cd, Pb, Se	Biró et al. (2005)
Penicillium sp. *Aspergillus* sp. *Aspergillus niger*	*Oryza* sp.	As	Guimarães et al. (2019)
Bacteria - fungi			
Brevibacillus sp. - *Glomus mosseae*	*Trifolium* sp.	Ni	Vivas et al. (2003a)
Brevibacillus sp. – AMF	*Trifolium* sp.	Cd	Vivas et al. (2004)
Brevibacillus agri – AMF	*Trifolium* sp.	Pb	Vivas et al. (2003b)

(contiuned)

Table 12.4 Metal(loid)s removal by plant/bacteria/fungi system from contaminated sites. ***Continued***

Bacteria/fungi	Plants	Contamination	References
Bacillus licheniformes *Bacillus thuringiensis* *Bacillus megaterium* *Bacillus subtilus* *Bacillus polymyxa* *Paenibacilus azotofixans* *Pseudomonas fluorescens* *Pseudomonas aeruginosa* *Pseudomonas brassicac-eraum* *Pseudomonas putida* *Pseudomonas protenges* - *Pisolithus tinctorius* *Entrophosphora colom-biana* *Glomus clarum* *Glomus etunicatum* *Glomus intraradices* *Glomus mosseae*	*Populus deltoids x Populus nigra* *Salix purpurea* subsp. *lambertiana*	As, Cd, Pb, Zn	Guarino et al. (2018)

increasing plant biomass (Zhang et al., 2010); *Rhizophagus irregularis* and *Funneliformis mossaeae* increased growth of *Helianthus annuus* reducing Cu, Cd, and Zn toxicity (Hassan et al., 2013); *Glomus intraradic*es enhanced the growth of *Helianthus annuus* and Ni phytoextraction (Ker and Charest, 2010); AM fungi alleviated As, Al, Cr, Zn, Cu, Cd, Pb, and Se toxicity in *Hordeum vulgare* (Biró et al., 2005). Guimarães et al. (2019) found that *Penicillium* sp., *Aspergillus* sp., and *Aspergillus niger* isolated from *Oryza* sp. promote As volatilization that is, they capable of producing of volatile As species (trimethylarsine, TMAs; mono- and dimethylarsine, MMAs, and DMAs) for 57.8%, 46.4%, and 5.2%, respectively and show high fungal biomass. Fungi, such as *Aspergillus glaucum, Candida humicola, Scorpulaipsis brevicaulis, Gliocladium roseum, Penicillium gladioli*, and *Fusarium* sp. are capable of As-volatilization that is, they biomethylate As into volatile forms (Lin, 2008). According to Čerňanský et al. (2009) and Urík et al. (2007) *Aspergillus clavatus* (18.8%–28%), *Neosartorya fischeri* (31.6%–36.7%), *Penicillium glabrum* (25.2%–26.2%), *Trichoderma viride* (4.0-9.3%) show different As volatilization efficiency.

Combined plant/bacterium/fungi system: Dual inoculation of bacteria and fungi from soils can be important in the metal(loid) plant tolerance and phytoremediation of contaminated sites (Turnau et al., 2006). Rhizosphere bacteria may improve mycorrhiza formation that promotes plant growth and tolerance to metal(loid)s. Selection procedure involves: (1) adapted bacteria and mycorrhizal fungi to heavy metal stress and their functional compatibility in terms of phytoremediation of polluted soils, (2) enhanced bacterial activities in the mycorrhizosphere, and (3) accumulation of heavy metals, avoiding further transport to plants (Turnau et al., 2006). Therefore, dual inoculation of *Glomus mosseae* and

Brevibacillus sp. from Ni-contaminated soils to *Trifolium* sp. increased Ni accumulation in plants and enhanced Ni phytoextraction from polluted soils (Vivas et al., 2003a). In addition, dual inoculation of *Brevibacillus* sp. and AM fungus to *Trifolim* sp. on the Cd-contaminated soils shows Cd accumulation in shoots indicating phytoextraction activity, probably because bacteria increased dehydrogenase, phosphatase, β-gluconase activities in the mycorrhizosphere improving microbial activities in order to promote plant growth on contaminated soils (Vivas et al., 2004). Dual inoculation of *Brevibacillus agri* and AMF to *Trifolium* sp. on Pb-contaminated soil lead to increased plant growth, and Pb accumulation in plant shoots showing that bacterial inoculation increased phytoextraction potential in plants inoculated with AMF (Vivas et al., 2003b). However, Hui et al. (2012) showed that Pb contamination may alter fungal richness and diversity in comparison to bacterial community in pine forest soil, probably due to the fact that fungi are more tolerant to Pb than bacteria (Rajapaksha et al., 2004). Thus, Pb decreased bacterial biomass, community structure and activity (Grandlic et al., 2006; Hu et al., 2007). Fungi, *Thelephora terrestris, Cortinarius* OTUs, and *Lactarius rufus* that colonize *Pinus sylvestris* roots show tolerance to high Pb content and high abundance in the polluted site (Krupa and Kozdroj, 2007; Hui et al., 2012). According to Guarino et al. (2018) inoculation of different bacteria and fungi to *Populus deltoids x Populus nigra* and *Salix purpurea* subsp. *lambertiana* significantly increase heavy metal concentrations in roots and enhance antioxidant enzymatic activity. Therefore, selected ectomycorrhizal fungi (*Pisolithus tinctorius*), and AMFs (*Entrophosphora colombiana, Glomus clarum, Glomus etunicatum, Glomus intraradices, Glomus mosseae*) and bacteria – *Bacillus licheniformes, Bacillus thuringiensis, Bacillus megaterium, Bacillus subtilus, Bacillus polymyxa, Paenibacilus azotofixans, Pseudomonas fluorescens, Pseudomonas aeruginosa, Pseudomonas brassicaceraum, Pseudomonas putida, Pseudomonas protenges*) are chosen for phytoremediation in the rhizosphere soil. Authors showed that after 12 and 36 months concentrations of heavy metals are significantly less in soil inoculated by microflora consortium than in control soil (As > Cd > Pb > Zn) and *Populus* and *Salix* show great removal efficiency. Furthermore, concentrations of all heavy metals were higher in roots than in aerial parts of both plants. Enzymatic activity of SOD, CAT, APX, GPX, PAL, GST in *Salix* and *Populus* inoculated with microflora consortium during 12 and 24 months increased, whereas MDA content did not change indicating plant adaptive response to stress. The optimization of the plant-microbe system to work synergistically, is important, for plant growth and tolerance, thus increasing the efficiency of phytoremediation of polluted soils (Guarino et al., 2018). Table 12.4.

12.5.2 Phytobial remediation toward radionuclides removal from contaminated sites

Radionuclide contamination presents a huge threat to human health and environment due to their application in nuclear energy generation, industrial, agricultural and medical activities (EPA, 2004a). Contamination with radionuclides can be due to their mining and milling process, nuclear explosion and accidental spills. According to IAEA (2010) hazardous radioactive waste is defined as any material that contains particles with radioactive levels that are not safe by national and international standards and with no further use.

The main radioactive elements released in the environment, such as cesium (Cs), strontium (Sr), uranium (U), plutonium (Pu), radium (Ra), technetium (Tc), yttrium (Y) have direct negative effects to human health and environment (De Filippis, 2015). Uranium (^{235}U) is one the most toxic radionuclides that is mostly generated during the manufacturing process of nuclear fuel, and byproduct of this process is depleted uranium (DU) which is used for many military applications (Bleise et al., 2003). ^{99}Tc is one of the fission products, such as ^{235}U and ^{239}Pu, released in the environment via nuclear weapons (Simonoff et al., 2007). These radionuclides are very toxic:

- ^{137}Cs is a β emitter with half-life of 30 years – 106 years, similar to K which can substitute for it in cells
- ^{90}Sr is a βγ emitter with half-life of 50 days – 29 years, can substitute Ca as its analogue in cells
- $^{234, 235,238, 239}$U is a αβ emitter with half-life of 69 years – 10^{9}years, absorbed by most cells
- $^{238,239,240, 242}$Pu is a α emitter with half-life of 88 years – 10^{5} years, highly reactive and accumulates especially in bones
- $^{223, 224, 226, 228}$Ra is a αβ emitter with half-life of 11 days – 10^{3}years, highly incorporated in all cells technetium (^{99}Tc) is radionuclide used in diagnostic nuclear medicine and cancer treatment, it is a βγ emitter with half-life of 4 days – 10^{6}years
- yttrium (^{90}Y) used in electronic industry and for treating lymphoma, it a β emitter with half-life of 2.7 days – 10^{7} days;
- iodine (^{131}I) used for treatment of thyroid cancer, it is a γβ emitter with half-life of 6 hours – 10^{7} years.

Radioactive waste must be stored in central storage/disposal facilities/deep geological repositories depending on waste categories: (1) low level waste (LLW) includes medical, industrial (paper, rags, glassware, clothing, filters), and nuclear fuel cycle waste with small amounts of short-lived radioactivity which require near-surface facilities or more complex engineered repositories, (2) intermediate level waste (ILW) includes reactor components, chemical sludge, sealed components from medicine and industry with high quantities of long-lived radionuclides, which require disposal in greater depths (10–100 m underground), and (3) high level waste (HLW) is produced by nuclear reactors containing fission products generated in the reactor core, such as nuclear fuel and the waste stream from the spent fuel, they require very deep geological repositories (several hundreds of meters) (De Filippis, 2015). According to ANSTO Australian Nuclear Science and Technology Organization, 2011 over 75% of all radioactive waste is already disposed. Radioactive waste from nuclear power facilities produces approximately 0.4 million tons per annum and around 340,000 tons of spent fuel is produced in power reactors. Furthermore, transportation of radioactive materials by roads, railways and ships also presents a great threat to public health and environment due to 'in-transit' accidents (De Filippis, 2015).

Engineering methods, such as encapsulation, size separation, soil washing, leaching with chelating agents, electro kinetics, ion exchange, etc. are used in cleanup for acute radioactive contamination (Malaviya and Singh, 2012; De Filippis, 2015). Biological remediation technology can be used efficiently for low level radioactive waste, including uranium contaminated soils (EPA US Environmental Protection Agency, 2004a; Malaviya and Singh, 2012; De Filippis, 2015). Cleanup of radionuclides in soils can be achieved by phytoremediation (plants) and bioremediation (bacteria and fungi) and their combination (plant-bacteria-fungi system) (Entry et al., 1999; Dushenkov, 2003; Malaviya and Singh, 2012; De Filippis, 2015; Ahsan et al., 2017). Radionuclides removal by plant/bacteria/fungi system from contaminated sites is given in Table 12.5.

Phytoremediation. Phytoremediation techniques that are used for cleanup of radionuclides from contaminated sites are: (1) phytoextraction (plants accumulate radionuclides from soil into shoots which were harvested), (2) rhizofiltration (plant roots can precipitate radionuclides from

Table 12.5 Radionuclides removal by plant/bacteria/fungi system from contaminated sites.

Plant/bacteria/fungi	Contaminants/ remediation	References
PLANTS		
Mellilotus, Sorghum, Trifolium	^{137}Cs phytoextraction	Rogers and Wiliams (1986)
Polytrichum, Festuca, Agrostis, Carex	^{137}Cs phytoextraction	Coughtrey et al. (1989)
Chenopodium, Brassica, Beta, Polygonum	^{137}Cs phytoextraction	Broadley and Willey (1997)
Pahlaris, Amaranthus, Brassica, Phaseoulus	^{137}Cs phytoextraction	Lasat et al. (1997); Lasat et al. (1998)
Vaccinium, Empetrum, Deschampsia	^{137}Cs phytoextraction	Bunzl et al. (1999)
Amaranthus, Helianthus, Brassica, Pisum	^{137}Cs phytoextraction	Dushenkov et al. (1999)
Helianthus, Populus	^{137}Cs phytoextraction	Vanek et al. (2001)
Salix	^{137}Cs phytoextraction	Dutton and Humphreys (2005)
Evergreen coniferous species	^{137}Cs phytoextraction	Yoshihara et al. (2013)
Triticum, Lolium sp.	^{99}Tc phytoextraction	Garten2 and Lomax, 1989
Helianthus, Brassica	^{137}Cs and ^{90}Sr rhizofiltration	Dushenkov et al. (1997)
Eichornia, Catharanthus	^{137}Cs and ^{90}Sr rhizofiltration	Prasad and Freitas (2003)
Helianthus, Brassica, Phaseoulus	^{238,235}U phytoextraction	Dushenkov et al. (1997)
Brassica, Chenopodium, Phaseolus	^{238,235}U rhizofiltration	Dushenkov et al. (1997)
Brassica juncea,Chenopodium amaranticolor	^{238,235}U rhizofiltration	Eapen et al. (2003)
Pinus sylvestris	^{238,235}U phytostabilization	Thyri et al. (2005)
Triticum aestivum, Brassica juncea, Zea mays, Pisum sativum	^{238,235}U phytostabilization	Huang et al. (1998)
Lolium perenne	^{238,235}U phytostabilization	Vanderhove et al. (2007)
Brassica juncea, Helianthus annuus	^{238,235}U phytostabilization	Shahandeh and Hossner (2002)
BACTERIA		
Arthrobacter, Rhodococcus, Nocardia, Deinococcus radiodurans	^{137}Cs, ^{99}Tc	Fredrickson et al. (2004)
Desulfovibrio desulfuricans, Schewanella putrefaciens, Geobacter sulfurreducens, Thiobacillus ferooxidans, Thiobacillus thioxidans	Tc	Simonoff et al. (2007)
Alcaligenese eutrophes, Pseudomonas fluorescens	Sr	Francis et al. (2000)
Arthobacter, Citrobacter	Sr	Dwivedi and Mathur (1995)
Rhizopus arrhizus, Citrobacter	Pu	

(contiunued)

Table 12.5 Radionuclides removal by plant/bacteria/fungi system from contaminated sites. *Continued*

Plant/bacteria/fungi	Contaminants/ remediation	References
mesophyllic Fe^{III} and sulfate-reducing bacteria *Shewanella puttrefaciens*, *Desulfovibrio vulgaris*	U	Lovley and Phillips (1992) Simonoff et al. (2007)
FUNGI		
Penicillium, Fusarium, Chrysosporium, Scopulariopsis, Hyalodendron, Verticillium, Aspergilus versicolor, A. niger, Hormoconis resinae, Cladosporium sphaerospermum, Cladosporium herbarum, Cladosporium cladosporioides, Alternaria alternate, Aurebasidium pullulans, Penicilium aurantiogriseu, Penicillium spinulosum, Acremonium strictum *Cladosporium sphaerospermum, C. herbarum, Hormoconis resinae, Alternaria alternate, Aureobasidium pullans,* *Rhizopus arrhizus, Aspergillus niger, Penicilium italicum, Penicilium acrysogenium*	Radionuclides	Zhdanova et al. (2004) Zhdanova et al. (2005) Zhdanova et al. (1995) White and Gadd (1990)
Cladosporium cladosporoides, Penicillium roseopurpureum	^{137}Cs	Zhdanova et al. (1991)
Alternaria alternate, Fusarium verticilioides	^{137}Cs	Mahmoud (2004)
Cortinarius praestans, Laccaria amethystine, Agaricus sp., *Lycoperdon* sp., *Inocybe longicystus, Xercomus badi*	^{137}Cs	Byrne (1988) Dighton and Horrill (1988)
Dematiceae	^{90}Sr	Zhdanova et al. (1990)
PLANT – BACTERIA		
Leptochloa fusca - Pantoea stewartii ASI11, *Enterobacter* sp. HU38, *Microbacterium arborescens*	U	Ahsan et al. (2017)
PLANT - FUNGI		
Brassica rapa, Solanum lycopersicum – Pseudosigmoidea ibarakiensis 1.2-2-1 *Veronaeopsis simplex* Y34, *Helminthosporium velutinum* 41-1	Cs	Diene et al. (2014)

Table 12.5 Radionuclides removal by plant/bacteria/fungi system from contaminated sites. *Continued*

Plant/bacteria/fungi	Contaminants/ remediation	References
*Paspalum notarum, Sorghum halpense Panicum virginiatum - Glomus mossaea*m *Glomus intraradices*	^{137}Cs, ^{90}Sr	Entry et al. (1999)
Soya, Pisum, Avena – *Glomus mossaea*m *Glomus intraradices*	^{137}Cs	Goncharova (2009)
Picea abies – AMF	^{134}Cs, ^{85}Sr	Riesen and Brunner (1996)
Sorghum × drummondii, Melilotus officinalis – AMF	^{137}Cs	Roger and Williams (1986)
Agrostis tenuis – AMF	^{137}Cs	Berreck and Haselwandter (2001)
Glycine max, Festuca ovina - Glomus mosseae	^{90}Sr	Jackson et al. (1973)
Sesbania rostrata - Azorrhizobium caulinodans ORS571, AMF GE	U	Ren et al. (2019)
Trifolium subterraneum - Glomus intraradices	U	Rufyikiri et al. (2004)
Cynodon dactylon – MF	U	Weiersbye et al. (1999)
Medicago truncalata, Lolium perenne – Glomus intraradices	U	Roos and Jakobsen (2008) Chen et al. (2008)

contaminated effluents), (3) phytostabilization (plants capture radionuclides *in situ* making them less harmless) (Dushenkov, 2003). Successful phytoremediation depends on bioavailability of radionuclides, soil properties, characteristics of radionuclides, plant species, the rate of uptake by plant roots, efficiency of translocation of radionuclides through vascular system from roots to shoots, plant's ability to accumulate and tolerate radionuclides (Dushenkov, 2003; Malaviya and Singh, 2012; De Filippis, 2015).

Plants that are effective in ^{137}Cs phytoextraction are: *Mellilotus, Sorghum, Trifolium* (Rogers and Wiliams, 1986); *Polytrichum, Festuca, Agrostis, Carex* (Coughtrey et al., 1989); *Chenopodium, Brassica, Beta, Polygonum* (Broadley and Willey, 1997); *Pahlaris, Amaranthus, Brassica, Phaseoulus* (Lasat et al., 1997; 1998); *Vaccinium, Empetrum, Deschampsia* (Bunzl et al., 1999); *Amaranthus, Helianthus, Brassica, Pisum* (Dushenkov et al., 1999); *Helianthus, Populus* (Vanek et al., 2001); *Salix* (Dutton and Humphreys, 2005); evergreen coniferous species after Fukushima nuclear power plant accident (Yoshihara et al., 2013). Plants that are effective in ^{99}Tc phytoextraction: in deciduous forests Tc accumulated (mostly in wood) (Garten et al., 1986); *Triticum* (92-95%) and *Lolium* sp. (62-78%) (Garten2 and

Lomax, 1989). Plants that are effective in ^{137}Cs and ^{90}Sr rhizofiltration are: *Helianthus* and *Brassica* (Dushenkov et al., 1997); *Eichornia* and *Catharanthus* (Prasad and Freitas, 2003). Plants that are effective in $^{238,235}U$ phytoextraction: *Helianthus, Brassica, Phaseoulus* (Dushenkov et al., 1997). Plants that are effective in $^{238,235}U$ rhizofiltration: *Brassica, Chenopodium, Phaseolus* (Dushenkov et al., 1997); *Brassica juncea, Chenopodium amaranticolor* (Eapen et al., 2003). According to Malaviya and Singh (2012) U from contaminated soils may be uptaken by adsorption onto root surface, or precipitation of uranium phosphate on/in root that can prevent U transport inside of plant tissue from roots to shoots. Uranium uptake is related to association of U, P, and Ca and that was confirmed by transmission electron microscopy and energy-dispersive spectroscopy (EDS) analysis (Laroche and Henner, 2003). Therefore, U predominantly accumulates in roots with its lower transportation in shoots of *Pinus sylvestris* (Thyri et al., 2005), *Triticum aestivum, Brassica juncea, Zea mays, Pisum sativum* (Huang et al., 1998), and *Lolium perenne* (Vanderhove et al., 2007). Shahandeh and Hossner (2002) reported that *Brassica juncea* and *Helianthus annuus* growing on soil contaminated with 600 mgU(VI) kg^{-1} accumulated U more in roots (600 mgU kg^{-1} and 6200 mgU kg^{-1}) and less in shoots (6.9 mgU kg^{-1} and 102 mgU kg^{-1}, respectively).

Bacteria: Microbial colonization of radioactive sites affects the solubility of radionuclides by biosorption, metabolite-dependent accumulation, enzymatic transformation, and mineralization by ligands (Simonoff et al., 2007). Generally, Chicote et al. (2004) found β-*Proteobacteria, Actinomycetes, Bacillus/Staphylococcus,* and fungi *Aspergillus* sp. attached to the pools that store nuclear materials. Similarly, Fredrickson et al. (2004) reported that *Arthrobacter, Rhodococcus, Nocardia,* and *Deinococcus radiodurans* were able to survive acute dozes of ionizing radiation (around 20 kGy) with ^{137}Cs and ^{99}Tc found in sediments formed from a waste tank leak. According to Thomas et al. (2000) *Pseudomonas aeruginosa* and *Pseudomonas putida* are capable to form radionuclide-citrate complexes. In addition, the microbial biomineralization of pollutants by phosphate is noted in *Escherichia coli*, in which U can precipitate in the form of uranyl-phosphate complex (Basnakova et al., 1998), whereas Francis et al. (2000) noted carbonate complex with Sr produced from *Alcaligenese eutrophes* and *Pseudomonas fluorescens*. Bacteria are capable of immobilizing Tc in the environment through enzymatic mechanisms or reduction to lower valency. Reduction of Tc at neutral pH is achieved by *Desulfovibrio desulfuricans* (sulfate-reducing bacteria), *Schewanella putrefaciens* (metal-reducing bacteria), *Geobacter sulfurreducens,* whereas under acidic pH, *Thiobacillus ferooxidans* and *Thiobacillusthioxidans* reduce Tc to lower valency (Simonoff et al., 2007). Sr can be biosorbed by *Arthobacter* or *Citrobacter* forming crystalline Sr - carbonate/insoluble phosphate complex (Dwivedi and Mathur, 1995). Pu removal is related with *Rhizopus arrhizus* which can at pH 4-9 lead to the reduction and biosorption of Pu, whereas $^{238,239}Pu$ can be 50% removed from solution by *Citrobacter* forming a metal-phosphate complex (Macaskie et al., 1994). Bacteria capable of reducing U are: thermophyllic bacterium, mesophyllic Fe^{III}, sulfate-reducing and fermentative bacteria (*Shewanella puttrefaciens, Desulfovibrio vulgaris*) (Lovley and Phillips, 1992; Simonoff et al., 2007). Reduction of U can vary with the amount of U-organic complex (acetate, malonate, oxalate, citrate, aromatic [Ganesh et al., 1997]).

Fungi: Fungi show long-term accumulation of radionuclides and radio-resistance (Dighton et al., 2008). It was reported that for 20 years, 2000 strains of 180 fungal species of 92 genera were isolated from the area around Chernobyl (Zhdanova et al., 2004, Zhdanova et al. 2005). There were noted resistant fungi with high frequency of 70%–80% (*Penicillium, Fusarium, Chrysosporium, Scopulariopsis, Hyalodendron, Verticillium*) and low to medium frequency of 10%–50% (*Aspergilus versicolor, A. niger, Hormoconis resinae, Cladosporium sphaerospermum, Cladosporium herbarum, Cladosporium cladosporioides, Alternaria alternate, Aurebasidium pullulans, Penicilium aurantiogriseu, Penicillium spinulosum, Acremonium strictum*). Saprophytic fungi have a high surface area of hyphae capable of absorbing nutrients and radionuclides from decomposed leaf litter that is, they can mineralize radionuclides and incorporate them into own biomass (Witkamp, 1968). It was shown that accumulation of ^{137}Cs into *Cladosporium cladosporoides* and *Penicillium roseopurpureum* can destroy these "hot radioactive particles" (<1147 Bq, γ ray) within 50–150 days (Zhdanova et al., 1991). Immobilization of radionuclides in fungal mycelial biomass leads to the radionuclides (Cs) loading into upper organic layer where they are resistant to leaching that is, they are retained in grasslands soil (Olsen et al., 1990) and spruce forest (Rommelt et al., 1990). *Rhizopus arrhizus, Aspergillus niger, Penicilium italicum, Penicilium acrysogenium* are capable of biosorption of thorium (Th) (White and Gadd, 1990), whereas great uptake capacity for ^{137}Cs have *Alternaria alternate* and *Fusarium verticilioides* (Mahmoud, 2004). It was noted that a large number of ectomycorrhizal fungi, such as *Cortinarius praestans, Laccaria amethystine, Agaricus* sp., *Lycoperdon* sp., *Inocybe longicystus, Xercomus badius* accumulated Cs in pre-Chernobyl accidents suggesting that these fungi act as long-term retainers of radionuclides (Byrne, 1988; Dighton and Horrill, 1988). However, some saprophytic fungi (*Mycena polygramma, Cyctoderma amianthinum, Mycena sanguinolenta*) show higher accumulation rates than ectomycorrhizal fungi (Clint et al., 1991). Melanized fungal species have been found to dominate in the area with radiation (Zhdanova et al., 1995). Melanin is pigment in the cell wall of fungus that acts as a receptor for radiation, i.e. this pigment has potential to change biochemical pathways in fungal cell when it is exposed to radiation (Zhdanova et al., 1995). Fungi isolated from the reactor in Chernobyl had melanin: *Cladosporium sphaerospermum, C. herbarum, Hormoconis resinae, Alternaria alternate, Aureobasidium pullans*. Zhdanova et al. (1990) show that dark pigmented *Dematiceae* accumulated more ^{90}Sr (20–1510 Bq g^{-1}) compared to light pigmented *Moniliaceae* (18–335 Bq g^{-1}).

Combined plant/bacterium/fungi system. The joint action of microorganisms (bacteria and fungi) and plants contributes to the efficient and successful soil remediation polluted with radionuclides. Endophytic bacterial inoculation of *Pantoea stewartii* ASI11, *Enterobacter* sp. HU38 and *Microbacterium arborescens* HU33 improved plant growth and chlorophyll content of *Leptochloa fusca* planted on artificial U contaminated soil (Ahsan et al., 2017). Results in this study show that roots accumulate more U (1758 mg kg^{-1}) than shoots indicating reduced U transport in the shoots of the plant. Multiple action of bacterial consortia enhances the production of organic compounds, that act as chelators increasing U uptake by roots. Thus, *L. fusca* and consortia of endophytic bacteria efficiently improved phytostabilization of U contaminated soils (Ahsan et al., 2017). Furthermore, dark septate endophytic fungi (DSE) form a symbiotic relationship with forest trees, and plant species which belong to families *Solanaceae* and *Brassicaceae* (Usuki and Narisawa, 2007; Narisawa et al., 2007). Thus, Diene et al. (2014) noted that inoculation of melanized DSE fungi *Pseudosigmoidea ibarakiensis* 1.2-2-1, *Veronaeopsis simplex* Y34, and *Helminthosporium velutinum* 41-1 to *Brassica rapa* and *Solanum lycopersicum* improved plant growth (82% and 122%, respectively) when plants grew on Cs contaminated soils. Also, Cs accumulation was higher in aboveground parts in *B. rapa* whereas in *S. lycopersicum* Cs accumulation was restricted from roots to shoots, and in both cases combination of DSE and plants enhanced phytoremediation of Cs polluted soils (Diene et al., 2014). AMF can increase accumulation of radionuclides, such as ^{137}Cs and ^{238}U due to their hyphal accumulation (Dighton et al., 2008). However, uptake of radionuclides by AMF is conflicting and there are differences between plant species. Entry et al. (1999) noted that inoculation of *Glomus mossaea* and *Glomus intraradices* to grasses (*Paspalum notarum, Sorghum halpense* and *Panicum virginiatum*) improved their biomass and enhanced accumulation of ^{137}Cs and ^{90}Sr indicating that combined plant-fungi system may present efficient strategy for remediation of radionuclides in contaminated soils. Goncharova (2009) noted that inoculation of mycorrhizal fungi *Glomus mossaea* and *Glomus intraradices* to *Soya, Pisum,* and *Avena* increased accumulation of ^{137}Cs in their tissue; however, 15–20 years are necessary for remediation, because substantial amount of the radionuclide has been leached below the root zone. According to Riesen and Brunner (1996) ectomycorrhizal fungus *Hebeloma crustuliniformein* inoculated to *Picea abies* accumulated ^{134}Cs and ^{85}Sr in plant roots, however the uptake and translocation rate of ^{85}Sr were smaller than those for ^{134}Cs, because ^{85}Sr is more readily adsorbed by mycelium or plant cells. *Sorghum × drummondii* and *Melilotus officinalis* slightly increased ^{137}Cs uptake and accumulation by AMF (Roger and Williams, 1986), whereas *Agrostis tenuis* inoculated with AMF decreased Cs accumulation in shoots (Berreck and Haselwandter, 2001). Similarly, *Glycine max* inoculated with *Glomus mosseae* increased the uptake of ^{90}Sr from soils whereas in *Festuca ovina* uptake of Cs decreased (Jackson et al., 1973) that is, radionuclides can be sequestrated in the mycorrhizal fungus or fungi may restrict transport of radionuclides to shoots. Triple symbiotic association of a legume plant species *Sesbania rostrata*, rhizobia *Azorrhizobium caulinodans* ORS571, and AMF GE show high U removal rate (50.5%–73.2%) improving phytoremediation efficiency of U-contaminated soils (Ren et al., 2019). Results in this study show that rhizobia and AMF enhance the growth of plant, phytochelatin synthase and production of organic acids (malic acid, succinic acid and citric acid) in root exudates (Ren et al., 2019). AMF can be used in phytostabilization of U in soils reducing U translocation from roots to shoots, because intraradical mycelium largely retains U within fungal structure (de Boulois et al., 2008; Malaviya and Singh, 2012; Davies et al., 2015). Inoculation of *Glomus intraradices* to *Trifolium subterraneum* showed that U concentrations in plants were 60–360 times higher in underground plant part than in aboveground plant part (uranium transfer factor was 0.008–0.02) and that was associated with formation of U-phosphate complexes by AMF (Rufyikiri et al., 2004). According to Weiersbye et al. (1999) uranium accumulation in mycorrhizal fungi vesicles decrease its further translocation from roots to shoots in *Cynodon dactylon*. U concentrations in shoots of non-inoculated plants of *Medicago trunculata* and *Lolium perenne* can be 20–30 times higher than in inoculated plants by *Glomus* intraradices (Roos and Jakobsen, 2008; Chen et al. 2008). Therefore, uranium immobilization in soils (adsorption/complex AMF – glycoproteins/complex AMF – polyphosphates) reduced its transport from underground plant part than in aboveground plant part and its entry in the food chain, thereby reducing environmental risk (Malaviya and Singh, 2012; Davies et al., 2015). Table 12.5.

12.5.3 Phytobial remediation toward chlorinated compounds removal from contaminated sites

POPs are synthetic chemicals that have broad application in industry, agriculture, public space, workspace, homes, etc., and due to lipophilic nature POPs are resistant to degradation under natural conditions, they are capable to accumulate in the environment having long-range dispersal potential and present a serious threat to human health and environment (EPA, 2004b; Weber, et al., 2011; Arslan et al., 2017). In addition, chlorinated solvents are hazardous chemicals that are slightly soluble in water occurring in the form of dense non-aqueous phase liquid (DNAPL) (EPA, 2004b). POPs named “dirty dozen” include chlorinated compounds that should be banned and regulated and due to their toxicity, human and environmental health deterioration they were listed in Stockholm Convention of POPs in 2001 (EPA, 2004b):

- Chlorinated pesticides: aldrin, dieldrin, chlordane, 2,2-bis(chlorophenyl)-1,1,1-trichloroethane (DDT), 1,1-dichloro-2,2-bis(chlorophenyl)ethane (DDD), 1,1-dichloro-2,2-bis(chlorophenyl) ethylene (DDE), endrin, heptachlor, mirex, toxaphene, hexachlorobenzene, chlordecone, lindane/ γ-hexachlorocyclohexane, α,β-hexachlorocyclohexane (HCH), endosulfan
- PCBs, polychlorinated dibenzo-p-dioxins (PCDDs), polychlorinated dibenzofutans (PCDFs)

Chlorinated solvents that are primarily used as metal cleaning agents and adhesives, such as trichloroethylene (TCE), perchloroethylene (PCE), carbon tetrachloride (CT), trichlorethanol, trichloracetic acid, vinyl chloride, chloroform, etc., and polychlorinated phenols that are used as bactericides, algicides, molluscides, fungicides, such as 3,4,6-trichlorphenol, pentachlorophenol (PCP) have been listed as EPA priority pollutants (http://oaspub.epa.gov/) and in priority list of hazardous substances (http://www.atsdr.cdc.gov./).

Chlorinated compounds as organic pollutants contain one or more chlorine atoms/one or more cycling ring structures/a lack of polar functional groups, and because of that they persist in the environment several years or decades, they are easily dissolved in fats and accumulated in food chain, and they can travel long distances in the environment (Weinberg, 2008). They are accumulated in plants, animals and humans (Sweetman et al., 2005). Human exposure to POPs causes cancer, cardiovascular diseases, endocrinological, neurological and immunological problems, learning disabilities, they affect pregnancy, cross the placenta, and after birth they can be found in breastfeeding (Sweetman et al., 2005; La Merril et al., 2012).

Remediation of contaminated soils with organic pollutants includes transport to artificial landfills, excavation, vitrification, vapor extraction, air sparging, pumping, chemical processes (ozonation, Fenton's reaction), or thermal processes (incineration) (EPA, 2004b). However, phytoremediation and bioremediation have been successfully used to reduce toxicity of persistent chlorinated compounds (EPA, 2004b; Van Aken and Doty, 2009; Lenoir et al., 2016).

Plants. Phytoremediation of chlorinated compounds is a technology that utilizes plants to transform, sequester, transport and extract contaminants that is, it includes rhizodegradation, phytodegradation, phytovolatization, and hydraulic control (Mc Cutcheon and Schnoor, 2003). Uptake and translocation of organic chemicals depend on octanol-water coefficient (log K_{ow}), acidity constant (pK_a), aqueous solubility (S_w), octanol solubility (S_o), concentration of pollutant and plant species (Burken and Schnoor, 1998; Admire et al., 2014). Moderately hydrophobic compounds have log K_{ow} values from 0.5–3.0 and they are efficiently taken up by plants and transported into plant tissue whereas if plant log K_{ow} values range from 3.0–8.3, plants are resilient to pollutant uptake and they in plant roots bind to lipid membranes (Chaudhry et al., 2002; Ajo-Franklin, et al., 2006). Pollutants enter the plant via simple diffusion through cell wall, into the xylem stream (Campos et al., 2008) and they can be released in the atmosphere through leaves evapotranspiration or they can be degraded (Sanderman, 1994). The metabolisms of organic pollutants in plants are characterized by three phases of the *green liver* model: *Phase I* includes reactions of oxidation, and hydrolysis of pollutants by plant-derived enzymes, such as cytochrome P-450 monooxygenases, peroxidases, flavin-dependent monooxygenases; *Phase II* includes conjugation by enzyme gluthatione-S-transferases and *Phase III* includes deposition and sequestration in vacuole (Sanderman, 1994). Plants that are efficient in phytoremediation of different chlorinated compounds (chlorinated solvents, pesticides, PCB, etc.) are: *Brassica napus, Zea mays, Coffea liberica, Oryza sativa, Glycine max, Cucurbita pepo,*

Daucus carota, Beta vulgaris, Solanum tuberosum, Spinacia oleracea, Lactuca sativa, Vigna unguiculata, Taraxacum, Lolium perenne, Phragmites australis, Ipomea batatas, Colocasia esculenta, Picea abies, Salix, Populus, etc. (Arslan et al., 2017).

Combined plant/bacterium system. Plant-associated bacteria (rhizobacteria and endophytic bacteria) are capable of degrading persistent chlorinated compounds (Glick, 2010; Mackova et al., 2009). In this relationship, bacteria possess catabolic genes responsible for activation of catabolic enzymes for the degradation, transformation and mineralization of pollutant mitigated plant toxicity (Sessitsch et al., 2005; Khan et al., 2013; Naveed et al., 2014). Rhizobacteria use a carbon source from organic pollutants for their metabolism whereas plants alleviate the proliferation and colonization of bacteria through production of root exudates, organic acid production, co-metabolite induction, and H^+/OH^- ion excretion (Khan et al., 2013). Furthermore, structural analogs in plants can act as inducers (terpenes, flavonoids, salicylic acid) enhancing bacterial colonization which can degrade organic pollutants (Tandlich et al., 2001; Arslan et al., 2017). In degradation of chlorinated compounds (pentachlorphenol, 2,4-D, PCBs) the following plant – associated rhizobacteria are included: *Triticum aestivum, Medicago sativa, Solanum nigrum, Zea mays, Hordeum aestivum, Panicum virogatum, Robinia pseudoacacia, Betula pendula, Fraxinus excelsior, Salix caprea, Pinus nigra, Platanus, Populus deltoids, Liquidambar styracula, Salix sachalinensis, Pinus taeda, Cupressocyparis leylandii, Pinus palustris – Rhodococcus sp. Arhobacter, Pseudomonas Burkholderia, Sphinomonas, Microbacterium, Bacillus, Achromobacter, Rhizobium, Gordonia, Mesorhizobium* (Strycharz and Newman, 2009; Strycharz and Newman, 2010; Arslan et al., 2017). Endophytic bacteria are capable to degrade organic pollutants in the rhizosphere and in the endosphere (root cortex, xylem) while plants provide nutrients to bacteria and facilitate their colonization in the root and the shoot (Sessitsch et al., 2005; Moore et al., 2006). In degradation of chlorinated compounds (chlorobenzoic acids, 2,4-D, TCE, hexachlorocyclohexane) the following plant-associated endophyte bacteria are included: *Lolium perenne, Cytisus striatus, Populus alba, Lupinus luteus – Pseudomonas, Burkholderia, Enterobacter* (Germaine et al., 2006; Arslan et al., 2017).

Combined plant/fungi system. Organic compounds can negatively affect the AMF development, such as germination, colonization, extraradical hyphal elongation, and sporulation (lindane – *Glomus* sp., Sainz et al., 2006; aldrin – *Rhizoglomus fasciculatum*, Twanabasu et al., 2013). Dissipation of organic pollutants presents the sum of all processes involved in the disappearance of the organic pollutants (adsorption onto roots, uptake by plants, degradation), whereas degradation implies only a break-down of organic pollutants into smaller compounds (Lenoir et al., 2016). Dissipation of organic pollutants can vary with AMF species, plant species, type and concentrations of pollutants, distance of pollutants from root and presence and number of other microorganisms with capability to degrade (Wang et al., 2011). Better dissipation of PCBs was shown in soils with mycorrhizal plant species than with non-mycorrhizal plants (74%, 58%, respectively, Lu et al., 2014). AMF can improve plant biomass in the presence of chlorinated compounds, such as pesticides (*Funneliformes mosseae – Oryza sativa; Rhizoglomus fasciculatum – Triticum aestivum*) (Zhang et al., 2006; Chhabra and Jalali, 2013), and PCB (*Funneliformes mosseae – Cucurbita pepo; Funneliformes caledonium – Lolium multiflorum; Funneliformes caledonium – Medicago sativa*) (Teng et al. 2008; Qin et al., 2014; Lu et al., 2014).

12.5.3.1 Plant/bacterium/fungi system in remediation of chlorinated solvents

Trichlorethylene (TCE) is a volatile organic compound (VOC) that has been widely used as an anesthetic, dry cleaning solvent and degreaser for electronics, textiles, and military weapons (Doherty, 2000). TCE is water-soluble with log K_{ow} of 2.42 and presents one of the most prevalent organic compounds in groundwater (McCarty, 2010). Human exposure to TCE causes headache, dizziness, confusion, unconsciousness, toxic effects in the liver and kidney, leads to neurological disorder amyotrophic lateral sclerosis and syndrome that resembles Parkinson's disease, and has negative effects on the central nervous system (Gash et al., 2008; USEPA, 2009).

TCE degradation pathway includes TCE transformation via oxidative metabolism under the control of plant oxidative enzymes that catalyze dechlorination of TCE (Newman et al., 1997; Shang et al., 2001). TCE is degraded by monooxygenase (TC-oxygen-P450/TCE epoxide formation) and the first major metabolite is chloral hydrate (CH) which can be further transformed to the trichloroacetic acid (TCAA) and trichloroethanol (TCEOH) (Lash et al., 2000). TCAA can be excreted or it is further decomposed to dichloroacetic acid (DCAA) with release of chloride and oxalic acid which are then mineralized to CO_2 (Shang et al., 2001). Plant glucosyltransferase/glutathione *S*-transferase - mediated conjugations of TCE have been incorporated into cell wall and large molecules (Newman et al., 1997).

TCE removal by plant/bacteria/fungi system from contaminated sites is given in Table 12.6. According to Godsy et al. (2003) rhizomicrobes of the mature tree of *Populus deltoides* (22 years old) were capable of degradation of TCE to vinyl chloride. Reductive dechlorination of TCE under anaerobic conditions is capable of using bacteria (*Clostridium* sp. DC-1, KYT-1, *Dehalobacter, Dehalococcoides, Desulforomonas, Desulfobacterium, Propionibacterium*, sp. HK-1, *Sulfutospirillum*) and fungi (*Trametes versicolor, Phanerochaete chrysosporium* ME-446) (Kang and Doty, 2014). Reductive dehalogenase PceA isolated from *Sulfurospirillum multivorans* catalyses the reductive dechlorination of TCE (John et al., 2009) by dehalorespiration releasing the halogen atoms and using chloroethenes as electron acceptors (Bommer et al., 2014).

Engineering endophyte/rhizobia/ plantsto improve phytoremediation of TCE. Methane monooxygenase OB3b from *Methylosinius trichosporium*, toluene dioxygenases and toluene monooxygenases from different bacteria have been reported to possess activity for TCE degradation (Tsien et al., 1989; Kang and Doty, 2014). Engineered transfer of *Pseudomonas putida* F1 genes encoding toluene dioxygenases (*todC1C2BA*) into *Escherichia coli* for TCE degradation have been reported (Kang and Doty, 2014). Toluene–oxidizing/TCE–degrading aerobic bacteria *Burkholderia cepacia* G4 requires substrates, such as toluene, phenol and benzene to induce degrading TCE enzymes (Snyder et al., 2000; Kim et al., 2010). Plants synthetize natural phenolic compounds that are capable of inducing catabolic genes in microorganisms and can be useful as inducers for TCE co-metabolic degradation (Suttinun et al., 2009). Leaf homogenates of *Populus* sp. induce expression of the gene encoding the enzyme toluene-*ortho*-monooxygenase (TOM) in *Burkholderia cepacia* G4 that degrade TCE to CO_2 and chloride ion (Kang and Doty, 2014). TOM of *B. cepacia* G4 consists of six genes (*tomA012345*) and three components: (1) 211-kDa hydroxylase with a catalytic oxygen-bridged binuclear Fe center, (2) 40-kDa reductase, and (3) 10.4-kDa protein involved in electron transfer between hydroxylase and reductase (Newman and Wackett, 1995). Inoculation of *B. cepacia* VM1468, which possesses the pTOM-Bu61 plasmid coding for TCE degradation to plant *Lupinus luteus* show reduced TCE toxicity (Weyens et al., 2010). Furthermore, rhizomediation of TCE by recombinant, root-colonizing *Pseudomonas fluorescens* 2-79 strain which expresses toluene *ortho*-monooxygenase (*tomA*$^+$) genes from *B. cepacia* PRI_{23} (TOM_{23C}) by using *Triticum* sp. showed decreased initial TCE concentration for 63% in 4 days (Yee et al., 1998). Shim et al. (2000) showed that recombinant bacteria for TCE degradation that actively expressed TOM from *B. cepacia* G4, such as wild type bacteria (PM2, PM4, Pb1, Pb2, Pb3, and Pb5) obtained from *Populus*, *Pseudomonas fluorescens* 2-79TOM obtained from *Triticum*, *Rhizobium* ATCC 1032D obtained from *Robinia pseudoacacia* and *Rhizobium* ATCC35645 obtained

Table 12.6 Trichlorethylene (TCE) removal by plant/bacteria/fungi system from contaminated sites.

Plant/bacteria/fungi	Contaminants	References
PLANTS		
Populus deltoids	TCE	Godsy et al. (2003)
BACTERIA		
Clostridium sp. DC-1, KYT-1, *Dehalobacter, Dehalococcoides, Desulforomonas, Desulfobacterium, Propionibacterium*, sp. HK-1, *Sulfurospirillum*	TCE	Kang and Doty (2014) John et al. (2009)
FUNGI		
Trametes versicolor, Phanerochaete chrysosporium ME-446	TCE	Kang and Doty (2014)
ENGINEERING BACTERIA/PLANTS		
Burkholderia cepacia G4 - *Populus* sp.	TCE	Kang and Doty (2014)
Burkholderia cepacia VM1468 – *Lupinus luteus*	TCE	Weyens et al. (2010)
Pseudomonas fluorescens 2-79 (express [a]TOM_{23C} from *B. cepacia* PRI_{23}) -*Triticum* sp.	TCE	Yee et al. (1998)
Recombinant bacteria expressed [a]TOM from *Burkholderia cepacia* G4 in the rhizosphere of *Populus canadensis* var. *eugeni* 'Imperial Carolina': -*Pseudomonas fluorescens* 2-79 TOM obtained from *Triticum* -*Rhizobium* ATCC 1032D -obtained from *Robinia pseudoacacia* -*Rhizobium* ATCC 35645 obtained from *Leucaena leucocephala*	TCE	Shim et al. (2000)

[a]*enzyme toluene-ortho-monooxygenase (TOM).*

from *Leucaena leucocephala* possess the ability to compete in the rhizosphere of *Populus canadensis* var. *eugeni* 'Imperial Carolina'; poplar recombinants from Pb3-1 (79%) and Pb5-1 (39%) from *Pseudomonas fluorescens* 2-79TOM were the most competitive and retained the ability to express TOM for 29 days.

12.5.3.2 Plant/bacterium/fungi system in remediation of organochlorine pesticides

Organochlorine pesticides have been widely used to control pests for crop protection and for use in protection against insect - diseases (malaria and typhus) (Arslan et al., 2017; Eevers et al., 2017; Tarla et al., 2020). They are fat - soluble, and hydrophobic with a carbon-chlorine (C-Cl) bond that resists hydrolysis (Arslan et al., 2017). Their persistence increases with the number of chlorine atoms, they have long half - life and high log K_{ow}: DDT (2-15 years and 6,2), aldrin (5 years, 5.52), dieldrin (5 years, 5.48), chlordane (1-3 years, 5.44-5.66), endrin (12-15 years, 4.71), heptachlor (2 years, 6.10), Mirex (10 years, 6.89), toxaphene (1-12 years, 4.77-6.64), chlordecone (50 years, 4.5), endosulfan (50 years, 3.5) (Arslan et al., 2017). They can be released in the atmosphere, soil, sediment, groundwater, and surface water, easily enter the food chain having toxic effect to plants, animals, and humans (Tarla et al., 2020). The soil and water contamination with organochlorine pesticides happens often due to spills in storage sheds and leaking containers. They can be transported to long distances, such as Arctic (Vorkamp and Riget, 2014). Human exposure to organochlorine pesticides causes cancer, cardiovascular disease, endocrine disruption and reproductive damages, they can be found in placenta, and mother's milk (Singh and Singh, 2017). In order to protect environment and human health, these chemicals were banned (obsolete chemicals), they are no longer useful or legal to use/approved in the world, except of DDT which can be used because of malaria in Africa and India (UNEP, 2005; UNEP, 2013).

Plants possess a high organochlorine pesticide-degrading potential. Chemicals can enter the plant roots through diffusion and be translocated via xylem to shoots where they can be accumulated and transformed (Eevers et al., 2017; Singh and Singh, 2017). According to Agbeve et al. (2013) all organochlorine pesticides have been found in anti-malarial plant *Cryptolepis sanguinolenta.* Organochlorine pesticide removal by plant/bacteria/fungi system from contaminated sites is given in Table 12.7. Plant species with high phytoremediation potential for organochlorine pesticides are as follows: DDT, DDD, DDE – *Achilea millefolim, Brassica juncea, Brassica napus, Cajanus cajan, Cucumis sativus, Cucurbita pepo, Daucus carota, Erigeron canadensis, Festuca arundacea, Glycine max, Helianthus annus, Hordeum vulgare, Juncus maritimus, Lolium, multiflorum, Lupinus albus, Medicago sativa, Myriophyllum, Oenothera biennis, Phaseolus vulgaris, Plantago lagopus, Raphanus sativus, Ricinus communis, Solanum lycopersicum, Sorghum bicolor, Taraxacum officinale, Trifolium incarnatum, Triticum vulgare, Vicia vilosa, Zea mays* (Gao et al., 2000; Lunney et al., 2004; White et al., 2005; Mikes et al., 2009; Moklyachuk et al., 2010; Bogdevich and Cadocinicov, 2010; Carvalho et al., 2011; Moklyachuk et al., 2012; Mitton et al., 2014; Rissato et al., 2015); Chlordane – *Cirsium arvense, Cucumis sativus, Cucurbita pepo, Lactuca sativa, Lupinus albus, Lycopersicum esculentum, Ricinus communis, Spinacia oleracea* (Mattina et al., 2003; Rissato et al., 2015); Aldrin, diedrin, endrin – *Cucurbita pepo, Eichornia crassipes, Ricinus comminis* (Matsumoto et al., 2009; Mercado-Borrayo et al., 2015; Rissato et al., 2015); HCH – *Chenopodium sp. Cucurbita pepo, Cytisus striatus, Daucus carota, Glycine max, Hordeum vulgare, Phaseoulus vulgaris, Raphanus sativus, Ricinus communis, Solanum nigrum, Sorghum bicolor, Triticum vulgare, Zea mays* (Calvelo Pereira et al., 2006;Mikes et al., 2009; Bogdevich and Cadocinicov, 2010; Moklyachuk et al., 2010; Rissato et al., 2015).

Biodegradation of DDT, its residues, aldrin, dieldrin, endosulfan, chlordane, HCH by bacteria has been studied: *Achromobacter, Agrobacterium, Alcaligenes, Bacillus, Cladosporium, Clostridium, Erwinia, Escherichia coli, Enterobacter, Pseudomonas, Stenotrophomonas, Streptomyces, Sphingobacterium, Stenotrophomonas, Xantomonas* (Aislabie et al., 1997; Xu et al., 2008; Fuentes et al., 2010; Narkhede et al., 2015; Pan et al., 2016). However, combined plant – bacterium system can play a significant role in plant-promoting capacities and pesticide-degrading processes: *Nicotiana tabacum*

Table 12.7 Organochlorine pesticides removal by plant/bacteria/fungi system from contaminated sites.

Plant/bacteria/fungi	Contaminants	References
PLANTS		
Achilea millefolim, Brassica juncea, Brassica napus, Cajanus cajan, Cucumis sativus, Cucurbita pepo, Daucus carota, Erigeron canadensis, Festuca arundacea, Glycine max, Helianthus annus, Hordeum vulgare, Juncus maritimus, Lolium, multiflorum, Lupinus albus, Medicago sativa, Myriophyllum, Oenothera biennis, Phaseolus vulgaris, Plantago lagopus, Raphanus sativus, Ricinus communis, Solanum lycopersicum, Sorghum bicolor, Taraxacum officinale, Trifolium incarnatum, Triticum vulgare, Vicia vilosa, Zea mays	DDT, DDD, DDE	Gao et al. (2000); Lunney et al. (2004); White et al. (2005); Mikes et al. (2009); Moklyachuk et al. (2010); Bogdevich and Cadocinicov (2010); Carvalho et al. (2011); Moklyachuk et al. (2012); Mitton et al. (2014); Rissato et al. (2015)
Cirsium arvense, Cucumis sativus, Cucurbita pepo, Lactuca sativa, Lupinus albus, Lycopersicum esculentum, Ricinus communis, Spinacia oleracea	Chlordane	Mattina et al. (2003); Rissato et al. (2015)
Cucurbita pepo, Eichornia crassipes, Ricinus comminis	Aldrin, Diedrin, Endrin	Matsumoto et al. (2009); Mercado-Borrayo et al. (2015); Rissato et al. (2015)
Chenopodium sp. Cucurbita pepo, Cytisus striatus, Daucus carota, Glycine max, Hordeum vulgare, Phaseoulus vulgaris, Raphanus sativus, Ricinus communis, Solanum nigrum, Sorghum bicolor, Triticum vulgare, Zea mays	HCH	Calvelo Pereira et al. (2006); Mikes et al. (2009); Bogdevich and Cadocinicov (2010); Moklyachuk et al. (2010); Rissato et al. (2015)
BACTERIA		
Achromobacter, Agrobacterium, Alcaligenes, Bacillus, Cladosporium, Clostridium, Erwinia, Escherichia coli, Enterobacter, Pseudomonas, Stenotrophomonas, Streptomyces, Sphingobacterium, Stenotrophomonas, Xantomonas	DDT,Aaldrin, Diedrin, Endosulfan, Chlordane, HCH	Aislabie et al. (1997); Xu et al. (2008); Fuentes et al. (2010); Narkhede et al. (2015); Pan et al. (2016)
FUNGI		
Phanerochaete chrysosporium, Saccharomyces cervisae, Trichoderma viridae	DDT, Aldrin, HCH	Bumpus et al. (1985); Subba-Rao and Alexander (1985); Arslan et al. (2017)
PLANT – BACTERIA		
Nicotiana tabacum – Arthrobacter	DDT, HCH	Wang et al. (2015)

Table 12.7 Organochlorine pesticides removal by plant/bacteria/fungi system from contaminated sites. *Continued*

Plant/bacteria/fungi	Contaminants	References
Cucurbita pepo – Enterobacter, Methylobacterium, Sphigomonas	DDE	Eevers et al. (2017)
Zea mays – Sphingomonas sp.	HCH	Abhilash et al. (2013)
Withania somnifera – Staphylococcus	HCH	Abhilash et al. (2011)
Cytisus striatus – Rhodococcus, Sphingomonas	HCH	Becerra-Castro et al. (2013)
PLANT - FUNGI		
Medicago sativa - Glomus etunicatum	DDT	Wu et al. (2008)
Salix alaxensis, Populus balsamifera – MF	Aldrin, Diedrin	Schnabel and White (2001)

DDD, 1,1-dichloro-2,2-bis(chlorophenyl)ethane; DDE, 1,1-dichloro-2,2-bis(chlorophenyl)ethylene; DDT, 2,2-bis(chlorophenyl)-1,1,1-trichloroethane; HCH, α,β-hexachlorocyclohexane.

– *Arthrobacter* (DDT, HCH) (Wang et al., 2015); *Cucurbita pepo – Enterobacter, Methylobacterium, Sphigomonas* (DDE) (Eevers et al., 2017); *Zea mays – Sphingomonas* sp. (HCH) (Abhilash et al., 2013); *Withania somnifera – Staphylococcus* (HCH) (Abhilash et al., 2011); *Cytisus striatus – Rhodococcus, Sphingomonas* (HCH) (Becerra-Castro et al., 2013).

Biodegradation of DDT, aldrin, and HCH occurs by ligninolytic fungi (*Phanerochaete chrysosporium, Saccharomyces cervisae, Trichoderma virid*ae) which can oxidize these pollutants to CO_2 by lignin-degrading enzyme systems (LiP, MnP, laccases) (Bumpus et al., 1985; Subba-Rao and Alexander, 1985; Arslan et al., 2017). According to Walter (1992) *Phanerochaete chrysosporium* was capable of DDT degrading for 80-90% during 35 - day incubation. Furthermore, it was found that combined AMF inoculation of plants and soil application of surfactant have high potential for decontamination of DDT (Wu et al., 2008). Therefore, inoculation of *Glomus etunicatum* to *Medicago sativa* with addition of surfactant Triton X-100, showed that accumulation of DDT increased in roots, but decreased in shoots (Wu et al., 2008). According to Schnabel and White (2001) mycorrhizal fungi were related with the uptake of aldrin and dieldrin in *Salix alaxensis*, but slightly associated with their uptake in *Populus balsamifera*.

12.5.3.3 Plant/bacterium/fungi system in remediation of polychlorinated biphenyls

PCB are chlorinated aromatic compounds with biphenyl molecule and chlorine atoms (1-10) that can attach *meta, ortho,* and *para* position allowing 209 different congeners (Vergani et al., 2017). They are low soluble in water, have long half–life (3 to 37 years) and high log K_{ow} (4.7–6.8) (Passatore et al., 2014). The more chlorinated PCB with log $K_{ow} > 6$, the higher the tendency it has to be associated with organic matter in the soils whereas in the atmosphere lower chlorinated PCB congeners dominate (Passatore et al., 2014). Mixture of PCB congeners had been used from 1930 to 1970 as lubricants, coolants in transformer oil, and plasticizers with different commercial names (Aroclor, Fenclor, Delor, Askarel, Santotherm, Clophen, Phenoclor) (Passatore et al., 2014; Sharma et al., 2018). PCBs contaminate environment due to their production, usage, improper disposal, transportation, and accidental spills and leaks and they can be found in Arctic and Antarctic (Van Aken et al., 2010; Passatore et al., 2014; Sharma et al., 2018). They accumulate in lipids in plants, animals and humans (Vergani et al., 2017). They have toxic effects on human health causing endocrine disruption, cancer, reproductive diseases, neurological and dermal abnormalities (ASTDR, 2000; Lauby-Secretan et al., 2013).

Plants are capable to uptake PCBs with log K_{ow} 4.5–8.2 (2-monochlorbiphenyl – decachlorbiphenyl, respectively) (Van Aken et al., 2010). Uptake of PCBs by plants could be mediated by the root exudates and they are further translocated in shoots through xylem sap (Greenwood et al., 2011). In hydroponic experiments with hybrid *Populus sp.* Liu and Schnoor (2008) it was shown that roots had higher chlorinated PCB congeners (mono to tetrachlorinated PCBs) than shoots (mono-, di-, and trichlorinated PCBs) (Liu and Schnoor, 2008). However, in the field experiment with Aroclor 1260, Zeeb et al. (2006) higher chlorinated PCBs congeners were found in shoots, such as tetrachloro – to hexaclorobiphenyls, even heptachloro – and nonachlorobiphenyls. Generally, metabolisms of PCBs in plants depend on plant species, degree of chlorination (mono- to tetrachlorinated PCBs), congener type and include oxidation of biphenyl core (Van Aken et al., 2010; Passatore et al., 2014; Vergani et al., 2017). In roots of *Solanum nigrum* hydroxylation activity produces biphenylol metabolites without changing the degree of chlorinated molecules (Rezek et al., 2007). Mono-chlorophenyls (-CB) (PCB2, PCB3, PCB4) were hydroxylated and transformed in monohydroxy-CB and dihydroxy-CB whereas PCB7, PCB8, PCB10, PCB11, PCB13 congeners were transformed in different monohydroxy-CB (Kucerova et al., 2000). In *Nicotiana tabacum* WSC-38, 6 congeners of di-CB were transformed in three types of metabolites, such as hydroxyl-CB, methoxy-CB and hydroxyl-methoxy-CB where *O*-methyltransferases were responsible for production of methylated compounds (Rezek et al., 2008). PCBs removal by plant/bacteria/fungi system from contaminated sites is given in Table 12.8. Plant species that can successfully uptake and metabolize PCBs are as follows:

- Field – *Pinus nigra, Fraxinus excelsior, Betula pendula, Robinia pseudoacacia* (Leight et al., 2006); *Medicago sativa* (Tu et al. 2011); *Cucurbita pepo, Festuca arundacea, Carex normalis* (Whitfield Aslund et al., 2007); *Chrysanthemum leucanthemum, Rumex crispa, Solidago canadensis* (Ficko et al., 2011a, Ficko et al., 2011b); *Zea mays. Helianthus annuus, Ppulus nigra x P. maximowiczii, Salix x smithiana* (Kacalkova and Tlustoš, 2011); *Ambrosia artemisifolia, Daucus carota, Polygonum persicaria, Setaria pumila, Somchus asper, Vicia cracca, Amaranthus retroflexus, Brassica nigra, Cirsium arvense, Cirsium vulgare, Echinichloa crusgalli, Lythrum salicaria, Solidago Canadensis, Symphyotrichium ericoides, S. novae-anlie, Barbarea vulgaris, Capsella bursa-pastoris, Chenopodium album, Chrrysanthemum leucanthemum, Echium vulgare, Medicago lupulina, Polygonum convolvulus, Sisymbrium officinale, Solanum nigrum, Trifolium pretense and Verbascum thapsus* (Ficko et al., 2010).
  - Greenhouse – *Oryza sativa* (Chen et al., 2014b); *Cucurbita pepo, Cucurbita maxima, Cucumis sativus, Cucumis melo, Daucus carota, Lycopersicum esculentum* (Campanela and Paul, 2000); *Festuca arundacea, Phalaris arundacea, Panicum variegatum, Panicum clandestinum, Medicago sativa, Coronilla varia, Lespedeza cuneata, Lathyrus sylvestris* (Dzantor et al., 2000); *Cucurbita pepo, Bidens cernua, Chenopodium album, Daucus carota, Plantago major, Rumex crispus* (Ficko et al., 2011b); *Brassica napus* (Javorska et al., 2009).

Table 12.8 Polychlorinated biphenyls (PCB)s removal by plant/bacteria/fungi system from contaminated sites.

Plant/bacteria/fungi	Contaminants	References
PLANTS		
FIELD		
Pinus nigra, Fraxinus excelsior, Betula pendula, Robinia pseudoacacia	PCBs	Leight et al. (2006)
Medicago sativa	PCBs	Tu et al. (2011)
Cucurbita pepo, Festuca arundacea, Carex normalis	PCBs	Whitfield Aslund et al. (2007)
Chrysanthemum leucanthemum, Rumex crispa, Solidago Canadensis	PCBs	Ficko et al. (2011a); Ficko et al. (2011b)
Zea mays. Helianthus annuus, Ppulus nigra x P. maximowiczii, Salix x smithiana	PCBs	Kacalkova and Tlustoš (2011)
Ambrosia artemisifolia, Daucus carota, Polygonum persicaria, Setaria pumila, Somchus asper, Vicia cracca, Amaranthus retroflexus, Brassica nigra, Cirsium arvense, Cirsium vulgare, Echinichloa crusgalli, Lythrum salicaria, Solidago Canadensis, Symphyotrichium ericoides, S. novae-anlie, Barbarea vulgaris, Capsella bursa-pastoris, Chenopodium album, Chrrysanthemum leucanthemum, Echium vulgare, Medicago lupulina, Polygonum convolvulus, Sisymbrium officinale, Solanum nigrum, Trifolium pretense, Verbascum Thapsus	PCBs	Ficko et al. (2010)
GREENHOUSE		
Oryza sativa	PCBs	Chen et al. (2014b)
Cucurbita pepo, Cucurbita maxima, Cucumis sativus, Cucumis melo, Daucus carota, Lycopersicum esculentum	PCBs	Campanela and Paul (2000)
Festuca arundacea, Phalaris arundacea, Panicum variegatum, Panicum clandestinum, Medicago sativa, Coronilla varia, Lespedeza cuneata, Lathyrus sylvestris	PCBs	Dzantor et al. (2000)
Cucurbita pepo, Bidens cernua, Chenopodium album, Daucus carota, Plantago major, Rumex crispus	PCBs	Ficko et al. (2011b)
Brassica napus	PCBs	Javorska et al. (2009)

(*continued*)

Table 12.8 Polychlorinated biphenyls (PCB)s removal by plant/bacteria/fungi system from contaminated sites. ***Continued***

Plant/bacteria/fungi	Contaminants	References
BACTERIA		
Dehalococcoides	PCBs	Furukawa and Fujihara (2008) Smidt et al. (2000)
Pseudomonas, Burkholderia, Sphigobium, Alcaligenes, Arthobacter, Achromobacter, Bacillus, Rodococcus	PCBs	Furukawa and Fujihara (2008) Vergani et al. (2017)
FUNGI		
Irpex lacteus BAFC 1168D, *Irpex lacteus* BAFC 1171, *Lenzites elegans* BAFC 2127, and *Pleurotus sajor-caju* LBM 105	PCBs	Sadanoski et al. (2018),
Phanerochaete chrysosporium, Trametes versicolor, Lentinus edodes, Phlebia brevispora, Irplex lacteus, Pycnoporus cinnabarinus, Phanerochaete magnolia, Pleurotus ostreatus	PCBs	Thomas et al. (1992) Chun et al. (2019)
PLANT – BACTERIA		
Medicago sativa – Rhizobium sp., *Rhodococcus*	PCBs	Teng et al. (2010)
Panicum virgatum – Burkholderia xenovorans LB400	PCBs	Liang et al., 2014
Nicotiana tabacum, Solanum nigrum – Pseudomomas sp. L15, *Pseudomonas* sp., JAB1, *Ochrobactrum* sp. KH6	PCBs	Kurzawova et al. (2012)
Brassica nigra - Ralstonia eutropha H850, *Rhodococcus* sp. ACS	PCBs	Singer et al. (2003a)
Festuca arundacea – Burkholderia xenovorans LB400	PCBs	Secher et al. (2013)
Citrus sinensis, Citrus reticulata, Pinus nigra, Hedera helix – Pseudomonas stutzeri	PCBs	Dudašova et al. (2012)
Mentha spicata – Arthobacter sp. B18	PCBs	Gilbert and Crowley (1997)
Salix caprea – Sphigobacterium mizutae, Burkholderia cepacia	PCBs	Ionescu et al. (2009)
Populus - Burkholderia cepacia	PCBs	Taghavi et al. (2005)
Pinus nigra, Salix caprea, Fraxinus excelsior, Robinia pseudoacacia – Rhodococcus	PCBs	Leight et al. (2006)

Table 12.8 Polychlorinated biphenyls (PCB)s removal by plant/bacteria/fungi system from contaminated sites. ***Continued***

Plant/bacteria/fungi	Contaminants	References
PLANT – FUNGI		
Lolium perenne – Glomus caledonium	PCBs	Lu et al. (2014)
Cucurbita pepo, Cucurbita moschata, Cucurbita maxima, Cucumis sativus – AMF	PCBs	Qin et al. (2014).
PLANT – BACTERIA – FUNGI		
Medicago sativa – Rhizobium sp. – *Glomus caledonium*	PCBs	Teng et al. (2010)
ENGINEERING BACTERIA/FUNGI/PLANTS		
Burkholderia xenovorans LB400 – Transgenic *Nicotiana tabacum*	PCB degradation	Mohammadi et al. (2007)
Commamonas testeteroni, Pseudomonas testeteroni - Transgenic *Nicotiana tabacum*	PCB degradation	Francova et al. (2003) Novakova et al. (2009)
B. xenovorans LB400 → *Pseudomonas fluorescens* F113 → F113pcb → plants	PCB degradation	Brazil et al. (1995)
B. xenovorans LB400 → *Pseudomonas fluorescens* F113 → *Sinorhizobium meliloti* (promoter nodbox4) → recombinant F113: 1180 → Salix	PCB degradation	Villacieros et al. (2005)
Sinorhizobium meliloti → plasmid (oxygenolitic *ortho*-dechlorinate gene) → *Medicago sativa*	PCB degradation	Toure et al. (2003)
Pseudomnas fluorescens→1GM strain with insertation of *bph* operon and *Pseudomnas fluorescens*→2GM strain with insertation of *bph* operon from *S. meliloti*→ *Salix viminalis x schwerini*	PCB degradation	de Carcer et al. (2007)
Phaenerochete chrysosporium →DNA of*Arabidopsis thaliana*	PCB degradation	Sonoki et al. (2007)

PCB degradation by bacteria occurs through anaerobic (reductive dechlorination of the biphenyl ring) and aerobic pathways (oxidative degradation of the biphenyl ring) that depends on the degree of chlorination of PCB congeners, the redox condition and bacterium species (Furukawa et al., 2010).

- Anaerobic pathway – reductive dehalogenation of PCBs congeners with four and more chlorine atoms is achieved by halorespiring bacteria, such as *Dehalococcoides,* which remove chlorines in *par-* and *meta* rather than in *ortho* position on the biphenyl structure (Furukawa and Fujihara, 2008). These bacteria use PCBs as electron acceptors for their metabolisms replacing chlorines with hydrogen atoms (Smidt et al., 2000). Anaerobic pathway of PCB degradation has been noted in flooded paddy fields and river sediments (Chen et al., 2014b).

- Aerobic pathway – oxidative degradation of low-chlorinated PCB congeners mediated by dioxygenases opens biphenyl ring and undergoes complete mineralization of the molecule (Furukawa and Fujihara, 2008). Different oxygenases codified catabolic enzymes that are expressed the *bph* gene cluster in *Pseudomonas, Burkholderia, Sphigobium, Alcaligenes, Arthobacter, Achromobacter, Bacillus, Rodococcus*, etc. (Furukawa and Fujihara, 2008; Vergani et al., 2017). Two clusters of genes are responsible for (1) the transformation of PCBs in chlorobenzoates and chlorinated aliphatic acids (biphenyl upper pathway), and (2) for further mineralization (biphenyl lower pathway) (Pieper and Seeger, 2008; Leewis et al., 2016). The biphenyl upper pathway involves seven genes that are grouped into one operon (*bph A1-A4, bph B, bph C, bph D*) (Furukava, 1994). A multicomponent dioxygenase *(bph A, bph E, bph F, bph G*), dehydrogenase (*bph B*), dioxygenase (*bph C*), and hydrolase (*bph D*) induce the insertion of two oxygen atoms into the aromatic ring, dehydrogenate the compound and cleave the dehydroxylated ring in *meta/ortho* position (Furukava, 1994). In the lower pathway, enzymes hydratase, dehydrogenase and aldolase encoded in one cluster (*bph H, bph I* and *bph J*, respectively) cleave the molecule into chlorobenzoate and 2-hydroxypenta-2,4-dienoate (Borja et al., 2005; Pieper and Seeger, 2008).

Plants are capable to release by roots some organic compounds that can act as co-metabolite/inducers used by microorganisms in PCB degradation that is, these compounds may induce PCB co-metabolism, increase bioavailability, uptake, transport, and enhance PCBs degradation: lignin - *Rhodocucoccus erythroplis* TA421; naringin - *Ralstonia eutrophus* H850; myricetin - *Burkholderia cepacia* LB400; coumarin – *Coynebacterium* sp. MB1; biphenyl – *Alcaligenes eutrophus* H850, *Pseudomonas putida* – LB400, *Rodococcus globenrlus* P6; terpenoides – *Arthrobacter* sp.B1B; carvone and limonene – *Pseudomonas stutzeri, Ralstonia eutrophus* H850, *Rhodococcus* sp. T104 (Fava et al., 2003; Jha et al., 2015).

Decline of PCB content was found in vegetated soils where rhizobacteria and endophytic bacteria assisted in PCB phytoremediation (Jha, et al., 2015; Vergani et al., 2017; Sharma et al., 2018).

- *Plant* – bacterium combinations that are exploted in PCB degradation: *Medicago sativa – Rhizobium* sp., *Rhodococcus* (Teng et al., 2010); *Panicum virgatum – Burkholderia xenovorans* LB400 (Liang et al., 2014); *Nicotiana tabacum, Solanum nigrum – Pseudomomas* sp. L15, *Pseudomonas* sp., JAB1, *Ochrobactrum* sp. KH6 (Kurzawova et al., 2012); *Brassica nigra - Ralstonia eutropha* H850, *Rhodococcus* sp. ACS (Singer et al., 2003a); *Festuca arundacea – Burkholderia xenovorans* LB400 (Secher et al., 2013); *Citrus sinensis, Citrus reticulata, Pinus nigra, Hedera helix – Pseudomonas stutzeri* (Dudašova et al., 2012); *Mentha spicata– Arthobacter* sp. B18 (Gilbert and Crowley, 1997); *Salix caprea – Sphigobacterium mizutae, Burkholderia cepacia* (Ionescu et al., 2009); *Populus - Burkholderia cepacia* (Taghavi et al., 2005); *Pinus nigra, Salix caprea, Fraxinus excelsior, Robinia pseudoacacia – Rhodococcus* (Leight et al., 2006).

PCB degradation pathway by fungi has not yet been described (Passatore et al., 2014; Vergani et al., 2017). Intracellular fungal enzymes, such as cytochrome 450 monooxygenases, dehydrogenases seem to be involved (ČvanČarova et al., 2012). Extracellular enzymes involved in degradation of lignin, such as peroxidases, MiP, LiP, laccases could also have degradation PCB abilities as well as secondary plant metabolites (terpenoids, flavonoids) (Singer et al., 2003b). According to Sadanoski et al. (2018), *Irpex lacteus* BAFC 1168D, *Irpex lacteus* BAFC 1171, *Lenzites elegans* BAFC 2127, and *Pleurotus sajor-caju* LBM 105 can be promising PCB degraders with high laccase secretion. Lignolitic *Phanerochaete chrysosporium* can mineralize tetra – and hexachloro PCB congeners (Thomas et al., 1992). White – rot fungi that can be efficient PCB degraders are: *Phanerochaete chrysosporium, Trametes versicolor, Lentinus edodes, Phlebia brevispora, Irplex lacteus, Pycnoporus cinnabarinus, Phanerochaete magnolia, Pleurotus ostreatus* (Chun et al., 2019). PCB degradation was found in the presence of combination of plants, bacteria and fungi that efficiently assist in PCB phytoremediation:

- *Plant* – fungi combinations that are exploited in PCB degradation are: *Lolium perenne – Glomus caledonium* (Lu et al. 2014); *Cucurbita pepo, Cucurbita moschata, Cucurbita maxima, Cucumis sativus* – AMF (Qin et al., 2014).
- *Plant* – bacterium - fungi combination that is exploited in PCB degradation: *Medicago sativa – Rhizobium* sp. – *Glomus caledonium* (Teng et al., 2010).

Engineering endophyte/rhizobia/ plantsto improve phytoremediation of PCBs. Engineering plants for phytoremediaton of PCBs is challenging due to a number of PCB congeners and it needs all multiple genes introduced for complete catabolism (Sylvestre et al., 2009). Construction of transgenic plants expressing PCB-degrading enzymes from bacteria requires enzymes that are a key step in the biphenyl degradation: initial enzyme (biphenyl 2,3-dioxygenase, BPDO) and final enzyme which catalyzes ring cleavage (2,3-dihydroxybiphenyl-1,2-dioxygenase, 2,3-DHBP) (Sylvestre et al., 2009). BPDO consists of the oxygenase (*BphAE*), the ferredoxin (*BphF*), and the ferredoxin reductase (*BphG*). It was reported that *bph* genes from *Burkholderia xenovorans* LB400 were successfully inserted in *Nicotiana tabacum* and each of the components (*BphAE, BphF* and *BphG*) can be produced individually in transgenic plants (Mohammadi et al., 2007). *bphC* genes responsible for 2,3-dihydroxybiphenyl ring cleavage were inserted in *Nicotiana tabacum* from the *Commamonas testeteroni* (Francova et al., 2003) and *Pseudomonas testeteroni* (Novakova et al., 2009). In bacterium that colonized roots of a number of plants, *Pseudomonas fluorescens* F113 was inserted *bph* genes from *B. xenovorans* LB400 and this recombinant bacterium F113pcb expressed the ability for PCB degradation (Brazil et al., 1995). Furthermore, *bph* operon from *B. xenovorans* LB400 was inserted into strain 113 under control promoter (nodbox4) from *Sinorhizobium meliloti* and new recombinant F113: 1180 expressed high PCB metabolizing rates in *Salix* rhizosphere Villacieros et al., 2005). In addition, nitrogen-fixing bacterium *S. meliloti* was transformed by introduction of PCB-degrading plasmid (oxygenolitic *ortho*-dechlorinate gene, *ohb*) and when *Medicago sativa* was inoculated with these transformed bacteria, PCB degradation was much more pronounced (Toure et al., 2003). Also, genetically modified strains of *P. fluorescens* (1GM strain with insertion of *bph* operon, and 2GM strain with insertion of *bph* operon from *S. meliloti*) that inoculated the roots of *Salix viminalis x schwerini* after 6 months showed increased PCB degradation (de Carcer et al., 2007). In rhizoengineering, metabolic mutants of *Arabidopsis thaliana* capable of releasing secondary metabolites, such as flavonoids can be used to promote growth of bacterium *Pseudomonas putida* PML2 that can efficiently degrade and remove PCBs for 90% (2-monochloro- and 4-monochlorobiphenyls) (Narasimhan et al., 2003). Finally, Sonoki et al. (2007) inserted genes that were responsible for the production of LiP, MnP and Lac enzymes from *Phaenerochete chrysosporium* in DNA of *Arabidopsis thaliana* to make this transgene plant degrade PCBs.

12.6 Conclusion

Environmental pollution caused by metal(loid)s, radionuclides, chlorinated solvents, pesticides, PCBs, petroleum hydrocarbons, polycyclic aromatic hydrocarbons, explosives, etc. released from industrial processes, mining, fuel combustion, waste disposal, agricultural practice, and accidental spills pose a global threat for environment and human health and demand urgent attention. In addition, POPs are resistant to degradation and have a long-range dispersal potential. Human exposure to heavy metals, metaloids and organic contaminants leads to cancer, cardiovascular diseases, endocrinological, neurological and immunological problems, behavior disabilities, they affect pregnancy, etc. and they have been listed in EPA Priority Pollutants and Priority List of Hazardous Substances.

Phytoremediation is a proven technology and a sustainable solution for cleanup of contaminated sites. Phytobial remediation employs plants with bacteria and fungi and may increase phytoremediation potential and revitalization of polluted sites. Bacteria/fungi assisted phytoremediation offers a cost-effective and public acceptable application at commercial scale. PGPB and mycorrhizal fungi live in symbiosis with plant host. The plants create habitats for rhizospheric/endophytic bacteria and ectomycorrhizal/endomycorrhizal fungi providing $O_{2,}$ organic C, sugars, and organic acids that improve aerobic metabolisms utilizing them as a source of energy or use them as an electron donor for anaerobic metabolisms; plants release a number of phenolic compounds that are structural analogs for some persistent organic compounds inducing biosynthesis of enzymes for their co-catabolisms; plants may release in root exudates some compounds that enhance pollutant mobility and bioavailabilty. However, bacteria and fungi benefit the plant through fixation of nitrogen, solubilization of phosphorus, release of siderophore, biosynthesis of phytochormones, and production of ACC deaminase stimulating plant growth, reduce disease, decrease metal(loid) toxicity and degrade persistent organic compounds.

Further research is needed for optimization of plants and their associated bacteria and fungi for phytoremediation, i.e. it is necessary to research plant-microbial interactions with regard to beneficial microbiome in contaminated rhizosphere. Novel insight should be given to competition-driven model, selection of effective plant host and microorganisms strain or consortia that are tolerant to pollutants and environmental stressors. Bioengineering and developed transgenic plants, bacteria and fungi with improved phytoremediation capabilities present advance knowledge in order to maximize phytoremediation efficiency. All these involve molecular data, ecological models, 'omics' technologies, such as genomics, transcriptomics, proteomics, metabolomics that can use as 'biological designers' of traits and identification of the most effective genes for microbe assisted phytoremediation. Particular attention should be given to potential risk that underlies transgenic biotechnology. However, biomolecular engineering has been developed to engineer microorganisms and plants creating great potential for enhanced removal of contaminants. Overall, biomolecular research can open exciting new insights into environmental protection and could be used in sustainable management programs in the coming years on the globe.

Acknowledgments

This research was funded by Ministry of Education, Science and Technological Development of the Republic of Serbia (No. 451-03-9/2021-14/ 200007).

References

Abhilash, P.C., Singh, B., Srivastava, P., Schaeffer, A., Singh, N., 2013. Remediation of lindane by *Jatropha curcas* L.: utilization of multipurpose species for rhizomediation. Biomass Bioenerg 51, 189–193.

Abhilash, P.C., Srivastava, S., Srivastava, P., Singh, B., Jafri, A., Singh, N., 2011. Influence of rhizospheric microbial inoculation and tolerant plant species on the rhizomediation of lindane. Environ. Exp. Bot. 74, 127–130.

Abou-Shanab, R.A.I., Angle, J.S., Delorme, T.A., Chaney, R.L., Berkum, P., Moavad, H., Ghanem, K., Ghozlan, H.A., 2003. Rhizobacterial effects on nickel extraction from soil and uptake by *Alyssum murale*. New Phytol 158 (1), 219–224.

Admire, B., Lian, B., Yalkowsky, S.H., 2014. Estimating the physicochemical properties of polyhalogenated aromatic and aliphatic compounds using UPPER: part 2. Aqueous solubility, octanol solubility and octanol-water partition coefficient. Chemosphere 119, 1441–1446.

Adriano-Anaya, M., Salvador-Figueroa, M., Ocampo, J.A., Garcia-Romera, I., 2005. Plant cell-wall degrading hydrolytic enzymes of *Gluconoacetobacter diazotrophicus*. Symbiosis 40 (3), 151–156.

Agbeve, S.K., Carboo, D., Duker-Eshun, G., Afful, S., Ofosu, P., 2013. Burden of organochlorine pesticide residues in the root of *Cryptolepis sanguinolenta*, antimalarial plant used in traditional medicine in Ghana. Eur. Chem. Bull. 2 (11), 936–941.

Ahmed, H., Holmstrom, S.J.M., 2014. Siderophores in environmental research: roles and applications. Microbiol. Biotechnol. 7 (3), 196–208.

Ahsan, M.T., Najam-ul-had, M., Idrees, M., Ullah, I., Afzal, M., 2017. Bacterial endophytes enhance phytostabilization in soils contaminated with uranium and lead. Int. J. Phytoremediation. 19 (10), 937–946.

Aislabie, J.M., Richards, N.K., Boul, H.L., 1997. Microbial degradation of DDT and its residues—a review. New Zeal. J. Agr. Res. 40 (2), 269–282.

Ajo-Franklin, J.B., Geller, J.T., Harris, J.M., 2006. A survey of the geophysical properties of chlorinated DNAPLs. J. Appl. Geophys. 59 (3), 177–189.

Ali, S.S., Vidhale, N.N., 2013. Bacterial siderophore and their application: a review. Int. J. Curr. Microbiol. Appl. Sci. 2 (12), 303–312.

ANSTO (Australian Nuclear Science and Technology Organization), 2011. Management of Radioactive Waste in Australia. Australian Government Printing Office, Melbourne.

Aranda, E., Scervino, J.M., Godoy, P., Reina, R., Ocampo, J.A., Wittich, R.M., Garcia-Romera, I., 2013. Role of arbiscular mycorrhizal fungus *Rhizophagus custos* in the dissipation of PAHs under roo-organ culture condition. Environ. Pollut. 181, 182–189.

Arora, P.K., Srivastava, A., Singh, V.P., 2010. Application of Monooxygenases in dehalogenatio, desulphurization, denitrification and hydroxylation of aromatic compounds. J. Bioremediat. Biodegrad. 1 (3), 1–8.

Arslan, M., Imran, A., Khan, Q.M., Afzal, M., 2017. Plant-bacteria partnership for the remediation of persistent organic pollutants. Environ. Sci. Pollut. Res. 24 (5), 4322–4336.

Asad, S.A., Farooq, M., Afzal, A., West, H., 2019. Integrated phytobial heavy metal remediation strategies for a sustainable clean environment – A review. Chemosphere 217, 925–941.

ASTDR (Agency for toxic Substances and Disease Registry), 2000. Toxicological Profile For Polychlorinated Biphenyls (PCBs). US Department of Health and Human Services, Public Health Service, Atlanta, Georgia. https://www.atsdr.cdc.gov/toxprofiles/tp.asp?id=142&tid=26.

Babu, A.G., Kim, G.D., Oh, B.T., 2013. Enhancement of heavy metal phytoremediation by *Alnus firma* with endophytic *Bacillus thuringiensis* GDB-1. J. Hazard. Mater. 250-251, 477–483.

Babu, A.G., Shea, P.J., Sudhakar, D., Jung, I.B., Oh, B.T., 2015. Potential use of *Pseudomonas koreensis* mining site soil. J. Environ. Manage. 151, 160–166.

Banerjee, A., Sharmar, R., Banerjee, U.C., 2002. The nitrile-degrading enzymes: current status and future prospects. Appl. Microbiol. Biotechnol. 60 (1–2), 33–44.

Barclay, M., Hart, A., Knowles, C.J., Meenssen, J.C.L., Tett, V.A., 1998. Biodegradation of metal cyanides by mixed and pure cultures of fungi. Enzyme Microbiol. Technol. 22 (4), 223–231.

Barriault, D., Vedadi, M., Powlowski, J., Sylvestre, M., 1999. *cis*-2,3-dihydro-2,3-dihydroxybiphenyl dehydrogenase and *cis*-1,2-dihydro-1,2-dihydroxynaphathalene dehydrogenase catalyze dehydrogenation of the same range of substrates. Biochem. Biophys. Res. Commun. 260 (1), 181–187.

Barriuso, J., Solano, B.R., Lucas, J.A., Lobo, A.P., Garcia-Villanaco, A., Monero, F.J.G., 2008. Ecology, genetic diversity and screening strategies of plant growth promoting rhizobacteria. In: Ahmad, I., Pichel, J., Hayat, S. (Eds.), Plant-Bacteria Interactions. Wiley-VCH, Weinheim, pp. 1–13.

Basheer, S., Kut, O.M., Prenosil, J.E., Bourne, J.R., 1993. Development of an enzyme membrane reactor for treatment of cyanide-containing wastewaters from the food industry. Biotechnol. Bioeng. 41 (4), 465–473.

Basnakova, G., Stephens, E.R., Thaller, M.C., Rossolini, G.M., Macaskie, L.E., 1998. The use of *Escherichia coli* bearing a *phoN* gene for the removal of uranium and nickel from aqueous flows. Appl. Microbiol. Biotechnol. 50 (2), 266–272.

Becerra-Castro, C., Prieto-Fernandez, A., Kidd, P.S., Weyens, N., Rodriguez-Garrido, B., Touceda-Gonzales, M., Acea, M.J., Vangronsveld, J., 2013. Improving performance of *Cytisus striatus* on substrates contaminated with hexachlorocyclohexane (HCH) isomers using bacterial inoculants: developing a phytoremediation strategy. Plant Soil 362, 247–260.

Beena, A.K., Geevarghese, P.I., 2010. A solvent tolerant thermostable protease from a psychrotrophic isolate obtained from pasteurized milk. Develop. Microbiol. Mol. Biol. 1, 113–119.

Belimov, A.A., Hontzeasb, N., Safronovaa, V.I., Demchinskayaa, S.V., Piluzzac, G., Bullittac, S., Glick, B.R., 2005. Cadmium-tolerant plant growth promoting bacteria associated with the roots of Indian mustard (*Brassica juncea* L. Czern.). Soil Biol. Biochem. 37, 241–250.

Bellion, M., Courbot, M., Jacob, C., Blaudez, D., Chalot, M., 2006. Extracellular and cellular mechanisms sustaining metal tolerance in ectomycorrhizal fungi. FEMS Microbiol. Lett. 254 (2), 173–181.

Berendsen, R.L., Pieterse, C.M., Bakker, P.A., 2012. The rhizosphere, microbiome and plant health. Trends Plant Sci 17 (8), 478–486.

Berg, G., Grube, M., Schloter, M., Smalla, K., 2014. Unraveling the plant microbiome: looking back and future perspectives. Front. Microbiol. 5, 148.

Berreck, M., Haselwandter, K., 2001. Effect of the arbuscular mycorrhizal symbiosis upon uptake of cesium and other cations by plants. Mycorrhiza 10, 275–280.

Biró, B., Posta, K., Füzy, A., Kadar, I., Nemeth, T., 2005. Mycorrhizal functioning as part of the survival mechanisms of barley (*Hordeum vulgare* L.) at long-term heavy metal stress. Acta Biol. Szegedien. 49 (1–2), 65–67.

Bleise, A., Danesi, P.R., Burkart, W., 2003. Properties, use and health effects of depleted uranium (DU): a general overview. J. Environ. Radioact. 64 (2–3), 93–112.

Bogdevich, O., Cadocinicov, O., 2010. Elimination of acute risks from obsolete pesticides in Moldavia. Phytoremediation experiment at a former pesticide storehouse. In: Kulakov, P.A., Pidlisnyuk, V. (Eds.), Aplication of Phytotechnologies for Clean Up Industrial, Agricultural and Wastewater Contamination. Springer Science + Buiness Media, B.V, Dordrect.

Bommer, M., Kunze, C., Fesseler, J., Schubert, T., Diekert, G., Dobbek, H., 2014. Structural basis for organohalide respiration. Science 346 (6208), 455–458.

Borja, J., Taleon, D.M., Auresenia, J., Gallardo, S., 2005. Polychlorinated biphenyls and their biodegradation. Process Biochem 40 (6), 1999–2013.

Braud, A., Geoffroy, V., Hoegy, F., Misli, G.L.A., Schalk, I.J., 2010. Presence of the siderophores pyoverine and pychelin in the extracellular medium reduces toxic metal accumulation in *Pseudomonas aeruginosa* and increases bacterial metal tolerance. Environ. Microbiol. Rep. 2 (3), 419–425.

Braud, A., Jezequel, K., Bazot, S., Lebeau, T., 2009. Enhanced phytoextraction of an agricultural Cr, Hg, and Pb-contaminated soil by bioaugmentation with siderophores producing bacteria. Chemosphere 74 (2), 280–286.

Brazil, G.M., Kenefick, L., Callanan, M., Haro, A., Delorenzo, V., Dowling, D.N., Ogara, F., 1995. Construction of a rhizosphere *Pseudomonas* with potential to degrade polychlorinated-biphenyls and detection of *bph* gene –expression in the rhizosphere. Appl. Environ. Microb. 61 (5), 1946–1952.

Broadley, M.R., Willey, N.J., 1997. Difference in root uptake of radiocesium by 30 plant taxa. Environ. Pollut. 97 (1–2), 2–11.

Bumpus, J.A., Tien, M., Wright, D., Aust, S.D., 1985. Oxidation of persistent environmental pollutants by a white rot fungus. Science 288 (4706), 1434–1436.

Bunzl, K., Albersa, B.P., Shimmacka, W., Rissanenb, K., Suomelab, M., Puhakainenb, M., Raholab, T., Steinnsc, E., 1999. Soil to plant uptake of fallout 137Cs by plants from boreal areas polluted by industrial emissions from smelters. Sci. Total Environ. 234, 213–221.

Burd, G.I., Dixon, D.G., Glick, B.R., 2000. Plant growth-promoting bacteria that decrease heavy metal toxicity in plants. Can. J. Microbiol. 46 (3), 237–245.

Burd, G.I., Dixon, D.G., Glick, B.R., 1998. A plant growrh-promoting bacterium that decreases nickel toxicity in plant seedlings. Appl. Environ. Microbiol. 64 (10), 3663–3668.

Burken, J.G., Shnoor, J.L., 1998. Predictive relationship for uptake of organic contaminants by hybrid poplar trees. Environ. Sci. Technol. 32 (21), 3379–3385.

Byrne, A.R., 1988. Radioactivity in fungi in Slovenia, Yugoslavia, following the Chernobyl accident. J. Environ. Radioactivity 6 (2), 177–183.

Calvelo Pereira, R., Camps-Arbestain, M., Rodriguez-Garrido, B., Macias, F., Monterroso, C., 2006. Behaviour of alpha-, beta-, gamma-, and delta-hexachlorocyclohexane in the soil – plant system of contaminated site. Environ. Pollut. 144 (1), 210–217.

Cameron, M.D., Aust, S.D., 1999. Degradation of chemicals by reactive radicals produced by cellobiose dehydrogenase from *Phanerochaete chrysosporium*. Arch. Biochem. Biophys. 367 (1), 115–121.

Cameron, M.D., Timofeevski, S., Aust, S.D., 2000. Enzymology of *Phanerochaete chrysosporicum* with respect to the degradation of recalcitrant compounds and xenobiotics. Appl. Microbiol. Biotechnol. 54 (6), 751–758.

Campanella, B., Paul, R., 2000. Presence in the rhizosphere and leaf extracts of zucchini (*Cucurbita pepo* L.) and melon (*Cucumis melo* L.) of molecules capable of increasing the apparent aqueous solubility of hydrophobic pollutants. Int. J. Phytoremediat. 2 (2), 145–158.

Campos, V., Merino, I., Casado, R., Gomez, L., 2008. Phytoremediation of organic pollutants. Span. J. Agric. Res. 6 (Special issue), 38–47.

Carvalho, P.N., Rodrigues, P.N.R., Evangelista, R., Basto, M.C.P., Vasconcelos, M.T.S.D., 2011. Can salt marsh plants influence levels and distributionof DDTs in estuarine areas? Estuar. Coast. Shelf Sci. 93 (4), 415–419.

Čerñanský, S., KolenČik, M., Ševe, J., Urik, M., Hiller, E., 2009. Fungal volatilization of univalent and pentavalent arsenic under laboratory conditions. Bioresour. Technol. 100 (2), 1037–1040.

Chaudhry, Q., Schroder, P., Werck-Reichhart, D., Grajeck, W., Marecik, R., 2002. Prospects and limitation of phytoremediaton for the removal of persistent pesticides in the environment. Environ. Sci. Pollut. Res. 9 (1), 4–17.

Chen, J., He, F., Zhang, X., Sun, X., Zheng, J., Zheng, J., 2014a. Heavy metal pollution decreases microbial abundance, diversity, and activity within particle-size fractions of a paddy soil. FEMS Microbiol. Ecol. 87 (1), 164–181.

Chen, B.D., Roos, P., Zhu, Y.G., Jakobsen, I., 2008. Arbuscular mycorrhizas contribute to phytostabilization of uranium in uranium mining tailings. J. Environ. Radioact. 99, 801–810.

Chen, C., Yu, C.N., Shen, C.F., Tang, X.J., Qin, Z.H., Yang, K., Hashimi, M.Z., Huang, R.L., Shi, H.X., 2014b. Paddy field-a natural sequential anaerobic-aerobic bioreactor for polychlorinated biphenyls transformation. Environ. Pollut. 190, 43–50.

Chhabra, M.L., Jalali, B.L., 2013. Impact of pesticides-mycorrhiza development of wheat interaction on growth. J. Biopestic. 5 (6), 129–132.

Chicote, E., Moreno, D., Garcia, A., Sarro, I., Lorenzo, P., Montero, F., 2004. Biofouling on the walls of a spent nuclear fuel pool with radioactive ultrapure water. Biofuling 20 (1), 35–42.

Christov, L.P., Prior, B.A., 1996. Repeated treatments with *Aureobasidium pullans* hemicellulases and alkali enhance biobleaching of sulphite pulps. Enzyme Microb. Biotechnol. 18, 244–250.
Chun, S.C., Muthu, M., Hasan, N., Tasneem, S., Gopal, J., 2019. Mycoremediation of PCBs by *Pleurotus ostreatus*: possibilities and prospects. Appl. Sci-Basel. 9, 4185.
Clint, G.M., Dighton, J., Rees, S., 1991. Influx of ^{137}Cs into hyphae of basidiomycete fungi. Mycol. Res. 95 (9), 1047–1051.
Colpaert, J.V., 1998. Biological interaction: the significance of root-microbial symbiosis for phytorestoration of metal-contaminated soils. In: Vangronsveld, J., Cunningham, S.D. (Eds.), Metal-contaminated soils: in situ inactivation and phytorestoration. Springer-Verlag, Berlin, p. 75.
Copnet, A., Bliard, C., Conteniente, J.P., Couturier, Y., 2001. Enzymatic degradation and deacetylation of native and acetylated starch-based extruded blends. Polym. Degrad. Stabil. 71, 203–212.
Cornelis, P., 2010. Iron uptake and metabolism in pseudomonas. Appl. Microbiol. Biotechnol. 86 (6), 1637–1645.
Coughtrey, P.J., Kirton, J.A., Mitchell, N.G., Morris, C., 1989. Transfer of radioactive cesium from soil to vegetation and comparasion with potassium in upland grasslands. Environ. Pollut. 62, 281–283.
ČvarČarova, M., Kresinova, Z., Filipova, A., Covino, S., Cajthami, T., 2012. Biodegradation of PCBs by lignolotitic fungi and characterization of the degradation products. Chemosphere 88 (11), 1317–1323.
Daft, M.J., Nicolson, T.H., 1974. Arbiscular mycorrhizas in plants colonizing coal wastes in Scotland. New Phytol 73, 1129.
Daniels, R., Vanderleyden, J., 2004. Quorum sensing and swarming migration in bacteria. FEMS Microbiol 28 (3), 261–289.
Davies, H.S., Cox, F., Robinson, C.H., Pittman, J.K., 2015. Radioactivity and the environment: technical approaches to understand the role of arbuscular mycorrhizal plants in radionuclide bioaccumulation. Front. Plant Sci. 6, 580.
De Boulois, H.D., Joner, E.J., Leyval, C., Jakobsen, I., Chen, B.D., Roos, P., Thiry, Y., Rufyikiri, G., Deelvaux, B., Declerck, S., 2008. Impact of arbuscular mycorrhizal fungi on uranium accumulation by plants. J. Environ. Radioact. 99 (5), 775–784.
De Carcer, D.A., Martin, M., Mackova, M., Macek, M., Macek, T., Karlson, U., Rivilla, R., 2007. The introduction of genetically modified microorganisms designed for rhizomediation induces changes on native bacteria in the rhizosphere but not in the surrounding soil. ISME J 1, 215–223.
De Filippis, L.F., 2015. Role of phytoremediation in radioactive waste treatment. In: Hakeem, K.R., Sabir, M., Öztürk, M., Mermut, A.R. (Eds.), Soil Remediation and Plants. Academic Press, Amsterdam, pp. 207–254.
Deguchi, T., Kakezawa, M., Nishida, T., 1997. Nylon degradation by lignin-degrading fungi. Appl. Environ. Microbiol. 63 (1), 329–331.
Deguchi, T., Kitaoka, Y., Kakezawa, M., Nishida, T., 1998. Purification and characterization of a nylon-degrading enzyme. Environ. Microbiol. 64 (4), 1366–1371.
Del Val, C., Barea, J.M., Azcon-Aguilar, C., 1999. Diversity of arbiscular mycorrhizal fungus populations in heavy metal-contaminated soils. Appl. Environ. Microbiol. 65 (2), 718.
Dell'Amico, E., Cavalca, L., Andreoni, V., 2008. Improvement of *Brassica napus* growth under cadmium stress by cadmium-resistant rhizobacteria. Soil Biol. Biochem. 40 (1), 74–84.
Delvasto, P., Ballester, A., Munoz, J.A., Gonzalez, F., Blazquez, M.I., Igual, J.M., 2009. Mobilization of phosphorous from iron ore by the bacterium *Burkholderia caribensis* FeGLO3. Miner. Eng. 22 (1), 1–9.
Diene, O., Sakagami, N., Narisawa, K., 2014. The role of dark septate endophytic fungal isolates in the accumulation of cesium by Chinese cabbage and tomato plants under contaminated environments. PLOS ONE 9 (10), e109233.
Dighton, J., Horrill, A.D., 1988. Radiocesium accumulation in the mycorhhizal fungi *Lactarius rufus* and *Inocybe longicystis*, in upland. Britain. Trans. Br. Mycol. Soc. 91 (2), 335–337.
Dighton, J., Tugay, T., Zhdanova, N., 2008. Fungi and ionizing radiation from radionuclides. FEMS Microbiol. Lett. 281 (2), 109–120.

Dimpka, C., Svatoš, A., Merten, D., Buchel, G., Kothe, E., 2008. Hydroxamate siderophores produced by *Streptomyces acidiscabies* E13, bind nickel and promote growth in cowpea (*Vigna unuiculata* L.) under nickel stress. Can. J. Microbiol. 54 (3), 163–172.

Dimpka, C.O., Merten, D., Svatoš, A., Buchel, G., Kothe, E., 2009. Siderophores mediate reduced and increased uptake of cadmium by *Streptomyces tendae* F4 and sunflower (*Helianthus annus*). J. Appl. Micribiol. 107 (5), 1687–1696.

Djurdjević, L., Mitrović, M., Pavlović, P., Gajić, G., Kostić, O., 2006. Phenolic acids as bioindicators of fly ash deposit revegetation. Arch. Environ. Contam. Toxicol. 50 (4), 488–495.

Doherty, R.E., 2000. A history of the production and use of carbon tetrachloride, tetrachlorethylene, trichloroethylene, and 1,1,1-trichlorethane in the United States: Part2.-Trichlorethylene and 1,1,1-trichlorethan. J. Environ. Foren. 1 (2), 83–93.

Dolphen, R., Thiravetyan, P., 2019. Reducing arsenic in rice grains by leonardite and arsenic-resistant endophytic bacteria. Chemosphere 223, 448–454.

Doombos, R.F., van Loon, L.C., Bakker, P.A.H.M., 2012. Impact of root exudates and plant defense signaling on bacterial communities in the rhizosphere. A review. Agron. Sustain. 32, 227–243.

Dua, M., Singh, A., Sethunathan, N., Johri, A., 2002. Biotecnology and bioremediation: success and limitations. Appl. Microbiol. Biot. 59 (2-3), 143–152.

Dudašova, H., LukaČova, L., Murinova, S., Dercova, K., 2012. Effects of plant terpenes on biodegradation of polychlorinated biphenyls (PCBs). Int. Biodeterior. Biodegrad. 69, 23–27.

Duran, N., Esposito, E., 2000. Potential applications of oxidative enzymes and phenoloxidase-like compounds in wastewater and soil treatment: a review. Appl. Catal.-B-Environ. 28 (2), 83–99.

Dushenkov, S., 2003. Trends in phytoremediation of radionuclides. Plant Soil 249, 167–175.

Dushenkov, S., Mikheev, A., Prokhnevsky, A., Ruchko, M., Sorochinsky, B., 1999. Phytoremediation of radiocesium-contaminated soil in the vicinity of Chernobyl, Ukraine. Environ. Sci. Technol. 33 (3), 469–475.

Dushenkov, S., Vasudev, D., Kapulnik, Y., Gleba, D., Fleisher, D., Ting, K.C., Ensley, B., 1997. Removal of uranium from water using terrestrial plants. Environ. Sci. Technol. 31 (12), 3468–3474.

Dutton, M.V., Humphreys, P.N., 2005. Assesing the potential of short rotation coppice (Src) for cleanup of radionuclide contaminated sites. Inter. J. Phytoremediation. 7 (4), 279–293.

Dwivedy, K.K., Mathur, A.K., 1995. Bioleaching—our experience. Hydrometallurgy 38 (1), 99–109.

Dzantor, E.K., Chekol, T., Vough, L.R., 2000. Feasibility of using forage grasses and legumes for phytoremediation of organic pollutants. J. Environ. Sci. Health A 35 (9), 1645–1661.

Eapen, S., Suseelan, K.N., Tivarekar, S., Kotwal, S.A., Mitra, R., 2003. Potential for rhizofiltration of uranium using hairy root cultures of *Brassica juncea* and *Chenopodium amaranticolor*. Environ. Res. 91 (2), 127–133.

Eevers, N., White, J.C., Vangronsveld, J., Weyens, N., 2017. Bio – and phytoremediation of pesticide–contaminated environments: a review. Adv. Bot. Res. 83, 277–318.

Entry, J.A., Watrud, L.S., Reeves, M., 1999. Accumulation of ^{137}Cs and ^{90}Sr from contaminated soil by three grass species inoculated with mycorrhizal fungi. Environ. Pollut. 104, 449–457.

EPA, 2004a. Radionuclide Biological Remediation Resource Guide. US Environmental Protection Agency, Chicago, Illinois.

EPA, 2004b. Phytoremediation Field Studies Database for Chlorinated Solvents, Pesticides, Explosives, and Metals. Office of Superfund Remediation and Technology Innovation. US Environmental Protection Agency, Washigton, DC. www.vlu-in.org.

Farwell, A.J., Vesely, S., Nero, V., Rodriguez, H., Shah, S., Dixon, D.G., Glick, B.R., 2006. The use of 877 transgenic canola (*Brassica napus*) and plant growth-promoting bacteria to enhance plant 878 biomass at a nickel-contaminated field site. Plant Soil 288, 309–318.

Fava, F., Bertin, L., Fedi, S., Zannon, D., 2003. Methyl-Beta-Cyclodextrin-enhanced solubilization and aerobic biodegradation of polychlorinated byphenyls in two aged-contaminated soils. Biotechnol. Bioeng. 81 (4), 381–390.

Ficko, S.A., Rutter, A., Zeeb, B.A., 2010. Potential for phytoextraction of PCBs from contaminated soils using weeds. Sci Total Environ 408 (16), 3469–3476.
Ficko, S.A., Rutter, A., Zeeb, B.A., 2011a. Effect of pumpkin root exudates on ex situ polychlorinated biphenyl (PCB) phytoextraction by pumpkin and weed species. Environ. Sci. Pollut. Res. 18, 1536–1543.
Ficko, S.A., Rutter, A., Zeeb, B.A., 2011b. Phytoextraction and uptake patterns of weathered polychlorinated biphenyl-contaminated soils using three perennial weed species. J. Environ. Qual. 40, 1870–1877.
Fox, B.G., Borneman, J.G., Wackett, L.P., Lipscomb, J.D., 1990. Haloalkene oxidation by the soluble methane monooxygenase from *Metylosinus trichosporium* OB3b: mechanistic and environmental implications. Biochemistry 29 (27), 6419–6427.
Francis, A.J., Dodge, C.J., Gillow, J.B., Papenguth, H.W., 2000. Biotransformation of uranium compounds in high ionic strength brine by a halophilic bacterium under denitrifying conditions. Environ. Sci. Technol. 34 (11), 2311–2317.
Francova, K., Sura, M., Macek, T., Szekers, M., Bancos, S., Demnerova, K., Sylvestre, M., Mackova, M., 2003. Preparation of plants containing bacterial enzyme for degradation of polychlorinated biphenyls. Fresen. Environ. Bull. 12, 309–313.
Fredrickson, J.K., Zachara, J.M., Balkwill, D.L., Kenney, D., Li, S.M.W., Kostandarithes, H.M., Daly, M.J., Romine, M.F., Brockman, F.J., 2004. Geomicrobiology of high-level nuclear waste-contaminated vadose sediments at the Hanford site, Washington state. Appl. Environ. Microbiol. 70 (7), 4230–4241.
Fuentes, M.S., Benimeli, C.S., Cuozzo, S.A., Amoroso, M.J., 2010. Isolation of pesticide-degrading actinomycetes from a contaminated site: Bacterial growth, removal and dechlorination of organochlorine pesticides. Int. Biodeter. Biodegr. 64 (6), 434–441.
Fujishige, N.A., Kapadia, N.N., Hirsh, A.M., 2006. A feeling for the microorganism: structure on a small scale. Biofilms on plant roots. Bot. J. Linn. Soc. 150 (1), 79–884.
Furukawa, K., 1994. Molecular and evolutionary relationship of PCB-degrading bacteria. Biodegradation 5 (3–4), 289–300.
Furukawa, K., 2010. Biochemical and genetic bases of microbial degradation of polychlororinated biphenyls (PCBs). J. Gen. Appl. Microbiol. 46 (6), 283–296.
Furukawa, K., Fujihara, H., 2008. Microbial degradation of polychlorinated biphenyls: biochemical and molecular features. J. Biosci. Bioeng. 105 (5), 443–449.
Gadd, G.M., 2010. Metals, minerals and microbes: geomicrobiology and bioremediation. Microbiol 156 (Pt3), 609–643.
Gadd, M.G., 1993. Interactions of fungi with toxic metals. New Phytol 124, 25.
Gajić, G., Djurdjević, L., Kostić, O., Jarić, S., Mitrović, M., Stevanović, B., Pavlović, P., 2016. Assessment of the phytoremediation potential and an adaptive response of *Festuca rubra* L. sown on fly ash deposits: native grass has a pivotal role in ecorestoration management. Ecol. Eng. 93, 250–261.
Gajić, G., Djurdjević, L., Kostić, O., Jarić, S., Mitrović, M., Stevanović, B., Mitrović, M., Pavlović, P., 2020b. Phytoremediation potential, photosinthetic and antioxidant response to arsenic-induced stress of *Dactylis glomerata* L. sown on the fly ash deposits. Plants-Basel 9 (5), 657.
Gajić, G., Đurđević, L., Kostić, O., Jarić, S., Mitrović, M., Pavlović, P., 2018. Ecological potential of plants for phytoremediation and ecorestoration of fly ash deposits and mine wastes. Front. Environ. Sci. 6, 124.
Gajić, G., Mitrović, M., Pavlović, P., 2019. Ecorestoration of fly ash deposits by native plant species at thermal power stations in Serbia. In: Pandey, V.C., Bauddh, K. (Eds.), Phytomanagement of Polluted Sites: Market Opportunities in Sustainable Phytoremediation. Elsevier, Amsterdam, Netherlands, pp. 113–177.
Gajić, G., Mitrović, M., Pavlović, P., 2020. Feasibility of *Festuca rubra* L. native grass in phytoremediation. In: Pandey, V.C., Singh, D.P. (Eds.), Phytoremediation Potential of Perennial Grasses. Elsevier, Amsterdam, Netherlands, pp. 115–164.
Gajić, G., Mitrović, M., Pavlović, P., Stevanović, B., Đurđević, L., Kostić, O., 2009. An assesment of the tolerance of *Ligustrum ovalifolium* Hassk. to traffic-generated Pb using physiological and biochemical markers. Ecotox. Environ. Safe. 72 (4), 1090–1101.

Gajić, G., Pavlović, P., 2018. The Role of Vascular Plants in the Phytoremediation of Fly Ash Deposits. In: Matichenkov, V. (Ed.), Phytoremediation: Methods, Management and Assessment. Nova Science Publishers Inc, New York, USA, pp. 151–236.

Gajić, G., Pavlović, P., Kostić, O., Jarić, S., Djurdjević, L., Pavlović, D., Mitrović, M., 2013. Ecophysiological and biochemical traits of three herbaceous plants growing of the disposed coal combustion fly ash of different weathering stage. Arch. Biol. Sci. 65 (1), 1651–1667.

Gamalero, E., Glick, B.R., 2012. Plant growth-promoting bacteria and metal phytoremediation. In: Anjum, N.A., Pereira, M.E., Ahmad, I., Duarte, C., Umar, S., Khan, N.A. (Eds.), Phytotechnologies. Tayloe and Francis, Boca Raton, Florida, USA, pp. 359–374.

Ganesh, R., Robinson, K.G., Reed, G.D., Sayler, G.S., 1997. Reduction of hexavalent uranium from organic complexes by sulfate- and iron-reducing bacteria. Appl. Environ. Microbiol. 63 (11), 4385–4391.

Gao, J., Garrison, A.W., Hoehamer, C., Mazur, C.S., Wolfe, N.L., 2000. Uptake and phytotransformation of o,p'-DDT and p,p'-DDT by axenically cultivated aquatic plants. J. Agric. Food Chem. 48 (12), 6121–6127.

Garten, C.T., Tucker, C.S., Walton, B.T., 1986. Environmental fate and distribution of technetium-99 in a deciduous forest ecosystem. J. Environ. Radioact. 3, 163–188.

Garten2, C.T., Lomax, R.D., 1989. Technettium-99 cycle in maple trees. Characterization of changes in chemical form. Health. Phys. 57 (2), 299–307.

Gash, D.M., Rutland, K., Hudson, N.L., Sullivan, P.G., Bing, G., Cass, W.A., Pandya, J.D., Liu, M., Choi, D.Y., Hunter, R.L., Gerhardt, G.A., Smith, C.D., Slevin, J.T., Prince, T.S., 2008. Trichlorethylene: parkinsonism and complex 1 mytochondrial neurotoxicity. Ann. Neurol. 63 (2), 184–192.

Germaine, K.J., Liu, X., Cabellos, G.G., Hogan, J.P., Ryan, D., Dowling, D.N., 2006. Bacterial endophyte-enhanced phytoremediation of the organochlorine herbicide 2,4-dichlorphenoxyacetic acid. FEMS Microbiol. Ecol. 57 (2), 302–310.

Gevao, B., Semple, K.T., Jones, K.C., 2000. Bound pesticide residues in soil: a review. Environ. Pollut. 108 (1), 3–12.

Gianfreda, L., Bollag, J.M., 2002. Isolated enzymes for the transformation and detoxification of organic pollutants. In: Burns, R.G., Dick, R. (Eds.), Enzymes in the Environment: Activity, Ecology and Applications. Marcel Dekker, NY, pp. 491–538.

Gianfreda, L., Nannipieri, P., 2001. Basic principles, agents and feasibility of bioremediation of soil polluted by organic compounds. Minerva Biotechnol 13 (1), 5–12.

Gianfreda, L., Rao, M.A., 2004. Potential of extracellular enzymes in remediation of polluted soils: a review. Enz. Microb. Tecnol. 35 (4), 339–354.

Gilbert, E.S., Crowley, D.E., 1997. Plant compounds that induce polychlorinated biphenyl biodegradation by *Arthrobacter* sp., strain B1B. Appl. Environ Microbiol. 63 (5), 1933–1938.

Glick, B.R., 2003. Phytoremediation: synergistic use of plants and bacteria to clean up the environment. Biotechnol. Adv. 21 (5), 383–393.

Glick, B.R., 2010. Using soil bacteria to facilitate phytoremediatio. Biotechnol. Adv. 28 (3), 367–374.

Glick, B.R., 2012. Plant growth-promoting bacteria: mechanisms and applications. Scientifica, 963401 2012.

Godsy, E.M., Warren, E., Paganelli, V., 2003. The role of microbial reductive dechlorination of TCE at a phytoremediation site. Int. J. Phytoremediat. 5 (1), 73–87.

Gonchareva, N.V., 2009. Availability of radiocesium in plant from soil: facts, mechanisms and modeling. Global NEST J 11 (3), 260–266.

Gonzáles – Chávez, M.C.A., Vangronsveld, J., Colpaert, J., Leyval, C., 2006. Arbiscular mycorrhizal fungi and heavy metals: tolerance mechamnisms and potential use in bioremediation. In: Prasad, M.N.V., Sajwan, K.S., Naidu, R. (Eds.), Trace Elements in the Environment (Biogeochemistry, Biotechnology, and Bioremediation). CRC Taylor and Francis, Boca Raton, pp. 211–234.

Gonzales, M.C., 2000. Arbiscular Mycorrhizal Fungi from As/Cu Polluted Soils, Contribution to Plant Tolerance and Importance of the External Mycelium. Reading University, UK Ph.D. thesis.

González-Chávez, C., D'Haen, J., Vangronsveld, J., Dodd, J.C., 2004. Copper sorption and accumulation by the extraradical mycelium of different *Glomus* spp. (arbuscular mycorrhizal fungi) isolated from the same polluted soil. Plant Soil 240, 287–297.

Grandlic, C.J., Geib, I., Pilon, R., Sandrin, T.R., 2006. Lead pollution in a large, prarie-pothole lake (Rush Lake, WI, USA): effects on abundance and community structure of indigenous sediment bacteria. Environ. Pollut 144, 119–126.

Gravel, V., Antoun, H., Tweddell, R.J., 2007. Growth stimulation and fruit yield improvement of greenhouse tomato plants by inoculation with *Pseudomonas putida* or *Trichoderma atroviride*: possible role of indole acetic acid (IAA). Soil Biol. Biochem. 39 (8), 1968–1977.

Grbović, F., Gajić, G., Branković, S., Simić, Z., Ćirić, A., Rakonjac, Lj., Pavlović, P., Topuzović, M., 2019. Allelopathic potential of selected woody species growing on fly-ash deposits. Arch. Biol. Sci. 71 (1), 83–94.

Grbović, F., Gajić, G., Branković, S., Simić, Z., Vuković, N., Pavlović, P., Topuzović, M., 2020. Complex effect of *Robinia pseudoacacia* L. and *Ailathus altissima* (Mill.) Swingle growing on asbestos deposits: allelopathy and biogeochemistry. J. Serbian Chem. Soc. 85 (1), 141–153.

Greenwood, S.J., Rutter, A., Zeeb, B.A., 2011. The absorption and translocation of polychlorinated biphenyl congeners by *Cucurbita pepo* spp. *pepo*. Environ. Sci. Technol. 45 (15), 6511–6516.

Guarino, C., Paura, B., Sciarrillo, R., 2018. Enhancing phytoextraction of HMs at real scale, by combining *Salicaceae* trees with microbial consortia. Front. Environ. Sci. 6, 137.

Guarino, C., Sciarrillo, R., 2017. Effectiveness of in situ application of an integrated phytoremediation system (IPS) by adding a selected blend of rhizosphere microbes to heavily multi-contaminated soils. Ecol. Eng. 99, 70–82.

Guimarães, L.H.S., Segura, F.R., Tonani, L., von-Zeska-Kress, M.R., Rodrigues, J.L., Calixto, L.A., Silva, F.F., Batista, B.L., 2019. Arsenic volatilization by *Aspergillus* sp., and *Penicillium* sp., isolated from rice rhizosphere as a promising eco-safe tool for arsenic mitigation. J. Environ. Manage. 237, 170–179.

Hassan, S.E., Hijri, M., St-Arnaud, M., 2013. Effects of arbuscular mycorrhizal fungi on trace metal uptake by sunflower plants grown on cadmium contaminated soil. New Biotechnol 30 (6), 780–787.

Hassan, Z., Ali, S., Rizwan, M., Ibrahim, M., Nafees, M., Waseem, M., 2017. Role of bioremediation agents (bacteria, fungi, and algae) in alleviating heavy metal toxicity. In: Kumar, V. (Ed.), Probiotics in Agroecosystem. Springer Nature, Singapore, pp. 517–537.

Heijden Van Der, M.G.A., Bardgett, R.D., Van Straalen, N.M., 2008. The unseen majority, soil microbes as drivers of plant diversity and productivity in terrestrial ecosystems. Ecol. Lett. 11 (3), 296–310.

Hermansyah, H., Wijanarko, A., Gozan, M., Arbianti, R., Utami, T.S., Kubo, M., Shibasaki-Kitakawa, N., Yonemoto, T., 2007. Consecutive reaction model for triglyceride hydrolosis using lipase. J. Teknol 2, 151–157.

Hider, R.C., Kong, X., 2010. Chemistry and biology of siderophores. Nat. Prod. Rep. 27 (5), 637–657.

Howard, G.T., 2002. Biodegradation of polyurethane: a revie. Int. Biodeter. Biodegr. 49 (4), 245–252.

Hu, Q., Qi, H.Y., Zeng, J.H., Zhang, H.X., 2007. Bacterial diversity in soils around a lead and zinc mine. J. Environ. Sci. 19 (1), 74–79.

Huang, J.W., Blaylock, M.J., Kapulnik, Y., Ensley, B.D., 1998. Phytoremediation of uranium contaminated soils: role of organic acids in triggering hyperaccumulation in plants. Environ. Sci. Technol. 32, 2004–2008.

Huang, X.D., El-Alawi, Y., Penrose, D.M., Glick, B.R., Greenberg, B.M., 2004. A multi-process phytoremediation system for removal of polycyclic aromatic hydrocarbons from contaminated soils. Environ. Pollut. 130 (3), 465–476.

Hui, N., Liu, X-X., Kurola, J., Mikola, J., Romantschuk, M., 2012. Lead (Pb) contamination alters richness and diversity of the fungal, but not the bacterial community in pine forest soil. Boreal Environ. Res. 17, 46–58.

IAEA (International Atomic Energy Agency), 2010. Managing radioactive waste. International Atomic Energy Agency, Vienna.

Ionescu, M., Beranova, K., Dudkova, V., Kochankova, L., Demnerova, K., Macek, T., Mackova, M., 2009. Isolation and characterization of different plant associated bacteria and their potential to degrade polychlorinated biphenyls. Int. Biodeter. Biodegr. 63 (6), 667–672.

Jackson, N.E., Miller, R.H., Franklin, R.E., 1973. The influence of vesicular-arbuscular mycorrhizae on uptake of ^{90}Sr from soil by soybeans. Soil Biol. Biochem. 5 (2), 205–212.
Javorska, H., Tlustos, P., Kaliszova, R., 2009. Degradation of polychlorinated biphenyls in the rhizosphere of rape, *Brassica napus* L. Bull. Environ. Contam. Toxicol. 82 (6), 727–731.
Jha, P., Panwar, J., Jha, P.N., 2015. Secondary plant metabolites and root exudates: guiding tools for polychlororinated biphenyl biodegradation. Int. J. Environ. Sci. Technnol. 12 (2), 789–802.
John, M., Rubick, R., Schmitz, RPH., Rakoczy, J., Schubert, T., Diekert, G., 2009. Retentive memory of bacteria: long-term regulation of dehalorespiraspiration in *Sulforospirillum multivorans*. J. Bacteriol. 191 (5), 1650–1655.
Joseph, B., Ramteke, P.W., Kumar, P.A., 2006. Studies on the enhanced production of extracellular lipase by *Staphylococcus epidermidis*. J. Gen. Appl. Microbiol. 52 (6), 315–320.
Kacalkova, L., Tlustoš, P., 2011. The uptake of persistent organic pollutants by plants. Cent. Eur. J. Biol. 6 (2), 223–235.
Kang, J.W., Doty, S.L., 2014. Cometabolic degradation of trichloroethylene by *Burkholderia cepacia* G4 with poplar leaf homogenate. Can. J. Microbiol. 60 (7), 487–490.
Karigar, C., Rao, S.S., 2011. Role of microbial enzymes in the bioremediation of pollutants: a review. Enzyme Res 2011, 805187.
Kassa-Laouar, M., Mechakra, A., Rodrigue, A., Meghnous, O., Bentellis, A., Rached, O., 2020. Antioxidative enzyme responses to antimony stress of *Serratia marcescens* – an endophytic bacteria of *Hedysarum pallidum* roots. Pol. J. Environ. Stud. 29 (1), 141–152.
Ker, K., Charest, C., 2010. Nickel remediation by AM-colonized sunflower. Mycorrhiza 20, 399–406.
Khan, A.G., Kuek, C., Chaudhry, T.M., Khoo, C.S., Hayes, W.J., 2000. Role of plants, mycorrhizae and phytochelators in heavy metal contaminated land remediation. Chemosphere 41, 197.
Khan, M.,U., Sessitdch, A., Harris, M., Fatima, K., Imran, A., Arslan, M., Shabir, G., Khan, Q.M., Afzal, M., 2015. Cr-resistant rhizo-and endophytic bacteria associated with *Prosopis juliflora* and their potential as phytoremediation enhancing agents in meta-degraded soils. Front. Plant Sci. 5, 755.
Khan, M.S., Zaidi, A., Wani, P.A., Oves, M., 2009. Role of plant growth promoting rhizobacteria in the remediation of metal contaminated soils. Environ. Chem. Lett. 7, 1–19.
Khan, S., Afzal, M., Iqbal, S., Khan, Q.M., 2013. Plant-bacteria parthership for the remediation of hydrocarbon contaminated soils. Chemosphere 90, 1317–1332.
Kim, S., Bae, W., Hwang, J., Park, J., 2010. Aerobic TCE degradation by encapsulated toluene-oxidizing bacteria, *Pseudomonas putida* and *Bacillus* spp. Water Sci. Technol. 62 (9), 1991–1997.
Kostić, O., Jarić, S., Gajić, G., Pavlović, D., Pavlović, M., Mitrović, M., Pavlović, P., 2018. Pedological properties and ecological implications of substrates derived 3 and 11 years after the revegetation of lignitefly ash disposal sites in Serbia. Catena 163, 78–88.
Kostić, O., Mitrović, M., Knežević, M., Jarić.S., G.a.j.i.ć., G., D.j.u.r.d.j.e.v.i.ć., L., P.a.v.l.o.v.i.ć., 2012. The potential of four woody species for the revegetation of fly ash deposits from the 'Nikola Tesla –A' thermoelectric plant (Obenovac, Serbia). Arch. Biol. Sci. 64 (1), 145–158.
Kothari, S.K., Marschner, H., George, E., 1990. Effect of VA mycorrhizal fungi and rhizosphere micro-organisms on root and shoot morphology, growth, and water relations of maize. New Phytol 116, 303–311.
Kramer, J., Özkaya, Ö., Kümmerli, R., 2020. Bacterial siderophores in community and host interactions. Nat. Rev. Microbiol. 18 (3), 152–163.
Krupa, P., Kozdroj, J., 2007. Ectomycorrhizal fungi and associated bacteria provide protection against heavy metals in inoculated pine (*Pinus sylvestris* L.) seedlings. Water Air Soil Pollut 182 (1), 83–90.
Kucerova, P., Mackova, M., Chroma, L., Burkhard, J., Triska, J., Demnerova, M.T., 2000. Metabolism of polychlororinated biphenyls by *Solanum nigrum* hairy root clone SNC-90 and analysis of transformation products. Plant Soil 225, 109–115.
Kuffner, M., Puschnreiter, M., Wieshammer, G., Gorfer, M., Sessitch, A., 2008. Rhizosphere bacteria affect growth and metal uptake of heavy metal accumulating willows. Plant Soil 304, 35–44.

Kurzawova, V., Stursa, P., Uhlik, O., Norkova, K., Strohalm, M., Lipov, J., Kochankova, L., Mackova, M., 2012. Plant-microorganism interactions in bioremediation of polychlorinated biphenyl-contaminated soil. New Biotechnol 30 (1), 15–22.

La Merill, M., Emond, C., Kim, M.J., Antignac, J-P., Le Bizec, B., Clement, K., Birnbaum, L.S., Barouki, R., 2012. Toxicological function of adipose tissue: focus on persistent organic pollutants. Environ. Health Perspect. 121 (2), 162–169.

Lampis, S., Ferrari, A., Cunha-Queda, A.C.F., Alvarenga, P., Di Gregorio, S., Vallini, G., 2009. Selenite resistant rhizobacteria stimulate SeO_3^{2-} phytoextraction by *Brassica juncea* in bioaugmented water-filtering artificial bed. Environ. Sci. Pollut. Res. 16, 663–670.

Lanfranco, L., Bolchi, A., Cesale, Ros, E., Ottonello, S., Bonfante, P., 2002. Differential expressionof a metallothionein gene during the presymbiotic vs. the symbiotic phase of an arbiscular mycorrhizal fungus. Plant Physiol 130 (1), 58.

Laroche, L., Henner, P., 2003. Chemical speciation of uranium in soil solution and bioavailability for root uptake in plants. In: Gobran, G.R., Lepp, N. (Eds.), 7th International Conference on the Biochemistry of Trace Elements, Uppsala, Sweden. SLU, Sweden, 1, June 15-19, 2003, pp. 168–169.

Lasat, M.M., 2002. Phytoextraction of toxic metals: a revie of biological mechanisms. J. Environ. Qual. 31 (1), 109–120.

Lasat, M.M., Fuhrmann, M., Ebbs, S.D., Cornish, J.E., Kochian, L.V., 1998. Phytoremediation of a radiocesium-contaminated soil: Evaluation of cesium-137 bioaccumulation in the shoots of three plant species. J. Environ. Qual. 27 (1), 165–169.

Lasat, M.M., Norvell, W.A., Kochian, L.V., 1997. Potential for phytoextraction of ^{137}Cs from contaminated soil. Plant Soil 195, 99–106.

Lash, L.H., Fisher, J.W., Lipscomb, J.C., Parker, J.C., 2000. Metabolism of trichloerethylene. Environ. Health Perspect. 108 (Suppl. 2), 177–200.

Lauby-Secretan, B., Loomis, D., Grosse, Y., Ghissassi, F., Bouvard, V., Benbrahim – Talla, L., Guha, N., Baan, R., Mattock, H., Straif, K., 2013. Carcinogenicity of polychlorobiphenyls and polybromoinated biphenyls. Lancet. Oncol. 14 (4), 287.

Leewis, M.C., Uhlik, O., Leight, M.B., 2016. Synergistic processing of biphenyl and benzoate: carbon flow through the bacterial community in polychlorinated – biphenyl-contaminated soil. Sci. Rep. 6, 22145.

Leight, M.B., Prouzova, P., Mackova, M., Macek, T., Nagle, D.P., Fletcher, J.S., 2006. Polychlorinated biphenyl (PCB)–degrading bacteria associated with trees in PCB-contaminated site. Appl. Environ. Microbiol. 72 (4), 2331–2342.

Lenoir, I., Sahraoui, L-H., Fontaine, J., 2016. Arbiscular mycorrhizal fungal-assisted phytoremediationof soil contaminated with persistent organic pollutants: a reviw. Eur. J. Soil Sci. 67 (5), 624–640.

Leung, M., 2004. Bioremediation: techniques for cleaning up a mess. J. Biotechnol. 2, 18–22.

Li, W.C., Ye, Z.H., Wong, M.H., 2010. Metal mobilization and production of short chain organic acids by rhizosphere bacteria associated with a Cd/Zn hyperaccumulating plant *Sedum alfredii*. Plant Soil 326, 453–467.

Liang, Y., Meggo, R., Hu, D., Schnoor, J.R., Mattes, T.E., 2014. Enhanced polychlorinated biphenyl removal in aswitchgrass rhizosphere by bioaugmentation with *Burkholderia xenovorans* LB400. Ecol. Eng. 71, 215–222.

Lin, Z.-Q., 2008. Volatilization. In: Jorgensen, S.E., Fath, B.D. (Eds.), Ecological Processes. Encyclopedia of Ecology. Elsevier, Oxford, pp. 3700–3705.

Liu, J.Y., Schnoor, J.L., 2008. Uptake and translocation of lesser-chlorinated polychlororinated biphenyls (PCBs) in whole hybride poplar plants hydroponic exposure. Chemosphere 73 (10), 1608–1616.

Lohman, R., Breivik, K., Dachs, J., Muir, D., 2007. Global fate of POPs: current knowledge and future research direction. Environ. Pollut. 150, 150–165.

Lovley, D.R., Phillips J.P., E, 1992. Bioremediation of uranium contamination with enzymatic uranium reduction. Environ. Sci. Technol. 26 (11), 2228–2234.

Lu, Y.F., Lu, M., Peng, F., 2014. Remediation of polychlorinated biphenyl-contaminated soil by using a combination of ryegrass, arbuscular mycorrhizal fungi and earthworms. Chemosphere 106, 44–50.

Luecking, A.D., Huang, W., Soderstrom-Schwarz, S., Kim, M., Weber, W.J.R., 2000. Relationship of soil organic matter characterization to organic contaminant sequestration and bioavailability. J. Environ. Qual. 29 (1), 317–323.

Lunney, A.I., Zeeb, B.A., Reimer, K.J., 2004. Uptake of weathered DDT in vascular plants: Potential for phytoremediation. Environ. Sci. Technol. 38 (22), 6147–6154.

Luo, S., Xu, T., Chen, L., Chen, J., Rao, C., Xiao, X., Wan, Y., Zeng, G., Long, F., Liu, C., Liu, Y., 2012. Endophyte-assisted promotion of biomass production and metal uptake of energy crop sweet sorghum by plant growt-promoting endophyte *Bacillus* sp. SLS18. Appl. Microbiol. Biotechnol. 93 (4), 1745–1753.

Lynch, J.M., 2002. Resilience of the rhizosphere to anthropogenic disturbance. Biodegradation 13 (1), 21–27.

Ma, Y., Prasad, M.N.V., Rajkumar, M., Freitas, H., 2011. Plant growth promoting rhizobacteria and endophytes accelerate phytoremediation of metalliferous soils. Biotechnol. Adv. 29 (2), 248–258.

Ma, Y., Rajkumar, M., Freitas, H., 2009a. Improvement of plant growth and nickel uptake by nickel resistant plant growh promoting bacteria. J. Hazard. Mater. 166 (2–3), 1154–1161.

Ma, Y., Rajkumar, M., Freitas, H., 2009b. Inoculation of plant growth promoting bacterium *Achromobacter xylosoxidans* strain Ax10 for the improvement of copper phytoextraction by *Brassica juncea*. J. Environ. Manage. 90, 831–837.

Macaskie, L.E., Jeong, B.C., Tolley, M.R., 1994. Enzymically accelerated biomineralization of heavy metals: application to the removal of americium and plutonium from aqueous flows. FEMS Microbiol. Rev. 14 (4), 351–367.

Mackova, M., Prouzova, P., Stursa, P., Ryslava, E., Uhlik, O., Beranova, K., Rezek, J., Kurzawova, V., Demnerova, K., Macek, T., 2009. Phyto/hizomediation studies using long-term PCB-contaminated soil. Environ. Sci. Pollut. Res. 16 (7), 817–829.

Madhaiyan, M., Pooguzhali, S., Sa, T., 2007. Metal tolerating methylotrophic bacteria reduces nickel and cadmium toxicity and promotes plant growrh of tomato (*Lycopersicum esculentum* L.). Chemosphere 69, 220–228.

Mahmmoud, Ya-G., 2004. Uptake of radionuclides by some fungi. Mycrobiology 32 (3), 110–114.

Mai, C., Schormann, W., Milstein, O., Huttermann, A., 2000. Enhanced stability of laccase in the presence of phenolic compounds. Appl. Microbiol. Biotechnol. 54 (4), 510–514.

Maiti, S.K., 2013. Ecorestoration of the Coalmine Degraded Lands. Springer, New Delhi, India, p. 361.

Malaviya, P., Singh, A., 2012. Phytoremediation strategies for remediation of uranium-contaminated environments: A review. Crit. Rev. Environ. Sci. Technol. 42 (24), 2575–2647.

Matsumoto, E., Kawanaka, Y., Yun, S.J., Oyaizu, H., 2009. Bioremediation of the organochlorine pesticides, dieldrin, and endrin and their occurrence in the environment. Appl. Microbiol. Biotechnol. 84 (2), 205–216.

Mattina, M.J.I., Lannucci-Berger, W., Musante, C., White, J., 2003. Concurrent plant uptake of heavy metals and persistent organic pollutants from soil. Environ. Pollut. 124 (3), 375–378.

McCarty, P.L., 2010. Groundwater contamination by chlorinated solvents: history, remediation technologies and strategies. In: Stroo, H.F., Ward, C.H. (Eds.), In situ Remediation of Chlorinated Solvent Plumes. Springer, New York, pp. 1–4.

McCutcheon, S.C., Schnoor, J.L., 2003. Phytoremediation. Transformation and Control of Contaminants. John Wiley and Sons, Inc, Publications, Hoboken, New Jersey, p. 987.

Mendes, R., Gabareva, P., Raaijmakers, J.M., 2013. The rhizosphere microbiome: significance of plant beneficial, plant pathogenic, and human pathogenic microorganisms. FEMS Microbiol. Rev. 37 (5), 634–663.

Mercado-Borrayo, B.M., Heydrich, S.M., Perez, I.R., Quiroz, M.H., Hill, C.P.L., 2015. Organophophorous and organochlorine pesticides bioaccumulation by *Eichornia crassipes* in irrigation canals in an urban agricultural system. Int. J. Phytoremediat. 17 (7), 701–708.

Mikes, O., Cupr, P., Stefan, T., Klanova, J., 2009. Uptake of polychlorinated biphenyls and organochlorine pesticides from soil and air into radishes (*Raphanus sativus*). Environ. Pollut. 157 (2), 488–496.

Mitrović, M., Pavlović, P., Lakušić, D., Stevanović, B., Djurdjevic, L., Kostić, O., Gajić, G., 2008. The potencial of *Festuca rubra* and *Calamagrostis epigejos* for the revegetation on fly ash deposits. Sci. Tot. Environ. 407 (1), 338–347.

Mitton, F.M., Miloranza, K.S.B., Gonzales, M., Shimabukoro, V.M., Monserrat, J.M., 2014. Assessment of tolerance and efficiency of crop species in the phytoremediation of DDT polluted soils. Ecol. Eng. 71 (2), 501–508.

Mohammadi, M., Chalavi, V., Novakova-Sura, M., Laliberte, J.F., Sylvestre, M., 2007. Expression of bacterial biphenyl-chlorobiphenyl dioxygenase genes in tobacco plants. Biotechnol. Bioeng. 97 (3), 496–505.

Moklyachuk, L., Goridska, I., Slobodenyuk, O., Petryshyna, V., 2010. Phytoremediation of soil polluted obsolete pesticides in Ukraine. In: Kulakov, P.A., Pidlisnyuk, V. (Eds.), Aplication of Phytotechnologies for Clean Up Industrial, Agricultural and Wastewater Contamination. Springer Science + Buiness Media, B.V, Dordrecht.

Moklyachuk, L., Petryshyna, V., Slobodenyuk, O., Zatsarinna, Y., 2012. Sustainable strategies of phytoremediation of the sites polluted with obsolete pesticides. In: Vitale, K. (Ed.), Environmental and Food Safety and security for South-East Europe. Springer Science + Buiness Media B.V, Dordrecht.

Moore, F.P., Barac, T., Borremans, B., Oeyen, L., Vangronsveld, J., van der Lelie, D., Campbell, C.D., Moore, E.R.B., 2006. Endophytic bacterial diversity in poplar trees growing on a BTEX-contaminated site: The characterization of isolates with potential to enhance phytoremediation. Syst. Appl. Microbiol. 29 (7), 539–556.

Nair, A., Juwarkar, A.A., Singh, S.K., 2007. Production and characterization of siderophores and its application in arsenic removal from contaminated soil. Water Air Soil Pollut 180 (4), 199–212.

Nakajima-Kambe, T., Shigeno-Akutsu, Y., Nomura, N., Onuma, F., Nakahara, T., 1999. Microbial degradation of polyurethane, polyester polyurethanes and polyether polyurethanes. Appl. Microbiol. Biotechnol. 51 (2), 134–140.

Nakamura, K., Tomita, T., Abe, N., Kamio, Y., 2001. Purification and characterization of an extra cellular poly(L-lacz+tic acid) depolymerase from a soil isolate, *Amycoatopsis* sp. strainK104-1. Appl. Environ. Microbiol. 67 (1), 345–353.

Narasimhan, K., Basheer, C., Bajic, V.B., Swarup, S., 2003. Enhancment of plant-microbe interactions using a rhizosphere metabolomics-driven approach and its application in the removal of polychlorinated biphenyls. Plant Physiol 132 (1), 146–153.

Narisawa, K., Hambleton, S., Currah, R.S., 2007. *Heteroconium chaetospira*, a dark septate root endophyte allied to the Herpotrichiellaceae (Chaetothyriales) obtained from some forest soil samples using bait plants. Mycoscience 48 (5), 274–281.

Narkhede, C.P., Patil, A.R., Koli, S., Syryawanshi, R., Wagh, N.D., Salunke, B.K., Patil, S.V., 2015. Studies on endosulfan degradation by local isolate *Pseudomonas aeruginosa*. Biocat. Agric. Biotechnol. 4 (2), 259–265.

Naveed, M., Mitter, B., Yousaf, S., Pastar, M., Afzal, M., Sessitsch, A., 2014. The endophyte *Enterobacter* sp. FD17: a maize growth enhancer selected based on rigorous testing of plant beneficial traits and colonization characteristics. Biol. Fertil. Soils 50, 249–262.

Newmann, L., Strand, S., Duffy, J., Ekuan, G., Raszaj, M., Shurtleff, B., Wilmoth, J., Heilman, P., Gordon, M., 1997. Uptake and biotransformation of trichloroethylene by hybride poplars. Environ. Sci. Technol. 31 (4), 1062–1067.

Newmann, L.M., Wacket, L.P., 1995. Purification and characterization of toluene 2-monooxygenase from *Burkholderia cepacia* G4. Biochemistry 34 (43), 14066–14076.

Novakova, M., Mackova, M., Chrastilova, Z., Viktorova, J., Szekers, M., Demnerova, K., Macek, T., 2009. Cloning the bacterial *bphC* gene into *Nicotiana tabacum* to improve the efficiency of PCB-phytoremediation. Biotechnol. Bioeng. 102 (1), 29–37.

Novotny, C., Vyas, B.R.M., Erbanova, P., Kubatova, A., Sasek, V., 1997. Removal of various PCBs by various white-rot fungi in liquid cultures. Folia Microbiol 42 (2), 136–140.

Olse, R.A., Joner, E., Bakken, L.R., 1990. Soil fungi and the fate of radiocesium in the soil ecosystem – a discussion of possible mechanisms involved in the radiocesium accumulation in fungi, and the role of fungi as a Cs-sink

in the soil. In: Desmet, G., Nassimbeni, P., Belli, M. (Eds.), Transfer of Radionuclides in Natural and Semi-Natural Environments. Elsevier, London, pp. 657–663.
Orlowska, E., Godzik, B., Turnau, K., 2012. Effect of different arbiscular fungal isolates on growth and arsenic accumulation in *Plantago lanceolata* L. Environ. Pollut. 168, 121–130.
Pal, C., Asiani, K., Arya, S., Rensing, C., Stekel, D.J., Larsson, D.G.J., Hobman, J.L., 2017. Metal resistance and its association with antibiotic resistance. Adv. Microb. Physiol. 70, 261–313.
Pan, X., Lin, D., Zheng, Y., Zhang, Q., Yin, Y., Cai, L., Fang, H., Yu, Y., 2016. Biodegradation of DDT by *Stenotrophomonas* sp. DDT-1: characterization and genome functional analysis. Sci. Rep. 6, 21332.
Pandey, N., Bhatt, N., 2016. Role of soil associated *Exiguobacterium* in reducing arsenic toxicity and promoting planth growth in *Vigna radiate*. Eur. J. Soil Biol. 75, 142–150.
Pandey, V.C., 2012. Invasive species based efficient green technology for phytoremediation of fly ash deposits. J. Geochem. Explor. 123, 13–18.
Pandey, V.C., 2013. Suitability of *Ricinus communis* L. cultivation for phytoremediation of fly ash disposal sites. Ecol. Eng. 57, 336–341.
Pandey, V.C., 2015. Assisted phytoremediation of fly ash dumps through naturally colonized plants. Ecol. Eng. 82, 1–5.
Pandey, V.C., 2020. Phytomanagement of Fly Ash. Elsevier, Amsterdam, p. 352.
Pandey, V.C., Bajpai, O., Sinhg, N., 2016. Plant regeneration potential in fly ash ecosystem. Urban For. Urban Gree 15, 40–44.
Pandey, V.C., Bauddh, K., 2019. Phytomanagement of Polluted Sites: Market Opportunities in Sustainable Phytoremediation. Elsevier, Amsterdam, Netherlands, p. 602.
Pandey, V.C., Pandey, D.N., Singh, N., 2015a. Sustainable phytoremediation based on naturally colonizing and economically valuable plants. J. Clean Prod. 86, 37–39.
Pandey, V.C., Prakash, P., Bajpai, O., Kumar, A., Sing, N., 2015b. Phytodiversity on fly ash deposits: evaluation of naturally colonized species for sustainable phytorestoration. Environ. Sci. Pollut. Res. 22 (4), 2776–2787.
Pandey, V.C., Singh, D.P., 2020. Phytoremediation Potential of Perennial Grasses. Elsevier, Amsterdam, Netherlands, p. 371.
Pandey, V.C., Singh, K., Singh, R.P., Singh, B., 2012. Naturally growing *Saccharum munja* L. On the fly ash lagoons: a potential ecological engineer for the revegetation and stabilization. Ecol. Eng. 40, 95–99.
Pandey, V.C., Singh, N., 2014. Fast green capping on coal fly ash basins through ecological engineering. Ecol. Eng. 73 (2), 671–675.
Pandey, V.C., Singh, N., Singh, R.P., Singh, D.P., 2014. Rhizomediation potential of spontaneous grown *Typha latifolia* onfly ash basins: study from the field. Ecol. Eng. 71 (2), 722–727.
Pandey, V.C., Singh, J.S., Singh, R.P., Singh, N., Yunus, M., 2011. Arsenic hazards in coal fly ash and its fate in Indian scenario. Resour. Conserv. Recy. 55 (9–10), 819–835.
Passatore, L., Rossetti, S., Juwarkar, A.A., Massacci, A., 2014. Phytoremediation and bioremediation of polychlorinated biphenyls (PCBs): state of knowledge and research perspectives. J. Hazard. Materials 278, 189–202.
Pavlović, P., Marković, M., Kostić, O., Sakan, S., Djordjević, D., Perović, V., Pavlović, D., Pavlović, M., Čakmak, D., Jarić, S., Paunović, M., Mitrović, M., 2019. Evaluation of potentially toxic element contamination in the riparian zone of the River Sava. Catena 174, 399–412.
Pavlović, P., Mitrović, M., Djurdjevic, L., 2004. An ecophysiological study of plants growing on the fly ash deposits from the "Nikola Tesla – A" thermal power station in Serbia. Environ. Manage. 33 (5), 654–663.
Pavlović, P., Mitrović, M., Đorđević, D., Sakan, S., Slobodnik, J., Liška, I., Csanyi, B., Jarić, S., Kostiž, O., Pavlović, D., Marinković, N., Tubić, B., Paunović, M., 2016. Assessment of the contamination of riparian soil and vegetation by trace metals - a Danube river case study. Sci. Tot. Environ 540, 396–409.
Pawlowska, T.E., Chaney, R.I., Chin, M., Charval, I., 2000. Effects of metal phytoextraction practices on the indigenous community of arbiscular mycorrhizal fungi at a metal-contaminated landfill. Appl. Environ. Microbiol. 66 (6), 2526.

Perry, R.D., Bobrov, A.G., Fetherson, J.D., 2015. The role of transition metal transporters for iron, zinc, manganese, and copper in the pathogenesis of *Yesinia pestis*. Metallomics 7 (6), 965–978.

Pieper, D.H., Seefer, M., 2008. Bacterial metabolism of polychlorinated biphenyls. J. Mol. Microbiol. Biotechnol. 15 (2–3), 121–138.

Pilon – Smits, E., 2005. Phytoremediation. Annu. Rev. Plant. Biol. 56 (1), 15–39.

Prasad, M.N.V., Favas, P.J.C., Maiti, S.K., 2018. Bio - Geotechnologies for Mine Site Rehabilitation. Elsevier, Amsterdam, Netherlands, p. 709.

Prasad, M.N.V., Freitas, H.M.O., 2003. Metal hyperaccumulation in plants – biodiversity prospecting for phytoremediation technology. Electron. J. Biotech. 6 (3), 285–305.

Prasad, M.N.V., Sajwan, K.S., Naidu, R., 2006. Trace Elements in the Environment (Biogeochemistry, Biotechnology, and Bioremediation). CRC Taylor and Francis, Boca Raton, p. 726.

Purchase, D., Miles, R.J., Yong, T.W.K., 1997. Cadmium uptake and nitrogen fixing ability in heavy metal-resistant laboratory and field strains of *Rhizobium leguminosarum* biovar trifoli. FEMS Microbiol. Ecol. 22 (1), 85–93.

Qin, H., Brookes, P.C., Xu, J., Feng, Y., 2014. Bacterial degradation of Aroclor 1242 in the mycorrhizosphere soils of zucchini (*Cucurbita pepo* L.) inoculated with arbuscular mycorrhizal fungi. Environ. Sci. Pollut. Res. Inter. 21 (22), 12790–12799.

Radnoti de Lipthay, J., Barkay, T., Vekova, J., Sorensen, S., 1999. Utilization of phenoxyacetic acid, by strains using either the *ortho* or *meta* cleavage of catechol during phenol degradation, after conjugal transfer of *tfdA*, the gene encoding a 2,4-dichlorophenoxyacetic acid/2-oxoglutarate dioxygenase. J. Appl. Microbiol. Biotechnol. 51, 207–214.

Rajapaksha, R.M.C.P., Took-Kapton, M.A., Baath, E., 2004. Metal toxicity affects fungal and bacterial activities in soil differently. Appl. Environ. Microbiol. 70 (5), 2966–2973.

Rajkumar, M., Nagendran, R., Lee, K.J., Lee, W.H., Kim, S.Z., 2006. Influence of plant growth promoting bacteria and Cr^{6+} on the growth of Indian mustard. Chemosphere 62 (5), 741–748.

Rajkumar, M., Sandhya, S., Prasad, M., Freitas, H., 2012. Perspectives of plant-associated microbes in heavy metal phytoremediation. Biotechnol. Adv. 30 (6), 1562–1574.

Rao, M.A., Scelza, R., Scotti, R., Gianfreda, L., 2010. Role of enzymes in the remediation of polluted environments. J. Soil Sci. Plant Nutr. 10 (3), 333–353.

Reed, M.I.E., Glick, B.R., 2005. Growth of canola (*Brassica napus*) in the presence of plant growrh-promoting bacteria and either copper or polycyclic aromatic hydrocarbons. Can. J. Microbiol. 51 (12), 1061–1069.

Reed, M.L.E., Warmer, B.G., Glick, B.R., 2005. Plant growth promoting bacteria facilitate the growth of the common reed *Phragmites australis* in the presence of copper or polycyclic aromatic hydrocarbons. Curr. Microbiol. 51 (6), 425–429.

Ren, C-G., Kong, C-C., Wang, S-X., Xie, Z-H., 2019. Enhanced phytoremediation of uranium-contaminated soils by arbuscular mycorrhiza and rhizobium. Chemosphere 217, 773–779.

Retamal-Morales, G., Menhert, M., Schwabe, R., Tischler, D., Zapata, C., Chavez, R., Schlomann, M., Levican, G., 2018. Detection of arsenic-binding siderophores in arsenic – tolerating actinobacteria by a modified CAS assay. Ecotox. Environ. Safe. 157, 176–181.

Rezek, J., Macek, T., Mackova, M., Trisha, J., 2007. Plant metabolites of polychlorinated biphenyls in hary root culture of black nightshade *Solanum nigrum* SNC-90. Chemosphere 69 (8), 1221–1227.

Rezek, J., Macek, T., Mackova, M., Triska, J., Ruzickova, K., 2008. Hydroxy-methoxy-PCBs: metabolites of polychlorinated biphenyls formed in vitro by tobacco cells. Environ. Sci. Technol. 42, 5746–5751.

Rezende, M.I., Barbosa, A.M., Vasconcelos, A.-F.D., Haddad, R., Dekker, F.H., 2005. Growth and production of laccases by the ligninolytic fungi, *Pleurotus ostreatus* and *Botryosphaeria rhodina*, cultured on basal medium containing the herbicide, Scepter® (imazaquin). J. Basic Microbiol. 45 (6), 460–469.

Riesen, T.K., Brunner, I., 1996. Effect of ectomycorrhizae and ammonium on ^{134}Cs and ^{85}Sr uptake into *Picea abies* seedlings. Environ. Pollut. 93 (1), 1–8.

Rilling, M.C., Antonovics, J., Caruso, T., Lehmann, A., Powell, J.R., Veresoglou, S.D., Verbruggen, E., 2015. Interchange of entire comminities: microbial community coalescens. Trends Ecol. Evol. 30 (8), 470–476.

Rilling, M.C., Wright, S., Allen, M.F., Field, C.B., 1999. Rise in carbon dioxide changes soil structure. Nature 400 (6745), 628.

Rissato, S.R., Galhiane, M.S., Fernandes, J.R., Gerenutti, M., Gomes, H.M., Ribeiro, R., Vinicius de Almeida, M., 2015. Evaluation of *Ricinus communis* L., for the phytoremediation of polluted soil with organochlorine pesticides. Biomed. Res. Int., ID549863.

Rixon, J.E., Ferreira, L.M., Ferreira, L.M.A., Durrant, A.J., Laurie, J.I., Hazlewood, G.P., Gilbert, H.J., 1992. Characterization of the gene celD and its encoded product 1,4-β-D-glucan glucohydrolase D from *Pseudomonas fluorescens* subsp. Cellulose. Biochem. J. 285 (3), 947–955.

Robinson, A.E., Lowe, J.E., Koh, E.I., Henderson, J.P., 2018. Urapathogenic enterobacteruse in the yersiniabactin metallophore system to acquire nickel. J. Biol. Chem. 293 (39), 14953–14961.

Rodriguez, H., Fraga, R., 1999. Phosphate solubilizing bacteria and their role in plant growth promotion. Biotechnol. Adv. 17 (4–5), 319–339.

Rodriguez, H., Gonzalez, T., Goire, I., Bashan, Y., 2004. Gluconic acid production and phosphate solubilization by the plant growth promoting bacterium *Azospirillium* spp. Naturwissenschaften 9 (11), 552–555.

Rogers, R.D., Williams, S.E., 1986. Vesicular-arbuscular mycorrhiza: Influence on plant uptake of cesium and cobalt. Soil Biol. Biochem. 18 (4), 371–376.

Rojjanateeranaj, P., Sangthong, C., Prapagdee, B., 2017. Enhanced cadmium phytoremediation of *Glycine max* L., through bioaugmentation of cadmium-resistant bacteria assisted by biostimulation. Chemosphere 185, 764–771.

Rommelt, R., Hiersche, L., Schaller, G., Wirth, E., 1990. Influence of soil fungi (Basidiomycetes) on the migration of Cs 134 + 137 and Sr90 in coniferous forest soils. In: Desmet, G., Nassimbeni, P., Belli, M. (Eds.), Transfer of Radionuclides in Natural and Semi-Natural Environments. Elsevier, London, pp. 152–160 Applied Science.

Roos, P., Jakobsen, I., 2008. Arbuscular mycorrhiza reduces phytoextraction of uranium, thorium, and other elements from phosphate rock. J. Environ. Radioact. 99 (5), 811–819.

Ross, S.M., Kaye, K.J., 1994. The meaning of metal toxicity in soil-plant systems. In: Ross, S.M. (Ed.), Toxic Metals in Soil-Plant Systems. Jon Willey and Sons, Chichester, UK.

Roy, M., Giri, A.K., Dutta, S., Mukherjee, P., 2015. Integrated phytobial remediation for sustainable management of arsenic in soil and water. Environ. Int. 75, 180–198.

Roy, M., Pandey, V.C., 2020. Role of microbes in grass-based phytoremediation. In: Pandey, V.C., Singh, D.P. (Eds.), Phytoremediation Potential of Perennial Grasses. Elsevier, Amsterdam, Netherlands, pp. 303–336.

Rufyikiri, G., Huysmans, L., Wannijna, J., Hees, M.V., Leyval, C., Jakobsen, I., 2004. Arbuscular mycorrhizal fungi can decrease the uptake of uranium by subterranean clover grown at high levels of uranium in soil. Environ. Pollut. 130 (3), 427–436.

Ryan, P.R., Delhaize, E., Jones, D.I., 2001. Function and mechanisms of organic anion exudation from plant roots. Annu. Rev. Plant. Physiol. Plant Mol. Biol. 52 (1), 527–560.

Sadanoski, M.A., Velazquez, J.E., Fonseca, M.I., Zapata, P.D., Levin, L.N., Villaba, L., 2018. Assesing the ability of white-rot fungi to tolerate polychlorinated biphenyls using predictive mycology. Mycology 9 (4), 239–249.

Sainz, M.J., Gonzales, P.B., Vilarino, A., 2006. Effects of hexachlorocyclohexane on rhizosphere fungal propagules and root colonization by arbuscular mycorrhizal fungi in *Plantago lanceolata*. Eur. J. Soil Sci. 57 (1), 83–90.

Saleh, S.X., Huang, D., Greenberg, B.M., Glick, B.R., 2004. Phytoremediation of persistent organic contaminants in the environment. In: Singh, A., Ward, O. (Eds.). Applied Bioremediation and Phytoremediation (Series: Soil Biology, 1. Springer-Verlag, Berlin, Heildeberg, pp. 115–134.

Salt, DE., Smith, RD., Raskin, I., 1998. Phytoremediation. Annu. Rev. Plant Biol. 49 (1), 643–668.

Sambadan, K., Kannan, K., Raman, N., 1992. Distribution of vesicular – arbiscular mycorrhizal fungi in heavy metal polluted soils of Tamil Nadu. J. Environ. Biol. India 13, 159–167.

Sandermann Jr., H., 1994. Higher plant metabolism of xenobiotics: the 'green liver' concept. Pharmacog. Genom. 4 (5), 225–241.
Santoyo, G., Moreno-Hagelsieb, G., Orozco-Mosqueda, M.C., Glick, B.R., 2016. Plant growth-promoting bacterial endophytes. Microb. Res. 183, 92–99.
Saravann, V.S., Madhaiyan, M., Thangaraju, M., 2007. Solubilization of zinc compounds by the diazotrophic plant growth promoting bactererium *Gluconacetpbacter diazoztophicus*. Chemosphere 66 (9), 1794–1798.
Schalk, I.J., Hannauer, M., Braud, A., 2011. New roles for bacterial siderophores in metal transport and tolerance. Environ. Microbiol. 13 (11), 2844–2854.
SchatVerklij, J.A.C., 1998. Biological interactions: the role for nonwoody plants in phytorestoration: possibilities to exploit adaptive heavy metal tolerance. In: Vangronsveld, J., Cunningham, S.D. (Eds.), Metal-Contaminated Soils: In Situ Inactivation and Phytorestoration. Springer-Verlag, Berlin, p. 51.
Schnabel, W.E., White, D.M., 2001. The effect of mycorrhizal fungi on the fate of aldrin: phytoremediation potential. Int. J. Phytoremediat. 3 (2), 221–241.
Scussler, A., Schwarzott, D., Walker, C., 2001. A new fungal phylum, the Glomeromycota: phylogeny and evolution. Mycol. Res. 105 (12), 1413.
Secher, C., Lollier, S., Jezequel, K., Cornu, J.Y., Amalric, L., Lebeau, T., 2013. Decontamination of polychlorinated biphenyls contaminated soil by phytoremediation-assisted bioaugmentation. Biodegradation 24, 549–562.
Segura, A., Rodriguez-Conde, S., Ramos, C., Ramos, J.L., 2009. Bacterial responses and interactions with plants during rhizomediation. Microb. Biotechnol. 2 (4), 452–464.
Selvi, A., Rajasekar, A., Theerthagiri, J., Ananthaselvam, A., Sathishkumar, K., Madhaven, J., Rahman, P.K.S.M., 2019. Integrated remediation processess toward heavy metal removal/recovery from various environments- a review. Front. Environ. Sci. 7, 66.
Sessitsch, A., Coenye, T., Sturz, A., Vandamme, P., Barka, E.A., Salles, J., Van Elsas, J., Faure, D., Reiter, B., Glick, B., 2005. *Burkholderia phytofirmas* sp. nov., a novel plant-associated bacterium with plant-beneficial properties. Int. J. Syst. Evol. Microbiol. 55, 1187–1192.
Shahandeh, H., Hossner, L.R., 2002. Role of soil properties in phytoaccumulation of uranium. Water Air Soil Pollut 141, 165–280.
Shang, T.Q., Doty, S.L., Wilson, A.M., Howald, W.N., Gordon, M.P., 2001. Trichlorethylene oxidative metabolism in plants: the trichlorethanol pathway. Phytochemistry 58 (7), 1055–1065.
Sharma, D., Sharma, B., Shukla, A.K., 2011. Biotechnological approach of microbial lipase: a review. Biotechnology 10 (1), 23–40.
Sharma, J.K., Gautam, R.K., Nanekar, S.V., Weber, R., Singh, B.K., Singh, S., Juwarkar, A.A., 2018. Advances and perspective in bioremediation of polychlororinated biphenyls contaminated soils. Environ. Sci. Pollut. Res. Int. 25 (17), 16355–16375.
Shim, H., Chauhan, S., Ryoo, D., Bowers, K., Thomas, S.M., Canada, K.A., Burken, J.G., Wood, T.K., 2000. Rhizosphere competitiveness of trichlorethylene-degrading poplar-colonizing recombinat bacteria. App. Environ. Microbiol. 66 (11), 4673–4678.
Shin, M., Shim, J., You, Y., Myung, H., Bang, K.S., Cho, M., Kamala-Kannan, S., Oh, B.T., 2012. Characterization of lead resistant endophytic Bacillus sp., MN3-4 and its potential for promoting lead accumulation in metal hyperaccumulator Alnus firma. J. Hazard. Mater. 199-200, 314–320.
Silver, S., Phung, L.T., 2009. Heavy metals, bacterial resistance. In: Scaechter, M. (Ed.), Encyclopedia of Microbiology. Elsevier, Oxford, pp. 220–227.
Simonoff, M., Sergeant, C., Poulain, S., Pravikoff, M.S., 2007. Microorganisms and migration of radionuclides in environment. C.R. Chimie 10 (10–11), 1092–1107.
Singer, A.C., Crowley, D.E., Thompson, IP., 2003a. Secondary plant metabolites in phytoremediation and biotransformation. Trends Biotechnol 21 (3), 123–130.
Singer, A.C., Smith, D., Jury, W.A., Hathuc, K., Crowley, D.E., 2003b. Impact of the plant rhizosphere and augmentation on remediation of polychlorinated biphenyls contaminated soil. Environ. Toxicol. Chem. 22 (9), 1998–2004.

Singh, C.J., 2002. Optimization of an extracellular protease of *Chrysosporium keratinophylum* and its potential in bioremediation of keratinic wastes. Mycopathologia 156 (3), 151–156.

Singh, T., Singh, D.K., 2017. Phytoremediation of organochlorine pesticides: review. Int. J. Phytoremediat. 19 (9), 834–843.

Smidt, H., Akkermans, A.D.L., van der OOst, J., de Vos, W.M., 2000. Halorespiring bacteria-molecular characterization and detection. Enzym. Microb. Technol. 27 (10), 812–820.

Smith, S.E., Read, D.J., 1997. Mycorrhizal Symbiosis, 2nd. Academic Press, London.

Snyder, R.A., Millward, J.D., Steffensen, W.S., 2000. Aquifer protest response and the potential for TCE bioremediation with *Burkholderia cepacia* G4 PRI. Microb. Ecol. 40 (3), 189–199.

Solano, B.R., Maicas, J.B., Monero, F.J.G., 2008. Physiological and molecular mechanisms of plant growth promoting rhizobacteria (PGPR). In: Ahmad, I., Pichel, J., Hayat, S. (Eds.), Plant-Bacteria Interactions. Wiley-VCH, Weinheim, pp. 41–52.

Sonoki, S., Fujihiro, S., Hisamatsu, S., 2007. Genetic engineering of plants for phytoremediation of polychlorinated biphenyls. In: Willey, N. (Ed.), Phytoremediation, Methods and Reviews. Humana Press, Totowa, New Jersey, pp. 3–13.

Spaepen, S., Vanderleyeden, J., Remans, R., 2007. Indole-3-acetic acid in microbial and microrganisms-plant signaling. FEMS Microbiol. Rev. 31 (4), 425–448.

Strycharz, S., Newman, L., 2009. Use of native plants for remediation of trichloroethylene: I. deciduous trees. Int. J. Phytoremediat. 11 (2), 150–170.

Strycharz, S., Newman, L., 2010. Use of native plants for remediation of trichloroethylene: II coniferous trees. Int. J. Phytoremediation 11 (2), 171–186.

Subba-Rao, R.V., Alexander, M., 1985. Bacterial and fungal cometabolism of 1,1-trichloro-2,2-bis(p-chlorophenyl) ethane (DDT) and its breakdown products. App. Environ. Microbiol. 49 (3), 509–516.

Sullivan, T.S., Gadd, G.M., 2019. Metal bioavailability and the soil microbiome. Adv. Agron. 155, 79–120.

Suttinun, O., Muller, R., Luepromchai, E., 2009. Trichlorethylene cometa-degradation by *Rhodococcus* sp. I4 induced with plant essential oil. Biodegradation 20 (2), 281–291.

Sweetman, A.J., Valle, M.D., Prevedouros, K., Jones, K.C., 2005. The role of soil organic carbon in the global cycling of persistent organic pollutants (POP)s: interpreting and modeling field data. Chemosphere 60 (7), 959–972.

Sylvestre, M., Macek, T., Mackova, M., 2009. Transgenic plants to improve rhiomediation of polychlorinated biphenyls (PCBs). Curr. Opin. Biotechnol. 20 (2), 242–247.

Sylvia, D.M., Fuhrmann, J.J., Hartel, P.G., Zuberer, D.A., 2005Principles and Applications of Soil Microbiology2nd edition. Prentice Hall, New Jersey, NyUpper Saddle River.

Taghavi, S., Barac, T., Greenberg, B., Borremans, B., Vangronsveld, J., van der Lelie, D., 2005. Horizontal gene transfer to endogenous endophytic bacteria from poplar improves phytoremediation of toluene. Appl. Environ. Microbiol. 71 (12), 8500–8505.

Tandlich, R., Brežna, B., Dercova, K., 2001. The effect of terpenes on the biodegradation of polychlororinated biphenyls by *Pseudomobas stutzeri*. Chemosphere 44 (7), 1547–1555.

Tangahu, B.V., Sheikh, Abdullah, S.R., Basri, H., Idris, M., Anuar, N., Mukhlisin, M., 2011. A review on heavy metals (As, Pb, Hg) uptake by plants through phytoremediation. Int. J. Chem. Eng. 2011, 939161.

Tarla, D.N., Erickson, L.E., Hettiarachchi, G.M., Amadi, S.I., Galkaduwa, M., Davis, L., Nurzhanova, A., Pidlisnyuk, V., 2020. Phytoremediation and bioremediation of pesticide-contaminated soil. Appl. Sci. 10 (4), 1217.

Tegli, S., Cerbonesch, M., Corsi, M., Bonnani, M., Bianchini, R., 2013. Water recycle as a must: decolorization of textile wastewaters by plant-associated fungi. J. Basic Microbiol. 54 (2), 120–132.

Teng, Y., Luo, Y., Sun, X., Tu, C., Xu, L., Liu, W., Li, Z., Christie, P., 2010. Influence of arbuscular mycorrhza and *Rhizobium* on phytoremediation by alfafa of an agricultural soil contaminated with weathered PCBs: a field study. Int. J. Phytoremediation. 12 (5), 516–533.

Teng, Y., Luo, Y.M., Gao, J., Li, Z., 2008. Combined remediation effects of arbuscular mycorrhizal fungi-legumes-rhizobium symbiosis on PCBs contaminated soils. Huan Jing Ke Xue 29, 2925–2930.

Teng, Y., Wang, X., Li, L., Li, Z., Luo, Y., 2015. Rhizobia and their bio-partners as novel drivers for functional remediation in contaminated soils. Front. Plant Sci. 6, 132.

Thiem, D., Zloch, M., Gadzala-Kopciuch, R., Szymanska, S., Baum, C., Hrynckiewicz, K., 2018. Cadmium-induced changes in the production of siderophores by a plant growth promoting strainof *Pseudomonas fulva*. J. Basic Microbiol. 58 (7), 623–632.

Thijs, S., Sillen, W., Rineau, F., Weyens, N., Vangronsveld, J., 2016. Towards an enhanced understanding of plant–microbiome interactions to improve phytoremediation: Engineering the Metaorganisms. Front. Microbiol. 7, 341.

Thiry, Y., Schmidt, P., Hees, M.V., Wannijn, J., Bree, P.V., Rufyikiri, G., Vandenhove, H., 2005. Uranium distribution and cycling of Scots pine (*Pinus sylvestris* L.) growing on a revegetated uranium-mining heap. J. Environ. Radioact. 81 (2–3), 201–219.

Thomas, D.R., Carswell, K.S., Ceorgiou, C., 1992. Mineralization of biphenyl and PCBs by the white-rot fungus *Phanerochaete chrysosporium*. Biotechnol. Bioeng. 40 (11), 1395–1402.

Thomas, R.A.P., Beswick, A.J., Basnakova, G., Moller, R., Macaskie, L.E., 2000. Growth of naturally occurring microbial isolates in metalcitrate medium and bioremediation of metal-citrate wastes. J. Chem. Technol. Biotechnol. 75, 187–195.

Toure, O., Chen, Y.Q., Dutta, S.K., 2003. *Sinorhizobium meliloti* electrotransporant containing *ortho*-dechlorination gene shows enhanced PCB dechlorination. Fresen. Environ. Bull. 12, 320–322.

Tripathi, M., Munor, H.P., Shouch, Y., Meyer, J.M., 2005. Isolation and functional characterization of siderophore producing lead and cadmium resistant *Pseudomonas putida* KNP9. Curr. Microbiol. 50 (5), 233–237.

Tsien, H.C., Brusseau, G.A., Hanson, R.S., Waclett, L.P., 1989. Biodegradation of trichloroethylene by *Methylosinus trichosporium* OB3b. Appl. Environ. Microbiol. 55 (12), 3155–3161.

Tu, C., Teng, Y., Luo, Y., Sun, X., Deng, S., Li, Z., Liu, W., Xu, Z., 2011. PCB removal, soil enzyme activities, and microbial community structures during the phytoremediation by alfafa in field soils. J. Soils Sediments 11 (4), 649–656.

Turnau, K., 1998. Heavy metal content and localization in mycorrhizal *Euphorbia cyparissias* from zinc wastes in southern Poland. Acta Soc. Bot. Poloniae 67 (1), 105.

Turnau, K., Jurkewicz, A., Lingua, G., Barea, J.M., Gianiazzi-Pearson, V., 2006. Role of Arbiscular mycorrhiza and associated microorganisms in phytoremediation of heavy metal-polluted sites. In: Prasad, M.N.V., Sajwan, K.S., Naidu, R. (Eds.), Trace Elements in the Environment (Biogeochemistry, Biotechnology, and Bioremediation). CRC Taylor and Francis, Boca Raton, pp. 235–252.

Turnau, K., Kottke, I., Oberwinkler, F., 1993. Element localization in mycorrhizal roots of *Pteridium aquilinum* L. Kuhn collected from experimental plots treated with cadmium dust. New Phytol 123, 313.

Turnau, K., Ryszka, P., Gianinazzi-Pearson, V., van Tuinen, D., 2001. Identification of arbiscular mycorrhizal fungi in soils and roots of plants colonizing zinc wastes insouthern Poland. Mycorrhiza 10, 169.

Twanabasu, B.R., Stevens, K.J., Venables, B.J., 2013. The effects of triclosan on spore germination and hyphal growth of the arbuscular mycorrhizal fungus *Glomus irregularis*. Sci. Total Environ. 455, 51–60.

Ullah, A., Heng, S., Munis, M.F.H., Fahad, S., Yang, X., 2015. Phytoremediation of heavy metals assisted by plant growth promoting (PGP) bacteria: a review. Environ. Exp. Bot. 117, 28–40.

UNEP, 2005. Riding the world of POPs: a guide on the Stockholms convention on persistent organic pollutants. https://www.pops.int/documents/guidance/beg_guide.pdf.

UNEP, 2013. The Hazardous Chemicals and Waste Convections; WHO: Rome, Italy; UNEP: Nairoby, Kenya;. FAO, Geneva, Switzerland, p. 4. https://www.pops.int/documents/background/hcwc.pdf.

Urík, M., Čerňanský, S., Ševe, J., ŠimonoviČova, A., Littera, P., 2007. Biovolatilization of arsenic by different fungal strains. Water Air Soil Pollut 186, 337–342.

USEPA (Air Toxics Web Site). Technical air pollution resources, 2009. https://www.epa.gov./ttn/atw/hlthef/tri-ethy.html.

Usuki, F., Narisawa, K., 2007. A mutualistic symbiosis between a dark septate endiphytic fungus *Heteroconium chaetospira* and nonmycorrhizal plant, Chinese cabbage. Mycologia 99 (2), 175–184.

Van Aken, B., Correa, P.A., Schnoor, J.L., 2010. Phytoremediation of polychlorinated biphenyls: new trends and promises. Environ. Sci. Technol. 44 (8), 2767–2776.

Van Aken, B., Doty, S.L., 2009. Transgenic plants and associated bacteria for phytoremediation of chlorinated compounds. Biotechnol. Genet. Eng. 26, 43–64.

Van der Heijden, M.G.A., Boller, T., Wiemken, A., Sanders, I.R., 1998. Different arbiscular mycorrhizal fungal species are potential determinants of plant community structure. Ecology 79 (6), 2082–2091.

Van Wyk, J.P.H., 1999. Saccharification of paper products by cellulose from *Penicillium funiculosum* and *Trichoderma reesei*. Biomass. Bioeng. 16, 239–242.

Vanderhove, H., Hees, M.V., Wannijn, J., Wouters, K., Wang, L., 2007. Can we predict uranium bioavailability based on soil parameters? Part.2. Soil solution uranium concentration is not a good bioavailability index. Environ. Pollut. 145 (2), 577–586.

Vanek, T., Soudek, P., Tykva, R., 2001. Study of radiophytoremediation. Minerv. Biotechnol. 13 (2), 117–121.

Vansuyt, G., Robin, A., Briat, J.F., Curie, C., Lemanceau, P., 2007. Iron acquisition from Fe-pyoverdine by *Arabidopsis thaliana*. Mol. Plant Microbe Interact. 20 (4), 441–447.

Verbon, E.H., Liberman, L.M., 2016. Beneficial microbes affect endogenous mechanisms controlling root development. Trends Plant Sci 21 (3), 218–229.

Vergani, L., Mapelli, F., Zanardini, E., Terzaghi, E., Di Guardo, A., Morosini, C., Raspa, G., Borin, S., 2017. Phyto-rhizomediation of polychlorinated biphenxl contaminated soils: an outlook on plant-microbe beneficial interaction. Sci. Total Environ. 575, 1395–1406.

Vessey, J.K., 2003. Plant growth promoting rhizobacteria as bifertilizers. Plant Soil 255, 571–586.

Villacieros, M., Whelan, C., Mackova, M., Molgaard, J., Sanchez-Cotreras, M., Lloret, J., de Carcer, D.A., Oruezabal, R.I., Bolanos, L., Macek, T., Karlson, U., Dowling, D.N., Martin, M., Rivilla, R., 2005. Polychlorinated biphenyl rhizomediation by *Pseudomonas fluorescens* F113 derivativies, using a *Sinorhizobium meliloti* nod system to drive *bph* gene expression. Appl. Environ. Microbiol. 71 (5), 2687–2694.

Vinichuk, M., Martensson, A., Ericsson, T., Rosen, K., 2013. Effect of arbiscular mycorrhizal (AM) fungi on 137Cs uptake by plants grown on different soils. J. Environ. Radioact. 115, 151–156.

Vivas, A., Azcon, R., Biró, Barca, J.M., Ruiz-Lozano, J.M., 2003b. Influence of bacterial strains isolated from lead-polluted soil and their interactions with arbuscular mycorrhizae on the growth of *Trifolium pretense* L. under lead toxicity. Can. J. Microbiol. 49, 577.

Vivas, A., Barea, JM., Azcon, R., 2004. Interactive effects of *Brevibacillus brevis* and *Glomus mosseae*, both from Cd contaminated soil, on plant growth, physiological mycorrhizal characteristic and soil enzymatic activities in Cd-spiked soil. Environ. Pollut. 134 (2), 275.

Vivas, A., Biró, Anton, A., Vörös, I., Barea, J.M., Azcón, R., 2003. Possibility of phytoremediation by co-inoculated Ni-tolerant mycorrhiza – bacterium strains. In: Simon, Szilagyi, L., György, M. (Eds.), Trace Elements in Food Chain. Bessenyei Publishers, Nyiregyhaza, Hungary, p. 76.

Vorkamp, K., Riget, F.F., 2014. A revie of new and currenz – use contaminants in the Arctic environment: evidence of long-range transport and indications of bioaccumulation. Chemosphere 111, 379–395.

Walter, M., 1992. Biodegradation of DDT and DDE. Lincoln University, New Zealand Msc Thesis.

Wang, W., Qui, Z., Tan, H., Cao, L., 2014. Siderophore production by actinobacteria. Biometals 27 (4), 623–631.

Wang, S., Zhang, S., Huang, H., Christie, P., 2011. Behavior of decabromodiphenyil ether (BDE-209) in soil: effects of rhizosphere and mycorrhizal colonization of ryegrass roots. Environ. Pollut. 159 (3), 749–753.

Wang, Y., Wang, C., Li, A., Gao, J., 2015. Biodegradation of pentachloronitrobenzene by *Arthrobacter nicotianae* DH19. Lett. Appl. Microbiol. 61 (4), 403–410.

Wani, P.A., Khan, M.S., Zaidi, A., 2008. Effect of metal-tolerant plant growth-promoting rhizobium on the perfomance of pea grown in metal-amended soil. Arch. Environ. Contam. Toxicol. 55 (1), 33–42.

Weber, R., Watson, A., Forter, M., Oliaei, F., 2011. Persistent organic pollutants and landfills- a review of past experiences and future challenges. Waste Manage. Res. 29 (1), 107–121.

Weiersbye, I.M., Stracker, C.J., Przybylowicz, W.J., 1999. Micro-PIXE mapping of elemental distribution in arbuscular mycorrhizal roots of the grass, *Cynodon dactylon*, from gold and uranium mine tailings. Nuc. Instrum. Meth. Phys. Res. B 158 (1–4), 335–343.

Weinberg, I., 2008. An NGO Guide to Persistent Organic Pollutants. International POPs Elimination Network. Stockholm, Sweden.
Weyens, N., Croes, S., Dupae, J., Newmann, L., van der Lelie, D., Carleer, R., Vangronsveld, J., 2010. Endophytic bacteria improve phytoremediation of Ni and TCE co-contamination. Environ. Pollut. 158 (7), 2422–2427.
White, C., Gadd, G.M., 1990. Biosorption of radionuclides by fungal biomass. J. Chem. Tech. Biotech. 49 (4), 331–343.
White, J.C., Parrish, Z., IsČeyen, M., Gent, M., Iannucci-Berger, W., Eitzer, B., Mattina, M., 2005. Uptake of weathered p,p'-DDE by plants species effective at accumulating soil elements. Microchem. J. 81 (1), 148–155.
Whitfield Aslund, M.L., Zeeb, B.A., Rutter, A., Reimer, K.J., 2007. In situ phytoextraction of polychlorinated biphenyl-(PCB) contaminated soil. Sci Total Environ 374 (1), 1–12.
Whiting, S.N., deSouza, M.P., Terry, N., 2001. Rhizosphere bacteria mobilize Zn for hyperaccumulation by *Thlaspi caerulescens*. Environ. Sci. Technol. 35 (15), 3144–3150.
Witkamp, M., 1968. Accumulation of ^{137}Cs by *Trichoderma viride* relative to ^{137}Cs in soil organic matter and soil solution. Soil Sci 106, 309–311.
Wright, S.F., Upadhyaya, A.A., 1998. A survey of soils for aggregate stability and glomain, a glycoprotein produced by hyphae of arbiscular mycorrhizal fungi. Plant Soil 198, 97.
Wu, C.H., Wood, T.K., Mulchandani, A., Chen, W., 2006. Engineering plant-microbe symbiosis for rhizomediation of heavy metals. Appl. Environ. Microbiol. 72 (2), 1129–1134.
Wu, N., Zhang, S., Huang, H., Shan, X., Christie, P., Wang, Y., 2008. DDT uptake by arbuscular mycorrhizal alfafa and depletion in soil as influenced by soil application of a non-ionic surfactant. Environ. Pollut. 151 (3), 569–579.
Xu, G., Li, Z.W., Wang, Y., Zhang, S., Yan, Y.C., 2008. Biodegradation of chlorpyrifos and 3,5,6-trichloro-2-pyridinol by a newly isolated *Paracoccus* sp. strain TRP. Int. Biodeter. Biodegr. 62, 51–56.
Yee, D.C., Maynard, J.A., Wood, T.K., 1998. Rhizomediation of trichlorethylene by recombinan, root-colonizing *Pseudomonas fluorescens* strain expressing toluene *ortho*-monooxygenase constitutively. Appl. Environ. Microbiol. 64 (1), 112–118.
Yoshihara, T., Matsumura, H., Hashida, S.N., Nagaoka, T., 2013. Radiocesium contamination of 20 wood species and the corresponding gamma-ray dose rates around the canopies 5 months after Fukushima nuclear power plant accident. J. Environ. Radioactiv. 115, 60–68.
Zaidi, S., Usmani, S., Singh, B.R., Musarrat, J., 2006. Significance of *Bacillus subtilis* strain SJ-101 as bioinoculant for concurrent plant growth promotion and nickel accumulation in *Brassica juncea*. Chemosphere 64 (6), 991–997.
Zeeb, B.A., Amphlett, J.S., Rutter, A., Reimer, K.J., 2006. Potential for phytoremediation of polychlororinated biphenyl-(PCB)-contaminated soil. Int. J Phytoremediat. 8 (3), 199–221.
Zhang, H.H., Tang, M., Chen, H., Zheng, C.L., Niu, Z.C., 2010. Effect of inoculation with AM fungi on lead uptake, translocation and stress alleviation of *Zea mays* L. seedlings planting in soil with increasing lead concentrations. Eur. J. Biol. 46 (5), 306–311.
Zhang, X.H., Zhu, Y.G., Lin, A.J., Chen, B.D., Smith, S.E., Smith, F.A., 2006. Arbuscular mycorrhizal fungi can alleviate the adverse effects of chlorothalonil on *Oryza sativa* L. Chemosphere 64 (10), 1627–1632.
Zhdanova, N.N., Lashko, T.N., Redchitz, T.I., Vasiliveskaya, A.I., Bosisyuk, L.G., Sinyavskaya, O.I., Gavrilyuk, V.I., Muzalev, P.N., 1991. Interaction of soil micromycetes with 'hot' particles in the model system. Microbiol. J 53 (4), 9–17.
Zhdanova, N.N., Tugay .m., T., Dighton, J., Zheltonozhsky, V., McDermott, P., 2004. Ionizing radiation attracts fungi. Mycol. Res. 108 (9), 1089–1096.
Zhdanova, N.N., Vasiliveskaya, A.I., Artyshkova, L.A., Sadnovikov, Y.u.S., Gavrilyik, VI., Dighton, J., 1995. Changes in the micromycete communities in soil in response to pollution by long-lived radionuclides emitted by in the Chernobyl accident. Mycol. Res. 98 (7), 789–795.
Zhdanova, N.N., Vasiliveskaya, A.I., Sadnovikov, Y.u.S., Artyshkova, L.A., 1990. The dynamics of micromycete complexes contaminated with soil radionuclides. Mikologia I Ditopatologiya 24 (6), 504–512.
Zhdanova, N.N., Zakharchenko, V.A., Haselwandter, K., 2005. Radionuclides and fungal communities. In: Dighton, J., White, J.F., Oudemnas, P. (Eds.), The Fungal Community: Its Organization and Role in the Ecosystem. CRC Press, Baton Rouge, pp. 759–768.

CHAPTER

13

Recent developments in phosphate-assisted phytoremediation of potentially toxic metal(loid)s-contaminated soils

Tariq Mehmood[a], Cheng Liu[a], Irshad Bibi[b], Mukkaram Ejaz[c], Anam Ashraf[d], Fasih U. Haider[e], Umair Riaz[f], Azhar Hussain[g], Sajid Husain[h], Mehak Shaz[i], Sumeera Asghar[j], M. Shahid[k], Nabeel Khan Niazi[b]

[a]*College of Environment, Hohai University Nanjing, China*

[b]*Institute of Soil and Environmental Sciences, University of Agriculture Faisalabad, Faisalabad, Pakistan*

[c]*School of Environmental and Municipal Engineering, Lanzhou Jiaotong University, Lanzhou, PR China*

[d]*School of Environment, Tsinghua University, Beijing, China*

[e]*College of Resources and Environmental Sciences, Gansu Agricultural University, Lanzhou, China*

[f]*Soil and Water Testing Laboratory for Research Bahawalpur, Pakistan*

[g]*Department of Soil Science, the Islamia University of Bahawalpur, Pakistan*

[h]*Hainan Key Laboratory for Sustainable Utilization of Tropical Bioresource, College of Tropical Crops, Hainan University, Haikou, Hainan, China*

[i]*Department of Environmental Sciences and Engineering, Government College University, Faisalabad, Pakistan*

[j]*Department College of Horticulture, China Agricultural University, Beijing, China*

[k]*Department of Environmental Sciences, COMSATS University Islamabad, Vehari, Pakistan*

13.1 Introduction

Potentially toxic metal(loid)s (PTMs; so-called heavy metals and metalloids), including arsenic (As), selenium (Se), cobalt (Co), copper (Cu), chromium (Cr), manganese (Mn), iron (Fe), zinc (Zn), mercury (Hg), cadmium (Cd), and lead (Pb) are considered to be a great threat to the environment, agriculture and human health (Asati et al., 2016; Liu et al., 2020). Generally, PTMs are classified into two groups that is, group I contain essential elements (Se, Co, Cu, Cr, Mn, Fe, and Zn), which are required as micronutrients by plants or humans. However, their high concentration becomes toxic for plants, animals, and humans (Kim et al., 2019). Group II of PTMs comprises non-essential elements (Hg, Cd, and Pb), and living organisms do not require these; therefore, their ingestion by humans/animals and

Assisted Phytoremediation. DOI: https://doi.org/10.1016/B978-0-12-822893-7.00014-8

uptake by crop plants can cause toxicity. The PTMs are ubiquitous in the Earth's environment; hence, suitable remediation strategies are required to reduce their environmental and human health risk. Particularly, food crops are grown in PTMs-contaminated soils irrigated with contaminated water or soils having deposition of sewage sludge have been identified unfit for human consumption (Raj and Maiti, 2020). Due to their severe health consequences, many PTMs, including Zn, As, Hg, Cu, Cd, and Cr in the soil-plant ecosystem, have been studied (Khan et al., 2020; Mehmood et al., 2017).

On the other hand, global food security issue is compelling the masses for extensive agriculture practices and increased livestock and poultry farming. Sewage sludge, wastewater irrigation and recycled products from organic waste of livestock and poultry (biosolids, biochar, compost, and manure) are being used in soil to enhance fertility to achieve better production (Antonious, 2018). In addition, soils are also used as dumping sites for extensive waste products originating from agriculture and industrial activities (Abdel-Shafy and Mansour, 2018). Dealing with land in such a way is widely considered a potential route of PTMs reaching the food chain, mainly via plant uptake and animal transfer. Different factors involved in the transportation of PTMs to food may include: (1) nature, source, and concentration of PTMs, (2) physicochemical properties of soil (soil texture, pH, cation exchange capacity, organic matter), (3) plant uptake and accumulation potential, (4) PTMs bioavailable fractions, and (5) metal absorption in dietary parts of grazing animals (Khan et al., 2020; Clemens and Ma, 2016).

The long persistence of these PTMs (particularly, PTMs with non-metabolic functions such as Pb, Cd, As and Hg) makes them potentially hazardous for living objects. The quality of edible crops depends on their PTMs contents and the PTMs concentration in food and diet of animal and human determine corresponding health effects. Specifically, cereals and legumes are more affected due to elevated PTMs accumulation. These nutritional enrich staple crops have high demand and consumption worldwide. The majority of these crops are being irrigated with contaminated water, particularly in developing countries where indusial waste water drains out and enter surface water without proper treatment. Therefore, in low-income countries, the food security attributed to PTMs contamination has worsened (Rizwan et al., 2016). Sharma et al. (2020) reported that elevated levels of PTMs in vegetables severely affected their consumption. They further stated that succulents and roots tubers are efficient hyperaccumulators of PTMs, so these vegetables contain above permissible levels of PTMs. Thus, consumption of these vegetables can be toxic to the consuming population. Subsequently, regular information circulation from health authorities has increased public awareness regarding the severe health effects of metal exposure, which has resulted in public avoidance of such PTMs enriched food items (Mehmood et al., 2020). For example, knowing about excessive concentration of Cd in food crops such as wheat, potato, and rice (Rai et al., 2019) and its bioaccumulation in the liver and kidney of animals, which makes them toxic for the human diet, has disrupted its overseas market (Kim et al., 2016). Hence, it becomes imperative to ensure that PTMs meet the regulatory requirements of foodstuffs. Food quality assurance has led to increased demand for research on PTMs in food, their sources, fate in the environment, and the development of their remediation strategies.

Unlike organic contaminants, PTMs cannot undergo microbial or chemical degradation because their soil persistence remains for several years after their introduction (Yazdankhah et al., 2018). Compared to all PTMs concentrations in soil, the bioavailable fraction of PTMs is more critical than potentially bioavailable or residual fractions (RESs) (Mehmood et al., 2017). The plant uptake and leaching of PTMs to groundwater can be effectively reduced using biological and chemical immobilization processes. Unlike other methods (e.g., soil excavation, photodegradation, oxidation), phytoremediation is a technology in which bioavailability of PTMs is increased, and the PTMs uptake by plants are facilitated (Niazi et al., 2017; Ali et al., 2013). Once plants uptake the PTMs and accumulate in aboveground biomass, then these plants are harvested, and their dried biomass is dumped in barren land. Phytoremediation has proven effective for PTMs removal from soil and water media. Moreover, efficiency could be enhanced by several organic and inorganic amendments (Mehmood et al., 2017; Niazi et al., 2017).

Many studies have suggested that phosphate application can modulate PTMs availability in soil (Niazi et al., 2017; Xu et al., 2020). The application of phosphate appeared to be a less disruptive and cost-effective approach than other amendments. Phosphate has diverse interactions with PTMs as it may mobilize and or immobilize them or may have no effect at all. The phosphate's behavior toward PTMs depends on several factors like the nature of metal, an initial concentration of phosphorus (P) in soil, and soil properties. For instance, P amendment significantly immobilized Pb and reduced its bioavailability in soil (Zeng et al., 2017). On contrary, Niazi et al. (2017) stated that the addition of P significantly enhanced As bioavailability and also helped plants to grow better under As toxicity. Regardless of several studies that have been conducted on the use of phosphate in the soil to remediate PTMs, there has been little discussion on the behavior of phosphate towards metal (im)mobilization in soil and its corresponding role in phytoremediation.

Research on the role of phosphate in strengthening phytoremediation has seen significant attention in the last decade. There is a substantial scientific interest in studying phosphate's impact on phytoremediation due to its complex interactions in soil, the importance for plant growth, and PTMs management in phytoremediation. Hence a systematic, detailed evaluation of various interactions, and working routes of phosphate in phytoremediation is needed for instructive evaluations of phosphate-assisted phytoremediation feasibility.

The current chapter provides a systematic, comprehensive, and advanced knowledge to the scientific community in the field of phosphate-assisted phytoremediation programs for PTMs-contaminated soils. The approaches in phosphate-assisted phytoremediation, characteristics of plants, phosphate-solubilizing microorganism (PSM), plant growth-promoting rhizobacteria (PGPR), which are beneficial for PTMs phytoremediation, are debated in detail in this chapter. The challenges and limitations in traditional PTMs phytoremediation and prospects of phosphate-assisted PTMs phytoremediation have also been discussed.

13.2 Phytoremediation

Phytoremediation is a cost-effective, environmentally-friendly, and sustainable strategy and relatively a better option than the other costly methods like soil removal and burial methods (Niazi et al., 2017; Pandey and Bajpai, 2019). In phytoremediation, specific plants are grown to remediate PTMs contaminated soils. These plants can effectively accumulate biotransform or volatilize PTMs from contaminated soils and water in rare cases.

The key features of plants that contribute to phytoremediation include:

1. PTMs tolerant
2. Rapid growth rate
3. Ability to accumulate a high concentration of PTMs in aboveground biomass
4. Prolific root system
5. High bioconcentration and translocation factors

13.2.1 Functions of plants in different phytoremediation approaches

Generally, most of the environmental PTMs remain inactive by natural attenuation or mineralized by various natural mechanisms. Plants also have the potential to reduce environmental PTMs as they could fix them in the rhizosphere and make them inactive and accumulate, mineralize and volatilize contaminants. The plant response towards PTMs depends on various factors such as morphological attributes of plant, nature of PTMs, media in which plant is grown, and plant-microbial interaction etc. (Mehmood et al., 2017; Rakshit et al., 2018) Fig. 13.1 showed important phytoremediation routes to treat PTMs contamination. The major types of phytoremediation that plant performs while treating environmental contaminants, particularly PTMs, are discussed here.

13.2.1.1 Phytoextraction

The other names of phytoextraction are phytoaccumulation, photoabsorption and phytosequestration. Major characteristics of both approaches are given in Table 13.1. In this method, plants accumulate PTMs in their tissues, like in hemofiltration plants withdraw PTMs from the wastewater system (Xiao et al., 2019). The essential PTMs are extracted or removed from contaminated soil, water, and sediments (Ali et al., 2013). This technique helps to treat various PTMs (Hg, Cd, Co, Cr, Cu, Ag, Pb, Mn, Mo, Ni, Pb, and Zn), metalloids (As and Se), radionuclides (Strontium-90, Caesium-137, uranium-234, and uranium-238), and non-PTMs that is, boron. These pollutants can be metabolized, mineralized, or volatilized by the plants, thus preventing the contaminant's high accumulation (Niazi et al., 2011,

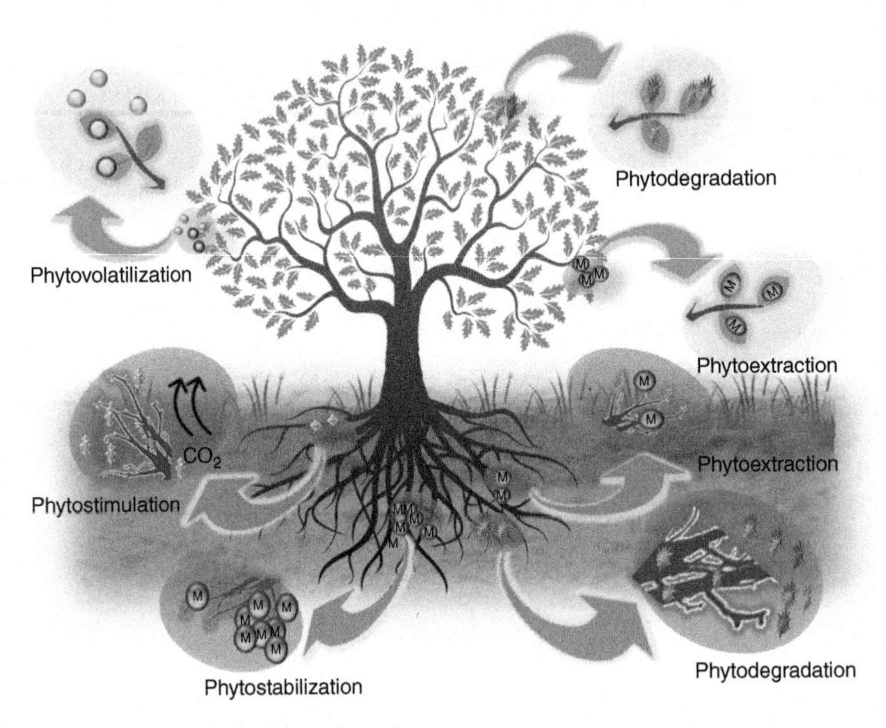

FIG. 13.1

A conceptual diagram showing various types of phytoremediation strategies.

Table 13.1 Important features of continuous vs induced phytoextraction.

Continuous phytoextraction	Assisted phytoextraction
Steady growing with low biomass yielding plants	Swift growing with high biomass plants
It extracts contaminants naturally	Chelators and organic acids enhance uptake
Systematic transports of PTMs from roots to shoots	Chemicals enhance the efficiency of translocation
High tolerance to PTMs concentration	Low tolerance
Environment friendly	Leaching down may enhance contamination
Plants play a role in naturally hyper accumulate PTMs	Plants play the role of metal excluders

PTM, potentially toxic metal(loid)s.

2012). Phytoextraction is further classified as continuous and chemically assisted or induced phytoextraction (Bhargava et al., 2012).

13.2.1.2 Phytodegradation or phytotransformation

This approach helps to biodegrade noxious ecological contaminants into simple harmless forms using plants and microbes. Plants have a natural capacity to uptake and degrade several pollutants. Various plant-microbial interactions arc also involved in the phytodegradation process (Camacho-González et al., 2019). Recently, entophytic bacteria have gained the attraction of scientists as these bacteria can increase plant growth and can help to produce high aboveground plant biomass (Pawlik et al., 2020). While in the phytotransformation process, plants transform the symphony of specific molecules by transforming toxic chemicals into non-toxic ones. The toxicity of the environment is either reduced or finished. Certain plant species such as Parrot's Feather, Hydrilla and Eurasian Watermilfoil are used for this sake (Ioannidis and Zouboulis, 2005; Singh et al., 2015). Periwinkle is another example of a plant that can convert toxic material into its non-toxic form (MORPHO, 2018). Studies have revealed that genetically modified organisms (transgenic organisms) are also beneficial for phytotransformation purposes.

13.2.1.3 Photovolatilization

This mechanism relates to the uptake of PTMs from growing media and volatilizes them into the atmosphere from areal parts via evapotranspiration (Wiszniewska et al., 2016). For the remediation of organic contaminants and the uptake of certain PTMs such as Hg, Se, and As, this technique could be used (Ali et al., 2013). Poplar trees showed excellent uptake of trichloroethylene (TCE) from the soil and can volatilize 90% of the total accumulated TEC (Legault et al., 2017). Guarino et al. (2020) reported excellent As phytovolatilization capacity of *Arundo donax* L. They stated *Arundo donax* L grown at 2.0, 10.0, 20.0 mg L^{-1} removed 75% As from system in via transpiration and only accumulate 0.15% As in tissue.

13.2.1.4 Phytostabilization

Phytostabilization or phytoimmobilization refers to holding or fixing PTMs in the soil by adsorption. It inhibits the bioavailability of PTMs and minimizes the risk of further contamination of the ecosystem by moving them through the soil to other components of the environment or by leaching them into the groundwater water table. Instead of plant tissues, phytostabilization primarily focuses on sequestering PTMs and other contaminants by precipitation within the rhizosphere (Radziemska et al., 2017).

13.2.1.5 Phytodesalination

The plants having the potential of extracting salt from soil or soil solution are called halophytes (salt-loving). Phytodesalination is a mechanism in which halophytes are used to eliminate extra salt from the soil which is mediated by root uptake. These plants can grow and survive at higher salt concentrations in soil. For instance, *Sesuvium portulacastrum* L. (an obligate halophyte) can accumulate 872 mg $plant^{-1}$ Na^+ ions (about 1 t ha^{-1}) in roots and aboveground biomass (Rabhi et al., 2010). Two American native vegetation that is, cattail (*Typha latifolia*) and alkaligrass (*Puccinellia nuttalliana*) showed substantial accumulation of salts compared to control. Particularly the accumulation of Cl^- was 120.14% and 94.47% time more in alkaligrass and cattail compared to corresponding control treatment, when treated with landfill leachate (Xu et al., 2019). Similarly in a hydroponic experiment *Ludwigia adscendens, Ipomoea aquatica and Alternanthera philoxeroides* grown at 7 dS m^{-1} showed promising phytodesalination capacity that is, 130, 105, and 80 kg Na^+ ha^{-1}, respectively. *A. philoxeroides* showed highest sodium accumulation (145.63 g kg^{-1}) in roots that other two plants (Islam et al., 2019).

13.2.2 Advantages of phytoremediation

Phytoremediation is given different names due to its link with plants that is, green remediation, botano-remediation, agroremediation, or vegetative remediation (Farraji et al., 2016). Phytoremediation is considered one of the best methods, which decrease environmental contamination through plant uptake along with their mediation techniques by microbiota, soil amendments, and agronomic practices (Niazi et al., 2011; Mehmood et al., 2017; Pandey and Souza-Alonso, 2019). The advantages and disadvantages of phytoremediation are presented in Table 13.2.

13.2.3 Limitation/challenges of phytoremediation

Unlike environmental toxins, PTMs do not experience chemical or microbial deterioration, and the overall accumulation of these PTMs in the soil continues even after their incorporation (Mehmood et al., 2021). However, by reducing the bioavailability of these PTMs through biological and chemical bonding, the mobilization of PTMs in the soil for the uptake by plant and leaching into the groundwater could be reduced (Bolan et al., 2014; Natasha et al., 2021).

Table 13.2 Pros and cons of phytoremediation technique. (McGrath and Zhao, 2003) (Farraji et al., 2016).

Advantages of phytoremediation	Disadvantages of phytoremediation
In situ and ex situ	It takes a long time to remediate an affected soil
Well suited for larger areas of soil	Very limited to shallow water
Cheap and cost-effective method	Slower than conventional methods
Easy to implement and publically accepted method	Not very effective for sites with higher contamination
Very low transmission of contaminants via air and water	Bioavailability and toxicity of the biodegraded product is unknown
Eco-friendly and pleasing for people	Contaminants can move to groundwater
Conserves natural resources	Influenced by soil type, soil conditions, and weather
Can be used for diverse organic and inorganic compounds	Plants accumulation so chances of pollution again

Table 13.3 Major challenges, limitations factor involved in phytoremediation of potentially toxic metal(loid)s.

PTMs stress	**Plants growth stopped under PTMs stress, decreasing the catabolism of PTMs (Harvey et al., 2002)**
Biotic and abiotic stress	The factors mentioned above can slow the growth process and may affect phytoremediation negatively (Soares et al., 2019). For instance, in Florida, NASA Kennedy Space Center failed to run the process of phytoremediation at a hydrocarbon burn-up facility due to several biotic and abiotic factors, i.e., weed competition and water stress situations. However, the results of various researches have shown that weeds can also be a source of phytoremediation. In contrast, in this NASA project, intensive weeds competition resulted in disturbing (Van Epps, 2006).
Contaminants' depth	Another factor that may affect phytoremediation is contaminants' depth; on average, plants can take up pollutants up to 50 cm depth (Pilon-Smits, 2005).
Time of Treatment	Dendroremediation showed great promise for deeper soil contamination and groundwater remediation (Doucette et al., 2003; Susarla et al., 2002). However, survival, establishment, and growth are significant problems in this technique. Trees take a long time to achieve enough biomass for useful phytoremediation. Besides, PTMs contained deep roots excavation is also a big challenge.
Spatial and temporal variability	Moreover, factors such as root architecture, root depth, soil morphology, nutrient composition pH, moisture availability, and microbiological actions commonly display considerable spatial and temporal differences (Nedunuri et al., 2000; Corwin et al., 2006) (Pilon-Smits, 2005).
Difficulties in field trials	Transgenic plants' implementation involves water storage to remove PTMs in conditions like a greenhouse and in vitro. However, their evaluation has become problematic in open fields due to inconsistency (Pilon-Smits and Freeman, 2006).

PTMs, potentially toxic metal(loid)s.

Plants that can accumulate higher PTMs concentration in their aboveground biomass are known as hyperaccumulators (Niazi et al., 2017). The effectiveness of phytoremediation depends on plant performance that is, how they uptake PTMs from the soil and store them in their tissues. Naturally, hyperaccumulators can accumulate substantial amounts of PTMs in aboveground biomass. Phytoremediation efficiency declined with time due to PTMs toxicity and biomass production reduction (Mehmood et al., 2017). Besides, PTMs accessibility, spatial-temporal variation of contaminants and plant attributes, competitive effects of weeds for nutrients, PTMs depths, and phytoremediation time are major challenges and limitations in the proper phytoremediation process (Table 13.3).

13.3 Phosphate-assisted phytoremediation

For successful assisted-phytoremediation, the first important factor is to increase PTMs concentration in soil solution. The next factor is the nature of PTMs (either easily water-soluble or not) available for plant uptake. The bioavailability of PTMs has been widely studied, and chelating agents have shown promising results in improving metal availability and enhance the nutrients status of soil (Adamo et al., 2018). Previous researches have evaluated soil-supplied chelating agent efficiency to increase PTMs and nutrients accessibility to plants. The inclusion of synthetic chelating agents like polyamino polycarboxylic acids has helped improve the deficiency of micronutrients in plants (McKnight and Rayborn, 2019). The affinity of chelating agents mainly depends on the type of soil and reactivity with concerned nutrient in soil. Most studies discussed use of chelating agents to recover iron deficiency and facilitating the PTMs uptake in plant (Lucena and Hernandez-Apaolaza, 2017).

Recently, inorganic phosphate compounds (apatite), lime, and organic amendments (biosolids) are being used to immobilize PTMs in soil (Lwin et al., 2018). Studies showed that phosphate fertilizer has promising remediation solutions for Pb-contaminated soils (Santos et al., 2020; Ren et al., 2020). However, phosphate fertilizers have also been recognized as the primary cause of metalloid soil contamination. Hence, using fertilizers, which are considered a sink for the immobilization of hazardous metalloids in the soils, has caused PTMs to contaminate the agricultural lands (Ma et al., 2020). However, the impact of phosphate in phytoremediation depends on the type of PTMs, phosphate dynamics in the soil, and microbial interactions (Seshadri et al., 2015).

As phytoremediation is concerned, phosphorus interacts differently with different PTMs and subsequently has a different impact on plant metal uptake efficiency (Seshadri et al., 2015). For phosphate deficiency management in soil, bulk quantities of fertilizers rich in phosphate are added to them. This induces abnormal deposition of phosphate in soils and is expected to conflict with the mobilization and absorption of PTMs in soils. Substantial studies have been conducted to delineate phosphate interactions with PTMs in various crops (Ren et al., 2020; Bolan et al., 2014; Niazi et al., 2017). Soils with elevated soil phosphate levels (naturally or man-made) must be observed for possible interaction with micronutrients. It has also been noted that the continued application of soluble phosphate-containing fertilizers to soils could lower Zn availability and induces Zn deficiency in plants (Alloway, 2009). Studies of nutrient accumulation in maize have shown that the patterns of phosphate and Zn uptake, translocation, and accumulation are very similar. The high phosphate level in the soil inhibits roots from shooting translocation of Zn and also could cause physiological inactivation of Zn inside the plants (Mousavi, 2011). More likely phosphate impacts on Zn deficiency in plants by following three ways: (1) rise in soil phosphate, dilute zinc concentration inside the plant, (2) The addition of phosphate fertilizer cause competition between Zn and other cations (e.g., Ca) in soil, and (3) phosphate anion-enhanced soil Zn adsorption. Nevertheless, the solubility of Zn could be improved by elevated fertilization rates of phosphate (Mousavi, 2011; Hafeez et al., 2013).

Arsenic and phosphate interaction depends on soil characteristics (Anawar et al., 2018). Both As and P belongs to group 5 elements and known chemically analogy to each other due to similar chemical speciation arsenate (AsV) and phosphate. In chemical reaction both can substitute each other. In chemical reaction (adsorption, dissolution reactions/precipitation both compete, however, metabolic substitution of As(v) to P cause serious toxicity and health issues in plant and or microorganisms (Strawn, 2018). It is assumed that the bioavailability of phosphate increase in soils having acidic pH is the consequence of increased translocation and absorption due to dihydrogen phosphate ion ($H_2PO_4^-$) (Malkanthi et al., 1995). Yang et al. (2017) stated phosphate rock is an ideal P source and it increased As removal efficiency of P. vittate. Phosphate addition increased boron (B) uptake in rapeseed; improved B uptake was further attributed to improved plant growth, water uptake efficiency, and calcium influence with B availability in the soil-plant system (Masood et al., 2019). Interaction between phosphorus and Cu was detected when high phosphate levels accentuated an acute Cu deficiency in citrus seedlings (Hardy, 2004). Contrary, Lin et al. (2018) found phosphate availability attributed to phosphate solubilizing bacterial (PSB) substantially upregulated translocation factors of Cu (i.e., leaf:root and stem:root) in *Wedelia trilobata,* which significantly increase Cu accumulation in *W. trilobata.*

Nevertheless, the solubility of Zn and Cu could be improved by elevated fertilization rates of phosphate. It is assumed that this contact takes place at the absorption site, probably with Cu's precipitation at the root surface. Phosphate application decreased the impact of the toxic Cu levels in some experiments. On the other hand, excess Cu could inhibit the absorption of phosphate and Fe. Phosphorus/Fe interaction existed in bush beans grown at either a surplus or insufficient soil phosphate level. In any case, there was a reduction in the absorption of Fe. Both maize and rice, grown on excess Cu soils, show extreme Fe chlorosis. Under these conditions, high fertilization with phosphate is always recommended. Interactions with phosphorus/Mn may evolve when soil Mn availability increases with higher soil phosphate levels. This could be due to phosphate impact, which increases soil acidity in some soil when it applied at higher concentration (Bolan et al., 2003). Pestana et al. (2018) stated that a moderate level of P application increases the Cd accumulation performance of *Egeria densa.* Wang et al. (2017) reported that P application switched RES of Cd into carbonate-binding fraction (CA), resulting in higher accumulation in *C. comosum*. A dose of 200 mg kg^{-1} of P reduced 10.15 mg kg^{-1} Cd in planted soil, which was significantly higher than control. Phosphate and molybdenum (Mo) relationship is subject to either the soil is by nature, acidic, or alkaline. Phosphate improves the Mo uptake for acidic soils, thus reducing the Mo uptake in alkaline soils (Bolan et al., 2003; Opala, 2017). Yu et al. (2020) reported application of sodium dihydrogen phosphate (SDP) and single superphosphate (SSP) reduce metal stress in *Polygonum pubescens* blume. Both SDP and SSP reduced >50.0% Mn concentration in leaves increased 71.8% and 135% biomass.

Besides it, phosphate fertilizers also promote plant growth and produce positive impacts on the soil microbial community. Phosphorus is categorized as one of the essential nutrients for the plant as required in an adequate amount by the plant for optimum growth and yield (Ashley et al., 2011). Phosphorus is necessary for various physiological functions in plants. Phosphorus plays an essential role in plant metabolism; it plays a significant role in cell division and the energy cycle. It is an integral part of cell components and helps plants in respiration and photosynthesis also supports roots development and elongation. Therefore, depending on the above-mentioned condition, phosphate fertilizers can assist the phytoremediation by immobilization and mobilization of PTMs and promoting plant growth.

To fully understand these factors, we provide a detailed discussion of phosphate characteristics and dynamics in soil and their role in phytoremediation.

13.4 Phosphorus dynamics in soil as an assisted phytoremediation agent

Phosphorus (P) is the 11th most abundant element in the Earth's crust. Phosphorous is highly reactive and in combination with other elements, it forms phosphate. Plants can only uptake phosphorus in the form of orthophosphate. However, organic and inorganic constituents in soil bind it and make it unavailable for plant uptake (Behera et al., 2014). Specific microorganisms can solubilize phosphate and make it more bioavailable for plant (Chibuike and Obiora, 2014; Girma, 2015). For instance, the use of PSM plays a significant role in solubilizing P containing insoluble inorganic and organic compounds (Yadav and Verma, 2012; Spohn et al., 2015). The inorganic phosphorus availability regulated by PSM is the second significant nutrient that could limit the development of plants after nitrogen (Kumar et al., 2008).

Microbes convert inorganic P forms into the organic P through immobilization and are then absorbed into the microbe's living cells. The factor influencing the P immobilization and mineralization is the same as the immobilization and mineralization of nitrogen, including moisture, aeration, and temperature (Silva et al., 2000). A typical soil has around 50% of the overall P in organic form. Therefore, soil organic P represents very significant components of the P cycle. The major organic P sources include phospholipids, nucleic acids, and phytin. Likewise, nitrogen (N), organic P is transformed into the inorganic P through mineralization (Jiang et al., 2021).

Therefore, understanding the microbial functions in phosphate-assisted phytoremediation is necessary. The following section gives a detailed description of the microbial role in the phosphate-assisted phytoremediation process.

13.5 Role of the microbial community in phosphate-assisted phytoremediation

This topic explains the role of microbes in P cycle to emphasize plant-PSM interaction processes and solubilization of phosphate (P solubilization) mechanism. Further investigation was by describing the efficacy of PSM in assisting the phytoremediation cycle.

13.5.1 Microbial-mediated phosphate solubilization

The first article on the solubilization of insoluble inorganic phosphate by microbial strains was published in 1948. The study had contributed to an improved understanding of phosphate solubilization (Rashid et al., 2012). Microorganisms that solubilize phosphates are ubiquitous in the environment (Prabhu et al., 2019).

Several fungal, actinobacteria, and bacterial strains have been investigated and identified for the phosphate-solubilizing potentials. The vast numbers of microbes that have been reported for having phosphate-solubilizers potentials belong to the Bacillus genus, proceeded by Pseudomonas (Awais et al., 2017). The genera Penicillium and Aspergillus are the most widespread among phosphate-solubilizing fungi, followed by Rhizoctonia and Trichoderma (Yaghoubian et al., 2019). In the past, scientists also reported novel genera and species of microorganisms solubilizing phosphate (Madhaiyan et al., 2015). Under the effect of abiotic stress that is, salinity, drought, and temperature, bacteria were documented to solubilize the soil (Vassilev et al., 2012). Cold-resistant *Mycobacterium spp., Mycoplasma spp., Acinetobacter spp., Pantoea spp.*, and *Pseudomonas spp.* have phosphate-solubilizing capabilities at lower temperature ranges between 4 °C – 16 °C (Prabhu et al., 2019; Pandey et al., 2006). The *Azospirillum spp.* and *Pseudomonas spp.* have the potential to solubilize the phosphate under extreme drought conditions (Arzanesh et al., 2011). Similarly, *Arthrobacter spp., Aerococcus spp., Bacillus spp., Pseudomonas spp., and Pantoea spp.* have significant phosphates solubilizing potential under salinity in the soil (Fig. 13.2) (Srinivasan et al., 2012).

13.5.2 Microbe-mediated phosphate solubilization mechanism

A broad series of mechanisms are involved in phosphate solubilization by microbes. Table 13.2 represents the various phosphate-solubilization mechanisms of microbes that is, siderophore production,

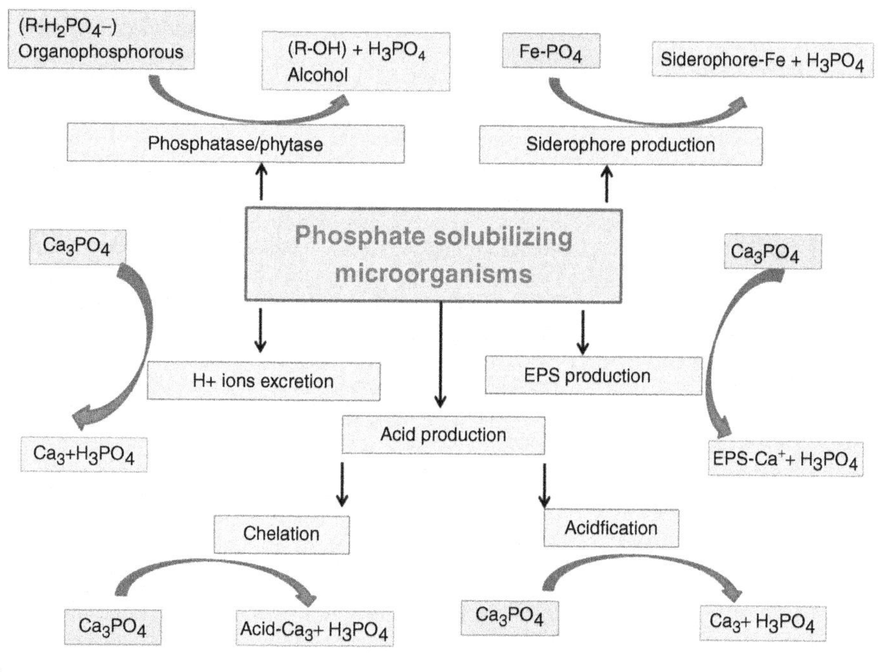

FIG. 13.2

Brief mechanism of microbial phosphate solubilization. Organophosphorous, iron-phosphate, and calcium phosphate are the parent materials that are solubilized into phosphate compounds (R-OH) + H_3PO_4-alcohol, siderophore-Fe + H_3PO_4, Ca_3 + H_3PO_4, EPS-Ca + H_3PO_4, and acid-Ca_3+ H_3PO_4 by the activity of microbial flora in the rhizosphere.

organic acid production, H^+ ions excretion, and exopolymeric compounds production, leading to the phytoremediation of PTMs in contaminated soils. Table 13.4 cover major functions and mechanisms contributing to direct and indirect microbial assisted phosphate solubilization. Likewise, a detailed description of direct and indirect mechanisms for the solubilization of phosphate compounds by microbes are mentioned below.

13.5.2.1 Direct effects

The production of organic acid is claimed to be the principal method of inorganic-phosphate solubilization (Prabhu et al., 2018). The substrate's acidification allows organic acids' capacity to solubilize the phosphate rock (Awais et al., 2017). Gulati et al. (2010) recorded a decrease in the soil pH due to the formation of organic acid during the growth of phosphate-solubilizing microbes in the aqueous solution. Phosphate-solubilizing microbes are observed to be generating the different acids which facilitate P solubilization process. In addition, the chelating compounds formation by microorganism are very effective in solubilization of complexed inorganic phosphate compounds. Besides, microbial biodegradation activities, production of inorganic acids, by sulfur-oxidizing and nitrifying bacteria and production of H^+ ions during assimilation of NH_4^+ also accelerate over all P solubility and mentalization (Table 13.4).

Table 13.4 Microbial direct and indirect mechanisms for the solubilization of phosphate.

Type of effect	Function of microorganism	Mechanism
Direct effect	Lower the pH	Phosphate-solubilizing microbes form organic acid, which reduces the pH and increases the solubilization of inorganic-phosphate (Gulati et al., 2010).
	Acidification	Several microorganisms, including phosphate-solubilizing microbes and acid-producing bacteria, produce mono, di and tricarboxylic and hydroxyl acids, i.e., acetic, citric, formic, glycolic, gluconic, lactic, 2-keto gluconic, oxalic, malic, and succinic acids in the aqueous medium (Rashid et al., 2012).
	Cation-chelation	Acid produces by microorganism, i.e., fulvic acid, humic acid, 2-keto-gluconic acid, are classified as active cation chelators such as Al, Fe, and Ca, facilitating solubilization of inorganic phosphate compounds (Singer et al., 2004).
	Microbial degradation activities	Microbial degradation of complex molecules produces fulvic acid and humic acids that enhance phosphates solubilization (Singer et al., 2004).
	Production of inorganic acids	Sulfur-oxidizing and nitrifying bacteria form inorganic acids which convert insoluble phosphates into soluble forms (Singer et al., 2004).
	H^+ ions production	Microorganisms excrete H^+ ions during assimilation of $NH_4^{+,}$ formation of carbonic acid (H_2CO_3) in the respiratory system, and the anions extrusion of organic acids that promote phosphate solubilization (Aken and Doty, 2009).
Indirect effect	Bind PTMs	Microorganisms under stress released exo-polysaccharides which has high affinity to bind PTMs, indirectly influence rhizosphere's metal-phosphates solubility (Ionescu and Belkin, 2009; Ochoa-Loza et al., 2001; Prabhu et al., 2018).
	Iron chelation	Microorganisms produced siderophores that cause iron chelation and significantly affect iron phosphate solubility in the soil (Gaonkar and Bhosle, 2013).
	Organic-phosphates solubilization	Several microbial enzymes, i.e., C-P lyase, phytase, phosphatase, phosphonatase, and phosphohydrolase, perform organic-phosphates solubilization in soil (Yan et al., 2020; Lwin et al., 2018).
	Hydrolyzation	The phytase enzyme activity is hydrolyzed myo-inositol-phosphate and phytic acid compounds. Phosphatase and C-P lyase hydrolyzes phosphonates ester bonds (phosphonoacetate, pyruvate, phosphoenol) and transforms phosphonates for assimilation into phosphate ions and hydrocarbon (Yahaghi et al., 2018).

PTMs, potentially toxic metal(loid)s.

13.5.2.2 Indirect effects

Exo-polysaccharides are extracellular enzymes released by the microbial population indirectly impact the phosphate availability in soil (Ionescu and Belkin, 2009). In addition, to the organic acid anions, microbial exo-polysaccharides have demonstrated tri-calcium-phosphate solubilization (Yi et al., 2008). A strong association was found between the rate of phosphate solubilization and the exo-polysaccharides concentration (Teixeira et al., 2014). Yet, the role of exo-polysaccharides of higher molecular weight in the phosphate solubilization from the rhizosphere still requires further investigations.

Further, many phosphate-solubilizing microbes showed siderophore secretion in the rhizosphere (Astriani et al., 2020). Yet, a direct association between the synthesis of siderophores and phosphate solubilization is still not well established (Zeng et al., 2020). Besides, microbial enzymes activities also contributes in solubilization of organic-phosphates and break down of ester bonds during hydrolysis of complex myo-inositol-phosphate and phytic acid compounds and transforms phosphonates for assimilation into phosphate ions and hydrocarbon (Yahaghi et al., 2018) (Table 13.4).

13.6 Phosphate effects on plant growth and potentially toxic metal(loid)s detoxification

Phytoremediation has demonstrated promising success in PTMs contaminated soil remediation. This method still includes some drawbacks, especially in the remediation of areas with higher PTMs concentration (Ahemad, 2015). In previous studies, the promotion of plant growth (Oves et al., 2013; He et al., 2010) and metal detoxification capacity of PSM strains have been identified (Misra et al., 2012). Table 13.5 and Fig. 13.3 represent the various microbial phosphate-solubilization mechanisms (i.e., siderophore production, organic acid production, H^+ ions excretion, and exopolymeric compounds production) that leads PTMs phytoremediation in soils.

PSM reduced the metal in soil via either the phytostabilization or phytoextraction process. The synthesis of organic acid, emission of siderophore, creation of indole-3-acetic acid (IAA), and 1-aminocyclopropane-1-carboxylate (ACC)-deaminase stimulation improved the phytoremediation potential of the plant. The stress-resistant phosphate-solubilizing microbes are considered potential bio-inoculants for the agricultural lands affected by stress (Prabhu et al., 2019). These studies show that microbes help plants mitigate abiotic stresses adverse effects (Madhaiyan et al., 2015). It has been stated that bacteria solubilize phosphate under the influence of abiotic stresses like drought, high, low, or moderate pH, salinity, and temperature (Vassilev et al., 2012).

Table 13.6 showed some PGPR, which support plants in stressed environment. Besides, stress-tolerant microbes modify the phytohormone ethylene by modulating the plant growth under stressful environments (Awais et al., 2017). Ethylene is a phytohormone under threat and affects seed germination, abscission of the leaves, senescence of the leaves, early maturation of the fruits, blooming and wilting of the flowers, initiation of base, elongation, and branching (Hussain et al., 2020). The precursor for ethylene is ACC. Microbes promoting plant growth under stressed conditions produce ACC deaminase enzyme that cleaves ACC. This results in a drop in ethylene and secondary plant growth promotion (Hussain et al., 2020). Table 13.2 showed some PGPR.

The role of bacteria that solubilizes phosphate in promoting plant growth and the detoxification of PTMs is significant. Similarly, the use of mineralization and solubilization properties of PSB is an eco-friendly and economical method that provides a substantial amount of solubilized P to plants and improves the plant's growth (Selvakumar et al., 2008).

In addition, PSB prevents the plant from phytopathogens by developing antifungal metabolites, phenazines, hydrogen cyanide (HCN), and antibiotics (Zaidi et al., 2009). Moreover, it also stimulates plant growth by N_2 fixation, siderophore development, phytohormone secretion, and reduction of ethylene concentration (Placek et al., 2016). Phosphate-solubilizing microbes that is, *Klebsiella spp.*,

Table 13.5 Microbial phosphate-solubilization mechanisms in different PTMs phosphate assisted phytoremediation process.

Microbes	PTMs	Phosphate solubilization mechanism	Host plant	Reference
Penicillium aculeatum PDR-4, *Trichoderma* spp., *Trichoderma* sp. PDR-16	As	Siderophore, phosphatase	Sunflower, sudan grass	(Babu et al., 2014; Govarthanan et al., 2018)
Rahnella spp., *Endophytic* bacteria spp., *Enterobacte*r spp., *Klebsiella* spp.	Cd	Siderophores, organic acids	Amaranthus spp., sunflower, mustrard	(Marques et al., 2013; Afzal et al., 2014; Jing et al., 2014; Yuan et al., 2017; Lin et al., 2018; Mahajan and Kaushal, 2018)
Enterobacter ludwigii L.,	Co	Organic acids,	Sunflower	(Arunakumara et al., 2014)
Pseudomonas aeruginosa L., *Rhizobacteria* spp., *Endophytic bacteria* spp.,	Cr	organic acid, EPS	Chick pea, prosopis	(Oves et al., 2013; Khan et al., 2015)
Pseudomonas spp. *Rhizobacteria* spp., *Paenibacillus polymyxa* L.,	Cu	Organic acid-production, siderophore -production	Maize, sunflower, trailing daisy	(He and Yang, 2007; Wenzel, 2009; Li and Ramakrishna, 2011; Lin et al., 2018)
Psychrobacter spp., *Pseudomonas* spp.,	Fe	Organic acids, Siderophore	Castor bean	(Ma et al., 2015)
Rhodococcus globerulus X80619, *Rhodococcus erythropolis* X79289, *Psychrobacter* spp., *Pseudomonas* spp.,	Ni	Organic acid-production, siderophore -production	Alyssum spp., castor bean	(Becerra-Castro et al., 2011; Ma et al., 2015; Hassan et al., 2019)
Penicillium aculeatum PDR-4, *Trichoderma* spp., *Trichoderma* sp. PDR-16, *Enterobacter* spp., *Klebsiella* spp., *Enterobacter ludwigii* L.,	Pb	Siderophore, phosphatase, organic acid	Sunflower, mustard	(Arunakumara et al., 2014; Babu et al., 2014; Jing et al., 2014; Yuan et al., 2017; Govarthanan et al., 2018; Zulfiqar et al., 2019)
Rahnella spp., *Rhizobacteria* spp., *Psychrobacter* spp., *Pseudomonas* spp., *Enterobacter* spp., *Klebsiella* spp., *Enterobacter ludwigii* L.,	Zn	Organic acid-production, siderophore -production	Mustard, sunflower, and castor bean	(Ma et al., 2011; He et al., 2013; Marques et al., 2013; Arunakumara et al., 2014; Jing et al., 2014; Ma et al., 2015)

PTMs, potentially toxic metal(loid)s.

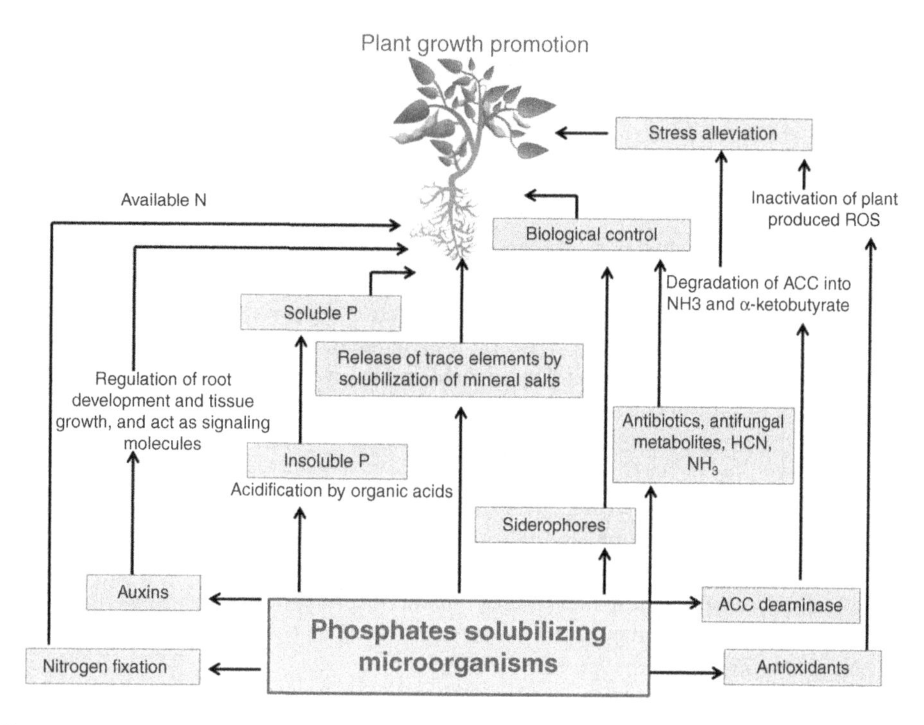

FIG. 13.3

The phosphate-solubilizing microbe's mechanisms facilitated the development of plant productivity and stimulated the PTMs contamination in the rhizosphere of the contaminated soils ACC, 1-aminocyclopropane-1-carboxylate; ROS, reactive oxygen species.

Table 13.6 Substances released by phosphate solubilizing bacteria for plant growth promotion.

PGPR	Plant growth promotion traits	Reference
Pseudimonas sp.	IAA, HCN	(Yogendra et al., 2013)
Bacillus Thuringiensis	IAA	(Majeed et al., 2015)
Acinetobactor haemolyticus RP19	IAA	(Misra et al., 2012)
Bacillus sp.	IAA, HCN	(Karuppiah, 2011)
Enterobacter asburiae	IAA, HCN, ammonia, siderophores	(Ahemad and Khan, 2010; Ahemad and Khan, 2010)
Burkholderia	IAA, siderophores, ACC deaminase	(Jiang et al., 2008)
Klebsiella sp.	Ammonia, HCN, IAA, siderophores	(Ahemad and Khan, 2010; Ahemad and Khan, 2011)
Pseudomonas putida	IAA, HCN, ammonia, siderophores	(Ahemad and Khan, 2011, 2010)
Enterobacter sp.	IAA, siderophores, ACC deaminase	(Kumar et al., 2008)

ACC, 1-aminocyclopropane-1-carbocyloate; HCN, hydrogen cynate; IAA indole-3-acetic acid; PGPR, plant growth-promoting rhizobacteria.

Enterobacter spp., and *Pseudomonas spp.* are useful in the bioremediation of PTMs and to influence phytoremediation through phytostabilization and phytoextraction in PTMs polluted soils (DalCorso et al., 2019).

13.7 Advantages and limitation of phosphate-assisted phytoremediation

Phosphate-assisted phytoremediation is generally based on plant-microbes association, which involves utilizing plants and microbes extract to remediate PTMs pollutants and decrease their soil bioavailability (Yan et al., 2020). Phosphate-assisted phytoremediation supports the plants to accumulate ionic compounds in the soil through their root network, even at small concentrations (Prabhu et al., 2019). Plants expand their root system into the soil matrix and create a rhizosphere ecosystem to absorb PTMs pollutants and modulate their bioavailability (Jacob et al., 2018), thereby enhancing soil fertility stabilization reclaiming the contaminated soil (DalCorso et al., 2019). Phosphate-assisted phytoremediation has several advantages, including (1) it is an economically feasible, solar-powered autotrophic system (Niazi et al., 2017), therefore easy to manage, low installation and maintenance costs, (2) eco-friendly and environmentally sustainable, it may limit the exposure of PTMs pollutants to the ecosystem (Farraji et al., 2016), (3) easily applicable over a wide area and can quickly be disposed of (Shah and Daverey, 2020), (4) avoid erosion and leaching of PTMs by stabilizing pollutants (Wuana and Okieimen, 2011), decreasing the hazard of contaminant distribution, and (5) enhance the soil fertility by the release of various organic matter into the soil (Yan et al., 2020).

Although having significant advantages of phosphate-assisted-phytoremediation, still there are certain limitations, which includes (1) phytoremediation is just limited to the depth and area which is occupied by roots (only rhizosphere of the soil) (Farraji et al., 2016), (2) this process is time-consuming and higher concentration of PTMs toxicity minimizes the remediation efficacy of plants (Rane et al., 2015), (3) having phosphate-plant based remediation systems, it is impossible to prevent the contaminates leaching into the subsurface water (without the absolute remediation of pollutants in the soil, still the problem of PTMs prevails in the contaminated soils) (Ahemad, 2015), (4) the toxicity of polluted land and the general soil conditions adversely affect the survival of the plants in contaminated soils (Palansooriya et al., 2020), (5) bioaccumulation of pollutants, especially PTMs in the plants results in consumer products i.e., cosmetics and foods that requires a proper disposal of the affected plant material (Lwin et al., 2018), and (6) as PTMs are absorbed, the metal is often attached to the soil organic matter that allows the plants inaccessible to the extracts (Grobelak and Napora, 2015). The advantages and drawbacks of this remediation technique rely mainly on the plant vulnerability as the key leading species into the treatment processes. In contrast, many of the drawbacks are related to the improper implementation of this sustainable environmental form of remediation. For instance, P is a non-renewable resource needed for crop production. Excessive application of P as amendment may discharge from agriculture lands and enter into freshwater. Soluble reactive phosphorus encourages the growth of algal and toxic cyanobacteria resulting in eutrophication and anoxic conditions, both of which are detrimental to human and animal health (Cao et al., 2016). While upland practices, ecological engineering showed promising recovering and recycling of P, ensuring food security while also reducing eutrophication (Roy, 2017).

In short, phytoremediation is a specific technique with both advantages and disadvantages; hence, this method's efficacy depends on the suitable application, proper selection, and modifications.

13.8 Conclusions and future outlooks

Many previous studies have shown conclusive evidence for the potential value of phytoremediation to treat environmental contaminants. Phytoremediation technology or technique is in its rising phase, and various amendments, e.g., phosphate, showed promising effects in removing PTMs from the soil. The particular future outlooks are as follows:

- Phosphate amendments have been proven effective in immobilizing PTMs in soils and plant growth, thereby enhancing phytoremediation efficiency in various polluted sites.
- Phytoremediation has proved not only the most economical and cost-effective method but also environment-friendly. Employing low-cost and budget-friendly plants can also improve air quality by removing carbon dioxide and particulate matter pollution.
- The limitation in phosphate-assisted PTMs phytoremediation may be addressed by using different amendments and best-suited plants.
- Phosphate amendment requires special assessment due to (1) its competition with other nutrients in the soil, (2) it can cause deficiency of other nutrients in plants, (3) it has dual effects on PTMs mobility (i.e., it can either mobilize or immobilize the PTMs in soil), (4) many P fertilizers contain other PTMs, and (5) P enhances PTMs ' mobilization and makes them more bioavailable to plant, PTMs mobilization increases their transport in soils and subsequent groundwater contamination.

  A substantial amount of phosphate remains inactive in soil and is not available for plants uptake or react with PTMs. Consequently, the role of phosphate-solubilizing microorganisms is essential in phosphate-assisted phytoremediation.
- The use of PSB in phosphate-assisted phytoremediation in PTMs contaminated soil has been found promising for plant growth under stress conditions (Ma et al., 2011). However, their level of effect on different plants differs, which depends on the bacterial strain, soil types, ecological condition, and plant species (Ahemad, 2015). The outcomes of different PTMs removal from soil using PSB s can be concluded as follows:
  1. Laboratories or greenhouse studies showed that PSB inoculation with plants from the *Brassicaseae spp.* can significantly accrue a higher concentration of PTMs in plant parts, especially in leaves.
  2. *Pseudomonas aeruginosa* L. was also studied for phytoremediation of PTMs. However, being a devious human pathogen could pose a risk to soil ecosystems, thus needing to be addressed (Walker and Bernal, 2004). Therefore, moral and legal concerns are important if they are to be used in fields as inoculants.
  3. Most PSB studies cover only a few PTMs that is, Cr, Cu, Ni, and Zn, while other toxicologically significant PTMs such as Pb, Cd, Hg, and As are the least known. Assessment of PSB potential in soil contaminated with these PTMs could uncover novel obstacles, observations, and problems and contribute to paving ways for more work in this direction.
  4. Such experiments included all methods for PTMs remediation, phytostabilization, and phytoextraction. The plants used in phytoremediation in severe PTMs contaminated soils could face severe physiological injury and become vulnerable to diseases and pests. Therefore, it is a prerequisite that plants used in phytoremediation, particularly in phytoextraction and phytostabilization, should have tolerance against phytopathogens and

diseases for effective performance. In addition, further development and application of PSB strains with additional characteristics that impart plant with resistance against different diseases would be a healthier alternative to metal phytoextraction.

Further studies are warranted to propose a multifunction approach of using P-assisted phytoremediation without disturbing other environmental stakeholders.

Acknowledgements

Dr Nabeel Khan Niazi and Dr Irshad Bibi are thankful to Higher Education Commission of Pakistan for financial support. Thanks are also extended to University of Agriculture Faisalabad, Pakistan.

Abbreviations

Ni	Nickel
Zn	Zinc
EDTA	Ethylenediamine tetraacetic acid
Cd	Cadmium
NTA	Nitrilotriacetic acid
Pb	Lead
DTPA	Diethylenetrinitrilo pentaacetic acid
Mg	Milligram
Kg	Kilogram
G	Gram
US	United States
As	Arsenic
Se	Selenium
Sr	Strontium
Cs	Caesium
U	Uranium
TNT	Trinitrotoluene
TCE	Tricholoroethylene
EPA	Environmental Protection Agency
GGTs	g-glutamyltranspPhosphate-solubilizing microorganism; Soil contamination; Toxicity; Sorption and desorptioneptidases
ATP	Adenosine triphosphate
CO_2	Carbon Dioxide
NASA	National Aeronautics and Space Administration
Cm	Centimeter

P	Phosphorus Phosphate
Fe	Iron
Al	Aluminium
Ca	Calcium
DCP	Dicalcium-phosphate
HAP	Hydroxyapatite
OCP	Octocalcium-phospahte
N	Nitrogen
C:P ratio	Carbon: Phosphorus ratio
ACC	1- Aminocyclopropane-1-carboxylate
H_3PO_4	Phosphoric Acid
NH4+	Ammonium
H_2CO_3	Carbonic Acid
NO_3	Nitrate
Mm	Millimetre
B	Boron
Cu	Copper
Mn	Manganese
Mo	Molybdenum
Cr	Chromium
Co	Cobalt
Hg	Mercury
PSM	Phosphate Solubilizing Microorganisms
Ppm	Parts per million
PSB	Phosphate Solubilizing Bacteria
Ga	Gallium
Ar	Argon
IAA	Indole- 3-acetic acid
HCN	Hydrogen cynate

References

Abdel-Shafy, H.I., Mansour, M.S., 2018. Solid waste issue: Sources, composition, disposal, recycling, and valorization. Egypt. J. Pet. 27 (4), 1275–1290.

Adamo, P., Agrelli, D., Zampella, M., 2018. Chemical speciation to assess bioavailability, bioaccessibility and geochemical forms of potentially toxic metals (PTMs) in polluted soilsEnvironmental Geochemistry. Elsevier, Amsterdam, The Netherlands.

Afzal, M., Khan, Q.M., Sessitsch, A., 2014. Endophytic bacteria: prospects and applications for the phytoremediation of organic pollutants. Chemosphere 117, 232–242.

Ahemad, M., 2015. Enhancing phytoremediation of chromium-stressed soils through plant-growth-promoting bacteria. J Genet. Eng. Biotechnol. 13 (1), 51–58.

Ahemad, M., Khan, M., 2010. Influence of selective herbicides on plant growth promoting traits of phosphate solubilizing Enterobacter asburiae strain PS 2. Res. J. Microbiol. 5 (9), 849–857.

Ahemad, M., Khan, M., 2011. Toxicological assessment of selective pesticides towards plant growth promoting activities of phosphate solubilizing Pseudomonas aeruginosa. Acta Microbiol. Immunol. Hung. 58 (3), 169–187.

Aken, B.V., Doty, S.L., 2009. Transgenic plants and associated bacteria for phytoremediation of chlorinated compounds. Biotechnol. Genet. Eng. Rev. 26 (1), 43–64.

Ali, H., Khan, E., Sajad, M.A., 2013. Phytoremediation of heavy metals—concepts and applications. Chemosphere 91 (7), 869–881.

Alloway, B., 2009. Soil factors associated with zinc deficiency in crops and humans. Environ. Geochem. Health 31 (5), 537–548.

Anawar, H.M., Rengel, Z., Damon, P., Tibbett, M., 2018. Arsenic-phosphorus interactions in the soil-plant-microbe system: Dynamics of uptake, suppression and toxicity to plants. Environ. Pollut. 233, 1003–1012.

Antonious, G.F., 2018. Biochar and animal manure impact on soil, crop yield and quality. Agricultural Waste. In: Aladjadjiyan, A (Ed.), Intec-Open Science Books. National Biomass Association, Bulgaria, Rijeka, Croatia.

Arunakumara, K., Walpola, B.C., Song, J.-S., Shin, M.-J., Lee, C.-J., Yoon, M.-H., 2014. Phytoextraction of heavy metals induced by bioaugmentation of a phosphate solubilizing bacterium. Korean J. Environ. Agric. 33 (3), 220–230.

Arzanesh, M.H., Alikhani, H., Khavazi, K., Rahimian, H., Miransari, M., 2011. Wheat (Triticum aestivum L.) growth enhancement by Azospirillum sp. under drought stress. World J. Microbiol. Biotechnol. 27 (2), 197–205.

Asati, A., Pichhode, M., Nikhil, K., 2016. Effect of heavy metals on plants: an overview. Int. J. Appl. Innov. Eng. Manage 5 (3), 2319–4847.

Ashley, K., Cordell, D., Mavinic, D., 2011. A brief history of phosphorus: from the philosopher's stone to nutrient recovery and reuse. Chemosphere 84 (6), 737–746.

Astriani, M., ZUBAIDAH, S., ABADI, A.L., SUARSINI, E., 2020. Pseudomonas plecoglossicida as a novel bacterium for phosphate solubilizing and indole-3-Sacetic acid-producing from soybean rhizospheric soils of East Java, Indonesia. Biodiversitas 21 (2), 578–586.

Awais, M., Tariq, M., Ali, A., Ali, Q., Khan, A., Tabassum, B., Nasir, I.A., Husnain, T., 2017. Isolation, characterization and inter-relationship of phosphate solubilizing bacteria from the rhizosphere of sugarcane and rice. Biocatal. Agric. Biotechnol. 11, 312–321.

Babu, A.G., Shim, J., Shea, P.J., Oh, B.-T., 2014. Penicillium aculeatum PDR-4 and Trichoderma sp. PDR-16 promote phytoremediation of mine tailing soil and bioenergy production with sorghum-sudangrass. Ecol. Eng. 69, 186–191.

Becerra-Castro, C., Prieto-Fernández, Á., Álvarez-Lopez, V., Monterroso, C., Cabello-Conejo, M., Acea, M., Kidd, P., 2011. Nickel solubilizing capacity and characterization of rhizobacteria isolated from hyperaccumulating and non-hyperaccumulating subspecies of Alyssum serpyllifolium. Int. J. Phytoremediation 13 (suppl1), 229–244.

Behera, B., Singdevsachan, S.K., Mishra, R., Dutta, S., Thatoi, H., 2014. Diversity, mechanism and biotechnology of phosphate solubilizing microorganism in mangrove—a review. Biocatal. Agric. Biotechnol. 3 (2), 97–110.

Bhargava, A., Carmona, F.F., Bhargava, M., Srivastava, S., 2012. Approaches for enhanced phytoextraction of heavy metals. J. Environ. Manage. 105, 103–120.

Bolan, N.S., Adriano, D.C., Naidu, R., 2003. Role of phosphorus in (im) mobilization and bioavailability of heavy metals in the soil-plant system. Rev. Environ. Contam. Toxicol. 117, 1–44.

Bolan, N., Kunhikrishnan, A., Thangarajan, R., Kumpiene, J., Park, J., Makino, T., Kirkham, M.B., Scheckel, K., 2014. Remediation of heavy metal (loid) s contaminated soils–to mobilize or to immobilize? J. Hazard. Mater. 266, 141–166.

Camacho-González, M.A., Quezada-Cruz, M., Cerón-Montes, G.I., Ramírez-Ayala, M.F., Hernández-Cruz, L.E., Garrido-Hernández, A., 2019. Synthesis and characterization of magnetic zinc-copper ferrites: Antibacterial activity, photodegradation study and heavy metals removal evaluation. Mater. Chem. Phys. 236, 121808.

Cao, X., Wang, Y., He, J., Luo, X., Zheng, Z., 2016. Phosphorus mobility among sediments, water and cyanobacteria enhanced by cyanobacteria blooms in eutrophic Lake Dianchi. Environ. Pollut. 219, 580–587.

Chibuike, G.U., Obiora, S.C., 2014. Heavy metal polluted soils: effect on plants and bioremediation methods. Appl. Environ. Soil Sci. 2014, 1–12.

Clemens, S., Ma, J.F., 2016. Toxic heavy metal and metalloid accumulation in crop plants and foods. Ann. Rev. Plant Biol. 67, 489–512.

Corwin, D.L., Hopmans, J., de Rooij, G.H., 2006. From field-to landscape-scale vadose zone processes: Scale issues, modeling, and monitoring. Vadose Zone J. 5 (1), 129–139.

DalCorso, G., Fasani, E., Manara, A., Visioli, G., Furini, A., 2019. Heavy metal pollutions: state of the art and innovation in phytoremediation. Int. J. Mol. Sci. 20 (14), 3412.

Doucette, W., Bugbee, B., Smith, S., Pajak, C., Ginn, J., 2003. Uptake, metabolism, and phytovolatilization of trichloroethylene by indigenous vegetation: impact of precipitation. Phytoremediation. John Wiley & Sons, Inc., New York, USA, pp. 561–588.

Farraji, H., Zaman, N.Q., Tajuddin, R., Faraji, H., 2016. Advantages and disadvantages of phytoremediation: a concise review. Int J Env Tech Sci 2, 69–75.

Gaonkar, T., Bhosle, S., 2013. Effect of metals on a siderophore producing bacterial isolate and its implications on microbial assisted bioremediation of metal contaminated soils. Chemosphere 93 (9), 1835–1843.

Girma, G., 2015. Microbial bioremediation of some heavy metals in soils: an updated review. Egypt. Acad. J. Biol. Sci. 7 (1), 29–45.

Govarthanan, M., Mythili, R., Selvankumar, T., Kamala-Kannan, S., Kim, H., 2018. Myco-phytoremediation of arsenic-and lead-contaminated soils by Helianthus annuus and wood rot fungi, Trichoderma sp. isolated from decayed wood. Ecotoxicol. Environ. Saf. 151, 279–284.

Grobelak, A., Napora, A., 2015. The chemophytostabilisation process of heavy metal polluted soil. PLoS One 10 (6), e0129538.

Guarino, F., Miranda, A., Castiglione, S., Cicatelli, A., 2020. Arsenic phytovolatilization and epigenetic modifications in Arundo donax L. assisted by a PGPR consortium. Chemosphere 251, 126310.

Gulati, A., Sharma, N., Vyas, P., Sood, S., Rahi, P., Pathania, V., Prasad, R., 2010. Organic acid production and plant growth promotion as a function of phosphate solubilization by Acinetobacter rhizosphaerae strain BIHB 723 isolated from the cold deserts of the trans-Himalayas. Arch. Microbiol. 192 (11), 975–983.

Hafeez, B., Khanif, Y., Saleem, M., 2013. Role of zinc in plant nutrition-a review. J. Exp. Agric. Int. 3 (2), 374–391.

Hardy, S., 2004. Growing Lemons in Australia–A Production Manual. NSW Department of Primary Industries, New South Wales, Australia.

Harvey, P.J., Campanella, B.F., Castro, P.M., Harms, H., Lichtfouse, E., Schäffner, A.R., Smrcek, S., Werck-Reichhart, D., 2002. Phytoremediation of polyaromatic hydrocarbons, anilines and phenols. Environ. Sci. Pollut. Res. 9 (1), 29–47.

Hassan, M.U., Chattha, M.U., Khan, I., Chattha, M.B., Aamer, M., Nawaz, M., Ali, A., Khan, M.A.U., Khan, T.A., 2019. Nickel toxicity in plants: reasons, toxic effects, tolerance mechanisms, and remediation possibilities—a review. Environ. Sci Pollut. Res. 26 (13), 12673–12688.

He, H., Kang, H., Ma, S., Bai, Y., Yang, X., 2010. High adsorption selectivity of ZnAl layered double hydroxides and the calcined materials toward phosphate. J. Colloid Interface Sci. 343 (1), 225–231.

He, H., Ye, Z., Yang, D., Yan, J., Xiao, L., Zhong, T., Yuan, M., Cai, X., Fang, Z., Jing, Y., 2013. Characterization of endophytic Rahnella sp. JN6 from Polygonum pubescens and its potential in promoting growth and Cd, Pb, Zn uptake by Brassica napus. Chemosphere 90 (6), 1960–1965.

He, Z.-l., Yang, X.-e., 2007. Role of soil rhizobacteria in phytoremediation of heavy metal contaminated soils. J. Zhejiang Univ. Sci. B 8 (3), 192–207.

Hussain, S., Huang, J., Zhu, C., Zhu, L., Cao, X., Hussain, S., Ashraf, M., Khaskheli, M.A., Kong, Y., Jin, Q., 2020. Pyridoxal 5′-phosphate enhances the growth and morpho-physiological characteristics of rice cultivars by mitigating the ethylene accumulation under salinity stress. Plant Physiol. Biochem. 154, 782–795.

Ioannidis, T.A., Zouboulis, A., 2005. Phytoremediation of lead-contaminated soils. Water Encyclopedia, 5. John Wiley & Sons, Inc, United States. pp. 381–385.

Ionescu, M., Belkin, S., 2009. Overproduction of exopolysaccharides by an Escherichia coli K-12 rpoS mutant in response to osmotic stress. Appl. Environ. Microbiol. 75 (2), 483–492.

Islam, M.S., Hosen, M.M.L., Uddin, M.N., 2019. Phytodesalination of saline water using Ipomoea aquatica, Alternanthera philoxeroides and Ludwigia adscendens. Int. J. Environ. Sci. Technol. 16 (2), 965–972.

Jacob, J.M., Karthik, C., Saratale, R.G., Kumar, S.S., Prabakar, D., Kadirvelu, K., Pugazhendhi, A., 2018. Biological approaches to tackle heavy metal pollution: a survey of literature. J. Environ. Manage. 217, 56–70.

Jiang, C.-y., Sheng, X.-f., Qian, M., Wang, Q.-y., 2008. Isolation and characterization of a heavy metal-resistant Burkholderia sp. from heavy metal-contaminated paddy field soil and its potential in promoting plant growth and heavy metal accumulation in metal-polluted soil. Chemosphere 72 (2), 157–164.

Jiang, F., Zhang, L., Zhou, J., George, T.S., Feng, G., 2021. Arbuscular mycorrhizal fungi enhance mineralization of organic phosphorus by carrying bacteria along their extraradical hyphae. New Phytol. 230 (1), 304–315.

Jing, Y.X., Yan, J.L., He, H.D., Yang, D.J., Xiao, L., Zhong, T., Yuan, M., Cai, X.D., Li, S.B., 2014. Characterization of bacteria in the rhizosphere soils of Polygonum pubescens and their potential in promoting growth and Cd, Pb, Zn uptake by Brassica napus. Int. J. Phytoremediation 16 (4), 321–333.

Karuppiah, P.a.R.S., 2011. Exploring the potential of chromium reducing Bacillus sp. and there plant growth promoting activities. J. Microbiol. Res. 1 (1), 17–23.

Khan, M.I., Cheema, S.A., Anum, S., Niazi, N.K., Azam, M., Bashir, S., Ashraf, I., Qadri, R., 2020. Phytoremediation of agricultural pollutants. Phytoremediation. In: Shmaefsky, B. (Ed.), Concepts and Strategies in Plant Sciences. Springer, Switzerland, pp. 381–385.

Khan, M.U., Sessitsch, A., Harris, M., Fatima, K., Imran, A., Arslan, M., Shabir, G., Khan, Q.M., Afzal, M., 2015. Cr-resistant rhizo-and endophytic bacteria associated with Prosopis juliflora and their potential as phytoremediation enhancing agents in metal-degraded soils. Front. Plant Sci. 5, 755.

Kim, D.-G., Kim, M., Shin, J.Y., Son, S.-W., 2016. Cadmium and lead in animal tissue (muscle, liver and kidney), cow milk and dairy products in Korea. Food Addit. Contam. Part B 9 (1), 33–37.

Kim, J.-J., Kim, Y.-S., Kumar, V., 2019. Heavy metal toxicity: An update of chelating therapeutic strategies. J. Trace Elements Med. Biol. 54, 226–231.

Kumar, K.V., Singh, N., Behl, H.M., Srivastava, S., 2008. Influence of plant growth promoting bacteria and its mutant on heavy metal toxicity in Brassica juncea grown in fly ash amended soil. Chemosphere 72 (4), 678–683.

Legault, E.K., James, C.A., Stewart, K., Muiznieks, I., Doty, S.L., Strand, S.E., 2017. A field trial of TCE phytoremediation by genetically modified poplars expressing cytochrome P450 2E1. Environ. Sci. Technol. 51 (11), 6090–6099.

Li, K., Ramakrishna, W., 2011. Effect of multiple metal resistant bacteria from contaminated lake sediments on metal accumulation and plant growth. J. Hazard. Mater. 189 (1-2), 531–539.

Lin, M., Jin, M., Xu, K., He, L., Cheng, D., 2018. Phosphate-solubilizing bacteria improve the phytoremediation efficiency of Wedelia trilobata for Cu-contaminated soil. Int. J. Phytoremediation 20 (8), 813–822.

Liu, C., Lu, J., Liu, J., Mehmood, T., Chen, W., 2020. Effects of lead (Pb) in stormwater runoff on the microbial characteristics and organics removal in bioretention systems. Chemosphere 253, 126721.

Lucena, J.J., Hernandez-Apaolaza, L., 2017. Iron nutrition in plants: an overview. Plant and Soil 418 (1–2), 1–4.

Lwin, C.S., Seo, B.-H., Kim, H.-U., Owens, G., Kim, K.-R., 2018. Application of soil amendments to contaminated soils for heavy metal immobilization and improved soil quality—a critical review. Soil Sci. Plant Nutr. 64 (2), 156–167.

Ma, J., Chen, Y., Antoniadis, V., Wang, K., Huang, Y., Tian, H., 2020. Assessment of heavy metal (loid) s contamination risk and grain nutritional quality in organic waste-amended soil. J. Hazard. Mater. 399, 123095.

Ma, Y., Prasad, M., Rajkumar, M., Freitas, H., 2011. Plant growth promoting rhizobacteria and endophytes accelerate phytoremediation of metalliferous soils. Biotechnol. Adv. 29 (2), 248–258.

Ma, Y., Rajkumar, M., Rocha, I., Oliveira, R.S., Freitas, H., 2015. Serpentine bacteria influence metal translocation and bioconcentration of Brassica juncea and Ricinus communis grown in multi-metal polluted soils. Front. Plant Sci. 5, 757.

Madhaiyan, M., Poonguzhali, S., Senthilkumar, M., Pragatheswari, D., Lee, J.-S., Lee, K.-C., 2015. Arachidicoccus rhizosphaerae gen. nov., sp. nov., a plant-growth-promoting bacterium in the family Chitinophagaceae isolated from rhizosphere soil. Int. J. Syst. Evol. Microbiol. 65 (2), 578–586.

Mahajan, P., Kaushal, J., 2018. Role of phytoremediation in reducing cadmium toxicity in soil and water. J. Toxicol. 2018, 1–16, 4864365.

Majeed, A., Abbasi, M.K., Hameed, S., Imran, A., Rahim, N., 2015. Isolation and characterization of plant growth-promoting rhizobacteria from wheat rhizosphere and their effect on plant growth promotion. Front. Microbiol. 6, 198.

Malkanthi, D.R., Moritsugu, M.Ramya, Yokoyama, K., 1995. Effects of low pH and Al on absorption and translocation of some essential nutrients in excised barley roots. Soil Sci. Plant Nutr. 41 (2), 253–262.

Marques, A.P., Moreira, H., Franco, A.R., Rangel, A.O., Castro, P.M., 2013. Inoculating Helianthus annuus (sunflower) grown in zinc and cadmium contaminated soils with plant growth promoting bacteria–Effects on phytoremediation strategies. Chemosphere 92 (1), 74–83.

Masood, S., Zhao, X.Q., Shen, R.F., 2019. Bacillus pumilus increases boron uptake and inhibits rapeseed growth under boron supply irrespective of phosphorus fertilization. AoB Plants 11 (4), plz036.

McGrath, S.P., Zhao, F.-J., 2003. Phytoextraction of metals and metalloids from contaminated soils. Curr. Opin. Biotechnol. 14 (3), 277–282.

McKnight, G.D., Rayborn, R.L., 2019. Aqueous organo liquid delivery systems containing dispersed organo polycarboxylate functionalities that improves efficiencies and properties of nitrogen sourcesNon-. Google Patents.

Mehmood, T., Bibi, I., Shahid, M., Niazi, N.K., Murtaza, B., Wang, H.L., Ok, Y.S., Sarkar, B., Javed, M.T., Murtaza, G., 2017. Effect of compost addition on arsenic uptake, morphological and physiological attributes of maize plants grown in contrasting soils. J. Geochem. Explor. 178, 83–91.

Mehmood, T., Liu, C., Niazi, N.K., Gaurav, G.K., Ashraf, A., Bibi, I., 2020. Compost-mediated arsenic phytoremediation, health risk assessment and economic feasibility using Zea mays L. in contrasting textured soils. Int. J. Phytoremediation, 1–12.

Mehmood, T., Lu, J., Liu, C., Gaurav, G.K., 2021. Organics removal and microbial interaction attributes of zeolite and ceramsite assisted bioretention system in copper-contaminated stormwater treatment. J Environ Manage. 292, 1–12, 112654.

Misra, N., Gupta, G., Jha, P.N., 2012. Assessment of mineral phosphate-solubilizing properties and molecular characterization of zinc-tolerant bacteria. J. Basic Mmicrobiol. 52 (5), 549–558.

Morpho, O., 2018. Phytoremediation potential of catharanthus roseus l. and effects. Pak. J. Bot 50 (4), 1323–1326.

Mousavi, S.R., 2011. Zinc in crop production and interaction with phosphorus. Aust. J. Basic Appl. Sci. 5 (9), 1503–1509.

Natasha, Bibi, I., Hussain, K., Amen, R., Masood Ul Hasan, I., Shahid, M., Bashir, S., Niazi, N.K., Mehmood, T., Asghar, H.N., Nawaz, M.F., Hussain, M.M., Ali, W., 2021. The potential of microbes and sulfate in reducing arsenic phytoaccumulation by maize (Zea mays L.) plants. Environ Geochem Health, 1–15.

Nedunuri, K., Govindaraju, R., Banks, M.K., Schwab, A., Chen, Z., 2000. Evaluation of phytoremediation for field-scale degradation of total petroleum hydrocarbons. J. Environ. Eng. 126 (6), 483–490.

Niazi, N.K., Bibi, I., Fatimah, A., Shahid, M., Javed, M.T., Wang, H., Ok, Y.S., Bashir, S., Murtaza, B., Saqib, Z.A., 2017. Phosphate-assisted phytoremediation of arsenic by Brassica napus and Brassica juncea: morphological and physiological response. Int. J. Phytoremediation 19 (7), 670–678.

Niazi, N.K., Singh, B., Van Zwieten, L., Kachenko, A.G., 2011. Phytoremediation potential of Pityrogramma calomelanos var. austroamericana and Pteris vittata L. grown at a highly variable arsenic contaminated site. Int. J. Phytoremediation 13 (9), 912–932.

Niazi, N.K., Singh, B., Van Zwieten, L., Kachenko, A.G., 2012. Phytoremediation of an arsenic-contaminated site using Pteris vittata L. and Pityrogramma calomelanos var. austroamericana: a long-term study. Environ. Sci. Pollut. Res. 19 (8), 3506–3515.

Ochoa-Loza, F.J., Artiola, J.F., Maier, R.M., 2001. Stability constants for the complexation of various metals with a rhamnolipid biosurfactant. J. Environ. Qual. 30 (2), 479–485.

Opala, P.A., 2017. Influence of Lime and phosphorus application rates on growth of maize in an acid soil. Adv. Agric. 2017, 7083206.

Oves, M., Khan, M.S., Zaidi, A., 2013. Chromium reducing and plant growth promoting novel strain Pseudomonas aeruginosa OSG41 enhance chickpea growth in chromium amended soils. Eur. J. Soil Biol. 56, 72–83.

Palansooriya, K.N., Shaheen, S.M., Chen, S.S., Tsang, D.C., Hashimoto, Y., Hou, D., Bolan, N.S., Rinklebe, J., Ok, Y.S., 2020. Soil amendments for immobilization of potentially toxic elements in contaminated soils: a critical review. Environ. Int. 134, 105046.

Pandey, A., Trivedi, P., Kumar, B., Palni, L.M.S., 2006. Characterization of a phosphate solubilizing and antagonistic strain of Pseudomonas putida (B0) isolated from a sub-alpine location in the Indian Central Himalaya. Curr. Microbiol. 53 (2), 102–107.

Pandey, V.C., Bajpai, O., 2019. Phytoremediation: from theory toward practice. In: Pandey, V.C., Bauddh, K. (Eds.), Phytomanagement of Polluted Sites. Elsevier, Netherlands, pp. 1–49.

Pandey, V.C., Souza-Alonso, P., 2019. Market opportunities in sustainable phytoremediation. In: Pandey, V.C., Bauddh, K. (Eds.), Phytomanagement of Polluted Sites. Elsevier, Netherlands, pp. 51–82.

Pawlik, M., Płociniczak, T., Thijs, S., Pintelon, I., Vangronsveld, J., Piotrowska-Seget, Z., 2020. Comparison of two inoculation methods of endophytic bacteria to enhance phytodegradation efficacy of an aged petroleum hydrocarbons polluted soil. Agronomy 10 (8), 1196.

Pestana, I.A., Meneguelli-Souza, A.C., Gomes, M.A.C., Almeida, M.G., Suzuki, M.S., Vitória, A.P., Souza, C.M.M., 2018. Effects of a combined use of macronutrients nitrate, ammonium, and phosphate on cadmium absorption by Egeria densa Planch. and its phytoremediation applicability. Aquat. Ecol. 52 (1), 51–64.

Pilon-Smits, E., 2005. Phytoremediation. Annu. Rev. Plant Biol. 56, 15–39.

Pilon-Smits, E.A., Freeman, J.L., 2006. Environmental cleanup using plants: biotechnological advances and ecological considerations. Front. Ecol. Environ. 4 (4), 203–210.

Placek, A., Grobelak, A., Kacprzak, M., 2016. Improving the phytoremediation of heavy metals contaminated soil by use of sewage sludge. Int. J. Phytoremediation 18 (6), 605–618.

Prabhu, N., Borkar, S., Garg, S., 2018. Phosphatesolubilization mechanisms in alkaliphilic bacteriumBacillus marisflavi FA7. Current Science 114 (4), 845–853.

Prabhu, N., Borkar, S., Garg, S., 2019. Phosphate solubilization by microorganisms: overview, mechanisms, applications and advancesAdvances in Biological Science Research. Elsevier, Amsterdam.

Rabhi, M., Ferchichi, S., Jouini, J., Hamrouni, M.H., Koyro, H.-W., Ranieri, A., Abdelly, C., Smaoui, A., 2010. Phytodesalination of a salt-affected soil with the halophyte Sesuvium portulacastrum L. to arrange in advance the requirements for the successful growth of a glycophytic crop. Bioresour. Technol. 101 (17), 6822–6828.

Radziemska, M., Vaverková, M.D., Baryła, A., 2017. Phytostabilization—management strategy for stabilizing trace elements in contaminated soils. Int. J. Environ. Res. Public Health 14 (9), 958.

Rai, P.K., Lee, S.S., Zhang, M., Tsang, Y.F., Kim, K.-H., 2019. Heavy metals in food crops: health risks, fate, mechanisms, and management. Environ. Int. 125, 365–385.

Raj, D., Maiti, S.K., 2020. Sources, bioaccumulation, health risks and remediation of potentially toxic metal (loid) s (As, Cd, Cr, Pb and Hg): an epitomized review. Environ. Monit. Assess. 192 (2), 108.

Rakshit, A., Sarkar, B., Abhilash, P., 2018. Soil Amendments for Sustainability: Challenges and Perspectives. CRC Press.

Rane, N.R., Chandanshive, V.V., Watharkar, A.D., Khandare, R.V., Patil, T.S., Pawar, P.K., Govindwar, S.P., 2015. Phytoremediation of sulfonated Remazol Red dye and textile effluents by Alternanthera philoxeroides: an anatomical, enzymatic and pilot scale study. Water Res. 83, 271–281.

Rashid, S., Charles, T.C., Glick, B.R., 2012. Isolation and characterization of new plant growth-promoting bacterial endophytes. Appl. Soil Ecol. 61, 217–224.

Ren, J., Zhao, Z., Ali, A., Guan, W., Xiao, R., Wang, J.J., Ma, S., Guo, D., Zhou, B., Zhang, Z., 2020. Characterization of phosphorus engineered biochar and its impact on immobilization of Cd and Pb from smelting contaminated soils. J. Soils Sediments 20 (8), 3041–3052.

Rizwan, M., Ali, S., Qayyum, M.F., Ibrahim, M., Zia-ur-Rehman, M., Abbas, T., Ok, Y.S., 2016. Mechanisms of biochar-mediated alleviation of toxicity of trace elements in plants: a critical review. Environ. Sci. Pollut. Res. 23 (3), 2230–2248.

Roy, E.D., 2017. Phosphorus recovery and recycling with ecological engineering: a review. Ecol. Eng. 98, 213–227.

Santos, M., Melo, V.F., Monte Serrat, B., Bonfleur, E., Araújo, E.M., Cherobim, V.F., 2020. Hybrid technologies for remediation of highly Pb contaminated soil: sewage sludge application and phytoremediation. Int. J. Phytoremediation, 1–8.

Selvakumar, G., Mohan, M., Kundu, S., Gupta, A., Joshi, P., Nazim, S., Gupta, H., 2008. Cold tolerance and plant growth promotion potential of Serratia marcescens strain SRM (MTCC 8708) isolated from flowers of summer squash (Cucurbita pepo). Lett. Appl. Microbiol. 46 (2), 171–175.

Seshadri, B., Bolan, N., Naidu, R., 2015. Rhizosphere-induced heavy metal (loid) transformation in relation to bioavailability and remediation. J. Soil Sci. Plant Nutr. 15 (2), 524–548.

Shah, V., Daverey, A., 2020. Phytoremediation: a multidisciplinary approach to clean up heavy metal contaminated soil. Environ. Technol. Innov. 18, 100774.

Silva, J., Evensen, C., Bowen, R., Kirby, R., Tsuji, G., Yost, R., 2000. Managing Fertilizer Nutrients to Protect the Environment and Human Health. University of Hawaii, Honolulu, USA.

Singer, A.C., Thompson, I.P., Bailey, M.J., 2004. The tritrophic trinity: a source of pollutant-degrading enzymes and its implications for phytoremediation. Curr. Opin. Microbiol. 7 (3), 239–244.

Singh, D., Vyas, P., Sahni, S., Sangwan, P., 2015. Phytoremediation: a biotechnological interventionApplied Environmental Biotechnology: Present Scenario and Future Trends. Springer.

Soares, M.M., FREITAS, C.D.M., OLIVEIRA, F.S.D., MESQUITA, H.C.D., Silva, T.S., Silva, D.V., 2019. Effects of competition and water deficiency on sunflower and weed growth. Rev. Caatinga 32 (2), 318–328.

Spohn, M., Treichel, N.S., Cormann, M., Schloter, M., Fischer, D., 2015. Distribution of phosphatase activity and various bacterial phyla in the rhizosphere of Hordeum vulgare L. depending on P availability. Soil Biol. Biochem. 89, 44–51.

Srinivasan, R., Yandigeri, M.S., Kashyap, S., Alagawadi, A.R., 2012. Effect of salt on survival and P-solubilization potential of phosphate solubilizing microorganisms from salt affected soils. Saudi J. Biol. Sci. 19 (4), 427–434.

Strawn, D.G., 2018. Review of interactions between phosphorus and arsenic in soils from four case studies. Geochem. Trans. 19 (1), 10.

Susarla, S., Medina, V.F., McCutcheon, S.C., 2002. Phytoremediation: an ecological solution to organic chemical contamination. Ecol. Eng. 18 (5), 647–658.

Teixeira, C., Almeida, C.M.R., da Silva, M.N., Bordalo, A.A., Mucha, A.P., 2014. Development of autochthonous microbial consortia for enhanced phytoremediation of salt-marsh sediments contaminated with cadmium. Sci. Total Environ. 493, 757–765.

Van Epps, A., 2006. Phytoremediation of Petroleum Hydrocarbons. Environmental Protection Agency, US.

Vassilev, N., Eichler-Löbermann, B., Vassileva, M., 2012. Stress-tolerant P-solubilizing microorganisms. Appl. Microbiol. Biotechnol. 95 (4), 851–859.

Walker, D.J., Bernal, M.P., 2004. The effects of copper and lead on growth and zinc accumulation of Thlaspi caerulescens J. and C. Presl: implications for phytoremediation of contaminated soils. Water Air Soil Pollut. 151 (1-4), 361–372.

Wang, Y., Zhu, C., Yang, H., Zhang, X., 2017. Phosphate fertilizer affected rhizospheric soils: speciation of cadmium and phytoremediation by Chlorophytum comosum. Environ. Sci. Pollut. Res. 24 (4), 3934–3939.

Wenzel, W.W., 2009. Rhizosphere processes and management in plant-assisted bioremediation (phytoremediation) of soils. Plant Soil 321 (1-2), 385–408.

Wiszniewska, A., Hanus-Fajerska, E., Muszynska, E., Ciarkowska, K., 2016. Natural Organic Amendments for Improved Phytoremediation of Polluted Soils: A Review of Recent Progress. Pedosphere 26 (1), 1–12.

Wuana, R.A., Okieimen, F.E., 2011. Heavy metals in contaminated soils: a review of sources, chemistry, risks and best available strategies for remediation. Isrn Ecol. 2011, 402647.

Xiao, R., Ali, A., Wang, P., Li, R., Tian, X., Zhang, Z., 2019. Comparison of the feasibility of different washing solutions for combined soil washing and phytoremediation for the detoxification of cadmium (Cd) and zinc (Zn) in contaminated soil. Chemosphere 230, 510–518.

Xu, X., Mao, X., Van Zwieten, L., Niazi, N.K., Lu, K., Bolan, N.S., Wang, H., 2020. Wetting-drying cycles during a rice-wheat crop rotation rapidly (im) mobilize recalcitrant soil phosphorus. J. Soils Sediments, 1–10.

Xu, Q., Sylvie, R., Qiuyan, Y., 2019. Phytodesalination of landfill leachate using Puccinellia nuttalliana and Typha latifolia. Int. J. Phytoremediation 21 (9), 831–839.

Yadav, B., Verma, A., 2012. Phosphate solubilization and mobilization in soil through microorganisms under arid ecosystems. In: Ali, M. (Ed.), The Functioning of Ecosystems, 93–108.

Yaghoubian, Y., Siadat, S.A., Telavat, M.R.M., Pirdashti, H., Yaghoubian, I., 2019. Bio-removal of cadmium from aqueous solutions by filamentous fungi: Trichoderma spp. and Piriformospora indica. Environ. Sci. Pollut. Res. 26 (8), 7863–7872.

Yahaghi, Z., Shirvani, M., Nourbakhsh, F., De La Pena, T.C., Pueyo, J.J., Talebi, M., 2018. Isolation and characterization of Pb-solubilizing bacteria and their effects on Pb uptake by Brassica juncea: implications for microbe-assisted phytoremediation. J. Microbiol. Biotechnol 28 (7), 1156–1167.

Yan, A., Wang, Y., Tan, S.N., Yusof, M.L.M., Ghosh, S., Chen, Z., 2020. Phytoremediation: a promising approach for revegetation of heavy metal-polluted land. Front. Plant Sci. 11, 1–15.

Yang, G.-M., Zhu, L.-J., Santos, J.A.G., Chen, Y., Li, G., Guan, D.-X., 2017. Effect of phosphate minerals on phytoremediation of arsenic contaminated groundwater using an arsenic-hyperaccumulator. Environ. Technol. Innov. 8, 366–372.

Yazdankhah, S., Skjerve, E., Wasteson, Y., 2018. Antimicrobial resistance due to the content of potentially toxic metals in soil and fertilizing products. Microb. Ecol. Health Dis. 29 (1), 1548248.

Yi, Y., Huang, W., Ge, Y., 2008. Exopolysaccharide: a novel important factor in the microbial dissolution of tricalcium phosphate. World J. Microbiol. Biotechnol. 24 (7), 1059–1065.

Yogendra, S., Ramteke, P., Shukla, P., 2013. Isolation and characterization of heavy metal resistant Pseudomonas spp. and their plant growth promoting activities. Adv. Appl. Sci. Res. 4 (1), 269–272.

Yu, F., Li, C., Dai, C., Liu, K., Li, Y., 2020. Phosphate: Coupling the functions of fertilization and passivation in phytoremediation of manganese-contaminated soil by Polygonum pubescens blume. Chemosphere 260, 127651.

Yuan, Z., Yi, H., Wang, T., Zhang, Y., Zhu, X., Yao, J., 2017. Application of phosphate solubilizing bacteria in immobilization of Pb and Cd in soil. Environ. Sci. Pollut. Res. 24 (27), 21877–21884.

Zaidi, A., Khan, M., Ahemad, M., Oves, M., 2009. Plant growth promotion by phosphate solubilizing bacteria. Acta Microbiol. Immunol. Hung. 56 (3), 263–284.

Zeng, G., Wan, J., Huang, D., Hu, L., Huang, C., Cheng, M., Xue, W., Gong, X., Wang, R., Jiang, D., 2017. Precipitation, adsorption and rhizosphere effect: the mechanisms for phosphate-induced Pb immobilization in soils—a review. J. Hazard. Mater. 339, 354–367.

Zulfiqar, U., Farooq, M., Hussain, S., Maqsood, M., Hussain, M., Ishfaq, M., Ahmad, M., Anjum, M.Z., 2019. Lead toxicity in plants: impacts and remediation. J. Environ. Manage. 250, 109557.

CHAPTER

14

Electrokinetic-assisted Phytoremediation

Luis Rodríguez[a], Virtudes Sánchez[a], Francisco J. López-Bellido[b]

[a]*Department of Chemical Engineering, School of Civil Engineering, University of Castilla-La Mancha, Ciudad Real (Spain)*

[b]*Department of Plant Production and Agricultural Technology, School of Agricultural Engineering, University of Castilla-La Mancha, Ciudad Real (Spain)*

14.1 Introduction

The advantages of phytoremediation as a green and environmentally sustainable technique to remediate polluted soils, waters and sludges have been widely reported and are the basis for the numerous research efforts carried out for more than thirty five years from the first work by Chaney (Chaney, 1983). Nevertheless, it can be said that phytoremediation has not yet reached a level of development enough to be considered as a common soil reclamation technique for its implementation in actual polluted sites (Sarwar *et al.*, 2017). However, this should not be seen as a problem but as an opportunity or a challenge; in this sense, it seems clear that a greater number of investigations must be carried out aimed at solving the main practical problems that phytoremediation still poses currently (Van Nevel *et al.*, 2007; Khalid *et al.*, 2017; Sarwar *et al.*, 2017; Ashraf *et al.*, 2019).

Amongst the different limitations that have been pointed out for the phytoremediation of polluted soils, the long time required for the cleanup has been one of the most cited (Ali, Khan and Sajad, 2013; Mahar *et al.*, 2016). Along with factors such as plant uptake capability and biomass development, the prolonged remediation time is mainly a consequence of the usual low or moderate availability of pollutants in soils. In fact, the term 'bioavailability' is one of the keywords most used in the research papers about phytoremediation of metals and it plays a key role in linking different research topics (Li et al., 2019). The availability and mobility of pollutants in soils is strongly determined by the interactions between them and the different minerals, chemical compounds, phases and microbiota present in soils. Toxic metals are mainly associated to five different geochemical soil fractions, i.e. water soluble, exchangeable and bound to carbonates, bound to Fe-Mn oxides, bound to organic matter and occluded in the mineral matrix (Bolan *et al.*, 2014). Only water soluble and exchangeable/carbonate fractions are considered as the readily available ones for plant uptake, while organic- and Fe-Mn oxides-bound metals can be partially solubilized by means of plant root exudates and other biochemical or chemical processes taking place naturally in polluted soils (Kumar Yadav *et al.*, 2018). Currently, there are a number of accepted chemical methods for the determination of metal availability in soils and sediments; they consist of both single extractions with different chemicals, such as $CaCl_2$, NH_4NO_3, NH_4OAc, DPTA, EDTA, and sequential extraction procedures, e.g. BCR and Tessier methods (Ruiz *et al.*, 2009; Bolan

Assisted Phytoremediation. DOI: https://doi.org/10.1016/B978-0-12-822893-7.00005-7

et al., 2014; Rodríguez, Gómez, *et al.*, 2016). Some chemical methods have also been used to assess the bioavailability of organic pollutants in soil and sediments ((Cui, Mayer and Gan, 2013; Ortega-Calvo *et al.*, 2015). Those methods can be classified into two groups: (i) extraction with cyclodextrin or Tenax (a compound-scavenging resin) which are considered as infinite sinks for organic pollutants, and (ii) the use of passive sampling materials (polyoxomethylene, polydimethylsiloxane, polyethylene, polyacrylate, etc.) for the determination of the free concentration of organics in the dissolved phase without significantly affecting the soil-water equilibrium. All of those types of extraction methods have been tried to correlate with biological assays (ecotoxicological ones) in order to assess their accuracy on determining the actual bioavailability of pollutants for plants, microorganisms and/or invertebrates (Menzies, Donn and Kopittke, 2007; Ortega-Calvo *et al.*, 2015; Rodríguez, Gómez, *et al.*, 2016).

Pollutants bioavailability is especially important when the phytoremediation experiments are conducted in the field or by using real soils with aged contamination; in that context, the assessment of the pollutant mobility is essential to determine the effectiveness of the technique since plants and rhizobacteria (in the case of rhizodegradation) are only capable to remove the available fraction of soil pollutants. Moreover, checking the availability of pollutants remaining after the decontamination process is the key point to evaluate its success and the potential mitigation of the environmental risk (Ortega-Calvo *et al.*, 2015). On the other hand, the objective of some of the assisted phytoremediation technologies reviewed in this book is just to increase pollutant availability and/or mobility in order to enhance the removal efficiencies and, thus, reducing the time necessary for the decontamination process. The most studied option for the mobilization of contaminants (and, more specifically, metals) from unavailable soil pools has been the use of synthetic chelating agents, e.g. ethylene diamine tetraacetic acid, ethylene diamine disuccinic acid or ethylene glycol tetraacetic acid (Evangelou, Ebel and Schaeffer, 2007; Sarwar *et al.*, 2017); alternatively, some low molecular weight organic acids (acetic acid, citric acid, malic acid and oxalic acid) and hydrochloric acid can also be effectively used because their chelating properties and their capability to decrease soil pH (Evangelou, Ebel and Schaeffer, 2007; Rodríguez, Alonso-Azcárate, *et al.*, 2016). However, despite of the demonstrated effectiveness of those chemicals for metal mobilization in soils, important environmental concerns such as plant toxicity and leaching and subsequent contamination of groundwater, should be carefully considered before their application in the field.

As an alternative to the use of chelating agents, the application of low intensity electric fields to the soil has been shown to be effective in the mobilization of metals and other organic pollutants due to the electrolysis of water, which generates redox and acid/base reactions, and the transport of charged species throughout the liquid phase of the soil (Virkutyte, Sillanpää and Latostenmaa, 2002; Yeung and Gu, 2011). The electrokinetic treatment has been studied in the last twenty five years for the removal of soil contaminants but, only more recently, it has been proposed to be merged with other remediation techniques in order to develop combined technologies that use the benefits and avoid the drawbacks of the individual ones. More specifically, the combination of electrokinetic remediation and phytoremediation, called as electrokinetic-assisted phytoremediation (EK-phytoremediation), has been studied from the early years of 2000s (O'Connor *et al.*, 2003), although there are still relatively few studies carried out, especially in the field of organic pollutants. As the main advantages of the EK-phytoremediation technology can be cited the following:

(i) The electric field is capable to increase the bioavailability of pollutants in soils and therefore enhancing plant uptake and/or degradation; moreover, the movement of pollutants in the soil can be adequately controlled by the electrode configuration avoiding the risk of leaching to groundwater.

(ii) The application of low-intensity electric current could improve the biomass reached by the plants.
(iii) These two previous effects would lead to an increase in the phytoremediation yields with the subsequent reduction in decontamination time.
(iv) It can be applied to both inorganic and organics pollutants.
(v) The combined technology may be considered as environmentally sustainable because the use of renewable resources, i.e. plants and solar energy (which could be applied as the source of the electric current); moreover, it is also a benign technology for soil and the environment.

Naturally, along with these listed advantages, some limitations should still be studied before the application of the EK-phytoremediation to real cases. These limitations will be analyzed in the subsequent sections along with the fundamentals of the technology and the revision of the most important knowledge developed in this field, based on the literature published to date. We will end this chapter summarizing the most relevant conclusions reached along with the issues that have not yet been sufficiently studied and should be addressed in future research works.

14.2 Fundamentals of electrokinetic-assisted phytoremediation

Electrokinetic remediation (EKR) of soils is an *in situ* technology consisting in the application of a direct electric potential to the polluted soil by means of several electrodes which act as anodes and cathodes; in the simplest case it is enough to use an anode and a cathode inserted in the soil (or within wells filled with electrolyte solutions) and connected to an external electric power source. The application of an electric current to the soil produces two types of simultaneous phenomena: (i) the electrochemical transport of contaminants (and other chemical species) throughout the soil from one electrode to another and (ii) a series of redox and acid/base reactions both on the surface of the electrodes and in the bulk soil. This technology allows the cleanup of soils by removing the contaminants in wells located close to the electrodes and, additionally, in the case of organics, by means of oxidation and/or volatilization. EKR is specially recommended for fine-grained soils with low permeability in which advection of pollutants and water is hindered and, therefore, other remediation methods, e.g. in situ chemical oxidation or soil flushing, cannot be used (Yeung and Gu, 2011). This method may be applied to inorganic (metals, metalloids, radionuclides) and organic pollutants as well as their combinations; moreover, both saturated and unsaturated soils may be treated (Virkutyte, Sillanpää and Latostenmaa, 2002; Reddy and Cameselle, 2009).

Another interesting feature of EKR is that it may be easily combined with other remediation treatments to overcome some of the drawbacks of the individual technologies. Among those, it can be mentioned permeable reactive barriers, bioremediation, chemical oxidation/reduction, phytoremediation, stabilization and thermal treatment (Yeung and Gu, 2011). Most of these combined methods take advantage of the pollutant fluxes electrochemically promoted to enhance the effectiveness of other treatments located between the electrode areas; the general idea is to create 'permeable treatment zones' in the whole soil matrix. This is also the main objective of the electrokinetic-assisted phytoremediation, i.e. to improve the processes carried out by the plants in phytoremediation by means of increasing the mobility and availability of pollutants under the electric field. For this, it is enough to interpose a vegetation cover or a certain number of plants in the area between the electrodes (Fig. 14.1). This combined technology allows the removal of pollutants not only by electrochemical mechanisms

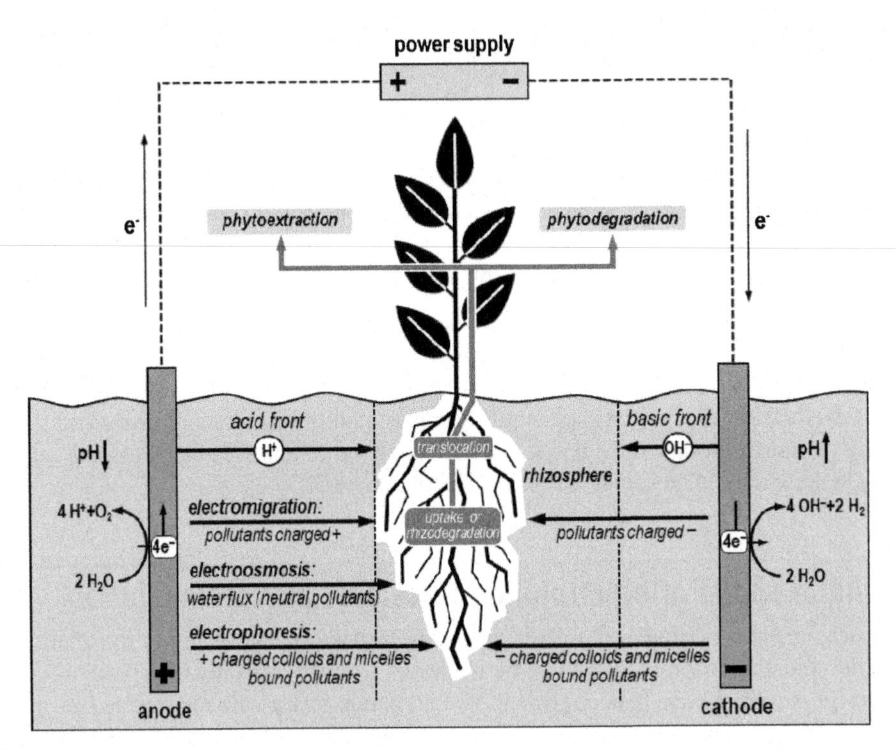

FIG. 14.1

Conceptual framework, mechanisms and electrochemical processes in the electrokinetic-assisted phytoremediation of polluted soils.

but also by plant uptake and accumulation (phytoextraction) or by degradation, both inside the plants (phytodegradation) or in the rhizosphere (rhizodegradation). Therefore, in most of practical cases, EK-phytoremediation avoids the removal of pollutants by pumping them from the electrode wells, with the subsequent benefit in terms of waste management.

14.2.1 Electrochemical reactions in EK-phytoremediation

The dominant redox reaction that takes place in the electrokinetic treatment occurs in the electrodes; it is the electrolysis of water, which generates oxygen gas and hydrogen ions (H^+) in the anode and hydrogen gas and hydroxyl ions (OH^-) in the cathode (Fig. 14.1):

$$Anode(oxidation): 2H_2O \rightarrow O_{2(g)} + 4H^+_{(aq)} + 4e^- \quad E^0 = -1.229V$$

$$Cathode(reduction): 4H_2O + 4e^- \rightarrow 2H_{2(g)} + 4OH^-_{(aq)} \quad E^0 = -0.828V$$

As a result of these electrochemical reactions, pH is decreased in the area surrounding the anode while it is increased near the cathode. Moreover, the formation and migration of two fronts in the soil profile can be detected: one acid, which is carried from the anode to the cathode, and another basic, which moves from the cathode to the anode; it is generally accepted that hydrogen ions moves at a rate of a least twice that of hydroxyl ions and other metal ions (Acar *et al.*, 1995; O'Connor *et al.*, 2003). The electrolysis of water has important effects on transport, transformation and degradation of pollutants during the electrochemical treatment. At the same time, it also leads to dynamic changes in the geochemical behavior of the soil, greatly influencing processes such as sorption-desorption, precipitation-dissolution and oxidation-reduction, and impacts other soil properties such as electrical conductivity or microbial activities (Cang *et al.*, 2011; Gill *et al.*, 2014); in turn, the presence of plants in EK-phytoremediation can significantly influence the extent of these processes.

14.2.2 Electrochemical fluxes in EK-phytoremediation

Let us now turn to the description of the electrochemical transport phenomena that take place in the soil under the action of an electric field. They are three: electromigration, electroosmosis and electrophoresis (Fig. 14.1). Electromigration is the transport of ions and other charged species present in the soil pore fluid to the electrode of opposite charge. Due to it, anions, e.g. chlorides, nitrates, or phosphates, move toward the anode while cations, e.g. heavy metals, move to the cathode. This type of transport is the most important for highly solubilized inorganic species, polar organic molecules, ionic micelles and colloidal electrolytes (Reddy and Cameselle, 2009). Electroosmosis is the advective flow of soil pore water due to electric gradient relative to the stationary charged solid surface; it is a result of the charging of the soil particles surface and the ionization of the water layer adjacent to it. As the soil particles surface is usually negatively charged, the counterions within the diffuse water double layer are cations and therefore, the net flux of water by electroosmosis is from the anode to the cathode. Electroosmosis is the dominant transport mechanism for neutral and non-ionic organic and inorganic species that are dissolved, suspended or emulsified within the soil pore water. Lastly, electrophoresis is the transport mechanism which causes the movement of charged particles of colloidal size and bound pollutants under the electric field relative to the pore fluid. Although it may be a dominant transport mechanism for biocolloids and micelles, it uses to be negligible in low-permeability soils as compared to electromigration and electroosmosis.

As it can be easily undertaken, the extent of those three electrochemical transport processes is a function of a lot of parameters related with the electric field (e.g. electric potential and current, application regime) and soil properties (e.g. porosity, pH, viscosity of the pore fluid, mineralogical composition, surface charge of the solid matrix). Moreover, the presence of plants adds one more factor to the already complex mechanisms of transport of chemical species and water in the soil.

14.3 Practical aspects of EK-phytoremediation

The content of this section, dedicated to the practical aspects of EK-phytoremediation, refers mainly to the experience accumulated in lab-scale tests, which represent the most part of all the works published to date.

14.3.1 Electric field type and application mode

The type of electric field and the mode of application are one of the first issues that must be considered in the EK-phytoremediation. Three different alternatives have been investigated (Table 14.1): (i) the use

Table 14.1 Summary of the main experimental parameters used in the electrokinetic-assisted phytoremediation tests carried out in polluted soils. Data from previous literature.

References	Plant species *(botanical name)*	Pollutants/ Source/ Experimental scale	Solubilizing agent	Electric field application			Electrode material[a] (+/–)	Electrode configuration *(Array type/ current direction[b]/ number/ position[c])*
				DC current	*AC current*	DC reversal polarity		
(O'Connor *et al.*, 2003)	*Lolium perenne*	Cu or Cd, As/ industrial/ laboratory		1.25 V cm^{-1} 24 h/ day			(+) graphite (–) stainless steel	1D/ H/ 2/ ends
(Lim, Salido and Butcher, 2004)	*Brassica juncea*	Pb, As/ industrial/ laboratory	EDTA	1.25 – 5 V cm^{-1} 1 h/ day			(+/–) Cu	1D/ H/ 2/ ends
(Zhou *et al.*, 2007)	*Lolium perenne*	Cu, Zn/ mining site/laboratory	EDTA and/or EDDS	1 V cm^{-1} 6 h/ day			(+/–) stainless steel	1D/ V/ 2/ ends
(Aboughalma, Bi and Schlaak, 2008)	*Solanum tuberosum*	Zn, Pb, Cu, Cd/ industrial/ laboratory		500 mA 24 h/ day	500 mA 24 h/ day		(+/–) Zn-coated Fe	1D/ H/ 2/ ends
(Bi *et al.*, 2011)	*Brassica napus* *Nicotiana tabacum*	Cd / spiked/ laboratory Cd, Zn, Pb/ industrial/ lab.			1 V cm^{-1} 24 h/ day	1 V cm^{-1}, 24 h/ day RP each 3 h	(+/–) carbon	1D/ H/ 2/ ends
(Cang *et al.*, 2011)	*Brassica juncea*	Cd, Cu, Pb, Zn/ industrial / laboratory		1 – 4 V cm^{-1} 8 h/ day			(+/–) graphite	1D/ H/ 2/ ends
(Cang *et al.*, 2012)	*Brassica juncea*	Cd, Cu, Pb, Zn/ industrial / laboratory		1 – 4 V cm^{-1} 8 h/ day			(+/–) graphite	1D/ H/ 2/ ends
(Putra, Ohkawa and Tanaka, 2013)	*Poa pratensis*	Pb/ spiked/ laboratory		50 mA 24 h/ day			(+) graphite (–) stainless steel	2D/ V/ 4 vertical anodes in rectangle, 1 horizontal cathode (top)
(Tahmasbian and Safari Sinegani, 2014)	*Helianthus annuus*	Cd/ pasture near mine/ laboratory	EDTA, cow manure, poultry manure	1.1 – 3.3 V cm^{-1} 1h/ day			(+) graphite (–) stainless steel	2D/ H/ 4 vertical anodes in square, 1 central cathode (plant)

(Chirakkara, Reddy and Cameselle, 2015)	*Avena sativa Helianthus annuus*	Pb, Cd, Cr, PAHs[†]/ spiked/ laboratory			1.2 V cm^{-1} 3 h/ day		(+/−) graphite	1D/ H/ 2/ ends
(Mao *et al.*, 2016)	*B. juncea**, *B. oleracea**, *Spinacia oleracea*	Pb, As, Cs/ spiked/ laboratory		1 V cm^{-1} 24h/ day			(+/−) graphite	1D/ H/ 2/ ends
(Tahmasbian and Safari Sinegani, 2016)	*Helianthus annuus*	Pb/ pasture near mine/ laboratory		1.1 – 3.3 Vcm^{-1} 1h/ day			(+) graphite (−) stainless steel	2D/ H/ 4 vertical anodes in square, 1 central cathode (plant)
(Acosta-Santoyo, Cameselle and Bustos, 2017)	*Lolium perenne*	Pb, Cr, Cd/ spiked/ laboratory		1 V cm^{-1} 8 h/ day	1 V cm^{-1} 8 h/ day	1 V cm^{-1}, 8 h/ day RP each 4 h	(+) graphite (−) graphite-Ti	1D/ H/ 2/ ends
(Xiao *et al.*, 2017)	*Sedum alfredii*	Cd/ industrial/ Laboratory	EDTA, pig manure, humic acid			1 V cm^{-1}, 24 h/ day RP each day	(+/−) graphite	1D/ H/ 2/ ends
(Luo, Wu, *et al.*, 2018)	*Eucalyptus globulus*	Cd, Pb, Cu/ industrial/ field trial		0.01-0.05 V cm^{-1} 6h/ day	0.01–0.05 V cm^{-1} 6 h/ day		(+/−) Ti	2D/ H/ rows with 4 vertical anodes and 4 horizontal cathodes

**Brassica juncea, Brassica oleracea;[†] naphthalene and phenanthrene*

[a](+) anode, (−) cathode

[b]H: horizontal, V: vertical

[c]ends: electrodes located at the ends of the pot/container

(*continued*)

Table 14.1 Cont'd.

References	Plant species (*botanical name*)	Pollutants/ Source/ Experimental scale	Solubilizing agent	Electric field application			Electrode material[a](+/–)	Electrode configuration (*Array type/ current direction[b]/ number/ position[c]*)
				DC current	*AC current*	*DC reversal polarity*		
(J. Luo, Cai, *et al.*, 2018)	*Eucalyptus globulus*	Cd, Pb, Cu/ industrial/ laboratory		0.1 – 0.5 V cm^{-1} 6 h/ day			(+) graphite (–) stainless steel	2D/ H/ 6 vertical anodes in circle, 1 central vertical cathode (plant)
(Jie Luo, Cai, *et al.*, 2018)	*Eucalyptus globulus*	Cd, Pb, Cu/ industrial/ mesocosm	EDTA	0.025 – 0.125 V cm^{-1}, 6 h/ day	0.025–0.125 V cm^{-1}, 6 h/ day		(+/–) stainless steel	1D/ V/ 1 horizontal anode (bottom), 1 vertical cathode (plant)
(Sánchez *et al.*, 2018)	*Zea mays*	Atrazine/ spiked/ laboratory				2 – 4 V cm^{-1}, 4h/ day RP each 2 h	(+/–) graphite	1D/ H/ 2/ ends
(Acosta-Santoyo *et al.*, 2019)	*Zea mays*	No pollutants/ laboratory		0.1 – 0.8 V cm^{-1} 24 h/ day			(+) $IrO_2.Ta_2O_5$ (–) Ti (+/–) Ti	1D/ H/ 2/ ends
		Petroleum hydrocarbons/ industr./ mesocosm, field trial		0.25 V cm^{-1} 8 h/ day			(+) $IrO_2.Ta_2O_5$ (–) Ti	2D/ H/ 6 vertical anodes in circle, 1 central vertical cathode
(Cameselle, Gouveia and Urréjola, 2019a)	*P. canariensis**,*Z. mays**, *B. rapa**, *L. perenne**	Cd, Cr, Pb/ spiked/ laboratory		0.3 – 1.3 V cm^{-1} 24 h/ day			(+/–) graphite	1D/ H/ 2/ ends
(Cameselle and Gouveia, 2019)	*Brassica rapa*	Pb, Cd, Cr, PAHs[‡]/ spiked/ laboratory		1 V cm^{-1}, 8 or 24 h/day	1 V cm^{-1},8 or 24 h/day	1 V cm^{-1}, 8 h/ day RP each 4 h	(+/–) graphite	1D/ H/ 2/ ends
(Luo *et al.*, 2019)	*Noccaea caerulescens*	Cd, Cu, Pb, Zn/ industrial/ laboratory		0.1 – 0.5 V cm^{-1} 4 h/ day			(+/–) stainless steel	1D/ V/ 1 horizontal anode (bottom), 1 horizontal cathode (top)
(Rocha *et al.*, 2019)	*Helianthus annuus*	Petroleum/ spiked/ laboratory	SDS (sodium dodecyl sulphate)	1 V cm^{-1} 24 h/ day		1 V cm^{-1}, 24 h/ day RP each day	(+/–) graphite	1D/ H/ 2/ ends

(Sánchez and López-Bellido, 2019a)	*Zea mays*	Atrazine/ spiked/ laboratory				2 – 4 V cm^{-1}, 4 h/ day RP each 2 h	(+/–) graphite	1D/ H/ 2/ ends
(Sánchez and López-Bellido, 2019b)	*Lolium perenne*	Atrazine/ spiked/ laboratory				1 V cm^{-1}, 6 or 24 h/ day RP each 2 h	(+/–) graphite	1D/ H/ 2/ ends
(Sánchez, López-Bellido, Rodrigo, *et al.*, 2020)	*Lolium perenne*	Atrazine/ spiked/ mesocosm		0.6 V cm^{-1} 24 h/ day			(+/–) graphite	1D/ H/ 2/ ends
(Sánchez, López-Bellido, Cañizares, *et al.*, 2020)	*Lolium perenne*	Atrazine/ spiked/ mesocosm				0.6 V cm^{-1}, 24 h/ day RP each 24 h	(+/–) graphite	1D/ H/ 2/ ends
(Siyar *et al.*, 2020)	*Vetiveria zizanoides*	As, Cd, Cu, Mo, Pb, Sb, Zn/industrial/ laboratory		1 – 2 V cm^{-1} 8 h/ day	2 V cm^{-1} 8 h/ day		(+/–) graphite	1D/ H/ 2/ ends
(Xu *et al.*, 2020a)	*Solanum nigrum*	Cd/ spiked/mesocosm				1 – 3 V cm^{-1}, 6, 10 or 14 h/ day, RP each 2 h	(+/–) graphite	2D/ H/ 2 anodes, 2 cathodes/ ends
(Xu *et al.*, 2020b)	*Solanum nigrum*	Cd/ spiked/mesocosm		1 V cm^{-1} 24 h/ day	1 V cm^{-1} 24 h/ day	1 V cm^{-1}, 24 h/ day RP each 3 h	(+/–) graphite (+/–) stainless steel	2D/ H/ 2 anodes, 2 cathodes/ ends

**Phalaris canariensis, Zea mays, Brassica rapa, Lolium perenne; ‡ anthracene and phenanthrene*
[a](+) anode, (−) cathode
[b]H: horizontal, V: vertical
[c]ends: electrodes located at the ends of the pot/container

of unidirectional electric current (DC); (ii) the use of DC current with periodic reversal of the electrode polarity (DC-RP); and (iii) the use of alternate electric current (AC).

The application of unidirectional DC current has been by far the most used in the EK-phytoremediation tests since it was the first proposed option and it is the only capable of causing stable electrochemical fluxes of water, ions and pollutants throughout the soil (Bi *et al.*, 2011; Cameselle, Chirakkara and Reddy, 2013; Acosta-Santoyo, Cameselle and Bustos, 2017). These fluxes are considered as the basis for the pollutant availability enhancement and, therefore, the essential mechanism to improve plant performance in EK-phytoremediation. However, the application of DC current can eventually cause marked decreases and increases in the pH of the soil areas close the anode and the cathode, respectively. The decrease in anode pH may be useful in the case of metal polluted soils because it contributes to the solubilization and mobilization of metals in some extent. However, the extreme pH values, along with the formation of the acid and basic fronts, have been shown to cause, in some cases, detrimental effects on plant growth and microorganisms, reducing the overall efficiency of EK-phytoremediation.

One way to avoid those extreme pH values is by means of inverting periodically the applied DC current; thus, in this 'reversal polarity' mode, electrodes act alternatively as anode and cathode allowing the maintenance of constant electrochemical fluxes of water and chemical species through the soil (Sánchez and López-Bellido, 2019a). It has been shown that the DC-RP mode does not cause significant pH changes and may significantly increase the uptake and/or degradation of pollutants due to their enhanced availability (Sánchez *et al.*, 2018; Rocha *et al.*, 2019; Sánchez, López-Bellido, Cañizares, *et al.*, 2020).

The third alternative studied for EK-phytoremediation is the application of AC current. This type of electric field results in a continuous variation of polarity which avoids the formation of acid and alkaline fronts (Chirakkara, Reddy and Cameselle, 2015). This fact, along with the periodic hyperpolarization-depolarization processes occurring in root cells and a potential higher movement of soil pore water, might explain the increase in plant growth that has been reported in several works (Bi *et al.*, 2011; Chirakkara, Reddy and Cameselle, 2015; Luo, Wu, *et al.*, 2018). However, since AC current does not cause a continuous transportation of contaminants either by electromigration or electroosmosis, it remains unclear if this type of current application mode is capable to increase phytoextraction and/or phytodegradation of pollutants (Bi *et al.*, 2011; Acosta-Santoyo, Cameselle and Bustos, 2017; Cameselle and Gouveia, 2019).

14.3.2 Electrode material

Another important issue in the practice of EK-phytoremediation is the material of which the electrodes are made. This is an important consideration which, however, has not been systematically investigated; most of the research works directly choose the electrode material without any previous analysis. Among other desirable features, electrodes used in EK-phytoremediation should be capable to keep their physical and chemical integrity in the experimental conditions; it would avoid deterioration and the release of any additional chemical species that could add complexity to the soil-plant system and interfere in the phytoremediation processes. This is especially important in the case of metallic anodes in which a potential release of dissolved metal species from the electrode oxidation could take place.

Although a great variety of electrode materials have been studied for the different electrochemical applications, the materials employed in EK-phytoremediation studies are basically three: graphite, stainless steel and titanium (Table 14.1). Copper and zinc-coated iron have been also used in one work each (Lim, Salido and Butcher, 2004; Aboughalma, Bi and Schlaak, 2008). Graphite has been the most used material because its inertness and relatively low cost; it can act as anode or cathode and the electrode shape includes rods (Cang *et al.*, 2009; Chirakkara, Reddy and Cameselle, 2015; Sánchez and López-Bellido, 2019a), plates (Bi *et al.*, 2011) and elongated sheets (Acosta-Santoyo, Cameselle and Bustos, 2017; Cameselle, Gouveia and Urréjola, 2019b). Stainless steel has been employed mainly for cathodes, in combination with graphite anodes, because it is cheaper and more easily available than graphite; in addition, it allows its use in the shape of a mesh (Zhou *et al.*, 2007; Putra, Ohkawa and Tanaka, 2013). Luo et al. (Luo and Wu, 2018) used reticulated titanium cylinders as anode and cathode while Acosta-Santoyo et al. (Acosta-Santoyo *et al.*, 2019) conducted a EK-phytoremediation study using two combinations of electrodes based on titanium for the decontamination of a soil polluted by hydrocarbons: (i) a Ti anode vs a Ti cathode, and (ii) an $IrO_2 \cdot Ta_2O_5$ coated titanium anode vs a Ti cathode. They suggested that, apart from their physicochemical resistance, titanium electrodes can generate superoxide and hydroxide anions that could improve uptake of water and oxygen by plants and degradation of organics. In any case, titanium is quite more expensive than graphite and stainless steel and the mentioned paper did not clearly demonstrate the advantages of using this type of material. In a very recent paper, Xu et al. (Xu *et al.*, 2020b) have carried out a comparative analysis of stainless steel and graphite electrodes concluding that their performance was basically the same.

14.3.3 Electrode configuration

The geometrical arrangement of electrodes in EK-phytoremediation is another practical aspect to consider. To date, the most used alternative has been to directly insert the electrodes into the soil but the immersion of them into compartments filled with water and/or electrolytes has also been used (Fig. 14.2 and Table 14.1). The first option is easier from a practical point of view; it consists solely of inserting the electrodes into the soil, ensuring their contact with the interstitial water to close the electrical circuit (Fig. 14.2a). As this configuration usually works quite good and it does not require especial containers for conducting the tests, it has been the preferred in most of the research works published until now.

However, some researchers have chosen to arrange the electrodes partially submerged in wells (Fig. 14.2b) (Putra, Ohkawa and Tanaka, 2013) or dedicated compartments in specially designed containers (O'Connor *et al.*, 2003; Chirakkara, Reddy and Cameselle, 2015; Rocha *et al.*, 2019; Siyar *et al.*, 2020). This arrangement has been widely used in the classical electrokinetic remediation of soils since the first objective of this technology is to transport the pollutants, dissolved in the pore liquid phase, from the bulk soil to the electrode areas, from which they must be removed by pumping and ex-situ treated (Virkutyte, Sillanpää and Latostenmaa, 2002). In EK-phytoremediation, the pumping of the liquid phase is not necessary because the pollutants are treated within the soil by means of the different capabilities of plant and/or rhizosphere microorganisms. However, despite the difficulty of designing and having enough special containers to carry out the EK-phytoremediation tests, they show a series of advantages that could make interesting its use, at least, with research purposes. The following can be cited: (i) the possibility of controlling the liquid level in the compartments or wells and, therefore, the height of the saturated soil area in the specimen; (ii) to make possible the measurement of the electrochemical fluxes, mainly electroosmosis; (iii) a better way of getting an homogeneous watering of the soil; (iv) to facilitate the addition of chemicals such as buffering solutions, chelating agents (for metal mobilization) or surfactants (for organics mobilization); (v) the possibility of simple replacement of electrodes in case of damage or failure.

Lastly, the location and geometrical arrangement of the electrodes can affect the effectiveness of EK-phytoremediation since it conditions the direction of the electrochemical fluxes within the soil

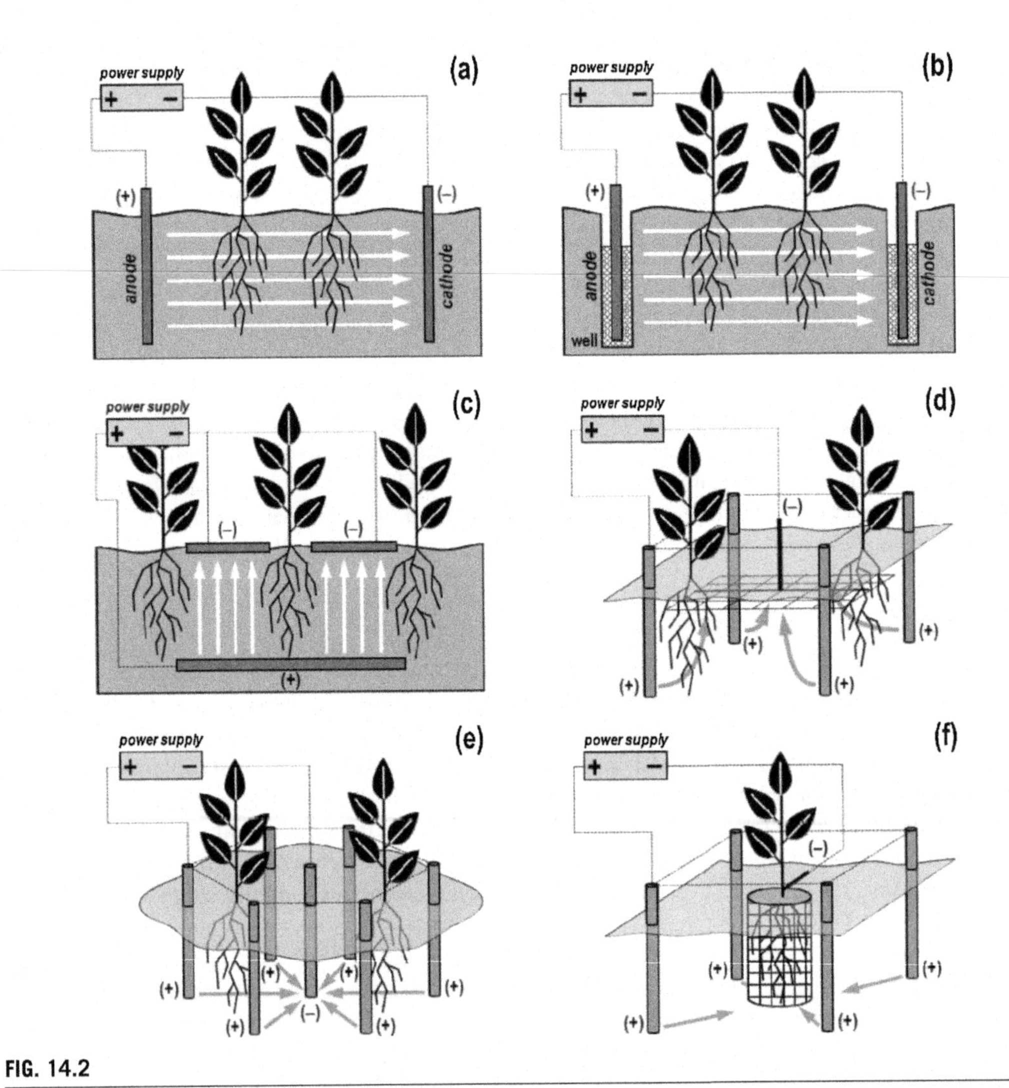

FIG. 14.2

Electrode configurations used in the electrokinetic-assisted phytoremediation tests carried out in polluted soils.

and, therefore, the area which could be potentially cleaned. In Table 14.1 are shown the electrode arrangements reported in the literature. The simplest geometrical configuration is to locate the electrodes (vertically inserted) in both ends of the containers used for the EK-phytoremediation tests (Fig. 14.2a); by this way, a one-dimensional (1D), horizontal and uniform electric field is generated. This arrangement, that has been used in most of the studies, allows treating the soil section located between the two electrodes, which usually comprises the root zone (Cameselle, Chirakkara and Reddy, 2013). The second option is to generate a 1D vertical electrical field by means of using an anode inserted horizontally in the soil zone that corresponds to the maximum depth reached by the pollution, together with a cathode located near the soil surface (Fig. 14.2c) (Luo *et al.*, 2019). This type of vertical arrangement has been proposed with two purposes: (i) to avoid the leaching of the chelating agents, e.g. EDTA or EDDS (ethylene diamine disuccinic acid), used in combination with EK-phytoremediation, and (ii) to promote the upward migration of metal complexes in the soil column (Zhou *et al.*, 2007). Despite of its advantages, the insertion of a horizontal electrode below the contaminated zone can be difficult in real cases. Two-dimensional (2D) electrode configurations have been also explored in some few works. Those configurations allow to get a more homogeneous cleanup of the polluted area and, moreover, as the electroosmotic flux goes from the anodes to the central cathode, it can act as a fence or barrier to avoid the spreading of the pollution. To date, all of 2D arrangements have included several anodes (from four to six) which are placed on the periphery of the soil area to be treated, while the cathode is located in the center of it (Fig. 14.2e). The anodes usually consisted of rods inserted vertically in the soil, but several types of cathodes have been used. Putra et al. (Putra, Ohkawa and Tanaka, 2013) used a stainless steel mesh placed in the surface of the soil (Fig. 14.2d), while Acosta-Santoyo et al. (Acosta-Santoyo *et al.*, 2019) employed a circular array with a central titanium rod. Another unusual option is to negatively charge the plant roots, which work as a cathode, by means of inserting an stainless steel needle in the lowest part of the stem (Fig. 14.2f) (Tahmasbian and Safari Sinegani, 2014, 2016) or placing small stainless steel sheets joined to the plant just above the soil surface (Luo, Wu, *et al.*, 2018). Cameselle and Gouveia (Cameselle and Gouveia, 2019) carried out a theoretical analysis of different electrode configurations concluding that the conventional 1D array (electrodes inserted in the soil side by side) is recommended for lab tests because the uniform electric field generated, while circular distributions are very appropriate when pursuing the concentration of pollutants in a central area, where plants should be placed.

14.4 Effects of EK-phytoremediation on soil properties and microbiota

As it has been pointed out above, the application of a DC electric field to a soil implies the development of several electrochemical processes which cause changes in some physical and chemical characteristics of the soil; those changes, in turn, can affect plant and microbial activities (Cameselle, Chirakkara and Reddy, 2013).

The most studied effect of the DC current is the acidification and, in a minor extent, the alkalinization of the anode and the cathode soil areas, respectively, because of the water electrolysis. The decrease of soil pH may be beneficial for pollutant availability and so, for the cleanup of the soil. In fact, since it is well known that most metals are more soluble at a low pH, it has been a desired objective in the electrokinetic remediation of soils polluted by metals (Yeung and Gu, 2011). However, sometimes, the very acidic pH values (< 5) found in the anode section have been shown to be detrimental for plant growth (O'Connor *et al.*, 2003; Aboughalma, Bi and Schlaak, 2008). On the other hand, the generation of OH^- ions in the cathode can immobilize metals by precipitation of the corresponding hydroxides. Hence, the introduction of weak acids (acetic or citric acid) in the cathode area has been studied as a mean of preventing alkalinization (Yeung and Gu, 2011). It has been reported that the capacity of electrokinetic treatment to change the soil pH is mainly dependent on the current applied (time and voltage gradient) and the buffering capacity of the soil (Cang *et al.*, 2011; Cameselle, Gouveia and Urréjola, 2019a). This latter characteristic would be the main responsible for avoiding changes in soil pH despite these do occur in the soil pore fluid. Sanchez et al. (Sánchez, López-Bellido, Rodrigo, *et al.*, 2020) found some punctual pH values of 2.5 and 11.5 in the pore water from the anode and the cathode soil sections, respectively; however, only variations of around 1 unit were registered for the soil pH.

In EK-phytoremediation, the electrical conductivity (EC) of soils is closely related with pH and the transportation of ions towards the electrodes (Cameselle and Gouveia, 2019; Cameselle, Gouveia and Urréjola, 2019a). Spatial and time variations of soil EC are usually included in the modelling of electrokinetic remediation processes, but this parameter has only been qualitatively evaluated in some EK-phytoremediation studies. Cang et al. (Cang *et al.*, 2012) found that the acidification of the anode zone, along with higher concentrations of solubilized anionic nutrients such as nitrate or phosphate, led to significant increases in soil EC when an unidirectional DC field was applied. However, higher EC values were found for the middle soil sections as compared to those of the electrode sections in EK-phytoremediation using DC current with reversal polarity (Xu *et al.*, 2020b). It was attributed to the continuous transport of nutrients by the electrochemical fluxes and the suction by plant roots. However, it has also been observed that, as the experiment progresses, the conductivity may be gradually decreased due to the absorption of nutrients and contaminants by the plants (Chirakkara, Reddy and Cameselle, 2015; Sánchez and López-Bellido, 2019b; Sánchez, López-Bellido, Rodrigo, *et al.*, 2020).

Although there are not many, some works on EKR and EK-phytoremediation reported not only changes in pH and EC, but also in other parameters related with soil fertility such as the concentration of available nutrients and dissolved organic carbon. Chen et al. (Chen *et al.*, 2006) found that after 60 h of EKR, the concentration of organic carbon did not changed significantly but the available nitrogen, phosphorus and potassium were increased by 0.44, 3.31, and 1.25-fold, respectively. That fact has been also reported by others authors who, in addition, suggested some hypothesis for the observed higher availability and mobility (Cang *et al.*, 2012; Zhou *et al.*, 2015). Thus, DC current application could transform some complex organic nitrogen species into simpler organic and inorganic nitrogen (nitrate and ammonium) by hydrolysis, while inorganic phosphorus species, e.g. calcium, aluminum or iron phosphates, would be partially solubilized by the low pH values reached in the anode. Moreover, H^+ ions generated in the anode may be exchanged by K^+ ions from the clay minerals increasing its availability in the soil. Lastly, negative nitrate and phosphate ions would be accumulated in the anode in a higher extent, while positive ammonium and potassium ions would move towards the cathode, being accumulated there. Zhou et al. (Zhou *et al.*, 2015) also found no significant variations in the organic carbon of the soil after the application of a low-voltage electric field (2 V cm^{-1}) during 10 days in an EKR test.

Among the soil parameters that can be affected by the application of the electric field, pH is the one that has the greatest influence on the viability of microbial communities. It has been shown that the application of a DC field with low intensity (0.3-1 mA cm^{-2}) had no overall effect on microorganisms (Gill *et al.*, 2014). Nevertheless, localized pH drops in the soil area close to the anode can cause stress responses such as increased rates of respiration and metabolic ability; it would be a result of the higher nutrients and carbon availabilities due to the low pH (Lear *et al.*, 2004). Moreover, the increase in soluble forms of metals and other pollutants might be one of the main factors reducing soil microbial activities in the soil anode area (Tahmasbian *et al.*, 2017). Therefore, it can be said that all the measures taken to avoid significant changes in soil pH may limit the negative effects of electrokinetic treatment on microbial communities. On the other hand, it has been shown that the application of low intensity DC current may have some positive effects on microbial activities such as increased substrate utilization (due to an increase in the availability of nutrients and organic compounds) and improved metabolism caused by direct (transfer of electrons from electrodes to microorganisms) and indirect (transfer of electrons through water hydrolysis) stimulation (Cang *et al.*, 2012; Sánchez and López-Bellido, 2019a). Lastly, it has been pointed out that an expected consequence of the electrokinetic remediation is the ohmic heating of the soil (Risco *et al.*, 2016). The increase in soil temperature caused by that heating could also adversely affect microbial growth; however, that effect has never been reported in the EK-phytoremediation literature, probably due to the low voltage gradients used.

On summary, in the previous paragraphs it has been shown that the application of a DC current and the consequent electrochemical processes induce important changes in several soil properties. However, those effects are only detected in a significant extent in the soil area near the electrodes, especially in the anode, with lesser effect elsewhere in the bulk soil. Therefore, it can be properly said that the application of the low intensity electric fields typical of EK-phytoremediation does not affect physico-chemical and biological properties of the soil to a greater extent than other technologies do. Moreover, the well-known capacity of plants to improve the soil properties related with its fertility and microbial viability, along with its ability to alleviate some of the changes carried out by the electrochemical processes, makes EK-phytoremediation as a feasible tool to remediate polluted soils without affecting soil health.

14.5 Effects of the electric current application on plant growth

The effect of the electric current application on the growth parameters of the plants, e.g. plant biomass (data from roots, shoots and/or total plant) and germination rate, is an issue usually discussed in the EK-phytoremediation papers although it is not usually the main object of the research. This type of studies should be carried out by including control tests in the experimental design, that is, by simultaneously using phytoremediation and EK-phytoremediation tests in the same experimental conditions (soil, pollution level, etc.). However, not all the works include those types of control treatments and, sometimes, comparisons between experiments with different values of pollutant concentration or voltage gradient are carried out; so, it is recommendable to follow a suitable experimental design in order to get conclusions adequately supported.

It should not also be forgotten that, among other aspects, EK-phytoremediation was initially proposed based on a former paper by Lemström (Lemström, 1904) regarding the influence of electricity in agriculture and horticulture. This author showed that plants exposed to an electric field grew greener and with higher biomass than those not subjected to the current. But, unfortunately, the results reported to date on EK-phytoremediation do not always support that hypothesis. From the revision of the existing literature on EK-phytoremediation, it can be found that at least eighteen different plant species have been investigated; those include both crop plants, such as maize (*Zea mays*) or sunflower (*Helianthus annuus*), and grasses, e.g. English ryegrass (*Lolium perenne,* the most used) or Kentucky bluegrass (*Poa pratensis*). As it can be also seen in Table 14.1, these plant species have been tested using a wide variety of pollutants and conditions of electric field application. Moreover, some of them have only been used in a single study and there are few the systematic studies regarding the effect of the different electric field parameters on plant growth. Nevertheless, some interesting trends can still be found from a detailed analysis of the literature.

In Fig. 14.3 are shown the data available in EK-phytoremediation literature regarding the increase or decrease found in plant biomass (roots, shoots or total plant), with respect to tests without electricity, as a function of the voltage gradient applied. The displayed data are not identified by plant species and different points may correspond with the same experiment, i.e. root and shoot biomass (data from total plant biomass have been only included when data from roots or shoots were not

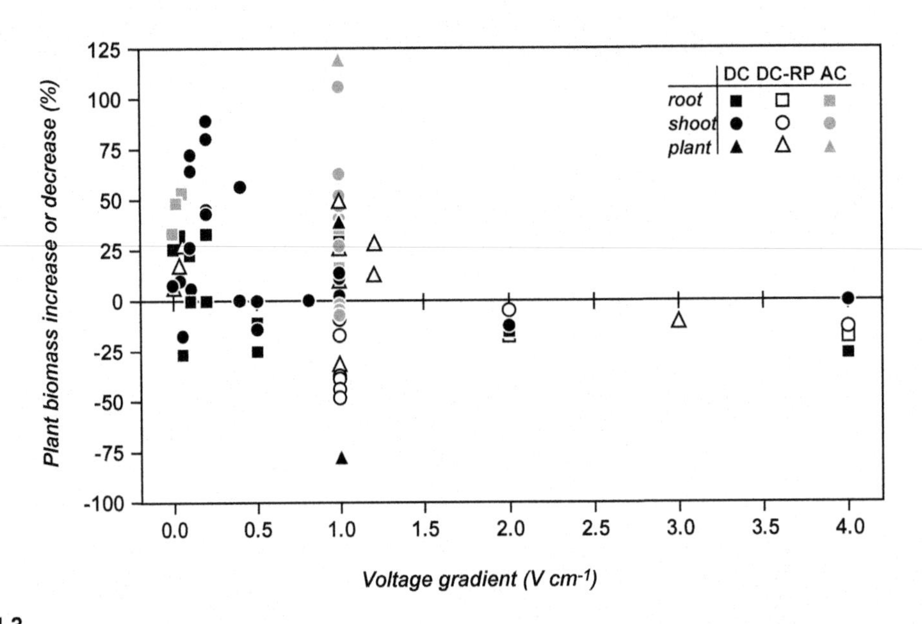

FIG. 14.3

Effect of voltage gradient on plant biomass in the electrokinetic-assisted phytoremediation of polluted soils. Data from previous literature.

available). It must be firstly mentioned that, regardless of the type of electric field (DC, AC or DC-RP), plant biomass always decreased when the voltage gradient was greater than 1.2 V cm^{-1}. Secondly, it clearly appears that the application of AC current (points in grey) had, in almost all cases, a beneficial effect on plant weight, while DC current, in any of its two application modes, affected it in a different way depending on plant species, electric parameters and/or pollution characteristics. Changes in plant biomass were in the ranges −78/+90%, −47/+50% and −6/+120% for the application of DC, DC-RP and AC electric fields, respectively. The highest increase (+120% of total plant weight) corresponded to the EK-phytoremediation of a soil polluted by metals and PAHs using English ryegrass and an AC current of 1 V cm^{-1} applied 4 h twice a day (4 h on, 8 h off), during 4 weeks (Acosta-Santoyo, Cameselle and Bustos, 2017). On the other hand, the highest decrease (−78% of the total plant biomass) was found by Rocha et al. (Rocha *et al.*, 2019) in an EK-phytoremediation test using a DC current of 1 V cm^{-1} applied 24 h a day during 20 days; the plant species used was sunflower and the soil was polluted with petroleum.

There is a practically unanimous agreement on the reasons for the different effect of AC and DC current on plant growth. These are mainly based on the alterations in the soil physicochemical properties caused by both types of electric current, which have been extensively discussed in the Section 14.4. Thus, unidirectional DC current led to marked soil pH changes, especially in the areas close the electrodes, due to the water electrolysis; those changes were higher as the voltage gradient was increased, with negatives consequences on the plant biomass. For instance, this fact has been highlighted by Cang et al. (Cang *et al.*, 2011) in an EK-phytoremediation pot experiment conducted with Indian mustard (*Brassica juncea*) using DC current with three different voltage gradients, i.e. 1, 2 and 4 V cm^{-1} (applied 8 h per day); they found that the higher the voltage gradient, the larger pH and electrical conductivity varied. As a result, root biomass increased for the 1 V cm^{-1} treatment and decreased gradually for higher voltage gradients. Similar results have been reported by Luo et al. (J. Luo, Cai, *et al.*, 2018; Luo, Wu, *et al.*, 2018) and Acosta-Santoyo et al. (Acosta-Santoyo *et al.*, 2019) for southern blue gum (*Eucalyptus globulus*) (in the remediation of a metal-polluted soil) and maize (in a soil polluted by petroleum hydrocarbons), respectively; these latter authors also found a gradually inhibition of germination with increasing voltage gradients. The toxicity caused by the higher availability of metals (the desired effect of the EK-phytoremediation) is another factor which should be considered in the inhibition of plant growth in the EK-phytoremediation of metal-polluted soils (O'Connor *et al.*, 2003; Aboughalma, Bi and Schlaak, 2008). However, other positive effects of DC current application could partially counteract the detrimental ones in the case of low or moderate voltage gradients. These are mainly the enhancement of the ionic transport, resulting in a higher availability of nutrients (nitrate, phosphate, calcium), and the activation of several enzymes that decrease oxidative stress and, in overall, strengthen plant metabolism and other physiological activities (Cang *et al.*, 2012; Putra, Ohkawa and Tanaka, 2013; J. Luo, Cai, *et al.*, 2018).

As previously described, the periodic change in the polarity of the DC electric field contributes to alleviate the soil pH changes although it may not always be enough to completely avoid detrimental effects on plants. Several authors have reported decreases of plant biomass (with respect to that of the plants grown without electricity) for DC application and increases for DC-RP using the same voltage gradient (Rocha *et al.*, 2019; Xu *et al.*, 2020b). Other times, the application of DC-RP improved the beneficial effect found also for unidirectional DC current (Acosta-Santoyo, Cameselle and Bustos, 2017). Finally, there were cases in which the use of DC-RP only achieved a reduction of the negative effect of DC current; Cameselle and Gouveia (Cameselle and Gouveia, 2019) found that the biomass production of turnip (*Brassica rapa*) decreased by 33% when a DC current of 1 V cm^{-1} was applied against a decrease of only 13% when DC-RP was used. In this last work, only the DC-RP treatment did not affect the germination rate while DC and AC current application decreased it between 5 and 10% with respect to the treatment without electricity.

Since the application of AC current do not cause changes in the soil pH, this is the main reason for not expecting negative effects on plant biomass by using this type of electric field. Regarding the beneficial effects usually reported in the literature, some researchers have argued the following three reasons: (i) a higher water movement in the soil due to the bidirectional electroosmotic flux induced by the AC current; (ii) the increase in the inflow of ions within the root cells caused by the periodic hyperpolarization and depolarization of the cell membranes; and (iii) the improvement in the metabolic activities of the plants due to higher activity of water ions at both extracellular and intracellular levels. Actually, those reasons were firstly hypothesized by Bi et al. (Bi *et al.*, 2011) and, later, other authors have repeated them without further discussion (Acosta-Santoyo, Cameselle and Bustos, 2017; Xu *et al.*, 2020b). In our opinion, the first point, i.e. the existence of a electroosmotic flux, seems difficult to understand considering that in the AC current usually applied in EK-phytoremediation the direction of the electric charge usually changes 50 times a minute; in such conditions, a relative higher mobility of water and/or ions (and pollutants) may be accepted but not the establishment of continuous electroosmotic nor electromigration fluxes (Chirakkara, Reddy and Cameselle, 2015).

Apart from the voltage gradient applied, it seems reasonable to expect that the effect of the electric field on plant biomass is also dependent on the application time. There are some few works in which different application times were tested in EK-phytoremediation. Cameselle and Gouveia

(Cameselle and Gouveia, 2019) used DC current (1 V cm^{-1}) for the EK-phytoremediation of a mixed contaminated soil (with metals and PAHs) with continuous (24 h/day) and periodic (8 h/day) application; they found that the reduction in the biomass suffered by *Brassica rapa* was decreased from 33%, for the continuous electric DC current, to only 7% when the electric field was applied 8 h a day. Sanchez et al. (Sánchez and López-Bellido, 2019b) reported a non-statistically significant decrease of 9.1% in the weight of English ryegrass plants grown in a soil polluted by atrazine when a DC-RP current was applied 6 h/day against a decrease of 32% when the current was continuously applied. Finally, Xu et al. (Xu *et al.*, 2020b) applied a DC-RP (1 V cm^{-1}) current to *Solanum nigrum* (a widespread weed) plants with three different daily application times, i.e. 6, 10 and 14 h a day; the higher biomass was found for the treatment of 10 h a day suggesting that could exist an optimal combination of voltage gradient and application time from the point of view of biomass production. This hypothesis can be further explored by graphing the increases or decreases on plant biomass reported in the literature vs the factor [voltage gradient x application time] expressed in V h cm^{-1} day^{-1}; those data are displayed in Fig. 14.4 (data from AC current have not been shown because they are all positive). As it can be seen, when the product of the voltage gradient by the application time is below 10 V h cm^{-1} day^{-1}, most of the data corresponded to increases in plant biomass; however, for higher values of that product, decreases in plant weight were predominant. So, that value of 10 V h cm^{-1} day^{-1} could be considered as an empirical threshold value for the product of the voltage gradient by the application time in order to avoid the detrimental effect of DC current in EK-phytoremediation studies. It additionally supports the fact that the physicochemical changes occurring in the soil as a result of DC field application are also function of both the voltage gradient and the time which it is applied (Cameselle and Gouveia, 2019).

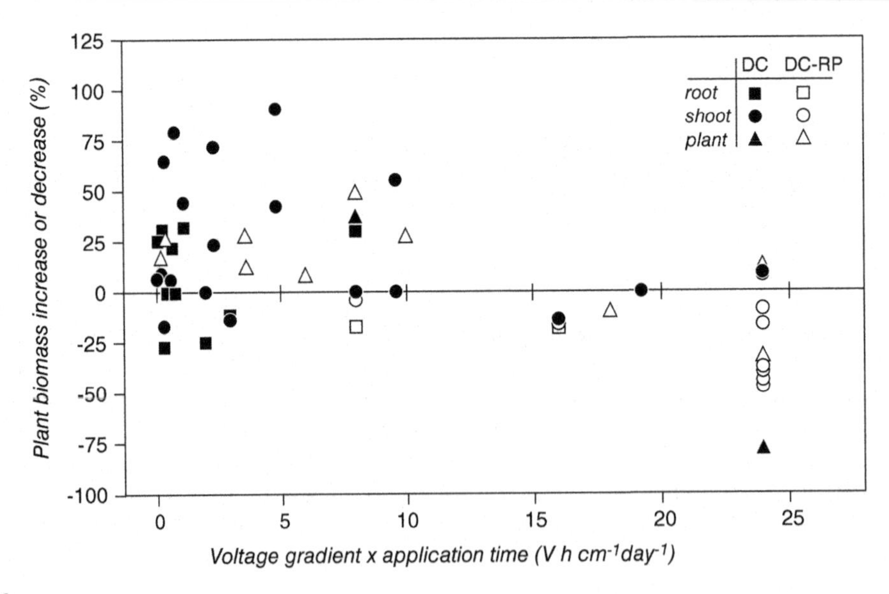

FIG. 14.4

Effect of voltage gradient and electric field application time on plant biomass in the electrokinetic-assisted phytoremediation of polluted soils. Data from previous literature.

14.6 EK-Phytoremediation of Metal-Polluted Soils

Electrokinetic remediation was firstly applied to soils polluted by metal(loid)s, both those which are present as cationic species, e.g. lead, cadmium, nickel or radionuclides such as uranium, and those which tend to form anionic species, e.g. arsenic, chromium or selenium (Cameselle, Chirakkara and Reddy, 2013). It may also be the reason for which, to date, EK-phytoremediation have been mainly focused on the enhancement of metal phytoextraction. The primary mechanisms of EKR for improving metal availability in soils are the partial solubilization and mobilization of metals by means of the acidification in the anode surrounding area (due to the generation of H^+ ions by water electrolysis) and the subsequent transport of cationic and anionic metal species to the cathode and anode, respectively, by electromigration (Virkutyte, Sillanpää and Latostenmaa, 2002; Reddy and Cameselle, 2009).

Cang et al. (Cang *et al.*, 2011) used several chemicals, i.e. DTPA, ammonium nitrate, a mixture of diluted acetic, lactic, citric, malic and formic acids (Rhizosphere-based method) and the first step of the sequential BCR extraction method (acetic acid 0.11 mol L^{-1}), to carry out single-step extractions of Cu, Cd, Pb and Zn after the application of the EK-phytoremediation by Indian mustard to an industrial polluted soil. They found that, under a DC electric field, the bioavailable Cd and Zn soil concentrations were increased from cathode to anode due to the formation of negatively charged complexes with soil organic acids, which were transported towards the anode by electromigration; however, available Cu increased from anode to cathode. Those results could not be well correlated with the concentrations of metals in the plant tissues showing that there is not always a good correspondence between the availability found in plants with that measured by chemical extraction methods (Ruiz, Alonso-Azcárate and Rodríguez, 2011; Bolan *et al.*, 2014). The application of a DC electric field of 2 V cm^{-1} led to an increase of approximately twice in the shoot Pb concentration of the plants grown in the anode region with respect to those in the control treatment. Cd and Cu shoot concentrations in the anode zone were also higher than those of the cathode one, which corresponded to the trend found for the available Cd but not for Cu. It should be considered that in this research the extraction methods were applied to soils after plant growth so that the concentrations of available metals could have been altered by the plant uptake; carrying out a control EKR test (using electric current but without plants) could have avoided this problem, leading to clearer conclusions regarding metal mobilization by the DC current. Similar results were reported by Tahmasbian et al. (Tahmasbian and Safari Sinegani, 2014) who found that the available Cd concentration was lower in the cathode than in the anode and related it with the Cd uptake by sunflower (which was negatively charged acting as a cathode). Chirakkara et al. (Chirakkara, Reddy and Cameselle, 2015) verified the higher exchangeable Cd soil concentrations caused by the application of an AC current by using a control EKR treatment; however, this higher availability of Cd was not further compared with its accumulation in the plant tissues. Some authors have found that, when a DC current with reversal polarity was applied, the concentration of DTPA-extractable cadmium was higher in the soil section located between the electrodes than in the anode or cathode soil sections (Xiao *et al.*, 2017; Xu *et al.*, 2020b); likewise, despite these authors did not use a control EKR test, the treatments with the higher DTPA-extractable soil Cd concentration corresponded with the larger plant Cd concentrations. From this paragraph, it follows that additional work should be done to clearly demonstrate the mobilization of metals in EK-phytoremediation by means of one-step or sequential extraction methods.

Regardless the issue of the metal availability assessment in EK-phytoremediation, from the literature it seems clear that, in general, the application of electric current really enhances metal uptake by plants. However, just like plant growth, the results must be carefully analyzed considering the conditions of electricity application, the pollutants and the plant species.

Starting with the application of unidirectional DC current, Cang et al. (Cang *et al.*, 2011) found that, compared with no current application, an electric field of 1 V cm^{-1} increased metal accumulation in the roots of Indian mustard but it had no significant effect on shoot concentration. They needed to apply 2 V cm^{-1} to obtain significant improvements in both shoot and root metal concentrations; however, when a voltage drop of 4 V cm^{-1} was applied, Cd, Cu and Zn shoot concentrations were lower than those of the test without electricity. In other EK-phytoremediation work reporting the use of the arboreal species *Eucalyptus globulus* in a field trial, Luo et al. (Luo, Wu, *et al.*, 2018) found that the plant concentrations of Pb, Cd and Cu followed an increasing trend as the voltage of a DC current was increased from 2 to 10 V (corresponding to voltage gradients of 0.01 to 0.05 V cm^{-1}). Similar trends between voltage gradient and plant metal uptake enhancement were reported by Luo et al. (Luo *et al.*, 2019) for the EK-phytoremediation of a soil polluted by Cd, Cu, Pb and Zn (applying voltage drops of 0.1, 0.2 and 0.5 V cm^{-1}) and Siyar et al. (Siyar *et al.*, 2020), who used Vetiver grass (*Vetiveria zizanoides*) and a DC voltage in the range of 1-2 V cm^{-1} for the EK-phytoremediation of a multi-contaminated (As, Cd, Cu, Pb, Sb, Zn) industrial soil; both authors also found an improvement in the translocation of metals from roots to shoots. Other works using unidirectional DC current and single values of voltage gradient have also reported increases in the metal accumulation by different plant species, e.g. Kentucky bluegrass (Putra, Ohkawa and Tanaka, 2013), Indian mustard and spinach (*Spinacia oleracea*) (Mao *et al.*, 2016), ryegrass (Acosta-Santoyo, Cameselle and Bustos, 2017) and *Solanum nigrum* (Xu *et al.*, 2020b). However, other results dependent on the specific metal have also been reported; for instance, Aboughalma et al. (Aboughalma, Bi and Schlaak, 2008) observed decreases in Zn, Pb, Cu and Cd shoot concentrations and Pb and Cd root concentrations, while the concentrations of Cu and Zn in roots were increased (applying a continuous unidirectional DC current of 500 mA to potato plants).

Following with EK-phytoremediation using DC-RP, Bi et al. (Bi *et al.*, 2011) found that Cd shoot and root concentrations in tobacco (*Nicotiana tabacum*) plants were significantly decreased (with respect to a control without electric field application) when a DC current of 1 V cm^{-1} was applied continuously with a change of polarity each 3 h; on the contrary, Cu, Zn and Pb root and shoot concentrations were significantly increased under the same conditions. However, in the same study, no significant effects were detected for rapeseed (*Brassica napus*) plants. Xu et al. (Xu *et al.*, 2020b), studying the EK-phytoremediation of Cd by *Solanum nigrum,* found that DC-RP current gave higher Cd plant concentrations than unidirectional DC (for a voltage gradient of 1 V cm^{-1}). These same authors reported that the higher the applied voltage gradient (in the range 1-3 V cm^{-1}), the larger the Cd concentrations in plants and roots of *Solanum nigrum*; however, the increase of the daily application time from 6 to 14 h day^{-1} did not lead to significant changes in the plant metal accumulation (Xu *et al.*, 2020a).

The application of AC current generally led to increases in the plant metal concentrations with respect to the experiments without electric current (Aboughalma, Bi and Schlaak, 2008; Bi *et al.*, 2011; Luo, Wu, *et al.*, 2018), and sometimes, results from AC current were also shown to improve those of DC current application in the same experimental conditions (Aboughalma, Bi and Schlaak, 2008; Bi *et al.*, 2011; Acosta-Santoyo, Cameselle and Bustos, 2017; Xu *et al.*, 2020b). Nevertheless, Chirakkara et al. (Chirakkara, Reddy and Cameselle, 2015), based on metal soil concentrations at the end of the experiments, concluded that AC current did not affect significantly the phytoextraction of metals by oat (*Avena sativa*) and sunflower. Luo et al. (Luo, Wu, *et al.*, 2018) reported that AC current increased the concentration of Cd, Pb and Cu in the tissues of *Eucalyptus globulus* as the applied voltage gradient was increased; however, those increases were equal or lower than those of DC current obtained in the same experiment. Xu et al. (Xu *et al.*, 2020b) found that root and shoot Cd concentrations in *Solanum nigrum* using AC current were higher than those of unidirectional DC current but equal or lower than those of DC-RP electric field application. Moreover, some researchers have found that AC current can promote the physiological mechanisms which carry out the translocation of metals from root to shoots; it has been described in the literature for Pb and Cu in *Eucalyptus globulus* (Luo, Wu, *et al.*, 2018) and for several metals in Vetiver grass (Siyar *et al.*, 2020). Lastly, it must be mentioned that, although in some cases there were not significant improvements in the plant metal concentrations, an enhancement in the phytoextraction yield (calculated as plant biomass by plant metal concentration) was registered. Those higher efficiencies were mainly caused by the increase in plant biomass with the application of AC current rather than from the rise of metal concentration in plants (Bi *et al.*, 2011; Xu *et al.*, 2020b).

The use of chelating and other mobilizing agents has been shown as a successful way of enhancing the effectiveness of both electrokinetic remediation and phytoremediation (Reddy and Cameselle, 2009; Bolan *et al.*, 2014), and therefore, there are some papers exploring this possibility for the EK-phytoremediation of metals. As it is well known, ethylene diamine tetraacetic acid (EDTA) has been the most widely used chelate in phytoremediation because of its large ability to produce soluble metal complexes and its relatively low cost (Kumar Yadav *et al.*, 2018); probably for this reason, it has been used in most of the papers reporting the combination of EK-phytoremediation and solubilization agents. Lim et al. (Lim, Salido and Butcher, 2004) found that the application of a DC current of 1.25–5 V cm^{-1} along with an EDTA addition of 0.5 mmol per kg of soil increased Pb concentrations in *Brassica juncea* by two to four fold as compared to those of using only EDTA. Zhou et al. (Zhou *et al.*, 2007) and Luo et al. (J. Luo, Cai, *et al.*, 2018) have investigated the combination of EDTA addition and a vertical DC electric field to avoid metal leaching. In both cases, the highest values of plant metal concentrations were achieved when electric current and EDTA were simultaneously applied; moreover, it was observed that the gravitational migration of the metals solubilized by EDTA to the bottom of the soil column was significantly reduced by the application of a vertical DC current. In addition to EDTA, Zhou et al., in the abovementioned study, also checked ethylene diamine disuccinic acid (EDDS) in the same conditions than those of EDTA, obtaining similar conclusions. Tahmasbian et al. (Tahmasbian and Safari Sinegani, 2014) studied the effect of some potential mobilizing agents, i.e. EDTA, cow manure and poultry manure extracts, and electricity on Cd and Pb phytoextraction from a mine soil using sunflower. In the case of Cd, the combined application of DC current and the amendments did not lead to any significant enhancement relative to individual agents; however, charging sunflower plants and using EDTA increased both Pb uptake and translocation by plants (Tahmasbian *et al.*, 2017). Lastly, Xiao et al. (Xiao *et al.*, 2017) reported a synergistic effect between humic acid and pig manure compost addition and the application of a DC current with reversal polarity in the phytoextraction of Cd by *Sedum alfredii*; however, no improvements were observed for only EDTA addition. The results included in this paragraph suggest the great potential of the combination of chelating agents and electric current to improve the effectiveness of metal phytoextraction but avoiding the environmental risk derived from the leaching of metal-complexes.

Three mechanisms have been proposed to explain the promotion of plant metal uptake by the application of an electric current (Luo, Wu, *et al.*, 2018): (i) the fluxes of metal ions by electromigration and/or electroosmosis both in horizontal and vertical directions (depending on the electrode arrangement), which causes metal redistribution in the soil; (ii) the changes in soil parameters, e.g. pH, electrical conductivity, microbial and enzymatic activities, density, etc., caused by electrochemical phenomena; and (iii) the changes in the permeability of root cell membranes which increase ion fluxes across them. As it has been seen, in case of applying DC current, it seems recommendable to use low to moderate voltage gradients (up to 2 V cm^{-1}) in order to increase plant metal uptake without causing damage in plant growth and metabolic activities. AC electric field is not able to produce a huge redistribution of metals near electrodes or promote changes in soil pH, but metal ions can still be driven to the rhizosphere of plants increasing their uptake and accumulation in a higher extent than phytoextraction (Xu *et al.*, 2020b). On the other hand, it seems demonstrated that AC current is capable to promote plant metabolism causing an improvement of metal translocation from root to shoots; this fact along with plant biomass enhancement, may lead in some cases to higher phytoextraction yields than those of using DC current.

14.7 EK-Phytoremediation of Organic Pollutants

Despite of both phytoremediation and electrokinetic remediation have demonstrated capability to treat soils contaminated by organic pollutants, the first paper regarding the use of EK-phytoremediation for this type of pollutants was not published until 2015 (Chirakkara, Reddy and Cameselle, 2015) and since then, some more have been published but, in any case, in a much lower number than those regarding metal polluted soils. To our knowledge, the only organic pollutants studied in EK-phytoremediation have been the pesticide atrazine, some PAHs (naphthalene, anthracene and phenanthrene) and petroleum hydrocarbons.

The application of EK-phytoremediation to PAHs has been studied using soils with mixed pollution, i.e. PAHs and metals. The two studies relative to these pollutants used a soil spiked in the laboratory with naphthalene or anthracene, phenanthrene, lead, cadmium and chromium. Chirakkara et al. (Chirakkara, Reddy and Cameselle, 2015) used sunflower and oat and AC current (around 1 V cm^{-1}, 3h per day); they found that naphthalene was completely removed from soil in all the treatments while phenanthrene removal was not improved by using either plants or electric current. In a more recent paper in which *Brassica rapa* was used together with AC, DC and DC-RP electric fields, Cameselle and Gouveia (Cameselle and Gouveia, 2019) concluded that the electric current application enhanced the PAHs (anthracene and phenanthrene) removal as compared to that of using only plants, being the highest efficiency found for the AC current. Neither of these works shows data from PAHs concentrations in the plant tissues, which makes difficult to understand the role of the plants in the PAHs removal; it is suggested that microbiological degradation would be the most relevant removal mechanism but volatilization and/or dissipation of PAHs could also occur in some extent.

Two studies have focused on the EK-phytoremediation of soils polluted by petroleum hydrocarbons. Acosta-Santoyo et al. (Acosta-Santoyo *et al.*, 2019) used maize in combination with DC current to remove hydrocarbons from a polluted soil from an oil refinery. They found that the removal rates varied in the anode, cathode and inter-electrode soil areas due to the redox reactions taking place; likewise, hydrocarbons were transported by the electroosmotic flux to the cathode, where the removal efficiency was the lowest. Hydrocarbons were only accumulated in the roots of maize but the data from EK-phytoremediation were not compared with others from any control test. Another recent work by Rocha et al. (Rocha *et al.*, 2019) used a soil spiked with petroleum to carry out EKR and EK-phytoremediation tests with sunflower and DC current, with and without using reversal polarity (in these experiments the surfactant sodium dodecyl sulfate was introduced in the electrode well in order to generate water-soluble charged species). The combined EK-phytoremediation technology increased the petroleum removal rates obtained by EKR in approximately 16-19%; it was attributed to the mobilization of the hydrocarbons by the electric field and the higher microbial degradation in the presence of plants. The highest removal efficiency was found for the reversal polarity strategy which allowed to maintain better experimental conditions for biological processes.

The removal of atrazine by EK-phytoremediation has been tested by our research group in pot and mesocosm experiments using maize and English ryegrass and using DC current with and without polarity change. In our first work, maize was tested for the remediation of atrazine-spiked soils using a DC electric field of 2 V cm^{-1} during 4 h per day and changing polarity every 2 h. Under these conditions, and despite of a slight decrease on plant biomass, the efficiency of atrazine removal was increased up to 36.5% with respect to the control experiment without electric current (Sánchez *et al.*, 2018). A more detailed study of the processes occurring in the soil and plants from the different sections of the pots (anode, cathode and inter-electrode ones) showed that atrazine transport between electrodes was enhanced by means of electromigration (flux of negatively charged species of atrazine towards the anode) and also by electroosmosis (flux of non-charged soluble species of atrazine towards the cathode). These opposite transport processes were periodically inverted by changing the polarity of the electrodes; as a result, atrazine plant accumulation was significantly higher in the inter-electrode section than in the electrode ones (Sánchez and López-Bellido, 2019a). Another EK-phytoremediation pot experiment was carried out using English ryegrass and different application times of a 1 V cm^{-1} DC current (6 and 24 h per day and changing electrode polarity each 2 h). The analysis of the atrazine soil concentration at different days throughout the experiment allowed to calculate the removal rate constant and the atrazine half-life; it was found that both parameters increased as the time of electric field application was increased (Sánchez and López-Bellido, 2019b). In that experiment, between 17 and 34% of the atrazine removed from the soil remained accumulated in the ryegrass tissues at its end, showing the relevant role of plants in the overall atrazine removal. Nevertheless, from our results, it seems than other mechanisms such as degradation within the plants and microbial degradation in the rhizosphere contributed in a significant extent to the observed atrazine removal. Lastly, a series of mesocosm EK-phytoremediation tests were conducted using mock-ups filled with 0.386 m^3 of soil an applying a 0.6 V cm^{-1} DC current, 24 h per day, with and without changing polarity (in the DC-RP test, the polarity of the electrodes was inverted each day). The strategy followed for the instrumentation and sampling of the mock-ups let us to obtain the spatial distribution and time course of atrazine (and its main metabolites) in both the soil and the interstitial water. In the experiment with unidirectional DC current, atrazine was mainly accumulated in the cathode section, showing the important contribution of the electroosmotic flux to the mobilization of atrazine (Sánchez, López-Bellido, Rodrigo, *et al.*, 2020); when the DC-RP strategy was used, the concentrations of remaining atrazine were similar in both electrode sections. Moreover, overall atrazine removal was approximately 15% higher in the DC-RP test than in the unidirectional DC one (Sánchez, López-Bellido, Cañizares, *et al.*, 2020). The results from the mesocosm experiments were compared with those coming from pot tests; it was concluded that it is difficult to achieve the necessary similarity (in terms of geometrical dimensions and electrical parameters) between different scales and, moreover, the extent of the processes for pollutant removal may be quite different. So, it seems difficult to adequately predict the effectiveness of the EK-phytoremediation technology in field conditions using data obtained from reduced-scale tests.

14.8 Learned lessons and future challenges

Previous sections have shown that electrokinetic-assisted phytoremediation is, in the current state of knowledge, a technology to be seriously considered for the remediation of soils polluted by metal(loid) s and organic pollutants. It takes advantage of the ability of electric fields to mobilize pollutants and to promote plant growth, not only for phytoremediation purposes, but also for the recovery of the physical, chemical, biological and aesthetical characteristics of the contaminated soils. It has been also demonstrated that EK-phytoremediation can, in the adequate experimental conditions, improve the pollutant removal efficiencies achieved by electrokinetic remediation and phytoremediation techniques applied in an individual way. In addition, and very important, using electric current does not cause phytoremediation to lose the status as 'green technology'; electrokinetic remediation does not involve 'adding a chemical to remove a chemical' because only electron transfer causes the electrochemical processes (Bejan and Bunce, 2015). On the other hand, electrokinetic remediation has shown to be powered by means of solar photovoltaic panels therefore becoming a more sustainable technology (Souza *et al.*, 2016). All of them are enough reasons to continue betting on this innovative technology for soil decontamination.

As it has been seen, much progress has been made in the last ten years in order to understand the effects of electrochemical reactions and fluxes on physicochemical and biological soil properties and plant growth. Nevertheless, there are still many questions and aspects that remain unclear and need more research efforts. If we consider the research lines suggested in 2013 in the complete review by Cameselle et al. (Cameselle, Chirakkara and Reddy, 2013), it is easy to realize that the majority of them have not been sufficiently addressed to date. These authors proposed to expand the number of plant species and pollutants tested in EK-phytoremediation experiments. Since then, more than ten new plant species have been used but, just like phytoremediation, it seems that the combination [soil-pollutant-plant species-electric field characteristics] are case-specific and require the experimentation at small scale in order to assess the potential efficiency of the technique. Regarding the number of studied pollutants, the greatest deficiencies are detected in the field of organic pollutants and the mixed contamination by organics and metal(loid)s. Another important issue (also pointed out in the abovementioned review) is the need to analyze in depth the changes in the geochemical partitioning of metals in soils caused by the electric field and their influence on the metal uptake by plants; the use of single and sequential methods for metal extraction along with an adequate experimental design including control tests (plants and no electricity, soil with electricity but no plants) are essential aspects to consider in the study of the EK-phytoremediation of metal-polluted soils. The determination of changes in nutrients availability and their effect on the promotion of plant biomass should be also considered in the EK-phytoremediation tests.

Regarding the parameters related with the electric field, very little has been done in relation with the influence of electrode materials and their geometrical arrangement. It would be interesting to check other materials different than graphite and stainless steel in order to assess whether the very probable cost increases may be offset by greater removal efficiencies, lower power consumptions or higher plant performances. On the other hand, the study of a higher number of geometrical electrode configurations could expand the potential applications of the technique. The most appropriate type of electric current to carry out EK-phytoremediation is also not an issue that is sufficiently clear to date. As it has been discussed, DC current is the only kind of electric field to cause significant changes in pollutants distribution and availability in soils; however, it has been also found that AC current is capable to improve in a significant extent both pollutant mobility and plant uptake, also causing less damage to plants than unidirectional DC current. More efforts should be done to design comparative experiments with both types of electric current and also considering the periodical change in the polarity of the electrodes which is an intermediate solution to promote electrochemical processes avoiding negative impacts in soil biological activity (plants and microbiota). In addition, experimentation with different combinations of voltage gradient and times of application would allow to better understand the joint influence of the parameters which control the overall efficiency of the remediation process, e.g. plant biomass, pollutants mobilization, degradation and/or accumulation of pollutants.

It is also evident that there are more knowledge gaps in the application of EK-phytoremediation to organic pollutants. Issues such as the role of rhizosphere microorganisms in the degradation of pollutants, the contribution of plant accumulation and/or degradation, the use of different surfactant materials for the treatment of non-polar organic pollutants, the effects of electric field on the organic pollutants availability and the expansion of the number and types of pollutants studied are some of the many aspect that should be addressed in the next years.

A last aspect to consider in future research is the scaling-up of the EK-phytoremediation technology. To our knowledge, only one work has been published on a field trial (Luo, Wu, *et al.*, 2018) and there are other six works reporting data from mesocosm tests (using more than 20 kg of soil) (Jie Luo, Cai, *et al.*, 2018; Acosta-Santoyo *et al.*, 2019; Sánchez, López-Bellido, Cañizares, *et al.*, 2020; Sánchez, López-Bellido, Rodrigo, *et al.*, 2020; Xu *et al.*, 2020b, 2020a); as it can be seen, all of these are very recent. Actually, these early works have not carried out a systematic study of extrapolation of laboratory-scale data to larger scales, but they have served to highlight the difficulties for achieving similar performances at different scales in terms of overall efficiency and homogeneity of treatment. This should be an important research line in the near future in order to obtain the necessary data, not only on the efficiency of the treatment in field conditions, but also for estimating treatment costs and conducting life-cycle analysis that will allow the comparison of EK-phytoremediation technology with others already available.

Acknowledgements

Authors acknowledge the Spanish Ministry of Science and Innovation for the funding of the project PID2019-107282RB-I00. They also thank to the Regional Government of Castilla-La Mancha and the European Social Fund for the PhD grant number PRE2014/8027 and the funding of the project SBPLY/19/180501/000254.

References

Aboughalma, H., Bi, R., Schlaak, M., 2008. Electrokinetic enhancement on phytoremediation in Zn, Pb, Cu and Cd contaminated soil using potato plants. Journal of Environmental Science and Health - Part A Toxic/Hazardous Substances and Environmental Engineering 43, 926–933. doi:10.1080/10934520801974459.

Acar, Y.B., et al., 1995. Electrokinetic remediation: Basics and technology status. Journal of Hazardous Materials 40 (2), 117–137. doi:10.1016/0304-3894(94)00066-P.

Acosta-Santoyo, G., et al., 2019. Analysis of the biological recovery of soils contaminated with hydrocarbons using an electrokinetic treatment. Journal of Hazardous Materials 371, 625–633. doi:10.1016/j.jhazmat.2019.03.015.

Acosta-Santoyo, G., Cameselle, C., Bustos, E., 2017. Electrokinetic – Enhanced ryegrass cultures in soils polluted with organic and inorganic compounds. Environmental Research, 158. doi:10.1016/j.envres.2017.06.004.

Ali, H., Khan, E., Sajad, M.A., 2013. Phytoremediation of heavy metals-Concepts and applications. Chemosphere. Elsevier Ltd 91 (7), 869–881. doi:10.1016/j.chemosphere.2013.01.075.

Ashraf, Sana, et al., 2019. Phytoremediation: Environmentally sustainable way for reclamation of heavy metal polluted soils. Ecotoxicology and Environmental Safety, Academic Press. 714–727. doi:10.1016/j.ecoenv.2019.02.068.

Bejan, D., Bunce, N.J., 2015. Acid mine drainage: electrochemical approaches to prevention and remediation of acidity and toxic metals. Journal of Applied Electrochemistry 45 (12), 1239–1254. doi:10.1007/s10800-015-0884-2.

Bi, R., et al., 2011. Influence of electrical fields (AC and DC) on phytoremediation of metal polluted soils with rapeseed (Brassica napus) and tobacco (Nicotiana tabacum. Chemosphere 83, 318–326. doi:10.1016/j.chemosphere.2010.12.052.

Bolan, N., et al., 2014. Remediation of heavy metal(loid)s contaminated soils – To mobilize or to immobilize? Journal of Hazardous Materials 266, 141–166. doi:10.1016/j.jhazmat.2013.12.018.

Cameselle, C., Chirakkara, R.A., Reddy, K.R., 2013. Electrokinetic-enhanced phytoremediation of soils: Status and opportunities. Chemosphere 93 (4). doi:10.1016/j.chemosphere.2013.06.029.

Cameselle, C., Gouveia, S., 2019. Phytoremediation of mixed contaminated soil enhanced with electric current. Journal of Hazardous Materials 361, 95–102. doi:10.1016/j.jhazmat.2018.08.062.

Cameselle, C., Gouveia, S., Urréjola, S., 2019a. Benefits of phytoremediation amended with DC electric field. Application to soils contaminated with heavy metals. Chemosphere 229, 481–488. doi:10.1016/j.chemosphere.2019.04.222.

Cameselle, C., Gouveia, S., Urréjola, S., 2019b. Sustainable Soil Remediation. Phytoremediation Amended with Electric Current. Lecture Notes in Civil Engineering, 51–61. doi:10.1007/978-981-13-7010-6_4.

Cang, L., et al., 2009. Effects of electrokinetic treatment of a heavy metal contaminated soil on soil enzyme activities. Journal of Hazardous Materials 172 (2–3), 1602–1607. doi:10.1016/j.jhazmat.2009.08.033.

Cang, L., et al., 2011. Effects of electrokinetic-assisted phytoremediation of a multiple-metal contaminated soil on soil metal bioavailability and uptake by Indian mustard. Separation and Purification Technology 79, 246–253. doi:10.1016/j.seppur.2011.02.016.

Cang, L., et al., 2012. Impact of electrokinetic-assisted phytoremediation of heavy metal contaminated soil on its physicochemical properties, enzymatic and microbial activities. Electrochimica Acta 86, 41–48. doi:10.1016/j.electacta.2012.04.112.

Chaney, R. L. (1983) "Plant Uptake of Inorganic Waste Constituents.," *Land Treat of Hazard Wastes*, pp. 50–76.

Chen, X., et al., 2006. Effects of electrokinetics on bioavailability of soil nutrients. Soil Science 171 (8), 638–647. doi:10.1097/01.ss.0000228038.57400.a8.

Chirakkara, R.A., Reddy, K.R., Cameselle, C., 2015. Electrokinetic Amendment in Phytoremediation of Mixed Contaminated Soil. Electrochimica Acta, 181. doi:10.1016/j.electacta.2015.01.025.

Cui, X., Mayer, P., Gan, J., 2013. Methods to assess bioavailability of hydrophobic organic contaminants: Principles, operations, and limitations. Environmental Pollution 172, 223–234. doi:10.1016/j.envpol.2012.09.013.

Edited by K.R. Reddy, C. Cameselle, 2009. Electrochemical Remediation Technologies for Polluted Soils, Sediments and Groundwater. In: Reddy, K.R., Cameselle, C. (Eds.), Electrochemical Remediation Technologies for Polluted Soils, Sediments and Groundwater. John Wiley & Sons, Inc, Hoboken, NJ, USA. doi:10.1002/9780470523650.

Evangelou, M.W.H., Ebel, M., Schaeffer, A, 2007. Chelate assisted phytoextraction of heavy metals from soil. Effect, mechanism, toxicity, and fate of chelating agents. Chemosphere 68 (6), 989–1003. doi:10.1016/j.chemosphere.2007.01.062.

Gill, R.T., et al., 2014. Electrokinetic-enhanced bioremediation of organic contaminants: A review of processes and environmental applications. Chemosphere 107, 31–42. doi:10.1016/j.chemosphere.2014.03.019.

Khalid, S., et al., 2017. A comparison of technologies for remediation of heavy metal contaminated soils. Journal of Geochemical Exploration 182, 247–268. doi:10.1016/j.gexplo.2016.11.021.

Kumar Yadav, K., et al., 2018. Mechanistic understanding and holistic approach of phytoremediation: A review on application and future prospects. Ecological Engineering 120, 274–298. doi:10.1016/j.ecoleng.2018.05.039.

Lear, G., et al., 2004. The effect of electrokinetics on soil microbial communities. Soil Biology and Biochemistry. doi:10.1016/j.soilbio.2004.04.032.

Lemström, S., 1904. Electricity in Agriculture and Horticulture. "The Electrician" Printing & Publishing Company, London, UK.

Li, C., Ji, X., Luo, X., 2019. Phytoremediation of heavy metal pollution: A bibliometric and scientometric analysis from 1989 to 2018. Int. J. Environ. Res. Public Health 16 (23), 4755. doi:10.3390/ijerph16234755.

Lim, J.M., Salido, A.L., Butcher, D.J., 2004. Phytoremediation of lead using Indian mustard (Brassica juncea) with EDTA and electrodics. Microchemical Journal 76, 3–9. doi:10.1016/j.microc.2003.10.002.

Luo, J., Wu, J., et al., 2018. A real scale phytoremediation of multi-metal contaminated e-waste recycling site with Eucalyptus globulus assisted by electrical fields. Chemosphere 201, 262–268. doi:10.1016/j.chemosphere.2018.03.018.

Luo, J., Cai, L., et al., 2018. Influence of direct and alternating current electric fields on efficiency promotion and leaching risk alleviation of chelator assisted phytoremediation. Ecotoxicology and Environmental Safety 149, 241–247. doi:10.1016/j.ecoenv.2017.12.005.

Luo, J., Cai, L., et al., 2018. The interactive effects between chelator and electric fields on the leaching risk of metals and the phytoremediation efficiency of Eucalyptus globulus. Journal of Cleaner Production 202, 830–837. doi:10.1016/j.jclepro.2018.08.130.

Luo, J., et al., 2019. Comparing the risk of metal leaching in phytoremediation using Noccaea caerulescens with or without electric field. Chemosphere 216, 661–668. doi:10.1016/j.chemosphere.2018.10.167.

Mahar, A., et al., 2016. Challenges and opportunities in the phytoremediation of heavy metals contaminated soils: A review. Ecotoxicology and Environmental Safety 126, 111–121. doi:10.1016/j.ecoenv.2015.12.023.

Mao, X., et al., 2016. Electro-kinetic remediation coupled with phytoremediation to remove lead, arsenic and cesium from contaminated paddy soil. Ecotoxicology and Environmental Safety 125, 16–24. doi:10.1016/j.ecoenv.2015.11.021.

Menzies, N.W., Donn, M.J., Kopittke, P.M., 2007. Evaluation of extractants for estimation of the phytoavailable trace metals in soils. Environmental Pollution 145 (1), 121–130. doi:10.1016/j.envpol.2006.03.021.

Van Nevel, L., et al., 2007. Phytoextraction of metals from soils: How far from practice? Environmental Pollution 150 (1), 34–40. doi:10.1016/j.envpol.2007.05.024.

O'Connor, C.S., et al., 2003. The combined use of electrokinetic remediation and phytoremediation to decontaminate metal-polluted soils: A laboratory-scale feasibility study. Environmental Monitoring and Assessment 84 (1–2), 141–158. doi:10.1023/A:1022851501118.

Ortega-Calvo, J.-J., et al., 2015. From Bioavailability Science to Regulation of Organic Chemicals. Environmental Science & Technology 49 (17), 10255–10264. doi:10.1021/acs.est.5b02412.

Putra, R.S., Ohkawa, Y., Tanaka, S., 2013. Application of EAPR system on the removal of lead from sandy soil and uptake by Kentucky bluegrass (Poa pratensis L. Separation and Purification Technology 102, 34–42. doi:10.1016/j.seppur.2012.09.025.

Risco, C., et al., 2016. Removal of oxyfluorfen from spiked soils using electrokinetic fences. Separation and Purification Technology 167, 55–62. doi:10.1016/j.seppur.2016.04.050.

Rocha, I.M.V.M.V, et al., 2019. Coupling electrokinetic remediation with phytoremediation for depolluting soil with petroleum and the use of electrochemical technologies for treating the effluent generated. Separation and Purification Technology. Elsevier 208, 194–200. doi:10.1016/j.seppur.2018.03.012.

Rodríguez, L., Gómez, R., et al., 2016. Chemical and plant tests to assess the viability of amendments to reduce metal availability in mine soils and tailings. Environmental Science and Pollution Research 23 (7), 6046–6054. doi:10.1007/s11356-015-4287-z.

Rodríguez, L., Alonso-Azcárate, J., et al., 2016. EDTA and hydrochloric acid effects on mercury accumulation by Lupinus albus. Environmental Science and Pollution Research 23 (24), 24739–24748. doi:10.1007/s11356-016-7680-3.

Ruiz, E., et al., 2009. Phytoextraction Of Metal Polluted Soils Around A Pb-Zn Mine by Crop Plants. International Journal of Phytoremediation 11 (4), 360–384. doi:10.1080/15226510802565568.

Ruiz, E., Alonso-Azcárate, J., Rodríguez, L., 2011. Lumbricus terrestris L. Activity increases the availability of metals and their accumulation in maize and barley. Environmental Pollution 159 (3), 722–728. doi:10.1016/j.envpol.2010.11.032.

Sánchez, V., et al., 2018. Can electrochemistry enhance the removal of organic pollutants by phytoremediation? Journal of Environmental Management 225, 280–287. doi:10.1016/j.jenvman.2018.07.086.

Sánchez, V., López-Bellido, F.J., et al., 2019a. Electrokinetic-assisted phytoremediation of atrazine: Differences between electrode and interelectrode soil sections. Separation and Purification Technology 211, 19–27. doi:10.1016/j.seppur.2018.09.064.

Sánchez, V., López-Bellido, F.J., et al., 2019b. Enhancing the removal of atrazine from soils by electrokinetic-assisted phytoremediation using ryegrass (Lolium perenne L. Chemosphere 232, 204–212. https://doi.org/10.1016/j.chemosphere.2019.05.216.

Sánchez, V., López-Bellido, F.J., Rodrigo, M.A., et al., 2020. A mesocosm study of electrokinetic-assisted phytoremediation of atrazine-polluted soils. Separation and Purification Technology 233, 116044. doi:10.1016/j.seppur.2019.116044.

Sánchez, V., López-Bellido, F.J., Cañizares, P., et al., 2020. Scaling up the electrokinetic-assisted phytoremediation of atrazine-polluted soils using reversal of electrode polarity: A mesocosm study. Journal of Environmental Management 255, 109806. doi:10.1016/j.jenvman.2019.109806.

Sarwar, N., et al., 2017. Phytoremediation strategies for soils contaminated with heavy metals: Modifications and future perspectives. Chemosphere, 710–721. doi:10.1016/j.chemosphere.2016.12.116.

Siyar, R., et al., 2020. Potential of Vetiver grass for the phytoremediation of a real multi-contaminated soil, assisted by electrokinetic. Chemosphere 246, 125802. doi:10.1016/j.chemosphere.2019.125802.

Souza, F.L., et al., 2016. Solar-powered electrokinetic remediation for the treatment of soil polluted with the herbicide 2,4-D. Electrochimica Acta 190, 371–377. doi:10.1016/j.electacta.2015.12.134.

Tahmasbian, I., et al., 2017. Application of manures to mitigate the harmful effects of electrokinetic remediation of heavy metals on soil microbial properties in polluted soils. Environmental Science and Pollution Research 24 (34), 26485–26496. doi:10.1007/s11356-017-0281-y.

Tahmasbian, I., Safari Sinegani, A.A., 2014. Chelate-assisted phytoextraction of cadmium from a mine soil by negatively charged sunflower. International Journal of Environmental Science and Technology 11 (3), 695–702. doi:10.1007/s13762-013-0394-x.

Tahmasbian, I., Safari Sinegani, A.A., 2016. Improving the efficiency of phytoremediation using electrically charged plant and chelating agents. Environmental Science and Pollution Research 23 (3), 2479–2486. doi:10.1007/s11356-015-5467-6.

Virkutyte, J., Sillanpää, M., Latostenmaa, P., 2002. Electrokinetic soil remediation - Critical overview. Science of the Total Environment 289, 97–121. doi:10.1016/S0048-9697(01)01027-0.

Xiao, W., et al., 2017. Enhancement of Cd phytoextraction by hyperaccumulator Sedum alfredii using electrical field and organic amendments. Environmental Science and Pollution Research 24 (5), 5060–5067. doi:10.1007/s11356-016-8277-6.

Xu, L., et al., 2020a. Optimal voltage and treatment time of electric field with assistant Solanum nigrum L. cadmium hyperaccumulation in soil. Chemosphere 253, 126575. doi:10.1016/j.chemosphere.2020.126575.

Xu, L., et al., 2020b. The effects of different electric fields and electrodes on Solanum nigrum L. Cd hyperaccumulation in soil. Chemosphere 246, 125666. doi:10.1016/j.chemosphere.2019.125666.

Yeung, A.T., Gu, Y.Y., 2011. A review on techniques to enhance electrochemical remediation of contaminated soils. Journal of Hazardous Materials 195, 11–29. doi:10.1016/j.jhazmat.2011.08.047.

Zhou, D.M., et al., 2007. Ryegrass uptake of soil Cu/Zn induced by EDTA/EDDS together with a vertical direct-current electrical field. Chemosphere 67, 1671–1676. doi:10.1016/j.chemosphere.2006.11.042.

Zhou, M., et al., 2015. Electrokinetic remediation of fluorine-contaminated soil and its impact on soil fertility. Environmental Science and Pollution Research 22 (21), 16907–16913. doi:10.1007/s11356-015-4909-5.

CHAPTER 15

Biosurfactant-assisted phytoremediation for a sustainable future

N.F. Islam[a], Rupshikha Patowary[b], Hemen Sarma[a]

[a]*Department of Botany, Nanda Nath Saikia College, Titabar, Assam, India*

[b]*Centre for the Environment, Indian Institute of Technology Guwahati, Guwahati, Assam, India*

15.1 Introduction

The industrial revolution, as well as the increasing human population, have had a negative impact on the environment.The establishment of numerous industries and the mass production of goods result in the creation of massive amounts of pollution. Further, the unscientific dumping of industrial waste has an adverse effect on the environment, jeopardising all living organisms. The major issue of the current age is soil contamination from a variety of organic and inorganic pollutants (Ahmad et al., 2017). Most of these pollutants are persistent and accumulate in the soil beyond the acceptable levels affecting soil environment and human health (Shayler et al., 2009; Liduino et al., 2018). Their extreme toxicity to soil microbiota and successive trophic levels necessitates their removal from the soil environment (Patra et al., 2016). Several remediation techniques have been adopted during the last few decades, either alone or in a combination of different processes, some of which are often difficult to implement, or of high cost (Das et al., 2017). Among the accepted technologies, bioremediation has gained priority owing to its low cost and favorable outcomes (Salam et al., 2017; Goswami et al., 2018).

The efficacy of phytoremediation can be enhanced by microbial biosurfactants. Plants release organic exudates enhancing soil microbial population. Biosurfactants released by the microbes mediate uptake of soil pollutants by plant roots (rhizoremediation). Once absorbed by the roots, the heavy metal(loid) can be transferred to the above-ground plant biomass and either lost via transpiration or stored in the plant cell's vacuole. These metals can be removed from the soil by harvesting and properly disposing of the plants.

Significant progress has been made in recent years in combining the plants with indigenous soil microbes (microbial-assisted phytoremediation) to remediate polluted soil (Kogbara et al., 2016). Biosurfactants are microbial products that improve the efficacy of phytoremediation. This is one of the most efficient methods for removing organic and inorganic pollutants from polluted soil (Zhang et al., 2010). Biosurfactant-assisted phytoremediation approaches have also shown better performance in remediation of soil co-contaminated with petroleum hydrocarbons, heavy metals (Liduino et al., 2018), and hazardous industrial wastes (Salam et al., 2017). This remediation technique has gained much attention in recent years owing to its efficiency, low-cost, easy availability and eco-friendly nature (Fig. 15.1).

Assisted Phytoremediation. DOI: https://doi.org/10.1016/B978-0-12-822893-7.00003-3

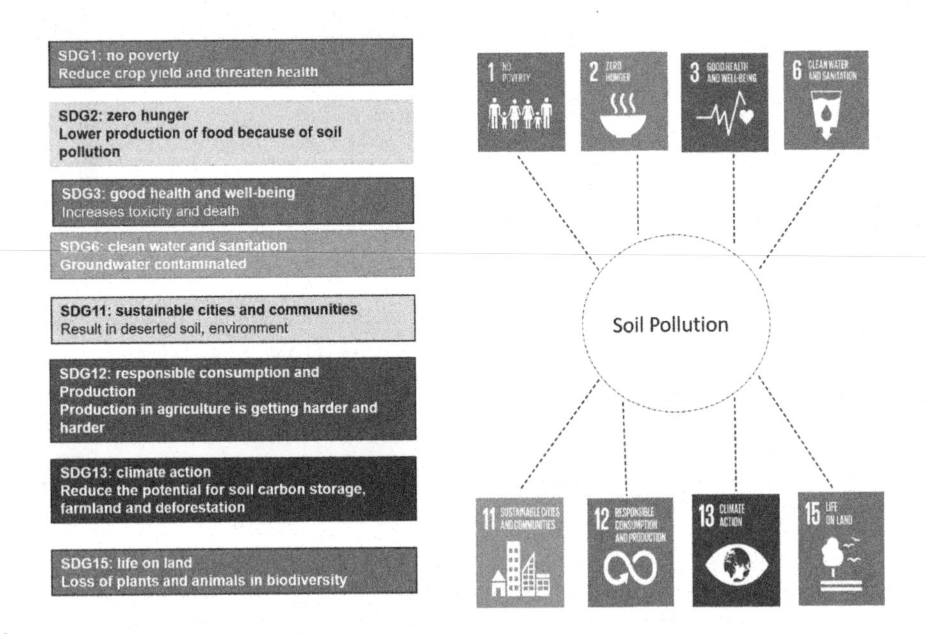

FIG. 15.1

Phytoremediation is a technique that utilizes plants and associated soil microbes to reduce the concentrations of contaminants in the soil as well as their toxic effects. Phytoremediation is widely recognized as a low-cost method of environmental restoration. These processes include a variety of techniques such as *in situ*, phytostabilization, phytodegradation, phytovolatilization, and phytoextraction.

It has been reported that the addition of biosurfactant or biosurfactant-producing microbes to a contaminated site aids in the faster degradation of organic contaminants (Wang et al., 2017) Thus, phytoremediation aided by biosurfactants has the potential to be an effective technology for contaminated soil remediation. Rhizobacteria-derived biosurfactants mediate and enhance bioavailability of complex organic and inorganic pollutants for biodegradation. Biosurfactants produced by rhizobacteria are released in the host rhizosphere, and form complexes with insoluble metals and facilitates desorption from soil particles, increasing their bioavailability (Lal et al., 2018). The metal chelating properties of biosurfactants improve plant efficiency in heavy metals contaminated soil remediation. Biosurfactants enhance the desorption of complex pollutants from soil and also promote their solubilization and mobilization through micelles formation (Moreira et al., 2011). These facilitate phytoextraction of contaminants from soil (Lal et al., 2018). Contaminated soil restoration is vital to a sustainable future, so that the ecosystem remains productive (Fig. 15.2). Numerous strategies have been adopted for achieving the United Nations Sustainable Development Goals (UN-SDGs). Based on existing research the development of sustainable biological methods could be one of the green options. In this section we focus on various aspects of biosurfactant-assisted phytoremediation and their role in the management of heavy metals and oil-polluted soils.

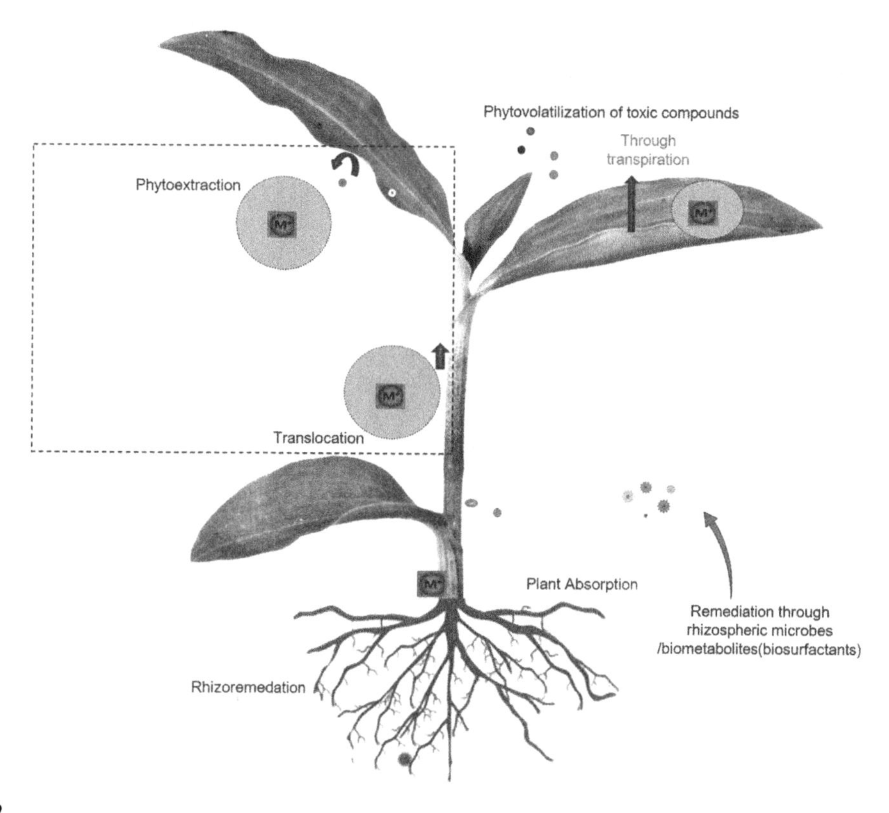

FIG. 15.2

The impact of soil pollution on the UN Sustainable Development Goals. Soil pollution has a negative impact on environmental sustainability, and it may impede the achievement of a number of United Nations Sustainable Development Goals (SDGs).

15.2 Soil inorganic and organic pollutants—source and concern

Global soil and water resources are heavily polluted with different organic and inorganic pollutants, resulting in deterioration of soil and food quality (Sarma, 2011). Soils contaminated with organic and inorganic pollutants are responsible for adverse ecosystem effects (Liduino et al., 2018). The pollutants, including polycyclic aromatic hydrocarbons (PAHs), pesticides, polychlorinated biphenyls (PCB), explosives, metals, metalloids and radionuclides are well reported in soils (Sarma et al., 2016; Sonowal et al., 2018; Ahmad et al., 2017). Heavy metals and hydrocarbons have significant environmental effects due to their persistence and adverse effects on living systems (Sarma et al., 2019a; Sarma et al., 2019b; Choden et al., 2020). Inorganic pollutants mainly consist of toxic metals, salts of phosphates, carbonates, sulfates, nitrates, inorganic ions, and various types of soil elements. Soil is the natural reservoir of inorganic compounds, derived from weathering of rocks, present in various anionic or cationic forms. Due to anthropogenic activities, these compounds are destabilized and emerge as contaminants of concern. Industrial and mining wastes add huge quantities of inorganic wastes mostly heavy metals, increasing soil toxicity, and posing threat to ecosystems. Industrial and agricultural activities release toxic heavy metals including cadmium (Cd), lead (Pb), mercury (Hg), and arsenic (As) into the environment (Kumar and Avery, 2016). Mining and metallurgical processes are important sources for the release of Cd, Pb, Zn, As, and Cu into the environment (Lee et al., 2001). Agricultural fertilizers and pesticides were known to release Ni, Cr, Zn, Cd, and Pb, some of which are persistent, with the immense possibility of entering the food chain (Lionetto et al., 2012). Besides these, other common inorganic contaminants are Cu, Zn, oxides of sulfur, nitrates, and phosphates. The inorganic pollutants may destabilize the physicochemical characteristics of soil, affecting the soil microbiota and plant growth (Shayler et al., 2009). Most of the inorganic pollutants are in the insoluble state in the soil, and are unavailable for uptake by plants (Lasat et al., 2002).

The source of organic pollutants in the environment is primarily anthropogenic (Nwoko, 2010). Organic pollutants include pesticides, polycyclic aromatic hydrocarbons (PAHs), azo dyes, chlorinated phenols, pharmaceuticals and various endocrine disrupting chemicals. They are released into the environment due to their application in agrochemicals, petrochemical industries, accidental oil spills, medicines etc. Most of these pollutants are toxic and persistent in nature, resulting in bioaccumulation and biomagnification, contributing to adverse health effects (Manohar et al., 2006). Persistent organic pollutants (POPs) like Dichlorodiphenyltrichloroethane (DDT), polychlorinated biphenyls (PCBs), bisphenol A (BPA), polychlorinated terphenyls (PCTs), polyaromatic hydrocarbon (PAHs) bioaccumulate in the food chain and disrupts the hormonal balance (Sarma and Prasad, 2015; Sarma and Prasad, 2016; Sarma and Prasad, 2019; Sarma et al., 2017; Vega et al., 2007). POPs affect the immune, endocrine and reproductive system, and also cause other serious health problems (Chaudhary & Shukla, 2019). Hydrophobic organic pollutants are more difficult to degrade due to their hydrophobicity. Organic pollutants are degraded via various biological and non-biological processes based on their physico-chemical properties. Phytovolatilization, phytoremediation, phytostimulation, and biosurfactant-assisted phytoremediation are commonly used for remediation of organic pollutants (Ali et al., 2013; Meagher, 2000). Bioventilation is another viable option for degradation of organic pollutants. The process involves promoting growth of native microorganisms for enhanced biodegradation (Yin et al., 2019). Microbial degradation is the most effective and economical process for conversion of toxic organic pollutants into less harmful compounds (Tripathi and Srivastava, 2011).

15.3 Biosurfactants—the 21st century's multifunctional biomolecules

Biosurfactants, also called green surfactants, are the most widely known molecule worldwide because of their wide range of applications. These are surface-active molecules that have shown prominence in various sectors, such as food, pharmaceutical, and agricultural sectors (Galabova et al., 2014). Due to their multifarious properties such as low toxicity, stability, foaming capacity, and faster degradability, they were the preferred alternative to their synthetic counterpart, the chemical surfactants. Biosurfactants are metabolites produced by a variety of microbes, such as bacteria, yeasts, and certain fungi and can be classified into various types such as glycolipid, fatty acids, and other polymers (Morita et al., 2016). Due to their hydrophilic head and a hydrophobic tail, these molecules have detergent characteristics, which reduce surface and interface tension (De Almeida et al., 2016). Among the different types of microbial biosurfactants, glycolipid biosurfactants were mostly commercialized including rhamnolipids, sophorolipids, and mannosylerythritol lipids. Surfactin derived from *Bacillus subtilis* is the widely used biosurfactant among lipopeptide biosurfactants. Similarly, sophorolipids, a glycolipid of long-chain fatty acids, are produced in high quantities from *Candida bombicola* (Jahan et al., 2020). Rhamnolipid biosurfactants have been widely applied in the remediation of polycyclic aromatic hydrocarbons (PAHs) contaminated soil owing to its environmental compatibility, and no phytotoxicity (Zhang et al., 2010). Rhamnolipid biosurfactants are one of the preferred groups of biosurfactants due to their minimal surface tension, higher emulsifying properties, and affinities for hydrophobic compounds (Aparna et al., 2012).

15.3.1 Biosurfactants producing microorganisms

Biosurfactants produced by diverse microbes ranging from bacteria to fungi are classified based on their chemical constituents and molecular mass. Several microbes with the capability to remediate contaminated soil were found to secrete biosurfactants that facilitate the uptake of organic and inorganic pollutants, thereby remediating the polluted site. The addition of biosurfactants to organic and inorganic contaminants increases the contaminant's solubility, which helps in biodegradation (Usman et al., 2016). The majority of the biosurfactants producing microbes investigated include *Bacillus* sp., *Pseudomonas* sp., *Arthrobacter* sp., *Acinetobacter* sp., *Rhodococcus* sp., *Serratia* sp., *Proteus* sp., *Halomonas* sp., *Micrococcus* sp. and *Rahnella* sp. (Saikia et al., 2012; Kotoky and Pandey, 2019). Biosurfactants produced by *Pseudomonas aeruginosa* BS2 promote solubility and mobility of Cd and Pb (Ullah et al., 2015). Rhamnolipids produced by *Pseudomonas aeruginosa* have been commercialized for bioremediation applications (Table 15.1). Through its chelating properties, the biosurfactant producing PGPR imparts metal resistance and promotes the sequestration pathway, making heavy metals phyto-available (Luo et al., 2011). Biosurfactant-producing bacterial strains with hydrocarbon-degrading potential have been successfully used in the environmental remediation of petroleum contaminants (Barin et al., 2014). Bacteria that live inside other organisms are known as endophytic bacteria. *Rahnella* sp., for example, is a potent biosurfactant producer that has been shown to improve phytoremediation of heavy metals like Cd, Zn, and Pb. Pb (He et al., 2013).

15.3.2 Classification of biosurfactants and their properties

Biosurfactants are microbe-derived surface-active molecules and are amphiphilic. These low molecular weight amphiphilic molecules have both hydrophilic and hydrophobic moieties having functional groups including glycolipids, lipopeptides, phospholipids, fatty acids, etc (Banat et al., 2010). They are secondary metabolites of microbes that remain attached to their cell surface or are liberated exterior to the cells (Dutta et al., 2020). Due to microbial origin, biosurfactants are less toxic, stable, and biodegradable. Biosurfactants reduce the surface and interfacial tension between immiscible liquids or between solid and liquid and are capable of reducing critical micelle concentration (CMC) (Singh and Rathore, 2019). High molecular mass biosurfactants have tended to stabilize oil-water emulsions, whereas low molecular weight biosurfactants can minimize interfacially and surface tensions (Geys et al., 2014). Based on their physical and chemical properties biosurfactants are ascribed as glycolipids (e.g. trehalolipids, sophorolipids, rhamnolipids, mannosylerythritol lipids, cellobiose lipids), lipoproteins and lipopeptides (e.g. lichenysin, surfactin, fengycin and iturin), phospholipids (e.g. corynomycolic acid), neutral lipids, fatty acid, and polymeric (e.g. emulsan, liposan, mannoprotein) biosurfactants

Table 15.1 Microbial biosurfactants used in phytoremediation of heavy metals and organic pollutants.

Biosurfactants	Producing Microorganisms	Applications	References
Rhamnolipids	*Pseudomonas putida*	Enhanced degradation of organic (alipathic and aromatic) pollutants	Rodrigues et al., 2006
Rhamnolipids	*Pseudomonas* sp. 2B	Enhanced bioremediation of petroleum hydrocarbons	Aparna et al., 2012
Rhamnolipids	*P. aeruginosa*	Remediation of petroleum hydrocarbons from oil contaminated soil	Chebbi et al., 2017
Rhamnolipids	*P. aeruginosa*	Remediation of Cu and Zn contaminated soil	Mulligan et al., 2001
Rhamnolipids	*Bacillus* Lz-2	Solubilization of polycyclic aromatic hydrocarbons (PAHs)	Li et al., 2015
Rhamnolipids	*P. sihuiensis*	Biodegradation of aliphatic and polycyclic aromatic hydrocarbons in seawater	Pereira et al., 2019
Surfactin	*Bacillus amyloliquefaciens*	Enhanced biodegradation of crude oil	Wang et al., 2020
Surfactin	*Bacillus subtilis*	Bioremediation of Cd, Cu and Zn contaminated soil	Mulligan et al., 1999
Surfactin and Iturins	*Bacillus methylotrophicus*	Biodegradation of aliphatic and polycyclic aromatic hydrocarbons in seawater	Pereira et al., 2019
Glycoprotein	*Aeribacillus pallidus* SL-1	Enhanced biodegradation of aliphatic and polycyclic aromatic hydrocarbons	Tao et al., 2020
Lipopeptide	*Paenibacillus dendritiformis* CN5	Biodegradation of pyrene	Bezza and Chirwa, 2017a
Lipopeptide	*Bacillus cereus* SPL-4	Promotes biodegradation of high molecular weight PAHs	Bezza and Chirwa, 2017b
Sophorolipid	*Torulopsis bombicola*	Bioremediation of Cu and Zn contaminated soil	Mulligan et al., 2001
Glycolipids	*Rhodococcus sp* TA6	Enhanced recovery of oil	Shavandi et al., 2011

(Sharma and Archana 2016; Mnif and Ghribi, 2016; Alahmad, 2015). Among these, the glycolipids biosurfactants are a highly studied group, including rhamnolipids, sophorolipids, trehalolipids, and mannosylerythritol lipids (Sarma et al., 2019). Glycolipids biosurfactants contain carbohydrate moieties attached to fatty-acid moieties (Das et al., 2020). Rhamnolipid (rhamnose linked to hydroxyl decanoic acid), a glycolipid produced by *Pseudomonas aeruginosa* and *Burkholderia* sp. can remove hydrophobic contaminants (Ghosh et al., 2008). Sophorolipids, a glycolipid biosurfactants, consist of sophorose carbohydrates linked to a chain of hydroxyl fatty acids (Gloor et al., 2004). Both *Torulopsis bombicola* and *T. apicola* produce hydrophobic sophorolipids along with lactones. Trehalolipids, a serpentine glycolipids, are produced by *Mycobacterium* sp. Lipopeptides biosurfactants consist of lipids linked to polypeptide chains (Kumar et al., 2015). Among lipopeptide biosurfactants, surfactin produced by *Bacillus* sp. is extensively studied (de Faria et al., 2011; Jacques, 2011). Among polymeric biosurfactants emulsan, liposan and mannoprotein have excellent emulsifying properties applied for emulsification of hydrocarbons in water (Randhawa et al., 2014).

15.4 Biosurfactants-assisted phytoremediation

15.4.1 Inorganic pollutants

Biosurfactants were described as excellent chelating agent owing to their biodegradability and minimal toxicity (Mekwichai et al., 2020). Biosurfactant mediated heavy metal bioremediation depends on i) complexation of the free form of metal present in the solution and ii) accumulation of metal with the biosurfactant under conditions of reduced interfacial tension at the solid-solution interface (Chakraborty and Das, 2014). Anionic biosurfactant forms non-ionic complexes with metal in the soil matrix. Cationic biosurfactants can competitively replace certain metal ions on the soil surface thereby enabling heavy metal remediation (Franzetti et al., 2014). Lipopeptide produced by marine sponge-associated bacteria (MSI 54) was found to show high affinity toward Pb, Hg, Mn, and Cd remediating 97.73% Pb, 75.5% Hg, 89.5% Mn, and 99.93% Cd, respectively, in 1,000 ppm of the respective metal solution at 2.0 × critical micelle concentration (CMC) (Ravindran et al., 2020). Enhanced removal efficacy of heavy metal from an oil-contaminated soil through batch washes with surfactin, rhamnolipid, and sophorolipid was reported by Mulligan et al., 1999. Several cases of pesticide remediation involving the use of biosurfactants have been reported in the literature. It has been shown that the presence of biosurfactant accelerates the degradation of chlorpyrifos (Singh et al., 2009). The addition of biosurfactants produced by *Bacillus subtilis* MTCC 1427 was found to enhance the biodegradation of endosulfan by 30-45% in both laboratory and soil conditions (Awasthi et al., 1999). Biosurfactant producing rhizobacteria helps the plants to tolerate heavy metal concentration by altering the plant metabolism (Welbaum et al., 2004). These rhizobacteria mediate numerous processes in metal contaminated soil including biosorption, biomineralization, bioaccumulation, and biotransformation (Karthik and Arulselvi, 2017; Ayangbenro and Babalola, 2017). Biosurfactants produced by rhizobacteria forms metal complexes in the rhizosphere soil interface for desorption of insoluble metals, increasing its bioavailability and facilitating metal uptake from the contaminated soil (Rajkumar et al., 2009). A wide number of biosurfactant-producing rhizobacteria having metal-reducing ability has been identified (Bacon and Hinton, 2011). Studies revealed the effectiveness of a bacteria-plant symbiotic system for the reclamation of heavy metal contaminated soil (Rajkumar et al., 2009; Rajkumar et al., 2012). *Azotobacter* bacteria and the plant *Lepidium sativum* were found to be effective in the remediation of Cr and Cd contaminated soils, according to the results of research (Sobariu et al., 2017). Similarly, biosurfactants produced by *Pseudochrobactrum lubricantis* and *Lysobacter novalis* have been reported to reduce Pb stress in plants (Yaseen et al., 2019). Further, *Cellulosimicrobium funkei* strain AR8 characterized from the rhizospheric region of *Phaseolus vulgaris* has been also found to improve the root length of plants in the presence of chromium stress (Karthik and Arulselvi, 2016). Findings provided by Abou-Shanab et al. (2003) revealed that rhizobacteria like *Microbacterium liquefaciens*, and *M. arabinogalactanolyticum* increases the availability of Ni in soils, and enhances its uptake by Ni-hyperaccumulator *Alyssum murale*.

Biosurfactant producing endophytic bacteria *Pseudomonas* sp. Lk9 significantly enhances metal uptake and growth of *Solanum nigrum*, a potent Cd-hyperaccumulator. This symbiotic association was reported to increase *Solanum nigrum* shoot biomass by 14% and Cd uptake of more than 46% mg kg^{-1} (Chen et al., 2014). Similarly, an association of biosurfactant producing *Pseudomonas koreensis* AGB-1 and *Miscanthus sinensis* was reported to enhance phytoremediation of Pb, Cd, As, Zn, and Cu with an increase in plant biomass by 54% (Babu et al., 2015). In another study, the association of biosurfactant producing *Bacillus* sp. J119 with tomato plant was reported to enhance Cd uptake with a significant increase in the plant biomass (Sheng et al., 2008) (Table 15.2).

Table 15.2 Examples of biosurfactant-assisted phytoremediation of contaminated soil in various parts of the world.

Types of Biosurfactant	Source/Origin	Plants sp.	Remediation efficiency	References
Rhamnolipid	Commercially procured	*Helianthus annuus* L.	Enhanced uptake of Zn	Vitor S. Liduino et al., 2018
Rhamnolipid	*Pseudomonas aeruginosa* L10	*Phragmites australis*	Biodegradation of C10-C26 *n*-alkanes and PAHs (including naphthalene, phenanthrene and pyrene), along with plant growth promoting effects.	Wu et al., 2018
Rhamnolipid	*Pseudomonas aeruginosa*	*Medicago sativa* L.	Enhanced phytoremediation efficiency of PAHs in association with arbuscular mycorrhizal fungi and aromatic hydrocarbon degrading bacteria (ARDB)	Zhang et al., 2010
Rhamnolipid and Tea Saponin	Commercially procured	Corn	Increase in the Cd uptake and corn biomass by 2.7 and 2.3-fold	Mekwichai et al., 2020
Sophorolipids	Commercially procured	*Medicago sativa and Bidens pilosa*	Increase in the root and shoot height by 17% and 11% respectively in *M. sativa* and *B. pilosa*, with increased uptake of Cd in *B. pilosa*	(Shah and Davery, 2021) (Shah and Davery, 2020)
Biosurfactant derived from Pseudomonas sp. SB	*Pseudomonas* sp. SB	Tall fescue and Rye grass	Increase in the bioavailability, and removal rate of DDT by 65.6% from contaminated soil	Wang et.al. 2017
Rhamnolipid	Commercially procured	*Helianthus annuus* L.	Enhanced desorption and biodegradation of PAH in slurry treated greenhouse planted soil	Volkerink et al., 2020
Rhamnolipid from *Pseudomonas aeruginosa* NY3	Commercially procured	*Cynodon dactylon* L. Pers. and *Panicum virgatum*	*Mycobacterium vanbaalenii* PYR-1 and biosurfactant assisted phytoremediation enhanced remediation of shooting range oils and high-molecular weight PAHs.	Wolf et al., 2020
Lipopeptide biosurfactant from *Brevi bacillus brevis* BAB-6437	Commercially procured	*Triticum aestivum* and *Capsicum annum*	Reduced toxicity of Cr and dye azulene in soil contaminated with textile effluent	Singh and Rathore, 2019
Biosurfactant derived from *Bacillus pumilus* 2A	Endophytic *Bacillus pumilus* isolated from synathropic plant-*Chelidonium majus* L.	*Sinapis alba, Lepidium sativum* and *Sorghum saccharatum*	High degradation potential for diesel oil and waste engine oil hydrocarbons, promoting growth of *Sinapis alba,* and enhanced seed germination of *Lepidium sativum*	Marchut-Mikolajczyk et al., 2018
Surfactant	*Bacillus flexus* S1I26 and *Paenibacillus* sp. S1I8	*Melia azadirachta*	Enhanced biodegradation of Benzo(*a*)pyrene	Kotoky and Pandey, 2019

15.4.2 Organic pollutants

Rhizobacteria communities that produce bio-surfactants have been shown to increase the solubility and bioavailability of organic pollutants by releasing the contaminants from soil particles and enhancing the degradation process (Arslan et al., 2017; Aslund and Zeeb, 2010). Biosurfactants increase the bioavailability of hydrophobic organic compounds and show prominence in remediating petroleum hydrocarbon contaminated soil (Marchut-Mikolajczyk et al., 2018). Biosurfactant-assisted remediation shows a greater extent of degradation of petroleum contaminants (Arslan et al., 2017). Biosurfactants are reported to enhance the bioavailability of hydrocarbon by i) emulsification of non-aqueous phase liquid contaminants, ii) enhancement of the apparent pollutants solubility, iii) facilitating transport of the pollutants from the solid phase, and iv) facilitating the absorption of microbes to soil particles occupied by the contaminant (Bustamante et al., 2012). Along with its efficacy against hydrocarbon contamination in soil, biosurfactant can facilitate hydrocarbon degradation in the aqueous phase assisting oil spill removal through the dispersal of contaminants in the aqueous phase and increasing the bioavailability of the hydrophobic substrate to microorganisms (Silva et al., 2014). It has been reported that plant endophytes can produce biosurfactants and degrade hydrocarbon from contaminated sites, and promote the growth of the plant (Marchut-Mikolajczyk et al., 2018). Rocha et al. (2011) reported that the presence of biosurfactant enhanced the degradation of linear and branched C11-C21 alkanes. Similarly, the presence of rhamnolipid was reported to enhance naphthalene degradation (Dasari et al., 2014). Increased bioavailability of anthracene and its ultimate metabolization by marine biosurfactant-producin strain *Bacillus circulansis* was reported (Das et al., 2008). Hydrocarbon contaminated sites if co-contaminated with toxic contaminants could inhibit the growth of hydrocarbon-degrading bacteria, thereby interfering with hydrocarbon degradation. The presence of rhamnolipid was found to reduce the toxicity of 4-chlorophenol (4-CP) or 2,4-dichlorophenol (2,4-DCP) to hydrocarbon-degrading bacteria, thereby facilitating diesel bioremediation (Chrzanowski et al., 2011). Biosurfactants bind with persistent organic pollutants (POPs) and facilitate their release from soil particles, promoting solubilization, improving bioavailability (Passatore et al., 2014). Biosurfactants reduce the toxic effect of persistent pollutants, promote soil enzymatic activities, and enhance plant growth under stress conditions (Sachdev and Cameotra, 2013). Biosurfactants produced by *Pseudomonas fluorescens* have been reported to contribute to the removal of total petroleum hydrocarbons (TPHs) (Gutiérrez et al., 2020). Endophytic bacteria *Pseudomonas aeruginosa* L10 has recently been attributed to have plant growth-promoting attributes, and produces biosurfactants that enhance emulsification of hydrocarbons, increasing the bioavailability for microbial degradation. *Pseudomonas aeruginosa* L10 produces rhamnolipid biosurfactants by *rhlABRI* gene cluster, which promotes hydrocarbon degradation (Wu et al., 2018). *Pseudomonas sp.* stimulates soil microbial communities by maintaining symbiotic relationships and enhances the degradation of petroleum hydrocarbons (Vinothini et al., 2015).

15.5 Significance of Biosurfactant in Phytoremediation

Plant-microbes interactions have always been proved instrumental for remediation of contaminated soil, as well as maintenance of healthy plant growth. The remediation capability of plants is enhanced by concomitant inoculation of biosurfactant-producing microbes/plant growth-promoting rhizobacteria. Phytoremediation in association with microorganisms imparts dual benefits of phytoremediation and microbial remediation of soils contaminated with organic pollutants (Wang et al., 2017). Plants provide a conducive environment for rhizospheric microorganisms and facilitate the remediation of soil contaminants. Rhizospheric microbes facilitate biotransformation, rhizodegradation, and volatilization of inorganic and organic pollutants (Sarma et al., 2019). They can mobilize metals through methylation, redox reactions, soil pH changes, the production and secretion of biosurfactants and siderophores, and thus reduce heavy metal toxicity (Shah and Daverey, 2020). Rhizobacterial biosurfactants have been extensively investigated in the removal of organic and inorganic pollutants from contaminated soil (Sriram et al., 2011). Biosurfactant producing rhizobacteria enhances plants tolerance to heavy metal stress, its accumulation and promotes plant growth (Wang et al., 2020). The beneficial plant-microbe interactions in the rhizospheric micro-environment promote detoxification of toxic wastes and enhance the phytoremediation potential of hyperaccumulators (Gleba et al., 1999). Biosurfactants have positive effects on the growth and activities of soil microbes and promote soil enzymatic activities (Shah and Daverey, 2020).

15.6 Conclusion

The increased number of industrial operations that release organic and inorganic contaminants into the environment jeopardizes the ecosystem. Heavy metals such as Zn, Cd, Cr, Pb, Hg, and Co are some of the pollutants of concern. Similarly, various organic pollutants, such as oil hydrocarbons, polyaromatic compounds, and pesticides, are categorised as emerging pollutants because they are harmful to all living organisms. These pollutants have been linked to a wide range of serious human diseases, many of which are associated with endocrine disorders. Their elimination is, therefore, a must in order to achieve good health and to achieve sustainable development goals of the United Nations. Phytoremediation is one of the low-cost, environmentally friendly options for removal of those pollutants. Many organic amendments were used to improve the effectiveness of this green technique. Biosurfactants are the biomolecules of choice that could be used to improve the efficacy of the phytoremediation process to repair and rejuvenate degraded and/or contaminated soil. As a result, the use of biosurfactants in phytoremediation is regarded as a green, long-term solution to pollution reduction.

Acknowledgments

Authors would like to acknowledge logistical support from the authorities of Nanda Nath Saikia College, India. This research did not receive any specific grant from funding agencies in the public, commercial, or not-for-profit sectors. We state that there is no competing financial interest.

References

Abou-Shanab, R.I., Delorme, T.A., Angle, J.S., Chaney, R.L., Ghanem, K., Moawad, H., Ghozlan, H.A., 2003. Phenotypic characterization of microbes in the rhizosphere of Alyssum murale. International Journal of Phytoremediation 5 (4), 367–379.

Ahmad, I., Imran, M., Hussain, M.B., Hussain, S., 2017. Remediation of organic and inorganic pollutants from soil: the role of plant-bacteria partnership. In: Naser, A.A. (Ed.), Chemical Pollution Control with Microorganisms. In: Chemical Pollution Control with Microorganisms, 2017. Nova Sci. Publisher, pp. 197–243.

Alahmad, K., 2015. The definition, preparation and application of Rhamnolipids as biosurfactants. Int. J. Nutr Food Sci. 4 (6), 613–623.

Ali, H., Khan, E., Sajad, M.A., 2013. Phytoremediation of heavy metals—concepts and applications. Chemosphere 91, 869–881.

Aparna, A., Srinikethan, G., Smitha, G., 2012. Production and characterization of biosurfactant produced by a novel Pseudomonas sp. 2B. Colloids Surf B Biointerfaces 95, 23–29.

Arslan, M., Imran, A., Khan, Q.M., et al., 2017. Plant–bacteria partnerships for the remediation of persistent organic pollutants. Environ Sci Pollut Res 24, 4322–4336.

Åslund, M.W., Zeeb, B., 2010. A Review of Recent Research Developments Into the Potential for Phytoextraction of Persistent Organic Pollutants (POPs) from Weathered, Contaminated Soil. Application of Phytotechnologies for Cleanup of Industrial, Agricultural, and Wastewater Contamination. in: NATO Security through Science Series C: Environmental Security. Springer, Dordrecht, pp. 35–59.

Awasthi, N., Kumar, A., Makkar, R., Cameotra, S.S., 1999. Biodegradation of soil-applied endosulfan in the presence of a biosurfactant. Journal of Environmental Science & Health Part B *34* (5), 793–803.

Ayangbenro, A.S., Babalola, O.O., 2017. A new strategy for heavy metal polluted environments: a review of microbial biosorbents. International journal of environmental research and public health *14* (1), 94.

Babu, A.G., Shea, P.J., Sudhakar, D., Jung, I.B., Oh, B.T., 2015. Potential use of Pseudomonas koreensis AGB-1 in association with Miscanthus sinensis to remediate heavy metal (loid)-contaminated mining site soil. J. Environ. Manag. 151, 160–166.

Bacon, C.W., Hinton, D.M., 2011. In planta reduction of maize seedling stalk lesions by the bacterial endophyte Bacillus mojavensis. Can J Microbiol 57, 485–492.

Banat, I.M., Franzetti, A., Gandolfi, I., Bestetti, G., Martinotti, M.G., Fracchia, L., Marchant, R., 2010. Microbial biosurfactants production, applications and future potential. Appl Microbiol Biotechnol 87, 427–444.

Barin, R., Talebi, M., Biria, D., 2014. Fast bioremediation of petroleum-contaminated soils by a consortium of biosurfactant/bioemulsifier producing bacteria. International Journal of Environmental Science and Technology 11, 1701–1710.

Bezza, F.A., Chirwa, E.M.N., 2017a. Pyrene biodegradation enhancement potential of lipopeptide biosurfactant produced by *Paenibacillus dendritiformis* CN5 strain. Journal of Hazardous Materials 321, 218–227.

Bezza, F.A., Chirwa, E.M.N., 2017b. The role of lipopeptide biosurfactant on microbial remediation of aged polycyclic aromatic hydrocarbons (PAHs)-contaminated soil. Chemical Engineering Journal 309, 563–576.

Bustamante, M., Duran, N., Diez, M.C., 2012. Biosurfactants are useful tools for the bioremediation of contaminated soil: a review. Journal of soil science and plant nutrition *12* (4), 667–687.

Chakraborty, J., Das, S., 2014. Biosurfactant-based bioremediation of toxic metals. In: Das, S. (Ed.), Microbial biodegradation and bioremediation. Elsevier, pp. 167–201. doi:10.1016/B978-0-12-800021-2.00007-8.

Chaudhary, T., Shukla, P., 2019. Bioinoculants for bioremediation applications and disease resistance:innovative perspectives. Indian Journal of Microbiology 59 (2), 129–136.

Chebbi, A., Hentati, D., Zaghden, H., Baccar, N., Rezgui, F., Chalbi, M., Sayadi, S., Chamkha, M., 2017. Polycyclic aromatic hydrocarbon degradation and biosurfactant production by a newly isolated *Pseudomonas* sp. strain from used motor oil-contaminated soil. International Biodeterioration & Biodegradation 122, 128–140.

Chen, L., Luo, S., Li, X., Wan, Y., Chen, J., Liu, C., 2014. Interaction of Cd-hyperaccumulator Solanum nigrum L. and functional endophyte Pseudomonas sp. Lk9 on soil heavy metals uptake. Soil Biol. Biochem. 68, 300–308.

Choden, D., Pokethitiyook, P., Poolpak, T., Kruatrachue, M., 2020. Phytoremediation of soil co-contaminated with zinc and crude oil using Ocimumgratissimum (L.) in association with Pseudomonasputida MU02. International Journal of Phytoremediation 23 (2), 181–189. doi:10.1080/15226514.2020.1803205.

Chrzanowski, Ł., Owsianiak, M., Szulc, A., Marecik, R., Piotrowska-Cyplik, A., Olejnik-Schmidt, A.K., Staniewski, J., Lisiecki, P., Ciesielczyk, F., Jesionowski, T., Heipieper, H.J., 2011. Interactions between rhamnolipid biosurfactants and toxic chlorinated phenols enhance biodegradation of a model hydrocarbon-rich effluent. International Biodeterioration & Biodegradation 65 (4), 605–611.

Das A.J., Ambust S., Kumar R. et al., 2020. Management of Petroleum Industry Waste Through Biosurfactant-Producing Bacteria: A Step Toward Sustainable Environment. In: G., Saxena, R.N., Bharagava (Eds.), Bioremediation of Industrial Waste for Environmental Safety. https://doi.org/10.1007/978-981-13-1891-7_8.

Das, A.J., Lal, S., Kumar, R., Verma, C., 2017. Bacterial biosurfactants can be an eco-friendly and advanced technology for remediation of heavy metals and co-contaminated soil. Int. J. Environ. Sci. Technol. 14, 1343–1354.

Das, P., Mukherjee, S., Sen, R., 2008. Improved bioavailability and biodegradation of a model polyaromatic hydrocarbon by a biosurfactant producing bacterium of marine origin. Chemosphere *72* (9), 1229–1234.

Dasari, S., Venkata Subbaiah, K.C., Wudayagiri, R., Valluru, L., 2014. Biosurfactant-mediated biodegradation of polycyclic aromatic hydrocarbons—naphthalene. Bioremediation Journal *18* (3), 258–265.

Datta, P., Tiwari, P., Pandey, L.M., 2020. Oil washing proficiency of biosurfactant produced by isolated Bacillus tequilensis MK 729017 from Assam reservoir soil. J. Pet. Sci. Eng. 195, 107612. doi:10.1016/j.petrol.2020.107612.

De Almeida, D.G., Soares Da Silva, R.D.C.F., Luna, J.M., Rufino, R.D., Santos, V.A., Banat, I.M., Sarubbo, L.A., 2016. Biosurfactants: promising molecules for petroleum biotechnology advances. Frontiers in microbiology *7*, 1718.

de Faria, A.F., Teodoro-Martinez, D.S., de Oliveira Barbosa, G.N., Vaz, B.G., ÍS, S.i.l.v.a., Garcia, J.S., et al., 2011. Production and structural characterization of surfactin (C14/Leu7) produced by *Bacillus subtilis* isolate LSFM-05 grown on raw glycerol from the biodiesel industry. Process Biochem 46 (10), 1951–1957.

Franzetti, A., Gandolfi, I., Fracchia, L., Van Hamme, J., Gkorezis, P., Marchant, R., Banat, I.M., 2014. Biosurfactant use in heavy metal removal from industrial effluents and contaminated sites. Biosurfactants: Production and utilization—Processes, technologies, and economics *159*, 361–370. doi:10.1201/b17599-20.

Galabova, D., Sotirova, A., Karpenko, E., Karpenko, O., 2014. Role of microbial surface-active compounds in environmental protection. In: M. Fanun, (Ed.), The role of colloidal systems in environmental protection. Amsterdam: Elsevier, pp. 41–83.

Geys, R., Soetaert, W., Bogaert, I.V., 2014. Biotechnological opportunities in biosurfactant production. Curr Opin Biotechnol 30, 66–72.

Ghosh, S., Kim, D., So, P., Blankschtein, D., 2008. Visualization and quantification of skin barrier perturbation induced by surfactant-humectant systems using two-photon fluorescence microscopy. Journal of cosmetic science 59 (4), 263–289.

Gleba, D., Borisjuk, N.V., Borisjuk, L.G., Kneer, R., Poulev, A., Skarzhinskaya, M., Dushenkov, S., Dushenkov, S., Logendra, S., Gleba, Y.Y., Raskin, I., May 1999. Use of plant roots for phytoremediation and molecular farming. Proc. Natl. Acad. Sci. U. S. A. 96 (11), 5973–5977. doi:10.1073/pnas.96.11.5973.

Gloor, M., Senger, B., Langenauer, M., Fluhr, J.W., 2004. On the course of the irritant reaction after irritation with sodium lauryl sulphate. Skin Res Technol 10 (3), 144–148.

Goswami, M., Chakraborty, P., Mukherjee, K., Mitra, G., Bhattacharyya, P., Dey, S., Tribedi, P., 2018. Bioaugmentation and biostimulation: a potential strategy for environmental remediation. J. Microbiol. Exp *6*, 223–231.

Gutiérrez, E.J., Abraham, M.R., Baltazar, J.C., Vazquez, G., Delgadillo, E., Tirado, D., 2020. Pseudomonas fluorescens: A Bioaugmentation Strategy for Oil-Contaminated and Nutrient-Poor Soil. Int. J. Environ. Res. Public Health 2020 (17), 6959. doi:10.3390/ijerph17196959.

He, H., Ye, Z., Yang, D., Yan, J., Xiao, L, Zhong, T, Yuan, M, Cai, X, 2013. Characterization of endophytic Rahnella sp. JN6 from Polygonum pubescens and its potential in promoting growth and Cd, Pb, Zn uptake by Brassica napus. Chemosphere 90 (6), 1960–1965. doi:10.1016/j.chemosphere.2012.10.057.

Jacques, P., 2011. Surfactin and other lipopeptides from *Bacillus* spp. In: Soberón-Chávez G. (Ed.), Biosurfactants. Microbiology Monographs, vol 20. Springer, Berlin, Heidelberg. https://doi.org/10.1007/978-3-642-14490-5_3.

Jahan, R., Bodratti, A.M., Tsianou, M., Alexandridis, P., 2020. Biosurfactants, natural 468 alternatives to synthetic surfactants: Physicochemical properties and applications. Adv. Colloid 469 Interface Sci. 275, 102061.

Karthik, C., Arulselvi, P.I., 2017. Biotoxic effect of chromium (VI) on plant growth-promoting traits of novel Cellulosimicrobium funkei strain AR8 isolated from Phaseolus vulgaris rhizosphere. Geomicrobiology Journal *34* (5), 434–442.

Kogbara, R.B., Ogar, I., Okparanma, R.N., Ayotamuno, J.M., 2016. Treatment of Petroleum Drill Cuttings Using Bioaugmentation and Biostimulation Supplemented with Phytoremediation. J. Environ. Sci. Health, Part. A 51 (9), 714–721. doi:10.1080/10934529.2016.1170437.

Kotoky, Pandey, 2019. Rhizosphere mediated biodegradation of benzo(A)pyrene by surfactin producing soil bacilli applied through *Meliaazadirachta* rhizosphere. International Journal of Phytoremediation. doi:10.1080/15226514.2019.1663486.

Kumar, A., Aery, N.C., 2016. Impact, metabolism, and toxicity of heavy metals in plantsPlant responses to Xenobiotics. Springer, Singapore, pp. 141–176.

Kumar, R., Das, A.J., Lal, S., 2015. Petroleum Hydrocarbon Stress Management in Soil Using Microorganisms and Their Products. In: Ram Chandra (Ed.), Environmental Waste Management. CRC Press Taylor and Francis, Boca Raton, p. 586.

Lal, S., Ratna, S., Said, O.B., Kumar, R, 2018. Biosurfactant and exopolysaccharide-assisted rhizobacterial technique for the remediation of heavy metal contaminated soil: An advancement in metal phytoremediation technology. Environmental Technology & Innovation 10, 243–263. doi:10.1016/j.eti.2018.02.011.

Lasat, M.M., 2002. Phytoextraction of toxic metals: a review of biological mechanisms. J Environ Qual 31, 109–120.

Lee, C.G., Hyo-Taek, C.M., Chae, J.C.G., 2001. Heavy metal contamination in the vicinity of the Daduk au–ag–Pb–Zn mine in Korea. Appl Geochem 16, 1377–1386.

Li, S., Pi, Y., Bao, M., Zhang, C., Zhao, D., Li, Y., Sun, P., Lu, J., 2015. Effect of rhamnolipid biosurfactant on solubilization of polycyclic aromatic hydrocarbons. Marine Pollution Bulletin 101, 219–225.

Liduino, V.S., Servulo, E.F., Oliveira, F.J., 2018. Biosurfactant-assisted phytoremediation of multi-contaminated industrial soil using sunflower (Helianthus annuus L.). J. Environ. Sci. Health, Part. A 53 (7), 609–616.

Lionetto, M.G., Calisi, A., Schettino, T., 2012. Earthworm biomarkers as tools for soil pollution assessment. In: Hernandez Soriano, M.C. (Ed.), Soil health and land use management. Intech Open, Croatia, pp. 305–332.

Luo, S.L., Wan, Y., Xiao, X., Guo, H., Chen, L., Xi, Q., Zeng, G., Liu, C., Chen, J., 2011. Isolation and characterization of endophytic bacterium LRE07 from cadmium hyperaccumulator Solanum nigrum L. and its potential for remediation. Appl Microbiol Biotechnol 89, 1637–1644.

Manohar, S., Jadia, C.D., Fulekar, M.H., 2006. Impact of ganesh idol immersion on water quality. In J Environ Protect 27, 216–220.

Marchut-Mikolajczyk, O., Drożdżyński, P., Pietrzyk, D., Antczak, T., 2018. Biosurfactant production and hydrocarbon degradation activity of endophytic bacteria isolated from Chelidonium majus L. Microb. Cell Factories 17, 171. doi:10.1186/s12934-018-1017-5.

Meagher, R.B., 2000. Phytoremediation of toxic elemental and organic pollutants. Curr Opin Plant Biol 3 (2), 153–162.

Mekwichai, P., Tongcumpou, C., Kittipongvises, S., Tuntiwiwattanapun, N., 2020. Simultaneous biosurfactant-assisted remediation and corn cultivation on cadmium-contaminated soil. Ecotoxicology and Environmental Safety 192 (2020), 110298 https://doi.org/10.1016/j.ecoenv.2020.110298.

Mnif, I., Ghribi, D., 2016. Glycolipid biosurfactants: main properties and potential applications in agriculture and food industry. J Sci Food Agric 96 (13), 4310–4320.

Moreira, I.T.A., Oliveira, O.M.C., Triguis, J.A., Santos, A.M.P., Queiroz, A.F.S., Martins, C.M.S., Silva, C.S., Jesus, R.S, 2011. Phytoremediation Using Rizophora Mangle L. in Mangrove Sediments Contaminated by Persistent Total Petroleum Hydrocarbons (TPH's). Microchem. J. 99 (2), 376–382. doi:10.1016/j.microc.2011.06.011.

Morita, T., Fukuoka, T., Imura, T., Kitamoto, D., 2016. Glycolipid Biosurfactants, Reference Module in Chemistry, Molecular Sciences and Chemical Engineering. Elsevier, https://doi.org/10.1016/B978-0-12-409547-2.11565-3, ISBN 9780124095472.

Mulligan, C.N., Yong, R.N., Gibbs, B.F., 1999. Removal of heavy metals from contaminated soil and sediments using the biosurfactant surfactin. Journal of Soil Contamination 8 (2), 231–254.

Mulligan, C.N., Yong, R.N., Gibbs, B.F., 2001. Heavy metal removal from sediments by biosurfactants. Journal of hazardous materials 85 (1–2), 111–125.

Nwoko, C.O., 2010. Trends in phytoremediation of toxic elemental and organic pollutants. Afr. J. Biotechnol. 9, 6010–6016. doi:10.5897/AJB09.061.

Passatore, L., Rossetti, S., Juwarkar, A.A., Massacci, A., 2014. Phytoremediation and bioremediation of polychlorinated biphenyls (PCBs): state of knowledge and research perspectives. J Hazard Mater 278, 189–202.

Patra, S., Mishra, P., Mahapatra, S.C., Mithun, S.K., 2016. Modelling impacts of chemical fertilizer on agricultural production: a case study on Hooghly district, West Bengal, India. Modeling Earth Systems and Environment *2* (4), 1–11.

Pereira, E., Napp, A.P., Allebrandt, S., Barbosa, R., Reuwsaat, J., Lopesa, W., Kmetzsch, L., Staats, C.C., Schrank, A., Dallegrave, A., Peralba, M.R., Passaglia, L.M.P., Bento, F.M., Vainstein, M.H, 2019. Biodegradation of aliphatic and polycyclic aromatic hydrocarbons in seawater by autochthonous microorganisms. International Biodeterioration & Biodegradation 145, 104789.

Rajkumar, M., Ae, N., Freitas, H., 2009. Endophytic bacteria and their potential to enhance heavy metal phytoextraction. Chemosphere 77, 153–160.

Rajkumar, M., Sandhya, S., Prasad, M. N.V., Freitas, H., 2012. Perspectives of plant associated microbes in heavy metal phytoremediation. Biotechnol Adv 30 (6), 1562–1574. doi:10.1016/j.biotechadv.2012.04.011.

Randhawa K.K.S., Rahman P.K., 2014. Rhamnolipid biosurfactants—past, present, and future scenario of global market. Frontiers in Microbiology 5, 454. https://www.frontiersin.org/article/10.3389/fmicb.2014.00454. doi:10.3389/fmicb.2014.00454.

Ravindran, A., Sajayan, A., Priyadharshini, G.B., Selvin, J., Kiran, GS., 2020. Revealing the efficacy of thermostable biosurfactant in heavy metal bioremediation and surface treatment in vegetables. Frontiers in Microbiology 11, 222.

Rocha, C.A., Pedregosa, A.M., Laborda, F., 2011. Biosurfactant-mediated biodegradation of straight and methyl-branched alkanes by Pseudomonas aeruginosa ATCC 55925. AMB express *1* (1), 1–10.

Rodrigues, L., Banat, I.M., Teixeira, J., Oliveira, R., 2006. Biosurfactants: potential applications in medicine. J Antimicrob Chemother 57 (4), 609–618. doi:10.1093/jac/dkl024.

Sachdev, D.P., Cameotra, SS., 2013. Biosurfactants in agriculture. Appl. Microbiol. Biotechnol. 97 (3), 1005–1016.

Saikia, R.R., Deka, S., Deka, M., Sarma, H., 2012. Optimization of environmental factors for improved production of rhamnolipid biosurfactant by Pseudomonas aeruginosa RS29 on glycerol. J. Basic Microbiol. 52 (4), 446–457. doi:10.1002/jobm.201100228.

Salam, J.A., Hatha, M.A.A., Das, N., 2017. Microbial-Enhanced Lindane Removal by Sugarcane (Saccharum Officinarum) in Doped SoilApplications in Phytoremediation and Bioaugmentation. J. Environ. Manage. 193, 394–399. doi:10.1016/j.jenvman.2017.02.006.

Sarma, H., Bustamante, K.L.T., Prasad, M.N.V., 2019. Biosurfactants for oil recovery from refinery sludge: Magnetic nanoparticles assisted purification. In: Prasad, M.N.V. (Ed.), Industrial and Municipal SludgeEmerging Concerns and Scope for Resource Recovery. Elsevier, pp. 107–132.

Sarma, H., Nava, A.R., Prasad, M.N.V., 2019. Mechanistic understanding and future prospect of microbe-enhanced phytoremediation of polycyclic aromatic hydrocarbons in soil. Environmental Technology & Innovation, 1864–2352. doi:10.1016/j.eti.2018.12.004.

Sarma, H., 2011. Metal Hyperaccumulation in Plants: A Review Focusing on Phytoremediation Technology. Journal of Environmental Science and Technology 4 (2), 118–138.

Sarma, H., Prasad, M.N.V., 2016. Phytomanagement of polycyclic aromatic hydrocarbons and heavy metals-contaminated sites in Assam, North Eastern State of India, for boosting bioeconomy. Bioremediation and Bioeconomy. In: Prasad, M.N.V. (Ed.), 1st. Elsevier Inc, USA, pp. 609–626.

Sarma, H., Prasad, M.N.V., 2015. Plant-microbe association-assisted removal of heavy metals and degradation of polycyclic aromatic hydrocarbons. In: Mukherjee S. (Ed.), petroleum geosciences: Indian contexts Springer Geology. Springer, Cham. https://doi.org/10.1007/978-3-319-03119-4_10.

Sarma, H., Prasad, M.N.V., 2019. Metabolic engineering of rhizobacteria associated with plants for remediation of toxic metals and metalloids. In: Prasad, M.N.V. (Ed.), Transgenic Plant Technology for Remediation of Toxic Metals and Metalloids 1st. Academic Press, pp. 299–311. https://doi.org/10.1016/B978-0-12-814389-6.00014-6.

Sarma, H., Sonowal, S., Prasad, M.N.V., 2019. Plant-microbiome assisted and biochar-amended remediation of heavy metals and polyaromatic compounds - a microcosmic study. Ecotoxicology and Environmental Safety 176 (30), 288–299.

Sarma, H., Islam, N.F., Prasad, M.N., 2017. Plant-microbial association in petroleum and gas exploration sites in the state of Assam, north-east India—significance for bioremediation. Environmental Science and Pollution Research 24 (9), 8744–8758. doi: 10.1007/s11356-017-8485-8. Epub 2017. PMID: 28213706.

Sarma, H., Deka, H., Saikia, R.R., 2012. Accumulation of heavy metals in selected medicinal plants. Reviews of Environ. Contam. Toxicol. 214, 63–86.

Sarma, H., Islam, N.F., Borgohain, P., Sarma, A., Prasad, M.N.V., 2016. Localization of polycyclic aromatic hydrocarbons and heavy metals in surface soil of Asia's oldest oil and gas drilling site in Assam, northeast India: Implications for the Bio economy, Emerging Contaminants 2 (3), pp. 119–127. https://doi.org/10.1016/j.emcon.2016.05.004.

Shah, V., Davery, A., 2021. Effects of sophorolipids augmentation on the plant growth and phytoremediation of heavy metal contaminated soil. J. Clean. Prod. 280, 124406. https://doi.org/10.1016/j.eti.2020.100774.

Shah, V., Daverey, A., 2020. Phytoremediation: A multidisciplinary approach to clean up heavy metal contaminated soil. Environmental Technology & Innovation 18, 100774. https://doi.org/10.1016/j.eti.2020.100774.

Sharma, D., Sarma, H., Hazarika, S., Islam, N.F., Prasad, M.N.V., 2018. Agro-Ecosystem Diversity in Petroleum and Natural Gas Explored Sites in Assam State, North-Eastern India: Socio-Economic Perspectives. Sustainable Agriculture Reviews 27 (37–60), 2210–4410.

Sharma, R.K., Archana, G., 2016. Cadmium minimization in food crops by cadmium resistant plant growth promoting rhizobacteria. Appl Soil Ecol 107, 66–78.

Shavandi, M., Mohebali, G., Haddadi, A., Shakarami, H., Nuhi, A., 2011. Emulsification potential of a newly isolated biosurfactant-producing bacterium, Rhodococcus sp. strain TA6. Colloids and Surfaces B: Biointerfaces 82, 477–482.

Shayler, H., McBride, M., Harrison, E., 2009. Sources and impacts of contaminants in soils. Cornell Waste Management Institute http://cwmi.css.cornell.edu/sourcesandimpacts.pdf.

Sheng, X., He, L., Wang, Q., Ye, H., Jiang, C., 2008. Effects of inoculation of biosurfactant-producing Bacillus sp. J119 on plant growth and cadmium uptake in a cadmium-amended soil. J. Hazard. Mater. 155, 17–22.

Silva, R.D.C.F., Almeida, D.G., Rufino, R.D., Luna, J.M., Santos, V.A., Sarubbo, L.A., 2014. Applications of biosurfactants in the petroleum industry and the remediation of oil spills. International journal of molecular sciences *15* (7), 12523–12542.

Singh, R., Rathore, D., 2019. Impact assessment of azulene and chromium on growth and metabolites of wheat and chilli cultivars under biosurfactant augmentation. Ecotoxicol. Environ. Saf. 186, 109789. doi:10.1016/j.ecoenv.2019.109789.

Singh, P.B., Sharma, S., Saini, H.S., Chadha, B.S., 2009. Biosurfactant production by Pseudomonas sp. and its role in aqueous phase partitioning and biodegradation of chlorpyrifos. Letters in applied microbiology *49* (3), 378–383.

Singh, R., Rathore, D., 2019. Impact assessment of azulene and chromium on growth and metabolites of wheat and chilli cultivars under biosurfactant augmentation. Ecotoxicol. Environ. Safety 186, 109789.

Sobariu, D.L., Fertu, D.I.T., Diaconu, M., Pavel, L.V., Hlihor, R.M., Drăgoi, E.N., Curteanu, S., Lenz, M., Corvini, P.F.X., Gavrilescu, M., 2017. Rhizobacteria and plant symbiosis in heavy metal uptake and its implications for soil bioremediation. New biotechnology *39*, 125–134.

Sonowal, S., Prasad, M.N.V., Sarma, H, 2018. C3 and C4 plants as potential phytoremediation and bioenergy crops for stabilization of crude oil spill laden soils – antioxidative stress responses. Tropical Plant Research 5 (3), 306–314 2018.

Sriram, M.I., Gayathiri, S., Gnanaselvi, U., Jenifer, P.S., Mohan Raj, S., Gurunathan, S., 2011. Novel lipopeptide biosurfactant produced by hydrocarbon degrading and heavy metal tolerant bacterium Escherichia fergusonii KLU01 as a potential tool for bioremediation. Bioresour. Technol. 102 (19), 9291–9295. doi:10.1016/j.biortech.2011.06.094.

Tao, W., Lin, J., Wang, W., Huang, H., Li, S., 2020. Biodegradation of aliphatic and polycyclic aromatic hydrocarbons by the thermophilic bioemulsifier-producing*Aeribacilluspallidus* strain SL-1. Ecotoxicology and Environmental Safety 189, 109994.

Tripathi, A., Srivastava, S.K., 2011. Ecofriendly Treatment of Azo Dyes: Biodecolorization using Bacterial Strains. Int. J. Biosci. Biochem. Bioinforma. 1 (1), 37–40.

Ullah, A., Heng, S., Munis, M.F.H., Fahad, S., Yang, X., 2015. Phytoremediation of heavy metals assisted by plant growth promoting (PGP) bacteria: a review. Environ Exp Bot 117, 28–40.

Usman, M.M., Dadrasnia, A., Lim, K.T., Mahmud, A.F., Ismail, S., 2016. Application of biosurfactants in environmental biotechnology, remediation of oil and heavy metal. AIMS Bioengineering *3* (3), 289–304.

Vega, F., Covelo, E., Andrade, M., 2007. Accidental organochlorine pesticide contamination of soil in Porrino. Spain. J environ quality 36 (1), 272–279.

Vinothini, C., Sudhakar, S., Ravikumar, R., 2015. Biodegradation of petroleum and crude oil by Pseudomonas putida and Bacillus cereus. Int J Curr Microbiol Appl Sci 4, 318–329.

Volkerink, S.N.J., Fernandez, J.L.G, Vila, J., 2020. Rhizosphere-enhanced biosurfactant action on slowly desorbing PAHs in contaminated soil. Sci. Total Environ. 720, 137608. https://doi.org/10.1016/j.scitotenv.2020.137608.

Wang, X., Cai, T., Wen, W., Ai, J., Ai, J., Zhang, Z., Zhu, L., George, S.C, 2020. Surfactin for enhanced removal of aromatic hydrocarbons during biodegradation of crude oil. Fuel 267, 117272.

Wang, B., Chu, C., Wei, H., Zhang, L., Ahmad, Z., Wu, S., Xie, B., 2020. Ameliorative effects of silicon fertilizer on soil bacterial community and pakchoi (Brassica chinensis L.) grown on soil contaminated with multiple heavy metals. Environ. Pollut. 267, 115411. doi:10.1016/j.envpol.2020.115411.

Wang, B., Wang, Q., Liu, W., Liu, X, Hou, J., Teng, Y., Luo, Y., Christie, P., 2017. Biosurfactant-producing microorganism *Pseudomonas* sp. SB assists the phytoremediation of DDT-contaminated soil by two grass species. Chemosphere 182, 137–142. doi:10.1016/j.envpol.2020.115411.

Welbaum, G.E., Sturz, A.V., Dong, Z., Nowak, J., 2004. Managing soil microorganisms to improve productivity of agro-ecosystems. Critical Reviews in Plant Sciences *23* (2), 175–193.

Wolf, D.C., Cryder, Z., Khoury, R., Carlan, C., Gan, Z., 2020. Bioremediation of PAH-contaminated shooting range soil using integrated approaches. Science of the Total Environment 726, 138440.

Wu, T., Xu, J., Xie, W., Yao, Z., Yang, H., Sun, C., Li, X., 2018. Pseudomonas aeruginosa L10: A Hydrocarbon-Degrading, Biosurfactant-Producing, and Plant-Growth-Promoting Endophytic Bacterium Isolated From a Reed (Phragmites australis). Frontiers in Microbiology 9, 1087.

Yaseen, R.Y., El-Aziz, S.A., Eissa, D.T., Abou-Shady, A.M., 2019. Application of biosurfactant producing microorganisms to remediate heavy metal pollution in El-Gabal El-Asfar area. Alex Sci Exch J 39, 17–34.

Yin, X., Sun, X., Yang, Y., Ding, H., 2019. In-situ bioremediation of soil pollution with electric heating temperature regulation bio-ventilation, IOP conference series: Earth and environmental science, IOP Publishing, 242, 042011. doi:10.1088/1755-1315/242/4/042011.

Zhang, J., Yin, R., Lin, X., Liu, W., Chen, R., Li, X., 2010. Interactive effect of biosurfactant and microorganism to enhance phytoremediation for removal of aged polycyclic aromatic hydrocarbons from contaminated soils. J. Health Sci. 56 (3), 257–266.

Index

Page numbers followed by "*f*" and "*t*" indicate, figures and tables respectively.

R

S

T

V

Z